Polymer Composites with Functionalized Nanoparticles

Related titles

Multifunctionality of Polymer Composites, (ISBN 978-0-323-26434-1)

Multifunctional Polymeric Nanocomposites Based on Cellulosic Reinforcements, (ISBN 978-0-323-44248-0)

Smart Polymers and their Applications, (ISBN 978-0-85709-695-1)

Polymer Hybrid Materials and Nanocomposites, (ISBN 978-0-12-813294-4)

Polyurethane Polymers: Composites and Nanocomposites, (ISBN 978-0-12-804065-2)

Micro and Nano Technologies

Polymer Composites with Functionalized Nanoparticles

Synthesis, Properties, and Applications

Edited by

Krzysztof Pielichowski

Department of Chemistry and Technology of Polymers,
Cracow University of Technology, Kraków, Poland

Tomasz M. Majka

Department of Chemistry and Technology of Polymers,
Cracow University of Technology, Kraków, Poland

ELSEVIER

Elsevier
Radarweg 29, PO Box 211, 1000 AE Amsterdam, Netherlands
The Boulevard, Langford Lane, Kidlington, Oxford OX5 1GB, United Kingdom
50 Hampshire Street, 5th Floor, Cambridge, MA 02139, United States

British Library Cataloguing-in-Publication Data
A catalogue record for this book is available from the British Library

Library of Congress Cataloging-in-Publication Data
A catalog record for this book is available from the Library of Congress

ISBN: 978-0-12-814064-2

Publisher: Matthew Deans
Acquisition Editor: Edward Payne
Editorial Project Manager: Peter Jardim
Production Project Manager: Debasish Ghosh
Cover Designer: Greg Harris

Typeset by MPS Limited, Chennai, India

Contents

List of Contributors xi

Preface xv

1. Synthesis Routes of Functionalized Nanoparticles 1

KATARÍNA MOSNÁČKOVÁ, JOZEF KOLLÁR, YI-SHEN HUANG, CHIH-FENG HUANG AND JAROSLAV MOSNÁČEK

1.1 Introduction 1

1.2 Functional Groups for Modification of Nanoparticles Surface 2

1.3 Modification of Nanoparticles by Polymer Chains 7

1.4 Modern Functionalized Nanoparticles for Advanced Applications 23

1.5 Conclusion 28

Acknowledgment 29

References 29

2. Design and Synthesis of Polymer Nanocomposites 47

E.I. AKPAN, X. SHEN, B. WETZEL AND K. FRIEDRICH

2.1 Polymer Nanocomposites 47

2.2 Design of Polymer Nanocomposites 50

2.3 Synthesis of Polymer Nanocomposites 66

2.4 Conclusion 74

Acknowledgments 75

References 75

3. Specific Interactions and Self-Organization in Polymer/Functionalized Nanoparticle Systems 85

OĞUZHAN OĞUZ, JAMAL SEYYED MONFARED ZANJANI, SERAP HAYAT SOYTAŞ AND YUSUF ZIYA MENCELOĞLU

3.1 Introduction 85

3.2 Synthesis, Modification, and Functionalization 88
3.3 Nanocomposites Containing Functionalized Nanoparticles 91
3.4 Self-Organization in Nanocomposites 103
3.5 Conclusion and Outlook 108
References 109

4. Polymer Composites With Functionalized Silica 119
VLADIMIR M. GUN'KO

4.1 Introduction 119
4.2 Structure of Unmodified and Modified Silicas as Fillers 122
4.3 Polymer Composites Filled by Silicas 125
4.4 Conclusion 141
References 141

5. Functionalized Clay-Containing Composites 149
TOMASZ M. MAJKA AND KRZYSZTOF PIELICHOWSKI

5.1 Introduction 149
5.2 Synthesis of Nanoparticles 150
5.3 Clays and Their Methods of Functionalization 151
5.4 Functionalized Clay-Containing Polymer Nanocomposites Formation 155
5.5 Types of Nanocomposites Structures 158
5.6 Functionalized Clays as Fire Retardants in Polymer Nanohybrids 159
5.7 Barrier Properties of Functionalized Layered Silicates 161
5.8 The Influence of Functional Clays on Aging of Polymer Composites 163
5.9 The Influence of Clay Functionalization on Human Health and the Environment 166
5.10 Concluding Remarks 170
References 171

6. Functionalized POSS-Based Hybrid Composites 179

HUI ZHOU, MING HUI CHUA AND JIANWEI XU

6.1 Introduction 179

6.2 Polyhedral Oligosilsesquioxanes–Based Hybrid Composites for Performance Enhancement of Materials 183

6.3 Conclusions and Perspectives 206

Acknowledgment 207

References 207

7. Polymer Composites With Functionalized Carbon Nanotube and Graphene 211

CHAOHUI WEI, BAMIDELE AKINWOLEMIWA, LINPO YU, DI HU AND GEORGE Z. CHEN

Abbreviations 211

7.1 Composites of Polymers With Carbon Nanomaterials 212

7.2 Functionalization of Carbon Nanomaterials and the Effects on Polymer Composites 216

7.3 Properties and Applications of Polymer Composites With Carbon Nanomaterials 228

7.4 Concluding Remarks 238

References 240

8. Polymer Composites With Metal Nanoparticles: Synthesis, Properties, and Applications 249

LAURA TAMAYO, HUMBERTO PALZA, JULIÁN BEJARANO AND PAULA A. ZAPATA

8.1 Introduction 249

8.2 Synthesis of Metal Nanoparticles 250

8.3 Functionalization of Silver and Gold Nanoparticles 253

8.4 Methods of Preparation of Metal/Polymer Nanocomposites 256

8.5 Properties of Metal/Polymer Nanocomposites 259

8.6 Conclusions 273

Acknowledgments 273

References 273

9. Polymer Nanocomposites With Decorated Metal Oxides 287

SERAP HAYAT SOYTAŞ, OĞUZHAN OĞUZ AND YUSUF ZIYA MENCELOĞLU

9.1 Introduction 287

9.2 Synthesis of Metal Oxide Nanoparticles 288

9.3 Surface Modification of Metal Oxide Nanoparticles 301

9.4 Polymer–Metal Oxide Nanocomposite Materials 304

9.5 Conclusion and Outlook 312

References 315

Further Reading 323

10. Optical Properties of Polymer Nanocomposites With Functionalized Nanoparticles 325

ANDREY L. STEPANOV

10.1 Introduction 325

10.2 Synthesis of Metal Nanoparticles in Polymethylmethacrylate by Ion Implantation 326

10.3 Fabrication of Viscous Polymer Composites With Metal Nanoparticles 338

10.4 Conclusion 352

Acknowledgments 353

References 353

Further Reading 355

11. Organic Conjugated Polymer-Based Functional Nanohybrids: Synthesis Methods, Mechanisms and Its Applications in Electrochemical Energy Storage Supercapacitors and Solar Cells 357

KAKARLA RAGHAVA REDDY, B. HEMAVATHI, GEETHA R. BALAKRISHNA, A.V. RAGHU, S. NAVEEN AND M.V. SHANKAR

11.1 Introduction 357

11.2 Organic Conjugated Polymers 358

11.3 Synthesis of Electrically Conducting Organic Conjugated Polymers 361

11.4 Conclusion 375

References 376

12. Medical Applications of Polymer/Functionalized Nanoparticle Systems 381

ASWATHY RAVINDRAN GIRIJA

12.1 Introduction 381

12.2 Functionalization and Bioconjugation of Nanoparticles 382

12.3 Medical Applications 387

12.4 Conclusion and Future Outlook 395

References 396

13. Thermal Decomposition of Polymer Nanocomposites With Functionalized Nanoparticles 405

KLAUDIA KRÓL-MORKISZ AND KINGA PIELICHOWSKA

13.1 Introduction 405

13.2 Effect of Functionalized Nanoparticles on the Polymer Nanocomposites' Stability 407

13.3 Decomposition of Polymer Nanocomposites With Functionalized Nanoparticles 408

13.4 Conclusions 429

Acknowledgments 430

References 430

14. Polymer Composites Containing Functionalized Nanoparticles and the Environment 437

JAMAL SEYYED MONFARED ZANJANI, OĞUZHAN OĞUZ, BURCU SANER OKAN, MEHMET YILDIZ AND YUSUF ZIYA MENCELOĞLU

14.1 Introduction 437

14.2 Toxicological and Exposure 438

14.3 Environmental Benefits and Application of Polymeric Nanocomposites 448
14.4 Conclusion and Outlook 455
References 456

15. Future Perspectives 467
TOMASZ M. MAJKA AND KRZYSZTOF PIELICHOWSKI

15.1 Trends in the Development of Functionalized Hybrid Materials 467

Index 471

List of Contributors

Bamidele Akinwolemiwa International Doctoral Innovation Centre, Faculty of Science and Engineering, University of Nottingham Ningbo China, Ningbo, P.R. China

E.I. Akpan Institute for Composite Materials, Kaiserslautern, Germany

Geetha R. Balakrishna Centre for Nano and Material Sciences, Jain University, Jain Global Campus, Bangalore, Karnataka, India

Julián Bejarano Departamento de Ingeniería Química y Biotecnología, Facultad de Ciencias Físicas y Matemáticas, Universidad de Chile, Las Palmeras, Casilla, Santiago, Chile; Departamento de Química Farmacológica y Toxicológica, Facultad de Ciencias Químicas y Farmacéuticas, Universidad de Chile, Sergio Livingstone, Independencia, Santiago, Chile

George Z. Chen Department of Chemical and Environmental Engineering, and Energy Engineering Research Group, Faculty of Science and Engineering, University of Nottingham Ningbo China, Ningbo, P.R. China; Department of Chemical and Environmental Engineering, and Advanced Materials Research Group, Faculty of Engineering, University of Nottingham, Nottingham, United Kingdom

Ming Hui Chua Institute of Materials Research and Engineering, Agency for Science, Technology and Research (A*STAR), Singapore

K. Friedrich Institute for Composite Materials, Kaiserslautern, Germany

Aswathy Ravindran Girija Future Industries Institute, University of South Australia, Mawson Lakes Campus, Mawson Lakes, SA, Australia

Vladimir M. Gun'ko Department of Amorphous and Structurally Ordered Oxides, Chuiko Institute of Surface Chemistry, Kyiv, Ukraine

B. Hemavathi Centre for Nano and Material Sciences, Jain University, Jain Global Campus, Bangalore, Karnataka, India

Di Hu Department of Chemical and Environmental Engineering, and Energy Engineering Research Group, Faculty of Science and Engineering, University of Nottingham Ningbo China, Ningbo, P.R. China

Chih-Feng Huang Department of Chemical Engineering, National Chung Hsing University, Taichung, Taiwan

Yi-Shen Huang Department of Chemical Engineering, National Chung Hsing University, Taichung, Taiwan

Jozef Kollár Polymer Institute, Slovak Academy of Sciences, Bratislava, Slovakia

Klaudia Król-Morkisz AGH University of Science and Technology, Faculty of Materials Science and Ceramics, Department of Biomaterials and Composites, Kraków, Poland

Tomasz M. Majka Department of Chemistry and Technology of Polymers, Cracow University of Technology, Kraków, Poland

Yusuf Ziya Menceloğlu Faculty of Engineering and Natural Sciences, Materials Science and Nano Engineering, Sabanci University, Tuzla, Istanbul, Turkey; Sabanci University Integrated Manufacturing Technologies Research and Application Center & Composite Technologies Center of Excellence, Istanbul, Turkey; Sabanci University Nanotechnology Research and Application Center (Sunum), Tuzla, Istanbul, Turkey

Jaroslav Mosnáček Polymer Institute, Slovak Academy of Sciences, Bratislava, Slovakia; Centre for Advanced Materials Application, Slovak Academy of Sciences, Bratislava, Slovakia

Katarína Mosnáčková Polymer Institute, Slovak Academy of Sciences, Bratislava, Slovakia

S. Naveen Department of Basic Sciences, School of Engineering and Technology, Jain University, Bangalore, Karnataka, India

Oğuzhan Oğuz Faculty of Engineering and Natural Sciences, Materials Science and Nano Engineering, Sabanci University, Tuzla, Istanbul, Turkey; Sabanci University Integrated Manufacturing Technologies Research and Application Center & Composite Technologies Center of Excellence, Istanbul, Turkey

Humberto Palza Departamento de Ingeniería Química y Biotecnología, Facultad de Ciencias Físicas y Matemáticas, Universidad de Chile, Las Palmeras, Casilla, Santiago, Chile

Kinga Pielichowska AGH University of Science and Technology, Faculty of Materials Science and Ceramics, Department of Biomaterials and Composites, Kraków, Poland

Krzysztof Pielichowski Department of Chemistry and Technology of Polymers, Cracow University of Technology, Kraków, Poland

A.V. Raghu Department of Basic Sciences, School of Engineering and Technology, Jain University, Bangalore, Karnataka, India

Kakarla Raghava Reddy The School of Chemical and Biomolecular Engineering, The University of Sydney, Sydney, NSW, Australia

Burcu Saner Okan Faculty of Engineering and Natural Sciences, Materials Science and Nano Engineering, Sabanci University, Tuzla, Istanbul, Turkey; Sabanci University Integrated Manufacturing Technologies Research and Application Center & Composite Technologies Center of Excellence, Istanbul, Turkey

Jamal Seyyed Monfared Zanjani Faculty of Engineering and Natural Sciences, Materials Science and Nano Engineering, Sabanci University, Tuzla, Istanbul, Turkey; Sabanci University Integrated Manufacturing Technologies Research and Application Center & Composite Technologies Center of Excellence, Istanbul, Turkey

M.V. Shankar Department of Materials Science & Nanotechnology, Yogi Vemana University, Kadapa, Andhra Pradesh, India

X. Shen Institute for Composite Materials, Kaiserslautern, Germany

Serap Hayat Soytaş Faculty of Engineering and Natural Sciences, Materials Science and Nano Engineering, Sabanci University, Tuzla, Istanbul, Turkey; Sabanci University Integrated Manufacturing Technologies Research and Application Center & Composite Technologies Center of Excellence, Istanbul, Turkey; Sabanci University Nanotechnology Research and Application Center (Sunum), Tuzla, Istanbul, Turkey

Andrey L. Stepanov Kazan Physical-Technical Institute, Russian Academy of Sciences, Kazan, Russian Federation; Kazan National Research Technological University, Kazan, Russian Federation

Laura Tamayo Departamento de Química, Facultad de Ciencias, Universidad de Chile, Las Palmeras, Casilla, Santiago, Chile

Chaohui Wei International Doctoral Innovation Centre, Faculty of Science and Engineering, University of Nottingham Ningbo China, Ningbo, P.R. China

B. Wetzel Institute for Composite Materials, Kaiserslautern, Germany

Jianwei Xu Institute of Materials Research and Engineering, Agency for Science, Technology and Research (A*STAR), Singapore

Mehmet Yildiz Faculty of Engineering and Natural Sciences, Materials Science and Nano Engineering, Sabanci University, Tuzla, Istanbul, Turkey; Sabanci University Integrated Manufacturing Technologies Research and Application Center & Composite Technologies Center of Excellence, Istanbul, Turkey

Linpo Yu Department of Chemical and Environmental Engineering, and Energy Engineering Research Group, Faculty of Science and Engineering, University of Nottingham Ningbo China, Ningbo, P.R. China

Jamal Seyyed Monfared Zanjani Faculty of Engineering and Natural Sciences, Materials Science and Nano Engineering, Sabanci University, Tuzla, Istanbul, Turkey; Sabanci University Integrated Manufacturing Technologies Research and Application Center & Composite Technologies Center of Excellence, Istanbul, Turkey

Paula A. Zapata Grupo Polímeros, Facultad de Química y Biología, Universidad de Santiago de Chile, USACH, Casilla 40, Correo 33, Santiago, Chile

Hui Zhou Institute of Materials Research and Engineering, Agency for Science, Technology and Research (A*STAR), Singapore

Preface

This book focuses on the synthesis, properties, and applications of polymer composites with functionalized nanoparticles. Attention is paid to the role of specific interactions in such systems, as well as to the modification routes of different types of nanoparticles, for example, carbon nanotubes (CNTs), layered silicates, and metal oxides.

Functionalization could be defined as an addition of chemical or physical functionality on the surface of (nano)particles. Some modification methods involve direct functionalization with functional ligand containing one reactive group able to attach onto the filler surface, and the other representing the required active surface functionality. Direct functionalization allows modification of the nanoparticles' surface in one single step. However, in some cases a competitive reaction with functionality, which is going to be introduced on the nanofillers' surface, can occur, and thus a two-step modification needs to be applied. Postfunctionalization via, for example, thiol–ene, Diels–Alder, or Huisgen reactions may be also applied to control the surface chemistry and energy.

Surface modification of nanofillers is generally realized during nanofillers synthesis in the presence of ligand, or in the case of prepared nanofillers by ligand exchange, ligand addition, formation of interdigitated layer, or by encapsulation.

Special attention has been given to functionalization of carbon, metal oxide, nanosilver and nanogold, or silicate-based nanomaterials. Functionalization of these nanomaterials is provided for several purposes, but the main goal is generally improvement of their dispersibility in both water and common organic solvents as well as in polymer matrices.

In Chapter 1, Synthesis Routes of Functionalized Nanoparticles, the synthetic approaches for functionalization of various nanoparticles for tailoring their surface properties are presented. Different methods of modification of nanoparticles surfaces by polymer chains are described in detail with some examples of functionalization of various types of nanoparticles to obtain advanced properties of the nanocomposites.

Chapter 2, Design and Synthesis of Polymer Nanocomposites, focuses on the design process and production techniques of polymer nanocomposite materials. In-depth analyses of design strategies, principles, and mechanisms that guide appropriate design are presented. The most commonly used synthesis techniques, including ultrasonication assisted solution mixing, shear mixing, three roll milling, ball milling, double-screw extrusion, as well as other ex situ and in situ processing methods, are discussed in the context of their applicability to different nanofillers and matrices, the process parameters, and the up-scalability for industrial applications.

In Chapter 3, Specific Interactions and Self-Organization in Polymer/Functionalized Nanoparticle Systems, specific interactions and self-organization in nanocomposites containing functionalized nanoparticles are discussed. Common techniques applied for the synthesis, modification, and functionalization of modified silica nanoparticles, layered silicates, and polyhedral oligomeric silsesquioxanes (POSS) are presented. The use of such nanoparticles in the preparation of polymer-based nanocomposites and their effects on the structure–property relationships of the resulting materials are described and exemplified by some representative studies reported in the literature. Moreover, a general overview on the self-organization phenomenon is provided based on the recent developments in the field of block copolymer–based nanocomposites.

Chapter 4, Polymer Composites With Functionalized Silica, deals with differently functionalized silica nanoparticles as fillers, both physical and chemical, for linear, branched, or cross-linked polymers. Moreover, in situ synthesized silica nanoparticles in polymeric media give another effective pathway to produce hybrid nanocomposites for various applications.

Chapter 5, Functionalized Clay-Containing Composites, is intended as an overview of the recent research on the use of functionalized layered silicates, dispersed at the nanometer scale in polymeric matrices, for improving composite properties, such as mechanical and barrier properties. Synthesis protocols, surface functionalization strategies, and structure–property relationships, including those on the barrier properties and aging/flammability behavior of clay-containing polymer nanocomposites, have been highlighted.

In Chapter 6, Functionalized POSS-Based Hybrid Composites, POSS particles functionalized through introduction of reactive organic groups to the vertex positions via various chemical methods are presented. The chapter summarizes the latest developments on POSS-based hybrid composites for materials performance enhancement in terms of thermal stability, mechanical strength, electric conductivity, antiatomic oxidation, antiphotodegradation, optical application, crystallinity, surface property, porosity, catalytic activity, biochemistry, and self-assembly.

In Chapter 7, Polymer Composites With Functionalized Carbon Nanotube and Graphene, the effects of carboxylic, carbonyl, sulfonic, amide, and halogen functional groups on the physicochemical properties of CNTs and graphene (GN) are discussed. An overview of fabrication of well-characterized polymer composites with functionalized CNTs and GN for energy conversion and storage, catalysis, and structural reinforcement is given, too.

Chapter 8, Polymer Composites With Metal Nanoparticles: Synthesis, Properties, and Applications, focuses on synthesis routes for obtaining silver and gold nanoparticles and their functionalization to improve nanoparticle dispersion in the polymer matrix. The main routes for nanocomposite preparation and their influence on the polymer composite properties are also shown. Special attention is paid to polymer nanocomposites designed for water treatment, textile materials, food packaging, and medical devices.

In Chapter 9, Polymer Nanocomposites With Decorated Metal Oxides, an overview of polymer nanocomposites containing functionalized metal oxide nanoparticles as inorganic fillers is presented. The main synthesis processes, surface functionalization of the metal oxide nanostructures, as well as the properties and potential applications of the nanocomposites are described. Polymer nanocomposites with functionalized metal oxide nanoparticles have received considerable attention as they are used in a wide range of applications in various fields including optoelectronic and electromagnetic devices, biomedicine, and fuel cell technology.

Chapter 10, Optical Properties of Polymer Nanocomposites With Functionalized Nanoparticles, focuses on the specific features of the metal nanoparticle fabrication in polymer matrix by applying ion implantation. The formation of silver nanoparticles in polymer carbonized by ion irradiation is also briefly reviewed and mechanisms of metal nanoparticle growth in viscous epoxy resin are discussed.

Then, polymer composites with functionalized nanoparticles for energy storage are discussed; the synthesis methodologies of nanostructures of conjugated polymers and their hybrid composites, as well as the formation mechanism of such nanoscale structures (from 0D to 3D) are demonstrated. Moreover, potential applications of this class of polymer composites as electrochemical supercapacitors and solar energy materials are outlined.

The following chapter 12 discusses the medical applications of polymeric nanoparticles highlighting the drug delivery and imaging modalities with polymeric nanoparticles. Recent progress in the understanding of the toxic behavior of various nanoparticles, used in polymer nanocomposites production, is shown. Functionalized nanoparticles' effects on living organs and their role on human life quality and the environment are also discussed.

The thermal decomposition behavior of selected polymer nanocomposites modified with functionalized nanoparticles is described in Chapter 13, Thermal Decomposition of Polymer Nanocomposites With Functionalized Nanoparticles. Surface modification significantly enhances the nanoadditive–polymer interactions and contact surface area and, in consequence, often improves the thermal properties of nanocomposites containing functionalized montmorillonite, hydroxyapatite, GN, CNTs, and titanium dioxide. Barrier effects and confined segmental motion of the polymer chains in the presence of nanostructured fillers are discussed as the reasons for observed thermal stabilization effects.

Possible harmful effects of nanomaterials, such as the toxicological effect on living organs, and other adverse environmental effects, are discussed in Chapter 14, Polymer Composites Containing Functionalized Nanoparticles and the Environment. The main theme focuses on the environmental effect of nanomaterials, nanotoxicology, and their novel and innovative environment-friendly applications. Solutions proposed to minimize the environmental consequences of nanoparticle use are also presented.

Chapter 15, Future Perspectives, discusses the main perspectives and trends in the field of polymer composites with functionalized nanoparticles. The role of advanced computational tools and modeling/simulation approaches, as well as supramolecular and interface chemistry, will contribute to the better understanding of next-generation multifunctional nanocomposite materials. An important issue that has to be addressed is the development of recycling technologies of composite materials containing functionalized nanoparticles.

We hope that this book will be a useful tool for scientists, academics, researchers, polymer science engineers, and those working in industries using nanomaterials. The book will also be useful to undergraduate and postgraduate students and hopefully an inspiration to young scientists to investigate the functionalization of nanomaterials in more detail.

Krzysztof Pielichowski, Tomasz M. Majka

1

Synthesis Routes of Functionalized Nanoparticles

Katarína Mosnáčková[1], Jozef Kollár[1], Yi-Shen Huang[2], Chih-Feng Huang[2], Jaroslav Mosnáček[1,3]

[1]POLYMER INSTITUTE, SLOVAK ACADEMY OF SCIENCES, BRATISLAVA, SLOVAKIA [2]DEPARTMENT OF CHEMICAL ENGINEERING, NATIONAL CHUNG HSING UNIVERSITY, TAICHUNG, TAIWAN [3]CENTRE FOR ADVANCED MATERIALS APPLICATION, SLOVAK ACADEMY OF SCIENCES, BRATISLAVA, SLOVAKIA

1.1 Introduction

Materials filled with nanoparticles (NPs) have attracted great interest due to their wide application in textiles, nanocomposites, biomedical and food industries, electronics, environment, and energy. Theoretically, Flory theories, density functional theories, and molecular dynamics methods can provide important directions thermodynamically and kinetically to predict the mixing stability of NPs and polymeric matrix. However, the precise calculations are still a challenge due to the difficulty in accessing the corresponding length and time scales [1–3]. To prepare materials with well-dispersed NPs for the practical applications, on the other prospect, surface functionalization is the key to stabilize NPs against aggregation in the polymer matrix. This chapter covers the functionalization of NPs surface with different functionalities and approaches. This process may provide several benefits such as improvement of NPs colloidal stability [4], their self-organization [5], compatibility with other phase [6,7], or incorporation of organic groups for specific interaction [8]. First, the surface modification includes simple functionalization of the NPs surface by low molecular weight compounds, which can directly improve stability and dispersibility of the NPs or can serve as introduction of functional groups needed for further postfunctionalization with low molecular weight compounds or polymers.

In the second part of this chapter the modification of NPs surface by polymers chains is reviewed. Controlled/living polymerizations [9–19], including ionic polymerizations, chain-growth condensation polymerization, and reversible-deactivation radical polymerizations (RDRPs), but partially also free radical polymerization (FRP) provide powerful tools for preparations of a variety of surface-functionalized NPs. Generally speaking, there are three major protocols for the surface functionalization, namely “grafting to,” “grafting from,” and “grafting through” methods. Accordingly, various polymerization techniques were successfully utilized for functionalization of NPs surface to improve the interfacial interactions between the NPs and polymeric matrices.

Polymer Composites with Functionalized Nanoparticles. DOI: https://doi.org/10.1016/B978-0-12-814064-2.00001-9

Due to the proper NPs dispersion, basic material properties can be improved and the final nanocomposites can be shifted toward various real-life applications with enhanced performance. The last part of this chapter is focused on the examples of specific properties and potential applications of the selected modern NPs, such as clays, Polyhedral Oligomeric Silsesquioxane (POSS), cellulose nanofibrils (CNFs), graphene, metal NPs , carbon nanotubes (CNTs), etc., functionalized by different strategies.

1.2 Functional Groups for Modification of Nanoparticles Surface

The common functionalization strategy can be defined as an addition of (new) chemical functionality on the surface of NPs. There are two main modification methods for attachment of a ligand and incorporation of functionality on NPs surface (Fig. 1–1). The first method involves direct functionalization with functional ligand containing one reactive group capable of attachment onto the NPs surface and the second one representing the required/targeted active surface functionality [20]. The second method consists of minimal two reaction steps. Initially, one reactive/functional group of bifunctional compound is reacted with NPs surface introducing a new functional group. Subsequently, the new functional group is used for postfunctionalization reaction to attach the ligand with required/targeted

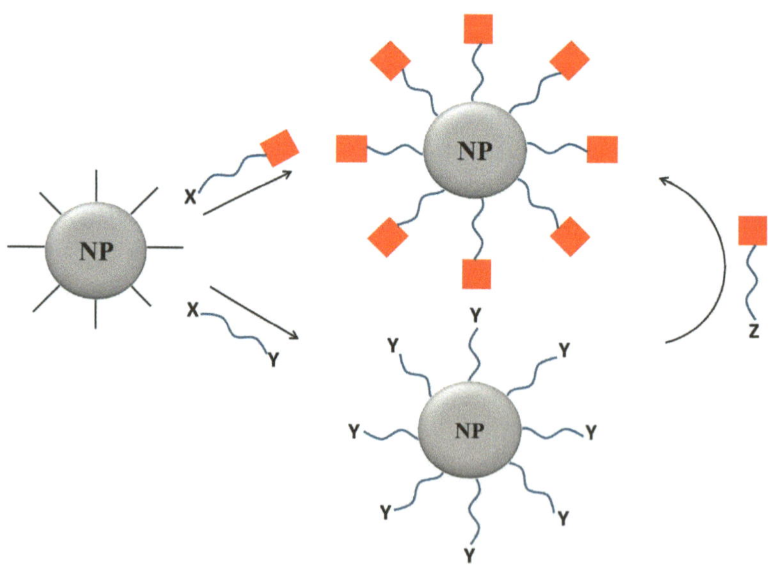

FIGURE 1–1 Two methods of nanoparticle surface modification: (top) direct functionalization with ligand containing reactive group; (bottom) two-step modification consisting of initial functionalization of the nanoparticle surface and subsequent postfunctionalization.

functionality on the NPs surface. Both strategies exhibit some advantages and limitations. Direct functionalization allows modification of NPs surface in one single step. On the other hand, in some cases a competitive reaction with functionality, which is going to be introduced on the NPs surface, can occur, and thus two-step modification needs to be applied. Another limitation of the direct functionalization approach relates with possible steric hindrance of the ligand, which can affect the surface grafting density. Postfunctionalization is a more versatile method and the main advantage of this technique is that plenty of coupling agents are commercially available [21].

Surface modification of NPs is generally realized during NPs synthesis in the presence of ligand, or in the case of prepared NPs by ligand exchange, ligand addition, formation of interdigitated layer, or by encapsulation. Ligand exchange technique is based on the point that the incoming ligand molecule binds more strongly to the inorganic NPs surface resulting in exchange of ligands [22,23]. The ligand addition method allows addition of ligand derivatives without removal of the original ligand [24]. Noncovalent interaction based on hydrophobic interaction between amphiphilic molecules or polymers leads to formation of an interdigitated bilayer [25,26]. Encapsulation can be generally accomplished by interaction of NPs stabilized with hydrophobic ligand with amphiphilic polymers [27].

1.2.1 Direct Functionalization of Nanoparticles Surface

1.2.1.1 Direct Functionalization of Noble Metals and Metal Oxides

Wide range of functional groups can be used for direct functionalization of NPs as depicted in Fig. 1–2. Modification of NPs surface by organosulfur derivatives is one of the most widely used and explored methods for noble metals, such as Au, Ag, Cu, Pt, etc. The surfaces of these metals exhibit strong affinity to thiols and disulfide groups. This chemisorption interaction leads to effective and quick surface modification or exchange of some other surfactants [28]. In addition, the concentration of organosulfur ligand can affect the size of NPs during their formation. Control over size of Au NPs from 2 to 5 nm was achieved by varying of dodecanethiol concentration [29]. Many authors reported the use of zwitterionic-based disulfide for stabilization of Au NPs. Resulting zwitterionic coated NPs provided sufficient colloidal stability in the pH region of 4–13 [30,31]. The character of the ligand determines overall properties of the NPs, especially the tendency to aggregate. Increased colloidal stability of Au NPs at higher tris–borate concentrations was obtained by modification of the Au NPs surface by mercaptopropionic acid [32].

Versatile route to stabilize NPs represents the addition of amine or ammonium containing ligands. Interaction between the amine groups and the surface of noble metal NPs is weaker as in the case of thiol groups, usually resulting in formation of larger NPs during the NPs preparation process. The spherical monodisperse magnetite NPs with the size around 5 nm can be however readily prepared in the presence of alkylamines as a ligand and do not undergo aggregation in various organic solvents [33]. Another method uses the adsorption of tetraalkylammonium halides for stabilization of noble metal NPs. Long chains of ammonium

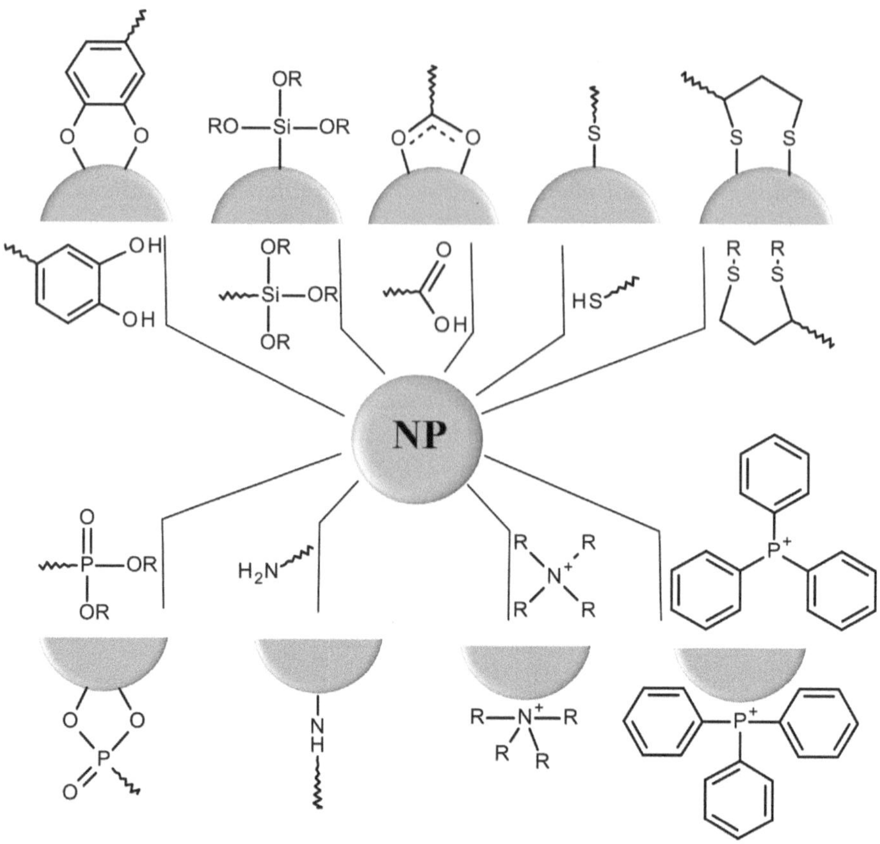

FIGURE 1–2 Schematic illustration of functional groups for direct functionalization of nanoparticle surface.

ions are often essential, because their amphiphilic character can improve the compatibility of NPs with particular medium [34].

A powerful method for surface modification of metal oxides NPs is coupling with alkoxysilane compounds. A great advantage of this technique is the possibility to utilize a variety of functional groups of organosilane derivatives for a subsequent postfunctionalization. Common examples represent silane derivatives with amine [35], carboxy [36], epoxy [37], or cyano [38] functionality. Functionalization can lead to cross-linking reactions through homocondensation forming a thin silica layer around the core with a fraction of the functional groups orientated outside of the NPs surface [39].

For some applications, like immobilization of proteins, the modification of NPs surface by phosphonate groups is preferable. Phosphonic acid shows stronger affinity to metal oxide NPs than the silyl group, especially under physiological conditions [40]. Moreover, the phosphonate groups do not undergo homocondensation as in the case of silane units. Several alkyl phosphonate derivatives can be effectively bound to iron oxide particles surface and improve their colloidal stability and biocompatibility [41]. Modification of NPs by phosphines, for example, triphenylphosphine, is easy to be achieved. However, contrary to the modification by phosphonates, the bond between the inorganic NP surface and electrodonating phosphine group is quite weak and undergoes dynamic binding–unbinding reaction. This process can be utilized in almost quantitative exchange with ligands containing strongly binding functional groups [42].

Another important technique for NPs modification benefits from coordination reaction between the carboxylic acid and surface of metal oxides. Oleic acid or other fatty acids are commonly used for the stabilization of colloidal solution [43]. Coordination bonds between the NPs' surface and carboxylates are weak and can be broken easily by increasing temperature or by ligand exchange reaction.

Polyalcohols bearing multiple alcohol groups are also able to bind various metal oxide NPs. Dopamine and ascorbic acid coordinate the surface of iron oxide NPs very effectively [44,45] and at the same time can introduce the functionality for following postmodification of NPs surface.

1.2.1.2 Functionalization of Carbon-Based Nanoparticles

Special attention should be given to functionalization of carbon-based nanomaterials, such as graphene or carbon nanotubes (CNTs). Functionalization of these nanomaterials is provided for several purposes, but the main goal is generally improvement of their dispersibility in both water and common organic solvents as well as in polymer matrices. The most chemical procedures are based on covalent or noncovalent functionalization of the aromatic structure or covalent attachment of functionalities on the oxidized surface of graphene or CNTs [46,47].

Covalent modification of the aromatic structure using diazonium salts consists of addition of radicals on the sp_2 carbon atoms. Surface structuring and functionalization at nanoscale via electrochemical grafting of diazonium compounds can be performed even under ambient conditions [48]. sp_2 carbon atoms also undergo reactions with dienophiles, for example, azomethine ylide. 1,3-Dipolar cycloaddition of paraformaldehyde and tert-butyloxycarbonyl (Boc) protected amino acid with graphene exfoliated sheets was successfully realized and used for special interaction between gold nanorods [49]. Another study reported functionalization of graphene sheets with alkylazides containing various alkyl chains. Modified graphene showed enhanced dispersibility in organic solvent such as acetone or toluene [50]. Nitrene addition of azidophenylalanine to exfoliated graphene provides another effective method for highly functionalized graphene nanosheets [51]. Another possible way for incorporation of functionalities on graphene surface is by substitution of metalized graphite. Reaction of graphene with lithium in the presence of ammonia results in formation of lithium–carbon bonds, which can be subsequently attacked with *n*-dodecyl iodide to yield dodecylated graphite [52]. Exfoliation of graphite and further functionalization can be achieved by fluorination of graphene sheets. Covalent functionalization of graphene was induced by CF_4 or CHF_3 plasma treatment [53]. Fluorine provides stable graphene nanosheets with stronger binding between agents and carbon, and the fluorinated graphite has attracted numerous studies for subsequent functionalization [54].

Covalent and noncovalent functionalization of graphene oxide (GO) with small organic derivatives has become the subject of intensive research. Carboxyl, hydroxyl, or epoxy groups represent a suitable locus for common organic synthesis. Functionalization using carboxyl groups present on GO or oxidized CNTs surface is generally realized via treatment with $SOCl_2$, followed by condensation with a nucleophile or through carbodiimide couplings with amino or hydroxy derivatives [55]. A similar approach via an acyl chloride intermediate was also used for the functionalization of GO with hydroxyl-terminated poly-(3-hexylthiophene) [56] or with alkylamine [57]. Similarly, the hydroxyl groups on GO surface can be functionalized using carbodiimide coupling chemistry. Alternatively, the hydroxyl groups can be modified through the reaction with *N,N*-dimethylacetamide dimethyl acetal [58] or with reaction with alkoxy or alkyltrichlorosilanes [59,60]. Epoxy group is another possible location for GO modification. The ring opening is generally attacked with nucleophile in the form of aliphatic or aromatic amine [61]. Another route of direct functionalization of GO and oxidized CNTs utilizes the reactive intermediates on the base of diazonium salts [62], carbenes [63], or nitrenes [64], similar to functionalization of surface of nonoxidized carbon-based NPs.

1.2.2 Postfunctionalization Reactions at Nanoparticles Surface

The amine functionality is very often associated with postfunctionalization of NPs surface, because the amine group is proper for further reactions with many derivatives containing various active species (Fig. 1–3). In addition, this technique can find the inspiration from bioconjugation chemistry [7,65]. Reaction between primary amines and terminal carboxylate in the presence of carbodiimides is widely used for linking enzymes or peptides on the NPs surface [66]. A versatile method to introduce a covalent amide bond between the NPs and (bio)molecules is also via reaction of amines on the NPs surface with *N*-hydroxysuccinimide terminated esters containing bioactive species [67]. A frequent approach for conjugation of biomolecules such as DNA or proteins uses the maleimide coupling based on transformation of amine to maleimide moiety in the first step and subsequent reaction with thiol derivatives [68]. The strategy can also be the opposite, when thiol group on the NPs surface can react with maleimide group at the (macro)molecule. When the NPs surface is modified with bifunctional ligand leading to free carboxylic acids on the NPs surface, the carboxylic acids can be used for postfunctionalization by condensation reaction with primary amines to yield amide bonds [69]. NPs with terminal aldehyde groups allow linking of amine containing (bio)molecules on their surface via Schiff base condensation [70]. In a similar manner, hydrazide is able to react with aldehyde groups and immobilize, for example, various antibodies [71].

Very high yield and selectivity between the functional groups attached on the NPs surface and functional ligand can be achieved by click chemistry [72,73]. The most common example represents catalyzed alkyne-azide cycloaddition giving a 1,2,3-triazole ring as a strong covalent bond between modified NPs surface and conjugate. This universal concept allows immobilization of numerous biological active molecules under mild conditions.

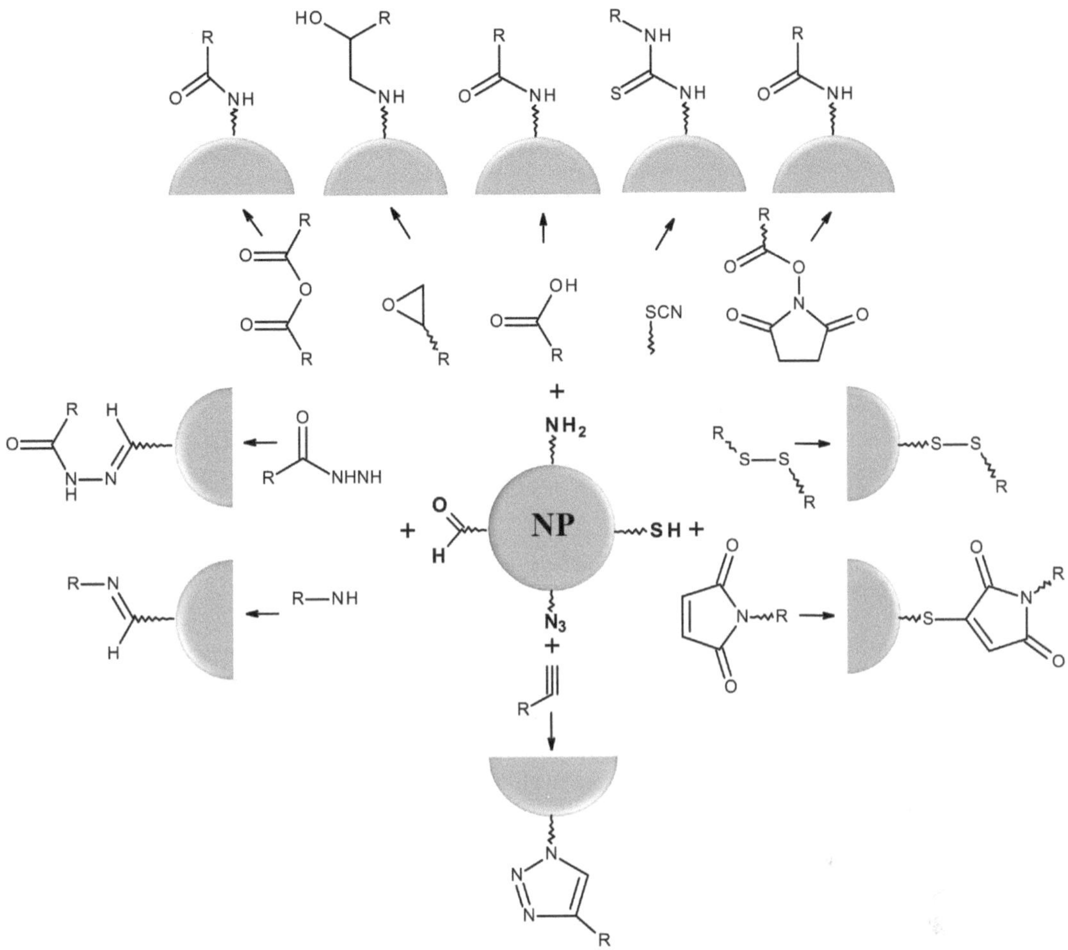

FIGURE 1–3 Strategies for postfunctionalization of nanoparticles using various functional groups.

1.3 Modification of Nanoparticles by Polymer Chains

The modification of NPs surface with polymer chains can lead to higher compatibility of the NPs with polymer matrix, [74] especially when the nature of the grafted polymer chains on the NPs surface is the same as or similar to the polymer matrix. The polymer chains bonded on the NPs surface interact with the matrix at the interface and affect the matrix performance. The polymer grafting enables to vary thickness of the polymer layer and structure, as well as density and thus the interphase ratio effectively.

1.3.1 Modification Using "Grafting to" Technique

The "grafting to" method involves the attachment of the polymer chains to reactive sites present on the NPs surface (Fig. 1–4A). This approach provides rather low grafting density,

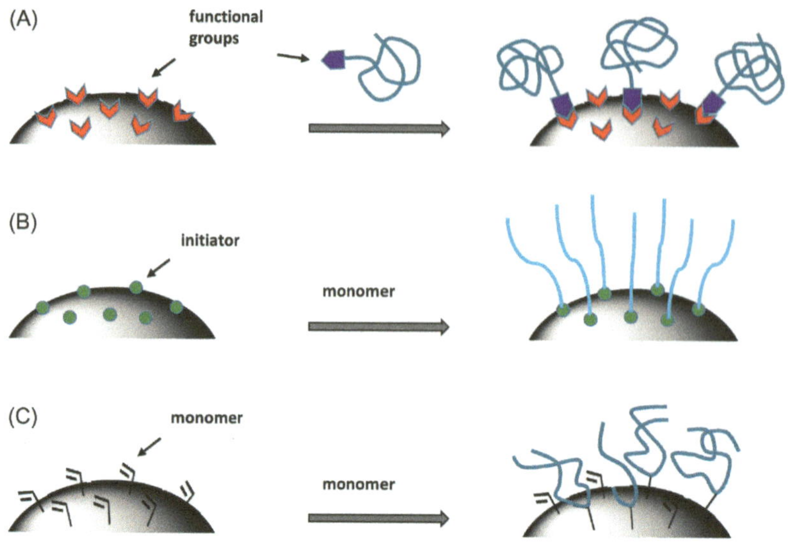

FIGURE 1–4 Schematic illustration of NPs surface modification using (A) "grafting to," (B) "grafting from," and (C) "grafting through" process.

due to steric hindrance having already attached polymer chains that prevents diffusion of the other polymer chains toward the free reactive sites on the NPs surface. On the other hand, functionalized polymer can be carefully characterized before the immobilization onto the surface, resulting in preparation of well-defined brushes [75]. Only polymer chains containing reactive functional groups can be used in this technique. Various commercial polymers with well-defined structure, possessing some functional groups available for reaction with the functional NPs surface, can be used. In general, many polymerization techniques, including cationic [76], anionic [77], ring-opening [78], or RDRP [79] can be applied for preparation of functional polymers for their subsequent grafting to the surface of NPs.

Generally, both direct functionalization and two-step modification using all functional groups described in the part 1.2 and shown in Fig. 1–2 and Fig. 1–3 can be applied for attachment of polymer chain by "grafting to" method. Some limitations can occur in preparation of functional polymers due to intolerance of different living polymerization techniques toward certain functional groups.

Covalent attachment of end-functionalized polymers can be performed frequently by thiols and disulfides for noble metal NPs or by silanes and phosphates for metal oxides. Monolayer-protected nanometer Au clusters based on thiolated poly(ethylene glycol) (PEG) was synthesized to improved core stabilization as well as to induce aqueous solubility of the NPs [80]. Polymer-stabilized Au NPs with relatively high grafting density have been prepared by absorption of thiol terminated polystyrene (PS) and (PEG) ligands [81–83]. Amphiphilic Au NPs were successfully prepared by "grafting to" method [84] using a mixture of poly(*N*-isopropylacrylamide) and PS chains at various compositions. In this case, the polymers were originally prepared by reversible addition-fragmentation transfer polymerization (RAFT) and the RAFT agent at the polymer chain end was reduced to thiol chain end groups, allowing the attachment of the polymers onto the surface of Au NPs formed in situ. The surface of Au NPs was efficiently modified also with hydrophilic nonionic polymers bearing terminal disulfide groups [85]. The resulting polymer monolayer allowed optimal water solubility of gold colloids as well as free diffusion of small molecules, with suppressed absorption of the large molecules such as proteins.

PEGylated silicon NPs were prepared by hydrosilylation of the Si–H terminated silicon NPs with alkene-terminated PEG in the presence of chloroplatinic acid catalyst [86]. PEGylated NPs can be stored as a solid material and then readily redispersed in good solvents for PEG, such as water, methanol, or chloroform. Another approach of silica NPs surface modification relies on the photochemistry of perfluorophenylazides (PFPA) [87]. The NPs were first functionalized with PFPA for introducing the photoactive group onto the surface. Irradiation of PFPA generated highly reactive perfluorophenylnitrenes that subsequently underwent C–H, N–H insertion, and C = C addition reactions with polymer chains. The similar methodology was used also for the preparation of gold NPs grafted with low density polyethylene (PE) [88]. Zhao and coworkers synthesized chain end–functionalized homopolymers by RAFT polymerization of vinyl monomers, such as methyl acrylate (MA), *n*-butyl acrylate (BA), *N,N*-dimethylacrylamide, *N*-isopropylacrylamide (NIPAM), *N*-acrylomorpholine, or methyl methacrylate (MMA). These polymers contained trimethoxysilane groups, which were used for "grafting to" silica NPs by coupling reaction between trimethoxysilane and hydroxyl groups on the NPs surface [89].

Development of various click chemistry system "grafting to" methods has become very efficient for the preparation of various grafted NPs. The combination of RDRP and click chemistry permits surface modification in a controlled manner. The alkyne-terminated trithiocarbonate transfer agent was used for RAFT polymerization to prepare alkyne chain end-functionalized PS and polyacrylamide (PAM). Subsequently, the alkyne chain end groups were used for silica NPs modification by click reaction with azide groups present on the silica NPs surface [90].

Tsubokawa reported surface grafting of polymers onto silica NPs in a solvent-free dry system [91]. In this study, amino functionalized silica NPs were coated with hyperbranched poly (amidoamine). Silica NPs coated with polymers with controlled molar mass were prepared by the termination of living macrocation with amino groups introduced onto silica surface [92]. Generally, the amino or *N*-phenylamino groups represent the suitable functional groups for "grafting to" techniques. It was shown that amino groups on silica can be used also for grafting of poly(isobutyl vinyl ether) or poly(2-methyl-2-oxazoline).

Polymer-coated cobalt NPs were synthesized by the thermolysis of dicobaltoctacarbonyl (CO_2CO_8) in the presence of end-functionalized PS [93]. Functionalities of the PS chains consisted of either a benzylamine or dioctylphosphine oxide groups. Such prepared ferromagnetic cobalt colloids afforded uniform particle size and long term colloidal stability [94–96]. Two different approaches were used in the study related to functionalized silica particles with poly(acrylic acid) (PAA) [97]. In the first approach, PAA was reacted with 1,1′-carbonyldiimidazole to give functionalized polymer. The resulting active carbonyl group reacted readily with 3-aminopropyl-functionalized silica NPs at low temperatures. The second approach used the magnesium chloride to catalyze the reaction between PAA and 3-glycidoxypropyl-functionalized silica NPs.

Neat or oxidized multi-walled carbon nanotubes (MWCNTs) were grafted by PS or PMMA chains significantly modifying their solubility and their suitability for nanocomposite applications. The grafting was obtained in situ during radical polymerizations, while MWCNTs reacted either directly with growing polymer radicals or with radicals formed by decomposition of radical initiator and subsequent coupling of radicals formed on the MWCNTs surface with growing polymer radicals [98]. However, reaction of the radicals formed on the MWCNTs surface with monomer providing "grafting from" method cannot be excluded. The achieved grafting efficiency was around 0.5% for the benzoyl peroxide initiating system, but reached 18% in the case of the potassium persulfate initiator [99,100].

1.3.2 Modification Using "Grafting From" Technique

The "grafting from" approach is based on polymer growth from the initiators bound on the NPs surface, therefore it is often called surface initiated (SI) polymerization (Fig. 1–4B). Controlled/living polymerization techniques including cationic, anionic, ring-opening, and RDRPs, such as nitroxide-mediated radical polymerization (NMP), atom transfer radical polymerization (ATRP), and RAFT can provide considerable control over polymer grafting density as well as dispersity and molar mass of the surface grafted polymer and thus over the polymer shell thickness. Using these techniques all the polymer chains grow simultaneously from the NPs surface and polymer chains with high degree of polymerization and densely packed on the NPs surface can be obtained. Additionally, the combination of these polymerization techniques applicable for polymer grafting was successfully used for modification of various surfaces leading to formation of multilayered core–shell NPs with well-defined structures as described below.

1.3.2.1 Cationic Polymerization

Covering of ultrafine silica with poly(ε-caprolactone) (PCL) and PS was successfully carried out via cationic ROP and graft polymerization initiated by benzylium perchlorate attached on the silica surface. Initiating sites on silica were formed through treatment with 4-(chloromethyl)phenyltrimethoxysilane followed by reaction with silver perchlorate in nitrobenzene. The polymer content grafted on silica was significantly affected by polymerization temperature due to accelerated chain transfer reactions of polymer cations at higher temperatures [101]. Grafting polymerization of isobutyl vinyl ether, *N*-vinylcarbazole, γ-butyrolactone, and 2,3-dihydrofuran was initiated by oxoaminium perchlorate groups attached onto silica surface. Oxoaminium perchlorate groups were introduced on the surface of silica by reaction of 4-hydroxy-2,2,6,6-tetramethylpiperidinyloxy radicals with acid anhydrides followed by chlorination with perchloric acid [102].

Jordan et al. [103] reported the preparation of end-functionalized self-assembled monolayered Au NPs. Modified Au NPs were prepared by reduction with lithium triethylborohydride and used for initiation of the living cationic ring opening polymerization of 2-ethyl-2-oxazoline. Polymerization was terminated by addition of *N,N*–di-*n*-octadecylamine resulting in well-defined amphiphilic core–shell NPs. The developed procedure can be used for a preparation of metal particles bearing various functional groups especially due to high purity and surfactant-free conditions.

Tsubokawa [104] described the grafting of poly(tetrahydrofuran) on carbon black (CB) bearing acid chloride groups via cationic polymerization under anhydrous ferric chloride catalysis. A similar approach was used for preparation of CB grafted with PS [105,106] and δ-valerolactone/ε-caprolactone/β-propiolactone [107] using $CO^+ClO_4^-$ as initiating species formed by reaction of surface tethered acyl chloride with $AgClO_4$. The authors extended the study on modification of carboxyl functionalized carbon fibers and were able to initiate the cationic polymerization of *N*-vinylcarbazole [108] and *N*-vinyl-2-pyrrolidone [109]. Initiating activity was associated with carboxyl functional group, which after blocking with bicarbonate/diazomethane became inactive. The yield of grafted polymer as a grafted ratio calculated from polymer loss after extraction with tetrahydrofuran was approximately 40%.

Metallocene coordination catalysts based on titanium catalytic systems were recently used for effective grafting of PE on CNTs. Tong et al. [110] modified the single-walled carbon nanotubes (SWCNTs by grafting with PE via in situ polymerization using Ziegler–Natta $MgCl_2/TiCl_4$ catalysis immobilized on the SWCNT surface. Park et al. [111,112] used for functionalization of MWCNTs with PE chains by grafting from method a Cp_2ZrCl_2 catalyst immobilized on the surface based on electronic interactions between CNT sidewalls and Cp ligands.

1.3.2.2 Anionic Polymerization

Oosterling et al. [113] modified silica NPs by grafting with PS and block copolymer poly(styrene-*co*-isoprene) via anionic living polymerization initiated by *tert*-butyllithium (*t*-BuLi) from the surface. First vinylbenzyltrichlorosilane was attached onto the NPs surface. The monomer was added to the system half an hour after addition of *t*-BuLi to ensure initiation of all double bonds at the surface. PS-grafted silica NPs were prepared also using 1,1-diphenylenethylene (DPE) as initiating precursor [114]. After its functionalization with alkyldimethylchlorosilane, the DPE was successfully grafted on the silica surface. Addition of *n*-BuLi activated silica-tethered initiator leading to anionic polymerization of styrene.

Joubert et al. [78] coated silica NPs via in situ anionic ring opening polymerization (AROP) of cyclic monomers such as ε-CL and L-lactide (L-LA) using tin, yttrium, and aluminum alkoxides catalysis. The polymerization was initiated by surface attached hydroxyl groups originated from bare silica or prehydrolyzed with (3-glycidoxypropyl)trimethoxysilane. Pretreatment with trimethoxysilane derivative in combination with tin alkoxide catalysis led to significant increase in grafting density resulting in poly(L-lactide) (PLLA) content up to 96 wt.%. Bentz et al. [115] applied the same approach for tuning the PCL graft density on nanosized silica particles based on initial treatment with (3-glycidoxypropyl)dimethylethoxysilane. Resulted PCL brushes length varied from 1 up to 84 nm depending on the molar mass of the grafted chains.

Schmidt et al. [116] described preparation of Fe_3O_4 magnetic NPs grafted with PCL brushes. The AROP was initiated by surface immobilized hydroxyls formed after chemisorption of glycolic acid. This strategy provided biodegradable, magnetic, and biocompatible core–shell hybrid particles with diameters varied between 5 and 15 nm. The same route of magnetic NPs functionalization was used for preparation of novel Fe_3O_4–PCL hybrids capped with carbazole tested in drug release as novel multifunctional delivery system with superparamagnetism and fluorescence properties [117].

Zhao et al. [118] grafted nanodiamonds via AROP of glycerol resulting in highly soluble NPs with hyperbranched polyglycerol. Prepared hybrids showed high stability and solubility in pure water as well as in buffer. This study was recently extended to conjugation of polyglycerol grafted nanodiamond with gadolinium (III) complex for preparation of novel magnetic resonance contrast agent partially soluble in physiological media [119]. Multihydroxyl polyglycerol was used for covalent attachment by grafting from quantum dots [120], MWCNTs [121], and Fe_3O_4 [122] surfaces due to its high biocompatibility, hydrophilicity, and possibility of further functionalization by fluorescent coupling applicable in the area of drug delivery and bioprobing.

Braun et al. [123] used *t*-BuLi and tetramethylethylenediamine to form initiating moieties on the CB for subsequent anionic polymerization of styrene. Viswanathan et al. [124] applied the activation of CNTs with *sec*-BuLi for in situ composite preparation by grafting of PS chains from untreated SWCNTs surface without disrupting the original conjugated structure. They showed that an anionic polymerization can be initiated by CNT carbanions produced by prior reaction with *sec*-BuLi. The molar mass of PS-grafted chains ranged from 1600 to 100,000 g/mol with narrow dispersities of about 1.02 but with low degrees of functionalization (approximately 10 wt.%). SWCNTs-bearing carbanions were also used for initiation of anionic polymerization of *tert*-butyl acrylate (*t*BA), MMA, and the block copolymerization of *t*BA with MMA. Chen et al. [125] found that the sensitivity of *t*BA and MMA to carbanion nature since contrary to *t*BA the homopolymerization of MMA did not occur. On the other hand the active polymeric *t*BA chains could further initiate the polymerization of MMA leading to the formation of diblock P*t*BA-*g*-PMMA copolymer chains.

For grafting of epoxide copolymers such as styrene oxide, chloromethyloxirane, and glycidyl phenylether with cyclic anhydrides, the potassium carboxylate (COOK) functionalized carbon fibers were used as initiating species [126]. This type of initiation was also applied for effective covering of CB with polyesters based on glycidyl methacrylate/succinic anhydride [127], styrene oxide/maleic anhydride [127], glycidyl methacrylate/phthalic anhydride [128], and β-propiolactones [129,130]. The initiating efficiency of alkali metal carboxylate groups increased in the order of $Cs > Rb > K > Na > Li$ [127].

Yang et al. [131] modified CB by grafting with PCL via AROP resulting in core–shell NPs with controlled crystallization orientation. Atomic force microscopy (AFM) and transmission electron microscopy (TEM) showed that grafted CB were oriented in the PCL matrix leading to lamella formation with a dendritic pattern. The similar dendritic lamellae structures were observed after modification of CB by grafting of tree-like block copolymer brushes based on poly(2-hydroxyethyl methacrylate) (PHEMA) grafted by PCL (PHEMA-*g*-PCL) prepared using combination of SI ATRP of 2-hydroxyethyl methacrylate (HEMA) and AROP of ε-CL [132].

Zheng et al. [133] reported preparation of PCL grafted on MWCNTs via AROP of ε-CL in the presence of tin octoate ($Sn(Oct)_2$) as a catalyst. Thanks to the PCL shell, the MWCNT-PCL nanohybrids showed good solubility/dispersibility in organic solvents. The biodegradation experiments showed that the PCL grafted on MWCNTs can be completely enzymatically degraded within 4 days. A similar approach was used for grafting of CNTs with PCL to prepare the vapor sensors for organic nonpolar and polar solvents. Here the ROP of ε-CL was mediated by trimethylaluminum anchored on the surface after reaction with hydroxyl functionalized MWCNTs activated by ultrasonication while the polymer shell with diameter approximately 13 nm in height was obtained [134]. Thicker polymer shell was obtained when hydroxyl groups were attached onto the CNTs surface using 2-hydroxyethyl benzocyclobutene (BCB-EO) via a [2 + 4] cycloaddition reaction as described by Priftis et al. [135]. They studied SI titanium-mediated coordination polymerization of L-LA, ε-CL, and *n*-hexyl isocyanate. The same type of modification based on benzocyclobutenes initiation was used for production of Janus type MWCNTs structures. Through the one-pot method the simultaneous growth of two different homopolymers such as PS/PMMA via ATRP and PCL/PLLA via AROP was achieved. Generation of binary polymer brushes, which can phase segregate under appropriate conditions, are applicable as stimuli-responsive systems [136]. The authors [137] used the mentioned initial functionalization with BCB-EO for MWCNTs modification by diblock copolymers via combination of AROP catalyzed by $Sn(Oct)_2$ and subsequent RDRP techniques. The hydroxyl-ending groups of PCL were transformed to isopropylbromide initiating species for subsequent ATRP of styrene or 2-dimethylaminoethyl methacrylate. This study was extended to modification of the MWCNTs by grafting with PEG brushes. The polymer contents grafted on the surface calculated from thermogravimetric analysis (TGA) were 98 wt.% and 90 wt.% for PCL and PEG, respectively. The height of the polymer coating determined by TEM reached 20 nm [138]. Sakellariou et al. [139] used covalently attached BCB-EO on the MWCNTs surface as precursor for generation of alkoxy and alkyllithium anions by reaction with triphenylmethane and *sec*-butyllithium. Using this approach the PEG and PS chains were grown from the MWCNTs surface via anionic polymerization technique. Polymer grafted nanocomposites MWCNTs-*g*-$(BCB\text{-}PEG)_n$ and MWCNTs-*g*-$(BCB\text{-}PS)_n$ contained a very high percentage of hairy polymer chains ($>$ 90 wt.%), while polymer shell of thickness up to 30 nm was obtained.

Chen et al. [140] modified the surface of oxidized MWCNTs by PLLA brushes with thickness of about 5 nm using AROP catalyzed by tin 2-ethylhexanoate ($Sn(Oct)_2$). Brzezinski et al. [141] studied thermal properties and crystallization behavior of enantiomeric polylactide (PLA) contained 0.5–1 wt.% of MWCNTs grafted with PLLA and/or poly(D-lactide) (PDLA) and showed that the stereocomplex formation was completely reversible with the absence of any homochiral crystallites. Diazonium coupling reactions of 4-hydroxymethylaniline for hydroxyl functionalization of SWCNTs were also used for subsequent grafting with PCL brushes [142]. The organic content determined by TGA reached the 63 wt.% leading to significant increase of dispersibility in chloroform. Lee et al. [143] modified the MWCNTs by combination of ROP of ε-CL/α-Cl-ε-CL and 1,3-dipolar cycloadditions. SI ROP of α-Cl-ε-CL led to formation of pendant chlorides, which subsequently reacted with sodium azides to produce reactive species for anchoring terminal alkynes, such as 2-propynyl benzoate or 1-hexyne.

Similarly, Xu et al. [144] described preparation of novel dendritic/treelike structured core–shell MWCNTs NPs through ROP of 3-ethyl-3-(hydroxymethyl)oxetane. Polymerization was initiated by hydroxyl functionalized MWCNTs and carried out under $BF_3 \cdot Et_2O$ catalysis.

1.3.2.3 Nitroxide Mediated Polymerization

NMP is the most used representative polymerization technique of so-called stable FRPs, which are based on reversible homolytic cleavage of a relatively weak covalent bond to generate a growing radical and a less reactive species (Fig. 1–5). These species should react reversibly only with growing radicals and should not react either among themselves or with monomers to form new growing radicals. In NMP the species reversibly reacting with growing radicals are based on nitroxides. In NMP technique either the source of radicals attached on the surface is combined with free nitroxide (Fig. 1–5, process A) or free source of radicals is combined with nitroxide covalently attached on the surface (Fig. 1–5, process B), even though process B should rather belong to the "grafting to" method, since the surface attached nitroxide serves for reversible reaction with polymer chains formed in bulk. In addition, the initiation can also be performed only by thermal decomposition of alkoxyamines covalently bound onto the surface without any additional source of radicals (Fig. 1–5, process C). Usually free (sacrificial) alkoxyamine and/or nitroxide is also used to decrease the local concentration of polymer radicals on the NPs surface and thus decrease the termination of radicals and improve the control over the polymerization process.

Silica NPs were grafted by poly(ethyl acrylate) (PEA), poly(*n*-butyl acrylate) (PBA), PS, and PBA-*b*-PS block copolymer brushes using silica-tethered diazo or peroxide initiators in the presence of *N*-tert-butyl-*N*-(1-diethylphosphono-2,2-dimethylpropyl) nitroxide (SG1)

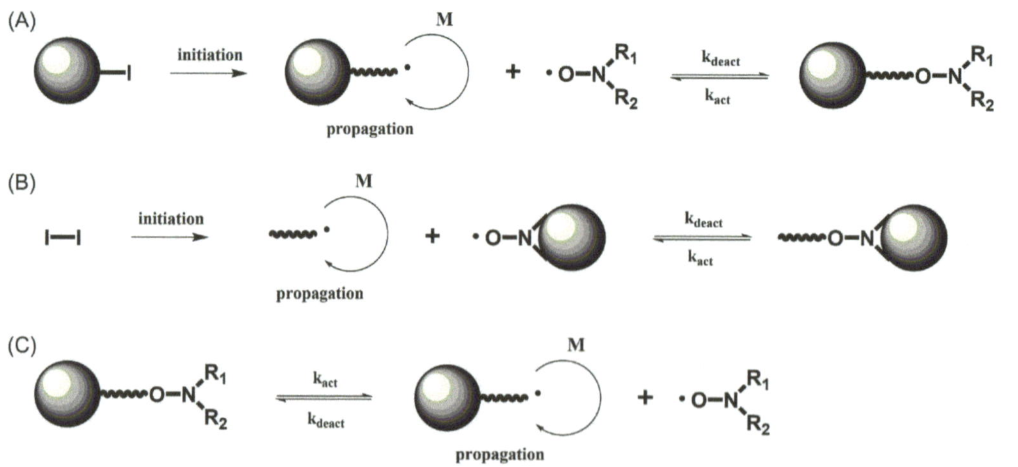

FIGURE 1–5 Schematic illustration of nanoparticles surface modifications by polymers using nitroxide mediated polymerization.

and/or its alkoxyderivative [145,146]. Similar modification of silica NPs was performed also using silica-tethered SG1-based alkoxyamine [147–149] in the presence of free alkoxyamine and/or nitroxide. Grafted PS chains with grafting density of 0.35 μmol/m^2 and controlled molar mass and dispersity up to 1.3 were obtained. Blomberg et al. [150] reported the preparation of poly(styrene-*co*-maleic anhydride) modified silica NPs using surface-bonded alkoxyamine derivative based on 2,2,5-trimethyl-4-phenyl-3-azahexane-3-nitroxide (TIPNO) The polymeric shell of the prepared hybrid NPs was subsequently crosslinked via interchain coupling reaction of maleic anhydride units with 2,2-(ethylenedioxy)bis(ethylamine).

PS was grafted by NMP also from monodisperse ZnS-SiO_2 particles [151]. Polymerization was carried out by 2,2,6,6-tetramethyl-1-piperidinyloxy (TEMPO) based alkoxyamine bonded to the NPS surface by silyloxy groups. NMP was performed in the presence of sacrificial alkoxyamine and polymer brushes of various molar masses ($M_n = 30–114$ kg/mol) and high grafting density (0.9 chains to nm^{-2}) were obtained. Matsuno et al. [152] described modification of magnetite NPs by direct grafting of PS and poly(3-vinylpyridine) from their surfaces using chemisorbed-based alkoxyamine bearing a phosphoric acid group in the presence of sacrificial alkoxyamine. The obtained polymer graft density on the Fe_3O_4 varied from 0.12 to 0.20 chains per nm^2, while the average size of the aggregates decreased from more than 300 nm before grafting down to about 100 nm for grafted NPs.

NMP was used also for grafting of PS chains from MWCNTs surface [153] using MWCNTs-supported alkoxyamine based on TEMPO. The obtained PS chains grafted from the MWCNTs surface were further used for chain extension polymerization of 4-vinylpyridine resulting in PS-*b*-poly(4-vinylpyridine)-grafted MWCNTs.

1.3.2.4 Atom Transfer Radical Polymerization

The basic mechanism of ATRP involves homolytic cleavage of, usually, alkyl halide bond R-X (or P_n-X) by a transition metal complex Mt^n/L (activator), while the corresponding higher oxidation state metal halide complex X-Mt^{n+1}/L (deactivator) and an alkyl radical $R^{\cdot}$ (or $P_n^{\cdot}$) are formed (Fig. 1–6). The radical can propagate with a monomer, terminate, or can be reversibly deactivated by X-Mt^{n+1}/L. In SI-ATRP the initiation is through the activation of alkyl halides bonded on the surface, while sacrificial alkyl halide initiator is usually also added to the system for the same purpose as described for the NMP technique. ATRP is currently the most widely used RDRP technique, also thanks to commercial availability of all necessary ATRP reagents including transition metal compounds, ligands, and alkyl halide initiators as well as thanks to the large scale of monomers polymerizable by this technique and wide range of conditions under which the polymerization can be performed. Therefore,

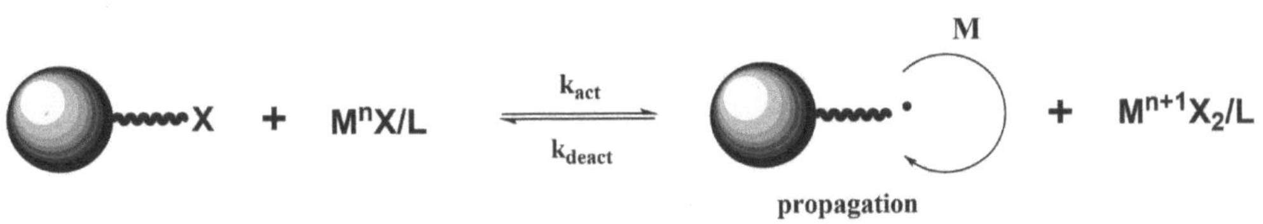

FIGURE 1–6 Schematic illustration of surface initiated atom transfer radical polymerization.

there is also a significantly higher number of works describing modification of NPs surface using ATRP compared with other RDRP techniques.

Huang and Wirth [154] used 1-trichlorosilyl-2-(*m-p*-chloromethylphenyl) ethane to immobilize ATRP initiator on the silica NPs and to form the surface-bound benzyl radicals by reaction with $Cu(bpy)_2Cl$ catalyst. The molar mass and dispersity PAM grafted from this initiator and determined after subsequent cleavage from the surface was 13−15 kg/mol and 1.5, respectively. Halogenated monosiloxane derivative of ATRP initiator was immobilized on the spherical silica NPs surface also by Patten et al. [155] who subsequently modified the NPs by PS and PMMA chains. They showed that the structure of the resulting polymer hybrids (grafting density, molar mass) can be controlled by setting of initial polymerization conditions. Harrak et al. [156] used two-step modification of silica NPs by ATRP initiator. Initially, the silica NPs were modified by mercaptopropyl triethoxysilane resulting in surface attached thiols. In the second step the thiols were transformed to bromide groups via reaction with 2-bromoisobutyryl bromide, which was used for SI-ATRP of styrene. Wang et al. [157] grafted the silica NPs with PMMA and blocks PMMA-*b*-PS via ATRP initiated by peroxide groups attached on the surface. Silica NPs were first treated with thionyl chloride to obtain the chlorosilyl groups and subsequently reacted with hydrogen peroxide resulting in peroxo-functionalized silica. Böttcher et al. [158] used a 2-chloro-2-phenylacetate initiator type immobilized on silica NPs for ATRP of styrene. It was shown that grafted PS layer bearing the Cl ending groups were still active and can be used for initiation of further chain growth. 2-Bromoisobutyrate-based siloxane is the most often used for immobilization of ATRP initiator on the various silica NPs surface [159]. It was used in numerous works for modification of the NPs by PS, PMA, and PBA [160−162]; various hydrophilic homopolymers as well as random or block copolymers of styrene/benzyl acrylate [163]; styrene/BA/MMA [164] and acrylonitrile [165]; and styrene/MMA [166]. The same immobilized initiator was used also for preparation of thermoresponsive silica hybrid NPs through grafting with poly(*N*-isopropylacrylamide) (PNIPAM) [167]. Hayashida et al. [168] prepared silica NPs grafted with poly(*n*-butyl methacrylate) (PBMA) via SI-ATRP, while the initiator was bonded on the NPs surface in two steps. First, phenyl amine derivative containing silyloxy groups was anchored on the surface and subsequently used for reaction p-(bromomethyl)benzyl 2-bromoisobutyrate. Polymerization was carried out in the absence of sacrificial initiator resulting in PBMA grafted silica NPs with dispersity as narrow as 1.1 and graft density of 0.54 chains/nm^2. Prepared core−shell hybrids were used for preparation of composites with poly(vinyl chloride).

Mori et al. [169] reported synthesis of hyperbranched polymer-silica hybrid NPs using covalently linked α-bromoester ATRP initiator. Self-condensing vinyl polymerization of an 2-(2-bromopropionyloxy)ethyl acrylate "inimer" (bearing both polymerizable acrylic and an initiating groups) was carried out using Cu^IBr/*N,N,N′,N″,N″*-pentamethyldiethylenetriamine (PMDETA) catalytic system in bulk. The obtained grafting density of the final polymer was approximately of 0.41 chains/nm^2. Copolymerization of inimer with *t*BA was performed for subsequent removal of *tert*-butyl group and forming of PAA copolymer for a variety of possible applications as biomaterials carriers, thermal reinforcement, and optical chemical sensing. Li et al. [170] used a bifunctional initiator anchored on the silica NPs surface for their modification by combination of ATRP and NMP resulting in mixed P*t*BA/PS brushes with possibility to produce amphiphilic hybrid NPs with PAA/PS brushes.

To prepare silica NPs with grafted high molar mass PS, Bombalski et al. [171] applied for grafting process an activator regenerated by electron transfer ATRP (ARGET-ATRP) technique, which uses only ppm amounts of metal catalyst in higher oxidation state in combination with reducing agent. Using 2-bromoisobutyrate-based silane initiator provided modification of SiO_2 NPs with grafting density up to 0.71 chains/nm^2 and PS chains with M_n varied from 1 to 14 kg/mol and dispersity in the range of 1.08–1.32.

Farmer et al. [172] grafted quantum nanodots/SiO_2 NPs with PMMA via SI-ATRP from attached 3-(2-bromopropionyloxy)propyl dimethylethoxysilane under $NiBr_2(PPh_3)_2$ catalysis.

ATRP technique was also used for preparation of core–shell hybrids by grafting of PMMA [173–175], PNIPAM [176], and poly(4-vinylpyridine) [177] from gold NPs. Generally, an ATRP initiator, such as 11,11′-dithiobis[1-(2-bromoisobutyryloxy)undecane] was attached on the gold surface allowing the grafting process from the gold surface. Kim [176] and Li [177] used attached initiator bearing the 2-bromoisobutyryl species by ligand exchange between disulfide initiator and citrates. Mandal et al. [173] modified the Au NPs first by 11-mercaptoundecanol and subsequently esterified with 2-bromoisobutyryl bromide. Ohno et al. [174] described the one-pot synthesis of initiator-anchored Au NPs by reduction of tetrachloroaurate with sodium borohydrate or reaction with citric acid trisodium salt [175] in the presence of disulfide initiator. Dong et al. [178] modified Au NPs with dual polymer layer via ATRP of dimethacrylate-based crosslinker resulting in inner polymer shell in the first step. In the second step outer linear PBA brushes were grown from the crosslinked inner shell via subsequent ATRP.

Li et al. [179] covered nanodiamond particles via ATRP of isobutyl methacrylate and *t*BA resulting in hydrophobic/hydrophilic ultradispersed diamond with calculated surface density of approximately 0.05 chains/nm^2. ATRP initiators based on hydroxylethyl-2-bromoisobutyrate or 2,2,2–trichloroethanol were covalently coupled on the NPs oxidized surface trough esterification reaction. Dahoumane et al. [180] prepared bioactive nanodiamond particles grafted by hairy protein brushes via ATRP of *tert*-butyl methacrylate (*t*BMA). Fluorescent nanodiamonds with attached ATRP initiator based on 1-bromoethylphenyl species were prepared via diazonium salt chemistry. An obtained P*t*BMA brushes were subsequently hydrolyzed to poly(methacrylic acid) for further coupling with bovine albumin. In similar way, the nanodiamond surface was grafted with poly(poly(ethylene glycol)methylether methacrylate (poly(PEGMA)) [181].

Gelbrich et al. [182] described preparation of thermoresponsive core–shell magnetic NPs based on Fe_3O_4 covered with covalently attached poly(2-methoxyethyl methacrylate) brushes using 2-bromoisobutyric acid anchored on the NPs surface as an initiator. Polymer modified NPs formed stable dispersion in methanol above critical solution temperature (>20°C). Yuan et al. [183] used the same anchored initiator for preparation of Fe_3O_4 NPs grafted with branched polymeric shell consisted of PHEMA-*g*-PCL block copolymer brushes via combination of ATRP and AROP. PCL chains were grown from the pendant hydroxyl groups of grafted PHEMA. It was shown that rate of degradation of PCL brushes depended on PCL length and could be adjustable as well as applicable in targeted delivery systems for anticancer therapy. Zhou et al. [184] prepared organic/inorganic hybrid magnetic NPs via SI ATRP of (2-dimethylamino)ethyl methacrylate (DMAEMA) and produced the novel pH-sensitive drug carriers with controllable release. The authors showed that ratio of drug release is higher at lower pH due to expanding conformation of grafted PDMAEMA brushes leading to easier drug diffusion. Vestal and Zhang [185] prepared $MnFe_2O_4$ magnetic core–shell NPs grafted with PS using ATRP initiated from 3-chloropropionic acid anchored on the NPs surface.

Qin et al. grafted the SWCNTs by PS [186] and PBMA [187] and Wu et al. [188] by PBA brushes via SI-ATRP using 2-hydroxyethyl 2-bromopropionate as an initiator covalently attached on the oxidized SWCNTs surface by esterification reaction. For better control of chain propagation and kinetics study, the sacrificial initiator methyl-2-bromopropionate was also used. The molecular characteristics of the cleaved PS ($M_n = 34$ kg/mol, dispersity < 1.30) were comparable with that for PS grown from the sacrificial initiator indicating living/controlled character of SI polymerization. The same covalently attached initiator was used by Baskaran et al. [189] and Shanmugharaj et al. [190] for grafting from MWCNTs surface to prepare homo and block copolymer brushes of PS, PMMA, PAN, and PS-*co*-PAN. The amount of grafted PS-*co*-PAN (22.6 wt.%) was significantly lower compared with PS (63.4 wt.%) and PAN (64.5 wt.%). The decrease in copolymer content was attributed to different radical reactivities and donor–acceptor interactions between acrylonitrile and styrene. Yao et al. [191] functionalized the SWCNTs with ATRP initiator using a two-step procedure involving first a 1,3-dipolar cycloaddition reaction using octanal and 4-hydroxyphenyl glycine resulting in phenol-functionalized CNTs, followed by an esterification with 2-bromiosobutyryl bromide. The SWCNTs served then for grafting of PMMA and P*t*BA chains. Similarly, Kong et al. modified MWCNT by ATRP initiator by reaction of hydroxyl functionalized MWCNTs with 2-bromoisobutyryl bromide. ATRP was subsequently used for preparation of core–shell hybrid nanostructure with PS [192] or amphiphilic PS-*b*-PAA [193] brushes. MWCNTs-PMMA were also prepared and used as macroinitiator for followed polymerization of HEMA resulting in MWCNTs grafted with PMMA-*b*-PHEMA nonpolar/polar layer [194]. Mosnacek et al. [195] functionalized CNTs by diazonium chemistry, that is, using 4-aminophenethyl alcohol in the presence of isoamyl nitrite, to introduce hydroxyl groups on the CNTs surface followed by esterification with 2-bromopropionyl bromide resulting in surface attached ATRP initiator. The CNTs were then modified with PS chains via SI-ATRP. Finally, the active PS chain ends were used for coupling with fluorescent nitroxides. The fluorescent labeling of PS grafted CNTs enabled visualization of the CNTs using confocal laser scanning microscopy and evaluation of their distribution in the nanocomposite.

Fang et al. [196] used the same approach of functionalization through the diazonium chemistry for modification of reduced GO with subsequent SI-ATRP of styrene. The same strategy but using 2-bromoisobutyryl bromide was applied also for grafting of 2-ethyl(phenylamino)ethyl methacrylate [197] as well as *t*BMA via ATRP mediated by Cu(0)/tris[2-(dimethylamino)ethyl]amine (Me_6TREN) [198]. Yang et al. [199] used for attachment of the ATRP initiator a different strategy, based on amidation of carboxyl functionalized GO with 1,3-diaminopropane and subsequent attachment of the initiator. The initiator functionalized

GO was subsequently used for grafting DMAEMA in the presence of sacrificial initiator. Goncalves et al. [200] used for surface attachment of 2-bromoisobutyryl bromide initiator on the GO surface by reaction with ethylene glycol reacted first with acyl chloride groups generated via initial chlorination. ATRP of MMA was carried out without addition of sacrificial initiator with relatively good control over molar mass and dispersity <1.1.

1.3.2.5 Reversible Addition-Fragmentation Chain Transfer Polymerization

RAFT polymerization has been shown to be a highly adaptable and widely applicable RDRP method for a wide range of monomers, including less activated monomers, using a broad range of initiation processes, solvents, and reaction temperatures. The main advantage of RAFT polymerization compared with the traditional ATRP is in the possibility to control the polymerization without using metal catalyst and better control in synthesis of water-soluble polymers [201]. In RAFT and other "degenerative transfer processes," conventional free radical initiators are used and control is assured by the presence of chain transfer agents (CTAs), which exchange a group/atom between all growing chains (Fig. 1–7). The steady state concentration of radicals is established via initiation and termination processes. RAFT mostly uses dithioesters, dithiocarbamates, trithiocarbonates, and xanthates as CTAs. General strategies for SI-RAFT polymerization can be achieved in two different ways: (1) covalent anchoring of free radical azoinitiator onto the surface in combination with free CTA [202–204], (2) surface CTA attachment followed by SI-RAFT initiated by free radical azoinitiator [79,201,205–217]. The CTA can be immobilized to the surface via either the Z group approach [205–209] or the R group approach [79,201,210–217] (Fig. 1–7). In the Z group approach, which is actually "grafting to" method, the stabilizing Z group of CTA is permanently fixed on the solid surface so that the polymer radicals always propagate in solution before they attached to surface via chain transfer reaction with immobilized CTA. These chain transfer reactions can be severely hampered due to steric hindrance of neighboring attached polymer chains leading to a lowered graft density. Whereas in the R group approach, the CTA is attached to the surface through R leaving group and the propagating radical is located at the terminal end of the polymer chain. An attachment via leaving and reinitiating R group always leads to detachment of CTA during the polymerization. This is an

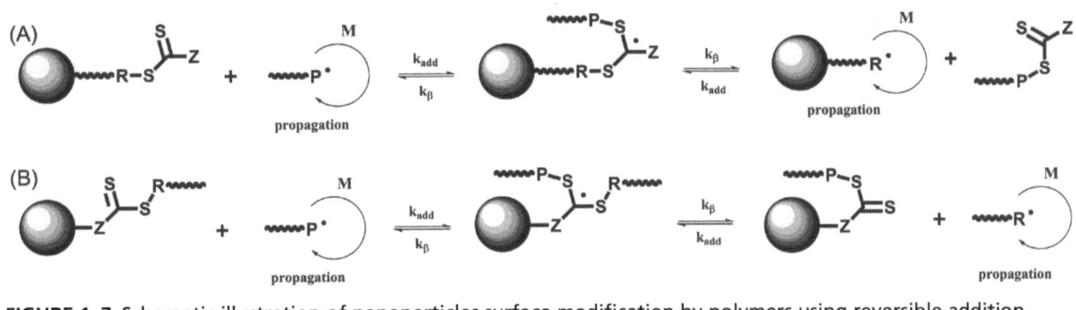

FIGURE 1–7 Schematic illustration of nanoparticles surface modification by polymers using reversible addition-fragmentation chain transfer polymerization: (A) R group approach and (B) Z group approach.

advantage over the Z group approach, especially for the preparation of polymer brushes with a high grafting density.

SI-RAFT polymerization technique was first applied for grafting from silica NPs [213]. Moraes at al. [218] prepared silica-g-poly(vinylbenzyl chloride) (PVBC) core–shell NPs with narrow dispersity. Polymerization of VBC using CTA-attached on silica NPs was carried out via thermal autoinitiaton of monomer or using conventional azoinitiator AIBN. Li and Benicewicz [210] modified silica NPs by combination of RAFT polymerization technique and click reactions of various functional alkynes. Grafting of poly(6-azidohexyl methacrylate) (PAHMA) and PAHMA-*b*-PMMA copolymer were performed via RAFT mediated by 4-cyano-pentanoic acid dithiobenzoate anchored on silica surface. Click reactions were applied for side-functionalization of grafted PAHMA with low molecular weight alkyne compounds or with alkyne terminated PS. This study demonstrated that combination of different techniques represents promising tool for advanced modification strategy of NP surfaces and possibility to achieving broad range of functionality and tailored properties. Core–shell NPs with silica core and PS shell were prepared by RAFT using silica NPs coated with trithiocarbonate or 2-butyric acid dithiobenzoate units as RAFT agents [219,220].

Baum and Brittain [203] reported as anchoring of initiator on the surface of silica particles using silane coupling agent. Thus grafting of PS, PMMA, poly(*N,N*-dimethylacrylamide) brushes and their copolymers were carried out from silica substrates using a azoundecyl-chlorosilane as surface immobilized initiator and 2-phenylprop-2-yl dithiobenzoate as a free CTA. Ohno et al. [221] modified monodisperse silica (SiPs) and Fe_3O_4 particles using newly designed RAFT agent having a triethoxysilane moiety, 6-(triethoxysilyl)-2-(((methylthio)car-bonothioyl)thio)-2-phenylacetate (EHT). SI-RAFT polymerization of styrene, MMA, NIPAM, and BA was carried out on EHT-modified SiPs/Fe_3O_4 in the presence of a free RAFT agent. Polymerization proceeded in a living manner resulting in coating with well-defined polymer layer of a target molar mass with a graft density as high as 0.3 chains/nm^2. RAFT technique has been used for the preparation of various other comb-like polymer brushes on silica [79,202,204–206,208–210,213,214,216] and Fe_3O_4 magnetic NPs [182,183,221].

Ability of trithiocarbonate groups to chemisorb on the Au surface was used for anchoring CTA onto Au NPs surface and subsequent modification of the NPs using RAFT polymerization of NIPAM [222].

Propanoic acid containing CTA was immobilized on the hydroxyl-functionalized CB [215,223] and used for grafting of PNIPAM and PAA. Similarly, CTA was immobilized onto surface of nanodiamond through reaction between the hydroxyl groups of nanodiamond and the carboxyl group of CTA [224,225]. Subsequently, SI-RAFT was used to polymerize PEG-containing methacrylates from the surface. The modified nanodiamonds loaded cisplatin exhibited enhanced cytotoxicity compared with neat cisplatin due to promotion of cellular uptake by the hydrophilic polymer shell.

CTA agent was covalently attached onto the oxidized MWCNTs surface by two-step synthesis, where first 2-hydroxyethyl-2′-bromoisobutyrate was reacted with carboxylic groups on the surface. Subsequently, bromides were used for attachment of CTA by reaction with PhC(S)SMgBr [223,224]. MWCNT were then grafted with PS, PMMA, PMMA-*b*-PS, and PNIPAM.

Similarly, molecularly imprinted polymer functionalized GO was also prepared [226]. Different approach of bonding of CTA agent, such as 4-(4-cyanopentanoic acid) dithiobenzoate includes reaction of carboxylic groups of oxidized MWCNTs with diethylene glycol and subsequent esterification with the CTA agent. MWCNTs modified by this approach were used for grafting of PS, poly(*N*-vinylcarbazole), or PHEMA [201,211,212,216,217,227,228]. The latest one was then hydrolyzed and formed PMAA chains were used for decoration of MWCNTs with Ag NPs. GO surface was modified by CTA agent by various approaches [226,229–232], including silanization, esterification, as well as click reaction. Subsequent SI-RAFT was used for grafting the GO surface by PS, PMA, P*t*BA, and PNIPAM.

1.3.2.6 Free Radical Polymerization

Contrary to previous techniques in the case of conventional FRP the polymer chains do not grow simultaneously from the surface, and therefore the control over the grafting density and thickness of the polymer shell is rather limited for this polymerization technique. FRP requires attachment of the thermally or photochemically unstable moieties, such as azo compounds or peroxides onto the NPs surface. It was used for the first time in a "grafting from" process by Dekking who performed SI polymerization from various clay surfaces [233,234]. In these studies, the 2,2′-azobisisobutyryamidine hydrochloride was absorbed on clay surface by means of an ion-exchange reaction and wide range of vinyl monomers were polymerized by subsequent thermal decomposition of the initiator. Later also Zhang et al. used additional azo derivatives for grafting of styrene and MMA from clay surface [235]. Nakatsuka [236] and later Tsubokawa [237] used immobilized 4,4′-azobis(4-cyanopentanoic acid) (ACPA) on silica glycidyl or amino functionalized silica. Prucker et al. [238–240] also reported the radical polymerization of styrene using covalently attached ACPA-derived initiator self-assembled in monolayers on silica surface. The amount of grafted polymer on the surface was about 25 g PS per g SiO_2. To prevent the formation of the free radicals during decomposition of azoinitiator and thus prevent formation of high fraction of free polymer chains, Czaun et al. [241] synthesized multifunctional azoinitiators, in which both formed radicals were covalently attached on the NPs surface.

1.3.3 Modification Using "Grafting-Through" Technique

An alternative approach for chemical linking of the growing polymer chains and forming assemblies of polymer chains tethered to surfaces represents "grafting-through" approach (Fig. 1–4C). In this method the polymerizable (monomer) moieties are covalently attached on the surface and are copolymerized with free monomers resulting in homogenous polymer covering [242]. The attached monomer layer can be either completely or partially transformed to (co)polymer chains, presenting a way to tune interfacial properties. The thickness of a surface-grafted polymer layer can be partially governed by adjusting the density of the surface-anchored monomers and the conditions of the polymerization reaction [243].

Chaimberg et al. [244] performed grafting of poly(*N*-vinyl pyrrolidone) to hydroxylated silica NPs using tethered vinyltriethoxysilane as a comonomer. This study was later extended by Browne [245] who grafted the silica NPs with poly(vinyl acetate), while the concentration of vinylsilane on the NPs surface was approximately 26 vinyl groups/nm^2 in comparison to Chaimberg who achieved approximately 9 vinyl groups/nm^2. Such high concentration of vinyl groups was explained by formation of multilayer coverage of the vinylsilane chains on the silica surface.

Laible and Hamann [246] reported using the same method for surface modification of silica, titanium dioxides, iron/zinc/chromium oxides, and quartz powders via free radical copolymerization of styrene with 3-methacryloxypropyltrimethoxysilane (MPS) covalently attached on the surfaces. MPS was used as comonomer also for modification of silica NPs by grafting of PMMA, [247,248] PMA [249], and P*t*BA [250].

Espirad et al. [95] used MPS for grafting the silica NPs via emulsion polymerization of ethyl acrylate initiated by potassium persulfate in the presence of nonionic surfactant. Similarly, Bourgeat-Lami and Lang [251] covalently bound PS to MPS functionalized silica NPs via dispersion polymerization in the presence of poly(*N*-vinyl pyrrolidone) used for efficient stabilization of NPs. Prepared NPs were monodisperse but larger (60−120 nm) compared with Barthet et al. [252] who prepared colloidal silica NPs up to 20 nm in diameter via free-radical copolymerization of vinyl monomers such as 4- or 2-vinylpyridine, MMA, styrene, and 2-(diethylamino)ethyl methacrylate. Henze et al. [253] studied the influence of reaction conditions on the grafting density as well as content of grafted polymer attached to the silica NPs. FRP of styrene was initiated by AIBN and carried out in the presence of MPS-capped silica NPs resulting in generation of self-assembled polymer monolayers. The authors reported how the temperature, polymerization time, initiator concentration, and grafting density of anchored silane linker bearing methacrylate group could affect the mechanism of polymer formation from the surface and concluded that polymerization process proceeded independently of the reaction conditions resulting in the same polymer monolayer thickness.

Wikberg et al. [254] activated silica in a similar manner through treatment with MPS and then grafted with zwitterionic polymers based on sulfoalkylbetaine. Polymerization of 3-(2-*N*-methacryloyloxyethyl)-*N*,*N*-dimethylammonio)propane sulfonate was carried out via FRP as well as RAFT mediated by sodium 2-(2-thiobenzoylsulfanyl propionylamino ethanesulfonate) used as CTA. Rong et al. [94] used for enhancing interfacial interaction of Al_2O_3 in nanocomposites initial modification of NPs with 3-aminopropyltrimethoxysilane for subsequent transformation to MPS monomer units. These were used for grafting of styrene and AM via FRP. The study was later extended to modification of silicon carbide [255] and silicon nitrite [256]. Prepared polymer−silicon nitrite NPs with active amide group of PAM were used for further modification with 4-diaminodiphenylsulfonate during the curing process to produce 3D-network structures covalently connected to the matrix [256].

Sugimoto et al. [257] used activated colloidal silica NPs functionalized with coupling agent based on 2-(methacryloyloxy)ethyl isocyanate for grafting of PMMA chains by FRP initiated with benzoyl peroxides. The prepared hybrid NPs exhibited enhanced heat resistance and surface hardness in contrast to lower volume contraction. Chinthamanipeta et al. [258] prepared silica NPs with well-defined PMMA brushes using RAFT copolymerization of MMA with MPS attached on the silica surface. Polymerization was mediated by 4-cyano-4-

(dodecylsulfanylthiocarbonyl)sulfanyl pentanoic acid as CTA agent. Sobani et al. [259] used also the RAFT polymerization for grafting of PS onto MPS modified silica NPs.

Chatterjee et al. [260] modified calcium carbonate ($CaCO_3$) NPs through grafting with PMMA [260,261] and PS [262,263] brushes to the surface using vinyltriethoxysilane anchored onto the $CaCO_3$ surface as a comonomer. Microemulsion polymerization was carried out in the presence of sodium dodecylsulfate as a hydrophobic surfactant. Core–shell NPs were used to reinforce the PP matrix resulting in enhanced rheological, mechanical, and thermal properties.

Ahmadian-Alan et al. [264] employed the ARGET ATRP technique for grafting of PS-*b*-PBA blocks into the nanoclay interlayers. The vinylbenzyltrimethylammonium chloride was selected as reactive surfactant due to its ability to participate in propagation reactions, facilitate monomer diffusion, and achieve exfoliated clay layers.

Qu et al. [265] described preparation of SWCNTs nanohybrids grafted with nylon 6 via AROP of free ε-caprolactam with ε-caprolactam attached on the SWCNT surface. ε-caprolactam was attached on the SWCNT surface by amidation with chlorides on the oxidized SWCNTs pretreated with thionyl chloride. TGA showed about one ε-caprolactam attached per 125 SWCNT carbons. Additionally, AROP catalyzed by well-defined metal-alkylidines was successfully implemented as an effective method for preparation of polymer brushes with well-defined structures. Liu et al. [266] applied this approach for preparation of SWCNTs-*g*-polynorbornene hybrids exhibited with enhanced solubility in organic solvents.

1.4 Modern Functionalized Nanoparticles for Advanced Applications

1.4.1 Functionalized Silicates: Clay and POSS

Clay minerals are normally layered silicate with less than 2 nm ultrafine grains and are commonly recognized as a first-generation nanomaterials. Since Kojima and coworkers discovered a great improvement to the thermal and mechanical properties of Nylon matrix [267], many efforts have demonstrated the hybridizations of a variety of commodity polymers with layered silicate. For example, Chang et al. presented a series of hybridizations, including (syndiotactic) PS/clay [268–271], PE/clay [272], polyimide/clay [273,274], PEG/clay [275,276], and polybenzoxazine/clay [277,278] through the manipulations of hydrogen bonding interactions and electrostatic interactions at the interfaces of clay and polymers. These results provide potential applications in low thermal expansion materials [273,274], lithium batteries [275,276], low surface free energy materials [277,278], and proton exchange membranes [279]. It is worth mentioning that Chang et al. discovered a novel nonfluorinated material with low surface free energy resulting from the strong intramolecular hydrogen bonding between the hydroxy groups of polybenzoxazines ($\gamma_{polybenzoxazine} = 16.4$ mJ/m^2; $\gamma_{teflon} = 21.4$ mJ/m^2) [280–282]. The same research group further presented an ultralow surface free energy ($\gamma_{polybenzoxazine/clay} = 12.7$ mJ/m^2) by synergistic the properties of polybenzoxazines and cetylpyridinium chloride (CPC) modified clay [278]. Recently, Gournis and coworkers reported the insertion of a water-soluble bisadduct fulleropyrrolidine (bis-C_{60}) derivative into the interlayer space of layered smectite clays through a common cation exchange reaction. The experiments, complemented by theoretical calculations, give insight into the formation process, structural details, and properties of the fullerene/clay nanocomposites. The reported materials constitute a new hybrid system and open new perspectives for the design and construction of novel C_{60}-based organic/clay hybrid materials [283].

POSS NPs possess well-defined 3-D cages composing of alternative Si–O bonds [$(RSiO_{1.5})_n$, $n = 8-12$] that were first discovered by Scott [284]. Due to the multisites of the peripheral *R* positions, we can easily tailor-make one or more functional groups on POSS to meet the demands of applications. Accordingly, numerous scientists have developed the preparations and studied the properties of polymer/POSS new generation nanocomposites. In addition to enhanced thermal stability and mechanical strength of the materials, the POSS-containing nanocomposites have been applied as sensors, in lithography, low dielectric constant materials, energy-storage materials, biomedical systems, coatings, and elastomers. Importantly, the existence of hydrogen bonding interactions between the siloxane (Si–O–Si) groups of the POSS and the hydroxyl (–OH) groups of polymers has been revealed [285,286]. Accordingly, Chang and coworkers have developed a series of POSS-containing nanocomposites with high homogeneity [287]. They further investigated the applications of high-performance materials [287], photoresists [288,289], and low dielectric constant materials [290]. POSS derivative-containing polymethacrylate-based photoresists for UV lithography demonstrated their high sensitivity, high contrast, and improved resolutions through the intermolecular hydrogen bonding interactions between the POSS moieties and the OH groups of polymethacrylate-based copolymers [289].

1.4.2 Functionalized Cellulose Nanofibrils

Celluloses are classified as one of the most abundant biomass of polysaccharides, which is also the major component of plants. Different from original cellulose, CNFs possess outstanding physical properties, such as high crystallinity, high aspect ratios, high tensile moduli (c. 145 GPa), high tensile strengths (c. 7 GPa), and high surface areas (c. 300 m^2/g). Accordingly, CNFs have been applied to be toughening agents, gas barriers, high-performance adsorbents, and electronics. Four major CNFs have been classified: micro/nanofibrillated cellulose (MFC/NFC) [291], nanocrystal cellulose/cellulose nanowhiskers (NCC/CNWs) [292], bacterial nanocellulose (BNC) [293], and TEMPO–oxidized cellulose nanofibers (TOCNs) [294,295]. Among them, the reported CNF-*graft*-polymer hybrids are still limited and challenging. Habibi and Dubois et al. conducted SI-AROP from NCC with *L*-LA, *DL*-LA, and ε-CL to obtain NCC-*g*-PLA [296], NCC-*g*-PDLA [297], and NCC-*g*-PCL [298,299] hybrids that enhanced the thermal and mechanical properties of the PLA, PLLA, and PCL matrices, respectively. Zhang and Thielemans et al. individually reported SI-ATRP from NCC with styrene to obtain a nanomaterial existed thermotropic/lyotropic liquid crystal states [300] and applied as an efficient adsorbent for removal of an aromatic pollutant [301]. Barros-Thimmons et al. conducted SI-ATRP from BNC membranes with acrylates to obtain

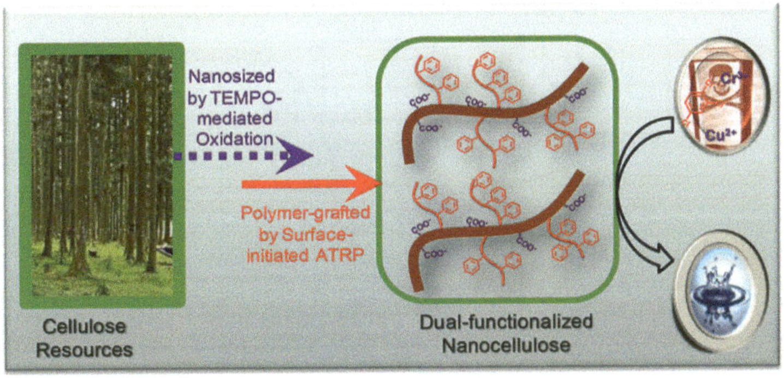

FIGURE 1–8 Modification of TOCNs to form a dual-functionalized adsorbent for the removal of organic pollutants and heavy metal ions [303,304]. *Reprinted from Huang C-F, Chen J-K, Tsai T-Y, Hsieh Y-A, Lin K-YA. Dual-functionalized cellulose nanofibrils prepared through TEMPO-mediated oxidation and surface-initiated ATRP. Polymer 2015;72:395–405. Copyright © 2015 with permission from Elsevier.*

novel nanostructured porous materials [302]. As shown in Fig. 1–8, we recently demonstrated a simple and facile approach for controlling the graft length and composition through SI-ATRP of styrene from the TOCN. The resulting dual-functionalized TOCN-*g*-PS composed of neutral carboxylic acid sodium salt groups and PS chains on the surfaces were highly efficient in removing an organic pollutant (e.g., 1,2,4-trichlorobenzene) and heavy metal ions (e.g., Cr^{3+} and Cu^{2+}) from aqueous media [303,304].

1.4.3 Functionalized Graphenes

Graphenes serve a very important role in contemporary nanotechnology due to their excellent properties, including high tensile modulus (c. 1 TPa), high tensile strength (c. 130 GPa), ultrahigh surface area (>2000 m^2/g), high electron mobility (c. 15,000 $cm^2/V/s$), and high thermal conductivity (c. 2000 W/m/K). In nanocomposite examples, Erogbogbo and coworkers demonstrated the infusion of graphene and quantum dots (GQDs) with dual acid and alcohol functional groups into an epoxy polymer matrix (see Fig. 1–9). The novel materials performed significant improvements: (1) a 2.6-fold increase in the toughness of the polymer resins, (2) a 2.25-fold increase in the tensile strength of the polymer resins compared with its native resins, (3) uniform loading at weight percentages as high as 10% of the polymer resin, (4) an 18% change to the max percentage increase in tensile strain compared with that of the neat polymer resin without GQDs, and (5) a 2.5-times increase in Young's modulus compared with that of the neat polymer resin; all while maintaining excellent optical properties of the composite formulation. These results indicated that properly functionalized graphene can enable high loading percentages, resulting in stronger and tougher nanocomposites [305]. The Chen group demonstrated an advance in which GO micropatterns can be firmly attached onto the surface of pristine graphene via a microcontact printing strategy through the interfacial of multiple $\pi-\pi$ interactions. They achieved GO-on-graphene hybrid materials that can be applied for heterogeneous functionalization from chemical/photoactive GO.

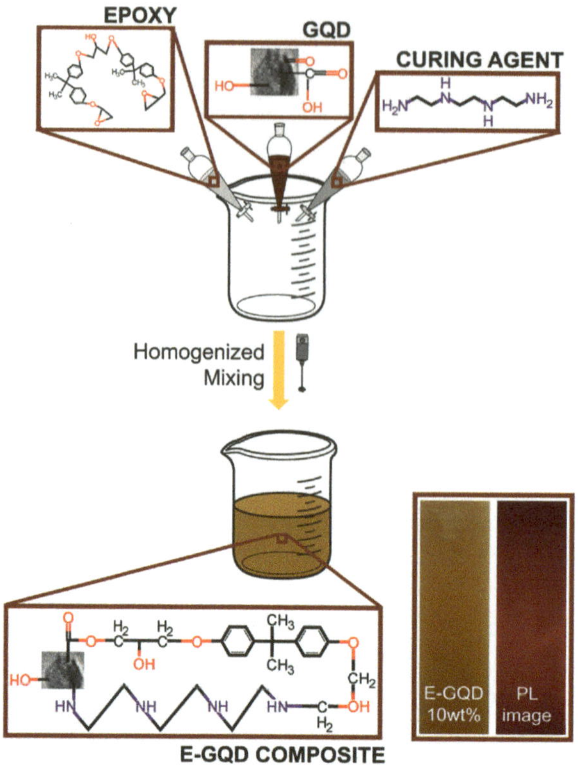

FIGURE 1–9 The incorporation of GQD into epoxy resin by chemical interactions [305]. *Reprinted with permission from Gobi N, Vijayakumar D, Keles O, Erogbogbo F. Infusion of graphene quantum dots to create stronger, tougher, and brighter polymer composites. ACS Omega 2017;2:4356–62. <https://pubs.acs.org/doi/full/10.1021/acsomega.6b00517> Copyright © 2017 American Chemical Society. Further permissions related to the material excerpted should be directed to the ACS.*

Thus different polymer brushes were introduced onto the photoactive sites of GO by self-initiated photografting and photopolymerization, offering the design of stimuli-responsive intelligent surfaces [306]. For the applications of energy storage, as shown in Fig. 1–10, Chang and coworkers exhibited a novel hybridization through the preparation of negatively charged Si NPs intercalated in positive charged graphene-modified protic ionic liquid (G-PIL) sheets by combining a self-assembly technique and a thermal process. The Si/G hybridized electrodes presented a capacity of 803 mAh/g in 100 cycles and $<0.25\%$ capacity decreasing per cycle before 150 cycles [307,308].

It was recently found that GO can be reduced in situ during ATRP when PMDETA is used as a ligand [309]. This ability of PMDETA to be involved in both the catalysis of ATRP and reducing of GO surface was utilized for tailoring the partial reduction of GO during its modification by SI-ATRP of various vinyl monomers. Thus conductivity of GO was controllably increased by only two orders to obtain materials with properties suitable for application in electrorheological suspensions [310,311].

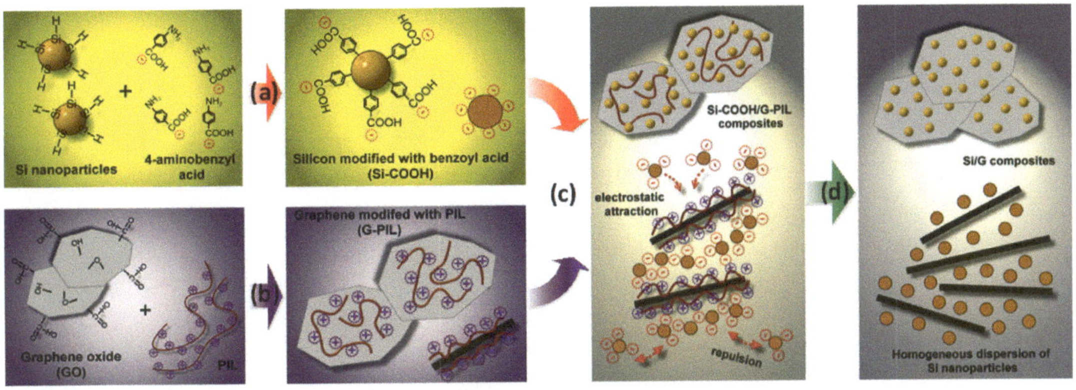

FIGURE 1–10 A Si/G nanocomposite fabricated by electrostatic attraction followed by thermal treatment for the applications of energy storage [307]. *Reprinted from Ye YS, Xie XL, Rick J, Chang FC, Hwang BJ. Improved anode materials for lithium-ion batteries comprise noncovalently bonded graphene and silicon nanoparticles. J Power Sources 2014;247:991–8. Crown copyright © 2013 with permission from Elsevier.*

1.4.4 Stimuli-Responsive Nanomaterials

Stimuli-responsive materials (SRMs) have recently shown their great impact for applications in sensors, shape memory materials, storage, catalysts, guest–host chemistry, and biomedical systems [312]. There are some commonly used external triggers to active the response of the SMRs including thermal, solvent, pH, light, electrical/magnetic field, mechanical strength, etc. In this part, we selectively discuss some work about the fabrications of SRMs since numerous review papers have been published. Through living polymerizations, Huang et al. synthesized a variety of NPs that possess thermal responsive compartment, including ZnS/PNIPAM hollow [313] and SiO_2@PNIPAM NPs [314], toward the applications of temperature sensors and biomaterials. In solvent-sensitive cases, Huang et al. fabricated different nanomaterials for sensing various polar environment, comprising poly(*N*-vinylpyrrolidone)-*b*-poly(*N*-vinylcarbazole) [315], PMMA brushes on silicon surfaces [316,317], and PS brushes on silicon surfaces [318].

Mosnacek et al. developed photoacutating materials based on nanocomposites consisting of thermoplastic elastomers (TPE), such as polystyrene-*b*-polyisoprene-*b*-polystyrene (SIS) [319,320] or poly(methyl methacrylate)-*b*-poly(butyl acrylate)-*b*-poly(methyl methacrylate) (PMMA-*b*-PBA-*b*-PMMA) [321,322] triblock copolymers, and covalently modified MWCNTs composites. For SIS matrix the MWCNTs were modified either with cholesteryl groups or with PS chains to obtain preferential interactions of MWCNTs with either polyisoprene or PS block, respectively. In the case of PMMA-*b*-PBA-*b*-PMMA, the MWCNTs were modified either with PBA chains or with PBA-*b*-PMMA block copolymer chains. In the later case both the modification of MWCNTs surface and synthesis of the triblock copolymer matrix were performed simultaneously (Fig. 1–11) resulting in significantly better dispersed MWCNTs in comparison with other described nanocomposites, and with preferential interactions of the MWCNTs with soft PBA phase. This combination provided materials with the best

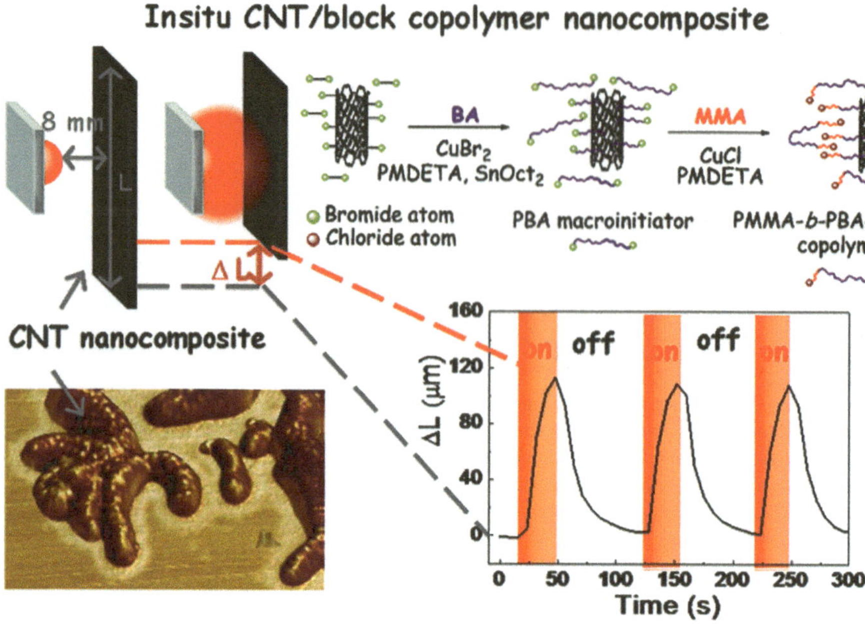

FIGURE 1–11 Schematic illustration of synthesis and properties of photoactuating nanocomposites based on CNTs grafted with poly(*n*-butyl acrylate)-*b*-poly(methyl methacrylate) block copolymer chains and poly(methyl methacrylate)-*b*-poly(*n*-butyl acrylate)-*b*-poly(methyl methacrylate) triblock copolymer matrix [321]. *Reprinted with permission from Ilčíková M, Mrlík M, Sedláček T, Šlouf M, Zhigunov A, Koynov K, Mosnáček J. Synthesis of photoactuating acrylic thermoplastic elastomers containing diblock copolymer-grafted carbon nanotubes. ACS Macro Lett 2014;3:999–1003. Copyright © 2014 American Chemical Society.*

photoactuation properties. The study was successfully extended also to photoactuating nanocomposites based on poly(dimethyl siloxane) as a matrix and GO covalently modified with PMMA or PBMA chains [323].

1.5 Conclusion

The aim of the functionalization of NPs surface, for their subsequent application in nanocomposites, is changing the surface nature to improve interactions of the NPs with polymer matrices. The chemical reactions leading to modification of NPs surface depend on the structure and reactivity of the NPs. In some cases the modification needs to be proceeded through several steps, where first some functional groups are created or introduced onto the NPs surface for their subsequent postfunctionalization. Typically thiol and disulfide functional groups are used for modification of noble metals surfaces, while silane, phosphonate, acidic, or polyol groups are commonly used for metal oxides surfaces. Surfaces of carbon-based NPs are commonly either oxidized before functionalization or directly functionalized using diazonium salts. Use of some functional groups can be limited due to their interference with the catalysts or growing polymer species during the polymerization process, when functional polymer is synthesized for subsequent "grafting to" process. Similarly, functional groups forming sufficiently stable bonds with the NPs should be selected for subsequent "grafting from" process to prevent cleavage during the SI polymerization. Thus suitability of the functional groups for the modification of NPs surface can depend not only on the chemical structure of the NPs but also on the method used for polymer synthesis and type modification process. This chapter reviews the most commonly used methods of the NPs surface modification from the point of view of a wide range of functional groups as well as various grafting techniques including "grafting to," "grafting from," and "grafting through" processes to help researchers choose the best modification method for their NPs/polymer system.

Acknowledgment

This work was supported by the Slovak Academy of Sciences and the Ministry of Science and Technology, Taiwan through the joint project (SAS-MOST JRP 2014-9 and MOST 106-2923-E-194-001, respectively) entitled "Synthesis of well defined novel copolymers by use of living polymerization methods and advanced chromatography technique." The authors also thank the Grant agency VEGA 2/0019/18, Slovak Research and Development Agency APVV through Grant APVV-15-0545 and APVV-14-0891, and Ministry of Science and Technology, Taiwan through project MOST 105-2628-E-005-003-MY3.

References

[1] Hooper JB, Schweizer KS, Desai TG, Koshy R, Keblinski P. Structure, surface excess and effective interactions in polymer nanocomposite melts and concentrated solutions. J Chem Phys 2004;121:6986–97.

[2] Hooper JB, Schweizer KS. Theory of phase separation in polymer nanocomposites. Macromolecules 2006;39:5133–42.

[3] Hooper JB, Schweizer KS. Contact aggregation, bridging, and steric stabilization in dense polymer-particle mixtures. Macromolecules 2005;38:8858–69.

[4] Sperling RA, Parak WJ. Surface modification, functionalization and bioconjugation of colloidal inorganic nanoparticles. Philos Trans R Soc A Math Phys Eng Sci 2010;368:1333–83.

[5] Kim B, Tripp SL, Wei A. Self-organization of large gold nanoparticle arrays. J Am Chem Soc 2001;123:7955–6.

[6] Jana NR, Earhart C, Ying JY. Synthesis of Water-Soluble and Functionalized Nanoparticles by Silica Coating. Chem Mater 2007;19:5074–82.

[7] Hao R, Xing R, Xu Z, Hou Y, Gao S, Sun S. Synthesis, Functionalization, and Biomedical Applications of Multifunctional Magnetic Nanoparticles. Adv Mater 2010;22:2729–42.

[8] Selvan ST, Yang Tan TT, Kee Yi D, Jana NR. Functional and multifunctional nanoparticles for bioimaging and biosensing. Langmuir 2010;26:11631–41.

[9] Ito S, Goseki R, Ishizone T, Hirao A. Synthesis of well-controlled graft polymers by Living anionic polymerization towards exact graft polymers. Polym Chem 2014;5:5523–34.

[10] Braunecker WA, Matyjaszewski K. Controlled/living radical polymerization: Features, developments, and perspectives. Prog Polym Sci 2007;32:93–146. Available from: https://doi.org/10.1016/j.progpolymsci.2006.11.002.

[11] Bielawski CW, Grubbs RH. Living ring-opening metathesis polymerization. Prog Polym Sci 2007;32:1–29.

[12] Sugi R, Yokoyama A, Furuyama T, Uchiyama M, Yokozawa T. Inductive effect-assisted chain-growth polycondensation. Synthetic development from para- to meta-substituted aromatic polyamides with low polydispersities. J Am Chem Soc 2005;127:10172–3.

[13] Yokozawa T, Ogawa M, Sekino A, Sugi R, Yokoyama A. Chain-growth polycondensation for well-defined aramide. Synthesis of unprecedented block copolymer containing aramide with low polydispersity. J Am Chem Soc 2002;124:15158–9.

[14] Hadjichristidis N, Pitsikalis M, Pispas S, Iatrou H. Polymers with complex architecture by living anionic polymerization. Chem Rev 2001;101:3747–92.

[15] Mecerreyes D, Jerome R, Dubois P. Novel macromolecular architectures based on aliphatic polyesters: Relevance of the "coordination-insertion" ring-opening polymerization. Adv Polym Sci 1999;147:1–59.

[16] Sawamoto M. Modern Cationic Vinyl Polymerization. Prog Polym Sci 1991;16:111–72.

[17] Huang C-F, Chen M-J, Lin C-H, Chiang Y-W. Synthesis of Well-Defined Poly(N-H Benzamide-co-N-Octyl Benzamide)s and the Study of their Blends with Nylon 6. Polymers (Basel) 2017;9:172.

[18] Huang CF, Aimi J, Lai KY. Synthesis of Novel m-Star Copolymers with Poly(N-Octyl Benzamide) and Poly(e-Caprolactone) Miktoarms through Chain-Growth Condensation Polymerization, Styrenics-Assisted Atom Transfer Radical Coupling, and Ring-Opening Polymerization. Macromol Rapid Commun 2017;38:1600607.

[19] Huang C-F, Ohta Y, Yokoyama A, Yokozawa T. Efficient Low-Temperature Atom Transfer Radical Coupling and Its Application to Synthesis of Well-Defined Symmetrical Polybenzamides. Macromolecules 2011;44:4140–8.

[20] Neouze M-A, Schubert U. Surface Modification and Functionalization of Metal and Metal Oxide Nanoparticles by Organic Ligands. Monatshefte Für Chemie - Chem Mon 2008;139:183–95.

[21] Mazur M, Barras A, Kuncser V, Galatanu A, Zaitzev V, Turcheniuk KV, et al. Iron oxide magnetic nanoparticles with versatile surface functions based on dopamine anchors. Nanoscale 2013;5:2692–702.

[22] Yang J, Ying JY. A general phase-transfer protocol for metal ions and its application in nanocrystal synthesis. Nat Mater 2009;8:683–9.

[23] Kango S, Kalia S, Celli A, Njuguna J, Habibi Y, Kumar R. Surface modification of inorganic nanoparticles for development of organic–inorganic nanocomposites—A review. Prog Polym Sci 2013;38:1232–61.

[24] Thanh NTK, Green LAW. Functionalisation of nanoparticles for biomedical applications. Nano Today 2010;5:213–30.

[25] Patil V, Mayya KS, Pradhan SD, Sastry M. Evidence for Novel Interdigitated Bilayer Formation of Fatty Acids during Three-Dimensional Self-Assembly on Silver Colloidal Particles. J Am Chem Soc 1997;119:9281–2.

[26] Duan H, Nie S. Cell-Penetrating Quantum Dots Based on Multivalent and Endosome-Disrupting Surface Coatings. J Am Chem Soc 2007;129:3333–8.

[27] Carion O, Mahler B, Pons T, Dubertret B. Synthesis, encapsulation, purification and coupling of single quantum dots in phospholipid micelles for their use in cellular and in vivo imaging. Nat Protoc 2007;2:2383–90.

[28] Woehrle GH, Brown LO, Hutchison JE. Thiol-Functionalized, 1.5-nm Gold Nanoparticles through Ligand Exchange Reactions: Scope and Mechanism of Ligand Exchange. J Am Chem Soc 2005;127:2172–83.

[29] Hostetler MJ, Wingate JE, Zhong C-J, Harris JE, Vachet RW, Clark MR, et al. Alkanethiolate Gold Cluster Molecules with Core Diameters from 1.5 to 5.2 nm: Core and Monolayer Properties as a Function of Core Size. Langmuir 1998;14:17–30.

[30] García KP, Zarschler K, Barbaro L, Barreto JA, O'Malley W, Spiccia L, et al. Zwitterionic-Coated "Stealth" Nanoparticles for Biomedical Applications: Recent Advances in Countering Biomolecular Corona Formation and Uptake by the Mononuclear Phagocyte System. Small 2014;10:2516–29.

[31] Zhan N, Palui G, Safi M, Ji X, Mattoussi H. Multidentate Zwitterionic Ligands Provide Compact and Highly Biocompatible Quantum Dots. J Am Chem Soc 2013;135:13786–95.

[32] Huang C-C, Chang H-T. Parameters for selective colorimetric sensing of mercury(ii) in aqueous solutions using mercaptopropionic acid-modified gold nanoparticles. Chem Commun 2007;1215–17.

[33] Vidal-Vidal J, Rivas J, López-Quintela MA. Synthesis of monodisperse maghemite nanoparticles by the microemulsion method. Colloids Surfaces A Physicochem Eng Asp 2006;288:44–51.

[34] Fan H, Gabaldon J, Brinker CJ, Jiang Y-B. Ordered nanocrystal/silica particles self-assembled from nanocrystal micelles and silicate. Chem Commun 2006;14:2323–5.

[35] Bini RA, Marques RFC, Santos FJ, Chaker JA, Jafelicci M. Synthesis and functionalization of magnetite nanoparticles with different amino-functional alkoxysilanes. J Magn Magn Mater 2012;324:534–9.

[36] Bagwe RP, Hilliard LR, Tan W. Surface Modification of Silica Nanoparticles to Reduce Aggregation and Nonspecific Binding. Langmuir 2006;22:4357–62.

[37] Khramov A, Balbyshev V, Voevodin N, Donley M. Nanostructured sol–gel derived conversion coatings based on epoxy- and amino-silanes. Prog Org Coatings 2003;47:207–13.

[38] Péré E, Cardy H, Latour V, Lacombe S. Low-temperature reaction of trialkoxysilanes on silica gel: a mild and controlled method for modifying silica surfaces. J Colloid Interface Sci 2005;281:410–16.

[39] De Palma R, Peeters S, Van Bael MJ, Van den Rul H, Bonroy K, Laureyn W, et al. Silane Ligand Exchange to Make Hydrophobic Superparamagnetic Nanoparticles Water-Dispersible. Chem Mater 2007;19:1821–31.

[40] Adden N, Gamble LJ, Castner DG, Hoffmann A, Gross G, Menzel H. Phosphonic Acid Monolayers for Binding of Bioactive Molecules to Titanium Surfaces. Langmuir 2006;22:8197–204.

[41] Sahoo Y, Pizem H, Fried T, Golodnitsky D, Burstein L, Sukenik CN, et al. Alkyl Phosphonate/Phosphate Coating on Magnetite Nanoparticles: A Comparison with Fatty Acids. Langmuir 2001;17:7907–11.

[42] Woehrle GH, Hutchison JE. Thiol-Functionalized Undecagold Clusters by Ligand Exchange: Synthesis, Mechanism, and Properties. Inorg Chem 2005;44:6149–58.

[43] Zhang L, He R, Gu H-C. Oleic acid coating on the monodisperse magnetite nanoparticles. Appl Surf Sci 2006;253:2611–17.

[44] Xu C, Xu K, Gu H, Zheng R, Liu H, Zhang X, et al. Dopamine as A Robust Anchor to Immobilize Functional Molecules on the Iron Oxide Shell of Magnetic Nanoparticles. J Am Chem Soc 2004;126:9938–9.

[45] Rajh T, Chen LX, Lukas K, Liu T, Thurnauer MC, Tiede DM. Surface Restructuring of Nanoparticles: An Efficient Route for Ligand – Metal Oxide Crosstalk. J Phys Chem B 2002;106:10543–52.

[46] Georgakilas V, Otyepka M, Bourlinos AB, Chandra V, Kim N, Kemp KC, et al. Functionalization of Graphene: Covalent and Non-Covalent Approaches, Derivatives and Applications. Chem Rev 2012;112:6156–214.

[47] Spitalsky Z, Danko M, Mosnacek J. Preparation of Functionalized Graphene Sheets. Curr Org Chem 2011;15:1133–50.

[48] Greenwood J, Phan TH, Fujita Y, Li Z, Ivasenko O, Vanderlinden W, et al. Covalent Modification of Graphene and Graphite Using Diazonium Chemistry: Tunable Grafting and Nanomanipulation. ACS Nano 2015;9:5520–35.

[49] Quintana M, Spyrou K, Grzelczak M, Browne WR, Rudolf P, Prato M. Functionalization of Graphene via 1,3-Dipolar Cycloaddition. ACS Nano 2010;4:3527–33.

[50] Vadukumpully S, Gupta J, Zhang Y, Xu GQ, Valiyaveettil S. Functionalization of surfactant wrapped graphenenanosheets with alkylazides for enhanced dispersibility. Nanoscale 2011;3:303–8.

[51] Strom TA, Dillon EP, Hamilton CE, Barron AR. Nitrene addition to exfoliated graphene: a one-step route to highly functionalized graphene. Chem Commun 2010;46:4097–9.

[52] Chattopadhyay J, Mukherjee A, Chakraborty S, Kang J, Loos PJ, Kelly KF, et al. Exfoliated soluble graphite. Carbon N Y 2009;47:2945–9.

[53] Chen M, Zhou H, Qiu C, Yang H, Yu F, Sun L. Layer-dependent fluorination and doping of graphene via plasma treatment. Nanotechnology 2012;23:115706.

[54] Bon SB, Valentini L, Verdejo R, Garcia Fierro JL, Peponi L, Lopez-Manchado MA, et al. Plasma Fluorination of Chemically Derived Graphene Sheets and Subsequent Modification With Butylamine. Chem Mater 2009;21:3433–8.

[55] Dreyer DR, Todd AD, Bielawski CW. Harnessing the chemistry of graphene oxide. Chem Soc Rev 2014;43:5288.

[56] Yu D, Yang Y, Durstock M, Baek J-B, Dai L. Soluble P3HT-Grafted Graphene for Efficient Bilayer – Heterojunction Photovoltaic Devices. ACS Nano 2010;4:5633–40.

[57] Jang J, Pham V, Rajagopalan B, Hur S, Chung J. Effects of the alkylamine functionalization of graphene oxide on the properties of polystyrene nanocomposites. Nanoscale Res Lett 2014;9:265.

[58] Collins WR, Lewandowski W, Schmois E, Walish J, Swager TM. Claisen Rearrangement of Graphite Oxide: A Route to Covalently Functionalized Graphenes. Angew Chemie Int Ed. 2011;50:8848–52.

[59] Ou X, Jiang L, Chen P, Zhu M, Hu W, Liu M, et al. Highly Stable Graphene-Based Multilayer Films Immobilized via Covalent Bonds and Their Applications in Organic Field-Effect Transistors. Adv Funct Mater 2013;23:2422–35.

[60] Ou J, Wang Y, Wang J, Liu S, Li Z, Yang S. Self-Assembly of Octadecyltrichlorosilane on Graphene Oxide and the Tribological Performances of the Resultant Film. J Phys Chem C 2011;115:10080–6.

[61] Bourlinos AB, Gournis D, Petridis D, Szabó T, Szeri A, Dékány I. Graphite Oxide: Chemical Reduction to Graphite and Surface Modification with Primary Aliphatic Amines and Amino Acids. Langmuir 2003;19:6050–5.

[62] Wei G, Yan M, Dong R, Wang D, Zhou X, Chen J, et al. Covalent Modification of Reduced Graphene Oxide by Means of Diazonium Chemistry and Use as a Drug-Delivery System. Chem - A Eur J 2012;18:14708–16.

[63] Chua CK, Ambrosi A, Pumera M. Introducing dichlorocarbene in graphene. Chem Commun 2012;48:5376.

[64] Xu LQ, Yee YK, Neoh K-G, Kang E-T, Fu GD. Cyclodextrin-functionalized graphene nanosheets, and their host-guest polymer nanohybrids. Polymer 2013;54:2264–71.

[65] Bohara RA, Thorat ND, Pawar SH. Role of functionalization: strategies to explore potential nano-bio applications of magnetic nanoparticles. RSC Adv 2016;6:43989–4012.

[66] Kang B, Opatz T, Landfester K, Wurm FR. Carbohydrate nanocarriers in biomedical applications: functionalization and construction. Chem Soc Rev 2015;44:8301–25.

[67] Kuhn SJ, Finch SK, Hallahan DE, Giorgio TD. Facile production of multivalent enzyme-nanoparticle conjugates. J Magn Magn Mater 2007;311:68–72.

[68] Huh Y-M, Jun Y, Song H-T, Kim S, Choi J, Lee J-H, et al. In Vivo Magnetic Resonance Detection of Cancer by Using Multifunctional Magnetic Nanocrystals. J Am Chem Soc 2005;127:12387–91.

[69] Bartczak D, Kanaras AG. Preparation of Peptide-Functionalized Gold Nanoparticles Using One Pot EDC/Sulfo-NHS Coupling. Langmuir 2011;27:10119–23.

[70] Zhou H-P, Xu C-H, Sun W, Yan C-H. Clean and Flexible Modification Strategy for Carboxyl/Aldehyde-Functionalized Upconversion Nanoparticles and Their Optical Applications. Adv Funct Mater 2009;19:3892–900.

[71] Kumar S, Aaron J, Sokolov K. Directional conjugation of antibodies to nanoparticles for synthesis of multiplexed optical contrast agents with both delivery and targeting moieties. Nat Protoc 2008;3:314–20.

[72] Kolb DM. Electrochemical Surface Science. Angew Chemie Int Ed 2001;40:1162–81.

[73] Li N, Binder WH. Click-chemistry for nanoparticle-modification. J Mater Chem 2011;21:16717–34.

[74] Tsubokawa N. Functionalization of carbon black by surface grafting of polymers. Prog Polym Sci 1992;17:417–70.

[75] Zdyrko B, Luzinov I. Polymer Brushes by the "Grafting to" Method. Macromol Rapid Commun 2011;32:859–69.

[76] Zhao B, Brittain WJ, Zhou W, Cheng SZD. Nanopattern Formation from Tethered PS- b -PMMA Brushes upon Treatment with Selective Solvents. J Am Chem Soc 2000;122:2407–8.

[77] Hübner E, Allgaier J, Meyer M, Stellbrink J, Pyckhout-Hintzen W, Richter D. Synthesis of Polymer/Silica Hybrid Nanoparticles Using Anionic Polymerization Techniques. Macromolecules 2010;43:856–67.

[78] Joubert M, Delaite C, Bourgeat-Lami E, Dumas P. Ring-opening polymerization of?-caprolactone andL-lactide from silica nanoparticles surface. J Polym Sci Part A Polym Chem 2004;42:1976–84.

[79] Li C, Benicewicz BC. Synthesis of Well-Defined Polymer Brushes Grafted onto Silica Nanoparticles via Surface Reversible Addition – Fragmentation Chain Transfer Polymerization. Macromolecules 2005;38:5929–36.

[80] Wuelfing WP, Gross SM, Miles DT, Murray RW. Nanometer Gold Clusters Protected by Surface-Bound Monolayers of Thiolated Poly(ethylene glycol) Polymer Electrolyte. J Am Chem Soc 1998;120:12696–7.

[81] Corbierre MK, Cameron NS, Lennox RB. Polymer-Stabilized Gold Nanoparticles with High Grafting Densities. Langmuir 2004;20:2867–73.

[82] Otsuka H, Akiyama Y, Nagasaki Y, Kataoka K. Quantitative and Reversible Lectin-Induced Association of Gold Nanoparticles Modified with α-Lactosyl-ω-mercapto-poly(ethylene glycol). J Am Chem Soc 2001;123:8226–30.

[83] Corbierre MK, Cameron NS, Sutton M, Mochrie SGJ, Lurio LB, Rühm A, et al. Polymer-Stabilized Gold Nanoparticles and Their Incorporation into Polymer Matrices. J Am Chem Soc 2001;123:10411–12.

[84] Shan J, Nuopponen M, Jiang H, Viitala T, Kauppinen E, Kontturi K, et al. Amphiphilic Gold Nanoparticles Grafted with Poly(N -isopropylacrylamide) and Polystyrene. Macromolecules 2005;38:2918–26.

[85] Mangeney C, Ferrage F, Aujard I, Marchi-Artzner V, Jullien L, Ouari O, et al. Synthesis and Properties of Water-Soluble Gold Colloids Covalently Derivatized with Neutral Polymer Monolayers. J Am Chem Soc 2002;124:5811–21.

[86] Sudeep PK, Page Z, Emrick T. PEGylated silicon nanoparticles: synthesis and characterization. Chem Commun 2008;14:6126–7.

[87] Gann JP, Yan M. A Versatile Method for Grafting Polymers on Nanoparticles. Langmuir 2008;24:5319–23.

[88] Schmid D, Handge UA, Gann JP, Yan M, Caseri W. Melt Elongation of Polymer Nanocomposites: A Method for the Controlled Production of Dichroic Films. Macromol Mater Eng 2008;293:471–8.

[89] Huang Y, Liu Q, Zhou X, Perrier S, Zhao Y. Synthesis of Silica Particles Grafted with Well-Defined Living Polymeric Chains by Combination of RAFT Polymerization and Coupling Reaction. Macromolecules 2009;42:5509–17.

[90] Ranjan R, Brittain WJ. Combination of Living Radical Polymerization and Click Chemistry for Surface Modification. Macromolecules 2007;40:6217–23.

[91] Tsubokawa N. Surface Grafting of Polymers onto Nanoparticles in a Solvent-Free Dry-System and Applications of Polymer-grafted Nanoparticles as Novel Functional Hybrid Materials. Polym J 2007;39:983–1000.

[92] Tsubokawa N, Yoshikawa S. Grafting of polymers with controlled molecular weight onto ultrafine silica surface. J Polym Sci Part A Polym Chem 1995;33:581–6.

[93] Korth BD, Keng P, Shim I, Bowles SE, Tang C, Kowalewski T, et al. Polymer-Coated Ferromagnetic Colloids from Well-Defined Macromolecular Surfactants and Assembly into Nanoparticle Chains. J Am Chem Soc 2006;128:6562–3.

[94] Rong MZ, Ji QL, Zhang MQ, Klaus F. Graft polymerization of vinyl monomers onto nanosized alumina particles. Eur Polym J 2002;38:1573–82.

[95] Espirad P, Guyot A. Poly(ethyl acrylate) latexes encapsulating nanoparticles of silica: 2. Grafting process onto silica. Polymer 1995;36:4391–5.

[96] Rong MZ, Ji QL, Zhang MQ, Friedrich K. Graft polymerization of vinyl monomers onto nanosized silicon carbide particles. Polym Polym Compos 2002;10:531–9.

[97] Heikkinen JJ, Heiskanen JP, Hormi OEO. Grafting of functionalized silica particles with poly(acrylic acid). Polym Adv Technol 2006;17:426–9.

[98] Jia Z, Wang Z, Xu C, Liang L, Wei B, Wu D, et al. Study on poly(methyl methacrylate)/carbon nanotube composites. Mater Sci Eng A 1999;271:395–400.

[99] Shaffer MSP, Koziol K. Polystyrene grafted multi-walled carbon nanotubes. Chem Commun 2002;2074–5.

[100] Park SJ, Cho MS, Lim ST, Choi HJ, Jhon MS. Synthesis and Dispersion Characteristics of Multi-Walled Carbon Nanotube Composites with Poly(methyl methacrylate) Prepared by In-Situ Bulk Polymerization. Macromol Rapid Commun 2003;24:1070–3.

[101] Tsubokawa N, Saitoh K, Shirai Y. Surface grafting of polymers onto ultrafine silica: cationic polymerization initiated by benzylium perchlorate groups introduced onto ultrafine silica surface. Polym Bull 1995;35:399–406.

[102] Tsubokawa N, Kimoto T, Endo T. Surface grafting of polymers onto ultrafine silica: Cationic polymerization initiated by oxoaminium perchlorate groups introduced onto ultrafine silica surface. Polym Bull 1994;33:187–94.

[103] Jordan R, West N, Ulman A, Chou Y-M, Nuyken O. Nanocomposites by Surface-Initiated Living Cationic Polymerization of 2-Oxazolines on Functionalized Gold Nanoparticles. Macromolecules 2001;34:1606–11.

[104] Tsubokawa N. Cationic grafting of poly(tetrahydrofuran) onto the surface of carbon black. Carbon N Y 1980;18:378–9.

[105] Tsubokawa N. Cationic grafting from carbon black. I. Graft polymerization of styrene using COCIO4 groups on the surface of carbon black. J Polym Sci Polym Lett Ed 1983;21:705–9.

[106] Tsubokawa N. Cationic grafting from carbon black. II. Polymerization of styrene initiated by COClO4 group on the surface of carbon black. J Polym Sci Polym Chem Ed 1984;22:1515–24.

[107] Tsubokawa N. Cationic grafting from carbon black. III. Grafting of polyesters from carbon black by ring-opening polymerization of lactones initiated by COClO4 groups on carbon black. J Appl Polym Sci 1985;30:2041–9.

[108] Tsubokawa N, Maruyama H, Sone Y. Carbon fiber as an initiator of cationic polymerization of N-vinylcarbazole. Polym Bull 1986;15:209–14.

[109] Tsubokawa N, Maruyama H, Sone Y. Cationic Polymerization of *N* -Vinylcarbazole and *N* -Vinyl-2-Pyrrolidone Initiated by Carboxyl Groups on Carbon Fibers. J Macromol Sci Part A - Chem 1988;25:171–82.

[110] Tong X, Liu C, Cheng H-M, Zhao H, Yang F, Zhang X. Surface modification of single-walled carbon nanotubes with polyethylene viain situ Ziegler-Natta polymerization. J Appl Polym Sci 2004;92:3697–700.

[111] Park S, Yoon SW, Lee K-B, Kim DJ, Jung YH, Do Y, et al. Carbon Nanotubes as a Ligand in Cp2ZrCl2-Based Ethylene Polymerization. Macromol Rapid Commun 2006;27:47–50.

[112] Park S, Yoon SW, Choi H, Lee JS, Cho WK, Kim J, et al. Pristine Multiwalled Carbon Nanotube/Polyethylene Nanocomposites by Immobilized Catalysts. Chem Mater 2008;20:4588–94.

[113] Oosterling MLCM, Sein A, Schouten AJ. Anionic grafting of polystyrene and poly(styrene-block-isoprene) onto microparticulate silica and glass slides. Polymer 1992;33:4394–400.

[114] Zhou Q, Wang S, Fan X, Advincula R, Mays J. Living Anionic Surface-Initiated Polymerization (LASIP) of a Polymer on Silica Nanoparticles. Langmuir 2002;18:3324–31.

[115] Bentz KC, Savin DA. Chain Dispersity Effects on Brush Properties of Surface-Grafted Polycaprolactone-Modified Silica Nanoparticles: Unique Scaling Behavior in the Concentrated Polymer Brush Regime. Macromolecules 2017;50:5565–73.

[116] Schmidt AM. The Synthesis of Magnetic Core-Shell Nanoparticles by Surface-Initiated Ring-Opening Polymerization of Caprolactone. Macromol Rapid Commun 2005;26:93–7.

[117] Zhou L, Yuan J, Yuan W, Zhou M, Wu S, Li Z, et al. Synthesis and characterization of multi-functional hybrid magnetite nanoparticles with biodegradability, superparamagnetism, and fluorescence. Mater Lett 2009;63:1567–70.

[118] Zhao L, Takimoto T, Ito M, Kitagawa N, Kimura T, Komatsu N. Chromatographic Separation of Highly Soluble Diamond Nanoparticles Prepared by Polyglycerol Grafting. Angew Chemie Int Ed 2011;50:1388–92.

[119] Zhao L, Shiino A, Qin H, Kimura T, Komatsu N. Synthesis, Characterization, and Magnetic Resonance Evaluation of Polyglycerol-Functionalized Detonation Nanodiamond Conjugated with Gadolinium(III) Complex. J Nanosci Nanotechnol 2015;15:1076–82.

[120] Zhou L, Gao C, Xu W, Wang X, Xu Y. Enhanced Biocompatibility and Biostability of CdTe Quantum Dots by Facile Surface-Initiated Dendritic Polymerization. Biomacromolecules 2009;10:1865–74.

[121] Zhou L, Gao C, Xu W. Efficient Grafting of Hyperbranched Polyglycerol from Hydroxyl-Functionalized Multiwalled Carbon Nanotubes by Surface-Initiated Anionic Ring-Opening Polymerization. Macromol Chem Phys 2009;210:1011–18.

[122] Wang L, Neoh KG, Kang ET, Shuter B, Wang S-C. Superparamagnetic Hyperbranched Polyglycerol-Grafted Fe3O4 Nanoparticles as a Novel Magnetic Resonance Imaging Contrast Agent: An In Vitro Assessment. Adv Funct Mater 2009;19:2615–22.

[123] Braun D, Kamprath A. Versuche zur pfropfung von polystyrol auf rußoberflächen. Die Angew Makromol Chemie 1984;120:1–41.

[124] Viswanathan G, Chakrapani N, Yang H, Wei B, Chung H, Cho K, et al. Single-Step in Situ Synthesis of Polymer-Grafted Single-Wall Nanotube Composites. J Am Chem Soc 2003;125:9258–9.

[125] Chen S, Chen D, Wu G. Grafting of Poly(tBA) and PtBA-b-PMMA onto the Surface of SWNTs Using Carbanions as the Initiator. Macromol Rapid Commun 2006;27:882–7.

[126] Tsubokawa N, Hamada H, Sone Y. Grafting of Polyesters from Carbon Fiber. Anionic Ring-Opening Copolymerization of Epoxides with Cyclic Acid Anhydrides Initiated by COOK Groups on the Surface of Carbon Fiber. Polym Plast Technol Eng 1989;28:201–14.

[127] Tsubokawa N, Yamada A, Sone Y. Grafting of Polyesters on Carbon Black V. Preparation of Polyester-Grafted Carbon Black with a Higher Grafting Ratio by the Copolymerization of Epoxide with Cyclic Acid Anhydrides Using COOK Groups on Carbon Black as the Initiator. Polym J 1984;16:333–40.

[128] Tsubokawa N, Yamada A, Sone Y. Grafting of polyester onto carbon black. Polym Bull 1983;10:63–9.

[129] Tsubokawa N, Funaki A, Hada Y, Sone Y. Grafting polyesters onto carbon black. I. Polymerization of β-propiolactone initiated by alkali metal carboxylate group on the surface of carbon black. J Polym Sci Polym Chem Ed 1982;20:3297–304.

[130] Tsubokawa N, Funaki A, Hada Y, Sone Y. Grafting of polyesters onto carbon black. Polym Bull 1982;7:589–96.

[131] Yang Q, Wang L, Xiang W, Zhou J, Deng L, Li J. Preparation of core-shell carbon black nanoparticles and their crystallization-induced orientation. Eur Polym J 2007;43:1718–23.

[132] Yang Q, Wang L, Huo J, Ding J, Xiang W. Novel comb-structured-polymer-grafted carbon black by surface-initiated atom transfer radical polymerization and ring-opening polymerization. J Appl Polym Sci 2010;117:824–7.

[133] Zeng H-L, Gao C, Yan D-Y. Poly(ε-caprolactone)-Functionalized Carbon Nanotubes and Their Biodegradation Properties. Adv Funct Mater 2006;16:812–18.

[134] Castro M, Lu J, Bruzaud S, Kumar B, Feller JF. Carbon nanotubes/poly(ε-caprolactone) composite vapour sensors. Carbon N Y 2009;47:1930–42.

[135] Priftis D, Petzetakis N, Sakellariou G, Pitsikalis M, Baskaran D, Mays JW, et al. Surface-Initiated Titanium-Mediated Coordination Polymerization from Catalyst-Functionalized Single and Multiwalled Carbon Nanotubes. Macromolecules 2009;42:3340–6.

[136] Priftis D, Sakellariou G, Baskaran D, Mays JW, Hadjichristidis N. Polymer grafted Janus multi-walled carbon nanotubes. Soft Matter 2009;5:4272–8.

[137] Priftis D, Sakellariou G, Mays JW, Hadjichristidis N. Novel diblock copolymer-grafted multiwalled carbon nanotubes via a combination of living and controlled/living surface polymerizations. J Polym Sci Part A Polym Chem 2010;48:1104–12.

[138] Priftis D, Sakellariou G, Hadjichristidis N, Penott EK, Lorenzo AT, Müller AJ. Surface modification of multiwalled carbon nanotubes with biocompatible polymers via ring opening and living anionic surface initiated polymerization. Kinetics and crystallization behavior. J Polym Sci Part A Polym Chem 2009;47:4379–90.

[139] Sakellariou G, Ji H, Mays JW, Baskaran D. Enhanced Polymer Grafting from Multiwalled Carbon Nanotubes through Living Anionic Surface-Initiated Polymerization. Chem Mater 2008;20:6217–30.

[140] Chen G-X, Kim H-S, Park BH, Yoon J-S. Synthesis of Poly(L-lactide)-Functionalized Multiwalled Carbon Nanotubes by Ring-Opening Polymerization. Macromol Chem Phys 2007;208:389–98.

[141] Brzeziński M, Bogusławska M, Ilčíková M, Mosnáček J, Biela T. Unusual Thermal Properties of Polylactides and Polylactide Stereocomplexes Containing Polylactide-Functionalized Multi-Walled Carbon Nanotubes. Macromolecules 2012;45:8714–21.

[142] Buffa F, Hu H, Resasco DE. Side-Wall Functionalization of Single-Walled Carbon Nanotubes with 4-Hydroxymethylaniline Followed by Polymerization of ε-Caprolactone. Macromolecules 2005;38:8258–63.

[143] Lee RS, Chen WH, Lin JH. Polymer-grafted multi-walled carbon nanotubes through surface-initiated ring-opening polymerization and click reaction. Polymer 2011;52:2180–8.

[144] Xu Y, Gao C, Kong H, Yan D, Jin YZ, Watts PCP. Growing Multihydroxyl Hyperbranched Polymers on the Surfaces of Carbon Nanotubes by in Situ Ring-Opening Polymerization. Macromolecules 2004;37:8846–53.

[145] Parvole J, Billon L, Montfort J. Formation of polyacrylate brushes on silica surfaces. Polym Int 2002;51:1111–16.

[146] Kasseh A, Ait-Kadi A, Riedl B, Pierson JF. Organic/inorganic hybrid composites prepared by polymerization compounding and controlled free radical polymerization. Polymer 2003;44:1367–75.

[147] Laruelle G, Parvole J, Francois J, Billon L. Block copolymer grafted-silica particles: a core/double shell hybrid inorganic/organic material. Polymer 2004;45:5013–20.

[148] Bartholome C, Beyou E, Bourgeat-Lami E, Chaumont P, Zydowicz N. Nitroxide-Mediated Polymerizations from Silica Nanoparticle Surfaces: "Graft from" Polymerization of Styrene Using a Triethoxysilyl-Terminated Alkoxyamine Initiator. Macromolecules 2003;36:7946–52.

[149] Bartholome C, Beyou E, Bourgeat-Lami E, Chaumont P, Zydowicz N. Nitroxide-mediated polymerization of styrene initiated from the surface of fumed silica. Comparison of two synthetic routes. Polymer 2005;46:8502–10.

[150] Blomberg S, Ostberg S, Harth E, Bosman AW, Van Horn B, Hawker CJ. Production of crosslinked, hollow nanoparticles by surface-initiated living free-radical polymerization. J Polym Sci Part A Polym Chem 2002;40:1309–20.

[151] Ladmiral V, Morinaga T, Ohno K, Fukuda T, Tsujii Y. Synthesis of monodisperse zinc sulfide particles grafted with concentrated polystyrene brush by surface-initiated nitroxide-mediated polymerization. Eur Polym J 2009;45:2788–96.

[152] Matsuno R, Yamamoto K, Otsuka H, Takahara A. Polystyrene- and Poly(3-vinylpyridine)-Grafted Magnetite Nanoparticles Prepared through Surface-Initiated Nitroxide-Mediated Radical Polymerization. Macromolecules 2004;37:2203–9.

[153] Zhao X-D, Fan X-H, Chen X-F, Chai C-P, Zhou Q-F. Surface modification of multiwalled carbon nanotubes via nitroxide-mediated radical polymerization. J Polym Sci Part A Polym Chem 2006;44:4656–67.

[154] Huang X, Wirth MJ. Surface Initiation of Living Radical Polymerization for Growth of Tethered Chains of Low Polydispersity. Macromolecules 1999;32:1694–6. Available from: https://doi.org/10.1021/MA981304X.

[155] von Werne T, Patten TE. Preparation of Structurally Well-Defined Polymer – Nanoparticle Hybrids with Controlled/Living Radical Polymerizations. J Am Chem Soc 1999;121:7409–10.

[156] Harrak A El, Carrot G, Oberdisse J, Eychenne-Baron C, Boué F. Surface – Atom Transfer Radical Polymerization from Silica Nanoparticles with Controlled Colloidal Stability. Macromolecules 2004;37:6376–84.

[157] Wang Y-P, Pei X-W, He X-Y, Yuan K. Synthesis of well-defined, polymer-grafted silica nanoparticles via reverse ATRP. Eur Polym J 2005;41:1326–32.

[158] Böttcher H, Hallensleben ML, Nuß S, Wurm H. ATRP grafting from silica surface to create first and second generation of grafts. Polym Bull 2000;44:223–9.

[159] Perruchot C, Khan MA, Kamitsi A, Armes SP, von Werne T, Patten TE. Synthesis of Well-Defined, Polymer-Grafted Silica Particles by Aqueous ATRP. Langmuir 2001;17:4479–81.

[160] Carrot G, Diamanti S, Manuszak M, Charleux B, Vairon J-P. Atom transfer radical polymerization ofn-butyl acrylate from silica nanoparticles. J Polym Sci Part A Polym Chem 2001;39:4294–301.

[161] Radhakrishnan B, Constable AN, Brittain WJ. A Novel Route to Organic-Inorganic Hybrid Nanomaterials. Macromol Rapid Commun 2008;29:1828–33.

[162] von Werne T, Patten TE. Atom Transfer Radical Polymerization from Nanoparticles: A Tool for the Preparation of Well-Defined Hybrid Nanostructures and for Understanding the Chemistry of Controlled/"Living" Radical Polymerizations from Surfaces. J Am Chem Soc 2001;123:7497–505.

[163] Pyun J, Matyjaszewski K, Kowalewski T, Savin D, Patterson G, Kickelbick G, et al. Synthesis of Well-Defined Block Copolymers Tethered to Polysilsesquioxane Nanoparticles and Their Nanoscale Morphology on Surfaces. J Am Chem Soc 2001;123:9445–6.

[164] Pyun J, Jia S, Kowalewski T, Patterson GD, Matyjaszewski K. Synthesis and Characterization of Organic/Inorganic Hybrid Nanoparticles: Kinetics of Surface-Initiated Atom Transfer Radical Polymerization and Morphology of Hybrid Nanoparticle Ultrathin Films. Macromolecules 2003;36:5094–104.

[165] Tang C, Bombalski L, Kruk M, Jaroniec M, Matyjaszewski K, Kowalewski T. Nanoporous Carbon Films from "Hairy" Polyacrylonitrile-Grafted Colloidal Silica Nanoparticles. Adv Mater 2008;20:1516–22.

[166] Zhang H, Lei X, Su Z, Liu P. A novel method of surface-initiate atom transfer radical polymerization of styrene from silica nanoparticles for preparation of monodispersed core-shell hybrid nanospheres. J Polym Res 2007;14:253–60.

[167] Zhang K, Ma J, Zhang B, Zhao S, Li Y, Xu Y, et al. Synthesis of thermoresponsive silica nanoparticle/PNIPAM hybrids by aqueous surface-initiated atom transfer radical polymerization. Mater Lett 2007;61:949–52.

[168] Hayashida K, Tanaka H, Watanabe O. Miscible blends of poly(butyl methacrylate) densely grafted on fumed silica with poly(vinyl chloride). Polymer 2009;50:6228–34.

[169] Mori H, Seng DC, Zhang M, Müller AHE. Hybrid Nanoparticles with Hyperbranched Polymer Shells via Self-Condensing Atom Transfer Radical Polymerization from Silica Surfaces. Langmuir 2002;18:3682–93.

[170] Li D, Sheng X, Zhao B. Environmentally Responsive "Hairy" Nanoparticles: Mixed Homopolymer Brushes on Silica Nanoparticles Synthesized by Living Radical Polymerization Techniques. J Am Chem Soc 2005;127:6248–56.

[171] Bombalski L, Dong H, Listak J, Matyjaszewsk K, Bockstaller MR. Null-Scattering Hybrid Particles Using Controlled Radical Polymerization. Adv Mater 2007;19:4486–90.

[172] Farmer SC, Patten TE. Photoluminescent Polymer/Quantum Dot Composite Nanoparticles. Chem Mater 2001;13:3920–6.

[173] Mandal TK, Fleming MS, Walt DR. Preparation of Polymer Coated Gold Nanoparticles by Surface-Confined Living Radical Polymerization at Ambient Temperature. Nano Lett 2002;2:3–7.

[174] Ohno K, Koh K-M, Tsujii Y, Fukuda T. Synthesis of Gold Nanoparticles Coated with Well-Defined, High-Density Polymer Brushes by Surface-Initiated Living Radical Polymerization. Macromolecules 2002;35:8989–93.

[175] Ohno K, Koh K, Tsujii Y, Fukuda T. Fabrication of Ordered Arrays of Gold Nanoparticles Coated with High-Density Polymer Brushes. Angew Chemie Int Ed 2003;42:2751–4.

[176] Kim DJ, Kang SM, Kong B, Kim W-J, Paik H, Choi H, et al. Formation of Thermoresponsive Gold Nanoparticle/PNIPAAm Hybrids by Surface-Initiated, Atom Transfer Radical Polymerization in Aqueous Media. Macromol Chem Phys 2005;206:1941–6.

[177] Li D, He Q, Cui Y, Li J. Fabrication of pH-Responsive Nanocomposites of Gold Nanoparticles/Poly(4-vinylpyridine). Chem Mater 2007;19:412–17.

[178] Dong H, Zhu M, Yoon JA, Gao H, Jin R, Matyjaszewski K. One-Pot Synthesis of Robust Core/Shell Gold Nanoparticles. J Am Chem Soc 2008;130:12852–3.

[179] Li L, Davidson JL, Lukehart CM. Surface functionalization of nanodiamond particles via atom transfer radical polymerization. Carbon N Y 2006;44:2308–15.

[180] Dahoumane SA, Nguyen MN, Thorel A, Boudou J-P, Chehimi MM, Mangeney C. Protein-Functionalized Hairy Diamond Nanoparticles. Langmuir 2009;25:9633–8.

[181] Zhang X, Fu C, Feng L, Ji Y, Tao L, Huang Q, et al. PEGylation and poly PEGylation of nanodiamond. Polymer 2012;53:3178–84.

[182] Gelbrich T, Feyen M, Schmidt AM. Magnetic Thermoresponsive Core–Shell Nanoparticles. Macromolecules 2006;39:3469–72.

[183] Yuan W, Yuan J, Zhou L, Wu S, Hong X. Fe3O4@poly(2-hydroxyethyl methacrylate)-graft-poly(ε-caprolactone) magnetic nanoparticles with branched brush polymeric shell. Polymer 2010;51:2540–7.

[184] Zhou L, Yuan J, Yuan W, Sui XS, Wu S, Li Z, et al. Synthesis, characterization, and controllable drug release of pH-sensitive hybrid magnetic nanoparticles. J Magn Magn Mater 2009;321:2799–804.

[185] Vestal CR, Zhang ZJ. Atom Transfer Radical Polymerization Synthesis and Magnetic Characterization of MnFe2O4/Polystyrene Core/Shell Nanoparticles. J Am Chem Soc 2002;124:14312–13.

[186] Qin S, Qin D, Ford WT, Resasco DE, Herrera JE. Functionalization of Single-Walled Carbon Nanotubes with Polystyrene via Grafting to and Grafting from Methods. Macromolecules 2004;37:752–7.

[187] Qin S, Qin D, Ford WT, Resasco DE, Herrera JE. Polymer Brushes on Single-Walled Carbon Nanotubes by Atom Transfer Radical Polymerization of *n* -Butyl Methacrylate. J Am Chem Soc 2004;126:170–6.

[188] Wu W, Tsarevsky NV, Hudson JL, Tour JM, Matyjaszewski K, Kowalewski T. "Hairy" Single-Walled Carbon Nanotubes Prepared by Atom Transfer Radical Polymerization. Small 2007;3:1803–10.

[189] Baskaran D, Mays JW, Bratcher MS. Polymer-Grafted Multiwalled Carbon Nanotubes through Surface-Initiated Polymerization. Angew Chemie Int Ed 2004;43:2138–42.

[190] Shanmugharaj AM, Bae JH, Nayak RR, Ryu SH. Preparation of poly(styrene-co-acrylonitrile)-grafted multiwalled carbon nanotubes via surface-initiated atom transfer radical polymerization. J Polym Sci Part A Polym Chem 2007;45:460–70.

[191] Yao Z, Braidy N, Botton GA, Adronov A. Polymerization from the Surface of Single-Walled Carbon Nanotubes – Preparation and Characterization of Nanocomposites. J Am Chem Soc 2003;125:16015–24.

[192] Kong H, Gao C, Yan D. Functionalization of Multiwalled Carbon Nanotubes by Atom Transfer Radical Polymerization and Defunctionalization of the Products. Macromolecules 2004;37:4022–30.

[193] Kong H, Gao C, Yan D. Constructing amphiphilic polymer brushes on the convex surfaces of multi-walled carbon nanotubes by in situ atom transfer radical polymerization. J Mater Chem 2004;14:1401–5.

[194] Kong H, Gao C, Yan D. Controlled Functionalization of Multiwalled Carbon Nanotubes by in Situ Atom Transfer Radical Polymerization. J Am Chem Soc 2004;126:412–13.

[195] Ilčíková M, Danko M, Doroshenko M, Best A, Mrlík M, Csomorová K, et al. Visualization of carbon nanotubes dispersion in composite by using confocal laser scanning microscopy. Eur Polym J 2016;79:187–97.

[196] Fang M, Wang K, Lu H, Yang Y, Nutt S, Guo TY, et al. Covalent polymer functionalization of graphene nanosheets and mechanical properties of composites. J Mater Chem 2009;19:7098–105.

[197] Wang D, Ye G, Wang X, Wang X. Graphene Functionalized with Azo Polymer Brushes: Surface-Initiated Polymerization and Photoresponsive Properties. Adv Mater 2011;23:1122–5.

[198] Chen X, Yuan L, Yang P, Hu J, Yang D. Covalent polymeric modification of graphene nanosheets via surface-initiated single-electron-transfer living radical polymerization. J Polym Sci Part A Polym Chem 2011;49:4977–86.

[199] Yang Y, Wang J, Zhang J, Liu J, Yang X, Zhao H. Exfoliated Graphite Oxide Decorated by PDMAEMA Chains and Polymer Particles. Langmuir 2009;25:11808–14.

[200] Gonçalves G, Marques PAAP, Barros-Timmons A, Bdkin I, Singh MK, Emami N, et al. Graphene oxide modified with PMMA via ATRP as a reinforcement filler. J Mater Chem 2010;20:9927–34.

[201] Pei X, Hao J, Liu W. Preparation and Characterization of Carbon Nanotubes-Polymer / Ag Hybrid Nanocomposites via Surface RAFT Polymerization. J Phys Chem C 2007;111:2947–52.

[202] Rotzoll R, Vana P. A bipedal silica-immobilized azo-initiator for surface-confined radical polymerizations. Aust J Chem 2009;62:1473–8.

[203] Baum M, Brittain WJ. Synthesis of Polymer Brushes on Silicate Substrates via Reversible Addition Fragmentation Chain Transfer Technique. Macromolecules 2002;35:610–15.

[204] Moraes J, Ohno K, Maschmeyer T, Perrier S. Synthesis of silica–polymer core–shell nanoparticles by reversible addition–fragmentation chain transfer polymerization. Chem Commun 2013;49:9077–88.

[205] Zhao Y, Perrier S. Synthesis of Well-Defined Homopolymer and Diblock Copolymer Grafted onto Silica Particles by Z-Supported RAFT Polymerization. Macromolecules 2006;39:8603–8.

[206] Perrier S, Takolpuckdee P, Mars CA. Reversible Addition – Fragmentation Chain Transfer Polymerization Mediated by a Solid Supported Chain Transfer Agent. Macromolecules 2005;38:6770–4.

[207] Stenzel MH, Zhang L, Huck WTS. Temperature-responsive glycopolymer brushes synthesized via RAFT polymerization using the Z-group approach. Macromol Rapid Commun 2006;27:1121–6.

[208] Zhao Y, Perrier S. Reversible Addition–Fragmentation Chain Transfer Graft Polymerization Mediated by Fumed Silica Supported Chain Transfer Agents. Macromolecules 2007;40:9116–24.

[209] Nguyen DH, Vana P. Silica-immobilized cumyl dithiobenzoate as mediating agent in reversible addition fragmentation chain transfer (RAFT) polymerization. Polym Adv Technol 2006;17:625–33.

[210] Li Y, Benicewicz BC. Functionalization of Silica Nanoparticles via the Combination of Surface-Initiated RAFT Polymerization and Click Reactions Functionalization of Silica Nanoparticles via the Combination of Surface-Initiated RAFT Polymerization and Click Reactions. Macromolecules 2008;41:7986–92.

[211] Xu G, Wu W-T, Wang Y, Pang W, Zhu Q, Wang P, et al. Constructing polymer brushes on multiwalled carbon nanotubes by in situ reversible addition fragmentation chain transfer polymerization. Polymer 2006;47:5909–18.

[212] Hong C-Y, You Y-Z, Pan C-Y. Synthesis of Water-Soluble Multiwalled Carbon Nanotubes with Grafted Temperature-Responsive Shells by Surface RAFT Polymerization. Chem Mater 2005;17:2247–54.

[213] Tsujii Y, Ejaz M, Sato K, Goto A, Fukuda T. Mechanism and Kinetics of RAFT-Mediated Graft Polymerization of Styrene on a Solid Surface. 1. Experimental Evidence of Surface Radical Migration. Macromolecules 2001;34:8872–8.

[214] Li C, Han J, Ryu CY, Benicewicz BC. A Versatile Method To Prepare RAFT Agent Anchored Substrates and the Preparation of PMMA Grafted Nanoparticles. Macromolecules 2006;39:3175–83.

[215] Yang Q, Wang L, Xiang W, Zhou J, Tan Q. hua. A temperature-responsive carbon black nanoparticle prepared by surface-induced reversible addition-fragmentation chain transfer polymerization. Polymer 2007;48:3444–51.

[216] Rotzoll R, Vana P. Synthesis of poly(methyl acrylate) loops grafted onto silica nanoparticles via reversible addition-fragmentation chain transfer polymerization. J Polym Sci Part A Polym Chem 2008;46:7656–66.

[217] Roghani-Mamaqani H, Haddadi-Asl V, Ghaderi-Ghahfarrokhi M, Sobhkhiz Z. Reverse atom transfer radical polymerization of methyl methacrylate in the presence of Azo-functionalized carbon nanotubes: a grafting from approach. Colloid Polym Sci 2014;292:2971–81.

[218] Moraes J, Ohno K, Gody G, Maschmeyer T, Perrier S. The synthesis of well-defined poly(vinylbenzyl chloride)-grafted nanoparticles via RAFT polymerization. Beilstein J Org Chem 2013;9:1226–34.

[219] Hong C-Y, Li X, Pan C-Y. Grafting polymer nanoshell onto the exterior surface of mesoporous silica nanoparticles via surface reversible addition-fragmentation chain transfer polymerization. Eur Polym J 2007;43:4114–22.

[220] Liu C-H, Pan C-Y. Grafting polystyrene onto silica nanoparticles via RAFT polymerization. Polymer 2007;48:3679–85.

[221] Ohno K, Ma Y, Huang Y, Mori C, Yahata Y, Tsujii Y, et al. Surface-Initiated Reversible Addition−Fragmentation Chain Transfer (RAFT) Polymerization from Fine Particles Functionalized with Trithiocarbonates. Macromolecules 2011;44:8944−53.

[222] Ebeling B, Vana P. RAFT-Polymers with Single and Multiple Trithiocarbonate Groups as Uniform Gold-Nanoparticle Coatings. Macromolecules 2013;46:4862−71.

[223] Ding W, Wang L. Synthesis of poly(acrylic acid) grafted carbon black and its application for sensing ethanol. J Polym Res 2014;21:425.

[224] Shi Y, Liu M, Wang K, Huang H, Wan Q, Tao L, et al. Direct surface PEGylation of nanodiamond via RAFT polymerization. Appl Surf Sci 2015;357:2147−53.

[225] Huynh VT, Pearson S, Noy J-M, Abboud A, Utama RH, Lu H, et al. Nanodiamonds with Surface Grafted Polymer Chains as Vehicles for Cell Imaging and Cisplatin Delivery: Enhancement of Cell Toxicity by POEGMEMA Coating. ACS Macro Lett 2013;2:246−50.

[226] Li Y, Li X, Dong C, Qi J, Han X. A graphene oxide-based molecularly imprinted polymer platform for detecting endocrine disrupting chemicals. Carbon N Y 2010;48:3427−33.

[227] Cui J, Wang W, YeZi Y, Liu C, Wang P. Functionalization of multiwalled carbon nanotubes by reversible addition fragmentation chain-transfer polymerization. Polymer 2004;45:8717−21.

[228] Zhang B, Wang J, Chen Y, Früchtl D, Yu B, Zhuang X, et al. Multiwalled carbon nanotubes covalently functionalized with poly(N-vinylcarbazole) via RAFT polymerization: Synthesis and nonliner optical properties. J Polym Sci Part A Polym Chem 2010;48:3161−8.

[229] Qiao M, Wu S, Wang Y, Ran Q. Brush-like block copolymer synthesized via RAFT polymerization for graphene oxide aqueous suspensions. RSC Adv 2017;7:4776−82.

[230] Yang Y, Song X, Yuan L, Li M, Liu J, Ji R, et al. Synthesis of PNIPAM polymer brushes on reduced graphene oxide based on click chemistry and RAFT polymerization. J Polym Sci Part A Polym Chem 2012;50:329−37.

[231] Jiang K, Ye C, Zhang P, Wang X, Zhao Y. One-Pot Controlled Synthesis of Homopolymers and Diblock Copolymers Grafted Graphene Oxide Using Couplable RAFT Agents. Macromolecules 2012;45:1346−55.

[232] Etmimi HM, Tonge MP, Sanderson RD. Synthesis and characterization of polystyrene-graphite nanocomposites via surface RAFT-mediated miniemulsion polymerization. J Polym Sci Part A Polym Chem 2011;49:1621−32.

[233] Dekking HGG. Propagation of vinyl polymers on clay surfaces. I. Preparation, structure, and decomposition of clay initiators. J Appl Polym Sci 1965;9:1641−51.

[234] Dekking HGG. Propagation of vinyl polymers on clay surfaces. II. Polymerization of monomers initiated by free radicals attached to clay. J Appl Polym Sci 1967;11:23−36.

[235] Zhang J, Yang Y, Zhao C, Zhao H. PS/PMMA mixed polymer brushes on the surface of clay layers: Preparation and application in polymer blends. J Polym Sci Part A Polym Chem 2007;45:5329−38.

[236] Nakatsuka T. Polyacrylate-graft silica gel as a support of lipase intersterifying triacylglycerol in organic solvent. J Appl Polym Sci 1987;34:2125−37.

[237] Tsubokawa N, Kogure A, Maruyama K, Sone Y, Shimomura M. Graft Polymerization of Vinyl Monomers from Inorganic Ultrafine Particles Initiated by Azo Groups Introduced onto the Surface. Polym J 1990;22:827−33.

[238] Prucker O, Rühe J. Synthesis of Poly(styrene) Monolayers Attached to High Surface Area Silica Gels through Self-Assembled Monolayers of Azo Initiators. Macromolecules 1998;31:592−601.

[239] Prucker O, Rühe J. Mechanism of Radical Chain Polymerizations Initiated by Azo Compounds Covalently Bound to the Surface of Spherical Particles. Macromolecules 1998;31:602−13.

[240] Prucker O, Rühe J. Polymer Layers through Self-Assembled Monolayers of Initiators. Langmuir 1998;14:6893–8.

[241] Czaun M, Hevesi L, Takafuji M, Ihara H. Novel Surface-Attachable Multifunctional Initiators: Synthesis, Grafting, and Polymerization in Aprotic and Protic Solvents. Macromolecules 2009;42:4539–46.

[242] Bialk M, Prucker O, Ruhe J. Grafting of polymers to solid surfaces by using immobilized methacrylates. Colloids Surfaces A Physicochem Eng Asp 2002;198–200:543–9.

[243] Datta P, Genzer J. "Grafting through" polymerization involving surface-bound monomers. J Polym Sci Part A Polym Chem 2016;54:263–74.

[244] Chaimberg M, Parnas R, Cohen Y. Graft polymerization of polyvinylpyrrolidone onto silica. J Appl Polym Sci 1989;37:2921–31.

[245] Browne T, Chaimberg M, Cohen Y. Graft polymerization of vinyl acetate onto silica. J Appl Polym Sci 1992;44:671–7.

[246] Laible R, Hamann K. Formation of chemically bound polymer layers on oxide surfaces and their role in colloidal stability. Adv Colloid Interface Sci 1980;13:65–99.

[247] Sunkara HB, Jethmalani JM, Ford WT. Composite of Colloidal Crystals of Silica in Poly(methyl methacrylate). Chem Mater 1994;6:362–4.

[248] Salami-Kalajahi M, Haddadi-Asl V, Rahimi-Razin S, Behboodi-Sadabad F, Roghani-Mamaqani H, Hemmati M. Investigating the effect of pristine and modified silica nanoparticles on the kinetics of methyl methacrylate polymerization. Chem Eng J 2011;174:368–75.

[249] Pu Z, Mark JE, Jethmalani JM, Ford WT. Mechanical properties of a poly(methyl acrylate) nanocomposite containing regularly-arranged silica particles. Polym Bull 1996;37:545–51.

[250] Sondi I, Fedynyshyn TH, Sinta R, Matijević E. Encapsulation of Nanosized Silica by in Situ Polymerization of tert-Butyl Acrylate Monomer†. Langmuir 2000;16:9031–4.

[251] Bourgeat-Lami E, Lang J. Encapsulation of Inorganic Particles by Dispersion Polymerization in Polar Media: 1. Silica Nanoparticles Encapsulated by Polystyrene. J Colloid Interface Sci 1998;197:293–308.

[252] Barthet C, Hickey AJ, Cairns DB, Armes SP. Synthesis of Novel Polymer-Silica Colloidal Nanocomposites via Free-Radical Polymerization of Vinyl Monomers. Adv Mater 1999;11:408–10.

[253] Henze M, Mädge D, Prucker O, Rühe J. "Grafting Through": Mechanistic Aspects of Radical Polymerization Reactions with Surface-Attached Monomers. Macromolecules 2014;47:2929–37.

[254] Wikberg E, Verhage JJ, Viklund C, Irgum K. Grafting of silica with sulfobetaine polymers via aqueous reversible addition fragmentation chain transfer polymerization and its use as a stationary phase in HILIC. J Sep Sci 2009;32:2008–16.

[255] Ji QL, Rong MZ, Zhang MQ, Klaus F. Graft Polymerization of Vinyl Monomers onto Nanosized Silicon Carbide Particles. Polym Polym Compos 2002;10:531–40.

[256] Rong MZ, Zhang MQ, Shi G, Ji QL, Wetzel B, Friedrich K. Graft polymerization onto inorganic nanoparticles and its effect on tribological performance improvement of polymer composites. Tribol Int 2003;36:697–707.

[257] Sugimoto H, Daimatsu K, Nakanishi E, Ogasawara Y, Yasumura T, Inomata K. Preparation and properties of poly(methylmethacrylate)–silica hybrid materials incorporating reactive silica nanoparticles. Polymer 2006;47:3754–9.

[258] Chinthamanipeta PS, Kobukata S, Nakata H, Shipp DA. Synthesis of poly(methyl methacrylate)–silica nanocomposites using methacrylate-functionalized silica nanoparticles and RAFT polymerization. Polymer 2008;49:5636–42.

[259] Sobani M, Haddadi-Asl V, Salami-Kalajahi M, Roghani-Mamaqani H, Mirshafiei-Langari S-A, Khezri K. "Grafting through" approach for synthesis of polystyrene/silica aerogel nanocomposites by in situ reversible addition-fragmentation chain transfer polymerization. J Sol-Gel Sci Technol 2013;66:337–44.

[260] Chatterjee A, Mishra S. Novel synthesis with an atomized microemulsion technique and characterization of nano-calcium carbonate (CaCO3)/poly(methyl methacrylate) core–shell nanoparticles. Particuology 2013;11:760–7.

[261] Chatterjee A, Mishra S. Rheological, thermal and mechanical properties of nano-calcium carbonate (CaCO3)/Poly(methyl methacrylate) (PMMA) core-shell nanoparticles reinforced polypropylene (PP) composites. Macromol Res 2013;21:474–83.

[262] Chatterjee A, Mishra S. Nano - Calcium carbonate (CaCO 3) / Polystyrene (PS) core - shell nanoparticle: It's effect on physical and mechanical properties of high impact polystyrene (HIPS). J Polym Res 2013;20:1–12.

[263] Mishra S, Chatterjee A, Singh R. Novel synthesis of nano-calcium carbonate (CaCO3)/polystyrene (PS) core-shell nanoparticles by atomized microemulsion technique and its effect on properties of polypropylene (PP) composites. Polym Adv Technol 2011;22:2571–82.

[264] Ahmadian-Alam L, Haddadi-Asl V, Roghani-Mamaqani H, Hatami L, Salami-Kalajahi M. Use of clay-anchored reactive modifier for the synthesis of poly (styrene-co-butyl acrylate)/clay nanocomposite via in situ AGET ATRP. J Polym Res 2012;19:9773.

[265] Qu L, Veca LM, Lin Y, Kitaygorodskiy A, Chen B, McCall AM, et al. Soluble Nylon-Functionalized Carbon Nanotubes from Anionic Ring-Opening Polymerization from Nanotube Surface. Macromolecules 2005;38:10328–31.

[266] Liu Y, Adronov A. Preparation and Utilization of Catalyst-Functionalized Single-Walled Carbon Nanotubes for Ring-Opening Metathesis Polymerization. Macromolecules 2004;37:4755–60.

[267] Kojima Y, Usuki A, Kawasumi M, Okada A, Fukushima Y, Kurauchi T, et al. Mechanical-Properties of Nylon 6-Clay Hybrid. J Mater Res 1993;8:1185–9.

[268] Fu HK, Huang CF, Huang JM, Chang FC. Studies on thermal properties of PS nanocomposites for the effect of intercalated agent with side groups. Polymer 2008;49:1305–11.

[269] Tseng CR, Wu JY, Lee HY, Chang FC. Preparation and characterization of polystyrene-clay nanocomposites by free-radical polymerization. J Appl Polym Sci 2002;85:1370–7.

[270] Wu HD, Tseng CR, Chang FC. Chain conformation and crystallization behavior of the syndiotactic polystyrene nanocomposites studied using Fourier transform infrared analysis. Macromolecules 2001;34:2992–9.

[271] Tseng CR, Wu JY, Lee HY, Chang FC. Preparation and crystallization behavior of syndiotactic polystyrene-clay nanocomposites. Polymer 2001;42:10063–70.

[272] Kuo SW, Huang WJ, Huang SB, Kao HC, Chang FC. Syntheses and characterizations of in situ blended metallocence polyethylene/clay nanocomposites. Polymer 2003;44:7709–19.

[273] Gu AJ, Kuo SW, Chang FC. Syntheses and properties of PI/clay hybrids. J Appl Polym Sci 2001;79:1902–10.

[274] Gu AJ, Chang FC. A novel preparation of polyimide/clay hybrid films with low coefficient of thermal expansion. J Appl Polym Sci 2001;79:289–94.

[275] Chen HW, Chiu CY, Chang FC. Conductivity enhancement mechanism of the poly(ethylene oxide)/modified-clay-LiClO4 systems. J Polym Sci Part B-Polymer Phys 2002;40:1342–53.

[276] Chen HW, Chang FC. The novel polymer electrolyte nanocomposite composed of poly(ethylene oxide), lithium triflate and mineral clay. Polymer 2001;42:9763–9.

[277] Yei DR, Fu HK, Chen WY, Chang FC. Synthesis of a novel benzoxazine monomer-intercalated montmorillonite and the curing kinetics of polybenzoxazine/clay hybrid nanocomposites. J Polym Sci Part B-Polymer Phys 2006;44:347–58.

[278] Fu H-K, Huang C-F, Kuo S-W, Lin H-C, Yei D-R, Chang F-C. Effect of an Organically Modified Nanoclay on Low-Surface-energy Materials of Polybenzoxazine. Macromol Rapid Commun 2008;29:1216–20.

[279] Ye YS, Cheng MY, Xie XL, Rick J, Huang YJ, Chang FC, et al. Alkali doped polyvinyl alcohol/graphene electrolyte for direct methanol alkaline fuel cells. J Power Sources 2013;239:424–32.

[280] Wang C-F, Su Y-C, Kuo S-W, Huang C-F, Sheen Y-C, Chang F-C. Low-surface-free-energy Materials Based on Polybenzoxazines. Angew Chemie-International Ed 2006;45:2248–51.

[281] Liao CS, Wu JS, Wang CF, Chang FC. Modification of polymer substrates with low surface free energy material by low-temperature cured polybenzoxazine. Macromol Rapid Commun 2008;29:52–6.

[282] Wang CF, Chiou SF, Ko FH, Chen JK, Chou CT, Huang CF, et al. Polybenzoxazine as a mold-release agent for nanoimprint lithography. Langmuir 2007;23:5868–71.

[283] Gournis D, Jankovic L, Maccallini E, Benne D, Rudolf P, Colomer JF, et al. Clay-fulleropyrrolidine nanocomposites. J Am Chem Soc 2006;128:6154–63.

[284] Scott DW. Thermal rearrangement of branched-chain methylpolysiloxanes. J Am Chem Soc 1946;68:356–8.

[285] Lee YJ, Kuo SW, Huang WJ, Lee HY, Chang FC. Miscibility, specific interactions, and self-assembly behavior of phenolic/polyhedral oligomeric silsesquioxane hybrids. J Polym Sci Part B-Polymer Phys 2004;42:1127–36.

[286] Xu HY, Kuo SW, Chang FC. Significant glass transition temperature increase based on polyhedral oligomeric silsequioxane (POSS) copolymer through hydrogen bonding. Polym Bull 2002;48:469–74.

[287] Kuo SW, Chang FC. POSS related polymer nanocomposites. Prog Polym Sci 2011;36:1649–96.

[288] Lin HM, Hseih KH, Chang FC. Characterization of negative-type photoresists containing polyhedral oligomeric silsesquioxane methacrylate. Microelectron Eng 2008;85:1624–8.

[289] Lin HM, Wu SY, Huang PY, Huang CF, Kuo SW, Chang FC. Polyhedral oligomeric silsesquioxane containing copolymers for negative-type photoresists. Macromol Rapid Commun 2006;27:1550–5.

[290] Wang YZ, Chen WY, Yang CC, Lin CL, Chang FC. Novel epoxy nanocomposite of low D-k introduced fluorine-containing POSS structure. J Polym Sci Part B-Polymer Phys 2007;45:502–10.

[291] Siro I, Plackett D. Microfibrillated cellulose and new nanocomposite materials: a review. Cellulose 2010;17:459–94.

[292] Eichhorn SJ. Cellulose nanowhiskers: promising materials for advanced applications. Soft Matter 2011;7:303–15.

[293] Nogi M, Iwamoto S, Nakagaito AN, Yano H. Optically Transparent Nanofiber Paper. Adv Mater 2009;21:1595–8.

[294] Isogai A, Kato Y. Preparation of polyglucuronic acid from cellulose by TEMPO-mediated oxidation. Cellulose 1998;5:153–64.

[295] Lin K-YA, Heish Y-T, Tsai T-Y, Huang C-F. TEMPO-oxidized Pulp as an Efficient and Recyclable Sorbent to Remove Paraquat from Water. Cellulose 2015;22:3261–74.

[296] Goffin A-L, Raquez J-M, Duquesne E, Siqueira G, Habibi Y, Dufresne A, et al. From Interfacial Ring-Opening Polymerization to Melt Processing of Cellulose Nanowhisker-Filled Polylactide-Based Nanocomposites. Biomacromolecules 2011;12:2456–65.

[297] Habibi Y, Aouadi S, Raquez J-M, Dubois P. Effects of interfacial stereocomplexation in cellulose nanocrystal-filled polylactide nanocomposites. Cellulose 2013;20:2877–85.

[298] Goffin A-L, Raquez J-M, Duquesne E, Siqueira G, Habibi Y, Dufresne A, et al. Poly(ε-caprolactone) based nanocomposites reinforced by surface-grafted cellulose nanowhiskers via extrusion processing: Morphology, rheology, and thermo-mechanical properties. Polymer 2011;52:1532–8.

[299] Habibi Y, Goffin A-L, Schiltz N, Duquesne E, Dubois P, Dufresne A. Bionanocomposites based on poly (ε-caprolactone)-grafted cellulose nanocrystals by ring-opening polymerization. J Mater Chem 2008;18:5002–10.

[300] Yi J, Xu Q, Zhang X, Zhang H. Chiral-nematic self-ordering of rodlike cellulose nanocrystals grafted with poly(styrene) in both thermotropic and lyotropic states. Polymer 2008;49:4406–12.

[301] Morandi G, Heath L, Thielemans W. Cellulose Nanocrystals Grafted with Polystyrene Chains through Surface-Initiated Atom Transfer Radical Polymerization (SI-ATRP). Langmuir 2009;25:8280–6.

[302] Lacerda PSS, Barros-Timmons AMMV, Freire CSR, Silvestre AJD, Neto CP. Nanostructured Composites Obtained by ATRP Sleeving of Bacterial Cellulose Nanofibers with Acrylate Polymers. Biomacromolecules 2013;14:2063–73.

[303] Tsai T-Y, Huang C-F. Data in Support of Dualfunctionalized Cellulose Nanofibrils Prepared through TEMPO-mediated Oxidation and Surface-initiated ATRP. Data Br 2015;3:195–200.

[304] Huang C-F, Chen J-K, Tsai T-Y, Hsieh Y-A, Lin K-YA. Dual-functionalized Cellulose Nanofibrils Prepared through TEMPO-mediated Oxidation and Surface-initiated ATRP. Polymer 2015;72:395–405.

[305] Gobi N, Vijayakumar D, Keles O, Erogbogbo F. Infusion of Graphene Quantum Dots to Create Stronger, Tougher, and Brighter Polymer Composites. ACS Omega 2017;2:4356–62.

[306] Xiao P, Qiu NX, Gu JC, Wang S, He J, Huang CF, et al. Direct supramolecular interacted graphene oxide assembly on graphene as an active and defect-free functional platform. Chem Commun 2017;53:1949–52.

[307] Ye YS, Xie XL, Rick J, Chang FC, Hwang BJ. Improved anode materials for lithium-ion batteries comprise non-covalently bonded graphene and silicon nanoparticles. J Power Sources 2014;247:991–8.

[308] Ye YS, Chen YN, Wang JS, Rick J, Huang YJ, Chang FC, et al. Versatile Grafting Approaches to Functionalizing Individually Dispersed Graphene Nanosheets Using RAFT Polymerization and Click Chemistry. Chem Mater 2012;24:2987–97.

[309] Ilčíková M, Mrlík M, Špitalský Z, Mičušík M, Csomorová K, Sasinková V, et al. A tertiary amine in two competitive processes: reduction of graphene oxide vs. catalysis of atom transfer radical polymerization. RSC Adv 2015;5:3370–6.

[310] Mrlík M, Ilčíková M, Plach T, Pavlínek V, Špitalsk Z, Mosnáček J. Graphene oxide reduction during surface-initiated atom transfer radical polymerization of glycidyl methacrylate: Controlling electro-responsive properties. Chem Eng J 2016;283:717–20.

[311] Mrlík M, Ilčíková M, Plachý T, Moučka R, Pavlínek V, Mosnáček J. Tunable electrorheological performance of silicone oil suspensions based on controllably reduced graphene oxide by surface initiated atom transfer radical polymerization of poly(glycidyl methacrylate). J Ind Eng Chem 2018;57:104–12.

[312] Stuart MAC, Huck WTS, Genzer J, Muller M, Ober C, Stamm M, et al. Emerging applications of stimuli-responsive polymer materials. Nat Mater 2010;9:101–13.

[313] Fu H-K, Kuo S-W, Huang C-F, Chang F-C, Lin H-C. Preparation of the Stimuli-responsive ZnS/PNIPAM Hollow Spheres. Polymer 2009;50:1246–50.

[314] Manivannan K, Huang YS, Huang BR, Huang CF, Chen JK. Real-Time Packing Behavior of Core-Shell Silica@Poly(N-isopropylacrylamide) Microspheres as Photonic Crystals for Visualizing in Thermal Sensing. Polymers (Basel) 2016;8:428.

[315] Huang C-F, Yoon JA, Matyjaszewski K. Synthesis of N-vinylcarbazole-N-vinylpyrrolidone Amphiphilic Block Copolymers by Xanthate-mediated Controlled Radical Polymerization. Can J Chem 2010;88:228–35.

[316] Chen J-K, Hsieh C-Y, Huang C-F, Li P-M. Characterization of Patterned Poly(methyl methacrylate) Brushes under Various Structures upon Solvent Immersion. J Colloid Interface Sci 2009;338:428–34.

[317] Chen J-K, Hsieh C-Y, Huang C-F, Li P-M, Kuo S-W, Chang F-C. Using Solvent Immersion to Fabricate Variably Patterned Poly(methyl methacrylate) Brushes on Silicon Surfaces. Macromolecules 2008;41:8729–36.

[318] Chen J-K, Wang J-H, Chang C-J, Huang C-F. Polarity-indicative Two-dimensional Periodic Concave Gratings of Tethered Polystyrene on Silicon Surfaces for Visualization in VOC Sensing. Sensors and Actuators B-Chemical 2013;188:1123–31.

[319] Ilčíková M, Mrlík M, Sedláček T, Chorvát D, Krupa I, Šlouf M, et al. Viscoelastic and photo-actuation studies of composites based on polystyrene-grafted carbon nanotubes and styrene-b-isoprene-b-styrene block copolymer. Polymer 2014;55:211–18.

[320] Ilčíková M, Mosnáček J, Mrlík M, Sedláček T, Csomorová K, Czaniková K, et al. Influence of surface modification of carbon nanotubes on interactions with polystyrene- *b* -polyisoprene- *b* -polystyrene matrix and its photo-actuation properties. Polym Adv Technol 2014;25:1293–300.

[321] Ilčíková M, Mrlík M, Sedláček T, Šlouf M, Zhigunov A, Koynov K, et al. Synthesis of Photoactuating Acrylic Thermoplastic Elastomers Containing Diblock Copolymer-Grafted Carbon Nanotubes. ACS Macro Lett 2014;3:999–1003.

[322] Ilčíková M, Mrlík M, Sedláček T, Doroshenko M, Koynov K, Danko M, et al. Tailoring of viscoelastic properties and light-induced actuation performance of triblock copolymer composites through surface modification of carbon nanotubes. Polymer 2015;72:368–77.

[323] Osicka J, Ilčíková M, Mrlik M, Minařík A, Pavlinek V, Mosnáček J. The Impact of Polymer Grafting from a Graphene Oxide Surface on Its Compatibility with a PDMS Matrix and the Light-Induced Actuation of the Composites. Polymers (Basel) 2017;9:264.

2

Design and Synthesis of Polymer Nanocomposites

E.I. Akpan, X. Shen, B. Wetzel, K. Friedrich
INSTITUTE FOR COMPOSITE MATERIALS, KAISERSLAUTERN, GERMANY

2.1 Polymer Nanocomposites

The science of fiber (particle) reinforced composites has shown that the size of filler materials has a profound effect on the resulting properties of the composite. This is because surface interactions with the matrix, adhesion, particle motion, dispersion, bonding, etc., depend a great deal on the size of the filler particle. Because some of these effects increase with decrease in filler size, they have more profound impact on the properties at the nanoscale. For example, nanoscale fillers possess very large surface to volume ratio and because properties like catalytic reactivity, electrical resistivity, adhesion, gas storage, and chemical reactivity depend on the nature of the interface, these properties change dramatically. On the other hand, effects such as quantum confinement, quantization of energy, molecular motion, and electromagnetic forces become very active at the nanoscale (Fig. 2–1). These effects give rise to increased intermolecular bonding, hydrogen bonding, van der Waals, hydrophobic effect, catalysis, magnetism, surface energy, etc. This prevalent increase in effects forms the basis for nanotechnology and nanostructured materials.

When a polymer matrix is combined with reinforcement materials having at least one of its dimensions in the nanoscale, a 3D structure is formed called polymer nanocomposite material [1–3]. Although elastomeric composites with nanosized fillers have been in use since 1959 [4], proper research on polymer nanocomposites began in 1990 when Toyota attempted to exfoliate clay nanofillers in nylon 6 [5–7]. The study demonstrated significant improvement in a wide range of mechanical properties and heat deflection temperature resulting in the use of the material in the fabrication of automobile tires. Following this study, extensive research in polymer nanocomposites and applications has been carried out globally. Today, polymer nanocomposites are used in several applications including packaging, electronic, electrical, structural, energy storage, automotive, aerospace, etc. The Tribology section of the Institute for Composite Materials, Kaiserslautern, Germany has also developed various kinds of tribology materials for automotive and calendar rolls applications

Polymer Composites with Functionalized Nanoparticles. DOI: https://doi.org/10.1016/B978-0-12-814064-2.00002-0

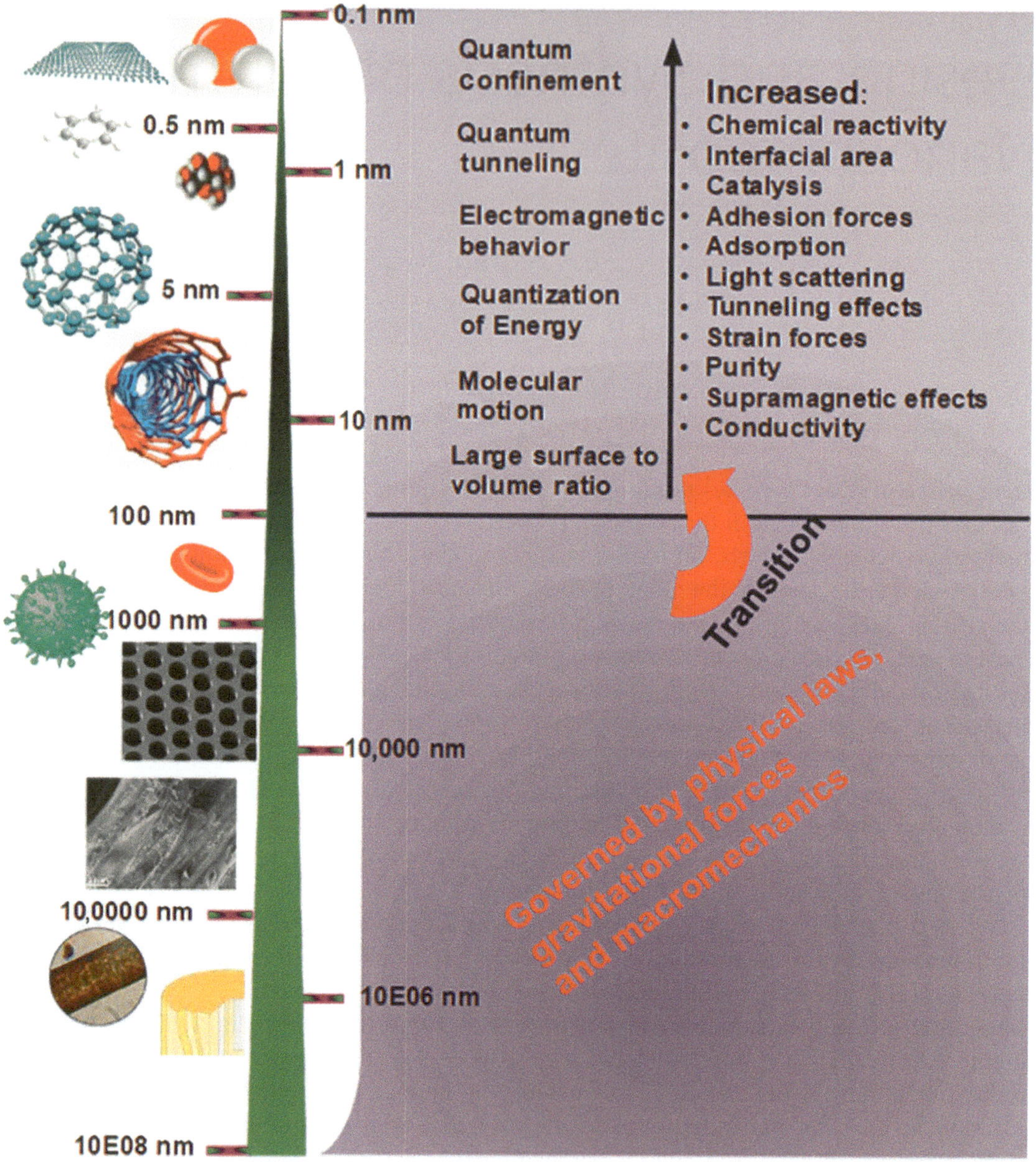

FIGURE 2–1 Peculiarities of the size scale.

from polymer nanocomposites [8–11]. In polymer nanocomposites some concepts are very fundamental and must be taken into consideration as they have direct effect on the properties of the resulting materials. These include size, shape, volume fraction, and state of dispersion of the nanofillers. These factors affect most properties of the nanocomposites [9,12–15].

2.1.1 Classification of Nanofillers

Nanofillers for polymer composite applications are mainly classified (Fig. 2–2) according to their dimensions [16–18]. These include:

1. one-dimensional nanofillers;
2. two-dimensional nanofillers; and
3. three-dimensional nanofillers.

2.1.1.1 One-Dimensional Nanofillers

One-dimensional (1D) nanofillers are those fillers that have one of their dimensions less than 100 nm [19]. They are usually in the form of sheets of one to a few nanometers thick to hundreds and thousands of nanometers long. Montmorillonite clay and nanographene platelets are common examples of 1D nanofillers. They are also found in the form of nanodisks [20], nanoprism, nanosheets, branched structures, nanoplates [21], and nanowalls [22–28]. Prominent examples include layered silicates, MMT [29], graphite nanoplatelets (GNP) [30], ZnO nanoplatelets [31], ZnO nanodiscs [32], carbon nanowall [33], amphiphilic graphene platelets [34], ZnO nanosheets [35], and Fe_3O_4 nanodiscs [36]. Most 1D nanofillers exhibit unique shape-dependent characteristics that make them useful in the formation of key components of nanodevices [28,37,38]. They are widely used in applications such as microelectronics, biosensors, sensors, biomedical, and coatings [17]. This is because of their excellent

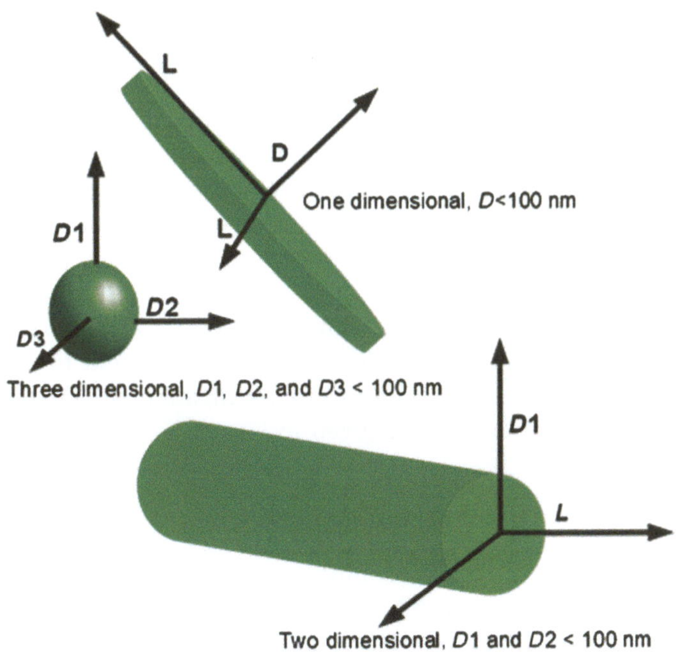

FIGURE 2–2 Classification of nanofillers.

electrical, magnetic, and optic properties [39]. 1D nanofillers are very useful in electrical and thermal applications because of their dimensionality.

2.1.1.2 Two-Dimensional Nanofillers

Nanofillers with two dimensions less than 100 nm are called two-dimensional (2D) nanofillers [19]. They are mostly in the shape of tubes, fibers, or filaments. Carbon nanotubes (CNTs), cellulose whiskers, boron nitride (BN) tubes, boron carbon nitride tubes, gold or silver nanotubes, 2D graphene, black phosphorus [40], and clay nanotubes are the most common examples of 2D nanofillers. Some 2D fillers used in the formulation of polymer nanocomposites are natural sepiolite clay fibers [41–44], cellulose fibers [45,46], nanotubes [47], sisal fibers [48–50], carbon fibers [51–54], gold nanowires [55,56], zinc oxide [57–60], titanium dioxide [61,62], silica [63–65], cerium dioxide [66–68], and copper oxide [69–72]. Others include molybdenum disulfide (MoS_2), hexagonal boron nitride (h-BN) [73], graphene [74], MoS_2 [75], and graphene oxide [76]. 2D nanofillers are very useful in energy [77], sensors [78], catalysis [79], photocatalysts, nanocontainers, nanoreactors [80], electronics, and optoelectronics. They are found to impart better flame-retardant property compared with 3D and 1D nanofillers [29]. They are also found to provide higher degree of reinforcement compared with 3D fillers [18,81].

2.1.1.3 Three-Dimensional Nanofillers

Three-dimensional (3D) nanofillers are relatively equiaxed particles having all three dimensions in the nanoscopic (nanometer) scale. They are usually in spherical [18] and cubical shapes, and are generally referred to as nanoparticles or isodimensional or zero dimensional nanoparticles. 3D nanofillers are also commonly called nanospheres, nanogranules, and nanocrystals. Examples of 3D nanofillers include nanosilica, nanoalumina, nanotitanium oxide, Polyhedral Oligomeric Silsesquioxane (POSS), nanomagnesium hydroxide, semiconductor nanoclusters, carbon black, silicon carbide, silica, and quantum dots (QDs) [17,82,83]. 3D nanofillers are very important in the formulation of polymer nanocomposites because of their inherent properties. Some of these nanoparticles (e.g., TiO_2, Ag, SiO_2, Fe_3O_4, and ZnO) possess good stability, high refractive index, hydrophilicity, ultraviolet (UV) resistance and excellent transparency to visible light, nontoxicity, high photocatalytic activity, and low cost [17]. When combined with suitable polymer matrix, they impart properties that enable them to be used in applications such as coatings [84–88], separation and purification [89–93], and biomedicine [94,95]. Some examples of nanofillers are given in Fig. 2–3.

2.2 Design of Polymer Nanocomposites

Development of high-performance nanocomposites requires a thorough design, material selection, formulation procedure, and fabrication. It involves the selection of a polymer as the matrix and the filler material, determination of the best process for production, consideration of end use requirements, and finally fabrication of the nanocomposite. In real

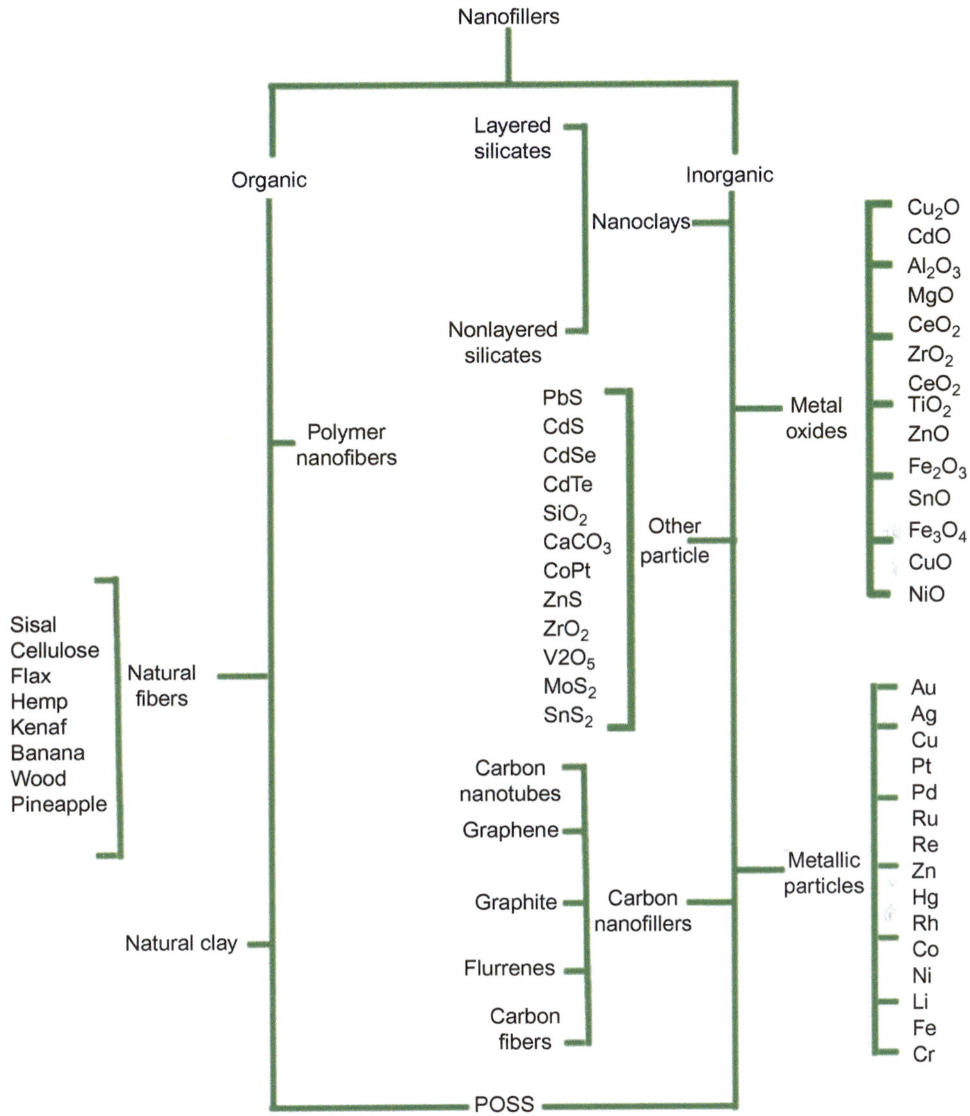

FIGURE 2–3 Examples of nanofillers.

applications, the designer is faced with the choice of polymers (thermoplastics, thermosetting, and elastomers) and filler materials (organic or inorganic) from a vast majority of available today. In choosing the polymer and the filler, the designer will have to take into consideration the compatibility of the polymer and the filler, adhesion and ease of incorporating the filler into the polymer, and dispersion (Fig. 2–4), making the design of polymer nanocomposite a rather complex procedure. To effectively formulate a desired polymer

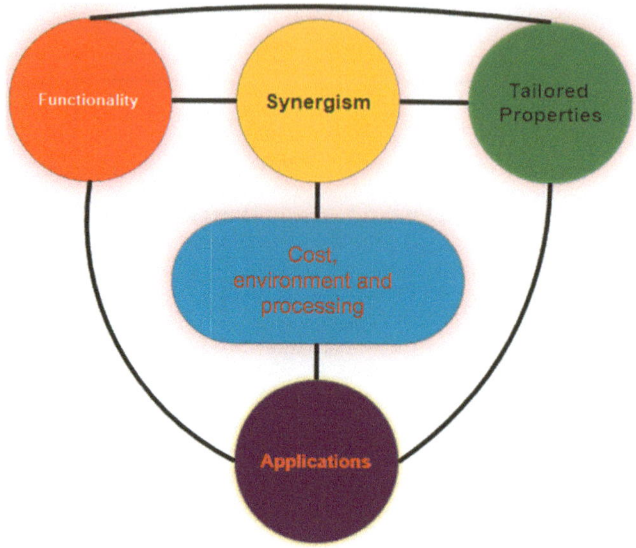

FIGURE 2–4 Polymer nanocomposite design variables.

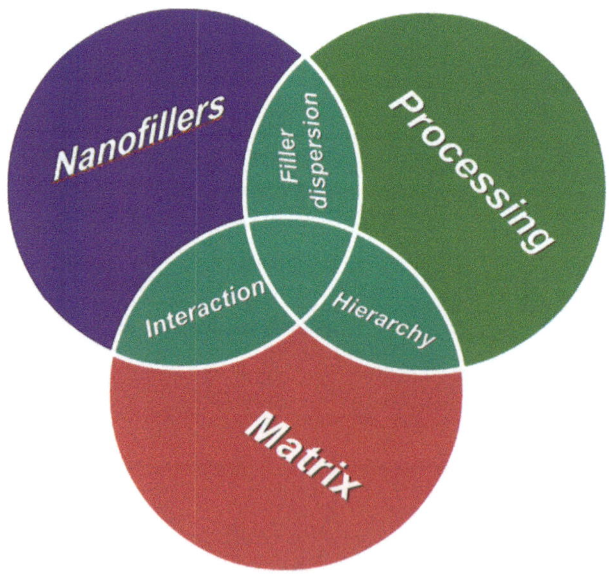

FIGURE 2–5 Design approach.

nanocomposite, the designer will have to consider some trade-offs including (1) applications; (2) required properties; (3) functionality; and (4) cost, environmental requirements, and processing (Fig. 2–5). These conditions constrain the design of the polymer nanocomposites to be based on rationality, tailored properties, and functionality.

2.2.1 Rationality-Based Design

Industrially, attention is usually paid to formulating materials for specific applications. In polymer nanocomposites, the use of nanofillers leads to large changes in the chemical, macromolecular, and crystalline structures, which in turn determines the bulk properties. Ability to predict how these structures affect the behavior of the composites enables the designer to create new materials with certain functionality. For example, knowing that the electrothermal actuation of polymer nanocomposites does not depend on the composition of the nanofiller, but only on the resultant macroscopic conductivity of the nanofiller, establishes a rationale for the selection of nanofiller, amount of nanofiller to be added, and the processing methods that will control the morphology. This type of design requires a fundamental understanding of physical and chemical properties and interactions in the chemical, macromolecular, and crystalline structures. Linking these structures with the bulk material properties is the heart of rational designs. This entails a very good knowledge of the chemistry and physics of the filler and the polymer materials.

A few examples of rational design of polymer nanocomposites are presented here. With knowledge of the critical role of interface structure of graphene/elastomer nanocomposites, a new strategy for constructing strong covalent interface has been developed. The rubber composites were reinforced with quinone-modified graphene. The modification was found to constrain the restacking of the nanosheets very effectively, leading to uniform dispersion and strong interfacial interaction [96]. Moreover, the knowledge of adhesive proteins in mussels was employed in the functionalization of TiO_2 nanowires to produce high–dielectric constant (High-κ) nanocomposites with solution to the large electrical mismatch and incompatibility between the TiO_2 nanowires and poly (vinylidene fluoride-co-hexafluoropropylene). The rational functionalization of nanofillers facilitated an excellent dispersion of the nanowires leading to a large discharged energy storage density [97]. Furthermore, a trade-off between gas permeability and selectivity has been the prevalent problem of conventional polymer membranes. Enhanced permeability with negligible losses has been achieved using zeolitic imidazolate nanoparticles in Matrimid [98] from the knowledge of free volume, molecular mobility, and polymer structure. In another study, deep understanding of the effects of filler fraction and thermomechanical properties of the polymer on the mechanical response of biomimetic cellulose nanofibril (CNF) nanocomposites was used to create predictive design space for rational design of covalent and noncovalent interactions of the CNF/polymer nanocomposites [99].

2.2.2 Functionality-Based Designs

Several structural, medical, electrical, and electronic applications possess vastly different functional requirements. In most cases, addition of nanofillers to a polymer matrix does more than improve on the properties of the polymer but imparts some other "unexpected" properties that can satisfy these requirements. This creates a situation where the nanofiller provides an extension of the composite functionality. By combining the filler and manipulating the parameter space, polymer nanocomposite research holds a vast potential to produce a wide range of functional materials with controllable properties. For example, knowledge of percolation properties of clay-filled polymers has given rise to the design of functional nanocomposites for sensor applications. Exploring filler abilities, the designer can design a nanocomposite material based on functionality. To illustrate functionality-based designs, a few of the nanocomposites that have been formulated with functional properties are presented here.

Functional polymer nanocomposites with antibacterial properties have been formulated using chitosan and metallic nanoparticles. Chitosan and Au/Ag nanoparticles were used to formulate functional nanocomposite with potentiometric sensing ability. The use of Ag nanoparticles enabled the detection of Cu^{++} ions. The same nanocomposite was also adapted for detection of NO_3 by replacing the Au nanoparticle with nanoclay. Another study reported the use of graphene oxide and poly(acryloyl-6-aminocaproic acid) to formulate hydrogels with self-healing capability [100]. Further studies realized self-healing composites based on reduced graphene and polyborosiloxane with sensing capabilities [101]. Functional nanocomposites for actuator applications have been made from CNTs and conducting shape memory polyurethanes [102]. Spray-coatable composite thin film with natural rubber latex, carbon fibers, and exfoliated graphite has been developed as a temperature sensor for accurate temperature measurement [103]. Bernabò et al. [104] used polyvinyl alcohol (PVA) and silver nanowires to realize pronounced polarization-dependent absorption properties that are suitable for innovative smart flexible films in packaging applications. To solve the problem of electroluminescence, ZnSe nanowires entangled into poly[2-methoxy-5-((2-ethylhexyl)oxy)-1,4-phenylenevinylene] was fabricated into a $p-n$ heterojunction for electroluminescence device [105]. CdSe/ZnS QDs embedded in an alternating triarylamine-functionalized fluorine-based copolymer are used as blue emitting layer in polymer light-emitting diodes [106].

Some particles have also been found to extend the functionalities of polymers so that they can be used as diodes in electronic equipment. Poly 9,9-dioctyl fluorene-alt-bithiophene and reduced graphene oxide were used to formulate a light-emitting diode with five times increase in efficiency [107]. Furthermore, a transparent conductive electrode was fabricated based on a silver nanowire percolation network modified with graphene oxide and embedded in polyurethane acrylate films. The study yielded fully stretchable white polymer light-emitting diodes [108]. In another study, blue light–emitting diode based on polyfluorene and poly(3,4-ethylenedioxythiophene):poly(styrenesulfonate)–titanium dioxide nanocomposite was fabricated [109]. Other authors also realized a hole injection and transport layer using a blend of poly(3,4-ethylenedioxythiophene):poly(styrene sulfonate) with SiO_2 and TiO_2 nanoparticles. The layer served as an organic light-emitting diode [110].

2.2.3 Tailored Property-Based Designs

The nanoscale dimensions of filler material in polymer nanocomposites induce some quantum effects [111], which make them possess an exceptional combination of properties when used as reinforcement in polymer matrix. This size influence can be compared with the role of chemical composition of the filler [112]. These effects are not obtainable in the microsized particle filled composites. For example addition of clay nanofiller to nylon 6 was found to exhibit a surprising substantial increase in heat distortion temperature, modulus, and strength without loss of impact strength and optical transparency [5,7]. This prevailing influence provides opportunity for several desirable material properties such as physical, chemical, optical, magnetic, etc., to be concurrently realized in one material by incorporating nanoscale fillers [113]. This concept enables polymer nanocomposites to be designed in search of tailored properties. Certain physical and chemical properties can be tailored using different type of fillers with appropriate particle size, particle shape, chemical nature [114], and process parameters [115]. A few examples show various advancements in this area.

Several studies have shown that processing and filler properties can be manipulated to develop a nanocomposite with controlled light scattering ability. A prime study in this respect involved the use of poly(methyl methacrylate) (PMMA) matrix and TiO_2 nanofillers to realize a "blue sky effect" [116]. The study modified the industrial method for synthesizing PMMA ("cell-cast" method) to fabricate a large-area nanocomposite plates with tuned optical diffusion. The PMMA matrix was then doped with commercial TiO_2. Further studies also reported tailored solar cells capability with enabled integrated photovoltaics using PMMA and Stokes-shift engineered cadmium selenide/cadmium(II) sulfide (CdSe/CdS) QDs. The ratio of the core diameter of CdS to the thickness of the shell (zinc sulfide) was modified such that the concentrators showed a large difference between the absorption and emission bands. The dots were then incorporated into the PMMA using a modified industrial method. Optical efficiency greater than 10% and an effective concentration factor of 4.4 were realized in the study [117]. Another study realized tailored refractive indices in the range of 1.52–1.65 as a function of TiO_2 content using hydrophilic poly(2-ethyl-2-oxazoline). The highly transparent nanocomposite was formulated with 44 wt.% of TiO_2 having its surface partially covered by an organic molecule. The nanocomposite was found to retain high transparency to visible light and also absorbs the UV radiation but with tailored refractive indices [118].

Other authors tailored polyvinyl phosphonic acid to realize a drug delivery agent and absorbents for the removal of organic contaminants by the addition silica-coated Fe_3O_4 nanoparticles [119]. Tailored mechanical properties and electric conductivity was also obtained from NbC nanocomposite thin films deposited on Si and polystyrene substrates. Required properties were obtained by adjusting the Nb flux to change the nanocrystal size and the percent of free-carbon phase in the films [120]. To gain excellent gas barrier properties, Scherillo et al. [121] tailored an interconnected network of self-assembled graphene oxide platelets in natural rubber. To fabricate microcantilevers with tunable sensor sensitivity, organically modified montmorillonite clay nanoparticles were dispersed in photoresist polymer SU-8. The resulting nanocomposite presented a good Young's modulus with acceptable thermal stability [122]. Talc filled polypropylene with tailored thermal, elastic, and electrical properties was also realized for applications in refrigerators jackets, infrared shielding, agricultural films, and packaging components [123]. Tailored mechanical properties with excellent biocompatibility have been achieved with hybrid reinforcement of PP with h-BN and nanohydroxyapatite [124].

Other important tailored property-based designs include simultaneous enhancement in mechanical and thermal properties achieved by incorporating tube-like natural halloysite in poly(tetrafluoroethylene) [125]; combined mechanical and thermal distortion realized with nylon six filled with montmorillonite intercalated with e-caprolactam [6]; superior, mechanical, and thermal properties of polylactic acid for automotive applications by addition of halloysite nanotubes modified with coaddition of a plasticizer [126]; temperature-responsive poly(N-cyclopropylacrylamide) polymers containing pyrene-bearing repeat units used to tailor the dispersion state of single-walled CNTs in water for synthesizing nanotube–polymer composites [127]; and enhancement in mechanical and gas barrier properties of graphene oxide filled natural rubber obtained without sacrificing the ultimate strain by coating the graphene oxide with bis(triethoxysilylpropyl)tetrasulfide [128].

2.2.4 Important Design Parameters

As noted in Section 2.2, the design of polymer nanocomposites involves several parameters depending on the product. Fig. 2–6 illustrates parameters that govern the design of polymer nanocomposites. Irrespective of the type of design under consideration, the designer must consider these parameters when formulating a new polymer nanocomposite material. Processing variables include process route, temperature, pressure, time, etc. On the other hand, filler parameters include filler shape, size, and type; surface nature of filler; filler volume or weight fraction; and treatment. Matrix parameters include type of polymer, surface nature and chemistry of the polymer, volume or weight fraction, and structure. All these variables can be condensed into three broad groups, namely aspect ratio (AR), interface, and orientation.

2.2.4.1 Aspect Ratio

Nanofillers are either 1D, 2D, or 3D in size. However, these fillers exist in different forms including, rods, fibers, tubes, platelets, wires, discs, flakes, spheroids, ellipsoids, etc. Despite the difference in shape and structure, nanofillers are generally characterized by 2D (the longest and shortest; the latter must always be less than 100 nm). Fig. 2–7 shows various shapes of nanofillers with these characteristic dimensions. These dimensions define two geometric features of the nanofillers, which are very important in the formation of polymer nanocomposites: (1) surface area and (2) AR.

Surface area is one of the distinguishing features between nanofillers and microfillers. It is the region responsible for "interaction" between the matrix and filler. Nanofillers possess extra-large surface area. When they are dispersed in polymers, the total area in contact with the polymer (interfacial area) is a direct result of the total surface area. This means that the larger the surface area, the larger the effect of the filler on the polymer. For simplicity in measurement, AR is generally used as a geometric definition of nanofiller. It is defined as the

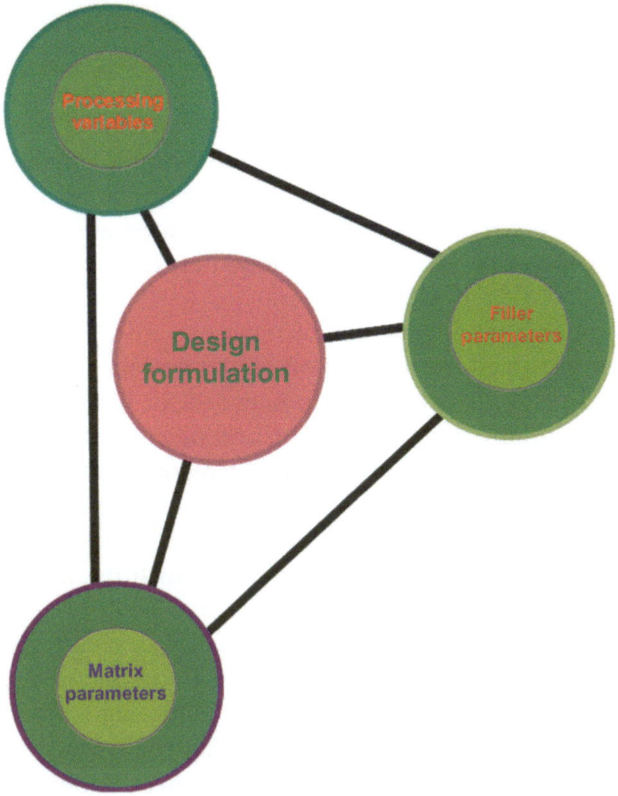

FIGURE 2–6 Parameters that govern the design of polymer nanocomposites.

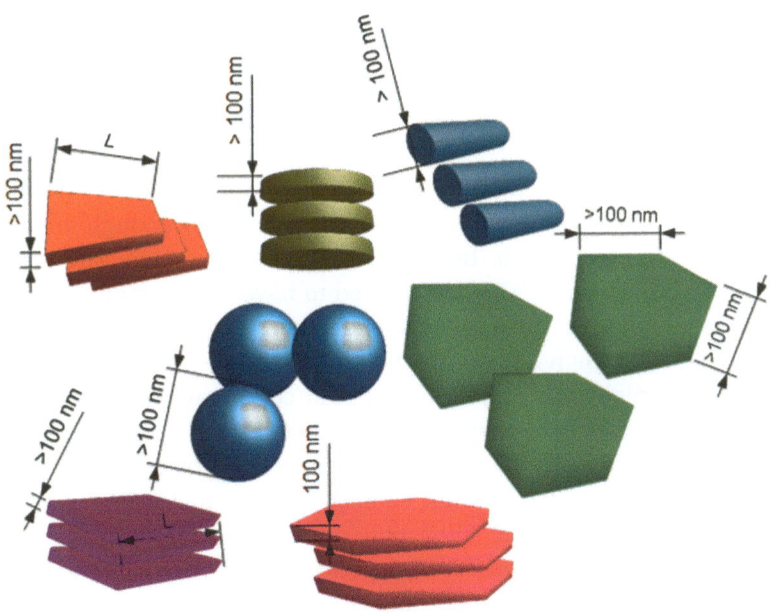

FIGURE 2–7 Various shapes of the nanofillers.

ratio of the longest dimension to the shortest dimension of the filler. Measurement for all possible shapes can be reduced to AR. The interfacial area per volume of filler is to a large extent dependent on the AR [129]. AR is found to play an important role in determination of physical and mechanical properties of polymer nanocomposites. For a successful formulation of polymer nanocomposites with desired properties, the material scientist must know and take into account the effect of AR of the filler.

A wide range of properties of polymer nanocomposites are a direct result of filler AR. Yao and coworkers [130] investigated the effect of AR on the percolation and dielectric properties of nanotube reinforced polyvinylidene fluoride intended for ferroelectric applications. They showed that dielectric constant, critical exponent, and percolation threshold increase as the AR increases until a critical value is reached. The increase is attributed to increase in total surface area of the nanotube. Similarly, Young's modulus of PMMA filled with CNTs intended for microelectromechanical systems was found to be dependent on AR. It was realized that there exists a critical value of the AR beyond which the modulus will be independent of the AR [131]. Thermal properties are also found to be dependent on the filler AR. In a study with clay reinforced nylon 6 nanocomposites, it was noted that modulus, strength, and heat distortion temperature of the nanocomposites decreased as the AR decreases [132]. Gojny et al. [133] also found a relationship between AR and electrical properties of CNT filled epoxy nanocomposites. They reported that AR of the filler is a crucial parameter for the realization of conductive nanocomposites at low filler contents. Highest electrical conductivity was obtained with relatively high AR. It was also noted that any treatment leading to a reduction of the AR (functionalization, ultrasonication) also led to increase in the percolation threshold. Another study based on epoxy and multiwalled CNTs showed higher electrical conductivity with higher AR. The authors also reported a percolation threshold in relation to the AR after which the epoxy shows conductivity [134]. A study on the effect of AR of multiwalled CNTs on flammability of polystyrene nanocomposites showed that larger ARs are more effective at reducing flammability. Larger ARs were also found to impart higher storage modulus and complex viscosity [135].

Another study showed that the thermal conductivity of multiwalled CNTs filled polyisobutene succinimide increased slightly with AR when the AR is high enough to keep the network structure but when the structure is broken the conductivity increases dramatically with increase in AR [136]. Nanoparticle AR was reported to have an effect on electrical conductivities of graphene/polyurethane and polyester/exfoliated graphite nanocomposites. Higher values were obtained when high AR nanoparticles were used [137,138]. Colonna and coworkers [139] also showed higher electrical and thermal conductivities with higher AR of GNP filled poly(butylene terephthalate) nanocomposites. Liang and Pearson reported negligible effect of AR on fracture toughness of some polymer nanocomposites [140].

In the case of dielectric properties, it has been shown that dielectric constant and loss factor of graphene and barium titanate filled poly(dimethyl siloxane) increase with fiber AR [141]. Dang and coworkers [142] observed an oscillating and nonmonotonic relationship between electrical percolation threshold and AR of CNTs in thermoplastic elastomer. Kalaitzidou and others [143] reported that as the AR of conductive fillers increases so does the critical concentration to induce bulk conductivity in the composite. This means that large AR particles maintain point-to-point contact at low concentrations allowing electron conduction and tunneling through and between particles leading to a decrease in percolation threshold.

Melt properties such as viscosity are also dependent on AR. In modified montmorillonite and synthetic silicate (chrysotile) nanotubes (SNT) filled polyimide nanocomposite, it was observed that relatively low AR SNT exhibits dramatic viscosity rise. This rise in viscosity is known to be consistent with formation of a percolation network structure. The authors also noted a large increase in modulus of MMT filled composites, which they attributed to the relatively high AR of the MMT filler. A study on vapor permeability of poly(ε-caprolactone) (PCL) clay nanocomposites by Messersmith and Giannelis [144] showed that water vapor permeability of the nanocomposites decreased as the AR of the platelets increased. It has also been shown that resistance-pressure sensitivity of multiwalled nanotubes methylvinyl silicone rubber nanocomposites are dependent on nanofiller AR. At low concentration of multiwalled nanotubes, low AR led to high resistance-pressure sensitivity whereas at high concentration of nanotubes high AR led to high resistance-pressure sensitivity [145]. Yang and coworkers [47] also found a relationship between filler AR and creep properties.

2.2.4.2 Interface

Mechanical, thermal, and electrical performances of polymer nanocomposites are largely dependent on the interaction between the nanofillers and the polymer matrix. For example, the strength and modulus of nanocomposites strongly depends on the effectiveness of stress transfer between matrix and fillers, which in turn is dependent on the interface adhesion. When fillers are dispersed in a polymer matrix, the filler surface area creates an "interaction zone" leading to some effects such as (1) altered behavior of the polymer, (2) changes in morphology, (3) changes in the space charge distribution, and (4) bond scattering. Nanofillers possess high surface area to volume ratio when compared with microparticle filled composites. This means that for the same volume concentration of the filler, nanocomposite possesses larger interfacial area ratio compared with microparticle filled composites. This means that for the same volume concentration of the filler, the nanocomposite possesses larger interfacial area. A large interfacial area implies also a large interaction zone and consequently a large altering of the behavior of the polymer and a large change in morphology. Consequent from the structure property relationship of materials, the nanocomposite will possess a large difference in property when compared with microcomposites. This underscores the importance of interfacial interaction in polymer nanocomposites. It is important to note that the nature of the interaction between the filler and the matrix determines the magnitude of the effects created by the presence of the filler. The interaction between the filler and the polymer is a complex phenomenon involving several interaction mechanisms, which are peculiar to the filler and matrix used. Figs. 2–8 and 2–9 show the nature of the interaction and the parameters involved.

These interaction mechanisms are peculiar to the polymer and the fillers used, depending on their type, source, size, shape, and preparation. Wettability is related to the ability of the

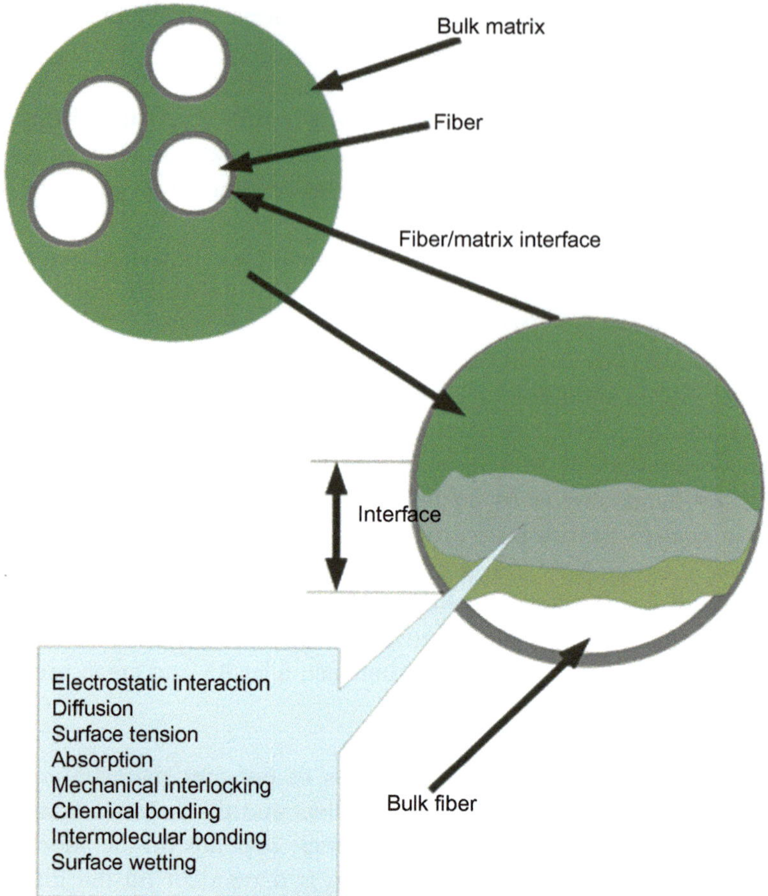

FIGURE 2–8 Simple model of interactions at the fiber–matrix interface.

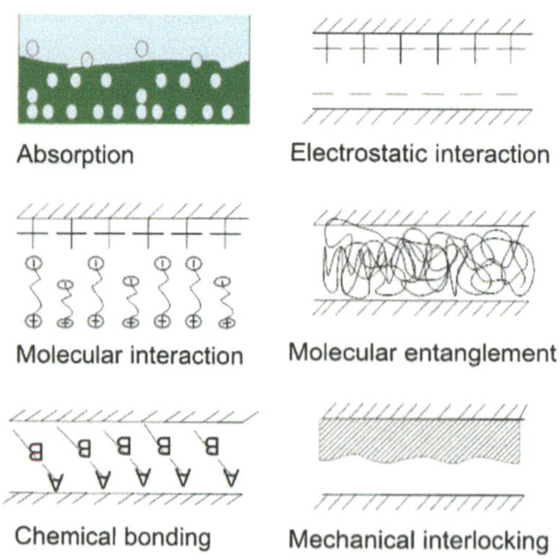

FIGURE 2–9 Schematic of the interaction mechanism at the fiber–matrix interface.

polymer to flow over the filler covering the surface completely and eliminating the presence of air. It is a product of polymer viscosity and critical surface energy. Mechanical locking depends on the roughness of the surface and contributes greatly to shear strength. Electrostatic interaction is the interaction between charges at the molecular level. It is very effective at very short separation distances. Nanocomposites possess very short separation distance between fillers and filler–matrix. This makes electrostatic interaction a very important phenomenon at the filler–matrix interface. Some polymers sometimes form covalent bonds with the filler. On the other hand, some polymers can diffuse into the surface of the filler. The variety of properties and phenomena exhibited by polymer nanocomposites compared with their base polymers are attributed to these interface features. The nature of the accumulated interaction arising from these features can be characterized as either strong or weak adhesion. Several models and theories have been formulated to explain the behavior of the polymer nanocomposite in relation to the interface. These models are in large part inaccurate because they are mostly not able to account for all the features at the fiber–matrix interface. To experimentally quantify the nature of this interfacial interaction between fillers and polymers, two methods are generally used: (1) adhesion strength and (2) work of adhesion [146]. Work of adhesion is usually measured using contact angle. The use of Young's equation enables estimates of surface energies and interfacial energies, which are then used to quantify the work of adhesion [147]. This method has a lot of problems including the lack of simplicity of the expression. On the other hand, adhesion strength of dispersed fiber composites is usually determined by single fiber fragmentation test. This involves embedding a single fiber in a polymer matrix and straining the resultant composite along the fiber axis in tensile mode. The fragmentation of the fiber into discrete lengths is then correlated with interfacial shear strength [148].

Although measurement of work of adhesion and adhesion strength are experimentally feasible, most researchers prefer direct determination of the effect of filler–matrix interaction on the properties of the composites. According to Fu and coworkers, composite strength and toughness are dependent on interfacial adhesion quality but not Young's modulus. This is because debonding is nearly impossible at small displacement and loads (where modulus is usually measured). This was noted in a study on silica particles filled nylon 6 polymer nanocomposites [149]. The study noted that surface treatment of silica nanoparticles with aminobutyric acid to improve the filler/matrix adhesion led to a large increase in tensile strength of the nanocomposites. Moreover, the modified composites showed good particle dispersion compared with the unmodified. Similar results were also obtained by Zhang and coworkers [150] using silica filled polypropylene and kaolin filled nylon 6,6 [151]. Thio and coworkers [152] also noted that tensile strength of the glass beads reinforced polypropylene increased with increasing interface adhesion. Similarly, higher interfacial strength was found to produce higher yield strength for $BaSO_4$ filled polypropylene [153]. Ou and coworkers [149] also showed that fracture toughness of silica particles filled nylon 6 polymer nanocomposites depends on the interfacial strength at the fiber–matrix interface. This was also noted with PP/$CaCO_3$ nanocomposites [154]. However, some authors [155–157] have noted that since crack growth is dominated by matrix failure and fiber breakage, interface debonding is

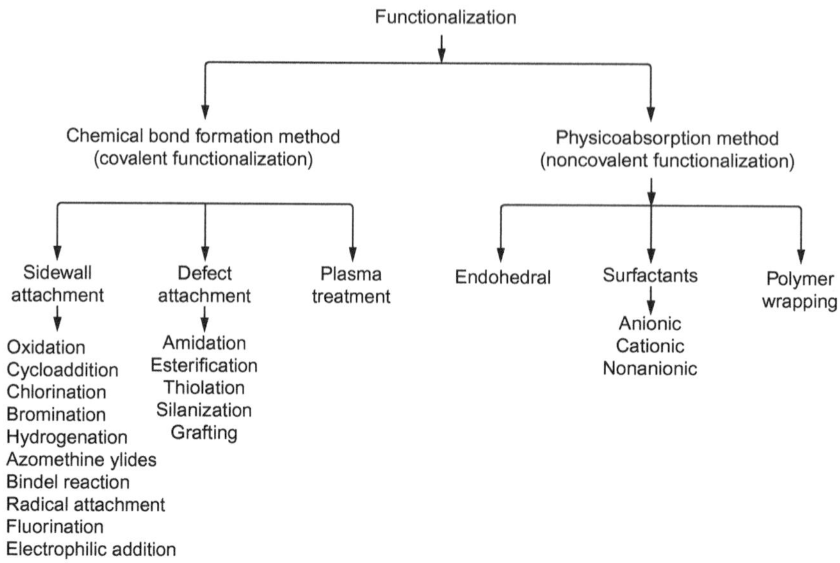

FIGURE 2–10 Functionalization strategies.

irrelevant to toughness. Moreover, several authors [154,157–159] also noted that tensile modulus is not affected by interfacial adhesion.

Most nanofillers used are chemically inert resulting in poor adhesion to the polymer matrix. For example CNTs present very poor interfacial interaction and they are very difficult to be dispersed in polymer matrix because of their strong electrostatic attraction [160,161]. Graphite is also reported to be chemically inert and mildly hydrophobic [162–164]. This was also noticed in graphene and graphene oxide [165,166]. Similarly, natural fillers are hydrophilic in nature and do not have good adhesion with thermoplastic polymers, which are naturally hydrophobic [167]. Fillers and polymers are usually modified (functionalized) to promote a good interfacial adhesion with each other.

Functionalization is generally classified as covalent (chemical) or noncovalent (physical) functionalization (Fig. 2–10). Covalent functionalization is based on linking of functional groups to the filler surface [168,169]. This can be done in the following two ways. The first is direct functionalization, that is, attaching the functional groups directly to the sidewalls of the fillers. Oxygen-containing groups such as carbonyl and carboxyl groups are then introduced directly onto the surface of the filler through bonding. This is generally achieved by reacting the filler with some molecules of high chemical reactivity [150,170–178]. The second is indirect functionalization, that is, attaching the functional groups to defects on the surface of the filler. Defects are deliberately created either by oxidative process with strong acids or with strong oxidants, which are later stabilized by bonding with functional groups [179–184]. Some drawbacks of covalent functionalization include generation of large number of defects and fragmentation of the fillers into smaller pieces. A novel covalent functionalization involves the use of plasma treatments. The surface of the filler are covalently attached with fluorine and then exposed to a polymerization initiator [185–188].

Noncovalent functionalization uses functional molecules or active species such as polymers (aromatic compounds) and surfactants as assembly mediators to functionalize the surface of fillers via hydrophobic, π–π stacking, electrostatic, and polymer wrapping interactions [168,189]. The assembled mediators then act as binding points for anchoring other molecules. Noncovalent functionalization methods are known to hamper dispersion [169]. Specific functionalization used for hollow tubular fillers involves the storing of molecules in the inner cavity of the filler through capillary effect [184]. This method is called endohedral functionalization.

Some authors have reported successful use of functionalization to improve on certain properties and dispersion [190–194]. Others have also reported deterioration of some properties with the use of these functionalization methods. Moreover, some recent research reports show the realization of some basic functional properties because of the use of functionalization. Details of different types of functionalization and their principles have been reviewed by several authors [195–201].

2.2.4.3 Orientation

Certain nano effects can be prompted in polymer nanocomposites by controlling the orientation of the nanofillers. For example, effective orientation of CNTs in polymer matrix can either trigger or annihilate van der Waals attraction and photoelastic response. On the other hand, some material properties are directional in nature such as thermal and electrical conductivity, diffusion, etc. Knowledge of these effects will help in the design of materials with functional properties. Although alignment of nanofillers presents a huge opportunity of property improvement in polymer nanocomposites, the means of aligning the fillers has been a major challenge in the past decade. Some successful methods of aligning fillers and the resulting properties are presented in this section.

SHEAR-INDUCED ALIGNMENT

Shear-induced alignment of nanofillers can be achieved by applying shear force or shear flow to the composite formulation. Some examples of the use of shear field to induce filler alignment are presented in this section. Costantino et al. [202] showed that nanoclay could be aligned in polymer matrix using shear-induced methods. The study reported the use of shear-controlled orientation in injection molding to align nanoclay in polypropylene matrix. Results showed that aligning the nanoclay led to higher degree of crystallinity against unaligned nanoclay. Tomer et al. [203] also reported the effect of filler alignment on the electric field breakdown strength and recoverable energy density of nanoclay filled polyethylene (PE). A commercial extrusion film blowing was used to induce the filler orientation. Nanocomposites with oriented fillers exhibited higher electric breakdown strength. Agarwal et al. [204] reported that alignment of nanofiber in heat flux direction resulted in a significant increase in thermal conductivity. Alignment of the carbon nanofibers in polycarbonate matrix was realized by extrusion. Galeski and Piorkowska [205] also noted that weak shear stress induces orientation of clay platelets and accelerates crystallization in nanocomposites. Application of shear-induced alignment to low-density PE filled with graphene nanoplatelets using single-screw melt extrusion was found to yield a highly ordered arrangement of graphene flakes resulting in a strong anisotropy of the material morphology [206]. Biaxial flow–induced alignment of clay particles was also realized in polypropylene/clay nanocomposite foams.

In another study, stretch-induced alignment was used to align single-walled CNTs in PVA solution using electrospinning. A surfactant stabilized nanotube suspension was injected into a stream of electrospun PVA solution giving rise to coagulation of the nanotubes through bridging flocculation. This led to the formulation of a dense interconnected network structure that when rewetted and stretched under tensile load leading to alignment of the nanofiller. A 400% improvement in tensile modulus was realized [207]. Badaire and coworkers [208] also employed stretch-induced alignment to gain preferred orientation of single-walled nanotubes in polyvinyl alcohol with improvement in electrical and thermal conductivity. Surfactant modified HiPco (high-pressure CO) conversion tubes were coagulated in a coflowing stream of polyvinyl alcohol. The fibers were dried and rewetted under tensile load to provide alignment of the fibers. Strain was also used to induce align CNTs in polyhydroxyaminoether [209]. Composites were fabricated by casting a suspension of CNTs in a solution of the polymer and chloroform. The cast composites were then uniaxially stretched at 100°C. Stretch-induced orientation has also been applied to PE/montmorillonite nanocomposite films. The 2D aligned inorganic nanoclays were found to provide strong improvement in dielectric breakdown [210]. Some other studies that deal with strain-induced alignment of nanofillers include Kelarakis et al. [211] and Ahir et al. [212] who also reported improvement in photomechanical actuation response in MCNT filled rubber composites due to the orientation. Seo and Weon [213] reported the use of equal channel angle extrusion to change the orientation of nanoclay in polyamide-6 nanocomposites.

FLOW-INDUCED ALIGNMENT

Flow-induced alignment of nanofillers is generally achieved by constraining the composite formulation to flow in a specified direction. The constrained flow induces an alignment of the nanofiller. Some successful examples are presented in this section. Flow-induced alignment has been used to align CNTs. Nanotubes alignment was obtained by injecting surfactants modified single-walled CNT dispersion in a coflowing stream of 5 wt.% of PVA to form a mesh. The flow-induced alignment led to preferential orientation of the nanotubes and improvement in electrical resistivity [214]. Kim and Macosko [215] also investigated flow-induced orientation of graphene sheets in polycarbonate resin. Injection and compression molding were used to obtain the orientation of the sheets. Percolation threshold is also found to be dependent on filler orientation. Using flow-induced orientation methods, Kalaitzidou and coworkers [143] realized that percolation threshold and electrical conductivity of GNP, carbon black in polypropylene matrix is dependent on the orientation of the nanofillers. A combination of electrospinning and in situ polymerization has been used to align CNTs in bulk polymer matrix [216]. Nanofiber scaffolds were first produced with aligned nanotubes by electrospinning, and incorporated into the PMMA matrix by in situ polymerization. Improvements in tensile strength and modulus were recorded because of aligning the fibers. The alignment was found to lead to very high electrical conductivity. Another study also show that unidirectional alignment of the clay layers parallel to the fiber axis of electrospun polyvinylidene fluoride gave rise to a highly laminated superstructure and imparted advanced mechanical properties [217].

ELECTRIC FIELD–INDUCED ALIGNMENT

In some cases, electric field is applied over a composite formulation before or after polymerization or cross linking to align nanofillers. Bower and coworkers [218] employed microwave plasma enhanced chemical vapor deposition to grow multiwalled CNTs in an electric field perpendicular to a silicon substrate. Their study realized a very good 1Dalignment. Another use of electric field–induced alignment of nanofillers was reported by Paineau et al. [219]. Polymer nanocomposite hydrogels were made from aqueous suspensions of natural swelling clays and subjected to high-frequency ac electric fields. The produced nanocomposite possesses uniformly aligned fibers with patterned platelet orientation. Multilayer graphene flakes have been aligned in epoxy matrix using external electric field [220]. The study showed that aligning the flakes led to improvement in thermal and electrical conductivity of the epoxy nanocomposites. The fracture toughness was also found to increase by 900%. Rozynek et al. [221] also noted that application of dc and ac electric field to clay reinforced polymer nanocomposites has a noticeable effect on filler orientation.

MAGNETIC FIELD–INDUCED ALIGNMENT

Magnetic field–induced alignment has been found very useful in the alignment of magnetic nanofillers. It involves the use of high magnetic field intensity to change the orientation of the fillers to a desired direction. The use of magnetic field to align fibers requires the use of very high magnetic field. This involve relatively high technical requirement, making the method less common. Magnetic field–induced alignment was applied to align graphene nanosheets in epoxy matrix incorporating Fe_3O_4 under a magnetic field. The composite, which was produced by coprecipitation, exhibited superparamagnetic behavior with high thermal conductivity in relation to nanosheet alignment [222]. Magnetic alignment was also applied to align single wall nanotube mat (bulky paper) in epoxy matrix [223]. However, this did not improve thermal conductivity as expected. Zhu and coworkers [224] fabricated a superior aligned nanopapers with combined thermal conductivity and mechanical strength using layered 2D BN nanosheets and 1D nanofibrillated cellulose. Fischer et al. [225] used filter deposition from suspension in strong magnetic fields to produce thick films of single wall CNTs with in-plane preferred orientation. This resulted in anisotropic electrical and thermal transport properties. Walters and coworkers [226] also used magnetic field alignment to produce highly aligned SWNT thin membranes. Magnetic alignment of nanofillers applied to epoxy-filled single-walled CNTs was shown to result in significant enhancement of thermal and electrical properties [227,228]. The composite was produced placing a mixture of the filler and matrix in a high magnetic field chamber.

2.3 Synthesis of Polymer Nanocomposites

In composite materials, the reinforcement phase needs to be uniformly distributed within the matrix to maximize reinforcing effect of fillers. Such uniform dispersion becomes increasingly difficult when the dimension of the filler is in the nanoscale. For nanofillers, the nature of dispersion is very much different from that of conventional microscale fillers such as carbon fibers and microspheres. First, the nanoscale dimension means that for a given volume fraction of fillers in the polymer matrix, a significantly larger amount of nanofillers is required compared with those microscale fillers. Fig. 2–11 depicts the amount of fillers required for different filler dimensions at a given volume fraction of 0.1 vol.% [169]. While microscale fillers are clearly discernible, nanofillers are extremely difficult to be distinguished from each other. This means that the interparticle distances between nanofillers are extremely close, leading to much larger van der Waals or electrostatic interactions between nanoparticles. In practice, therefore, it is extremely difficult, if not entirely impossible, to insert closely spaced but distinctly separated nanofillers within a polymer matrix.

Second, such a tendency to agglomeration can be more severe for anisotropic nanofillers such as nanotubes, nanowires, and nanoplatelets, which have 1D or 2D in nanometer scale while the rest can be in micrometer scale. Such anisotropic structures usually result in extremely high ARs, making them prone to severe restacking or reentanglement after proper dispersion. For example, monolayer graphene is a typical nanofiller that has a thickness of only one atomic layer (~ 0.34 nm) while it can be as large as a few to tens of micrometers in lateral dimensions, resulting in an AR of $\sim 10^4$. Such a high AR can lead to the stacking due to the strong $\pi-\pi$ interactions between huge graphene basal planes in favor of reducing the total surface energy.

Third, the large number of nanofillers creates immense amount of surface area, which in turn leads to large interface or interphase area with polymer matrices. The above characteristics suggest that the proper dispersion of nanofillers in polymer matrices are not trivial but is vital to the overall performance of nanocomposites. Therefore, the key issue in the synthesis of nanocomposites is the dispersion of nanofillers, the aim of which is to incorporate individual nanofillers or their small bundles into the polymer matrix and stabilize them to avoid

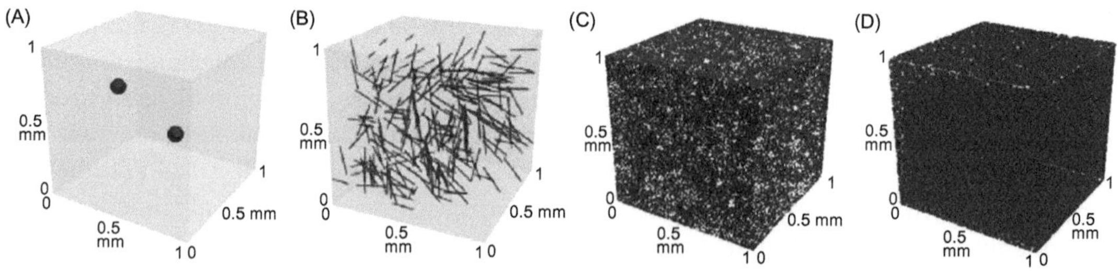

FIGURE 2–11 Distribution of 0.1 vol% micro and nanofillers in the same 1 mm^3. (A) Al_2O_3 microparticle; (B) carbon fiber; (C) GNP; and (D) CNTs. *Reprinted with permission from Ma PC, Siddiqui NA, Marom G, Kim JK, Compos Part A Appl Sci Manuf 2010;41: 1345–1367. Copyright 2010, Elsevier Ltd.*

further agglomeration during the course of synthesis. This requires the application of external energy of different forms to overcome the energy barrier within the nanofiller bundles. In the following sections, the common synthesis techniques are discussed with emphasis on their advantages and disadvantages.

2.3.1 Ultrasonication-Assisted Solution Mixing

Solution mixing is the most straightforward fabrication technique for polymer nanocomposites [229–233]. In this process, nanofillers are first solved in certain solvents followed by the simple mixing of polymer solutions and nanofiller dispersions. The dispersion of nanofillers in polymer matrices is facilitated by the ultrasonication, which applies the ultrasonic energy to the nanofiller bundles in solvent or polymer solutions. The general principle is that the bundles or agglomerates of nanofillers are agitated when ultrasound passes through in the form of high-frequency waves. These waves produce higher energy than the interaction energy between nanofillers in agglomerates, which leads to the separation of outer layers from the agglomerates so that smaller bundles form. These smaller bundles of nanofillers progressively exfoliate into even smaller ones as sonication time increases, and eventually turn into a monodispersed state where individual nanofillers are completely separated from each other in the polymer matrix. After proper dispersion, the solvent is evaporated to obtain the nanocomposites. The polymerization is usually done at elevated temperature at which monomers or prepolymers are polymerized in situ at the presence of nanofillers. Such an in situ polymerization process provides the possibility of chemical reactions between nanofillers and polymer matrices to enhance the interfacial interactions [234].

Due to its facile fabrication process, ultrasonication-assisted solution mixing is suitable for fast screening of new nanomaterials for nanocomposite applications. For example, the very first graphene-based nanocomposite was prepared using this technique [232]. The desired amount of GO sheets were mixed with polystyrene (PS) in *N,N*-dimethylformamide and sonicated for 1 hour. The GO was reduced to rGO before removing the solvent via filtration to make rGO/PS nanocomposites, which showed extraordinary mechanical and electrical properties. rGO was thus proved to be a promising nanofiller and later used for various polymer matrices for different applications.

To achieve homogeneous dispersion of nanofillers, both the sonication power and time are important parameters to control. In terms of sonication power, there are primarily two types of sonicators commonly used in the laboratory, bath sonicator and probe sonicator (Fig. 2–12). The bath sonicator has a smaller power of usually less than 100 W, while the probe sonicator produces a higher power commonly in the range of 100–1500 W with tunable power amplitude. The chosen sonication power should be based on the types of nanofillers and their dispersion states. High sonication power produced by the probe sonicator may not be damaging to the 3D nanofillers such as nanospheres, but can lead to serious fragmentation of 1D (e.g., nanosheets) and 2D (e.g., nanotubes) counterparts because of their large ARs. For example, various studies suggest that CNTs can suffer from severe damage to ultrasound evidenced by the Raman spectroscopy [235], where D band increases

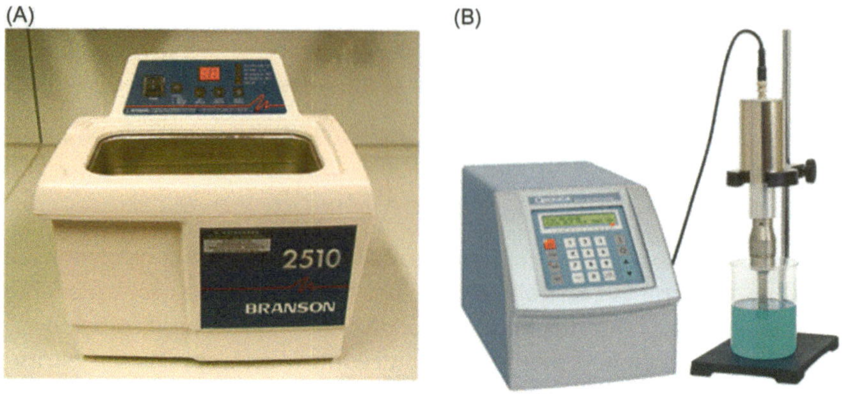

FIGURE 2–12 Photographs of (A) bath sonicator [236] and (B) probe sonicator [237].

significantly after sonication. This indicates an increasing number of defects due to the fragmentation of long CNTs into shorter ones, resulting in a reduced AR, which is deleterious to both the electrical and mechanical properties of the nanocomposites. Therefore, bath sonication with prolonged period is often used for the dispersion of 1D or 2D nanofillers. Nevertheless, probe sonications operated in short intervals with controlled temperature can be more effective and efficient if severe agglomerations occur.

The limitation of ultrasonication is that the solutions containing nanofillers/polymers need to be of low viscosity. However, most polymers are solids or rather viscous, and hence must be dissolved or diluted in solvent. The removal of solvents, particularly those with high boiling point, after the dispersion of nanofillers can be problematic. If the solvent is not completely removed, the quality of the final nanocomposites cannot be guaranteed. This should be an important consideration when choosing ultrasonication as the technique for nanocomposite synthesis.

2.3.2 Shear Mixing

While ultrasonication is a versatile method to disperse nanofillers in the laboratory, it always causes certain level of damages to the nanofillers and is not scalable for industrial applications [238]. Alternatively, shear mixing is less intensive and has the potential to be scaled up for industrial productions. The most common and easiest shear mixing setup is the magnetic stir. The stirrer rotating within the nanofiller solution generates shear forces, which drive the separation of agglomerates of nanofillers. Some level of dispersion can be achieved by extensive stirring of nanofillers/polymers solutions, but it is commonly observed that the reagglomerations easily occur after some time of dispersion. This is because the shear force generated with a magnetic stirrer is very weak, limited by its shape and rotating speed. Therefore, higher shear forces are needed to generate higher shear energies and thus more effectively disperse the nanofillers, which can be achieved by a high-speed shear mixer.

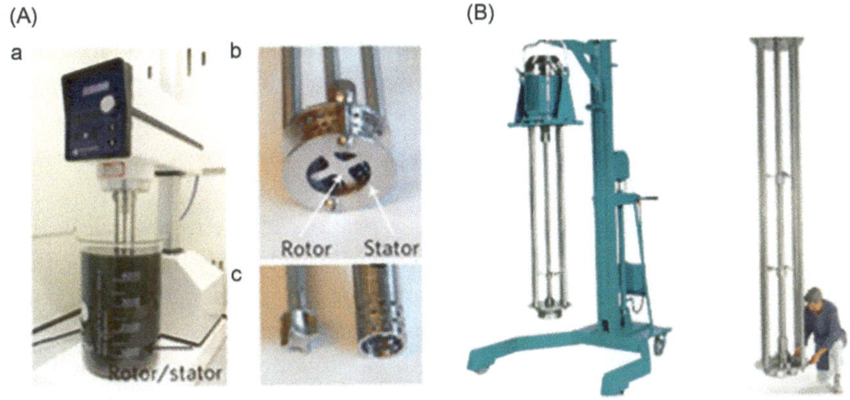

FIGURE 2–13 Photographs of (A) lab-scale [238] and (B) industry-scale shear mixer [247]. *(A) Reprinted with permission from K.R. Paton, E. Varrla, C. Backes, R.J. Smith, U. Khan, A. O'Neill, et al., Nat Mater 13, 2014, 624–630. Copyright 2014, Springer Nature.*

Fig. 2–13 shows a laboratory- and industry-scale high-shear mixer. The core component of a high-shear mixer is the workhead including a rotor blade and a stator. The rotor blades inside the workhead are capable of rotating at a high speed, suctioning nanofillers solutions from surroundings into the center of the workhead. Within the workhead, the nanofillers are redistributed to the periphery due to the centrifugal force, and the high-shear force between the rotor blades and the inner wall of the stator is exerted on agglomerates, effectively breaking them apart. Apparently, such mechanical forces are far less powerful compared with the high-energy ultrasonication, and are therefore conventionally only used to disperse agglomerates with weakly bonded nanoparticles such as silica nanospheres [239]. Intuitively, it may be considered as unsuitable for dispersing high-AR nanofillers such as layered nanosheets because the interlayer interaction energy is much higher compared with loosely bound agglomerates of spherical nanoparticles. It is therefore expected that shear mixing may be too weak to disperse a wide range of emerging layered materials such as graphene [238,240], BN [241], transition metal dichalcogenides (TMDs) [242], MXenes [242,243], and transition metal oxides [244]. However, recent advances in exfoliating various layered materials from their bulk counterparts suggest that shear mixing is a scalable alternative to the ultrasonication in effectively dispersing some of the layered materials [245]. The exfoliation process is dominated by the shear-induced interlayer sliding [246], with a minimum shear rate required depending on the types of materials and solvents. It is suggested that the shear rate required for the exfoliation is minimized when the surface energy of solvent matches that of nanofillers [238]. Therefore, it is important to choose the right solvent to achieve the most efficient dispersion.

To summarize, shear mixing is a dispersion technique that involves the application of shear energies to the nanofillers to separate them in solvents or polymer solutions. It is not only suitable for dispersing agglomerates of loosely bound nanoparticles, but also effective to exfoliate and disperse some layered nanosheets. The selection of solvents is an important consideration to minimize the shear energy required for different types of nanosheets. It usually does not cause damage to the nanofillers due to the low intensity of shear energies, and can be scaled up for large-quantity production for industrial applications. Like ultrasonication, the nanofillers need to be dispersed in solvents or low-viscosity polymer solutions.

2.3.3 Three Roll Milling

Three roll milling, also known as calendering, is a process involving the application of shear forces between rollers to mix and disperse materials with relatively high viscosities. This process is widely used in the industry to blend inks, color pigments, coatings, cosmetics, adhesives, and pastes. The machine used for the process, known as the three roll mill or calender, is composed of three adjacent cylindrical rollers that rotate at different speeds (Fig. 2–14A,B). Adjacent rollers rotate in the opposite direction to generate shear forces (Fig. 2–14C). Both the speed of rollers and the gaps between adjacent rollers are adjustable, generating a wide range of controllable shear forces. The composites containing polymer and nanofillers are fed into the gap between the first and second rollers. Then, the materials are transferred to the second gap through the rotating second roller. The second gap is usually narrower than the first one and the speed of the third roller is much higher, enabling higher shear force and thus allowing finer dispersion of nanofillers within the polymers. The final materials are scraped off from the third roller and repeatedly fed into the first gap for several times to further disperse the nanofillers.

One of the unique advantages of three roll milling is that the material can have a relatively high viscosity because of the high-shear force generated between the narrow gaps, which can be as thin as a few micrometers. This allows the use of high nanofiller content with little or no solvent involved. Moreover, both the distance between adjacent rollers and the speed of each roller can be fully controlled and maintained, and thus the consistency of

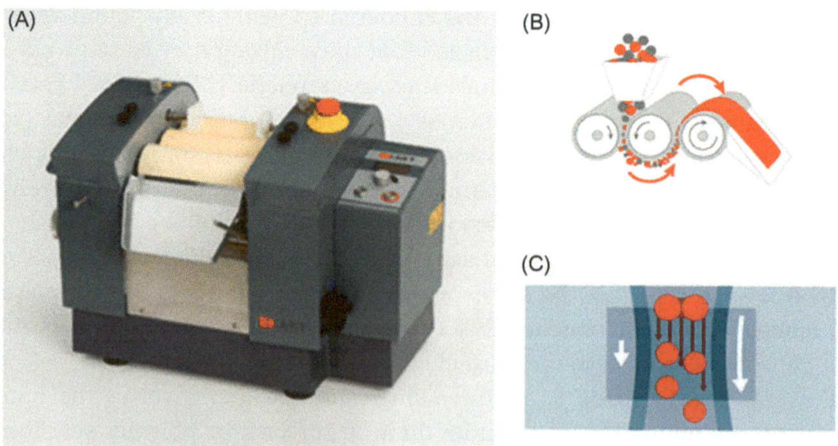

FIGURE 2–14 (A) Photograph of a laboratory three roll mill; (B) schematic of the typical three roll mill process; and (C) schematic of the shear force generated between adjacent rollers because of the different rotating direction and speed [248].

the dispersion in terms of size distribution and homogeneity of nanofillers can be guaranteed. Therefore, it has been used to disperse a range of nanofillers including nanoclays [249–251], CNTs [252–255], and GNPs [256–258]. The shear force applied to these nanofiller bundles can also partially align them along the shear direction.

However, it should be noted that these nanofillers all have anisotropic geometries, with at least one dimension in the micrometer scale. This means that three roll milling may not be suitable for dispersing spherical nanofillers that have all three dimensions at the nanometer scale. The reason behind is that the gap between the adjacent rollers is at least 1 μm, much higher than all the three dimensions of typical nanospheres, which are only a few to tens of nanometers. The disparity between dimensions may only lead to the breakage of agglomerates into smaller bundles, but not necessarily monodispersed particles. Another limitation for three roll milling is that viscous materials are required to be fed into the rollers, meaning that nanofillers can only be dispersed in liquid monomers or oligomers of some thermosets. Thermoplastics such as PE, polystyrene, and polypropylene are not suitable matrices for this technique.

2.3.4 Ball Milling

Ball milling is a technique traditionally used in the metallurgy and mineral industries that aims to grind or blend powders [259]. The rigid balls, typically made of high-hardness materials such as ceramics, flint pebbles, and stainless steels, collide with each other together with powders within a closed chamber. The collision of two balls exerts high shear or compressive forces to the powders at the interface, refining the powders with decreasing particle sizes. By the same token, such high-shear and compressive forces can also be useful to the dispersion of nanofillers. A variety of nanofillers, including silica nanoparticles [260], graphene nanosheets [261–264], CNTs [265], and BNs [266,267] have been dispersed in both thermoplastic and thermosetting matrices using the ball milling.

There are two major advantages of this technique compared with the others mentioned previously. First, it can achieve in situ exfoliation, dispersion, and functionalization of 2D nanosheets such as graphene and BNs [263,268]. The high-shear force generated by the collision of the balls facilitates the exfoliation of some 2D nanosheets from their counterpart [269–272]. While such forces are adequate to exfoliate nanosheets with only weak van der Waals bonding such as graphene and MoS_2 [270,271], it is impossible to exfoliate BN whose interlayer bonding is slightly ionic in nature [273]. In such cases, certain chemicals are needed to assist the mechanical exfoliation by functionalizing the nanosheets to reduce the interlayer bonding strength. For example, NaOH aqueous solution was used as an additive during the ball milling of BN nanosheets, which effectively reduced the shear force required for the exfoliation due to the cutting of BN sheets as a result of the reaction between OH ions and bulk BN [268]. Moreover, this method also offered the possibility of selectively functionalizing the edges of nanosheets without damaging the crystal structure of the basal planes of BN nanosheets. Similar results were also achieved by using dry ice to selectively functionalize the graphene nanosheets during the ball milling process [274]. These functionalized groups in turn facilitate the dispersion of 2D nanosheets within the solvent or polymer solutions [268], resulting in nanocomposites with excellent mechanical [262], thermal [267], electrical [264], and dielectric properties [265,266], and unique permeability [268]. Second, the materials used in the ball milling can be either in the wet or dry state. This means that this technique not only can be used to disperse nanofillers in liquid monomers of thermosets and soluble thermoplastics, but also allows the dispersion in solid thermoplastic matrices. For example, PMMA [260], PE [261,275], and polyphenylene sulfide [267,276] in the form of pellets or powders were blended with different types of nanofillers through ball millings, followed by the curing of composites using hot pressing or injection molding. The dispersion of nanofillers in these thermoplastic matrices would not be possible by using the abovementioned techniques such as ultrasonication, shear mixing, and three roll milling, which require the matrix either in a viscous state or dissolved in solvents.

2.3.5 Double-Screw Extrusion

Double-screw extrusion is a commonly used technique for fabricating thermoplastic-based nanocomposites [277]. Compared with the solution-based techniques, the main advantage of double-screw extrusion is the environmentally benign processing conditions without any solvent involved. Solid thermoplastic pellets or powders are either first compounded at high temperatures with nanofillers to produce the master batch [278], or fed with nanofillers together into the hopper by gravimetric dosing [279]. The dispersion of nanofillers is achieved by the high-shear forces exerted to the master batch due to the high-speed rotations of the twin screws.

Nanofillers with different dimensionalities in nanoscale, including 3D nanosilica [36], 2D CNTs [279], and 1D GNP [280], can be effectively dispersed using this technique. A number of material and processing parameters are considered critical for the effective dispersion of nanofillers, including filler content, processing temperature, and rotation speed. High filler content is always desired for nanocomposites to achieve optimum properties. Double-screw extrusion is a rather effective technique to disperse nanofillers at high loadings. For example, effective dispersion of CNTs of up to 15 wt% in PLA was achieved [279]. Nevertheless, such high filler content caused high torques of the extruder due to the high viscosity. Therefore, lower filler content is more practical in real applications. The temperature is another important parameter affecting the dispersion of nanofillers. The higher temperature can reduce the viscosity of the molten polymer, facilitating the incorporation of larger amount of nanofillers. However, a too high temperature can degrade polymer matrices or nanofillers, leading to deteriorated properties of final nanocomposites. In particular, some functionalized nanofillers can decompose at lower temperature than the typical processing temperature of 190–220°C due to the low thermal stability of the functional groups. This means that a trade-off must be made by either reducing the processing temperature or choosing more thermally resistant functional groups when functionalized nanofillers are used for such process [281]. The rotation speed of the screw mainly affects the shear forces applied, with higher rotation speed normally leading to higher shear stress. Such higher shear stress can translate into better dispersion of nanofillers with fewer agglomerates.

While double-screw extrusion is widely used for industrial applications of thermoplastics, there are a few issues need to be considered for the synthesis of nanocomposites containing functionalized nanoparticles. First, as mentioned previously, the processing temperature should be lower than the decomposition temperature of functional groups. If such a temperature is too low for the melting and processing of polymer matrix, alternative functionalization should be considered. Second, monodispersing of nanofillers is almost impossible using this technique due to the large physical gap between the screws, which is much larger than the nanoscale dimensions of nanofillers, with common presences of large agglomerates. While occasional agglomerates are sometimes acceptable, monodispersing is highly desirable for certain applications. In such circumstances, other synthesis techniques such as solution-based ultrasonication and ball milling should be used prior to or after double-screw extrusion to achieve satisfactory results [282,283].

Table 2–1 summarizes the common synthesis techniques for nanocomposites, which intends to provide a general guideline for the selection of proper techniques depending on the purpose of application, types of fillers and matrices, and the production scale.

Table 2–1 Summary of Common Techniques for the Fabrication of Nanocomposites

Technique	Suitable Filler	Suitable Matrix	Solvent	Scalability	Controlling Factors
Ultrasonication-assisted solution mixing	All types	Liquid or viscous monomers or oligomers of thermosets	Required	Only for laboratory	Sonication power and time
Shear mixing	Nanosheets such as graphene, BN, TMDs, MXenes	Liquid or viscous monomers or oligomers of thermosets	Required	Can be scaled for industrial production	Shapes of the rotor blades, rotating speed and time
Three roll milling	Nanosheets and nanotubes or nanowires	Liquid or viscous monomers or oligomers of thermosets	Not required	Can be scaled for industrial production	Speed of roller, gap between adjacent roller
Ball milling	All types	Liquid or solid thermoplastics and thermosets	Not required	Can be scaled for industrial production	Time of milling, ball size, rotating speed, ball/nanofiller ratio
Double-screw extrusion	All types	Solid thermoplastics	Not required	Can be scaled for industrial production	Processing temperature, screw configuration, rotation speed of the screw

2.3.6 Other Techniques

The above techniques are not comprehensive methods for the synthesis of nanocomposites, but rather most widely used in the laboratory and industry to achieve uniform dispersion of different types of nanofillers in polymer matrices. These processes usually involve the fabrication and functionalization of nanofillers in an external step followed by the incorporation and dispersion within monomers or resins, which are considered as ex situ processes [114]. Besides ex situ methods, another important synthesis strategy is the in situ process where nanofillers are synthesized from molecular precursors during the composite processing rather than used as received [114]. The in situ strategy includes chemical in situ methods and physical in situ methods [284]. Chemical in situ methods refer to processes involving chemical reactions in solutions where nanofillers are synthesized from precursors. A typical type of nanofiller that can be synthesized using this technique is the metal oxide nanoparticle. The nanocomposites can be fabricated through either a one-pot hydrothermal process [285] or a two-step sol–gel process [286]. Physical in situ methods are developed to synthesize inorganic/polymer core/shell nanocomposite particles [287]. Precursors in the gas phase are transformed into inorganic nanoparticles by plasma treatment, with subsequent organic compounds condensation to coat inorganic nanoparticles with polymer shells [114]. The in situ process, as well as other techniques, such as emulsion polymerization and pulverization, are either too specialized for a certain type of nanofiller or more general for polymer composites with fillers of different length scales. Therefore, these techniques are not the focus of this chapter. Those who are interested in these techniques can refer to a number of comprehensive reviews that discuss them in detail [114,288,289]. It is also common practice to combine these techniques to disperse nanofillers more efficiently.

2.4 Conclusion

To design polymer nanocomposites with the required properties, it is necessary to understand the exact influence of filler particle size, shape, orientation, dispersion, compatibility with matrix and volume/weight fraction, and how to control them. Moreover, understanding the relationship between these factors is required for successful design of polymer nanocomposites. Applications, required properties, functionality, cost, environmental requirements, and processing are the major trade-offs that should guide the design of polymer nanocomposites. Ability to predict how chemical and macromolecular changes, and changes in the crystalline structures of polymers, will affect the overall behavior of the composites is a major requirement in the creation of new polymer composite materials with certain functionality. To achieve tailored and multifunctional properties, a rational model of the design space is required. The design space, not necessarily a mathematical model, must be constructed with understanding of appropriate influences of filler, matrix and process parameters. In the synthesis of polymer nanocomposites, uniform dispersion of the nanofillers is the most important consideration, which should be taken into account when choosing the synthesis technique. Most physical, chemical, and mechanical properties of polymer nanocomposites depend on the interfacial interaction between filler and matrix. A good understanding of the interaction mechanism is the key to improved properties.

Acknowledgments

E.I.A. and X.S. were funded by the Alexander von Humboldt Foundation under Georg Forster Research Fellowship for postdoctoral researchers and Humboldt Research Fellowship for postdoctoral researchers, respectively.

References

[1] Rogers-Hayden T, Pidgeon N. Public Underst Sci 2007;16:345–64.
[2] Luo JJ, Daniel IM. Compos Sci Technol 2003;63:1607–16.
[3] Pomogailo AD. Polym Sci Ser C 2006;48:85–111.
[4] Bueche AM. J Polym Sci 1957;25:139–49.
[5] Kojima Y, Usuki A, Kawasumi M, Okada A, Fukushima Y, Kurauchi T, et al. J Mater Res 1993;8:1185–9.
[6] Kojima Y, Usuki A, Kawasumi M, Okada A, Kurauchi T, Kamigaito O. J Polym Sci Part A Polym Chem 1993;31:1755–8.
[7] Kojima Y, Usuki A, Kawasumi M, Okada A, Kurauchi T, Kamigaito O. J Appl Polym Sci 1993;49:1259–64.
[8] Chang L, Zhang Z, Ye L, Friedrich K. Wear 2007;262:699–706.
[9] Wetzel B, Haupeiit F, Friedrich K, Zhang MQIU, Rong MINZHI. Polym Eng Sci 2002;42:1919–27.
[10] Zhang MQ, Rong MZ, Friedrich K. In: Mai Y-W, Yu Z-Z, editors. Polym nanocomposites. Woodhead Publishing; 2006. p. 540–77.
[11] Chang L, Zhang Z, Ye L, Friedrich K. In: Friedrich K, Schlarb AK, editors. Tribol polym nanocomposites. 2nd ed. Oxford: Butterworth-Heinemann; 2013. p. 49–89.
[12] Wetzel B, Rosso P, Haupert F, Friedrich K. Eng Fract Mech 2006;73:2375–98.
[13] Deng S, Ye L, Friedrich K. J Mater Sci 2007;42:2766–74.
[14] Vassileva E, Friedrich K. J Appl Polym Sci 2003;89:3774–85.
[15] Zhi Rong M, Qiu Zhang M, Liu H, Zeng H, Wetzel B, Friedrich K. Ind Lubr Tribol 2001;53:72–7.
[16] Salavati-niasari M, Ghanbari D. in: Reddy B. (Ed.), Adv Divers Ind Appl Nanocomposites 2011;501–21.
[17] Vengatesan MR, Mittal V. Spherical fibrous fill compos. Wiley-VCH Verlag GmbH & Co. KGaA; 2016. p. 1–38.
[18] Bhattacharya M. Materials 2016;9:1–35.
[19] Raquel Verdejo LJR, Mar Bernal M, Francisco MAL-M, Tapiador J. Cell Polym 2011;30:45–62.
[20] Schmidt FP, Ditlbacher H, Hohenester U, Hohenau A, Hofer F, Krenn JR. Nano Lett 2012;12:5780–3.
[21] Nieto A, Lahiri D, Agarwal A. Carbon NY 2012;50:4068–77.
[22] Tiwari JN, Tiwari RN, Kim KS. Prog Mater Sci 2012;57:724–803.
[23] Nayak BB, Behera D, Mishra BK. J Am Ceram Soc 2010;93:3080–3.
[24] Dong X, Ji X, Jing J, Li M, Li J, Yang W. J Phys Chem C 2010;114:2070–4.
[25] Mann AKP, Skrabalak SE. Chem Mater 2011;23:1017–22.
[26] Siril PF, Ramos L, Beaunier P, Archirel P, Etcheberry A, Remita H. Chem Mater 2009;21:5170–5.

[27] Vizireanu S, Stoica SD, Luculescu C, Nistor LC, Mitu B, Dinescu G. Plasma Sources Sci Technol 2010;19:34016.

[28] Jung S, Oh E, Lee K, Yang Y, Park CG, Park W, et al. Cryst Growth Des 2008;8(1):265–9.

[29] Isitman NA, Dogan M, Bayramli E, Kaynak C. Polym Degrad Stab 2012;97:1285–96.

[30] Li B, Zhong WH. J Mater Sci 2011;46:5595–614.

[31] Mani GK, Rayappan JBB. RSC Adv 2014;4:64075–84.

[32] Umar A, Hahn YB. Nanotechnology 2006;17:2174–80.

[33] Liu X, Wen N, Wang X, Zheng Y. Nanomater Nanotechnol 2015;5:1–7.

[34] Shen J, Hu Y, Li C, Qin C, Ye M. Small 2009;5:82–5.

[35] Bai W, Zhu X, Zhu Z, Chu J. Appl Surf Sci 2008;254:6483–8.

[36] Zou H, Wu S, Shen J. Chem Rev 2008;108:3893–957.

[37] Bae S, Kim H, Lee Y, Xu X, Park J-S, Zheng Y, et al. NatNanotechnology 2010;5:574–8.

[38] Kim KS, Zhao Y, Jang H, Lee SY, Kim JM, Kim KS, et al. Nature 2009;457:706–10.

[39] Li BL, Setyawati MI, Chen L, Xie J, Ariga K, Lim CT, et al. ACS Appl Mater Interfaces 2017;9:15286–96.

[40] Passaglia E, Cicogna F, Lorenzetti G, Legnaioli S, Caporali M, Serrano-Ruiz M, et al. RSC Adv 2016;6:53777–83.

[41] Gupta P, Bera M, Maji PK. Polym Adv Technol 2017;28:1428–37.

[42] Ruiz-Hitzky E, Darder M, Alcantara ACS, Wicklein B, Aranda P. Functional polymer composites with nanoclays. The Royal Society of Chemistry; 2017. p. 1–53.

[43] Bilotti E, Fischer HR, Peijs T. J Appl Polym Sci 2008;107:1116–23.

[44] Can MF, Avdan L, Bedeloglu AC. Polym Compos 2015;36:2334–42.

[45] Fahmy TYA, Mobarak F, Fahmy Y, Fadl MH, El-Sakhawy M. Wood Sci Technol 2006;40:77–86.

[46] Fahmy TYA, Mobarak F. Carbohydr Polym 2008;72:751–5.

[47] Yang J, Zhang Z, Friedrich K, Schlarb AK. Macromol Rapid Commun 2007;28:955–61.

[48] Ibrahim ID, Jamiru T, Sadiku ER, Kupolati WK, Agwuncha SC. J Nanotechnol 2016;2016:9–11.

[49] Lee KY, Bharadia P, Blaker JJ, Bismarck A. Compos Part A Appl Sci Manuf 2012;43:2065–74.

[50] Garcia de Rodriguez NL, Thielemans W, Dufresne A. Cellulose 2006;13:261–70.

[51] Pozegic TR, Anguita JV, Hamerton I, Jayawardena KDGI, Chen J-S, Stolojan V, et al. Sci Rep 2016;6:37334.

[52] Xu Y, Van Hoa S. Compos Sci Technol 2008;68:854–61.

[53] Ulus H, Şahin ÖS, Avcı A. Fibers Polym 2015;16:2627–35.

[54] Ye G. Int J Electrochem Sci 2017;12:8467–76.

[55] Lonjon A, Laffont L, Demont P, Dantras E, Lacabanne C. J Phys D: Appl Phys 2010;43:34. Available from: https://doi.org/10.1088/0022-3727/43/34/345401.

[56] Kim YS, Yu BK, Kim JW, Suh YH, Kim DY, Kim WB. J Mater Chem A 2013;1:5015–20.

[57] Septiani NLW, Yuliarto B, Nugraha, Dipojono HK. Appl Phys A Mater Sci Process 2017;123:1–9.

[58] Saranya M, Ramachandran R, Wang F. J Sci Adv Mater Devices 2016;1:454–60.

[59] Alswata AA, Bin Ahmad M, Al-Hada NM, Kamari HM, Bin Hussein MZ, Ibrahim NA. Results Phys 2017;7:723–31.

[60] Ma J, Zhu W, Tian Y, Wang Z. Nanosc Res Lett 2016;11:200.

[61] Lu X, Chao D, Chen J, Zhang W, Wei Y. Mater Lett 2006;60:2851–4.

[62] Dhandapani S, Nayak SK, Mohanty S. J Vinyl Addit Technol 2016;22:529–38.

[63] D'Amato M, Dorigato A, Fambri L, Pegoretti A. Express Polym Lett 2012;6:954–64.

[64] Fambri L, Dabrowska I, Ceccato R, Pegoretti A. Polymers (Basel) 2017;9:41.

[65] Ślosarczyk A, Barełkowski M, Niemier S, Jakubowska P. J Sol-Gel Sci Technol 2015;76:227–32.

[66] Konduru NV, Jimenez RJ, Swami A, Friend S, Castranova V, Demokritou P, et al. Part Fibre Toxicol 2015;12:31.

[67] Shehata N, Gaballah S, Samir E, Hamed A, Saad M. Nanomaterials 2016;6:102. Available from: https://doi.org/10.3390/nano6060102.

[68] Shehata N, Samir E, Gaballah S, Hamed A, Elrasheedy A. Sensors (Switzerland) 2016;16:1371. Available from: https://doi.org/10.3390/s16091371.

[69] Sunny AT, Vijayan P, Adhikari R, Mathew S, Thomas S. Phys Chem Chem Phys 2016;18:19655–67.

[70] Ungur G, Hrůza J. Fibers Polym 2015;16:621–8.

[71] El-nahhal IM, Zourab SM, Kodeh FS, Selmane M, Genois I, Babonneau F. Int Nano Lett 2012;2:1–5.

[72] Arena A, Scandurra G, Ciof C. Sensors 2017;17:2198.

[73] Ribeiro H, Trigueiro JPC, Silva WM, Woellner CF, Owuor PS, Cristian Chipara A, et al. ACS Appl Mater Interfaces 2017. Available from: https://doi.org/10.1021/acsami.7b09945.

[74] Wang X, Song M. Nanomater Energy 2013;2:265–78.

[75] Wang X, Xing W, Feng X, Song L, Hu Y. Polym Rev 2017;57:440–66.

[76] Paszkiewicz S, Pawelec I, Szymczyk A, Rosłaniec Z. Polish J Chem Technol 2015;17:74–81.

[77] Wan J, Kaplan AF, Zheng J, Han X, Chen Y, Weadock NJ, et al. J Mater Chem A 2014;2:6051–7.

[78] Shahjamali MM, Salvador M, Bosman M, Ginger DS, Xue C. J Phys Chem C 2014;118:12459–68.

[79] Sanjeeva Rao K, Senthilnathan J, Ting J-M, Yoshimura M. Nanoscale 2014;6:12758–68.

[80] Pradhan S, Lach R, Le HH, Grellmann W, Radusch H-J, Adhikari R. ISRN Polym Sci 2013;2013:1–9.

[81] Brune DA, Bicerano J. Polymer (Guildf) 2002;43:369–87.

[82] Stephen R, Thomas S. Rubber nanocomposites. John Wiley & Sons, Ltd; 2010. p. 1–19.

[83] Kumar AP, Depan D, Singh Tomer N, Singh RP. Prog Polym Sci 2009;34:479–515.

[84] Macyk W, Szaciłowski K, Stochel G, Buchalska M, Kuncewicz J, èabuz P. CoordChem Rev 2010;254:2687–701.

[85] Paz Y. Appl Catal B Environ 2010;99:448–60.

[86] Su W, Wang S, Wang X, Fu X. J Weng, Surf Coatings Technol 2010;205:465–9.

[87] Wang N, Fu W, Zhang J, Li X, Fang Q. Prog Org Coatings 2015;89:114–22.

[88] Olad A, Nosrati R. Prog Org Coatings 2013;76:113–18.

[89] María Arsuaga J, Sotto A, del Rosario G, Martínez A, Molina S, Teli SB, et al. J MembSci 2013;428:131–41.

[90] Zhao S, Yan W, Shi M, Wang Z, Wang J, Wang S. J Memb Sci 2015;478:105–16.

[91] Huang J, Zhang K, Wang K, Xie Z, Ladewig B, Wang H. J Memb Sci 2012;423–424:362–70.

[92] Sotto A, Boromand A, Balta S, Kim J, Van der Bruggen B. J Mater Chem 2011;21:10311.

[93] Maximous N, Nakhla G, Wan W, Wong K. J Memb Sci 2009;341:67–75.

[94] Di Carlo G, Curulli A, Toro RG, Bianchini C, De Caro T, Padeletti G, et al. Langmuir 2012;28:5471–9.

[95] Matos AC, Marques CF, Pinto RV, Ribeiro IAC, Gonçalves LM, Vaz MA, et al. Int J Pharm 2015;490:200–8.

[96] Yang Z, Liu J, Liao R, Yang G, Wu X, Tang Z, et al. Compos Sci Technol 2016;132:68–75.

[97] G Wang, X Huang, F Liu, J Zhang and P Jiang, In: 2016 IEEE international conference on dielectrics (ICD), Montpellier, 2016, p. 864–7. https://doi.org/10.1109/ICD.2016.7547753.

[98] Song Q, Nataraj SK, Roussenova MV, Tan JC, Hughes DJ, Li W, et al. Energy Environ Sci 2012;5:8359.

[99] Benítez AJ, Lossada F, Zhu B, Rudolph T, Walther A. Biomacromolecules 2016;17:2417–26.

[100] Cong HP, Wang P, Yu SH. Chem Mater 2013;25:3357–62.

[101] D'Elia E, Barg S, Ni N, Rocha VG, Saiz E. Adv Mater 2015;27:4788–94.

[102] Goo NS, Paik IH, Jung YC, Cho JW. Smart Mater Struct 2006;15:1476–82.

[103] Sauerbrunn E, Chen Y, Didion J, Yu M, Smela E, Bruck HA. Phys Status Solidi 2015;212:2239–45.

[104] Bernabò M, Pucci A, Ramanitra HH, Ruggeri G. Materials (Basel) 2010;3:1461–77.

[105] Chen GH, Ho SJ, Chen HS. J Phys Chem C 2014;118:25816–22.

[106] Mucur SP, Lenkeit D, Kanelidis I, San SE, Holder E, Tekin E. J Polym Sci Part B Polym Phys 2014;52:147–56.

[107] Singh JP, Saha U, Jaiswal R, Anand RS, Srivastava A, Goswami TH. J Nanoparticle Res 2014;16:2693.

[108] Liang J, Li L, Tong K, Ren Z, Hu W, Niu X, et al. ACS Nano 2014;8:1590–600.

[109] Gupta N, Grover R, Mehta DS, Saxena K. Displays 2015;39:104–8.

[110] Riedel B, Shen Y, Hauss J, Aichholz M, Tang X, Lemmer U, et al. Adv Mater 2011;23:740–5.

[111] Kumar A, Ali SF, Arockiarajan A. Smart Polym Nanocompos 2017;43–75.

[112] Koo JH. Polymer nanocomposites processing, characterisation and applications. New York: McGraw-Hill; 2006.

[113] Noll A, Friedrich K, Burkhart T, Breuer U. Polym Compos 2013;34:1405–12.

[114] Hanemann T, Szabó DV. Polymer-nanoparticle composites: from synthesis to modern applications. Materials 2010;3:3468–517. Available from: https://doi.org/10.3390/ma3063468.

[115] Dubas ST, Schlenoff JB. Macromolecules 1999;32:8153–60.

[116] Colombo A, Tassone F, Santolini F, Contiello N, Gambirasio A, Simonutti R. J Mater Chem C 2013;1:2927.

[117] Meinardi F, Colombo A, Velizhanin KA, Simonutti R, Lorenzon M, Beverina L, et al. Nat Photonics 2014;8:392–9.

[118] Colombo A, Tassone F, Mauri M, Salerno D, Delaney JK, Palmer MR, et al. RSC Adv 2012;2:6628.

[119] Sengel SB, Sahiner N. Eur Polym J 2016;75:264–75.

[120] Yate L, Coy LE, Wang G, Beltrán M, Díaz-Barriga E, Saucedo EM, et al. RSC Adv 2014;4:61355–62.

[121] Scherillo G, Lavorgna M, Buonocore GG, Zhan YH, Xia HS, Mensitieri G, et al. ACS Appl Mater Interfaces 2014;6:2230–4.

[122] Chen H, Ojijo V, Cele H, Joubert T, Suprakas SR, Land K. Proc SPIE - Int Soc Opt Eng 2014;9257:1–11.

[123] Saq'an S, Zihlif AM, Ragosta G. J Thermoplast Compos Mater 2008;21:457–67.

[124] Chan KW, Wong HM, Yeung KWK, Tjong SC. Materials (Basel) 2015;8:992–1008.

[125] Gamini S, Vasu V, Bose S. Mater Res Express 2017;4:45301.

[126] Notta-Cuvier D, Murariu M, Odent J, Delille R, Bouzouita A, Raquez JM, et al. Macromol Mater Eng 2015;300:684–98.

[127] Etika KC, Jochum FD, Cox MA, Schattling P, Theato P, Grunlan JC. Macromolecules 2010;43:9447–53.

[128] Wu J, Huang G, Li H, Wu S, Liu Y, Zheng J. Polym (United Kingdom) 2013;54:1930–7.

[129] Gulotty R, Castellino M, Jagdale P, Tagliaferro A, Balandin AA. ACS Nano 2013;7:5114–21.

[130] Yao SH, Dang ZM, Jiang MJ, Xu HP, Bai J. Appl Phys Lett 2007;91:50.

[131] Ashrafi B, Hubert P, Vengallatore S. Nanotechnology 2006;17:4895–903.

[132] Weon JI, Sue HJ. Polymer (Guildf) 2005;46:6325–34.

[133] Gojny FH, Wichmann MHG, Fiedler B, Kinloch IA, Bauhofer W, Windle AH, et al. Polymer (Guildf) 2006;47:2036–45.

[134] Ayatollahi MR, Shadlou S, Shokrieh MM, Chitsazzadeh M. PolymTest 2011;30:548–56.

[135] Cipiriano BH, Kashiwagi T, Raghavan SR, Yang Y, Grulke EA, Yamamoto K, et al. Polymer (Guildf) 2007;48:6086–96.

[136] Yang Y, Grulke EA, Zhang ZG, Wu G. J Appl Phys 2006;99 056109, 1–4.

[137] Kim H, Macosko CW. Macromolecules 2008;3317–27.

[138] Kim H, Miura Y, MacOsko CW. Chem Mater 2010;22:3441–50.

[139] Colonna S, Bernal MM, Gavoci G, Gomez J, Novara C, Saracco G, et al. Mater Des 2017;119:124–32.

[140] Liang YL, Pearson RA. Polymer (Guildf) 2010;51:4880–90.

[141] Wang Z, Nelson JK, Miao J, Linhardt RJ, Schadler LS, Hillborg H, et al. IEEE Trans Dielectr Electr Insul 2012;19:960–7.

[142] Dang ZM, Shehzad K, Zha JW, Hussain T, Jun N, Bai J. Jpn J Appl Phys 2011;50.

[143] Kalaitzidou K, Fukushima H, Drzal LT. Materials (Basel) 2010;3:1089–103.

[144] Messersmith PB, Giannelis EP. J Polym Sci Part A Polym Chem 1995;33:1047–57.

[145] Jiang B, Liu C, Zhang C, Wang B, Wang Z. Compos Part B Eng 2007;38:24–34.

[146] Fu SY, Feng XQ, Lauke B, Mai YW. Compos Part B Eng 2008;39:933–61.

[147] Packham DE. Int J Adhes Adhes 1996;16:121–8.

[148] Harding PH, Berg JC. J Adhes Sci Technol 1997;11:37–41.

[149] Ou Y, Yang F, Yu Z-Z. J Polym Sci Part B Polym Phys 1998;36:789–95.

[150] Zhang L, Li C, Huang R. J Polym Sci Part B Polym Phys 2004;42:1656–62.

[151] Buggy M, Bradley G, Sullivan A. Compos Part A Appl Sci Manuf 2005;36:437–42.

[152] Thio YS, Argon AS, Cohen RE. Polymer (Guildf) 2004;45:3139–47.

[153] Wang K, Wu J, Ye L, Zeng H. Compos Part A Appl Sci. Manuf 2003;34:1199–205.

[154] Levita G, Marchetti A, Lazzeri A. Polym Compos 1989;10:39–43.

[155] Spanoudakis J, Young RJ. J Mater Sci 1984;19:473–86.

[156] Moloney AC, Kausch HH, Kaiser T, Beer HR. J Mater Sci 1987;22:381–93.

[157] Roulin-Moloney AC, Cantwell WJ, Kausch HH. Polym Compos 1987;8:314–23.

[158] Dekkers MEJ, Heikens D. J Appl Polym Sci 1983;28:3809–15.

[159] Dibenedetto AT, Wambach AD. Int J Polym Mater Polym Biomater 1972;1:159–73.

[160] Sahoo NG, Rana S, Cho JW, Li L, Chan SH. Prog Polym Sci 2010;35:837–67.

[161] Szleifer I, Yerushalmi-Rozen R. Polymer (Guildf) 2005;46:7803–18.

[162] Li ZF, Blum FD, Bertino MF, Kim CS. Sensors Actuat B Chem 2013;183:419–27.

[163] Xu K, Heath JR. Nat Mater 2013;12:872–3.

[164] Shih C-J, Strano MS, Blankschtein D. Nat Mater 2013;12:866–9.

[165] Nikkhah SJ, Moghbeli MR, Hashemianzadeh SM. J Mol Model 2015;21:1–12.

[166] Fu Y, Liu L, Zhang J. ACS Appl. Mater Interfaces 2014;6:14069–75.

[167] Akpan EI, Adeosun SO, Lawal GI, Balogun SA, Chen XD. J Nat Fibers 2016;13:103–24.

[168] Karimi M, Solati N, Amiri M, Mirshekari H, Mohamed E, Taheri M, et al. Expert Opin Drug Deliv 2015;12:1071–87.

[169] Ma PC, Siddiqui NA, Marom G, Kim JK. Compos Part A ApplSci Manuf 2010;41:1345–67.

[170] Kelly KF, Chiang IW, Mickelson ET, Hauge RH, Margrave JL, Wang X, et al. Chem Phys Lett 1999;313:445–50.

[171] Touhara H. Carbon Alloy Nov Concepts to Dev Carbon Sci Technol 2003;114:485–98.

[172] Stevens JL, Huang AY, Peng H, Chiang IW, Khabashesku VN, Margrave JL. Nano Lett 2003;3:331–6.

[173] Chen J, Hamon MA, Hu H, Chen Y, Rao AM, Eklund PC, et al. Science 1998;282:95–8.

[174] Hu H, Zhao B, Hamon MA, Kamaras K, Itkis ME, Haddon RC. J Am Chem Soc 2003;125:14893–900.

[175] Holzinger M, Steinmetz J, Samaille D, Glerup M, Paillet M, Bernier P, et al. Carbon N Y 2004;42:941–7.

[176] Unger E, Graham A, Kreupl F, Liebau M, Hoenlein W. Curr Appl Phys 2002;2:107–11.

[177] Kim KS, Bae DJ, Kim JR, Park KA, Lim SC, Kim JJ, et al. Adv Mater 2002;14:1818–21.

[178] Tagmatarchis N, Prato M. J Mater Chem 2004;14:437–9.

[179] Saito T, Matsushigem K, Tanaka K. Physica B 2002;323:280–3.

[180] Yu R, Chen L, Liu Q, Lin J, Tan K-L, Ng SC, et al. Chem Mater 1998;10:718–22.

[181] Sham ML, Kim JK. Carbon N Y 2006;44:768–77.

[182] Ma PC, Kim J-K, Tang BZ. Carbon N Y 2006;44:3232–8.

[183] Wang SC, Chang KS, Yuan CJ. Electrochim Acta 2009;54:4937–43.

[184] Ávila-Orta CA, Cruz-Delgado VJ, Neira-Velázquez MG, Hernández-Hernández E, Méndez-Padilla MG, Medellín-Rodríguez FJ. Carbon N Y 2009;47:1916–21.

[185] Valentini L, Puglia D, Armentano I, Kenny JM. Chem Phys Lett 2005;403:385–9.

[186] Valentini L, Puglia D, Carniato F, Boccaleri E, Marchese L, Kenny JM. Compos Sci Technol 2008;68:1008–14.

[187] Tseng CH, Wang CC, Chen CY. Chem Mater 2007;19:308–15.

[188] Kalita G, Adhikari S, Aryal HR, Afre R, Soga T, Sharon M, et al. Curr Appl Phys 2009;9:346–51.

[189] He H, Pham-Huy LA, Dramou P, Xiao D, Zuo P, Pham-Huy C. Biomed Res. Int. 2013;. Available from: https://doi.org/10.1155/2013/578290.

[190] Rong MZ, Zhang MQ, Zheng YX, Zeng HM, Walter R, Friedrich K. JMater Sci Lett 2000;19:1159–61.

[191] Rong MZ, Zhang MQ, Pan SL, Lehmann B, Friedrich K. Polym Int 2004;53:176–83.

[192] Wu CL, Zhang MQ, Rong MZ, Lehmann B, Friedrich K. Plast Rubber Compos 2004;33:71–6.

[193] Zhang Qiu M, Zhi M, Li S, Wetzel B, Friedrich K. 2002;253:1086–93.

[194] Ruan WH, Zhang MQ, Rong MZ, Friedrich K. Polym Polym Compos 2004;12:257–67.

[195] Punetha VD, Rana S, Yoo HJ, Chaurasia A, McLeskey JT, Ramasamy MS, et al. Prog Polym Sci 2017;67:1–47.

[196] Eyley S, Thielemans W. Nanoscale 2014;6:7764–79.

[197] Habibi Y, Aouadi S, Raquez JM, Dubois P. Cellulose 2013;20:2877–85.

[198] Brand J, Pecastaings G, Sèbe G. Carbohydr Polym 2017;169:189–97.

[199] Ferreira FV, Pinheiro IF, Gouveia RF, Thim GP, Lona LMF. Polym Compos 2017;1–21.

[200] Wei L, Agarwal UP, Hirth KC, Matuana LM, Sabo RC, Stark NM. Carbohydr. Polym 2017;169:108–16.

[201] Lin N, Huang J, Dufresne A. Nanoscale 2012;4:3274.

[202] Costantino A, Pettarin V, Viana J, Pontes A, Pouzada A, Frontini P. Proc Mater Sci 2012;1:34–43.

[203] Tomer V, Polizos G, Randall CA, Manias E. J Appl Phys 2011;109:1–11.

[204] Agarwal S, Khan MMK, Gupta RK. Polym Eng Sci 2008;48:2474–81.

[205] Galeski A, Piorkowska E. Polimery/Polymers 2007;52:323–8.

[206] Gaska K, Kádár R, Rybak A, Siwek A, Gubanski S. Polymers (Basel) 2017;9.

[207] Vigolo B, Poulin P, Lucas M, Launois P, Bernier P. Appl Phys Lett 2002;81:1210–12.

[208] Badaire S, Pichot V, Zakri C, Poulin P, Launois P, Vavro J, et al. J Appl Phys 2004;96:7509–13.

[209] Jin L, Bower C, Zhou O. Appl Phys Lett 1998;73(1197):1–4.

[210] Li B, Xidas PI, Triantafyllidis KS, Manias E. Appl Phys.Lett 2017;73(111):1–5 82906.

[211] Kelarakis A, Yoon K, Sics I, Somani RH, Hsiao BS, Chu B. Polymer (Guildf) 2005;46:5103–17.

[212] Ahir SV, Terentjev EM. Nat Mater 2005;4:491–5.

[213] Seo YR, Weon J-I. J Korean Phys Soc 2013;63:114–19.

[214] Vigolo B. Science 2000;290:1331–4.

[215] Kim H, Macosko CW. Polymer (Guildf) 2009;50:3797–809.

[216] Chang ZJ, Zhao X, Zhang QH, Chen DJ. Express PolymLett 2010;4:47–53.

[217] Yoon K, Kelarakis A. J Nanomater 2014;2014. Available from: https://doi.org/10.1155/2014/367671.

[218] Bower C, Zhu W, Jin S, Zhou O. Appl Phys Lett 2000;77:830–2.

[219] Paineau E, Dozov I, Bihannic I, Baravian C, Krapf MEM, Philippe AM, et al. ACS Appl Mater Interfaces 2012;4:4296–301.

[220] Wu S, Ladani RB, Zhang J, Bafekrpour E, Ghorbani K, Mouritz AP, et al. Carbon N Y 2015;94:607–18.

[221] Rozynek Z, de SM, Silva L, Fossum JO, da Silva GJ, de Azevedo EN, et al. Appl Clay Sci 2014;96:67–72.

[222] Yan H, Tang Y, Long W, Li Y. J Mater Sci 2014;49:5256–64.

[223] Gonnet P, Liang Z, Choi ES, Kadambala RS, Zhang C, Brooks JS, et al. CurrAppl Phys 2006;6:119–22.

[224] Zhu H, Li Y, Fang Z, Xu J, Cao F, Wan J, et al. ACS Nano 2014;8:3606–13.

[225] Fischer JE, Zhou W, Vavro J, Llaguno MC, Guthy C, Haggenmueller R, et al. J Appl Phys 2003;93:2157–63.

[226] Walters DA, Casavant MJ, Qin XC, Huffman CB, Boul PJ, Ericson LM, et al. Chem Phys Lett 2001;338:14–20.

[227] Choi ES, Brooks JS, Eaton DL, Al-Haik MS, Hussaini MY, Garmestani H, et al. J Appl Phys 2003;94:6034–9.

[228] Hone J, Llaguno MC, Nemes NM, Johnson AT, Fischer JE, Walters DA, et al. Appl Phys Lett 2000;77:666–8.

[229] Yousefi N, Sun X, Lin X, Shen X, Jia J, Zhang B, et al. Adv Mater 2014;26:5480–7.

[230] Yousefi N, Gudarzi MM, Zheng Q, Lin X, Shen X, Jia J, et al. Compos Part A Appl Sci Manuf 2013;49:42–50.

[231] Jang J, Bae J, Yoon S-H. J Mater Chem 2003;13:676–81.

[232] Stankovich S, Dikin DA, Dommett GHB, Kohlhaas KM, Zimney EJ, Stach EA, et al. Nature 2006;442:282–6.

[233] Shen X, Wang Z, Wu Y, Liu X, He Y-B, Kim J-K. Nano Lett 2016;16:3585–93.

[234] Yousefi N, Lin X, Zheng Q, Shen X, Pothnis JR, Jia J, et al. Carbon N Y 2013;59:406–17.

[235] Lu KL, Lago RM, Chen YK, Green MLH, Harris PJF, Tsang SC. Carbon N Y 1996;34:814–16.

[236] M Sęk. Bath sonicator device, Branson 2510. <https://commons.wikimedia.org/wiki/File:Bath_sonicator_device_Branson_2510.jpg>; 2017 [accessed 08.12.17].

[237] QSONICA. Q500 Sonicator. <https://www.sonicator.com/products/q500-sonicator.html>; 2017 [accessed 08.12.17].

[238] Paton KR, Varrla E, Backes C, Smith RJ, Khan U, O'Neill A, et al. Nat Mater 2014;13:624–30.

[239] R Wengeler, H Nirschl, J Colloid Interface Sci 306 (2007) 262–273.

[240] Hernandez Y, Nicolosi V, Lotya M, Blighe FM, Sun Z, De S, et al. Nat Nanotechnol 2008;3:563–8.

[241] Zhi C, Bando Y, Tang C, Kuwahara H, Golberg D. Adv Mater 2009;21:2889–93.

[242] Coleman JN, Lotya M, O'Neill A, Bergin SD, King PJ, Khan U, et al. Science 2011;331:568–71.

[243] Shahzad F, Alhabeb M, Hatter CB, Anasori B, Man Hong S, Koo CM, et al. Science 2016;353:1137–40.

[244] Hanlon D, Backes C, Higgins TM, Hughes M, O'Neill A, King P, et al. Chem Mater 2014;26:1751–63.

[245] Backes C, Higgins TM, Kelly A, Boland C, Harvey A, Hanlon D, et al. Chem Mater 2017;29:243–55.

[246] Chen X, Dobson JF, Raston CL. Chem Commun 2012;48:3703.

[247] Silverson. Multipurpose Batch Mixer. <http://www.silverson.com/us/products/batch-mixers/mixer-options> [accessed 08.12.17].

[248] EXAKT. Three roll mill. <https://www.exakt.de/en/products/three-roll-mills/technology.html>; 2017 [accessed 08.12.17].

[249] Dalir H, Farahani RD, Nhim V, Samson B, Lévesque M, Therriault D. Langmuir 2012;28:791–803.

[250] Park J-J, Lee J-Y. Mater Chem Phys 2013;141:776–80.

[251] Kothmann MH, Ziadeh M, Bakis G, Rios de Anda A, Breu J, Altstädt V. J Mater Sci 2015;50:4845–59.

[252] Gojny FH, Wichmann MHG, Köpke U, Fiedler B, Schulte K. Compos Sci Technol 2004;64:2363–71.

[253] Thostenson ET, Chou T-W. Carbon N Y 2006;44:3022–9.

[254] Souri H, Nam IW, Lee HK. Compos Sci Technol 2015;121:41–8.

[255] Viets C, Kaysser S, Schulte K. Compos Part B Eng 2014;65:80–8.

[256] Ahmadi-Moghadam B, Taheri F. J Mater Sci 2014;49:6180–90.

[257] Chandrasekaran S, Sato N, Tölle F, Mülhaupt R, Fiedler B, Schulte K. Compos Sci Technol 2014;97:90–9.

[258] Li Y, Zhang H, Bilotti E, Peijs T. MRS Adv 2016;1:1389–94.

[259] Zhang DL. Prog Mater Sci 2004;49:537–60.

[260] Castrillo PD, Olmos D, Amador DR, González-Benito J. J Colloid Interface Sci 2007;308:318–24.

[261] Gu J, Li N, Tian L, Lv Z, Zhang Q. RSC Adv 2015;5:36334–9.

[262] Tang L-C, Wan Y-J, Yan D, Pei Y-B, Zhao L, Li YB, et al. Carbon NY 2013;60:16–27.

[263] Wu H, Zhao W, Hu H, Chen G. J Mater Chem 2011;21:8626.

[264] Jiang X, Drzal LT. J Appl Polym Sci 2012;124:525–35.

[265] Gupta TK, Singh BP, Mathur RB, Dhakate SR. Nanoscale 2014;6:842–51.

[266] Donnay M, Tzavalas S, Logakis E. Compos Sci. Technol 2015;110:152–8.

[267] Gu J, Guo Y, Yang X, Liang C, Geng W, Tang L, et al. Compos Part A Appl Sci Manuf 2017;95:267–73.

[268] Lee D, Lee B, Park KH, Ryu HJ, Jeon S, Hong SH. Nano Lett 2015;15:1238–44.

[269] Brent JR, Savjani N, O'Brien P. Prog Mater Sci 2017;89:411–78.

[270] Yao Y, Lin Z, Li Z, Song X, Moon K-S, Wong C. J Mater Chem 2012;22:13494.

[271] Buzaglo M, Bar IP, Varenik M, Shunak L, Pevzner S, Regev O. Adv Mater 2017;29:1603528.

[272] Teng C, Xie D, Wang J, Yang Z, Ren G, Zhu Y. Adv Funct Mater 2017;27:1700240.

[273] Lin Y, Connell JW. Nanoscale 2012;4:6908.

[274] Jeon I-Y, Shin Y-R, Sohn G-J, Choi H-J, Bae S-Y, Mahmood J, et al. Proc Natl Acad Sci 2012;109:5588–93.

[275] Gu J, Guo Y, Lv Z, Geng W, Zhang Q. Compos Part A Appl Sci Manuf 2015;78:95–101.

[276] Gu J, Xie C, Li H, Dang J, Geng W, Zhang Q. Polym Compos 2014;35:1087–92.

[277] Fawaz J, Mittal V. Synthesis techniques for polymer nanocomposites. Weinheim: Wiley-VCH Verlag GmbH & Co. KGaA; 2014. p. 1–30.

[278] Wu CL, Zhang MQ, Rong MZ, Friedrich K. Compos Sci Technol 2002;62:1327–40.

[279] Villmow T, Pötschke P, Pegel S, Häussler L, Kretzschmar B. Polymer (Guildf) 2008;49:3500–9.

[280] Kim I-H, Jeong YG. J Polym Sci Part B Polym Phys 2010;48:850–8.

[281] Mittal V, editor. Optimization of polymer nanocomposite properties. Weinheim: Wiley-VCH Verlag GmbH & Co. KGaA; 2010.

[282] Moniruzzaman M, Du F, Romero N, Winey KI. Polymer (Guildf) 2006;47:293–8.

[283] Isayev AI, Kumar R, Lewis TM. Polymer (Guildf) 2009;50:250–60.

[284] Caseri WR. Mater Sci Technol 2006;22:807–17.

[285] Cao Z, Jiang W, Ye X, Gong X. J Magn Magn Mater 2008;320:1499–502.

[286] Xiong M, Zhou S, Wu L, Wang B, Yang L. Polymer (Guildf) 2004;45:8127–38.

[287] Vollath D, Szabó DV. J Nanoparticle Res 1999;1:235–42.

[288] Potts JR, Dreyer DR, Bielawski CW, Ruoff RS. Polymer (Guildf) 2011;52:5–25.

[289] Zou H, Wu S, Shen J. Chem Rev 2008;108:3893–957.

3

Specific Interactions and Self-Organization in Polymer/Functionalized Nanoparticle Systems

Oğuzhan Oğuz[1,2], Jamal Seyyed Monfared Zanjani[1,2], Serap Hayat Soytaş[1,2,3], Yusuf Ziya Menceloğlu[1,2,3]

[1]FACULTY OF ENGINEERING AND NATURAL SCIENCES, MATERIALS SCIENCE AND NANO ENGINEERING, SABANCI UNIVERSITY, TUZLA, ISTANBUL, TURKEY [2]SABANCI UNIVERSITY INTEGRATED MANUFACTURING TECHNOLOGIES RESEARCH AND APPLICATION CENTER & COMPOSITE TECHNOLOGIES CENTER OF EXCELLENCE, ISTANBUL, TURKEY [3]SABANCI UNIVERSITY NANOTECHNOLOGY RESEARCH AND APPLICATION CENTER (SUNUM), TUZLA, ISTANBUL, TURKEY

3.1 Introduction

Polymeric nanocomposites have been a focus of attention particularly because of their enhanced physicochemical, thermal, mechanical, and other engineering properties, enabling their use in a wide range of applications, such as automotive, aerospace, biomedical, sporting goods, and others [1–3]. These complex materials are generally prepared by the incorporation of inorganic fillers into polymer matrices using various processing techniques such as melt-compounding [4], solution blending [5], liquid exfoliation [6], solvent exchange [7], in situ polymerization, and so forth [8]. To date, a fairly large number of polymeric nanocomposites have been produced using different polymer matrices (amorphous or semicrystalline) and inorganic fillers such as nanoclays [9], nanotubes [10,11], nanofibers [12,13], and fumed silica nanospheres [14]. Due to the availability of various polymers and inorganic fillers, the associated functional properties of the nanocomposites can be tailored by modular design approaches [1–8,15–17] suggesting a high level of control over the structure and morphology. The structure–morphology–property behavior of the nanocomposites mainly depends on the characteristic features of the polymer matrix, the shape, size, that is, aspect ratio, surface area, concentration, and distribution of the filler along with the strength of polymer filler interactions.

Polymer Composites with Functionalized Nanoparticles. DOI: https://doi.org/10.1016/B978-0-12-814064-2.00003-2

One may note that the term "specific" is still ambiguous when it comes to defining interactions in polymer nanocomposites. Nevertheless, hydrogen bonding, hydrophobic, and hydrophilic interactions may be considered as commonly accepted examples of specific interactions that have been comprehensively reported in the literature. It should be also noted that it is impossible to completely cover all aspects of those interactions in a book chapter because of the voluminous literature on these subjects comprising many different theoretical approaches and experimental observations. Besides, it is well known that such specific interactions may be adjusted by various design and processing strategies to achieve advanced functional properties in polymer nanocomposites. Therefore, it is important to note that, in terms of specific interactions, instead of detailed descriptions of hydrogen bonding, hydrophilic and hydrophobic interactions, only some practical examples of polymer nanocomposites involving these well-known specific interactions will be discussed in this chapter. The interested reader is referred to the related studies provided in the References for more detailed discussions on these subjects [18–22].

Tuning specific interactions basically aims to achieve an excellent integration between organic and inorganic phases that allows for obtaining a strong polymer–filler interaction in polymer nanocomposites. In this respect, surface modification of inorganic nanoparticles (NPs) has been considered as a promising approach in the literature [23–27]. In general, surface-modified NP reinforced polymer-based materials effectively combine functional properties of polymers and unique characteristic features of NPs [23,28,29]. However, one of the main challenges in polymer nanocomposites is to achieve a homogeneous dispersion of NPs in polymer matrices as the NPs, in general, have a strong tendency to undergo aggregation [30,31]. Surface modification methods comprising the use of polymer surfactant molecules or other modification agents allow for achieving a strong repulsion between the NPs. For this reason, it is of particular interest to functionalize the NPs before incorporating them into polymers in an effort to produce high performance nanocomposite materials.

Over the years, polymer-based nanocomposites containing functionalized NPs have received substantial academic and industrial interest. Numerous studies reporting the preparation, processing, characterization, and structure–property relationships of those materials are available in the literature. Considering the high number of studies on nanocomposites containing functionalized NPs, it should be emphasized that it is impossible to completely describe this field in a book chapter. Therefore, this chapter will focus on several nanocomposite systems representing the field. Among the numerous polymer/functionalized NP systems in the literature, the nanocomposites containing (1) modified silica NPs, (2) organically modified layered silicates, and (3) polyhedral oligomeric silsesquioxane (POSS) hybrids may be listed as the most commonly investigated ones, which have received great attention in recent years as they have been utilized in a wide range of applications. In this chapter, various examples representing these three subdivisions will be reported. More detailed discussions on specific themes can be referred from relevant studies given in the References.

As described above, dispersion of inorganic NPs in a polymer matrix, interfacial interactions, and thus structure–property relationships of polymer nanocomposites containing functionalized NPs may be controlled by tuning specific interactions. However, morphological control is also significant in such systems especially for exploiting the full property potential of nanocomposites [32–34]. Although polymer–filler interactions render the distribution of NPs in polymer (homopolymer) matrices at a certain level, resulting nanocomposites lack for a precise control over the morphology. For instance, it is unlikely to align NPs in polymer matrices solely by using interfacial interactions in polymer nanocomposites. In contrast, block copolymers with well-defined self-assembled nanodomains may suggest a path toward solving this particular problem as the immiscibility of covalently bonded different blocks of copolymers induces the ordered phase formation in the neat materials [32,33,35–38]. The use of block copolymers has therefore become of particular interest to organic NPs [33,35,39–57] that allows for the facile production of nanocomposites with a high level of control over the (1) morphology, (2) domain size and shape, and (3) local environment. In addition, block copolymers also suggest (1) size and shape correspondence between polymer domains and NPs, (2) favorable and selective polymer–NP interactions, (3) NP alignment and domain orientation, and thus, (4) well-defined global (composite) morphology. For this reason, it should be emphasized that block copolymers have many advantages over homopolymers in terms of being matrix materials in nanocomposites.

Recently, nanocomposites based on block copolymers have received great attention as an emerging research field and remarkable efforts have been performed to investigate fundamental issues related to their preparation, characterization, and structure–property relationships from both theoretical and experimental aspects that have been summarized in several review articles available in the literature [32,33,35,39,46,49,50,52,53,56,58–75]. Thanks to the research in this field, considerable progress has been achieved in the production of block copolymer-based nanocomposites without sacrificing the precise control over the NP distribution and location in ordered domains, and thus, the morphology. However, it still remains a challenge to develop facile techniques for cost-effective continuous processing of such nanocomposite materials with controllable morphologies and functional properties at industrial scale, which is of particular interest to numerous technological applications required to maintain the maximum precision during the operation. Besides, controlling self-organization mechanisms in such nanocomposites dictates a solid understanding of both individual and combined effects of various interrelated factors that have not been clearly demonstrated yet. This is mainly due to the fact that even small changes in design parameters of block copolymer-based nanocomposites may lead to significant variations in their final properties. Depending on the design principles and characteristic features of both block copolymers and NPs of interest, self-organization mechanisms in block copolymer-based nanocomposites pave the way for achieving unique morphologies comprising hierarchical structures along the multiple length scales. The overall change in free energy arising from the total change in both enthalpy and entropy of block copolymer-based nanocomposites plays a key role on the final morphology. Enthalpic interactions between the polymer and the particle mainly depend on their surface chemistries, whereas entropic interactions are dependent on the dimensional features of the NPs and host polymer domains [33,76]. For instance, the surface chemistry of NPs can be adjusted by grafting appropriate polymeric ligands onto the particle surfaces that may be highlighted as an effective modification technique for tuning the enthalpic interactions [77–79]. The polymer–particle interactions are directly affected by the chemical properties of ligands and grafting density, which can be described as the number of ligands per unit NP surface area [79,80]. This mainly forms the basis of the selectivity in block copolymer-based nanocomposites. In accordance with the nature of polymer–particle interactions, NPs may be either selective or nonselective toward interacting with a specific block copolymer domain. In general, selective particles intend to localize in particular block copolymer domains, whereas nonselective particles tend to localize at the interfaces accompanied by different blocks [33,81]. The selectivity directly influences the self-organization mechanisms in block copolymer-based nanocomposites. In case of an AB type diblock copolymer nanocomposite, the selectivity (δ_γ) of the NP toward a particular block copolymer domain may be defined as $\delta_\gamma = (\gamma_{A\text{-}NP} - \gamma_{B\text{-}NP})$ where $\gamma_{A\text{-}NP}$ and $\gamma_{B\text{-}NP}$ represent the interfacial tensions between different segments of the copolymer (A and B) and the NP. A negative value of δ_γ implies that a particle is selective to the domains [33,81]. In addition, it has been clearly demonstrated that the location of the NPs and the overall morphology in such nanocomposites are also influenced by the presence of the small molecules as well as the characteristic features of the block copolymers and the NPs [82–84]. Therefore, it is important to note that each nanocomposite system must be particularly investigated by a series of characterization techniques to clearly understand the key parameters governing the self-organization mechanism along the multiple length scales.

On this basis, here we intend to provide a general overview of the current literature on the specific interactions and self-organization in polymer-based nanocomposites containing functionalized NPs. With this aim, we firstly discuss the synthesis, modification, and functionalization of NPs, and then focus on the nanocomposites containing functionalized NPs based on several representative systems. Afterward, we also address the self-organization mechanism in such materials.

3.2 Synthesis, Modification, and Functionalization

Numerous physical and chemical techniques may be applied to synthesize NPs with different elemental composition, size, shape, and properties [23,85]. In general, the physical approach comprises vapor deposition and relies on the principle of subdividing bulk precursors into NPs, whereas the chemical one includes reduction of metal ions into atoms in the presence of stabilizing agents that is followed by the aggregation of atoms in a controllable way [23,86].

In general, NPs have remarkable functional properties, such as mechanical, magnetic, electrical, and optical ones, that are significantly different than those of their bulk counterparts. This is basically due to increasing proportion of surface atoms with decreasing size of the material, which increases the reactivity and allows for fundamental catalytic processes [23,63]. As a direct result of their attractive properties, NPs can be used in numerous applications, such as drug delivery, immunotherapy, nanomedicine, smart textiles, catalysis, waste water treatment, paints, and coatings. Recently, Kango et al. [23] have summarized synthesis, properties, and applications of various NPs, such as TiO_2, ZnO, Al_2O_3, and SiO_2, in a comprehensive review article. Among these NPs, synthesis, properties, and applications of SiO_2 NPs may be highlighted as an example since they are also commonly employed in polymer nanocomposites. To synthesis SiO_2 NPs, sol–gel chemistry, flame synthesis and water-in-oil microemulsion processes may be applied. For instance, we applied the sol–gel chemistry to synthesize the SiO_2 NPs further used as reinforcing agents in various thermoplastic polyurethane–urea elastomers [87–90]. In our investigation, we synthesized the colloidal silica NPs with three different particle sizes by the reaction of tetraethylortosilicate (TEOS) and dilute aqueous ammonium hydroxide (NH_4OH) in isopropanol (IPA) at 25°C, using Stöber method. The sol–gel reactions were carried out in 100-mL glass reactors with low to moderate mixing speeds on a magnetic stirrer. Initially, IPA and aqueous ammonia were mixed and ultrasonicated for 20 minutes, and then TEOS was quickly poured into the reactor to initiate the reaction. Silica sols of various particle size were prepared at molar ratios of $[TEOS]/[NH_3]_{aq} = 0.8$, 1.7, and 2.8. TEM images of the silica NPs with average particle diameters of 18, 50, and 175 nm were representatively provided in Fig. 3–1. The NPs obtained by such techniques are commonly used in drug delivery, tissue engineering, and biosensing applications. The SiO_2 NPs can also be employed as carriers for antimicrobial applications [91–99].

Surface modification and functionalization of NPs has received particular attention as an effective way of dispersing NPs in various media, such as polymer matrices. Using proper modification agents on NPs' surfaces also enables NPs to easily adapt themselves to their environment. The importance of surface modification can be understood by examining polymer nanocomposites comprising functionalized NPs, which are well integrated into polymer matrices via proper surface modification techniques, as it basically improves interfacial interactions between NPs and polymer matrices in such systems that, in general, result in an enhancement in functional properties of the final nanocomposite materials.

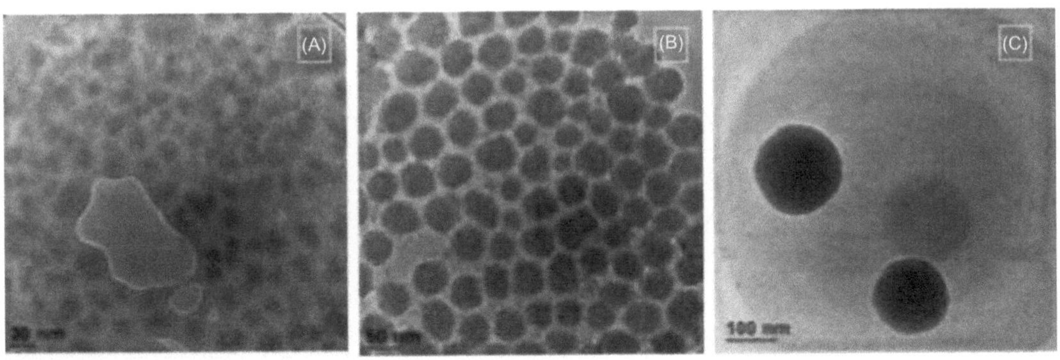

FIGURE 3–1 TEM images of SiO_2 nanoparticles prepared at molar ratios of $[TEOS]/[NH_3]_{aq}$ of (A) 2.8, (B) 1.7, and (C) 0.8. Reprinted from [88], Copyright 2013, with permission from Elsevier. *Adapted from Malay O, Oguz O, Kosak C, Yilgor E, Yilgor I, Menceloglu YZ. Polyurethaneurea–silica nanocomposites: preparation and investigation of the structure–property behavior. Polymer 2013;54(20):5313.*

Various strategies, such as chemical treatments, polymer grafting, ligand exchange, adsorption of polymeric dispersants, and in situ surface modification, may be applied to modify NPs. Among those strategies, chemical treatments and polymer grafting can be considered as commonly applied surface modification methods. The absorption of silane coupling agents may be considered as an example of chemical treatments for the surface modification of NPs. This is a facile technique to disperse NPs in various liquid media. Plueddemann and his coworkers [100] has comprehensively reported the use of silane coupling agents in polymers. Upon this pioneering study, remarkable efforts have been performed to produce NPs functionalized with silane coupling agents toward solving the problems related to the incompatibility between NPs and polymers [23]. This is of particular interest to achieve a homogeneous dispersion of NPs in a polymer matrix, and to improve the interfacial interactions, and thus, the final properties of that nanocomposite material. Surface modification of a NP using a silane coupling agent is representatively shown in Fig. 3–2.

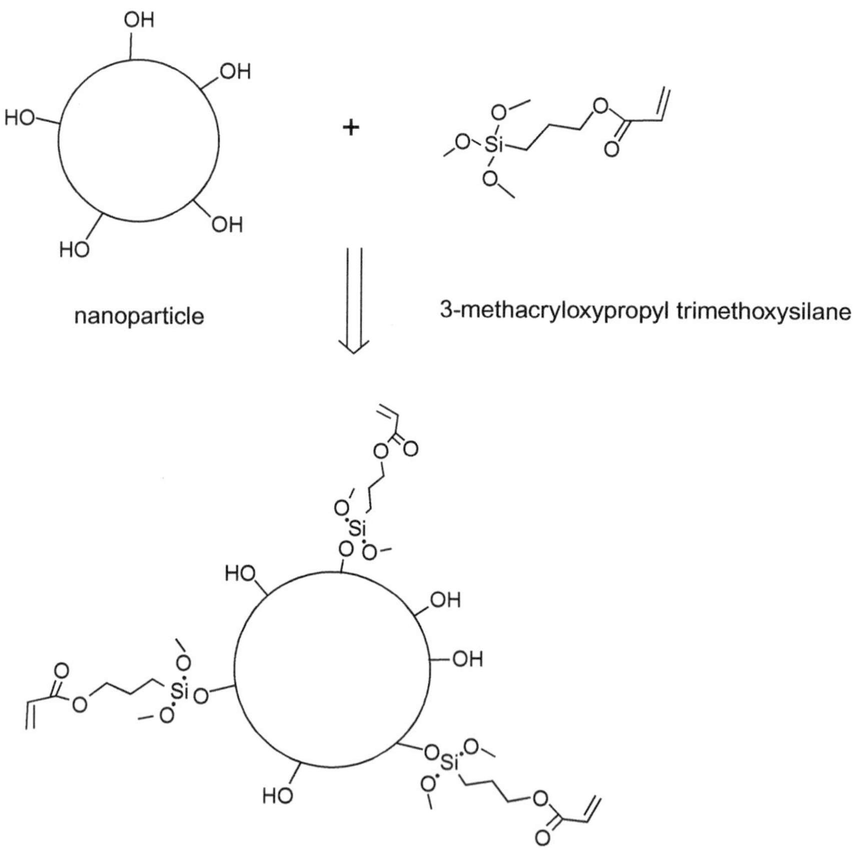

FIGURE 3–2 Modification of a nanoparticle with 3-methacryloxypropyl trimethoxysilane. Reprinted from [23], Copyright 2013, with permission from Elsevier. *Adapted from Kango S, Kalia S, Celli A, Njuguna J, Habibi Y, Kumar R. Surface modification of inorganic nanoparticles for development of organic–inorganic nanocomposites—a review. Prog Polym Sci 2013;38(8):1235.*

In this example, 3-methacryloxypropyl trimethoxysilane is used as the silane coupling agent. As illustrated in Fig. 3–2, only –OH groups cover the surface of the unmodified NP whereas the surface of the silane-modified NP is covered with 3-methacryloxypropyl trimethoxysilane groups. The modified NPs display different functional properties and better dispersion in polymer matrices compared with the unmodified counterparts [23]. In addition, metal alkoxides, epoxides, and alkyl or aryl isocyanates can also be used to produce modified NPs [23].

Another surface modification technique is grafting of polymers onto the surfaces of NPs, which is generally applied for improving chemical functionality and tailoring topological properties of the NPs. As polymeric species (oligomers) and/or monomers are intrinsically low molecular weight compounds, they can easily penetrate into NP aggregates and react with the active sites on the surfaces of NPs. Such macromolecular chains can partially fill the interstitial volume inside the aggregates that enables the NPs to be further separated from each other. To date, two different approaches have been demonstrated to covalently graft macromolecular chains onto NP surfaces. One of them is referred to as the "grafting to" strategy, which is based on the reaction of end-functionalized polymers with a proper surface, and the other one is known as the "grafting from" method, which mainly comprises polymer chains grown from an initiator-terminated self-assembled monolayer [23].

An overview on other techniques for the surface modification of NPs, such as ligand exchange, adsorption of polymeric dispersants, and in situ surface modification, can be found in the review article published by Kango et al. [23].

3.3 Nanocomposites Containing Functionalized Nanoparticles

3.3.1 Nanocomposites Reinforced by Modified Silica Nanoparticles

Numerous studies on the preparation and characterization of polymer nanocomposites containing surface modified SiO_2 NPs as reinforcing fillers are available in the literature [101–116]. For instance, Kang et al. [108] reported preparation and characterization of epoxy composites filled with functionalized nanosilica particles that are obtained via sol–gel process and then modified either by substituting surface silanol groups into epoxide ring (S–epoxide), amine (S–NH_2) or isocyanate (S–NCO) groups or by calcinating them to remove surface silanol groups (CS). They have demonstrated that surface modified silica NPs can be chemically reacted with epoxy matrix. The materials filled with CS or S–NCO show weak polymer–filler interfaces and aggregated NPs in the matrix. In contrast, S–epoxide and S–NH_2 with smaller and more reactive substituents display a better dispersion of silica NPs in epoxy matrix, which is mainly attributed to the better resin wettability.

Epoxy/silica nanocomposites have been widely utilized in various applications, such as electrical cable insulations and high performance capacitors, in the field of dielectrics. The effect of surface properties of silica NPs on the dielectric behavior of epoxy/silica nanocomposites has been reported by Cheng et al. [102]. They have mainly suggested that the coupling agent (γ-aminopropyl-triethoxysilane) enables the silica NPs to act cooperatively with the epoxy matrix via prolonging the polymer molecules, which results in an enhanced flexibility and dielectric response. Many different studies focusing on the preparation, characterization, and applications of epoxy-based nanocomposites containing modified silica NPs are also available in the literature. Another example is the enhancement of mechanical properties of fiber reinforced composites based on epoxy resins containing industrial surface-modified silica NPs published by Sprenger [113]. It is demonstrated that combining classical toughening strategies with the silica NP modification is the next step to further increase the performance of fiber-reinforced systems.

Surface-functionalized silica NPs have also been used in acrylate-based polymers, such as poly(methylmethacrylate) (PMMA) [103,105,106,110]. For instance, Liu et al. [110] reported the preparation of PMMA–silica nanocomposites via copolymerization of methylmethacrylate and allylglycidylether functionalized silica NPs. In this approach, no alkoxysilane coupling agents and sol–gel reactions were employed. Instead, it introduced covalent bonds between PMMA and silica particles. Transparent PMMA-based nanocomposite films with very high silica contents, superior thermal stability, low surface roughness, and high hardness were successfully obtained.

Poly(dimethylsiloxane) (PDMS) has also been utilized as the matrix of elastomeric nanocomposites containing surface-modified silica NPs. For example, Yuan and Mark [115] reported the reinforcement of PDMS networks by both preformed and in situ synthesized silica fillers with various sizes, size distributions, and modified surfaces. They demonstrated that silica particles modified with hydrophobic agents were likely to provide optimal enhancements as compared with the in situ precipitated ones. In another study, Ogbole et al. [112] investigated the effect of surface-treated fumed silica on the transport properties of O_2 and N_2 gases through PDMS-based nanocomposite membranes. It was found that the O_2 flux through the membranes increased with time, whereas N_2 flux decreased with time before attaining steady state.

Surface functionalized silica NPs have also been incorporated into various polyurethanes [101,104,107,111]. For instance, Dolatzadeh et al. [104] investigated the effect of organosilane-modified SiO_2 on the electrochemical properties of PU-SiO_2 nanocomposite coatings. They found that the incorporation of the NPs with more hydrophobic groups and longer hydrophobic chains on the surface significantly improves the interfacial interactions at the polymer/filler interfaces resulting in a better corrosion performance. Jang et al. [107] reported the preparation and characterization of shape memory nanocomposites based on polyurethanes cross-linked by surface-modified silica particles. An overall reaction scheme demonstrated in this study is reproduced in Fig. 3–3. The PU-based nanocomposites containing 2 wt.% SiO_2 displayed over 99% shape fixity and shape recovery up to the fourth thermomechanical cycle.

In another study, Theppradit et al. [114] reported the surface modification of silica particles and its effects on curing and mechanical properties of the natural rubber composites. The particles were modified with modified with methyl, vinyl, or aminopropyl groups using methyltriethoxysilane, vinyltriethoxysilane, or aminopropyltrimethoxysilane, respectively.

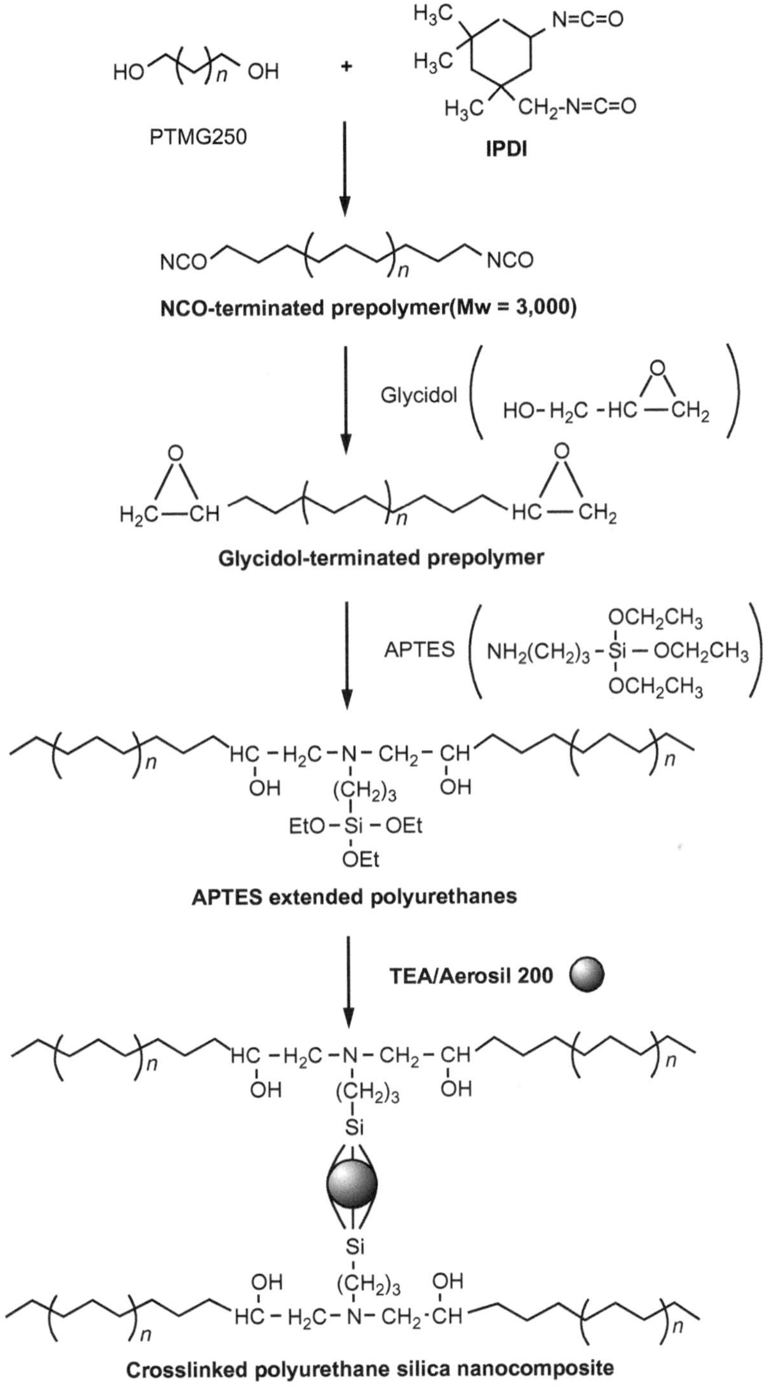

FIGURE 3–3 An overall reaction scheme for preparing the polyurethane–silica nanocomposites. The zig-zag lines are used as general symbol for the polymer chains formed in the different steps. Reprinted from [107], Copyright 2009, with permission from The Royal Society of Chemistry. *Adapted from Jang MK, Hartwig A, Kim BK. Shape memory polyurethanes cross-linked by surface modified silica particles. J Mater Chem 2009;19(8):1167.*

The modified particles displayed enhanced hydrophobicity as compared with the unmodified SiO_2. The final composite vulcanizate materials showed a good rubber–filler compatibility, and thus, an enhanced cure, mechanical, thermal, and dynamic mechanical properties due to the presence of well-dispersed modified SiO_2.

As another example, Zhou et al. [116] demonstrated the preparation of robust, self-healing, superamphiphobic fabrics via two-step coating of fluoro-containing polymer (poly (vinylidene fluoride-hexafluoropropylene), PVDF-HFP), fluoroalkyl silane (FAS), and modified silica NPs. Fig. 3–4 shows the chemical structures of FAS and PVDF (Fig. 3–4A) along with a schematic of the preparation method (Fig. 3–4B), the superrepellency of the coated polyester fabric to liquid droplets (Fig. 3–4C), the stability of the liquid repellency (Fig. 3–4D), and the dependence of contact and sliding angles on the surface tension of the liquids (Fig. 3–4E). Macroscopic photographs of the sphere-like droplets of blue-colored water, red-colored hexadecane, and clear soybean oil on the coated fabric are representatively shown in Fig. 3–4C. The contact angles of 172 degree, 165 degree, and 160 degree to water, soybean oil, and hexadecane, respectively, were obtained from the measurements performed on the coated fabric. The sliding angles of 2 degree, 5 degree, and 7 degree for water, soybean oil, and hexadecane were also recorded to measure the contact angle hysteresis. To demonstrate the stability of the liquid repellent behavior, the liquid droplets were kept on the fabric for an hour. As shown in Fig. 3–4D, the contact angles were not reduced by increasing time within an hour. The repellency of the coated fabric was also investigated using a series of organic solvents with different surface tensions. The dependence of the

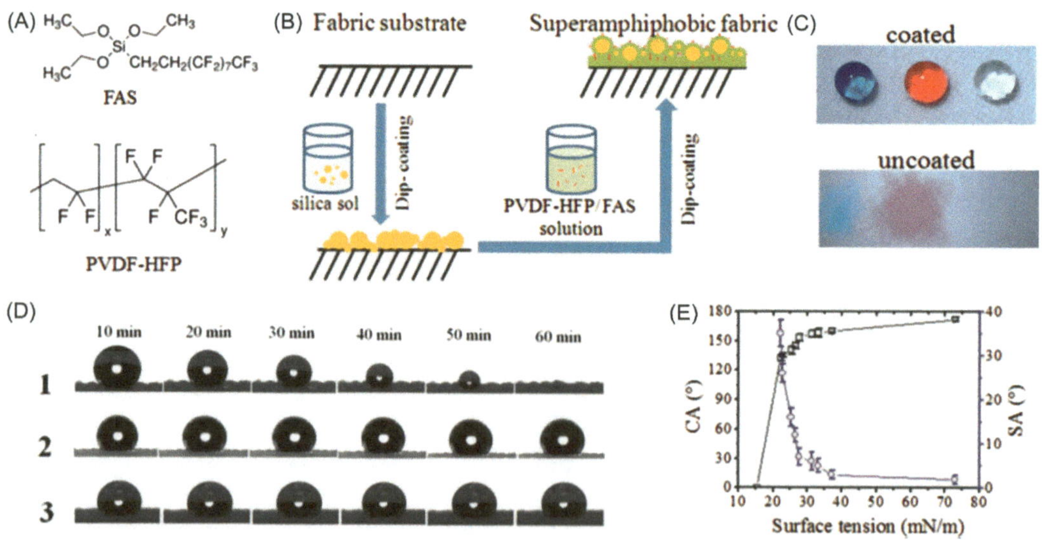

FIGURE 3–4 (A) Chemical structure of FAS and PVDF-HFP. (B) Coating procedure for superamphiphobic fabrics. (C) Photograph of blue-colored water, red-colored hexadecane, and clear soybean oil on the coated and uncoated polyester fabrics (10 μL for each drop; the small amount of dye used, oil blue in water and oil red in hexadecane, had no influence on the contact angles). (D) Variation of liquid drops with time: (1) water, (2) soybean oil, and (3) hexadecane. (E) Dependency of CA and SA on the surface tension of liquids. Reprinted from [116], Copyright 2013, with permission from John Wiley & Sons, Inc. *Adapted from Zhou H, Wang H, Niu H, Gestos A, Lin T. Robust, self-healing superamphiphobic fabrics prepared by two-step coating of fluoro-containing polymer, fluoroalkyl silane, and modified silica nanoparticles. Adv Func Mater 2013;23(13):1665.*

contact and sliding angles on the surface tension of the liquids is given in Fig. 3–4E. As clearly indicated by the results reproduced in Fig. 3–4, it was found that the coated fabric is superrepellent to both water and oil fluids with a surface tension larger than 27.5 mN/m that is a typical characteristic of superamphiphobicity [116].

Most recently, Kang et al. [109] investigated the dispersion of p-aminophenyl trimethoxysilane (APhTMS) modified silica NPs in polyamide-imide (PAI) films for improved thermal and mechanical properties. The preparation protocol is schematically presented in Fig. 3–5. In this study, the authors reported that the thermal stability and mechanical properties of the PAI-based nanocomposites were enhanced by increasing content of the modified silica NPs up to 12% by weight. In particular, 27.17% increase was recorded in the tensile strength, whereas 42.37% increase was achieved in the elongation as compared with those of pure PAI films [109].

Numerous studies on nanocomposites containing surface modified silica NPs and different polymer matrices are also available in the literature. It may be concluded that the surface modification of SiO_2 NPs suggest a modular strategy to tune filler distribution, polymer–filler interactions, interfaces, and thus, overall morphology and properties of the final polymer-based nanocomposite materials.

3.3.2 Nanocomposites Reinforced by Organically Modified Layered Silicates

In general, layered silicates have lateral dimension in centimeters, micron-sized in-plane dimension of individual layers, and thickness of single nanosized platelets. It is rather challenging to disperse the layered silicates in a polymer matrix, which is mainly due to the strong intralayer covalent bonds within the individual sheets comprising the clay. For this reason, it is essential to modify the layered silicates for using them as reinforcing fillers in polymer nanocomposites. Over the years, various methods [117] have been proposed to disperse various layered silicates based on two basic modification routes referred to as physical adsorption and chemical modification. A typical method comprising clay modification and polymer intercalation for the preparation of polymer-layered silicate nanocomposites is schematically shown in Fig. 3–6 [117].

However, different techniques, such as melt mixing and in situ polymerization, can also be used to add layered silicates into a polymer matrix. In such nanocomposites, it is possible to achieve remarkable enhancement in various functional properties, such as stiffness, strength, and permeability, which are basically related to dimensional features (e.g., high aspect ratio) and surface properties (e.g., high surface area) of the layered silicate employed. As a result of their advanced functional properties, the polymer-layered silicate nanocomposites can be utilized in a wide range of applications ranging from robust, heat resistant automotive components to high performance optical, electronic, and medical devices.

In general, smectite type of layered silicates are used in polymer nanocomposites. They have a layer thickness around 1 nm. With respect to the clay type, the lateral dimension can vary between 300 Å and several microns leading to an aspect ratio >1000. This group

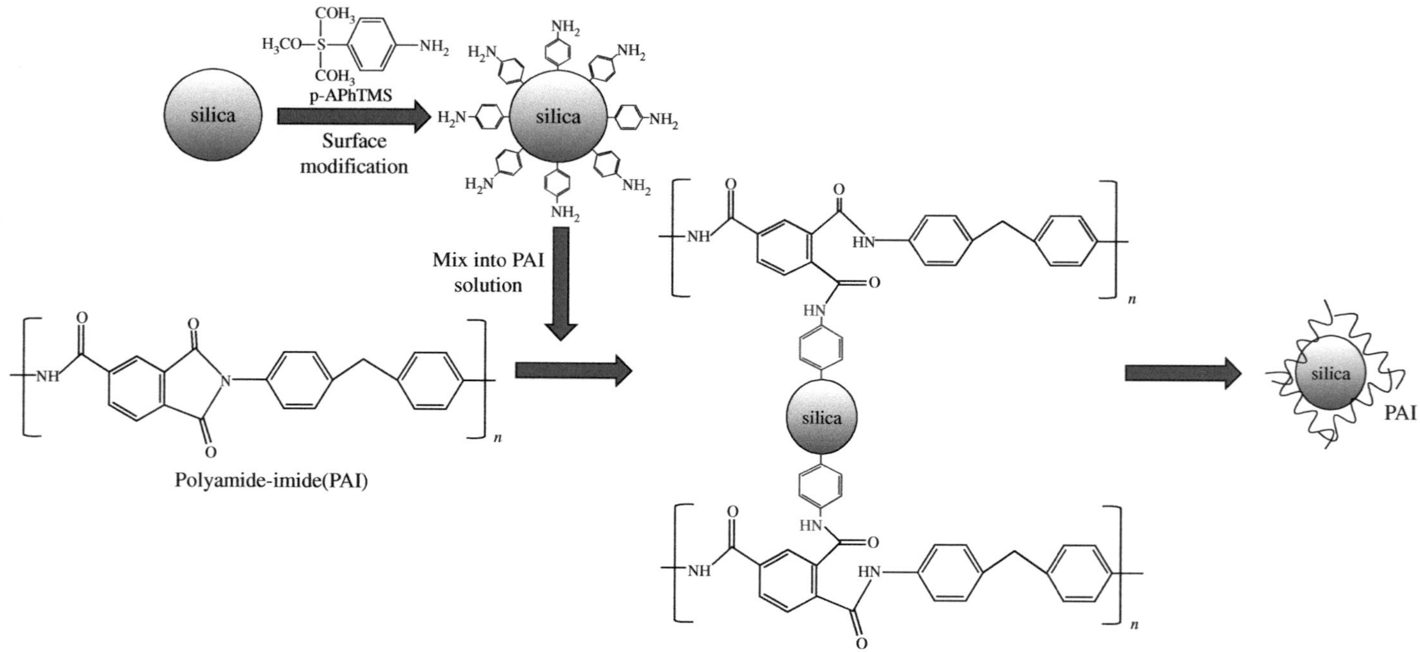

FIGURE 3–5 Schematic illustration of the APhTMS modified silica nanoparticles and APhTMS-modified silica/PAI film. Reprinted from [109] Copyright 2017, with permission from Elsevier. *Adapted from Kang T, Lee JH, Oh SG. Dispersion of surface-modified silica nanoparticles in polyamide-imide (PAI) films for enhanced mechanical and thermal properties. J Indus Eng Chem 2017;46:295.*

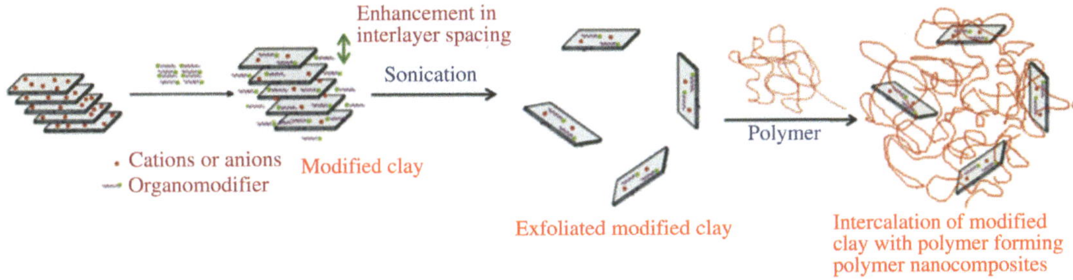

FIGURE 3–6 Schematic diagram showing clay modification and intercalation of polymer to form polymer nanocomposites. Reprinted from [117] Copyright 2015, with permission from Elsevier. *Adapted from Kotal M, Bhowmick AK. Polymer nanocomposites from modified clays: recent advances and challenges. Prog Polym Sci 2015;51:129.*

includes montmorillonite (MMT), saponite, laponite, sepiolite, hectorite, rectorite, bentonite, and vermiculite, which have an hydrophilic character and are miscible with hydrophilic polymers such as poly(ethylene oxide) (PEO). To incorporate them into various engineering polymers, they must have an organophilic character, which can be achieved by the cation exchange with various cationic surfactants comprising primary, secondary, tertiary, and quaternary alkylammonium or alkylphosphonium salts.

Among the smectite layered silicates, MMT is the most commonly used one in polymer nanocomposites due to its availability, well-documented intercalation/exfoliation chemistry, high surface area, and reactivity. MMT is a soft smectite layered silicate including plate-like particles with an average diameter of about 1 μm. It is available either in natural or synthetic forms. Its quaternary ammonium salt modified natural forms are referred to as Cloisites, which mainly show a hydrophobic character because of the enhanced interlayer spacing as a result of the presence of long alkyl chains. This makes them compatible with a polymer matrix.

Over the years, different types of Cloisites, such as Cloisite 6A [118], Cloisite 10A [118], Cloisite 15A [119], Cloisite 20A [120], Cloisite 30B [121], have been used in the preparation of nanocomposites containing modified layered silicates. Most recently, Kotal and Bhowmick have published a review article involving a comprehensive list of the studies on the nanocomposites containing different types of Cloisites as well as other modified layered silicates [117]. For instance, Maji et al. have reported the influence of the Cloisite 30B incorporation on the structure–property relationships of in situ polymerized hyperbranched polyurethane (HBP)-based nanocomposites. Briefly, an exfoliated structure was observed in all the nanocomposites investigated. Moreover, 100% increase in tensile strength, 120% increase in storage modulus, 200% increase in peel strength, and ~twofold increase in lap shear strength were achieved by the addition of 8 phr Cloisite 30B into polyurethane matrix, which also resulted in a significant enhancement in the thermal stability. In this study, the effect of the HBP functionality was also investigated. It has been reported that the HBP with higher functionality displayed better improvement in the final properties [122]. They further investigated physicomechanical properties of Cloisite 30B reinforced nanocomposites based on polyurethanes consisting of poly(propylene glycol) and toluene diisocyanate, and cured

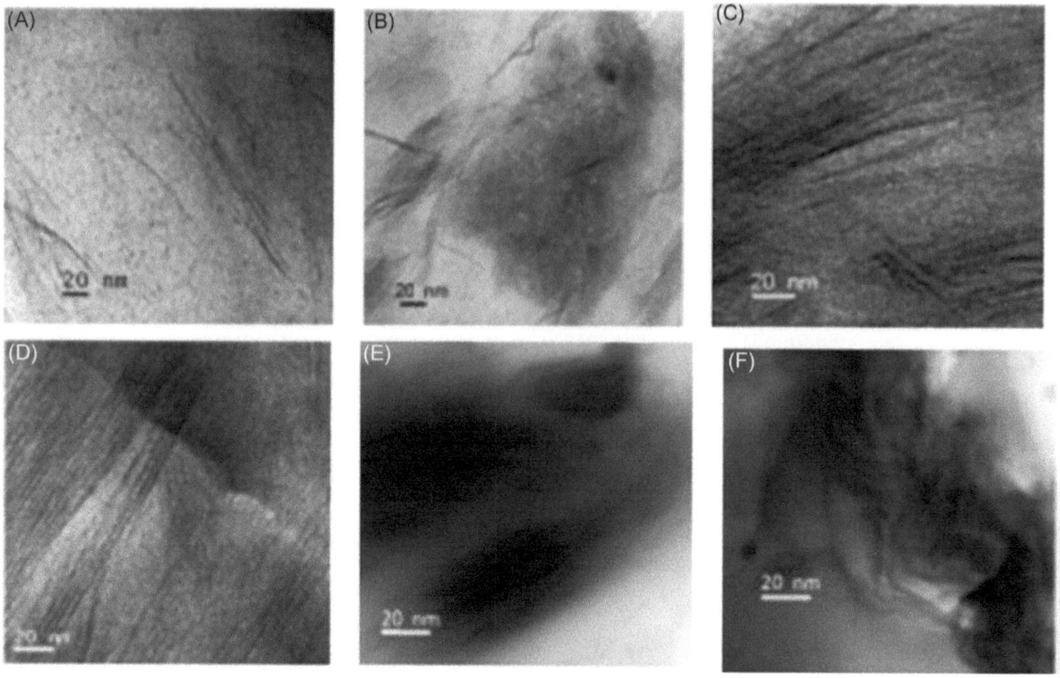

FIGURE 3–7 TEM photographs of different polyurethanes with varying clay loadings: (A) PUC2, (B) PUC4, (C) PUC8, (D) PUC16, (E) PU-Gy8, and (F) PU-TMP8. Reprinted from [123] Copyright 2012, with permission from John Wiley & Sons Inc. *Adapted from Maji PK, Bhowmick AK. Efficacy of clay content and microstructure of curing agents on the structure–property relationship of new-generation polyurethane nanocomposites. Polym Adv Technol 2012;23(9):1316.*

either by third generation hyperbranched (HB3) polyesterpolyol (HBP), trimethylolpropane (TMP), or glycerol (Gy). As shown in Fig. 3–7, the best exfoliated morphology was obtained in the nanocomposite based on HB3 cured PU containing 8 wt.% Cloisite 30B. It has been reported that 140% increase in tensile strength and 23% increase in storage modulus were achieved by the incorporation of 8 wt.% Cloisite 30B. The onset temperature of degradation was 5°C higher than that of the pure polymer. In line with this, 9°C increase in the temperature at which maximum degradation takes place was also observed [123].

In addition, it was found that the nanocomposites based on HB3 cured polyurethanes displayed better final properties in comparison with those of polyurethanes cured HBP and TMP, which is suggested to be due to the higher number of end hydroxyl functional groups taking part in cross-linking reactions leading to more ionic interaction between the polymer and the filler. Apart from the modified MMT, other surface modified layered silicates are also used for the nanocomposite preparation. For example, Ray et al. [124,125] investigated the structure–property relationships of poly(butylene succinate) (PBS)-based nanocomposites reinforced by saponite clay modified with quaternary hexadecyl tri-n-butylphosphonium bromide. Melt intercalation technique was used to prepare these nanocomposites. Results suggested that the final properties were significantly affected by a flocculated structure. In comparison with neat PBS, enhanced tensile and dynamic mechanical properties and decreased O_2 gas permeability were also noted.

Wheeler et al. [126] studied the modification of another smectite clay, laponite, using mono- and trifunctional alkoxy silanes. They employed 3-aminopropyltrimethoxy silane and aminopropyldimethylethoxy silane to covalently functionalize the laponite and reported that the trifunctional alkoxy silanes linked the clay sheets together whereas the monofunctional ones allowed for the formation of a flat monolayer on the edge of the clay plates. The basal spacing and surface area of the clay were increased by the modification with trifunctional alkoxy silanes, which makes the materials nondispersible in water, whereas the one with the monofunctional alkoxy silanes showed less effect on these properties. For this reason, the use of the monofunctional alkoxy silane modified laponite is suggested for nanocomposite production [126].

To achieve uniform dispersion of laponites in a polymer matrix, dual modification strategies have also been suggested in the literature. For instance, Mishra et al. [127] performed dual modification of laponite, using cetyltrimethylammonium bromide (CTAB) as an ionic modifier and octyl trimethoxy silane and 3-aminopropyl triethoxy silane as covalent modification agents, for the production of thermoplastic polyurethane (TPU)-based nanocomposites via both ex situ and in situ techniques. The ex situ technique is reported to be more effective for the exfoliation of laponite in polymer matrices. Mondal et al. [128] also prepared TPU-based nanocomposites containing dual-modified laponite fillers. In accordance with the viscoelastic properties and thermomechanical and dimensional stabilities of the nanocomposites, they have suggested that dual-modified laponite is more effective as a reinforcing agent.

Overall, it might be concluded that the modification is of particular interest to tune the surface properties of various layered silicates to make them available to use for designing high performance nanocomposites based on different polymers.

3.3.3 Nanocomposites Containing Polyhedral Oligomeric Silsesquioxane Hybrids

Numerous reactive organic groups can be used to functionalize POSS feedstocks, which makes them available to be used in polymers via facile techniques, such as grafting, copolymerization or blending. Utilizing POSS nanocluster cages in a polymeric system may be a convenient framework to produce functional materials with enhanced final properties, such as superior thermal stability, oxidation resistance, and reduced flammability. Hence, POSS containing polymeric systems have received particular attention over the years. In line with the versatility of POSS chemistry [129], remarkable efforts have performed to produce polymer/POSS nanocomposites with various structural and morphological features (Fig. 3–8) regarding to the number of POSS functional groups [130]. Some of those are representatively given in this section. However, it might be noted that two comprehensive review articles [129,130] recently published by Chordes et al. [129] and Kuo and Chang [130] are highly

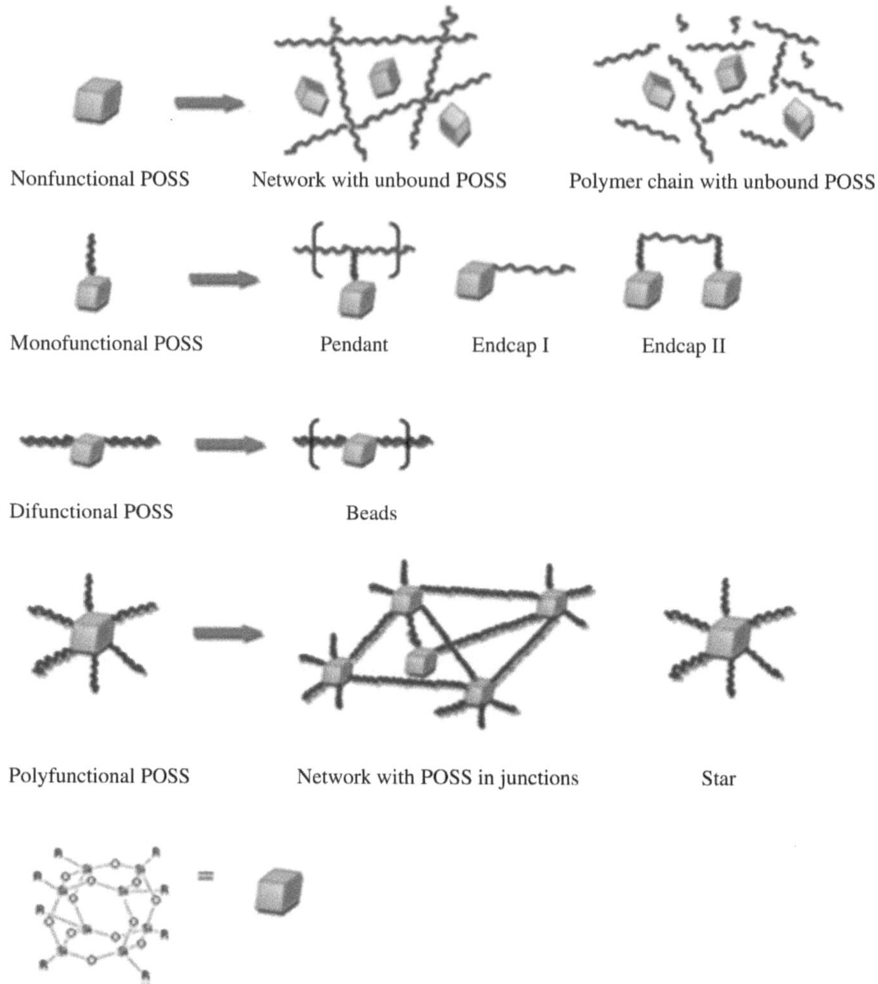

FIGURE 3–8 Polymer/POSS architectures. Reprinted from [130], Copyright 2011, with permission from Elsevier. *Adapted from Kuo S-W, Chang F-C. POSS related polymer nanocomposites. Prog Polym Sci 2011;36(12):1658.*

recommended for interested readers who would like to have more detailed discussions on POSS chemistry and POSS containing polymeric systems.

Over the years, various design approaches have suggested to produce polyolefin/POSS nanocomposites. Joshi et al. [131] demonstrated the preparation of high density polyethylene-based nanocomposites containing octamethyl-POSS (HDPE/OM-POSS) via melt mixing technique. According to the rheological results obtained in this study, it was reported that the POSS particles acted as a lubricant at low filler concentrations (0.25–0.5 wt.%) as they reduced the complex viscosity in these nanocomposites. In contrast, the complex viscosity was increased by the incorporation of POSS particles at higher

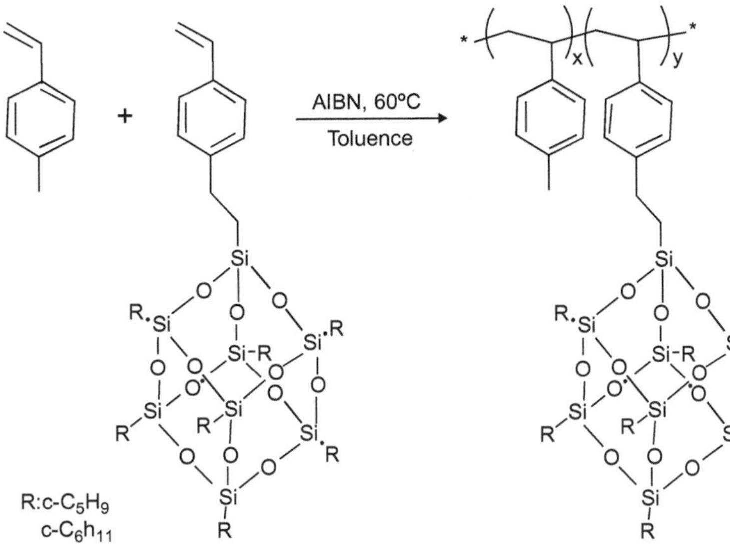

FIGURE 3–9 Polymerization of PS–POSS copolymers. Reprinted from [130], Copyright 2011, with permission from Elsevier. *Adapted from Kuo S-W, Chang F-C. POSS related polymer nanocomposites. Prog Polym Sci 2011;36(12):1661.*

concentrations, where they inclined to form agglomerates. However, the particles were miscible with the matrix at low concentrations. Applying the Kissinger method based on nonisothermal crystallization behavior of the materials, it was also demonstrated that the activation energy for the transport of the polymer segments to the growing surface was not significantly affected by the addition of POSS particles [132]. Tabuani and coworkers [133–136] investigated the morphological features and thermal properties of melt-blended PP/POSS hybrids with respect to the alkyl substituents (R = ethyl, isobutyl, isooctyl) used in the POSS structure. They reported that the filler distribution and polymer–filler interactions were significantly affected by the alkyl group length during the cooling from the melt.

In another study, free radical copolymerization was used to prepare linear thermoplastic hybrid materials containing large POSS groups pendent to a polystyrene (PS) backbone (Fig. 3–9) [137]. It was reported that the thermal properties of the PS were dramatically influenced by the presence of pendent groups and interchain and/or intrachain POSS-POSS interactions, which also affected the solubility.

POSS is also used as a filler in polyacrylates. For instance, Kopesky et al. [138] investigated poly(methylmetacrylate) (PMMA)-based nanocomposites containing both tethered and untethered POSS moieties. They demonstrated that the plateau modulus was decreased by the incorporation of tethered-POSS as compared with that of the pristine PMMA. Besides, the zero-shear viscosity is notably increased by the addition of untethered POSS, mainly because of the interactions between the POSS units.

On the other hand, Hirai and coworkers [139,140] produced POSS units block copolymers by anionic living polymerization technique. As shown in Fig. 3–10, various self-assembled

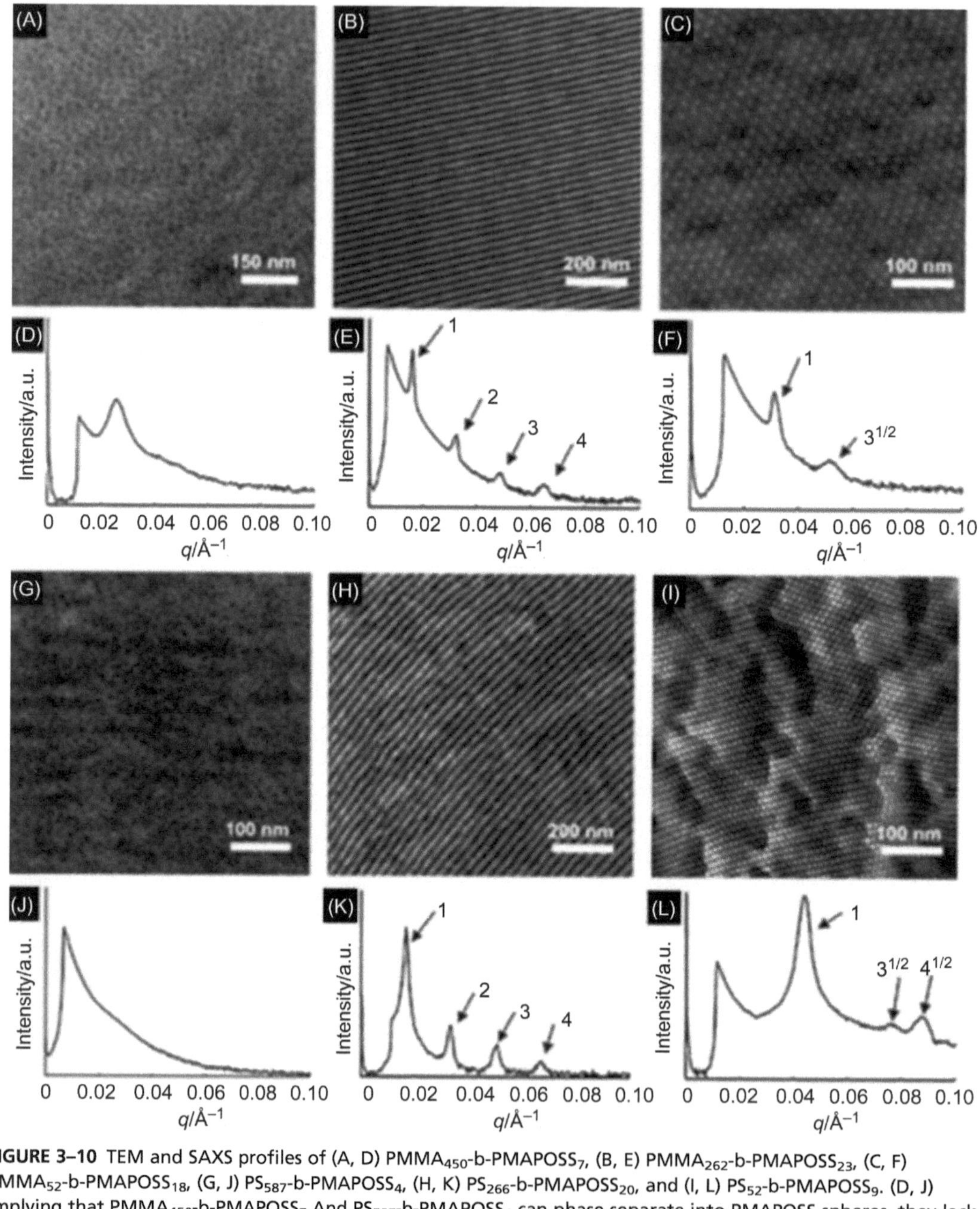

FIGURE 3–10 TEM and SAXS profiles of (A, D) $PMMA_{450}$-b-$PMAPOSS_7$, (B, E) $PMMA_{262}$-b-$PMAPOSS_{23}$, (C, F) $PMMA_{52}$-b-$PMAPOSS_{18}$, (G, J) PS_{587}-b-$PMAPOSS_4$, (H, K) PS_{266}-b-$PMAPOSS_{20}$, and (I, L) PS_{52}-b-$PMAPOSS_9$. (D, J) Implying that $PMMA_{450}$-b-$PMAPOSS_7$ And PS_{587}-b-$PMAPOSS_4$ can phase separate into PMAPOSS spheres, they lack long range order. (E, K) Characteristic ratio of 1:2:3:4, implying well-defined lamellar morphologies for $PMMA_{262}$-b-$PMAPOSS_{23}$ and PS_{266}-b-$PMAPOSS_{20}$. (F, L) Characteristic ratio of 1:31/2:41/2, suggesting hexagonally packed cylinders for $PMMA_{52}$-b-$PMAPOSS_{18}$ and PS_{52}-b-$PMAPOSS_9$. Reprinted from [140], Copyright 2009, with permission from American Chemical Society. *Adapted from Hirai T, Leolukman M, Jin S, Goseki R, Ishida Y, Kakimoto M-a, Hayakawa T, Ree M, Gopalan P. Hierarchical self-assembled structures from POSS-containing block copolymers synthesized by living anionic polymerization. Macromolecules 2009;42(22):8839.*

structures, such as spherical, cylindrical, lamellar morphologies, were successfully obtained in these diblock copolymers notated as PMMA-b-PMAPOSS and PS-b-PMAPOSS.

Another interesting example is the use of POSS along with PEO, which is utilized as an efficient polymeric electrolyte in lithium-ion batteries due to its ability to solvate lithium ions. To increase room temperature conductivity of PEO-based electrolytes, one of the strategies is to form comb- or hairy rod-like side chains via attaching PEO macromolecules with short chain length. It is possible to obtain a chain-length dependent increase in the glass transition temperature (T_g) and suppression of crystallization by grafting PEO onto POSS that is mainly due to the high ratio of chain ends in such dendrimers or star-shaped polymers. Moreover, this strategy allows for obtaining a lower melt viscosity and greater free volume compared with those of linear PEO macromolecules with a comparable molecular weight. As such PEO-POSS hybrids display relatively low T_g and do not crystallize, it is possible to utilize them as low temperature solvents in lithium-ion batteries [141–150]. POSS units are also incorporated into various polymers, such as polyesters [151], polyamides [152], polyimides [153], polyurethanes [154], epoxies [155], and polybenzoxazines [156] in an effort to produce advanced functional materials that can be utilized in a wide range of applications relying on tunable thermal, mechanical, electrical, and surface properties [130].

3.3.4 Other Systems

Many different nanocomposites representing polymer/functionalized NP systems are available in the literature. For instance, PMMA-based nanocomposites containing functionalized TiO_2 NPs were produced by an in situ radical polymerization of MMA in a toluene solution of surface modified TiO_2 NPs [157]. Functionalized alumina NPs were also dispersed in PMMA. Ritzhaupt-Kleissl et al. [158] studied the effect of surface modification of alumina NPs on its optical properties. Silane treated alumina NPs in PMMA displayed better optical transmittance properties in both visible and infrared regions in comparison with unmodified alumina NPs. Surface treated ZnO NPs were incorporated into polypropylene via extrusion [159]. Kim et al. [160] produced high energy density poly(vinylidene fluoride-co-hexafluoro propylene)-based nanocomposites containing phosphonic acid surface-modified $BaTiO_3$ NPs to investigate the dielectric permittivity and electric breakdown strength as a function of the volume fraction of NPs. It might be noted that various NPs can be properly functionalized and then incorporated into different polymer matrices in an effort to produce advanced nanocomposites with functional properties [3,23,40,117,130].

3.4 Self-Organization in Nanocomposites

It is well known that "top-down" strategies can be applied to tune the spatial organization of NPs at a few hundreds of nanometers. However, it still remains a challenge to organize NPs with a precise control over a single particle and tune interparticle distance to be less than 10 nm [161–163]. In contrast, "bottom-up" techniques based on various self-assembly processes suggest a convenient framework to overcome such limitations, particularly for tuning the coupling issue in NP clusters by monitoring ordering of the particles, and thus, the interparticle distance with a single particle precision. In addition, "bottom-up" strategies allow for the incorporation of other functional building blocks into nanocomposites to produce advanced multifunctional materials with tunable final properties.

Numerous functional building blocks with well-defined structures, such as liquid crystals, small organic species, supramolecules, suggest a toolbox for designing novel materials that can be readily used in conjunction with NPs to achieve exceptional physical, chemical, thermal, mechanical, electrical, and optical properties. Recent developments in polymer science and technology, particularly in the field of block copolymers (BCPs), provide a practical guideline to manipulate the self-assembled building blocks along the multiple length scales. Combining unique functional properties of inorganic NPs with such supramolecular assemblies offers a tremendous potential to produce multifunctional nanocomposite materials (Fig. 3–11), which have become of particular interest in various advanced technological applications [162].

Two main approaches have been proposed to design BCP microdomains containing NP assemblies. One of them is the in situ synthesis of NPs within BCP microdomains and the other one is the assembling of preformed NPs in BCP microdomains of interest [35,50]. The in situ synthesis can be performed by the reduction of NP precursors preferably inserted into one of the BCP microdomains. In this respect, Wiesner and coworkers proposed a facile method based on sol–gel chemistry mainly referred as the "dispersion route" to produce

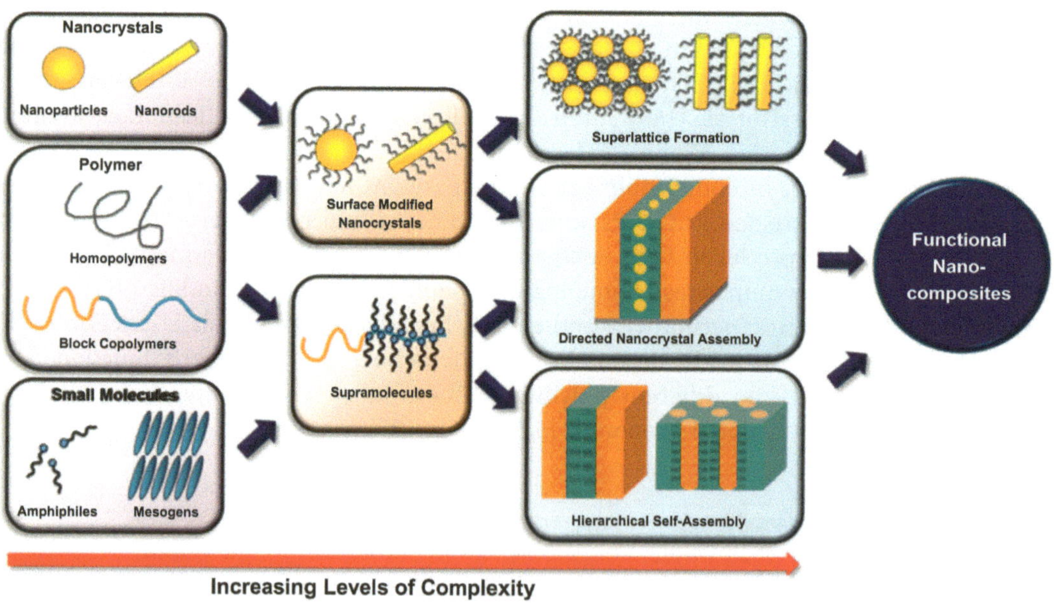

FIGURE 3–11 Hierarchical coassemblies of nanocrystals, polymers, and small molecules toward functional nanocomposites. Reprinted from [162] Copyright 2013, with permission from The Royal Society of Chemistry. *Adapted from Kao J, Thorkelsson K, Bai P, Rancatore BJ, Xu T. Toward functional nanocomposites: taking the best of nanoparticles, polymers, and small molecules. Chem Soc Rev 2013;42(7):656.*

inorganic particles within the PEO blocks of a poly(isoprene-b-ethylene oxide) (PI-b-PEO) block copolymer [164–167]. Apart from this, another in situ synthesis method referred to as the "immersion route" is also available in the literature. In this technique, a solid BCP film is immersed into a solution involving the metal precursor, which paves the way for the diffusion of the precursor into the swollen BCP microdomains. The NP formation in the microdomains is successfully completed by immersing the film containing metal salts into the reducing agent solution. Two main aspects of the in situ techniques render them highly applicable. Firstly, such techniques allow for achieving a homogeneous dispersion of NPs in the BCP matrix. Secondly, they can be applied to increase the volume fraction of NPs as it may be achieved solely by repeating the precursor loading and reduction steps [162].

On the other hand, the spatial organization of various preformed NPs with different sizes, shapes, and compositions can also be controlled by using BCPs. This may provide an alternative way of producing functional nanocomposites for new-generation nanodevices that may not be prepared by in situ techniques. However, it should be emphasized that balancing interactions between the BCP segments and the ligands attached to the NPs with a maximum precision is of particular importance for the coassembly of NPs and the BCPs, which is also dependent on the size effect of the NPs on the BCP chain configuration [40,168–173]. For instance, surface functionalized NPs can selectively interact with certain blocks of a BCP. Loading solid NPs into BCPs drives polymer chains to be stretched around the NPs. This mainly results in a loss of conformational entropy that increases with particle size [174]. An increase in particle size (if the particles become bigger than BCP domains) can make the BCP domains incapable of hosting the particles, which may result in a macrophase separation. Such effects may cause significant changes in BCP morphology, which can be referred to as size-induced phase transitions. In addition to the particle size, shape, volume fraction, and surface properties of NPs can also induce such morphological transitions in BCPs. For this reason, it has become of particular interest to construct phase/morphology diagrams of nanocomposites for a better understanding of structure–property correlations, which is extremely important in a wide range of applications. Recently, Garcia et al. [175] demonstrated the morphology space of nanocomposites based on poly(isoprene-block-ethylene oxide) (PI-b-PEO) diblock copolymers as structure directing agents for aluminosilicate NPs prepared from (3-glycidyloxypropyl)trimethoxysilane (GLYMO) and aluminum(III) sec-butoxide. The results obtained from the structural investigations of over 60 nanocomposite compositions were reported. A ternary composition map summarizing those results is representatively reproduced in Fig. 3–12. More detailed discussions can be found in their publication [175].

The shape of NPs is also a significant parameter in such nanocomposite systems. In particular, the anisotropy in shape must be considered for a precise control over the key mechanisms behind the spatial organization of NPs. This aspect has recently become of particular interest to experimentalists upon the pioneering work published by Deshmukh et al. focusing on the use of a BCP matrix to align gold nanorods parallel to a substrate [176].

Assemblies of NPs may be directed in BCP-based supramolecules, which can be fabricated by attaching small molecules to polymer backbones using secondary interactions, such as hydrogen bonding and ionic interactions. In such systems, polymer–NP interactions can

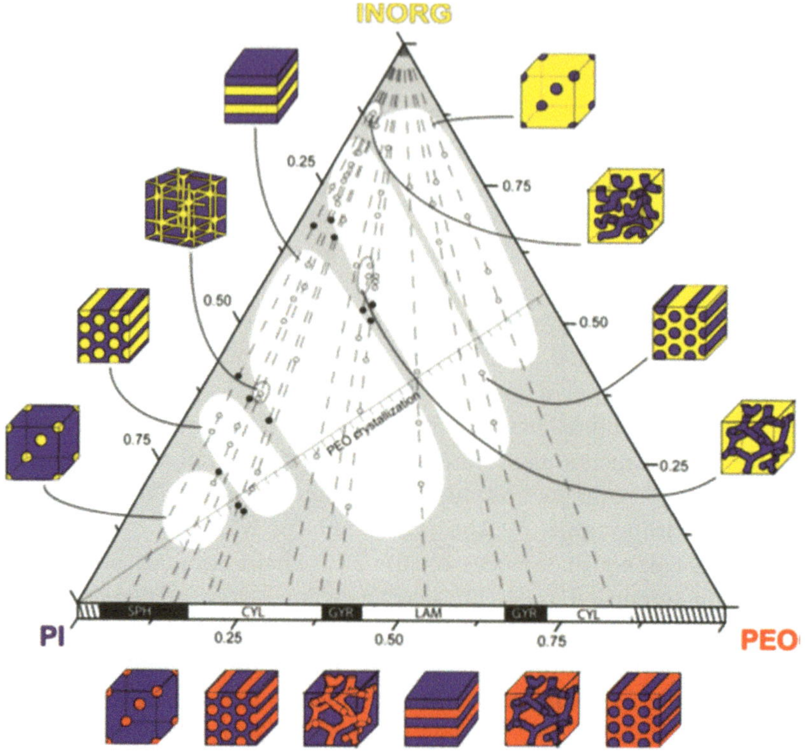

FIGURE 3–12 Ternary diagram mapping out the morphologies found for various compositions (in weight fractions) of the system PI-b-PEO plus aluminosilicate sol nanoparticles derived from (3-glycidyloxypropyl)trimethoxysilane (GLYMO) and aluminum(III) sec-butoxide ($Al(O^{s}Bu)_3$). At the bottom of the diagram are shown schematics of the morphologies found for the pure PI-b-PEO diblock copolymer. Hatched areas along the PI-PEO axis indicate areas where no data are available. Each white region within the diagram is labeled with a schematic representing the morphology of the composites formed. The yellow (light) regions in these schematic morphologies on the right and left are a representation of the PEO + inorganic domains. Closed dark points on a gray background indicate areas where biphasic behavior is observed. Reprinted from [175] Copyright 2009, with permission from American Chemical Society. *Adapted from Garcia BC, Kamperman M, Ulrich R, Jain A, Gruner SM, Wiesner U. Morphology diagram of a diblock copolymer – aluminosilicate nanoparticle system. Chem Mater 2009; 21(22):5403.*

be effectively tuned as attaching small molecules favorably interacting with the NP ligands to the polymer side chains simply eliminates the need to conduct NP surface chemistry. Utilizing secondary interactions also eliminates the need to modify polymer chains [162]. In addition, various external stimuli can be used to alter the strength of the noncovalent linkage between the BCP and small molecules. Hence, it should be emphasized that modular strategies relying on secondary interactions suggest a path for tuning the supramolecular structure and the spatial organization of NPs solely by applying external fields. Following such strategies, one can provide a precise control over the structural, dimensional, and spatial

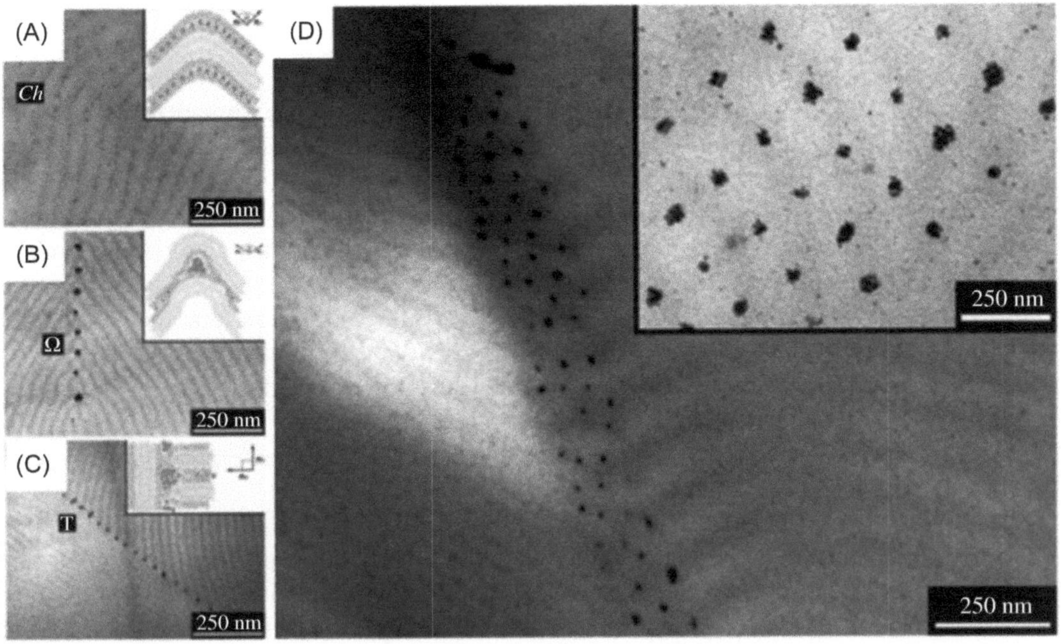

FIGURE 3–13 (A) Electron micrograph of the unstained PS-b-PEP/AuSPS depicting a chevron grain boundary structure (θ≈30°). Particles were uniformly distributed in the PS microdomains. (B) Electron micrograph of the stained PS-b-PEP/AuSPS depicting an omega-type grain boundary (Ω). The Au NPs segregated to the high-energy regions across the boundary, giving rise to the formation of the periodic array of NPs aggregates. (C) Electron micrograph of the stained PS-b-PEP/AuSPS depicting a T-junction grain boundary. The periodic array of aggregates reveals that particle accumulation occurred within the high-energy T-junctions across the boundary. Insets of panels (A–C) illustrate the particle distribution within the boundary region. Reprinted from [178] Copyright 2006, with permission from American Chemical Society. *Adapted from Listak J, Bockstaller MR. Stabilization of grain boundary morphologies in lamellar block copolymer/nanoparticle blends. Macromolecules 2006;39(17):5822-5.*

parameters along the multiple length scales to produce multifunctional stimuli-responsive nanocomposites [65,177].

The spatial arrangement of NPs in bulk can also be controlled by exploiting the defects in BCPs. Listak and Bockstaller [178] demonstrated that polymer-coated NPs inclined to be segregated at high-energy tilt and twist grain boundary structures in amorphous lamellar BCP-based nanocomposites containing PS-capped Au NPs (Fig. 3–13). They showed that the distribution of the NPs at small tilt angles (chevron grain boundary) was equal to that near the center of the particular domains in the BCP (Fig. 3–13A).

On the contrary, the NPs were found to preferably swell the high elastic energy regions along the grain boundary at larger tilt angles such as omega (Fig. 3–13B) or T-junction (Fig. 3–13C) grain boundary structures rendering the stabilization of the energetically unfavorable grain boundary structures through the relaxation of chains in the grain boundary region to maximize the conformational entropy of the chains located nearby those defects [162,178].

3.5 Conclusion and Outlook

Nanocomposites containing functionalized NPs form an important subdivision of polymer-based nanocomposites. In this chapter, specific interactions and self-organization in polymer/functionalized NP systems are discussed based on several representative studies. It may be concluded that surface modification of inorganic NPs is of particular interest to tailor the specific interactions in such materials. Polymer/functionalized NP systems combine unique properties of organic and inorganic building blocks within the same material that makes them promising alternatives to conventional materials used in various fields. Such nanocomposites can be utilized in a wide range of applications such as high performance chemicals, thermal and electrical insulation materials, smart coatings, artificial membranes, adsorbents, biomaterials, medical devices, imaging, telecommunication, and information technologies. To successfully develop these nanocomposites, dispersion stability of and compatibility of NPs with polymer matrices should be significantly improved, which can be achieved by various surface modification techniques.

In parallel with the remarkable efforts in synthesis, modification and functionalization of inorganic NPs, recent developments in polymer science suggest novel strategies toward producing high performance nanocomposites with advanced functional properties as the next-generation materials that can be implemented in emerging fields of science and technology. In particular, modular design approaches relying on the use of block copolymers in the production of multifunctional nanocomposites offer practical solutions to overcome main limitations, such as controlling the spatial organization of NPs in polymer matrices, in the field of conventional (homo) polymer-based nanocomposites. In this chapter, some of the recent advances in the field of block copolymer-based nanocomposites are highlighted. It is important to note that self-assembly processes in block copolymers pave the way for the morphological control in the resulting nanocomposites that may be the key to the realization of an envisioned nanoworld promising energy efficiency, sustainable production, and maximum performance. Critical design principles required to provide a precise control over the spatial organization of NPs in block copolymer matrices are discussed. In addition, common methods for producing block copolymer-based nanocomposites are summarized. Basically, techniques relying on the in situ synthesis allow for producing nanocomposites with high volume fractions of the NPs that can be utilized in various applications, such as photocatalysis, dye-sensitized solar cells, and metamaterials fabrication. Methods comprising the use of preformed NPs pave the way to produce multidimensional NP arrays in block copolymer-based nanocomposites. In this context, tuning particle size, shape, and surface chemistry is of significance to tailor the thermodynamic driving forces taking place during the self-assembly processes. Along these lines, particular emphasis has to be devoted to clearly understand the kinetics of the coassemblies to further optimize critical design principles and processing parameters to provide a precise morphological control over the systems, not only at equilibrium, but also kinetically trapped states. Considering the versatility of the nanocomposites together with the characteristic features of their organic and inorganic components, it is obvious that many other challenges can be highlighted. It is also obvious that such complex systems will bring new challenges into this interdisciplinary research field in the future. However, it should be emphasized that recent developments in computational tools, small molecule synthesis, supramolecular chemistry, and other interrelated experimental techniques undoubtedly look promising for the sustainable mass production of next-generation multifunctional nanocomposite materials with advanced physical properties. It is our hope that this chapter provides a practical guideline for the readers and scientifically inspires them to further advance this interdisciplinary research field.

References

[1] Jancar J, Douglas J, Starr FW, Kumar S, Cassagnau P, Lesser A, et al. Current issues in research on structure–property relationships in polymer nanocomposites. Polymer (Guildf) 2010;51(15):3321–43.

[2] Schaefer DW, Justice RS. How nano are nanocomposites? Macromolecules 2007;40(24):8501–17.

[3] Zou H, Wu S, Shen J. Polymer/silica nanocomposites: preparation, characterization, properties, and applications. Chem Rev 2008;108(9):3893–957.

[4] Dan CH, Lee MH, Kim YD, Min BH, Kim JH. Effect of clay modifiers on the morphology and physical properties of thermoplastic polyurethane/clay nanocomposites. Polymer (Guildf) 2006;47(19):6718–30.

[5] Zha W, Han CD, Moon HC, Han SH, Lee DH, Kim JK. Exfoliation of organoclay nanocomposites based on polystyrene-block-polyisoprene-block-poly (2-vinylpyridine) copolymer: solution blending versus melt blending. Polymer (Guildf) 2010;51(4):936–52.

[6] Coleman JN, Lotya M, O'Neill A, Bergin SD, King PJ, Khan U, et al. Two-dimensional nanosheets produced by liquid exfoliation of layered materials. Science 2011;331(6017):568–71.

[7] Liff SM, Kumar N, McKinley GH. High-performance elastomeric nanocomposites via solvent-exchange processing. Nat Mater 2007;6(1):76–83.

[8] Saeed K, Park S-Y, Lee H-J, Baek J-B, Huh W-S. Preparation of electrospun nanofibers of carbon nanotube/polycaprolactone nanocomposite. Polymer (Guildf) 2006;47(23):8019–25.

[9] Sinha Ray S, Okamoto M. Polymer/layered silicate nanocomposites: a review from preparation to processing. Prog Polym Sci 2003;28(11):1539–641.

[10] Koerner H, Price G, Pearce NA, Alexander M, Vaia RA. Remotely actuated polymer nanocomposites—stress-recovery of carbon-nanotube-filled thermoplastic elastomers. Nat Mater 2004;3(2):115–20.

[11] Nie M, Kalyon DM, Fisher FT. Interfacial load transfer in polymer/carbon nanotube nanocomposites with a nanohybrid shish kebab modification. ACS Appl Mater Interfaces 2014;6(17):14886–93.

[12] Jonoobi M, Harun J, Mathew AP, Oksman K. Mechanical properties of cellulose nanofiber (CNF) reinforced polylactic acid (PLA) prepared by twin screw extrusion. Compos Sci Technol 2010;70 (12):1742–7.

[13] Kuang Q, Zhang D, Yu JC, Chang Y-W, Yue M, Hou Y, et al. Toward record-high stiffness in polyurethane nanocomposites using aramid nanofibers. J Phys Chem C 2015;119(49):27467–77.

[14] Yilgor E, Eynur T, Kosak C, Bilgin S, Yilgor I, Malay O, et al. Fumed silica filled poly (dimethylsiloxane-urea) segmented copolymers: preparation and properties. Polymer (Guildf) 2011;52(19):4189–98.

[15] Bonderer LJ, Studart AR, Gauckler LJ. Bioinspired design and assembly of platelet reinforced polymer films. Science 2008;319(5866):1069–73.

[16] Demirörs AF, Courty D, Libanori R, Studart AR. Periodically microstructured composite films made by electric-and magnetic-directed colloidal assembly. Proc Natl Acad Sci 2016;113(17):4623–8.

[17] Erb RM, Libanori R, Rothfuchs N, Studart AR. Composites reinforced in three dimensions by using low magnetic fields. Science 2012;335(6065):199–204.

[18] Ben-Naim AY. Hydrophobic interactions. New York: Springer Science & Business Media; 2012.

[19] Burrows AD, Mingos DMP. Supramolecular assembly via hydrogen bonds I. Berlin Heidelberg: Springer-Verlag; 2004.

[20] Coleman MM, Painter PC, Graf JF. Specific interactions and the miscibility of polymer blends. Lancaster, Pennsylvania: CRC Press; 1995.

[21] Israelachvili JN. Intermolecular and surface forces. Academic Press; 2011.

[22] Jeffrey GA, Saenger W. Hydrogen bonding in biological structures. Springer Science & Business Media; 2012.

[23] Kango S, Kalia S, Celli A, Njuguna J, Habibi Y, Kumar R. Surface modification of inorganic nanoparticles for development of organic–inorganic nanocomposites—a review. Prog Polym Sci 2013;38 (8):1232–61.

[24] Peng Q, Lai DM, Kang E, Neoh K. Preparation of polymer – silicon (100) hybrids via interface-initiated reversible addition-fragmentation chain-transfer (RAFT) polymerization. Macromolecules 2006;39 (16):5577–82.

[25] Skaff H, Emrick T. Reversible addition fragmentation chain transfer (RAFT) polymerization from unprotected cadmium selenide nanoparticles. Angew Chem Int Ed 2004;43(40):5383–6.

[26] Taniguchi Y, Ogawa M, Gang W, Saitoh H, Fujiki K, Yamauchi T, et al. Preparation of hyperfunctional carbon black by grafting of hyperbranched polyester onto the surface. Mater Chem Phys 2008;108 (2):397–402.

[27] Walter R, Friedrich K, Privalko V, Savadori A. On modulus and fracture toughness of rigid particulate filled high density polyethylene. J Adhes 1997;64(1–4):87–109.

[28] Breiner J, Mark J. Preparation, structure, growth mechanisms and properties of siloxane composites containing silica, titania or mixed silica–titania phases. Polymer (Guildf) 1998;39(22):5483–93.

[29] Rong MZ, Zhang MQ, Zheng YX, Zeng HM, Friedrich K. Improvement of tensile properties of nano-SiO 2/PP composites in relation to percolation mechanism. Polymer (Guildf) 2001;42(7):3301–4.

[30] Kruenate J, Tongpool R, Panyathanmaporn T, Kongrat P. Optical and mechanical properties of polypropylene modified by metal oxides. Surf Interface Anal 2004;36(8):1044–7.

[31] Yang H, Zhang Q, Guo M, Wang C, Du R, Fu Q. Study on the phase structures and toughening mechanism in PP/EPDM/SiO 2 ternary composites. Polymer (Guildf) 2006;47(6):2106–15.

[32] Förster S, Plantenberg T. From self-organizing polymers to nanohybrid and biomaterials. Angew Chem Int Ed 2002;41(5):688–714.

[33] Sarkar B, Alexandridis P. Block copolymer–nanoparticle composites: structure, functional properties, and processing. Prog Polym Sci 2015;40:33–62.

[34] Xie X-L, Mai Y-W, Zhou X-P. Dispersion and alignment of carbon nanotubes in polymer matrix: a review. Mater Sci Eng R Rep 2005;49(4):89–112.

[35] Bockstaller MR, Mickiewicz RA, Thomas EL. Block copolymer nanocomposites: perspectives for tailored functional materials. Adv Mater 2005;17(11):1331–49.

[36] Mackay ME, Tuteja A, Duxbury PM, Hawker CJ, Van Horn B, Guan Z, et al. General strategies for nanoparticle dispersion. Science 2006;311(5768):1740–3.

[37] Pozzo DC, Hollabaugh KR, Walker LM. Rheology and phase behavior of copolymer-templated nanocomposite materials. J Rheol 2005;49(3):759–82.

[38] Spontak RJ, Alexandridis P. Advances in self-ordering macromolecules and nanostructure design. Curr Opin Colloid Interface Sci 1999;4(2):140–6.

[39] Balazs AC. Modeling self-assembly and phase behavior in complex mixtures. Annu Rev Phys Chem 2007;58:211–33.

[40] Balazs AC, Emrick T, Russell TP. Nanoparticle polymer composites: where two small worlds meet. Science 2006;314(5802):1107–10.

[41] Cao G, Liu D. Template-based synthesis of nanorod, nanowire, and nanotube arrays. Adv Colloid Interface Sci 2008;136(1):45–64.

[42] Caseri W. Nanocomposites of polymers and metals or semiconductors: historical background and optical properties. Macromol Rapid Commun 2000;21(11):705–22.

[43] Castelvetro V, De Vita C. Nanostructured hybrid materials from aqueous polymer dispersions. Adv Colloid Interface Sci 2004;108:167–85.

[44] Ciebien J, Clay R, Sohn B, Cohen R. Brief review of metal nanoclusters in block copolymer films. New J Chem 1998;22(7):685–91.

[45] Cohen R. Block copolymers as templates for functional materials. Curr Opin Solid State Mater Sci 1999;4(6):587–90.

[46] Darling S. Directing the self-assembly of block copolymers. Prog Polym Sci 2007;32(10):1152–204.

[47] Förster S, Antonietti M. Amphiphilic block copolymers in structure-controlled nanomaterial hybrids. Adv Mater 1998;10(3):195–217.

[48] Grubbs RB. Hybrid metal–polymer composites from functional block copolymers. J Polym Sci A Polym Chem 2005;43(19):4323–36.

[49] Hamley I. Nanostructure fabrication using block copolymers. Nanotechnology 2003;14(10):R39.

[50] Haryono A, Binder WH. Controlled arrangement of nanoparticle arrays in block-copolymer domains. Small 2006;2(5):600–11.

[51] Hulteen JC, Martin CR. A general template-based method for the preparation of nanomaterials. J Mater Chem 1997;7(7):1075–87.

[52] Lazzari M, López-Quintela MA. Block copolymers as a tool for nanomaterial fabrication. Adv Mater 2003;15(19):1583–94.

[53] Matsushita Y. Creation of hierarchically ordered nanophase structures in block polymers having various competing interactions. Macromolecules 2007;40(4):771–6.

[54] Mayer AB. Colloidal metal nanoparticles dispersed in amphiphilic polymers. Polym Adv Technol 2001;12(1–2):96–106.

[55] Park C, Yoon J, Thomas EL. Enabling nanotechnology with self assembled block copolymer patterns. Polymer (Guildf) 2003;44(22):6725–60.

[56] Smart T, Lomas H, Massignani M, Flores-Merino MV, Perez LR, Battaglia G. Block copolymer nanostructures. Nano Today 2008;3(3):38–46.

[57] Yoon J, Lee W, Thomas EL. Self-assembly of block copolymers for photonic-bandgap materials. MRS Bull 2005;30(10):721–6.

[58] Biswas A, Bayer IS, Biris AS, Wang T, Dervishi E, Faupel F. Advances in top–down and bottom–up surface nanofabrication: techniques, applications & future prospects. Adv Colloid Interface Sci 2012;170(1):2–27.

[59] Bombalski L, Listak J, Bockstaller MR. Structure, properties, and opportunities of block copolymer/particle nanocomposites. Annual Review of Nano Research, World Scientific; 2006. p. 295–336.

[60] Chiu C-W, Lin J-J. Self-assembly behavior of polymer-assisted clays. Prog Polym Sci 2012;37(3):406–44.

[61] Crosby AJ, Lee JY. Polymer nanocomposites: the "nano" effect on mechanical properties. Polym Rev 2007;47(2):217–29.

[62] Ganesan V, Ellison CJ, Pryamitsyn V. Mean-field models of structure and dispersion of polymer-nanoparticle mixtures. Soft Matter 2010;6(17):4010–25.

[63] Hanemann T, Szabó DV. Polymer-nanoparticle composites: from synthesis to modern applications. Materials 2010;3(6):3468–517.

[64] Haraguchi K. Nanocomposite hydrogels. Curr Opin Solid State Mater Sci 2007;11(3):47–54.

[65] Hsu L, Weder C, Rowan SJ. Stimuli-responsive, mechanically-adaptive polymer nanocomposites. J Mater Chem 2011;21(9):2812–22.

[66] Ji Y, Marshall JE, Terentjev EM. Nanoparticle-liquid crystalline elastomer composites. Polymers 2012;4 (1):316–40.

[67] Kumar SK, Krishnamoorti R. Nanocomposites: structure, phase behavior, and properties. Annu Rev Chem Biomol Eng 2010;1:37–58.

[68] Langner KM, Sevink G. Mesoscale modeling of block copolymer nanocomposites. Soft Matter 2012;8 (19):5102–18.

[69] Pyun J. Nanocomposite materials from functional polymers and magnetic colloids. Polym Rev 2007;47 (2):231–63.

[70] Schacher FH, Rupar PA, Manners I. Functional block copolymers: nanostructured materials with emerging applications. Angew Chem Int Ed 2012;51(32):7898–921.

[71] Schadler LS, Kumar SK, Benicewicz BC, Lewis SL, Harton SE. Designed interfaces in polymer nanocomposites: a fundamental viewpoint. MRS Bull 2007;32(4):335–40.

[72] Shoichet MS. Polymer scaffolds for biomaterials applications. Macromolecules 2009;43(2):581–91.

[73] Thomas A. Functional materials: from hard to soft porous frameworks. Angew Chem Int Ed 2010;49 (45):8328–44.

[74] Tjong SC. Structural and mechanical properties of polymer nanocomposites. Mater Sci Eng R Rep 2006;53(3):73–197.

[75] Yan L-T, Xie X-M. Computational modeling and simulation of nanoparticle self-assembly in polymeric systems: structures, properties and external field effects. Prog Polym Sci 2013;38(2):369–405.

[76] Morgan AB, Wilkie CA. Flame retardant polymer nanocomposites. Hoboken, New Jersey: John Wiley & Sons; 2007.

[77] Chiu JJ, Kim BJ, Yi G-R, Bang J, Kramer EJ, Pine DJ. Distribution of nanoparticles in lamellar domains of block copolymers. Macromolecules 2007;40(9):3361–5.

[78] Kim BJ, Bang J, Hawker CJ, Chiu JJ, Pine DJ, Jang SG, et al. Creating surfactant nanoparticles for block copolymer composites through surface chemistry. Langmuir 2007;23(25):12693–703.

[79] Kim BJ, Bang J, Hawker CJ, Kramer EJ. Effect of a real chain density on the location of polymer-modified gold nanoparticles in a block copolymer template. Macromolecules 2006;39(12):4108–14.

[80] Kim BJ, Fredrickson GH, Kramer EJ. Effect of polymer ligand molecular weight on polymer-coated nanoparticle location in block copolymers. Macromolecules 2008;41(2):436–47.

[81] Pryamitsyn V, Ganesan V. Strong segregation theory of block copolymer – nanoparticle composites. Macromolecules 2006;39(24):8499–510.

[82] Sarkar B, Alexandridis P. Self-assembled block copolymer–nanoparticle hybrids: interplay between enthalpy and entropy. Langmuir 2012;28(45):15975–86.

[83] Yaklin MA, Duxbury PM, Mackay ME. Control of nanoparticle dispersion in thin polymer films. Soft Matter 2008;4(12):2441–7.

[84] Zhao Y, Thorkelsson K, Mastroianni AJ, Schilling T, Luther JM, Rancatore BJ, et al. Small-molecule-directed nanoparticle assembly towards stimuli-responsive nanocomposites. Nat Mater 2009;8 (12):979–85.

[85] Masala O, Seshadri R. Synthesis routes for large volumes of nanoparticles. Annu Rev Mater Res 2004;34:41–81.

[86] Chen HM, Liu R-S. Architecture of metallic nanostructures: synthesis strategy and specific applications. J Phys Chem C 2011;115(9):3513–27.

[87] Koutsoumpis S, Raftopoulos KN, Oguz O, Papadakis C, Menceloglu YZ, Pissis P. Dynamic glass transition of the rigid amorphous fraction in polyurethane-urea/SiO 2 nanocomposites. Soft Matter 2017;13 (26):4580–90.

[88] Malay O, Oguz O, Kosak C, Yilgor E, Yilgor I, Menceloglu YZ. Polyurethaneurea–silica nanocomposites: preparation and investigation of the structure–property behavior. Polymer (Guildf) 2013;54(20):5310–20.

[89] Malay O, Yilgor I, Menceloglu YZ. Effects of solvent on TEOS hydrolysis kinetics and silica particle size under basic conditions. J Sol-Gel Sci Technol 2013;67(2):351–61.

[90] Oguz O, Simsek E, Soz CK, Heinz OK, Yilgor E, Yilgor I, et al. Effect of filler content on the structure-property behavior of poly (ethylene oxide) based polyurethaneurea-silica nanocomposites. Polym Eng Sci 2018;58(7):1097–107.

[91] Anglin EJ, Cheng L, Freeman WR, Sailor MJ. Porous silicon in drug delivery devices and materials. Adv Drug Deliv Rev 2008;60(11):1266–77.

[92] Aubert T, Grasset F, Mornet S, Duguet E, Cador O, Cordier S, et al. Functional silica nanoparticles synthesized by water-in-oil microemulsion processes. J Colloid Interface Sci 2010;341(2):201–8.

[93] Jia H, Hou W, Wei L, Xu B, Liu X. The structures and antibacterial properties of nano-SiO 2 supported silver/zinc–silver materials. Dent Mater 2008;24(2):244–9.

[94] Rahman I, Vejayakumaran P, Sipaut C, Ismail J, Chee C. Size-dependent physicochemical and optical properties of silica nanoparticles. Mater Chem Phys 2009;114(1):328–32.

[95] Rahman IA, Padavettan V. Synthesis of silica nanoparticles by sol-gel: size-dependent properties, surface modification, and applications in silica-polymer nanocomposites—a review. J Nanomater 2012;2012:8.

[96] Rossi LM, Shi L, Quina FH, Rosenzweig Z. Stöber synthesis of monodispersed luminescent silica nanoparticles for bioanalytical assays. Langmuir 2005;21(10):4277–80.

[97] Shekar S, Sander M, Riehl RC, Smith AJ, Braumann A, Kraft M. Modelling the flame synthesis of silica nanoparticles from tetraethoxysilane. Chem Eng Sci 2012;70:54–66.

[98] Slowing II, Trewyn BG, Giri S, Lin VY. Mesoporous silica nanoparticles for drug delivery and biosensing applications. Adv Funct Mater 2007;17(8):1225–36.

[99] Zawrah M, El-Kheshen A, Abd-Elaal HM. Facile and economic synthesis of silica nanoparticles. J Ovonic Res 2009;5(5):129–33.

[100] Plueddemann EP. Chemistry of silane coupling agents, silane coupling agents. Boston, MA: Springer; 1991. p. 31–54.

[101] Chen G, Zhou S, Gu G, Wu L. Modification of colloidal silica on the mechanical properties of acrylic based polyurethane/silica composites. Colloids Surf A Physicochem Eng Asp 2007;296(1):29–36.

[102] Cheng L, Zheng L, Li G, Zeng J, Yin Q. Influence of particle surface properties on the dielectric behavior of silica/epoxy nanocomposites. Phys B Condens Matter 2008;403(17):2584–9.

[103] Chinthamanipeta PS, Kobukata S, Nakata H, Shipp DA. Synthesis of poly (methyl methacrylate)–silica nanocomposites using methacrylate-functionalized silica nanoparticles and RAFT polymerization. Polymer (Guildf) 2008;49(26):5636–42.

[104] Dolatzadeh F, Moradian S, Jalili MM. Influence of various surface treated silica nanoparticles on the electrochemical properties of SiO 2/polyurethane nanocoatings. Corros Sci 2011;53(12):4248–57.

[105] Etienne S, Becker C, Ruch D, Grignard B, Cartigny G, Detrembleur C, et al. Effects of incorporation of modified silica nanoparticles on the mechanical and thermal properties of PMMA. J Therm Anal Calorim 2007;87(1):101–4.

[106] Hong R, Fu H, Zhang Y, Liu L, Wang J, Li H, et al. Surface-modified silica nanoparticles for reinforcement of PMMA. J Appl Polym Sci 2007;105(4):2176–84.

[107] Jang MK, Hartwig A, Kim BK. Shape memory polyurethanes cross-linked by surface modified silica particles. J Mater Chem 2009;19(8):1166–72.

[108] Kang S, Hong SI, Choe CR, Park M, Rim S, Kim J. Preparation and characterization of epoxy composites filled with functionalized nanosilica particles obtained via sol–gel process. Polymer (Guildf) 2001;42(3):879–87.

[109] Kang T, Lee JH, Oh S-G. Dispersion of surface-modified silica nanoparticles in polyamide-imide (PAI) films for enhanced mechanical and thermal properties. J Ind Eng Chem 2017;46:289–97.

[110] Liu Y-L, Hsu C-Y, Hsu K-Y. Poly (methylmethacrylate)-silica nanocomposites films from surface-functionalized silica nanoparticles. Polymer (Guildf) 2005;46(6):1851–6.

[111] Luo Z, Hong R, Xie H, Feng W. One-step synthesis of functional silica nanoparticles for reinforcement of polyurethane coatings. Powder Technol 2012;218:23–30.

[112] Ogbole EO, Lou J, Ilias S, Desmane V. Influence of surface-treated SiO 2 on the transport behavior of O 2 and N 2 through polydimethylsiloxane nanocomposite membrane. Sep Purif Technol 2017;175:358–64.

[113] Sprenger S. Improving mechanical properties of fiber-reinforced composites based on epoxy resins containing industrial surface-modified silica nanoparticles: review and outlook. J Compos Mater 2015;49(1):53–63.

[114] Theppradit T, Prasassarakich P, Poompradub S. Surface modification of silica particles and its effects on cure and mechanical properties of the natural rubber composites. Mater Chem Phys 2014;148 (3):940–8.

[115] Yuan QW, Mark JE. Reinforcement of poly (dimethylsiloxane) networks by blended and in-situ generated silica fillers having various sizes, size distributions, and modified surfaces. Macromol Chem Phys 1999;200(1):206–20.

[116] Zhou H, Wang H, Niu H, Gestos A, Lin T. Robust, self-healing superamphiphobic fabrics prepared by two-step coating of fluoro-containing polymer, fluoroalkyl silane, and modified silica nanoparticles. Adv Funct Mater 2013;23(13):1664–70.

[117] Kotal M, Bhowmick AK. Polymer nanocomposites from modified clays: recent advances and challenges. Prog Polym Sci 2015;51:127–87.

[118] Li X, Kang T, Cho WJ, Lee JK, Ha CS. Preparation and characterization of poly (butyleneterephthalate)/organoclay nanocomposites. Macromol Rapid Commun 2001;22(16):1306–12.

[119] Bhowmick AK, Bhattacharya M, Mitra S. Exfoliation of nanolayer assemblies for improved natural rubber properties: methods and theory. J Elastom Plast 2010;42(6):517–37.

[120] Ganguly A, Bhowmick AK, Li Y. Insights into montmorillonite nanoclay based ex situ nanocomposites from SEBS and modified SEBS by small-angle X-ray scattering and modulated DSC studies. Macromolecules 2008;41(16):6246–53.

[121] Maji PK, Das NK, Bhowmick AK. Preparation and properties of polyurethane nanocomposites of novel architecture as advanced barrier materials. Polymer (Guildf) 2010;51(5):1100–10.

[122] Maji PK, Guchhait PK, Bhowmick AK. Effect of the microstructure of a hyperbranched polymer and nanoclay loading on the morphology and properties of novel polyurethane nanocomposites. ACS Appl Mater Interfaces 2008;1(2):289–300.

[123] Maji PK, Bhowmick AK. Efficacy of clay content and microstructure of curing agents on the structure–property relationship of new-generation polyurethane nanocomposites. Polym Adv Technol 2012;23(9):1311–20.

[124] Okamoto K, Sinha Ray S, Okamoto M. New poly (butylene succinate)/layered silicate nanocomposites. II. Effect of organically modified layered silicates on structure, properties, melt rheology, and biodegradability. J Polym Sci B Polym Phys 2003;41(24):3160–72.

[125] Sinha Ray S, Okamoto K, Okamoto M. Structure – property relationship in biodegradable poly (butylene succinate)/layered silicate nanocomposites. Macromolecules 2003;36(7):2355–67.

[126] Wheeler PA, Wang J, Baker J, Mathias LJ. Synthesis and characterization of covalently functionalized laponite clay. Chem Mater 2005;17(11):3012–18.

[127] Mishra AK, Chattopadhyay S, Rajamohanan P, Nando GB. Effect of tethering on the structure-property relationship of TPU-dual modified Laponite clay nanocomposites prepared by ex-situ and in-situ techniques. Polymer (Guildf) 2011;52(4):1071–83.

[128] Mondal M, Chattopadhyay PK, Chattopadhyay S, Setua DK. Thermal and morphological analysis of thermoplastic polyurethane–clay nanocomposites: comparison of efficacy of dual modified laponite vs. commercial montmorillonites. Thermochim Acta 2010;510(1):185–94.

[129] Cordes DB, Lickiss PD, Rataboul F. Recent developments in the chemistry of cubic polyhedral oligosilsesquioxanes. Chem Rev 2010;110(4):2081–173.

[130] Kuo S-W, Chang F-C. POSS related polymer nanocomposites. Prog Polym Sci 2011;36(12):1649–96.

[131] Joshi M, Butola B, Simon G, Kukaleva N. Rheological and viscoelastic behavior of HDPE/octamethyl-POSS nanocomposites. Macromolecules 2006;39(5):1839–49.

[132] Joshi M, Butola B. Isothermal crystallization of HDPE/octamethyl polyhedral oligomeric silsesquioxane nanocomposites: role of POSS as a nanofiller. J Appl Polym Sci 2007;105(2):978–85.

[133] Baldi F, Bignotti F, Fina A, Tabuani D, Ricco T. Mechanical characterization of polyhedral oligomeric silsesquioxane/polypropylene blends. J Appl Polym Sci 2007;105(2):935–43.

[134] Fina A, Abbenhuis HC, Tabuani D, Camino G. Metal functionalized POSS as fire retardants in polypropylene. Polym Degrad Stab 2006;91(10):2275–81.

[135] Fina A, Tabuani D, Frache A, Camino G. Polypropylene–polyhedral oligomeric silsesquioxanes (POSS) nanocomposites. Polymer (Guildf) 2005;46(19):7855–66.

[136] Pracella M, Chionna D, Fina A, Tabuani D, Frache A, Camino G. Polypropylene-POSS nanocomposites: morphology and crystallization behaviour. Macromol Symp. Weinheim, Verlag: Wiley Online Library; 2006. p. 59–67.

[137] Haddad TS, Lichtenhan JD. Hybrid organic – inorganic thermoplastics: styryl-based polyhedral oligomeric silsesquioxane polymers. Macromolecules 1996;29(22):7302–4.

[138] Kopesky ET, Boyes SG, Treat N, Cohen RE, McKinley GH. Thermorheological properties near the glass transition of oligomeric poly (methyl methacrylate) blended with acrylic polyhedral oligomeric silsesquioxane nanocages. Rheol Acta 2006;45(6):971–81.

[139] Hirai T, Leolukman M, Hayakawa T, Kakimoto M-A, Gopalan P. Hierarchical nanostructures of organosilicate nanosheets within self-organized block copolymer films. Macromolecules 2008;41(13):4558–60.

[140] Hirai T, Leolukman M, Jin S, Goseki R, Ishida Y, Kakimoto M-A, et al. Hierarchical self-assembled structures from POSS-containing block copolymers synthesized by living anionic polymerization. Macromolecules 2009;42(22):8835–43.

[141] Chiu C-Y, Chen H-W, Kuo S-W, Huang C-F, Chang F-C. Investigating the effect of miscibility on the ionic conductivity of LiClO4/PEO/PCL ternary blends. Macromolecules 2004;37(22):8424–30.

[142] Chiu C-Y, Hsu W-H, Yen Y-J, Kuo S-W, Chang F-C. Miscibility behavior and interaction mechanism of polymer electrolytes comprising LiClO4 and MPEG-b lock-PCL copolymers. Macromolecules 2005;38(15):6640–7.

[143] Huang J, Li X, Lin T, He C, Yi Mya K, Xiao Y, et al. Inclusion complex formation between α, γ-cyclodextrins and organic–inorganic star-shaped poly (ethylene glycol) from an octafunctional silsesquioxane core. J Polym Sci B Polym Phys 2004;42(7):1173–80.

[144] Lee Y-J, Huang J-M, Kuo S-W, Chang F-C. Low-dielectric, nanoporous polyimide films prepared from PEO–POSS nanoparticles. Polymer (Guildf) 2005;46(23):10056–65.

[145] Maitra P, Wunder SL. Oligomeric poly (ethylene oxide)-functionalized silsesquioxanes: interfacial effects on Tg, Tm, and ΔHm. Chem Mater 2002;14(11):4494–7.

[146] Maitra P, Wunder SL. POSS based electrolytes for rechargeable lithium batteries. Electrochem Solid-State Lett 2004;7(4):A88–92.

[147] Markovic E, Ginic-Markovic M, Clarke S, Matisons J, Hussain M, Simon GP. Poly (ethylene glycol)-octafunctionalized polyhedral oligomeric silsesquioxane: synthesis and thermal analysis. Macromolecules 2007;40(8):2694–701.

[148] Yen Y-C, Cheng C-C, Kuo S-W, Chang F-C. A new poly (amide urethane) solid state electrolyte containing supramolecular structure. Macromolecules 2010;43(5):2634–7.

[149] Zhang H, Kulkarni S, Wunder SL. Polyethylene glycol functionalized polyoctahedral silsesquioxanes as electrolytes for lithium batteries. J Electrochem Soc 2006;153(2):A239–48.

[150] Zhang H, Kulkarni S, Wunder SL. Blends of POSS – PEO (n = 4) 8 and high molecular weight poly (ethylene oxide) as solid polymer electrolytes for lithium batteries. J Phys Chem B 2007;111 (14):3583–90.

[151] Kai W, Hua L, Dong T, Pan P, Zhu B, Inoue Y. Polyhedral oligomeric silsesquioxane-and fullerene-end-capped poly (ε-caprolactone). Macromol Chem Phys 2008;209(12):1191–7.

[152] Baldi F, Bignotti F, Ricco L, Monticelli O, Ricco T. Mechanical and structural characterization of POSS-modified polyamide 6. J Appl Polym Sci 2006;100(4):3409–14.

[153] Ye YS, Chen WY, Wang YZ. Synthesis and properties of low-dielectric-constant polyimides with introduced reactive fluorine polyhedral oligomeric silsesquioxanes. J Polym Sci A Polym Chem 2006;44 (18):5391–402.

[154] Janowski B, Pielichowski K. Thermo (oxidative) stability of novel polyurethane/POSS nanohybrid elastomers. Thermochim Acta 2008;478(1):51–3.

[155] Strachota A, Kroutilová I, Kovářová J, Matějka L. Epoxy networks reinforced with polyhedral oligomeric silsesquioxanes (POSS). Thermomechanical properties. Macromolecules 2004;37(25):9457–64.

[156] Nair CR. Advances in addition-cure phenolic resins. Prog Polym Sci 2004;29(5):401–98.

[157] Džunuzović E, Jeremić K, Nedeljković JM. In situ radical polymerization of methyl methacrylate in a solution of surface modified TiO 2 and nanoparticles. Eur Polym J 2007;43(9):3719–26.

[158] Ritzhaupt-Kleissl E, Boehm J, Hausselt J, Hanemann T. Thermoplastic polymer nanocomposites for applications in optical devices. Mater Sci Eng C 2006;26(5):1067–71.

[159] Zhao H, Li RK. A study on the photo-degradation of zinc oxide (ZnO) filled polypropylene nanocomposites. Polymer (Guildf) 2006;47(9):3207–17.

[160] Kim P, Doss NM, Tillotson JP, Hotchkiss PJ, Pan M-J, Marder SR, et al. High energy density nanocomposites based on surface-modified BaTiO3 and a ferroelectric polymer. ACS Nano 2009;3(9):2581–92.

[161] Cheng JY, Ross C, Chan VH, Thomas EL, Lammertink RG, Vancso GJ. Formation of a cobalt magnetic dot array via block copolymer lithography. Adv Mater 2001;13(15):1174–8.

[162] Kao J, Thorkelsson K, Bai P, Rancatore BJ, Xu T. Toward functional nanocomposites: taking the best of nanoparticles, polymers, and small molecules. Chem Soc Rev 2013;42(7):2654–78.

[163] Ling XY, Phang IY, Maijenburg W, Schönherr H, Reinhoudt DN, Vancso GJ, et al. Free-standing 3 D supramolecular hybrid particle structures. Angew Chem Int Ed 2009;48(5):983–7.

[164] Arora H, Du P, Tan KW, Hyun JK, Grazul J, Xin HL, et al. Block copolymer self-assembly–directed single-crystal homo-and heteroepitaxial nanostructures. Science 2010;330(6001):214–19.

[165] Du P, Li M, Douki K, Li X, Garcia CB, Jain A, et al. Additive-driven phase-selective chemistry in block copolymer thin films: the convergence of top–down and bottom–up approaches. Adv Mater 2004;16 (12):953–7.

[166] Simon PF, Ulrich R, Spiess HW, Wiesner U. Block copolymer – ceramic hybrid materials from organically modified ceramic precursors. Chem Mater 2001;13(10):3464–86.

[167] Templin M, Franck A, Du Chesne A, Leist H, Zhang Y, Ulrich R, et al. Organically modified aluminosilicate mesostructures from block copolymer phases. Science 1997;278(5344):1795–8.

[168] Bockstaller MR, Lapetnikov Y, Margel S, Thomas EL. Size-selective organization of enthalpic compatibilized nanocrystals in ternary block copolymer/particle mixtures. J Am Chem Soc 2003;125(18):5276–7.

[169] Chiu JJ, Kim BJ, Kramer EJ, Pine DJ. Control of nanoparticle location in block copolymers. J Am Chem Soc 2005;127(14):5036–7.

[170] Lee JY, Shou Z, Balazs AC. Predicting the morphologies of confined copolymer/nanoparticle mixtures. Macromolecules 2003;36(20):7730–9.

[171] Lee J-Y, Thompson RB, Jasnow D, Balazs AC. Entropically driven formation of hierarchically ordered nanocomposites. Phys Rev Lett 2002;89(15):155503.

[172] Lin Y, Böker A, He J, Sill K, Xiang H, Abetz C, et al. Self-directed self-assembly of nanoparticle/copolymer mixtures. Nature 2005;434(7029):55–9.

[173] Warren SC, Messina LC, Slaughter LS, Kamperman M, Zhou Q, Gruner SM, et al. Ordered mesoporous materials from metal nanoparticle–block copolymer self-assembly. Science 2008;320(5884):1748–52.

[174] Thompson RB, Ginzburg VV, Matsen MW, Balazs AC. Predicting the mesophases of copolymer-nanoparticle composites. Science 2001;292(5526):2469–72.

[175] Garcia BC, Kamperman M, Ulrich R, Jain A, Gruner SM, Wiesner U. Morphology diagram of a diblock copolymer – aluminosilicate nanoparticle system. Chem Mater 2009;21(22):5397–405.

[176] Deshmukh RD, Liu Y, Composto RJ. Two-dimensional confinement of nanorods in block copolymer domains. Nano Lett 2007;7(12):3662–8.

[177] Capadona JR, Shanmuganathan K, Tyler DJ, Rowan SJ, Weder C. Stimuli-responsive polymer nanocomposites inspired by the sea cucumber dermis. Science 2008;319(5868):1370–4.

[178] Listak J, Bockstaller MR. Stabilization of grain boundary morphologies in lamellar block copolymer/nanoparticle blends. Macromolecules 2006;39(17):5820–5.

4

Polymer Composites With Functionalized Silica

Vladimir M. Gun'ko

DEPARTMENT OF AMORPHOUS AND STRUCTURALLY ORDERED OXIDES, CHUIKO INSTITUTE OF SURFACE CHEMISTRY, KYIV, UKRAINE

4.1 Introduction

Properties of filled polymers strongly differ from those of pristine polymers due to effects of various fillers, and can be controlled by changes in filler characteristics and amounts [1–11]. Hybrid nanostructured composites (HNCs) are especially effective materials including a phase with one, two, or three dimensions of less than 100 nm or repeated structures with short (nanoscaled) distances between fragments of different phases. There are numerous fillers, for example, fumed metal or metalloid oxides (FMOs), carbons (graphene, graphene oxide (GO), carbon nanotubes (CNTs), carbon black, and activated carbons), metal (Au, Ag, etc.), and others. Addition of functionalized nanoparticles (FNPs) to polymers can strongly enhance the performance in comparison with nonfunctionalized NPs. It is important that FNPs are most effective for good NP dispersion in polymers. Chemical nature, particulate morphology, type of NP bonding to polymers, particle distribution in polymers, and other characteristics of fillers and related HNCs play an important role allowing one to control the material properties in applications. Silica-based fillers are mostly used for several reasons [12–25]. First, silica particle size distribution (PaSD), porosity or nonporosity of particles, and chemical structure of a surface can be easily changed upon filler synthesis alone (with subsequent modification) or in situ in parallel to polymerization or cross-linking in/with a polymeric matrix. Second, silica is characterized by relative chemical inertness, thermal stability, amorphism, mechanical strength, low cost, and other properties useful for fillers and HNCs. Easiness and simplicity of various chemical functionalization of silica are very important [12–16,26], for example, for partial or complete hydrophobization or other functionalization of a surface by coupling silanes, element-organic, or organic compounds. Silanes are preferable because the Si–O–Si bonds formed are more stable than Si–O–M (M = Al, Ti, Zr, C, etc.) [18,26]. Silanes can have one, two, or three reactive groups (e.g., Cl in Cl_nSiR_{4-n}, where R=CH_3 in trimethylchlorosilane (TMCS), dimethyldichlorosilane (DMS), methyltrichlorosilane, or R with C=C, COOH,

Polymer Composites with Functionalized Nanoparticles. DOI: https://doi.org/10.1016/B978-0-12-814064-2.00004-4

COH, CO, CNH_2, CSH, etc.), methoxysilanes $(CH_3O)_nSiR_{4-n}$ (methyltriethoxysilane, phenyltrimethoxysilane, 3-glycidoxypropyltrimethoxysilane, etc.), hexamethyldisilazane (HMDS, etc.), which interact with surface silanols directly or after hydrolysis (to form reactive SiOH in silanes). Their reactions with surface silanols give stable $Si-O-SiR_3$ bonds. For two or three reactive bonds in silanes, lateral cross-linking reaction (to form Si–O–Si between neighboring surface functionalities) results in enhanced functionalization (e.g., hydrophobization increased due to better cross-linking with increasing reaction temperature) and can cause cross-linking of adjacent NPs in aggregates (rather a negative effect). Reactions of some silanes (e.g., HMDS) with silanols can occur at room temperature. In other cases, the reactions catalyzed (e.g., by amines) can occur at low temperatures. However, the silica reactions with chlorosilanes alone need increased temperature. There is another pathway of the silica hydrophobization dealing with poly(dimethyl siloxane) (PDMS) in the presence of dimethyl carbonate (DMC). This type of modification provides better lyophilic property of fillers, but interactions between surface functionalities (short methyl groups or long PDMS fragments) and polymers are relatively weak due to only van der Waals interactions. Therefore, the functionalities length (e.g., PDMS fragments) can play an important role to control the properties (especially mechanic) of HNC with hydrophobized silicas [1–25].

To produce HNC with stronger bonds between FNP and polymers, functionalities should include polar/reactive bonds (CO, COH, COOH, CN, NH, NH_2, CN SH, SCN, PH, for example, in 3-aminopropyltriethoxysilane, 3-aminopropylmethyldimethoxysilane, 3-aminopropyldimethylmethoxysilane, mercaptopropyltriethoxysilane, 3-aminopropyltrie thoxysilane, 3-isocyanatopropyltriethoxysilane, bis(triethoxysilylpropyl)tetrasulfane, *n*-butyldimethyl(dimethylamino)silane, *n*-octadecyldimethyl(dimethylamino)silane, 3-mercapto propyltrimethoxysilane, 3-cyanopropyltrimethoxysilane, 3-thiocyanopropyltriethoxysilane, or with complex amino groups $(CH_3O)_3Si(CH_2)_3N(H)(CH_2)_2NH_2$, $(CH_3O)_3Si(CH_2)_3N[(CH_2)_2NH_2]_2$, etc.). Polar FNP groups can form strong intermolecular bonds with polar groups of polymers due to strong electrostatic interactions and hydrogen bonds. Reactive FNP functionalities can participate in chemical reactions with polymers (or monomers upon in situ polymerization), for example, condensation reactions of hydroxyls, or other reactions of C=O, COOH, NH_2, etc. with appropriate groups in polymers. Some groups (e.g., OH) can be also used in cross-linking with polymers by cross-linking agents (e.g., glutaraldehyde). Filler functionalities with C=C bonds (vinyltriethoxysilane, vinyltrimethoxysilane, methacryloxymethyl-triethoxysilane, 3-methacryloyloxypropyltrimethoxysilane, methacryloyloxymethylenemethyldiethoxysilane, etc.) can take part in polymerization with unsaturated bonds of the organic phase that strongly enhance the mechanical properties of HNC. Thus, there are many factors (morphology, PaSD, chemical nature of functionalities and polymers, stage of filler introduction before or after polymerization, type of binding of filler functionalities to polymers, functionalities length, etc.) affecting interactions between filler particles and polymers, and, therefore, the HNC properties [1–25].

Various silicas can be used as fillers: (1) fumed silicas (A-50, A-100, A-300, A-500, labels values show the specific surface area, S_{BET}, corresponding to average diameters

$d \approx 6/(\rho_0 S_{BET})$ (ρ_0 is the true density of NP) from 50 nm to 5 nm, respectively); (2) fume silica ($d \approx 150$ nm); (3) thermal silica wetted and then dried at high temperature ($d \approx 10$ μm); (4) precipitated silica (broad PaSD); (5) silica gels ($d \approx 0.01-1.0$ mm); (6) mesoporous ordered silicas (various MCM, SBA, etc.); and (7) in situ synthesized nanosilicas in polymer/monomer media ($d \sim 10$ nm), as well as silica in complex fillers. Silicas can be pretreated using mechanochemical activation (MCA), compaction by wetting–drying, hydrothermal treatment, heating, stirring, etc., that is "geometrical" modification. Chemical modification of silica (i.e., surface functionalization) results in increasing size of primary particles and possible formation of larger clusters of NP cross-linked by functionalities, which remain in filled polymers. Nanosilica can be also used as a modifier of larger filler particles, for example, CNT, GO, and metals.

Nanosilicas ($d < 100$ nm) can be prepared by several methods: pyrogenic synthesis at high temperature, flame spraying, sol–gel, precipitation, aerosol, etc. [12–15]. Depending on the method and composition, FMO can be completely amorphous (fumed silica) or crystalline with contributions of amorphous phase(s) (silica/titania, alumina/silica, etc.). FMO prepared at high temperatures are spherical-like and less aggregated than oxides prepared using other methods (especially in aqueous media) [12–26]. NPs are characterized by greater activity with decreasing size. Therefore, the smaller the FMO NP (larger S_{BET}), the stronger the NP aggregation. Active NPs form aggregates and agglomerates of aggregates, and these secondary particles are responsible for textural porosity of FMO that causes difficulties in dispersing NP in polymer matrices [27–31].

HNCs produced using a combination of functional polymers with inorganic nanostructured fillers (having size-dependent physical and chemical properties) are the focus of numerous investigations using various synthesis techniques including blending, sol–gel, emulsion polymerization, photopolymerization, metallosupramolecular and coordination approaches, intercalation, microwave-assisted and electrochemical synthesis, synthetic methods based on surface grafting, and self-assembly and block copolymer-mediated synthetic strategies [5]. Therefore, the design versatility of HNCs with various structural, compositional, and architectural characteristics creates new prospects for numerous applications of the materials in biomedicine, sensing, environmental remediation, energy, construction, automotive and coating technologies, catalysis, optoelectronics, etc. [5].

HNCs can be very interesting from a practical point of view because they can possess favorable properties of both components and demonstrate new properties not characteristic to the components alone due to synergetic effects [1–6]. These effects are maximal for HNCs because of the greatest contribution of interfaces as an area for polymer–filler interactions and, therefore, have specific or unique properties. To prepare HNCs with well-defined morphologies, block copolymer–directed nanoparticle hybrids present a promising approach [6]. The main characteristics for successful formation of stable nanostructures should be at or close to the thermodynamic equilibrium. Recent developments based on the application of di- and triblock polymers are of importance for progress in HNC, as well as for deeper understanding of the mechanisms of the HNC formation including silica FNP fillers.

4.2 Structure of Unmodified and Modified Silicas as Fillers

Flame ($H_2/O_2/N_2$) synthesis of FMO at 1000–1300°C using $SiCl_4$ or $SiCl_nR_{4-n}$ or SiR_4 (R is OCH_3, CH_3, etc.) as precursors leads to certain general characteristics many of which are related to their nanoparticulate morphology and the absence of pores in NPs [12–26]. Silica NP scale corresponds to that of macromolecules (Fig. 4–1) that provide multipoint interactions between a particle and a macromolecule in the HNC. There is silica fume (microsilica), consisting of spherical-like particles of 150 nm in average size, as a byproduct of silicon and ferrosilicon alloy production, with the primary particle size distribution (PPSD) broader than that of FMO. Typically, FMOs are composed of roughly spherical-like nonporous NPs (NPNPs) of 5–100 nm in size determining large S_{BET} values.

FMOs with NPNP forming aggregates (<1 μm in size) and less dense agglomerates of aggregates (>1 μm) are characterized by a small portion of the solid fraction in the powder ~2%–5%. The nanoparticulate morphology of FMO provides these materials with characteristics useful for fillers of polymers, drug carriers, pigments (titania-based FMO), thickeners, etc. due to easy distribution of NPNPs in polymeric matrices [12–20]. The morphological features enhance an interfacial phenomena role in the HNC characteristics applied in different media. Other silicas (silica gels, thermosilica, precipitated silica, in situ synthesized silica, etc.) are composed of NPNPs of ca. 10 nm in size but much more tightly organized in porous micro- or macroparticles. Surface silanols are the main adsorption and reactive sites for all silicas. Therefore, fumed silica could be analyzed as a representative of silica materials.

There is a simple ratio between the S_{BET} value and average diameter (d) of an NPNP: $S_{BET} \approx 6/(\rho_0 d)$, where ρ_0 is the true density of the material. The primary nanoparticles are characterized by a certain size distribution becoming broader with decreasing S_{BET} value (Fig. 4–2). The pore volume value (V_p) increases with increasing S_{BET} value due to stronger aggregation of smaller nanoparticles. In the powder, individual (nonaggregated) NPNPs (fraction at $d = 50–100$ nm) can be observed for A-50 [26] or OX-50 (Degussa, Evonik) [22], but individual particles of silicas (or other FMO [26]) possessing great S_{BET} value are typically not observed due to their strong aggregation. These morphological features can affect the behavior of adsorbates bound to different FMO or the characteristics of polymers filled by FMO. Applications of FMO in HNC could give significant advantages in comparison with

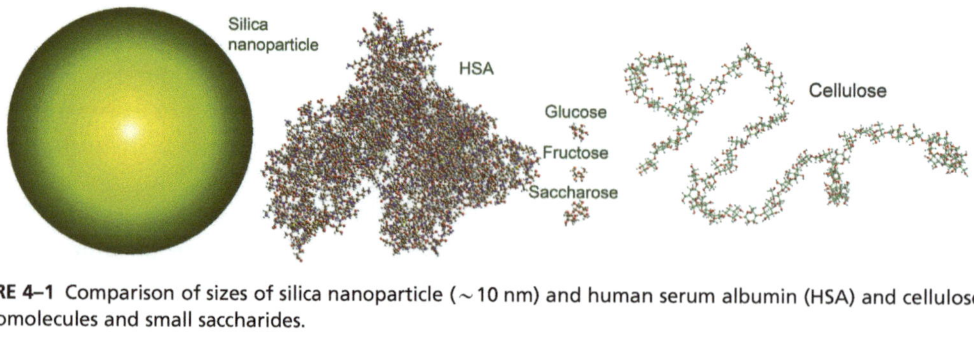

FIGURE 4–1 Comparison of sizes of silica nanoparticle (~10 nm) and human serum albumin (HSA) and cellulose macromolecules and small saccharides.

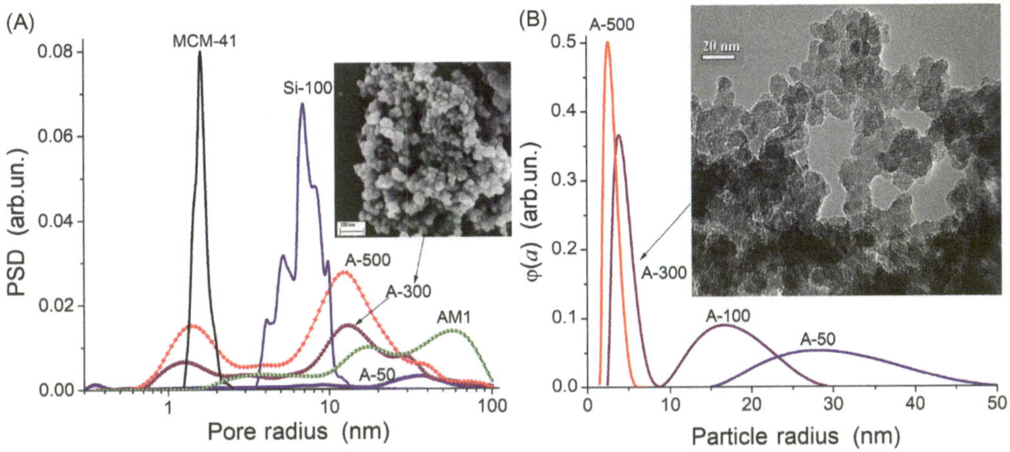

FIGURE 4–2 Pore size distributions, PSD (A) and particle size distributions, PaSD (B) of fumed silicas (initial A-50, A-300, A-500, and AM1 as A-300 hydrophobized using DMS) estimated from nitrogen adsorption data using self-consistent regularization (SCR) procedure and pore model as random voids between spherical nanoparticles aggregated; PSD for mesoporous ordered MCM-41 and silica gel Si-100 (Merck) are shown. Inserts with (A) SEM and (B) TEM images correspond to A-300 (recalculated here using methods described in [32] and adapted from [26,33,34]).

materials composed of much larger particles due to high activity of mobile nanoparticles characterized by a significant S_{BET} value (up to 500 m^2/g corresponding to $d = 5.5$ nm for A-500).

The empty volume, which is large as $V_{em} = 1/\rho_b - 1/\rho_0 = 10–25$ cm^3/g (where $\rho_b = 0.04–0.13$ g/cm^3 for various nanosilicas, and $\rho_0 = 2.2$ g/cm^3 is the true density of silica NPNP), can be about 95% for the initial nanooxide powders. For example, $\rho_b = 0.045$, 0.1, and 0.25 g/cm^3 for initial and compacted nanosilicas A-300 that correspond to $V_{em} = 21.8$, 9.5, and 3.5 cm^3/g. However, the V_p values (from nitrogen adsorption at relative pressure $p/p_0 \approx 0.99$) are much lower (0.524 cm^3/g at $\rho_b = 0.045$ g/cm^3 and 1.4 cm^3/g at $\rho_b = 0.25$ g/cm^3) [26]. The V_p value is much lower for the initial A-300 than that for the dense one because nitrogen cannot effectively fill broad macropores [35] and this filling increases due to the compaction of the powder [33,34,36]. Therefore, water amounts adsorbed onto initial A-300 from air is low (1–4 wt.%) [12–18]. Hydrophobic AM1, prepared by complete functionalization of fumed silica A-300 by DMS, is characterized by lower intensity of the pore size distribution (PSD) than hydrophilic A-300 (Fig. 4–3) because grafting of functionalities and their lateral cross-linking lead to a decrease in the pore volume and specific surface area (since size of modified NPNP increases) [34,37].

Despite NPNP hydrophobization by DMS in AM1, residual free silanols are observed in the infrared (IR) spectra (Fig. 4–4) because not all the OH groups of hydrolyzed DMS can participate in the lateral cross-linking reaction due to structural factors [26]. However, the amounts of free silanols and water adsorbed from air are much smaller for AM1 than that for unmodified A-300 (Fig. 4–4).

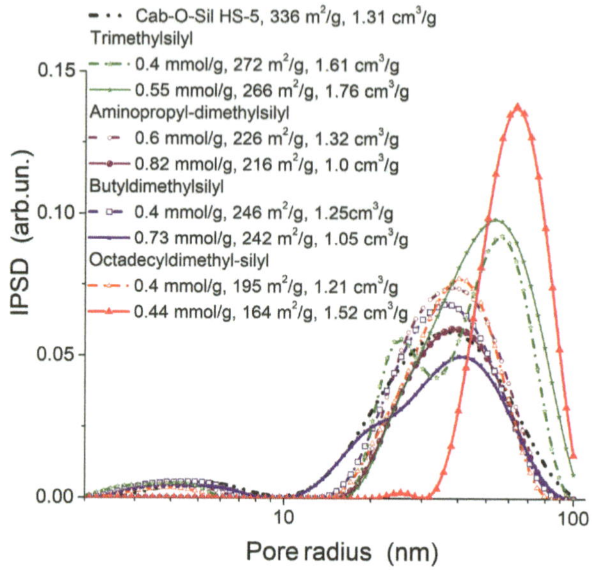

FIGURE 4–3 Incremental PSD for unmodified and modified Cab-O-Sil HS-5 with trimethylsilyl, aminopropyl-dimethylsilyl, butyldimethylsilyl, and octadecyldimethyl-silyl groups (their content, S_{BET} and V_p values are in Figure); IPSDs are recalculated here using nitrogen adsorption–desorption isotherms [37] and DFT method described in [26,32].

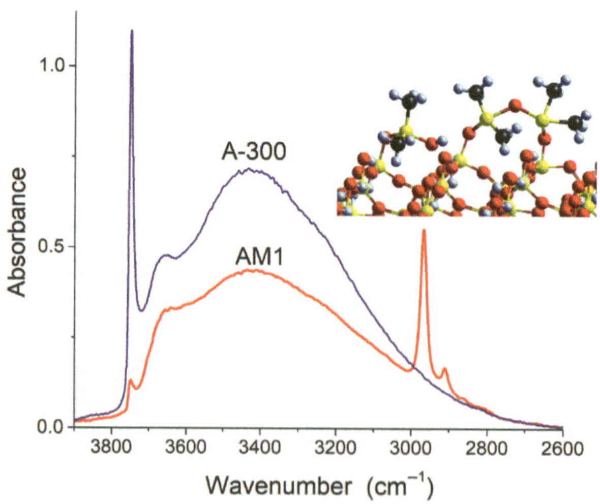

FIGURE 4–4 IR spectra of initial nanosilica A-300 and hydrophobic AM1, based on A-300 modified by DMS. *Adapted from Gun'ko VM, Pakhlov EM, Goncharuk OV, Andriyko LS, Marynin AI, Ukrainets AI, et al. Influence of hydrophobization of fumed oxides on interactions with polar and nonpolar adsorbates. Appl Surf Sci 2017;423:855–868, with permission, Copyright 2017, Elsevier.*

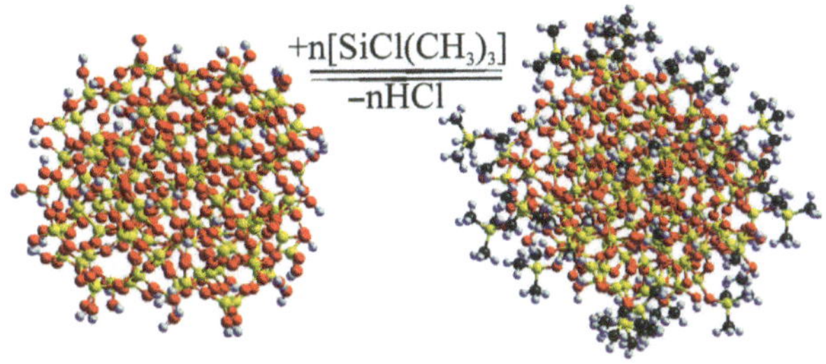

FIGURE 4–5 Modification of nanosilica surface by trimethylsilyl groups due to reaction with TMCS.

Silanes can react with a silica surface to form a functional layer of a thickness depending on silane type. Fig. 4–5 shows a molecular model with unmodified nanoparticle interacting with TMCS resulting in the formation of trimethylsilyl groups at a silica surface. Other silanes with one reactive group X can interact with silica in a similar way, but under different temperatures due to variation in the reactivity of the Si–X bonds (X = Cl, OR, etc.) [12–14,38–46].

For two (e.g., DMS as for AM1) or three (e.g., methyltrichlorosilane) reactive groups in silane molecules, an additional aspect appears because of the possibility of lateral cross-linking of grafted functionalities. The degree of lateral cross-linking (e.g., due to condensation reaction of silanols formed upon hydrolysis of the Si–X groups) depends strongly on temperature, structural features of silica, type and amount of silane, presence of water, or a catalyst. Thus, the number of reactive groups and modification reaction conditions strongly affect the characteristics and properties of modified silicas, for example, a number of residual OH groups, amounts of water adsorbed in air, zeta potential, and surface charge density in aqueous media [47]. Clearly these structural features of modified silicas can affect the properties of filled polymers in HNCs.

4.3 Polymer Composites Filled by Silicas

Synthetic strategies for HNC preparation can be based on appropriate classification on whether the inorganic and polymer components formed in situ or ex situ [48]. Four types of strategies were identified and described, referring to certain examples: (1) ex situ formation of the components and subsequent attachment or integration, either by covalent or noncovalent bonding; (2) in situ polymerization in the presence of ex situ formed inorganic nanoparticles; (3) in situ precipitation of inorganic components on or in polymer structures; and (4) strategies in which both polymer and inorganic component are simultaneously formed in situ [48]. HNC reinforcement depends on the nanostructure morphology, defects, dispersion of FNP in the polymer matrix, and the cross-linking density of the polymer or/and polymer and FNP functionalities. Inorganic nanomaterials are better reinforcing agents than carbon-based nanomaterials.

HNC can be prepared using sol–gel technique under different conditions [49]. Various methods and strategies of in situ synthesis of fillers based on the sol–gel chemistry include the hydrolysis and condensation reactions of such compounds as metal alkoxides or alkoxysilanes. The respective inorganic NPs can be synthesized under soft conditions due to high reactivity of silanes or metal alkoxides. The synthesis of silica (and titania) particles for their in situ dispersion in polymers is one of most effective methods to prepare HNCs with good mechanical and other properties. For the synthesis several methods can be used to create NP fillers: (1) by posthydrolysis with the method generally based on two subsequent processes: Mixing of alkoxysilane (or metal alkoxide) with molten polymers; and hydrolysis–condensation of the sample in hot water; (2) due to the presence or addition of water during the processing step; (3) by incorporation of a prehydrolyzed solution of inorganic precursors into the organic reaction media (e.g., a solution of monomers) or yet uncross-linking polymers. Clearly, the PaSD of the inorganic phase in HNC depends on the synthetic method, but generally observed at the nanoscale (~10 nm). For phase separation according to hydrolysis–condensation and miscibility with polymers, larger filler domains (NPNP aggregates with several particles) can be obtained. A derivative method was used to synthesize HNCs during the formation of chemical bonds between the polymer and inorganic phases. The hydrolysis/condensation reactions of silanes (or metal alkoxides) are fast, as the kinetics last only a few minutes with increasing temperature (~ 200°C). Consequently, the formation of HNCs using in situ FNP synthesis can be investigated from a one-step continuous process. The reactive extrusion process showed that the conversion of the hydrolysis–condensation reactions never exceeded 0.9. A reactive process with the addition of water in the extruder allowed the complete hydrolysis–condensation (conversion >0.97) inside the extruder. The particle size was coarser ($d \sim 200$ nm) compared with the subsequent processing method ($d \sim 10$ nm). However, better control of the NP size can be expected at controlled conditions. Another important aspect of the potential of in situ synthesizing FNP by judiciously selecting the precursors (phosphorylated and fluorinated silica as a filler), for example, diethylphosphatotriethoxysilane (SiP) having a catalytic effect on the hydrolysis–condensation during the polyamide-processing step. The methods for the in situ synthesis of HNC are good ways to prepare materials with good optical, mechanical reinforcement, fire retardancy, bioactive, and other properties. These properties can be related either to the chemical structure and size of FNP or to the incorporation of specific functionality at the nanoscale (e.g., phosphorous from the SiP in a polyamide matrix). These examples are illustrations of the wide field of investigations that can be opened by coupling sol–gel chemistry and molten polymer processing [49].

4.3.1 Hydrophobic Nonpolar Polymers

The mechanical properties of polymers are frequently poor and fillers can provide overcoming of this limitation. Silica powder was introduced into polypropylene (PP) to enhance the mechanical properties [50]. However, hydrophilic silica distribution in hydrophobic polymer is difficult. Therefore, styrene-ethylene/butylene-styrene (SEBS) triblock copolymer and its grafted maleic anhydride (SEBS-g-MA) were used as silica/PP compatibilizers. The compatibilizer addition enhances interfacial bonding and smoothens the transfer of the stresses between filler particles and polymers [50].

Nanocomposites were prepared by dispersing silica nano/microspheres (SNS, 49–282 nm, 0%–30%) into PP and two types of functionalized PP to clarify the effects of matrix–filler interfacial adhesion forces on the mechanical properties of PP-based HNCs [31]. There is good affinity between silanols and hydroxyls in functionalized PP; therefore, poly(5-hexen-1-ol-co-propylene) with 1.3 mol% and 6.4 mol% of hydroxyl groups were used as functionalized PP (PPOH). Tensile tests revealed that mechanical properties of PPOH composites were better than those with PP. PPOH HNCs maintained high ductility, whereas the ductility of PP HNCs were reduced by dispersing SNS. FTIR and XPS indicated the formation of hydrogen bonds between PPOH and silanols attributed to the high ductility [31].

Polysiloxane elastomers $[SiRR{-}O{-}]_x$ can be used as high-performance materials, but only when reinforced with relatively large amounts of filler [51] such as silica NP prepared in the gas phase (fumed NP) or in aqueous solution (precipitated or sol–gel NP).

4.3.2 Composites With Polar Polymers

The morphology and properties of HNC depend on the size, nature, and strength of the interface that are easily achieved in the sol–gel (in situ) processes. Amine-cured epoxy–silica HNCs have better compatibility due to strong hydrogen bonds between amino groups and silanols. The coupling agents can improve compatibility through chemical binding organic and inorganic structures [52].

Among HNCs, epoxy resins (ERs) are very important and widely used as thermosetting polymers [53], whose properties can be effectively modified by addition of reactive silanes, polysiloxanes, silsesquioxanes, silica, montmorillonite, and other fillers. Distinct advantages of HNCs with 3D nanofillers over microcomposites are due to performance improvement using relatively low NP amount. Nanosilica is the most widely used filler to improve the ER properties. Two approaches of the sol–gel process to prepare epoxy–silica hybrids can be used: (1) directly mixing ER with functional silanes, for example, tetraethoxysilane (TEOS); (2) introducing trialkoxysilyl groups into an organic elastomer with organofunctional silane prior to the sol–gel reactions. The ER properties can be improved in their HNCs by modifications with reactive silanes, silicon containing monomers and polymers, and silica-based fillers, enabling improvement of their mechanical properties, thermal and flame resistance, as well as providing corrosion and antimicrobial protection [52]. Due to the strong tendency of nanofillers to aggregate, their surface functionalization is an effective way to afford possibilities for compatibility between FNPs and polymers. The modification role of silica NPs in ER-based HNCs was more effective than that of silica microfiller due to large specific surface area and active groups on NP surfaces. Similar behavior was shown for the toughness of ER-based HNCs with silica NP and silica microfiller as showed in SEM images [54]. Epoxy–AMS (nanosilica modified with allyl glycidyl ether) HNCs exhibited enhanced thermal and mechanical properties and reduced moisture absorption and gas permeability.

Nanosilica functionalized with allyl ether and hydroxyl groups was applied for the modification of the ER properties and other polymers [55,56]. Epoxy-functionalized polystyrene/silica NP (10 ± 5 nm in size) core–shell HNC particles [up to 70 nm in size due to silica NP clusters (with 4–5 NPs) as cores] were prepared by the postaddition of glycidyl methacrylate (GMA) via emulsion polymerization. The outermost shell of the obtained multilayered core–shell particles was made up of poly(glycidyl methacrylate) (PGMA). A semicontinuous process involving the dropwise addition of GMA was used to avoid demulsification of the emulsion system. The binding efficiency was usually high (ca. 90%), indicating strong adhesion between the silica core and polymer shell. Other main factors of polymerization conditions included the amounts of sodium dodecyl sulfonate ($\sim$2%) and silica (3%–7%) [57]. HNCs with polyamide (PA), poly(ethylene terephthalate) (PET), and other polymers and nanosilicas (0.5–6 wt.%), having various surface functionalities, show unique and enhanced properties in comparison with unfilled PA, PET, etc., such as better barrier, mechanical, and thermal properties, and in case of polyamide also increased chemical resistance. Significant influences on the properties of HNCs include the content and size of spherical silica particles, the number of introduced functional groups, the presence of a compatibilizer, and also processing conditions [1–11,55–57].

HNCs with interpenetrating network (IPN) was obtained from diglycidyl ether of bisphenol-A, poly(oxypropylene)diamine (D2000, $M \approx 1970$), and TEOS [58]. The sol–gel hydrolytic condensation of TEOS was catalyzed with p-toluenesulfonic acid monohydrate. The inorganic phase caused reinforcement of the rubbery material. Bonding of silica–siloxane structures to the epoxy network was possible by condensation reactions of silanols with the C–OH groups that formed during reaction of epoxy groups with amines. The poly(oxypropylene) chain of D2000 ensured a good solubilization of the silica leading to transparent gels, which formed during TEOS polycondensation, in contrast to the base catalyzed process, for example, by benzyldimethylamine.

Composites with 1–7 wt.% silica were synthesized using bisphenol-A and epichlorohydrin, modified with 3-thiopropyltriethoxysilane (TPTES). The reactive coupling agent TPTES was introduced to the epoxy resin to obtain high adhesive strength material for the engineering industry. It was expected that after hydrolysis the multiple reactive functional silanols of TPTES could react with the epoxy resin to increase the shear strength and toughness of HNC [59].

Polyamide 6/silica HNCs were synthesized by in situ ring-opening anionic polymerization of ε-caprolactam in the presence of sodium caprolactamate as a catalyst and caprolactam-functionalized silica as an initiator [60]. The initiator precursor, isocyanate-functionalized silica, was prepared by directly reacting nanosilica with excess toluene 2,4-diisocyanate. This polymerization was found to occur in a highly efficient manner at relatively low reaction temperature (170°C) and short reaction times (6 h). FTIR spectroscopy was utilized to follow the introduction and consumption of isocyanate groups on the silica surface. Thermogravimetric analysis indicated that the polyamide 6 was successfully grown at the silica surface. Transmission electron microscopy images of polymer-functionalized silica showed fine dispersion of silica NP of 20–40 nm in size.

Silica NPs functionalized with vinyl isocyanate were UV cured to synthesize polyurethane–silica HNC tested for mechanical, dynamic mechanical, cross-link density, contact angle, and XPS. The modified silica provided composites with cross-links that were pinned at the surfaces and that effectively reinforce the materials mechanically as well as thermally [61]. The effect of silica particles in UV-cured polyurethane–silica HNCs seems to depend greatly on whether the silica surface provides HNCs with reaction sites or not. Without providing reaction sites, silica particles disturb the cure kinetics of urethane acrylate, aside from the particle reinforcement, giving rise to the decreased tensile strength below the unfilled one. Regarding modulus, increase by filler addition seems offset by the decreased extent of cure. If silica had surface reaction sites then it provided the composite with additional cross-links pinned on the surface. This contributed to the enhanced mechanical and thermal stability of the HNCs [61].

Synthesis of polymer composites filled with silica without the use of coupling agents could be considered as an advantageous route [62], for example, due to lower cost. However, many of the HNC characteristics could be worse than that of HNC with FNP, especially because of the use of low amounts of functionalized fillers [63–71]. Additionally, good dispersion of FNP in HNC can introduce new physical properties and novel behaviors (e.g., fire resistance or flame retardancy, and accelerated biodegradability) absent in the unfilled ones.

Such factors as good nanofiller dispersion in a polymer matrix, a high aspect ratio of filler particles, and surface functionalization of fillers to improve NP dispersion and to react with polymers to provide strong interfacial bonding are of importance. Considerable attention has been paid to HNC, especially with silica nanoclusters formed in situ in a matrix via a sol–gel process [63]. Some HNCs with 3D silica NPs were intensively studied [62–69]. A linear polymer matrix is considered to be entirely confined in nanovolumes, and the enhancement of HNC properties is expected if an average interparticle distance, L, is close to or less than the dimensions of the unperturbed macromolecular random coils (radius of gyration r_g is typically of the order of 10 nm for many polymers). According to [71], the substantial effect of 3D NP $\sim$10 nm in size on the dynamics and properties of a polymer matrix may be expected when their volume content is not less than 10%–20%. Really, various anomalies of glass transition dynamics (GTD) were observed, for example, at silica content of 10–50 wt.% [63–69]. Monomolecular polymer layer bound to a silica surface is characterized by multimodal dynamics within the glass transition range ΔT_g extending from 80°C to 230°C and a high apparent activation energy (80–560 kJ/mol). Small additives of 3D NP ($\sim$0.25 wt.% only) may strongly affect GTD and elastic properties at $L >> r_g$. This effect was observed for polyurethane-poly(2-hydroxyethyl methacrylate) (PU-PHEMA) semi-IPN composites with silica NPs. PU-PHEMA semi-IPNs are of interest as biomedical and damping materials. Polyurethane materials are extensively used in blood contacting applications and organ and tissue engineering. Although PUs are relatively biocompatible materials, they are also known to be prone to biodegradation, stress induced degradation, and surface cracking. Creation of IPN or semi-IPN is one of the most powerful approaches to improve PU biocompatibility, mechanical properties, and resistance to degradation. PHEMA is well known as a polymer with good biocompatibility but with unstable properties due to great hygroscopicity.

Semi-IPN creates an opportunity to synthesize PHEMA-based composites that retain PHEMA biocompatibility and acquire better stability and mechanical properties of PU. Semi-IPN based on PU-PHEMA were synthesized by a set of methods [64–69]. SAXS study showed that the semi-IPNs were two-phase systems with incomplete phase separation and had a nanoheterogeneous structure with characteristic quasi-periodicities of 6–9 and 22–27 nm due to spinodal decomposition. DMA analysis of the semi-IPN showed two broad mechanical loss peaks located at -55 to 50°C and 50–180°C associated with PU and PHEMA glass transitions, respectively. Discrete CRS analysis of glass transition dynamics in a series of the PU-PHEMA semi-IPN with regularly varied composition clarified an origin of the extraordinarily broad PU and PHEMA glass transitions in the networks. The complicated creep rate spectra of the semi-IPNs consisted of a few partly overlapping peaks and indicated the presence of several segmental dynamics modes, that is, the pronounced dynamic heterogeneity within an each glass transition range. HNCs with potential biomedical and technical significance based on the PU-PHEMA network were doped by small additives of silica NP with unmodified and functionalized by 3-aminopropylmethylsilyl groups (APMS, $-NH_2$ cover) or 3-methacryloylpropylsilyl groups (MAPS, $-CH{=}CH_2$ cover) [64–69]. AFM images (Fig. 4–6) demonstrate the presence of NP clusters in HNCs.

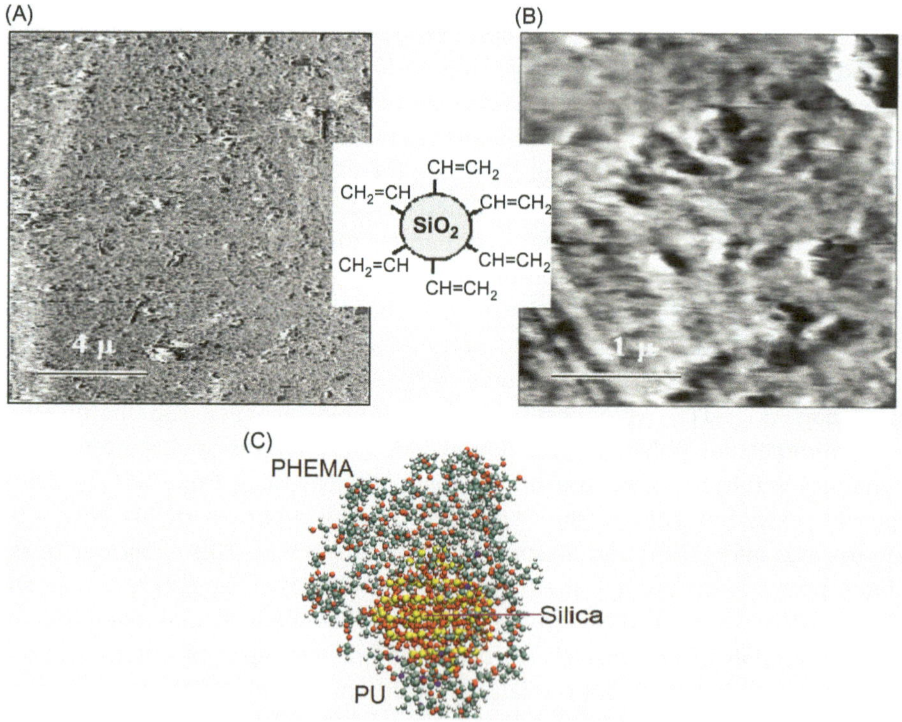

FIGURE 4–6 AFM images (contact mode) of 83PU-17PHEMA-based HNC film with (A and B) 3 wt.% MAPS-A-300 and (C) molecular model. *Reproduced from Bershtein VA, Gun'ko VM, Karabanova LV, Sukhanova TE, Yakushev PN, Egorova LM, et al. Polyurethane–poly(2-hydroxyethyl methacrylate) semi-IPN–nanooxide composites. RSC Adv 2013;3:14560, with permission from The Royal Society of Chemistry.*

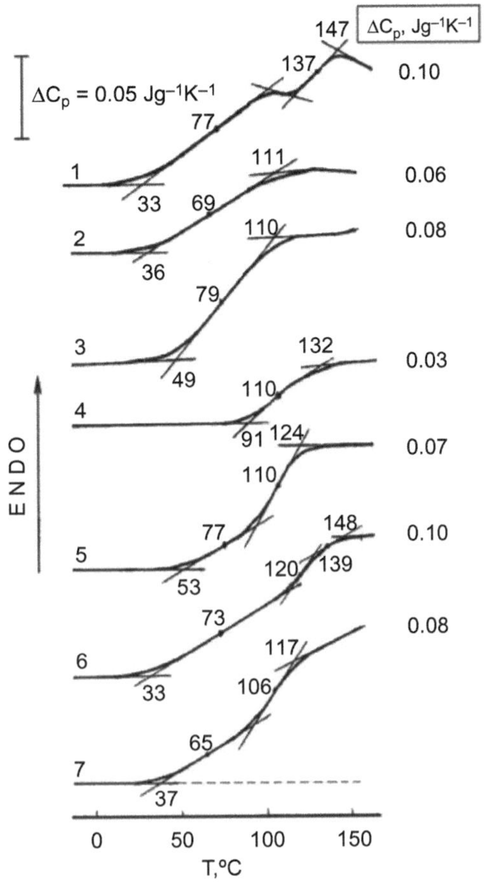

FIGURE 4–7 DSC curves obtained in the region of PHEMA glass transition for the (1) 83PU-17PHEMA network, and for HNC with silica (2, 4, 6) 0.25 wt.% or (3, 5, 7) 3 wt.%. Curves 2 and 3 relate to initial, as-prepared oxide particles; curves 4 and 5 to oxide surface with MAPS cover, and curves 6 and 7 to oxide surface with APMS groups. The T_g, T_g', T_g'', and ΔC_p values are indicated. *Reproduced from Bershtein VA, Gun'ko VM, Karabanova LV, Sukhanova TE, Yakushev PN, Egorova LM, et al. Polyurethane–poly(2-hydroxyethyl methacrylate) semi-IPN–nanooxide composites. RSC Adv 2013;3:14560, with permission from The Royal Society of Chemistry.*

Unlike one-stage transition with $T_g = 85°C$ and transition width $\Delta T_g = 37°C$ for the neat PHEMA, a strongly broadened, two-step PHEMA glass transition extending from 33 to 147°C, at $T_{g1} = 77°C$ and $T_{g2} = 137°C$ (Fig. 4–7, curve 1), was observed for the neat 83PU-17PHEMA network; that supposes the pronounced dynamic heterogeneity within its glass transition range. The second higher temperature ΔC_p step was caused by covalent hybridization of a part of the PHEMA segments with the PU network junctions. Impeding dynamics in the attached PHEMA segments could be caused by increasing motional event scale and making conformational transitions more difficult. Thus, the PU-PHEMA network must be considered just as the HNC semi-IPN one.

DSC data show (Fig. 4–7) a substantial effect of NP filler on the characteristics of PHEMA glass transition in the 83PU-17PHEMA network. It depends on the type of NPs, their content, and the type of functionalization. The following changes in the course of the DSC curves are observed: the total disappearance of the second ΔC_p step; a sharp narrowing of the glass transition range; a large increase in T_g and especially T_g' values; and changing the ΔC_p values in two opposite directions. The disappearance of the higher temperature ΔC_p step in some DSC curves corresponds to suppression of the mobility of PHEMA segments covalently anchored to the network or located close to the anchoring points.

For HNCs (Fig. 4–7), the most impressive DSC result is the quite different impact of 0.25 wt.% NP on the PHEMA glass transition; this distinctly demonstrates the connection of dynamics with the type of functional groups at the NP surface, that is, with the character of interfacial interactions (curves 2, 4, and 6). Really, the -OH and APMS covers could provide only hydrogen bonding of nanosilica with PHEMA. As a result, only suppression of the second ΔC_p step (curve 2) or even negligibly small changes in the DSC curve (APMS cover, curve 6) are observed. In contrast, it may be assumed with a large certainty that copolymerization between $-CH{=}CH_2$ cover of NPs and HEMA occurs, and covalent bonding of a part of PHEMA segments with these NPs takes place. This results in T_g' displacement from 33°C to 91°C; increase in T_g from 77 to 110°C, and threefold reduction of transition width ΔT_g and ΔC_p step (Fig. 4–7, curve 4). More complicated behavior of glass transition, as estimated by DSC, is observed for HNC with 3 wt.% silica (curves 3, 5, and 7).

High-resolution laser-interferometric creep rate spectroscopy (CRS) was used to analyze the segmental dynamics and heterogeneity of HNCs at 20–160°C covering the PHEMA glass transition range. The CR spectra (Fig. 4–8) of the neat matrix and HNC have the

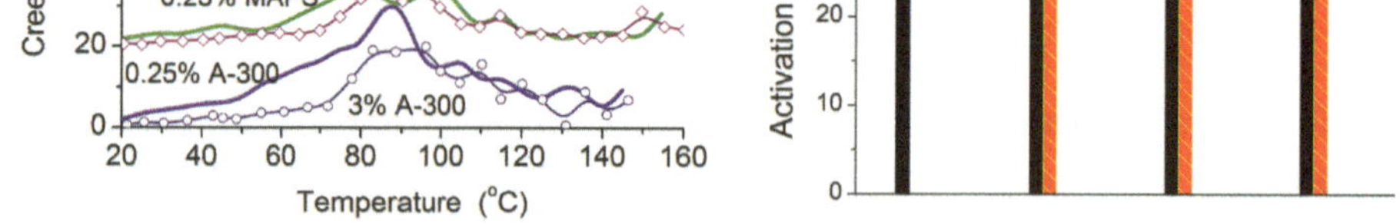

FIGURE 4–8 Creep rate spectra obtained at tensile stress of 0.3 MPa in the temperature region of PHEMA glass transition for the 83PU-17PHEMA network and HNCs containing 0.25 or 3 wt.% of (A) silica and with different surface functional groups; and corresponding activation energies for PU-PHEMA and HNCs with (B) silica fillers calculated using the initial rise of the creep rate to the first maximum. Creep rate spectra started at about zero creep rate at 20°C but were displaced by 20 units regarding one another along the ordinate axis. *Reproduced from Bershtein VA, Gun'ko VM, Karabanova LV, Sukhanova TE, Yakushev PN, Egorova LM, et al. Polyurethane–poly(2-hydroxyethyl methacrylate) semi-IPN–nanooxide composites. RSC Adv 2013;3:14560, with permission from The Royal Society of Chemistry.*

complicated contours in the 40–150°C range including a few partly overlapping peaks, the basic one at ca. 100°C (for the neat matrix) and the lesser peaks at 60–80°C, 90°C, 110–120°C, and 140–150°C. This indicates the multimodal dynamics (strong dynamic heterogeneity) within or close to the PHEMA glass transition in the HNC [68]. An influence of small amounts of silica FNP is observed and the spectral contours depend on a type of oxide, its content, and the surface structure (Fig.4–8). The basic impact of NPs is a strong suppression of segmental dynamics, namely, a sharp decrease in the height of the main CR peak. Secondly, the small CR peak at 140°C is suppressed or displaced to 150°C for HNCs. Third, some increasing creep rates (enhanced mobility) at moderate temperatures of 40–80°C simultaneously arose at 0.25 wt.% oxides. CRS measurements revealed the differences in the glass transition dynamics for HNCs with different types of NP functionalities. For the HNCs, the maximal suppression of the dynamics at ~90–160°C is observed for NP-MAPS when the largest displacement of the onset of a sharp acceleration of creep toward higher temperatures occurs. Really, at 3 wt.% silica NP, a low level of creep rate retains up to 160°C (Fig. 4–8A). Lesser effects are observed in the cases of the untreated silica surface (–OH cover) and, especially, for the NH_2-functionalized surface. In addition, at 0.25 wt.% silica NP some opposite effect of accelerating dynamics (increasing creep rates and diminution of the activation energy, Fig. 4–8B) is observed at ~40–80°C, that is, in the temperature region between β- and α-transitions of PHEMA (Fig. 4–8A).

The latter effect is caused, obviously, by some loosening molecular packing, with a partial or total collapse, locally, of intermolecular cooperativity of segmental motions in the HNCs ("nanoconfined geometries" effect). This leads to diminution of the activation energy, most substantial at the minimal amount of the MAPS-A-300 filler (Fig. 4–8B).

There is a considerable influence of NP on the modulus of elasticity $E(T)$ (Fig. 4–9). The effects are temperature-dependent, and the HNC modules are typically higher than those for neat polymer matrix. However, at some temperatures they may be close to the magnitude of that of the neat network. On the whole, the E_n/E_0 ratio varies from ~1 to 2.5. The role of interfacial interactions in HNC for modulus behavior is observed. For $CH = CH_2$ modified NP with PHEMA/silica covalent bonding, a considerable rise of E_n/E_0 ratio up to 1.5–2.0 is observed at high temperatures (60–140°C), with the maximal value at ~120°C, that is, in the temperature region of PHEMA glass transition (curve 5).

For NH_2 cover of NP, the modulus increased only slightly at high temperatures, whereas $E_n/E_0 = 2.0–2.5$ at 20°C and 60–70°C (curve 4) at temperatures rather close to the PU glass transition temperatures. This may be due to chemical reactions between isocyanate groups of the pristine mixture of reagents and APMS cover that inevitably proceed during preparing the HNCs. Due to this reaction, covalent bonding between NPs and the forming PU network occurs promoting increasing modulus of HNCs at moderate temperatures.

Epoxy ternary blends were prepared by mixing diglycidyl ether bisphenol-A, cycloaliphatic epoxy, and novolac epoxy [72]. Spherical silica particles were added into epoxy blends. The thermal properties of the epoxy composites were characterized using a thermomechanical analyzer, DSC, and DMA. It was found that the storage modulus of the epoxy was increased in the presence of silica. However, the coefficient of thermal expansion and

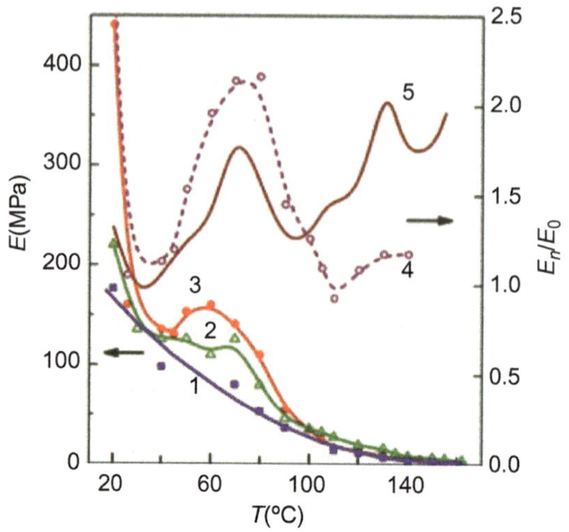

FIGURE 4–9 Modulus of elasticity E versus temperature dependencies obtained at tensile stress of 0.3 MPa for the 83PU-17PHEMA network, E_0 (1), and for HNC, E_n (2, 3). Lines 4 and 5 indicate the E_n/E_0 ratio. Composites with 3% silica NP functionalized with APMS (3, 4) or MAPS groups (2, 5). *Reproduced from Bershtein VA, Gun'ko VM, Karabanova LV, Sukhanova TE, Yakushev PN, Egorova LM, et al. Polyurethane–poly(2-hydroxyethyl methacrylate) semi-IPN–nanooxide composites. RSC Adv 2013;3:14560, with permission from The Royal Society of Chemistry.*

the glass transition temperature of the epoxy ternary blends was reduced. This was due to the expansion of the epoxy matrix constrained in the presence of silica fillers. The UV/Vis spectroscopy results demonstrated that the percentage of transmittance of epoxy was decreased by the incorporation of silica particles.

Dielectric HNCs need a large interfacial area between polymers and NP fillers [73]. The interface can create sites for trapping electrons leading to increased dielectric breakdown strength (DBS). NPs with a bimodal population of covalently anchored molecules were created using ligand engineering. Electrically active short molecules (oligothiophene or ferrocene) and matrix compatible long PGMA chains comprise the bimodal brush. The DBS was evaluated from recessed samples, and dielectric spectroscopy was used to study the dielectric constant and loss as a function of frequency. The DBS and permittivity increased with only 2 wt.% filler loading while the dielectric loss remained comparable with the reference epoxy. The long, epoxy compatible PGMA brush ensured good dispersion and allowed study of the effect of the electroactive layer on the breakdown strength. The DBS increased in the bimodal brush particle HNC.

HNCs can be used as semipermeable membranes for water desalination technologies such as the reverse osmosis process and emerging forward osmosis process [74]. Structural and chemical properties of the membranes determine water flux, salt rejection, fouling resistance, and chemical stability, which greatly impact energy consumption and costs in the osmosis separation processes. Significant progress has been made in the development of high-performance polymer and polymer composite membranes for desalination applications.

Polyelectrolyte membranes showing proton conductivity at moderate levels of relative humidity (RH) and temperatures are essential for the development of polyelectrolyte membrane fuel cells [75]. Monomethoxy oligoethylene glycol methacrylate derived polymer brush functionalized silica NPs are presented as a humidifying nanofiller to create Nafion HNC membranes. This HNC showed improved proton conductivity at moderate RH levels and temperatures. Polymer-brush functionalized nanoparticles (PBFNP) prepared using surface-initiated atom transfer radical polymerization and dispersed in the Nafion resin solution (Nafion/FNPPB) were created with solution casting. Composite membranes were also prepared with bare silica NP. Spectroscopic measurements confirmed the presence of polymer brushes in the final membranes and demonstrated increased water uptake with Nafion/PBFNP. Electrochemical impedance showed that 1 wt.% of FNP is sufficient to achieve HNC with superior proton conductivities at ambient and moderately high temperatures over the entire RH range. The membrane with 1 wt.% of FNPPB showed ~11 times higher proton conductivity at 20% RH and 25°C. At the same temperature but at 80% RH, the proton conductivity of the Nafion/PBFNP membrane was ~4 times higher than that of Nafion.

A continuum-based elastic micromechanics model was developed for silica NP/polyimide HNC with various treatments [76]. The silica (α-quartz) NPs were spherical-like of 1.2 nm in size. The polymer models, having an amorphous molecular structure, were based on a polyimide from 3,3′,4,4′-biphenyltetracarboxylic dianhydride (BPDA) and 1,3-bis(4-aminophenoxy)benzene (APB) monomers. The model used the molecular structures of the NPs, polyimide, and interfacial regions, which were determined using a molecular modeling method involving coarse-grained and reverse mapping techniques. The micromechanics model includes an effective interface between the polyimide and NP with properties and dimensions determined using the results of molecular dynamics simulations. It was shown that the model could be used to predict the elastic properties of silica NP/polyimide HNC for a large range of NP radii from 1 nm to 1 μm. For silica NP radii above 100 nm, the predicted properties are equal to those predicted using the standard Mori–Tanaka micromechanical approach, which does not incorporate the molecular structure. It was also shown that the specific silica NP/polyimide interface conditions had a significant effect on the HNC mechanical properties for nanoparticle radii below 100 nm.

Submicron silica particles surface-capped with diglycidyl ether of bisphenol-A were dispersed in a solution of epoxy resin, hardener, and acetone [77]. The resulting suspension was then spin-coated onto a surface of austenitic stainless steel AISI 316 L and cured at temperature, generating a thick silica/epoxy coating 300 nm. An epoxy coating without nanosilica was also prepared as a reference. The coatings were further functionalized with fluoroalkylsilane (FAS17) for an additional improvement of the nonwetting properties. The mechanical properties of epoxy coatings filled with different combinations of silica particles 30 nm, 200 nm, and 600 nm were compared and characterized using scratch-resistance tests. The effects of incorporating the silica particles on the surface characteristics of the epoxy-coated steel were investigated with contact-angle and surface-energy evaluations. The surface morphology of the coatings was characterized with SEM. The results indicated that the silica particles significantly improved the microstructure of the coating matrix that was reflected in increased damage resistance, reduced degree of delamination, and induced hydrophobicity. The formation of microsized silica agglomerates was observed due to the epoxy–matrix curing process. The significantly increased hydrophobicity of silica/epoxy coatings of 200 nm and (30 + 600) nm was due to appropriate combination of micro- and nanoroughness created by particle agglomerates embedded into the matrix [77,78].

Structure and temperature-responsive anion exchange properties of amino-functionalized mesoporous silica coated with temperature-responsive copolymer, poly(N-isopropylacrylamide-co-acrylamide) (P(NIPAM-co-Am)) were studied [79]. HNCs containing 0, 10, or 20 wt.% of Am were synthesized. From TG results, it was found that the amounts of copolymer immobilized on the mesoporous silica were 1.6–2.6 wt.%. XRD patterns revealed that the HNC structures were hexagonal and almost the same as those of the original mesoporous silica without polymer. At low temperature the methyl orange (MO) anions adsorbed and desorbed reversibly with changing pH of the solution, while at high temperature the MO anions did not. Even if the amount of added Am increased up to 20 wt.%, the composite structure (MCM type) was hexagonal. There was no influence on the porous property of the mesoporous silica, because the amount of immobilized polymer was very small at 1.6–2.6 wt.%. The transition temperatures for Am0-MS, Am10-MS, and Am20-MS were 30–35°C, 40–42°C, and 47–50°C, respectively, and it was found that the transition temperature of HNCs shifted to the higher temperature side with increasing amount of Am.

HNCs show unique properties combining the advantages of polymers and such fillers as mesoporous silicas (MCM-41, MCM-48, MCM-50, and SBA-15) [80]. To accommodate macromolecules, the nanopores led to unusually intimate interactions between the polymer and the inorganic phase, and some unusual properties can be observed, when compared with nonporous fillers. Whereas many review articles have been devoted to polymer/nonporous nanofiller HNCs, few review articles focus on polymer/mesoporous silica HNCs [80].

Mesoporous silica (MCM-41) was sulfonated (S-MCM-41) and incorporated into sulfonated polyether sulfone (SPES) matrix to prepare composite membranes [81]. Various concentrations of S-MCM-41 were used (1, 2, 5, 10, and 20 wt.%). FTIR and XRD of MCM-41 and S-MCM-41 confirmed the chemical and structural properties; AFM and UTM were used to analyze the morphology and mechanical properties of the HNCs. The water uptake and ionic conductivity of the composite membranes increases with MCM content in the membrane. Mechanical stability of the membrane increased with MCM content. The membranes showed good mechanical and physicochemical properties that can be utilized as polymer electrolyte membranes for fuel cells and other separation applications.

4.3.3 Complex Fillers

Not only functionalized FMO alone but also hybrid fillers (e.g., nanosilica and GO) can be used in composites [82]. A multicomponent curing agent for epoxy resin with GO core and poly(amidoamine) (PAMAM) dendrimer-grafted silica nanoparticles was synthesized for thermal stabilization of epoxy resins due to dendritic aminofunctionalization of the filler [82]. Hyperbranched PAMAM-grafted silica nanoparticles (silica dendrimer, SD) were attached at a surface of GO to prepare the HNC and curing agent (silica dendrimer-attached GO, GSD). Then, epoxy resin was treated in the presence of different amounts of GSD and the final products were compared with ethylenediamine-cured epoxy resin (E) in their char contents. Successful functionalization of silica NP and attachment of SD at the surface of GO were analyzed using a number of methods.

The characteristics (stiffness, rigidity) of poly(methyl methacrylate) filled by multiwall carbon nanotubes (MWCNTs) strongly improved if functionalized silica nanoparticles are bonded to the MWCNT surface [83].

Effects of silane and silica (in situ synthesized) enrichment of carbon fibers (CF, 7 μm and density of 1.76 g/cm^3) on interfacial properties of methylphenylsilicone resin (MPSR) composites were investigated [84]. CFs were oxidized, grafted with 3-aminopropyltriethoxysilane (APS), and then modified with silica NP prepared by the sol–gel polymerization of TEOS. Chemical structures of CF were characterized by confirming the successful grafting. SEM showed a uniform distribution of silica NP on the CF surface. The interlaminar shear strength (ILSS) and impact toughness of silanized CF (CF-siloxane) composites were 12.05% and 7.46% higher than those of untreated composites. However, ILSS and impact toughness of the hybrid fiber (CF/Si) composites obtained from the hydrolysis of different concentrations TEOS improved significantly, especially for grafting silica enrichment with the TEOS concentration of 0.05 mol/L increasing 45.6% in ILSS and 29.6% in impact properties. Moreover, the hydrothermal aging resistance was also improved, and functionalization processes did not decrease fiber tensile strength [84].

A composite polymer electrolyte based on an ionic liquid, Li salt, and polymeric ionic liquid (PIL) that included PIL-functionalized mesoporous silica nanoplates (PIL-SNP) was created using a solution-casting method [85,86]. A PIL-SNP content (8 wt.%) optimized composite polymer electrolyte had high ionic conductivity, a great Li ion transference number, and good interfacial stability characteristics superior to those of a plain PIL-based electrolyte or incorporating PIL-SNP. The incorporation of SNP allowed the creation of a continuous and interconnected conductive network, and weakened the anionic and cationic interactions of the IL and Li salts, thereby improving ionic mobility and charge carrier concentration. The PIL-SNP effectively improves the electrochemical stability and interfacial stability of the polymer electrolyte with the Li metal electrode and anode electrode, due to its large surface area and the high affinity resulting from porous structures and grafted PIL. Li/$LiFePO_4$ cells with composite polymer electrolyte and PIL-SNP showed significantly improved discharged capacities and cyclabilities, compared to cells with plain PIL-based polymer electrolyte, displaying a higher average specific capacities at 40°C and 60°C, approximating to 228% and 152% enhancements after 100 cycles, respectively.

Water-dispersible composite silica nanospheres covalently anchored with gold nanoparticles (AuNPs) possessing thermotunable spatial distributions at the outer periphery of thermoresponsive poly(N-isopropylacrylamide) (PNIPAM) brushes were prepared [87]. Starting from initiator functionalized silica NP, surface-initiated atom transfer radical polymerization (ATRP) of N-isopropylacrylamide (NIPAM) afforded hybrid silica NP coated with PNIPAM brushes. The substitution reaction of halogen terminal groups of grafted PNIPAM chains with sodium azide and subsequent click reaction with 1,2-dithiolane-3-pentanoic acid-N-propargylamide afforded hybrid silica nanoparticles coated with 1,2-dithiolane end-capped PNIPAM brushes. AuNPs were then covalently anchored to the outer periphery of hybrid silica nanoparticles by utilizing strong chemisorption of surface-attached dithiolane moieties to the AuNPs. Dynamic laser light scattering measurements revealed that thermo-sensitive PNIPAM brushes at the surface of hybrid silica NPs showed reversible thermoinduced collapse/swelling transitions, leading to the facile thermomodulation of spatial distributions of AuNPs covalently attached at the periphery of composite silica NPs and thermoreversible surface plasmon absorption band shift. The reported strategy of covalent assembly of AuNPs into well-defined composite nanospheres possessing thermotunable characteristics could be further exploited for colorimetric temperature sensing and responsive SERS detection purposes [87].

4.3.4 Biomedical Applications of Composites

Various HNCs could be used for biomedical applications such as tissue engineering, drug delivery, and cellular therapies [88]. Due to the strong interactions between polymer and NPs, a range of property combinations can be engineered to mimic native tissue structure and properties. Such natural and synthetic polymers as starch, cellulose, alginate, chitosan, collagen, gelatin, and fibrin, poly(vinyl alcohol), poly(ethylene glycol), poly(caprolactone), poly(lactic-co-glycolic acid), and poly(glycerol sebacate) can be used to design HNCs for biomedical applications. NPs with ceramics, polymers, metal oxides, and carbon-based nanomaterials can be incorporated into the polymeric network to obtain desired property combinations [88].

Ultrasound micromolding technology was used in processing biodegradable polymers with minimum material loss [28]. This method is appropriate for creation of biomedical microdevices. The use of silica NP is appropriate to improve the properties of composites. However, homogeneous dispersion of particles is crucial. The use of micro- and nanosized silica particles as fillers of a biodegradable polyester (with 1,9-nonanodiol and azelaic acid) demonstrated that HNCs could include up to 6 wt.% of silica. No degradation occurred if particles were functionalized with a compatibilizer (3-aminopropyltriethoxysilane).

Silanization of a silica filler with methacryloxy silanes can significantly improve mechanical properties (compression strength, angular flexural strength, flexural strength, and elasticity modulus) and monomer conversion percentage due to the chemical bonding between the silanized filler and the organic matrix, although it is not affected by the depth of cure and polymerization shrinkage in HNCs [89]. Since the release of monomers from restorative materials can lead to toxic effects on the patient, high degree of conversion can be very important to improve the biocompatibility of the dental composites. The silanization effect causing good mechanical properties of the composites can be protected lifelong in oral cavities. Silica fillers were prepared using colloidal silica solution, Ludox® HS-40, and silanized using 3-methacryloxypropyl-trimethoxysilane (MPTMS) in acidic media. SEM images, surface area, and particle sizes showed that agglomeration tendencies of silanized fillers were lower in comparison with nonsilanized fillers. The curing depth and degree of polymerization percentage of composites showed that structure and mechanical properties of composites could be strongly affected by silanization due to the polymer–filler interface bonding formed by silanes. Despite these findings, silica silanization was not affected by depth of cure and polymerization shrinkage values [89].

Effects of silica NP aggregation on the reflectance and crystallinity of dental composite materials were analyzed using two types of silica NP (ca. 10 nm in size) that is, Aerosil 200 nonfunctionalized and Aerosil DT4 functionalized with 3-methacryloxypropyl-trimethoxysilane [90]. NPs were dispersed in a monomer mix composed of urethane dimethacrylate and ethylene glycol dimethacrylate at 80:20 mass ratio. The particle size of silica and their aggregation were determined using SEM and TEM showing that Aerosil DT4 formed dense aggregates of 1 μm and larger in size. Aerosil 200 showed a particle gel-like structure. The functionalization degree of Aerosil DT4 was 7.57 wt.%. The composites were evaluated using DSC to determine their crystallinity. The composite reinforced by Aerosil DT4 showed lower crystallinity than the system with Aerosil 200 due to stronger interactions of polymers with functionalized Aerosil DT4. The aggregation effect of silica NP on the optical properties was determined by reflectance analysis. The Aerosil 200 composite showed a lower degree of nanoparticle aggregation and higher reflectance than the system with Aerosil DT4 [90].

Nanoparticles can penetrate into cells and their surface functionalization could offer a way to increase their sensitivity when detecting target molecules [91]. It increases the potential for flexibility in particle design, their therapeutic function, and variation possibilities in diagnostics. Mesoporous silica NPs have attractive physical and chemical characteristics such as particle morphology, controllable pore size, and great surface area and pore volume. Additionally, the surface functionalization of NPs is relatively straightforward, which enables optimization of the interactions between the particles and biosystem. Some of the properties of silica NP fillers can be used in composites prepared for biomedical applications [12–18].

Microbial fuel cells (MFCs) can be conveniently used for wastewater treatment and bioelectricity production [92]. Their efficiency is strongly influenced by the physicochemical properties of an ion-conducting membrane. Nafion-based composite membranes were prepared with mesoscale SBA-15 silica functionalized with SO_3H groups. The membranes were tested as alternative separators in MFC for wastewater treatment and the performances were compared with that of standard Nafion 117. Prolonged (3 months) MFC operation showed that the composite membrane with 5 wt.% of SBA-15 functionalized with 10 mol% of SO_3H gave maximum power density of 380 mW m^3 that was three times better than that of Nafion after 90 days of operation. The same membrane offered an effective chemical oxygen demand removal after 14 days (more than 95%), an impressive coulombic efficiency of 34%, and high resistance to biofouling. The use of silica-based SO_3H functionalized fillers could be a powerful strategy to improve the performance of Nafion™ membranes in MFCs.

Mechanoresponsive polymeric scaffolds, which can alter cellular functions, were created using silica NP (A-200) functionalized with OH (initial) and NH_2 (amino-silanized with aminopropyltrimethoxysilane) [93]. These silicas were mixed with biocompatible polyhedral oligomeric silsesquioxanes-poly(carbonate-urea) urethane (POSS-PCU) at various contents to produce scaffolds. XPS and mechanical testing demonstrated that bulk mechanical properties (Young's modulus) were modified without changing the scaffold's surface chemistry. Mechanical testing showed significant change in bulk properties of POSS-PCU scaffolds at low NP content of 1% ($P < 0.01$). Scaffolds modified with NH_2-silica showed significantly better bulk mechanical properties in comparison with that modified with initial silica. Enhanced cell adhesion, proliferation, and collagen production over 14 days were observed on scaffolds with good bulk mechanical properties (NH_2-silica) compared with that unmodified or with initial silica ($P < 0.05$) during in vitro analysis. Thus, an effective method of production of mechanoresponsive polymeric scaffolds to customize cellular responses for biomaterial applications could be based on the use of functionalized silica NPs [93].

Mesoporous silica NPs are among the most promising nanovectors for controlled release of active compounds, and a SBA-15-based delivery system was studied [94]. The release of the active compound was controlled by specific pore/wall interactions and can be triggered by an external stimulus, such as simple addition of water to the media. Due to its reversibility in the presence of water, imine bond formation between amino groups and aldehydes was proposed as a valuable approach to reach this scope. The amino groups were provided by aminofunctionalization of SBA-15 particles (<1 μm in size) with aminopropyltrietoxysilane. Natural antimicrobial vanillin was as a model bioactive compound. Prepared particles were embedded into biodegradable poly-ε-caprolactone (PCL) films. Release kinetics of vanillin showed that the diffusion of vanillin molecules from the filled films was delayed by about 20% and 75% with respect to films containing free vanillin, in water and in ethanol media, respectively. The most significant feature of the system was the stepwise behavior, which showed that the release of vanillin bound to the NH_2 groups of functionalized SBA-15 was triggered only after the addition of water. These findings give an insight into possible application of pristine and functionalized mesoporous silicas to develop polymer films for active responsive packaging materials [94].

The performance of polymer-based restorative dental composite materials can be strongly improved using appropriate nanofillers [95]. Long-term performance of composites is vulnerable due to its polymerization shrinkage and associated shrinkage stresses and also degradation of the filler/matrix interface. Development of low shrinkage resin systems that hinder the volumetric shrinkage and enhance the quality of the filler/matrix interface is of importance. A variety of coupling agents and techniques are currently available for surface treatment of inorganic fillers used in composites. However, for dental composites, there are still some concerns related to the best coupling agent at the interface that guarantee long-term stability and crack resistance of these restorations. The surface modification of fillers has a strong potential, more than just a bond between two different phases. It can be considered as a medium to compensate for the deficits of dental composites. Application of long chain hydrophobic polymers at the particle surface of composite materials not only improves hydrolytic stability and uniform dispersion of fillers, but also reduces the polymerization shrinkage and associated stresses, which would be very advantageous for restorative dental composites [95].

The main characteristics (textural, surface chemistry, PaSD, etc.) of modified silica fillers depend on type, condition and time of modification and treatments [96–100]. Characteristics of fillers and polymer matrices determine the ways and conditions of composite preparations [56,101–117], as well as expected applications [118–128]. The characteristics of composites with FMO can change over large ranges depending on the properties of media in which these composites are used [129–135]. The mechanical properties are especially important [136–141], and these properties are reflected in dynamics of the polymers under different conditions [142–150]. The effects can be enhanced if instead of pure silica complex FMO or silica modified by grafted oxide nanoparticles are used [140,141]. Physical properties of composites [151–153] or efficiency of their bioapplications [154,155] depend strongly on both structure of fillers and polymer matrix.

4.4 Conclusion

Silicas both amorphous and crystalline, pristine or functionalized, with nonporous or porous nano- or microparticles are the best fillers for a variety of polymers. Differently functionalized silica nanoparticles can be well distributed in the polymer matrices that provide good properties of composites with linear, branched, cross-linked, polar/nonpolar, ionic, and other polymers. There are a large number of modifiers of a silica surface, and among them silanes are the best to create stable functionalities since stable Si–O–Si bonds can be formed. Some functionalities can be cross-linked with polymers or take part in polymerization reactions, which is of importance for good mechanical properties of the materials. In situ synthesized silica nanoparticles in polymeric (or a monomer mix) media give another effective pathway to produce good hybrid nanocomposites for various applications.

References

[1] Kickelbick G, editor. Hybrid materials. Synthesis, characterization, and applications. Weinheim: Wiley-VCH; 2007.

[2] Ajayan PM, Schadler LS, Braun PV. Nanocomposite science and technology. Weinheim: Wiley-VCH; 2003.

[3] Wypych G. Handbook of fillers. Toronto: ChemTec Publishing; 1999.

[4] Mittal V, editor. Surface modification of nanoparticle and natural fiber fillers. 1st ed. Weinheim: Wiley-VCH; 2015.

[5] Kim C-S, Randow C, Sano T, editors. Hybrid and hierarchical composite materials. Switzerland: Springer; 2015.

[6] Hoheisel TN, Hur K, Wiesner UB. Block copolymer–nanoparticle hybrid self-assembly. Prog Polym Sci 2015;40:3–32.

[7] Balazs C, Emrick T, Russell TP. Nanoparticle polymer composites: where two small worlds meet. Science 2006;314:1107–10.

[8] Zou H, Wu S, Shen J. Polymer/silica nanocomposites: preparation, characterization, properties, and applications. Chem Rev 2008;108:3893–957.

[9] Schmidt G, Malwitz MM. Properties of polymer—nanoparticle composites. Curr Opin Colloid Interface Sci 2003;8:103–8.

[10] Kumar MSS, Raju NMS, Sampath PS, Jayakumari LS. Effects of nanomaterials on polymer composites - an expatiate view. Rev Adv Mater Sci 2014;38:40–54.

[11] Rothon RN, Roger N. Particulate fillers for polymers. Shrewsbury: Rapra; 2001.

[12] Iler RK. The chemistry of silica. Chichester: Wiley; 1979.

[13] Bergna HE, Roberts WO, editors. Colloidal silica: fundamentals and applications. Boca Raton: CRC Press; 2006.

[14] Vansant EF, Van Der Voort P, Vrancken KC. Characterization and chemical modification of the silica surface. Amsterdam: Elsevier; 1995.

[15] Legrand AP, editor. The surface properties of silicas. New York: Wiley; 1998.

[16] Dabrowski A, Tertykh VA, editors. Adsorption on new and modified inorganic sorbents, studies in surface science and catalysis, vol. 99. Amsterdam: Elsevier; 1996.

[17] Blitz JP, Gun'ko VM, editors. Surface chemistry in biomedical and environmental science, NATO science series II: mathematics, physics and chemistry, vol. 228. Dordrecht: Springer; 2006.

[18] Chuiko AA, editor. Chemistry of silica surface. Kiev: UkrINTEI; 2001 (in Russian).

[19] Auner N, Weis J, editors. Oganosilicon chemistry VI. Weinheim: Wiley-VCH; 2005.

[20] Ullmann's encyclopedia of industrial chemistry. Weinheim: Wiley-VCH, 2008.

[21] Büchel KH, Moretto H-H, Woditsch P. Industrial inorganic chemistry. Weinheim: Wiley-VCH; 2000.

[22] Basic characteristics of Aerosil fumed silica. 4th ed. Tech. Bull. Fine Particles 11. Hanau: Evonik Industries; 2014. Aerosil®—Fumed Silica. Technical Overview. Hanau: Evonik Industries; 2015. <http://www.aerosil.com/product/aerosil/en/products/hydrophobic-fumed-silica/Pages/default.aspx>.

[23] Cabot Corp. <http://www.cabotcorp.com/solutions/products-plus/fumed-metal-oxides/hydrophobic>.

[24] Wacker Chemie Ag. <https://www.wacker.com/cms/en/products/brands_2/hdk/hdk.jsp>.

[25] Hongwu International Group Ltd. <http://www.hwnanomaterial.com/The-Difference-Between-Hydrophilic-Silica-and-Hydrophobic-Silica-Nanoparticles_p255.html>.

[26] Gun'ko VM, Turov VV. Nuclear magnetic resonance studies of interfacial phenomena. Boca Raton: CRC Press; 2013.

[27] Krishnamoorti R. Strategies for dispersing nanoparticles in polymers. Mater Res Bull 2007;32:341–7.

[28] Díaz A, Casas MT, Puiggalí J. Dispersion of functionalized silica micro- and nanoparticles into poly(nonamethylene azelate) by ultrasonic micro-molding. Appl Sci 2015;5:1252–71.

[29] Bagwe RP, Hilliard LR, Tan W. Surface modification of silica nanoparticles to reduce aggregation and nonspecific binding. Langmuir 2006;22:4357–62.

[30] Oberdisse J. Aggregation of colloidal nanoparticles in polymer matrices. Soft Matter 2006;2:29–36.

[31] Watanabe R, Kunioka M, Mizukado J, Suda H, Hagihara H. Highly ductile polypropylene-based nanocomposites by dispersing monodisperse silica nanospheres in functionalized polypropylene containing hydroxyl groups. Polymer 2016;99:63–71.

[32] Gun'ko VM. Composite materials: textural characteristics. Appl Surf Sci 2014;307:444–54.

[33] Gun'ko VM, Turov VV, Zarko VI, Pakhlov EM, Charmas B, Skubiszewska-Zięba J. Influence of structural organization of silicas on interfacial phenomena. Colloids Surf A: Physicochem Eng Aspects 2016;492:230–41.

[34] Gun'ko VM, Turov VV, Zarko VI, Goncharuk OV, Pahklov EM, Skubiszewska-Zięba J, et al. Interfacial phenomena at a surface of individual and complex fumed nanooxides. Adv Colloid Interface Sci 2016;235:108–89.

[35] Adamson AW, Gast AP. Physical chemistry of surface. 6th ed. New York: Wiley; 1997.

[36] Gun'ko VM, Voronin EF, Nosach LV, Turov VV, Wang Z, Vasilenko AP, et al. Structural, textural and adsorption characteristics of nanosilica mechanochemically activated in different media. J Colloid Interface Sci 2011;355:300–11.

[37] Gun'ko VM, Sheeran DJ, Augustine SM, Blitz JP. Structural and energetic characteristics of silicas modified by organosilicon compounds. J Colloid Interface Sci 2002;249:123–33.

[38] Gun'ko VM, Pakhlov EM, Goncharuk OV, Andriyko LS, Marynin AI, Ukrainets AI, et al. Influence of hydrophobization of fumed oxides on interactions with polar and nonpolar adsorbates. Appl Surf Sci 2017;423:855–68.

[39] Gun'ko VM, Mironyuk IF, Zarko VI, Voronin EF, Turov VV, Pakhlov EM, et al. Morphology and surface properties of fumed silicas. J Colloid Interface Sci 2005;289:427–45.

[40] Shpak AP, Gorbyk PP, editors. Nanomaterials and supramolecular structures. Dordrecht: Springer; 2010.

[41] Blitz JP, Gun'ko VM. Kinetics and mechanism of chemical reactions at silica surfaces. In: Somasundaran P, editor. Encyclopedia of surface and colloid science. 3rd ed. Boca Raton: CRC Press; 2015. p. 3491–501.

[42] Blitz JP, Gun'ko VM, Samala R, Lawrence BA. Mixed bifunctional surface-modified silicas using tethered aminofunctional silane catalysts. Colloids Surf A: Physicochem Eng Aspects 2014;462:1–8.

[43] Frei R, Blitz JP, Gun'ko VM, Frost BE, Sergeev VS. Kinetics and computational studies of an aminosilane reaction with a silsesquioxane silanol. J Phys Chem A 2009;113:6612–19.

[44] Blitz JP, Christensen JM, Gun'ko VM. Silica surface modification reactions with aluminum and boron alkyls and (alkyl) chlorides: reactivities and surface nanostructures. J Nanosci Nanotechnol 2008;8:660–6.

[45] Blitz JP, Diebel RE, Deakyne CA, Christensen JM, Gun'ko VM. Experimental and computational studies of trialkylaluminum and alkylaluminum chloride reactions with silica. J Phys Chem B 2005;109:5667–77.

[46] Gun'ko VM. Interfacial phenomena: effects of confined space and structure of adsorbents on the behavior of polar and nonpolar adsorbates at low temperatures. Curr Phys Chem 2015;5:137–72.

[47] Gun'ko VM, Voronin EF, Pakhlov EM, Zarko VI, Turov VV, Guzenko NV, et al. Features of fumed silica coverage with silanes having three or two groups reacting with the surface. Colloids Surf A: Physicochem Eng Aspects 2000;166:187–201.

[48] Hood MA, Mari M, Muñoz-Espí R. Synthetic strategies in the preparation of polymer/inorganic hybrid nanoparticles. Materials 2014;7:4057–87.

[49] Bounor-Legaré V, Cassagnau P. In situ synthesis of organic-inorganic hybrids or nanocomposites from sol-gel chemistry in molten polymers. Prog Polym Sci 2014;39:1473–97.

[50] El-Midany AA, Ibrahim SS. Interfacial role of compatabilizers to improve mechanical properties of silica–polypropylene composites. Physicochem Probl Miner Process 2011;46:295–305.

[51] Paul DR, Mark JE. Fillers for polysiloxane ("silicone") elastomers. Prog Polym Sci 2010;35:893–901.

[52] Chruściel JJ, Leśniak E. Modification of epoxy resins with functional silanes, polysiloxanes, silsesquioxanes, silica and silicates. Prog Polym Sci 2015;41:67–121.

[53] Sarkar B, Alexandridis P. Block copolymer-nanoparticle composites: structure, functional properties, and processing. Prog Polym Sci 2015;40:33–62.

[54] Li HY, Zhang ZS, Ma XF, Hu M, Wang XY, Fan PF. Synthesis and characterization of epoxy resin modified with nanoSiO_2 and glycidoxypropyltrimethoxysilane. Surf Coat Technol 2007;201:5269–72.

[55] Liu YL, Hsu CY, Wang ML, Chen HS. A novel approach of chemical functionalization on nano-scaled silica particles. Nanotechnology 2003;14:813–19.

[56] Liu YL, Hsu CY, Hsu KY. Poly(methylmethacrylate)–silica nanocomposites films from surface-functionalized silica nanoparticles. Polymer 2005;46:1851–6.

[57] Zeng Z, Yu J, Guo Z. Preparation of epoxy-functionalized polystyrene/silica core–shell composite nanoparticles. J Polym Sci: Part A: Polym Chem 2004;42:2253–62.

[58] Matějka L, Dušek K, Pleštil J, Kříž J, Lednický F. Formation and structure of the epoxy–silica hybrids. Polymer 1999;40:171–81.

[59] Wang Z, Suo J, Li J. Synthesis and characterization of epoxy resin modified with thiopropyl triethoxysilane. J Appl Polym Sci 2009;114:2388–94.

[60] Yang M, Gao Y, He JP, Li HM. Preparation of polyamide 6/silica nanocomposites from silica surface initiated ring-opening anionic polymerization. eXPRESS Polym Lett 2007;1:433–42.

[61] Jang MK, Lee SK, Kim BK. Polyurethane nano-composite with functionalized silica particle. Compos Interfaces 2008;15:549–59.

[62] Qiao H, Chao M, Hui D, Liu J, Zheng J, Lei W, et al. Enhanced interfacial interaction and excellent performance of silica/epoxy group-functionalized styrene-butadiene rubber (SBR) nanocomposites without any coupling agent. Compos Part B 2017;114:356–64.

[63] Ke YG, Stroeve P. Polymer-layered silicate and silica nanocomposites. Amsterdam: Elsevier; 2005.

[64] Bershtein V, Gun'ko V, Egorova L, Guzenko N, Pakhlov E, Ryzhov V, et al. Well-defined silica core–poly (vinyl pyrrolidone) shell nanoparticles: interactions and multi-modal glass transition dynamics at interfaces. Polymer 2009;50:860–71.

[65] Bershtein VA, Gun'ko VM, Karabanova LV, Sukhanova TE, Yakushev PN, Egorova LM, et al. Hybrid polyurethane-poly(2-hydroxyethyl methacrylate) semi-ipn-silica nanocomposites: interfacial interactions and glass transition dynamics. J Macromol Sci Part B: Phys 2010;49:18–32.

[66] Bershtein VA, Gun'ko VM, Egorova LM, Guzenko NV, Pakhlov EM, Ryzhov VA, et al. Well-defined oxide core-polymer shell nanoparticles: interfacial interactions, peculiar dynamics and transitions in polymer nanolayers. Langmuir 2010;26:10968–79.

[67] Bershtein V, Gun'ko V, Egorova L, Guzenko N, Pakhlov E, Ryzhov V, et al. Well-defined oxide core-poly (vinyl pyrrolidone) shell nanoparticles: interactions and multi-modal glass transition dynamics at interfaces. Macromol Symp 2010;296:541–9.

[68] Bershtein VA, Gun'ko VM, Egorova LM, Wang Z, Illsley M, Voronin EF, et al. Dynamics, thermal behaviour and elastic properties of thin films of poly(vinyl alcohol) nanocomposites. RSC Adv 2012;2:1424–31.

[69] Bershtein VA, Gun'ko VM, Karabanova LV, Sukhanova TE, Yakushev PN, Egorova LM, et al. Polyurethane–poly(2-hydroxyethyl methacrylate) semi-IPN–nanooxide composites. RSC Adv 2013;3:14560–70.

[70] Giannelis EP, Krishnamoorti R, Manias E. Adv Polym Sci 1999;138:107.

[71] Hu Y, Shenderova OA, Hu Z, Padgett CW, Brenner DW. Rep Progr Phys 2006;69:1847.

[72] Chin LW, Hung HL, Shyang CW. Thermo-mechanical and light transmittance of silica diffusant filled epoxy composites. J Phys Sci 2010;21:93–107.

[73] Virtanen S, Krentz TM, Nelson JK, Schadler LS, Bell M, Benicewicz B, et al. Dielectric breakdown strength of epoxy bimodal-polymer-brush-grafted core functionalized silica nanocomposites. IEEE Trans Dielectr Electr Insul 2014;21:563–70.

[74] Li D, Yan Y, Wang H. Recent advances in polymer and polymer composite membranes for reverse and forward osmosis processes. Prog Polym Sci 2016;61:104–55.

[75] Farrukh A, Ashraf F, Kaltbeitzel A, Ling X, Wagner M, Ghaffar A, et al. Polymer-brush functionalized SiO_2 nanoparticles based Nafion nanocomposites: A novel avenue to low-humidity proton conducting membranes. Polym Chem 2015;6:5782–9.

[76] Odegard GM, Clancy TC, Gates TS. Modeling of the mechanical properties of nanoparticle/polymer composites. Polymer 2005;46:553–62.

[77] Conradi M, Intihar G, Zorko M. Mechanical and wetting properties of nanosilica/epoxy-coated stainless steel. Mater Technol 2015;49:613–18.

[78] Conradi M. Nanosilica-reinforced polymer composites. Mater Technol 2013;47:285–93.

[79] Murakami K, Sato Y, Inoue Y. Synthesis of P(NIPAM-co-Am)/mesoporous silica composites and their temperature-responsive anion exchange. J Mat Sci Chem Eng 2015;3:7–15.

[80] Wei L, Hu N, Zhang Y. Synthesis of polymer–mesoporous silica nanocomposites. Materials 2010;3:4066–79.

[81] Gahlot S, Sharma PP, Kulshrestha V. Synthesis and characterization of functionalized silica/SPES composite membranes. AIP Conf Proc 2015;1665 050042,1-4.

[82] Azimi R, Roghani-Mamaqani H, Gholipour-Mahmoudalilou M. Grafting poly (amidoamine) dendrimer-modified silica nanoparticles to graphene oxide for preparation of a composite and curing agent for epoxy resin. Polymer 2017;. Available from: https://doi.org/10.1016/j.polymer.2017.08.037.

[83] Olek M, Kempa K, Jurga S, Giersig M. Nanomechanical properties of silica coated multiwall carbon nanotubes–poly (methyl methacrylate) composites. NSTI-Nanotech 2005;2:167–70.

[84] Wu G, Chen L, Liu L. Effects of silanization and silica enrichment of carbon fibers on interfacial properties of methylphenylsilicone resin composites. Compos Part A 2017;98:159–65.

[85] Wang S, Shi QX, Ye YS, Wang Y, Peng HY, Xue ZG, et al. Polymeric ionic liquid-functionalized mesoporous silica nanoplates: a new high-performance composite polymer electrolyte for lithium batteries. Electrochim Acta 2017;245:1010–22.

[86] Wang S, Shi QX, Ye YS, Xue Y, Wang Y, Peng HY, et al. Nano Energy 2017;33:110–23.

[87] Wu T, Zhang Q, Hu J, Zhang G, Liu S. Composite silica nanospheres covalently anchored with gold nanoparticles at the outer periphery of thermoresponsive polymer brushes. J Mater Chem 2012;22:5155–63.

[88] Tang Z, He C, Tian H, Ding J, Hsiao BS, Chu B, et al. Polymeric nanostructured materials for biomedical applications. Prog Polym Sci 2016;60:86–128.

[89] Aydinoğlu A, Yoruç ABH. Effects of silane-modified fillers on properties of dental composite resin. Mater Sci Eng C 2017;79:382–9.

[90] Rodríguez-Quirós HA, Casanova-Yepes HF. Effect of the functionalization of silica nanoparticles as a reinforcing agent on dental composite materials. Rev Fac Ing Univ Antioquia 2015;75:36–44.

[91] Bergman L. Influence of surface functionalization on the behavior of silica nanoparticles in biological systems [Thesis]. Åbo, Finland: Åbo Akademi University; 2014.

[92] Angioni S, Millia L, Bruni G, Tealdi C, Mustarelli P, Quartarone E. Improving the performances of Nafion™-based membranes for microbial fuel cells with silica-based, organically-functionalized mesostructured fillers. J Power Sources 2016;334:120–7.

[93] Griffin M, Nayyer L, Butler PE, Palgrave RG, Seifalian AM, Kalaskar DM. Development of mechano-responsive polymeric scaffolds using functionalized silica nano-fillers for the control of cellular functions. Nanomedicine 2016;12:1725–33.

[94] Stanzione M, Gargiulo N, Caputo D, Liguori B, Cerruti P, Amendola E, et al. Peculiarities of vanillin release from amino-functionalized mesoporous silica embedded into biodegradable composites. Eur Polym J 2017;89:88–100.

[95] Amdjadi P, Ghasemi A, Najafi F, Nojehdehian H. Pivotal role of filler/matrix interface in dental composites: review. Biomed Res 2017;28:1054–65.

[96] Blitz JP, Little C, editors. Fundamental and applied aspects of chemically modified surfaces. Cambridge: Royal Soc Chem; 1999.

[97] Gun'ko VM, Zarko VI, Voronin EF, Turov VV, Mironyuk IF. Gerashchenko II, et al, Impact of some organics on structural and adsorptive characteristics of fumed silica in different media. Langmuir 2002;18:581–96.

[98] Gun'ko VM, Mironyuk IF, Zarko VI, Turov VV, Voronin EF, Pakhlov EM, et al. Fumed silicas possessing different morphology and hydrophilicity. J Colloid Interface Sci 2001;242:90–103.

[99] Gun'ko VM, Turov VV, Myronyuk IF, Goncharuk OV, Pakhlov EM, Bezruka NA, et al. Interfacial phenomena at a surface of partially silylated nanosilica. J Colloid Interface Sci 2014;434:28–39.

[100] Leboda R, Gun'ko VM, Skubiszewska-Zięba J, Gierak A, Oleszczuk P. Properties of thin polythylene glycol layers on the surface of silica gel and pyrocarbon/silica gel. Effect of topography and morphology of carbon deposit. Mater Chem Phys 2001;70:25–37.

[101] Sanchez C, Laine RM, Yang S, Brinker CJ. Organic/inorganic hybrid materials, vol. 726. Warrendale, PA: Materials Research Society; 2002.

[102] Nakatani AI, Hjelm RJ, Gerspacher M, Krishnamoorti R. Filled and nanocomposite polymer materials, vol. 661. Warrendale, PA: Materials Research Society; 2001.

[103] Raaijmakers MJT, Benes NE. Current trends in interfacial polymerization chemistry. Prog Polym Sci 2016;63:86–142.

[104] Chakkalakal GL. Homopolymer and block copolymer composites based on silica nanoparticles coated with polymeric single or double shells synthesized by atom transfer radical polymerization [Thesis]. Kiel: Christian-Albrechts-Universität; 2011.

[105] Li J, Yu K, Qian K, Cao H, Lu X, Sun J. The situ preparation of silica nanoparticles on the surface of functionalized graphene nanoplatelet. Nanoscale Res Lett 2014;9:172 (1-9).

[106] Osei-Prempeh G, Ingles J, Keffer M, Dunlap D, Thomas G, Davari A. Influence of the polymer surface charge on the synthesis and properties of polymer – silica composites. Ind Eng Chem Res 2015;54:11295–301.

[107] Suffner J, Schechner G, Sieger H, Hahn H. In situ coating of silica nanoparticles with acrylate-based polymers. Chem Vap Depos 2007;12:459–64.

[108] Park EJ, Kim WS, Hwang HS, Park C, Lim KT. PMMA encapsulated silica particles by dispersion polymerization in supercritical carbon dioxide. Macromol Symp 2007;249–250:196–201.

[109] Monteil V, Stumbaum J, Thomann R, Mecking S. Silica/polyethylene nanocomposite particles from catalytic emulsion polymerization. Macromolecules 2006;39:2056–62.

[110] Shimada T, Aoki K, Shinoda Y, Nakamura T, Tokunaga N, Inagaki S, et al. Functionalization on silica gel with allylsilanes. A new method of covalent attachment of organic functional groups on silica gel. J Am Chem Soc 2003;16:4688–9.

[111] Gu S, Kondo T, Konno M. Preparation of silica-polystyrene core-shell particles up to micron sizes. J Colloid Interface Sci 2004;272:314–20.

[112] Rhee SH, Choi JY. Preparation of a bioactive poly(methyl methacrylate)/silica nanocomposite. J Am Ceram Soc 2002;85:1318–20.

[113] Kang S, Hong SI, Choe CR, Park M, Rim S, Kim J. Preparation and characterization of epoxy composites filled with functionalized nanosilica particles obtained via sol-gel process. Polymer 2001;42:879–87.

[114] Campos JM, Lourenc JP, Cramail H, Ribeiro MR. Nanostructured silica materials in olefin polymerisation: From catalytic behaviour to polymer characteristics. Prog Polym Sci 2012;37:1764–804.

[115] von der Ehe C, Weber C, Gottschaldt M, Schubert US. Immobilized glycopolymers: synthesis, methods and applications. Prog Polym Sci 2016;57:64–102.

[116] Claesson EM, Philipse AP. Thiol-functionalized silica colloids, grains, and membranes for irreversible adsorption of metal(oxide) nanoparticles. Colloids Surf A: Physicochem Eng Aspects 2007;297:46–54.

[117] Genest A, Portinha D, Fleury E. The aza-Michael reaction as an alternative strategy to generate advanced silicon-based (macro)molecules and materials. Prog Polym Sci 2017;72:61–110.

[118] Hanemann T, Szabó DV. Polymer-nanoparticle composites: from synthesis to modern applications. Materials 2010;3:3468–517.

[119] Rahman IA, Padavettan V. Synthesis of silica nanoparticles by sol-gel: size-dependent properties, surface modification, and applications in silica-polymer nanocomposites – a review. J Nanomater 2012;2012 132424(1-15).

[120] Dian J, Konečný M, Broncová G, Kronďák M, Matolínová I. Electrochemical fabrication and characterization of porous silicon/polypyrrole composites and chemical sensing of organic vapors. Int J Electrochem Sci 2013;8:1559–72.

[121] Gupta S, Ramamurthy PC, Madras G. Future scope of silicone polymer based functionalized nanocomposites for device packaging: a mini review. J Chem Eng Process Technol 2015;6 1000213(1–7).

[122] Xu X, Asher SA. Synthesis and utilization of monodisperse hollow polymeric particles in photonic crystals. J Am Chem Soc 2004;126:7940–5.

[123] Stamatopoulou C, Klonos P, Bondaruk O, Koutsoumpis S, Gun'ko V, Pissis P, et al. Hydrophilic nanocomposites based on polyurethane/poly(2-hydroxyethyl methacrylate) semi-IPNs and modified/unmodified nanosilica for biomedical applications. J Polym Sci B: Polym Phys 2014;52:397–408.

[124] García N, Corrales T, Guzmán J, Tiemblo P. Understanding the role of nanosilica particle surface in the thermal degradation of nanosilica- poly(methyl methacrylate) solution-blended nanocomposites: from low to high silica concentration. Polym Degrad Stab 2007;92:635–43.

[125] Ye S, Azarnoush S, Smith IR, Cramer NB, Stansbury JW, Bowman CN. Using hyperbranched oligomer functionalized glass fillers to reduce shrinkage stress. Dent Mater 2012;28:1004–11.

[126] Avella M, Bondioli F, Cannillo V, Pace E, Errico ME, Ferrari AM, et al. Poly(ε-caprolactone)-based nanocomposites: Influence of compatibilization on properties of poly(ε- caprolactone)–silica nanocomposites. Compos Sci Technol 2006;66:886–94.

[127] Karger-Kocsis J, Fakirov S, editors. Nano- and micro-mechanics of polymer blends and composites. Munich: Hanser Publishers; 2009.

[128] Wu CL, Zhang MQ, Rong MZ, Friedrich K. Silica nanoparticles filled polypropylene: effects of particle surface treatment, matrix ductility and particle species on mechanical performance of the composites. Compos Sci Technol 2005;65:635–45.

[129] Cai LF, Huang XB, Rong MZ, Ruan WH, Zhang MQ. Effect of grafted polymeric foaming agent on the structure and properties of nano-silica/polypropylene composites. Polymer 2006;47:7043–50.

[130] Bansal A, Li C, Yang H, Benicewicz BC, Kumar SK, Schadler LS. Glass transition behavior of polystyrene filled with surface modified silica nanoparticles. PMSE Prepr 2005;92:260–1.

[131] Hsieh TH, Ho KS, Bi X, Han YK, Chen ZL, Hsu CH, et al. Synthesis and electromagnetic properties of polyaniline-coated silica/maghemite nanoparticles. Eur Polym J 2009;45:613–20.

[132] Gun'ko VM, Voronin EF, Zarko VI, Pakhlov EM, Chuiko AA. Modification of some oxides by organic and organosilicon compounds. J Adhes Sci Technol 1997;11:627–53.

[133] Sulym I, Goncharuk O, Sternik D, Terpilowski K, Derylo-Marczewska A, Borysenko MV, et al. Nanooxide/polymer composites with silica@PDMS and ceria-zirconia-silica@PDMS: textural, morphological and hydrophilic/hydrophobic features. Nanoscale Res Lett 2017;12:152.

[134] Gun'ko VM, Leboda R, Skubiszewska-Zięba J, Goncharuk EV, Nychiporuk YM, Zarko VI, et al. Influence of different treatments on characteristics of nanooxide powders alone or with adsorbed polar polymers or proteins. Powder Technol 2008;187:146–58.

[135] Gun'ko VM, Borysenko MV, Pissis P, Spanoudaki A, Shinyashiki N, Sulim IY, et al. Polydimethylsiloxane at the interfaces of fumed silica and zirconia/fumed silica. Appl Surf Sci 2007;253:7143–56.

[136] Pedrazzoli D, Pegoretti A, Kalaitzidou K. Interfacial interactions in silica-reinforced polypropylene nanocomposites and their impact on the mechanical properties. Polym Compos 2016;37:2018–26.

[137] Uygunoglu T, Gunes I, Brostow W. Physical and mechanical properties of polymer composites with high content of wastes including boron. Mater Res 2015;18:1188–96.

[138] Rutz BH. Improvement of mechanical properties of polymeric composites [Thesis]. Washington: University of Washington; 2014.

[139] Cho J, Joshi MS, Sun CT. Effect of inclusion size on mechanical properties of polymeric composites with micro and nano particles. Compos Sci Technol 2006;66:1941–52.

[140] Haeri SZ, Asghari M, Ramezanzadeh B. Enhancement of the mechanical properties of an epoxy composite through inclusion of graphene oxide nanosheets functionalized with silica nanoparticles through one and two steps sol-gel routes. Prog Org Coat 2017;111:1–12.

[141] Bian J, Wang ZJ, Lin HL, Zhou X, Xiao WQ, Zhao XW. Thermal and mechanical properties of polypropylene nanocomposites reinforced with nano-SiO_2 functionalized graphene oxide. Compos Part A 2017;97:120–7.

[142] Casalini R, Roland CM. Local and global dynamics in polypropylene glycol/silica composites. Macromolecules 2016;49:3919–24.

[143] Gun'ko VM, Zarko VI, Goncharuk EV, Andriyko LS, Turov VV, Nychiporuk YM, et al. TSDC spectroscopy of relaxational and interfacial phenomena. Adv Colloid Interface Sci 2007;131:1–89.

[144] Klonos P, Kulyk K, Borysenko M, Gun'ko V, Kyritsis A, Pissis P. Effects of molecular weight below the entanglement threshold on interfacial nanoparticles/polymer dynamics. Macromolecules 2016;49:9457–73.

[145] Klonos P, Sulym IYa, Kyriakos K, Vangelidis I, Zidropoulos S, Sternik D, et al. Interfacial phenomena in core–shell nanocomposites of PDMS adsorbed onto low specific surface area fumed silica nanooxides: effects of surface modifications. Polymer 2015;68:158–67.

[146] Klonos P, Sulym IYa, Borysenko MV, Gun'ko VM, Kripotou S, Kyritsis A, et al. Interfacial interactions and complex segmental dynamics in systems based on silica-polydimethylsiloxane core-shell nanoparticles: dielectric and thermal study. Polymer 2015;58:9–21.

[147] Sulym IYa, Klonos P, Borysenko MV, Pissis P, Gun'ko VM. Dielectric and thermal studies of segmental dynamic in silica/PDMS and silica/titania/PDMS nanocomposites. J Appl Polym Sci 2014;131:41154.

[148] Klonos P, Pissis P, Gun'ko VM, Kyritsis A, Guzenko NV, Pakhlov EM, et al. Interaction of poly(ethylene glycol) with fumed silica and alumina/silica/titania. Colloids Surf A: Physicochem Eng Aspects 2010;360:220–31.

[149] Galaburda MV, Klonos P, Gun'ko VM, Bogatyrov VM, Borysenko MV, Pissis P. Dielectric properties and thermal destruction of poly(dimethylsiloxane)/Fe_2O_3/SiO_2 nanocomposites. Appl Surf Sci 2014;305:67–76.

[150] Klonos P, Kaprinis S, Zarko VI, Peoglos V, Pakhlov EM, Pissis P, et al. Thermal and dielectric studies of PEG/C/AST nanocomposites. J Appl Pol Sci 2013;128:1601–15.

[151] Roy M, Nelson JK, Maccrone RK, Schadler LS, Reed CW, Keefe R, et al. Polymer nanocomposite dielectrics – the role of the interface. IEEE Trans Dielectr Electr Insul 2005;12:629–43.

[152] Meng H, Li G. A review of stimuli-responsive shape memory polymer composites. Polymer 2013;54:2199–221.

[153] Cangialosi D, Alegría A, Colmenero J. Effect of nanostructure on the thermal glass transition and physical aging in polymer materials. Prog Polym Sci 2016;54–55:128–47.

[154] Carrow JK, Gaharwar AK. Bioinspired polymeric nanocomposites for regenerative medicine. Macromol Chem Phys 2014;216:248–64.

[155] Lam SJ, Wong EHH, Boyer C, Qiao G. Antimicrobial polymeric nanoparticles. Prog Polym Sci 2017;. Available from: https://doi.org/10.1016/j.progpolymsci.2017.07.007.

5

Functionalized Clay-Containing Composites

Tomasz M. Majka, Krzysztof Pielichowski
DEPARTMENT OF CHEMISTRY AND TECHNOLOGY OF POLYMERS, CRACOW UNIVERSITY OF TECHNOLOGY, KRAKÓW, POLAND

5.1 Introduction

The mystery of nanotechnology has been progressively disclosed over the last three decades. The nanometer scale of 1–100 nm (as often defined in the literature) is a range between the micro and molecular dimensions. The sciences in these two dimensional ranges have already been well explored by chemists, physicists, and materials science experts. The real interest in nanoscience is based on new materials' properties and functions that occur at that length scale, and on wide possibilities in the design of various nanodevices [1–5].

Different materials with synergistic properties are generally chosen to produce composites with tailored physical, chemical, and mechanical properties. The great interest and demand generated for these macro- and microcomposites in scientific, technological, and structural applications over the last 60 years have finally led to the limits of optimizing composite properties, because the properties usually achieved involve compromises. To overcome these drawbacks from macro- and microscale reinforcements, nanoscale fillers, where the filler is less than 100 nm in at least one dimension, provided a radical alternative [3–10].

A plethora of polymeric nanomaterials have been developed using different types of nanofillers to achieve desired properties, for example, nanoclays are used for the enhancement of mechanical, barrier, and fire properties; carbon nanotubes and synthetic fibers are applied to improve tensile strength and conductivity; metal oxide nanoparticles usually impart antimicrobial properties, etc. There has been a great interest in studying the polymer hybrids with layered silicates after the successful preparation of a polyamide 6/clay system by the Toyota research team in the latter part of the 1980s and the beginning of the 1990s. Up to now, different methods have been elaborated for the preparation of polymer/clay nanocomposites, leading to incorporation through the layered silicates at the molecular level in the polymer matrix [2,11–15].

Functionalized clay-containing polymer nanocomposites exhibit not only markedly improved mechanical properties, but also significant enhancements in thermal, optical, and physicochemical properties, when compared with the pure polymer or conventional

Polymer Composites with Functionalized Nanoparticles. DOI: https://doi.org/10.1016/B978-0-12-814064-2.00005-6

micro- or macrocomposites. The potential to achieve these properties' improvements indicates the importance of these materials; however, it is still necessary to understand how the incorporation of clay influences the fundamental relationships between structure, processing, and properties. This knowledge will help to tailor and/or manipulate the properties of interest depending on the end application [2,11–15].

The wide and growing application of flammable materials in engineering and construction applications demands further research to improve the fire retardant properties. Functionalized clay-containing polymer nanocomposites are showing great promise as effective fire retardants for different types of materials. Layered silicates have been successfully used as additive fire retardants, such as sodium metasilicate pentahydrate, which has been found to delay ignition and reduce weight loss on ignited foams. Moreover, aluminosilicates have been used to reinforce carbon-fiber composites, creating inflammable materials [16–19].

In this chapter we describe protocols of clay functionalization, clay-containing nanocomposites formation and processing, examples of experimental results, and postulated interpretations of those results including the role of specific interactions between clays and other modifiers on flammability and aging in polymer/clay nanocomposites. The last section deals with the influence of modified layered silicates on human health.

5.2 Synthesis of Nanoparticles

The most widely used methods for synthesis of nanofillers are [1–7]:

- Chemical vapor deposition. In this technique, the substrate is exposed to a volatile precursor, which then reacts or decomposes on the surface of the substrate and is then obtained as the desired deposit. Materials are collected in various forms such as monocrystalline, polycrystalline, amorphous, and epitaxial.
- High-energy ball milling process. This method is used for the synthesis of nanoparticles of three types:
  - Mechanical alloying—mixtures of powders are milled together and then material transfer takes place to obtain homogenous alloy;
  - Mechanical milling—powders with uniform composition are milled together as no material transfer occurs;
  - Mechanochemical synthesis—chemical reaction between the powders takes place during the milling process at low temperature, which is far from equilibrium conditions.
- Microwave synthesis. Nonaqueous sol–gel technique is widely used for the synthesis of metallic nanoparticles. The size of the nanoparticles can be controlled by choosing the appropriate reaction conditions including temperature, reactant amount, and the reaction time.
- Polymerized complex method. In this technique, metal ions are first chelated to form complexes and are then polymerized to form a gel. Among other chemical processes, this

method is the most suitable due to homogenous dispersion of cations in the polymer network.

- Sol–gel synthesis. This method involves the production of quite high ultrapure materials at atomic scale and offers the advantage of tailoring the composition. Sol–gel synthesis is the most viable method for the production of homogenous alloys and composites in an efficient and cost-effective manner.
- Supercritical hydrothermal synthesis. This process is used mainly for the synthesis of metal oxide nanoparticles. During supercritical hydrothermal synthesis, water is used as a solvent due to its high dielectric constant.

5.3 Clays and Their Methods of Functionalization

Clays used in the production of the nanocomposites are hydrous aluminum silicates with a layered structure, with cations loosely bonded between sheets of silicon and oxygen. The clay montmorillonite (MMT) has been found to be the most suitable, but other clays as well as natural and manmade micas are also used. These alternatives include lithium fluorohectorite, synthetic fluorinated mica, and other types of layered silicates [2,8–14].

Silicates are frequently incorporated to, for example, superabsorbent polymers to improve their properties as superior properties have been found for the superabsorbent polymer hybrid materials compared with their clay-free counterparts. However, studies of the effect of silicates on the residual monomer content in hydrogels are rather scarce [15–19].

Silicate minerals belong to the nanoplatelet-type of structured filler particles (NSFPs) used in polymer composite (nano)technology. There are two types of silicates found in nature: Expanding and nonexpanding clays. The expanding clays are phyllosillicates, smectite, and MMT, and the nonexpanding clays are talc, mica, and kaolin. To prepare polymer nanohybrids, clay selection is made on the basis of its moderate surface charge, that is, cation exchange capacity (CEC) and layer morphology. The most commonly used layered silicates in the field of nanocomposites are MMT, hectorite, and saponite. MMT has the CEC 110 meqiv/100 g, hectorite 120 meqiv/100 g, and saponite 86.6 meqiv/100 g. These hydrophilic clays are miscible with hydrophilic polymers only, such as poly(ethylene oxide) (PEO) and poly(vinyl alcohol), whereby their interactions with nonpolar polymers are thermodynamically not favorable [2,5–10,15–24].

Layered silicates crystal structure is made of tetrahedral sheets, in which a silicon atom is surrounded by four oxygen atoms and octahedral sheets with a metal atom like aluminum or magnesium surrounded by eight oxygen atoms. Both the tetrahedral and octahedral sheets are fused together by sharing the oxygen atoms, whereby unshared oxygen atoms are present in hydroxyl functional form. Two main arrangements of tetrahedral and octahedral sheets fused together create a one-layered structure of clay. One tetrahedral fused to one octahedral (1:1) is known as kaolin with the general composition of $Al_2Si_2O_5(OH)_5$ and the base layer thickness of $\sim$0.7 nm. The crystal lattice consisting of one octahedral sheet sandwiched between two tetrahedral sheets (2:1) with the total thickness of 0.94 nm is known as phyllosilicate. The 2:1 phyllosilicate layers, electrostatically neutral with no interlayer ion present, are known as pyrophyllite. Due to the presence of no interlayer ions, the layers cannot expand in water. When silicon atoms in tetrahedral sheets are substituted by aluminum atoms, the 2:1 structure is called mica. Then, the negative charge induced by this substitution is balanced by the insertion of potassium cations between layers. Due to the equal size of potassium cation and the hole created by Si/Al tetrahedral sheets, the presence of potassium cation creates no interlayer gap. Consequently, the 2:1 layers are held together strongly and the swelling or exfoliation of layers is impossible. When in neutral pyrophyllite, the aluminum cations in the octahedral layers are partially substituted by divalent magnesium or iron cations, a new clay group called smectites is formed. The negative charge created by this substitution is balanced by interlayer counter sodium or calcium ions. The charge created on the layers is not locally constant and varies from layer to layer. An average charge value is considered for the layers, which is determined by CEC measurement. Due to the unequal size of interlayer cations (with the holes of tetrahedral sheets), the presence of interlayer cations causes an interlayer spacing. The layers stay very close together with a regular gap between them. This gap is called an interlayer or gallery. The thickness of the repeated units in a regular multilayer structure containing one layer and one gallery space is called d-spacing (d_{001}) or basal spacing. The d-spacing of clays can be measured or calculated from their X-ray diffraction patterns. The interlayer dimension is also dependent on the clay nature and swelling or hydration degree of interlayer cations. For instance, the interlayer distance is about 1 nm for MMT (Fig. 5–1) [2,5–10,15–27].

The electrostatic and van der Waals forces holding these layers together are relatively weak. The interlayer distance varies depending on the charge density on the layers, interlayer cation radius, and its degree of hydration. Because of the interlayer spacing and weak interlayer forces, the cations present between the layers could be hydrated in aqueous solutions, known as clay swelling, during which an increase of interlayer space occurs. The charge density on the silicate layers is different in various clay groups with different structures. Due to

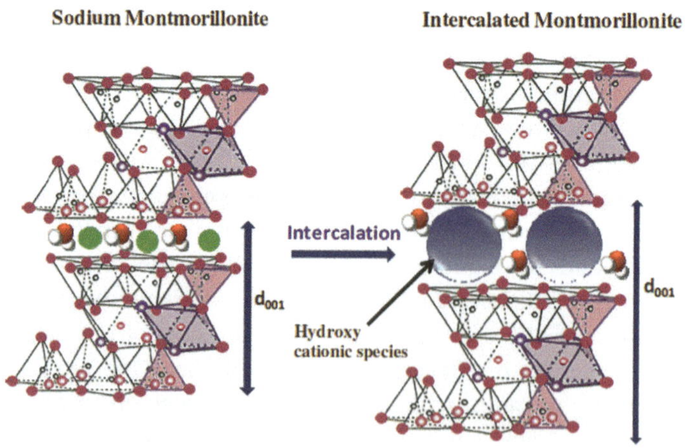

FIGURE 5–1 Schematic representation of intercalation process of montmorillonite.

the interlayer spacing and weak forces between the layers, other molecules could be intercalated between layers, leading to the expansion of layered lattice and, finally, to the separation of individual layers. The unique separation behavior of smectite minerals, which is responsible for the high aspect ratio of this clay group, makes them very efficient reinforcing fillers for polymeric matrices [15–27].

The single layer thickness is around 1 nm, and the lateral dimensions of these layers may vary from 30 nm to several microns or larger, depending on the source and preparation method of the clay. The clay lateral dimensions and its ability to disperse into separated layers determine the aspect ratio of clay, which plays an important role in polymer/clay interfacial interactions. Separation of individual clay layers having about 1 nm thickness and displaying an well-ordered structure beneficially affects polymer properties [24–34].

5.3.1 Clay Functionalization

Without layered silicates functionalization, preparation of polymer/clay nanocomposites by physical mixing of both components showing good dispersion of clay layers within the polymer matrix is practically impossible. To increase the compatibility of layered silicates with hydrophobic polymers, it is necessary to make these silicates organophilic. This is surface functionalization, which causes the reduction of surface energy of clay layers and matches their surface polarity with the matrix polarity. The organoclays with lowered surface energy are more compatible with polymer matrices and polymer molecules are able to intercalate within their interlayer space or galleries under well-defined experimental conditions [24–34].

The surface functionalization of layered silicates could be achieved by exchanging the cation present in the clay layers with the cationic surfactants such as quaternary alkylammonium or alkylphosphonium ions. In addition to the surface functionalization and increasing the hydrophobicity of mineral layers, the insertion of organosalt cations into the galleries causes an increase in the interlayer spacing, which promotes the penetration of macromolecules into the galleries during the nanocomposites' preparation [24–27].

The ion exchange process is carried out in aqueous solution with the hydrated interlayer cations. The affinity of monovalent sodium cations to hydration promotes the ion exchange process and increases the efficiency of organic functionalization of mineral layers. Layered silicates containing divalent counter cations such as calcium could not be easily hydrated and therefore their replacement by ion exchange process is not so efficient. Therefore, in these minerals, the divalent cations should be ion exchanged with easily exchangeable sodium or potassium cations prior to functionalization by onium cations [34–41].

The efficiency of organic functionalization by ion exchange process in the increasing of basal spacing and consequently the exfoliation of layers and formation of stable nanocomposite systems depends also on the surface charge of silicate layers. The surface charge density of these layers depends on the nature of the clay and its preparation conditions. When the charge density of mineral layer is as high as c. 1 equiv/mol for mica, the electrostatic forces between layers are also high. Therefore the interlayer cations could not be swollen and consequently the ion exchange process could not be carried out. The medium charge densities (0.5–0.8 equiv/mol), such as in vermiculite, could make a high potential for hydration of cations. However, the divalent cations present in galleries of this layered silicate make the swelling process difficult due to the lower affinity of divalent cations for hydration. The lower charge densities (0.25–0.5 equiv/mol) such as these in MMT correspond to the weak electrostatic forces between silicate layers [24–27,36–44]. Moreover, the presence of potassium or sodium cations between the MMT layers makes this kind of silicate more suitable for swelling, organic functionalization, and exfoliation. The structure of the organoclays and their basal spacing depends also on the molecular size of organic onium and its configuration in the gallery after functionalization [24–27].

It is clear that the charge density of mineral layers, the efficiency of the ion exchange process, and the spatial characteristics of the onium cation are effective on the configuration of onium cations in the modified silicates. The organic tallow of onium laid up parallel to the silicate layer could be obtained with lower charge densities. By increasing the charge density of silicate layers, bilayer, or even trilayer structures with increased gallery spaces could be formed (Fig. 5–2). Finally, with the higher charge densities, a larger interlayer gap is

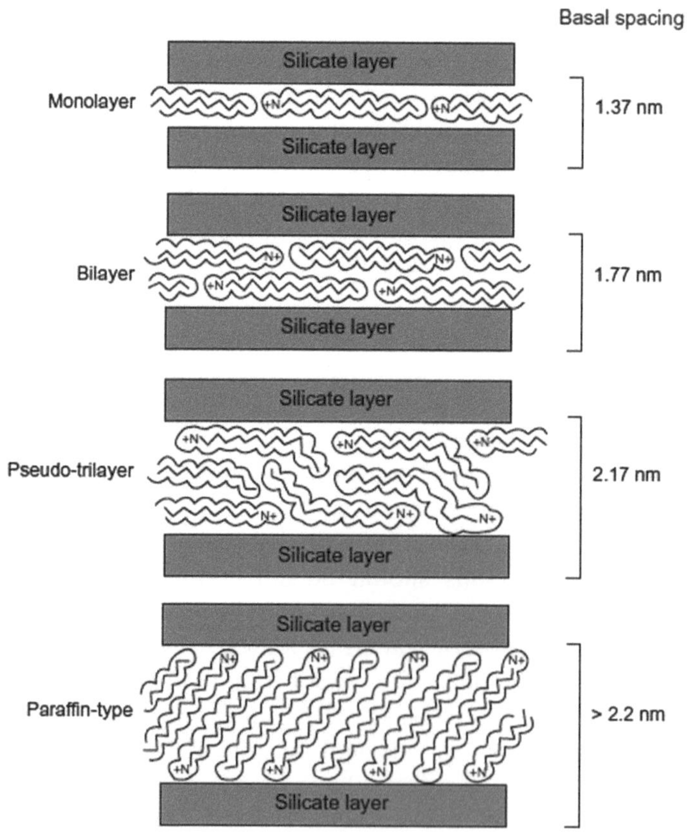

FIGURE 5–2 Possible arrangements of long-chain alkylammonium ions in the interlayer space of clay [30].

obtained by the formation of a paraffin structure. Also an increase in temperature could lead to the structures with larger basal spacing by increasing the mobility of onium molecules into the galleries. However, at a given temperature two parameters determine the equilibrium layer spacing—the charge density of layers, driving the packing and number of onium chains into the interlayer space, and the size or chain length of organic tallow in the onium compound [39–44].

5.3.2 Silicates Functionalization for Fire Retardant Applications

Recently, Majka et al. have proposed a clay functionalization technique to reduce the flammability of PA6/clay nanocomposites, utilizing adsorption of a phosphorous-containing compound (disodium H-phosphonate, DHP) on the clay surface [26].

For the adsorption of a stabilizer on the sodium MMT the following procedure was used. The clay and DHP ratio has been optimized and selected based on mechanical properties improvements, as well as flame retardant effects. The stabilizing compound DHP (0.75 wt.% of the total composition) was dissolved in distilled water. Then MMT (3 wt.% of the total composition) was added, the mixture was left for 24 hours at room temperature. Then the temperature was increased to 50°C and the mixture was stirred for 5 hours. The dispersion was left to cool down and was stirred for further 19 hours with a magnetic stirrer. The next step included evaporation of the solvent using an evaporator and drying in an oven at 100°C for 48 hours. In a further step, the samples were frozen in liquid nitrogen and freeze-dried over 7 days. The adsorption of the DHP on organomodified MMT required the use of a slightly modified procedure [26].

Surface functionalization of MMT by DHP, investigated by BET method, led to an increase in pore sizes and pore volumes for MMT-DHP and OMMT-DHP systems confirming physical interaction between clay and the stabilizer (Fig. 5–3). The morphology of polyamide six composites was shown via XRD analysis to be of intercalated type. Physical mixing of DHP with other components, clays, and PA6, during high-temperature processing did not give such high overall thermal stabilization effect (OSE) values as adsorption of stabilizer on organoclay surface. Surface modified MMT protects the polymeric material against the degradation within the temperature range used (at 100°C–150°C against thermal aging) and at the polymer processing temperatures (230°C–250°C). Samples containing OMMT/DHP and OMMT-DHP showed the best results in cone test, in particular in reducing peak heat release rate (PHRR). The addition of the DHP reduced the PHRR of the composite by 53% in the case of OMMT/DHP/PA6 and 59% in the case of OMMT-DHP/PA6, compared with the neat PA6 [26].

5.4 Functionalized Clay-Containing Polymer Nanocomposites Formation

Although studies on layered silicates/polymer intercalation can be traced back to before the 1970s, these developments should not be taken into account in the history of clay-containing polymer nanocomposites as the work did not result in substantial improvement in the

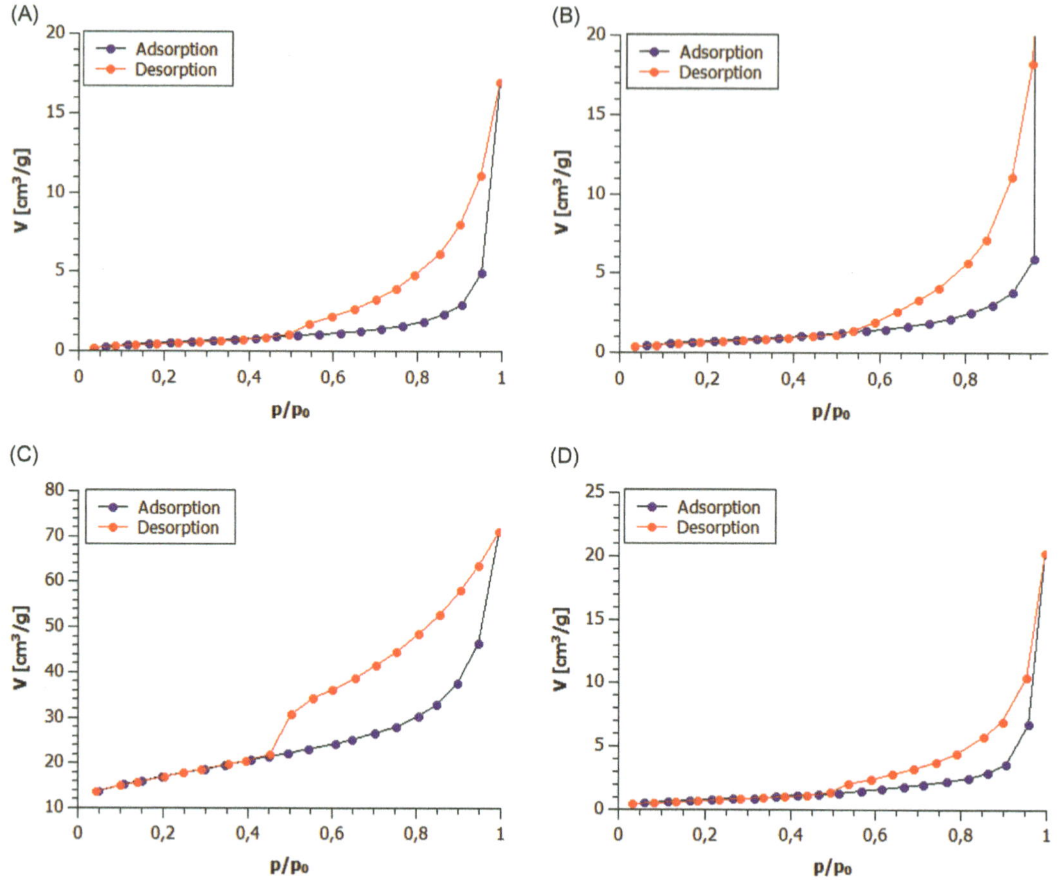

FIGURE 5–3 BET adsorption isotherms for (A) OMMT; (B) OMMT-DHP; (C) MMT; and (D) MMT-DHP [26].

physical and engineering properties of polymeric materials. At present, development has been widened into almost every engineering polymer including polyethylene (PE), poly(ethylene oxide) (PEO), poly(ethylene terephthalate) (PET), polyimide (PI), polylactide (PLA), poly(methyl methacrylate) (PMMA), polypropylene (PP), polystyrene (PS), polyurethane (PU), poly(vinyl chloride) (PVC), poly(4-vinylpyridine) (PVP), acrylonitrile-butadiene-styrene (ABS) copolymer, ethylene-vinyl acetate copolymer (EVA), polyacrylonitrile (PAN), polycarbonate (PC), polycaprolactone (PCL), epoxy resin, phenolic resin, and various rubbers [1–8,11–16,22–27,32–38,45–48].

To obtain polymer nanocomposites four main preparation methods are utilized.

1. Sol–gel synthesis

   In this technique, the clay minerals are synthesized within the macromolecular matrix, using an aqueous solution (or gel) containing the polymer and the silicate building blocks. As precursors for the clay silica sol, magnesium hydroxide sol, and lithium fluoride are used. During the process, the polymer aids the nucleation and growth of the inorganic host crystals and gets trapped within the layers as they grow. Although this technique has the potential of promoting the dispersion of the silicate layers in a one-step process, without needing the presence of the onium ion, it presents serious disadvantages. First of all, the synthesis of clay minerals generally requires high temperatures, at which the polymer matrices could decompose. Another problem is the aggregation tendency of the growing layered silicates. Therefore this technique is far less developed for layered silicates [1–8,11–16,22–27,32–38,45–48].

2. Solution Intercalation

   Following this technique, the layered silicate is exfoliated into single layers using a solvent in which the polymer (or prepolymer) is soluble. It is well known that such layered silicates, owing to the weak forces that stack the layers together, could be easily dispersed in an adequate solvent,for example, water, acetone, chloroform, or tetrahydrofurane. After the modified silicate has swollen in the solvent, the polymer is added to the solution and intercalates between the clay layers. The final step consists of removing the solvent by vaporization, usually under vacuum. Upon solvent removal the sheets reassemble, sandwiching the matrix to form a nanocomposite structure. This technique may be applied to fabricate nanocomposites obtained through the emulsion polymerization, where the layered silicate is dispersed in the aqueous phase. The major advantage of this method is that intercalated nanocomposites based on polymers with low polarity could be synthesized. However, the solvent approach is difficult to apply in industry owing to problems associated with the use of large amounts of solvents. However, this method found broad application in the electrospinning technique by which polymer nanofibers with controlled geometries can be produced [1–8,11–16,22–27,32–38,45–48].

3. In Situ Intercalative Polymerization

   In situ intercalative polymerization technique was the first method used to synthesize clay-containing polymer nanocomposites based on PA6. In this technique, the modified layered silicate is swollen by a liquid monomer or a monomer solution. The monomer migrates into the galleries of the layered silicate, so that the polymerization reaction could occur between the intercalated sheets. The reaction could be initiated either by heat or radiation, by the diffusion of a suitable initiator or by an organic initiator or catalyst fixed through cationic exchange inside the interlayer before the swelling step by the monomer. Polymerization produces long-chain macromolecules within the clay galleries [1–8,11–16,22–27,32–38,45–51].

4. Direct Melt Intercalation

   This technique consists of blending the layered silicate with the polymer matrix in the molten state. Under such conditions—if the layer surfaces are sufficiently compatible with the chosen polymer—the polymer could crawl into the interlayer space and form either an intercalated or an exfoliated nanocomposite [1–8,11–16,22–27,32–38,45–51]. This technique is vastly applied on industrial scale using the polymer processing methods, especially double-screw extrusion.

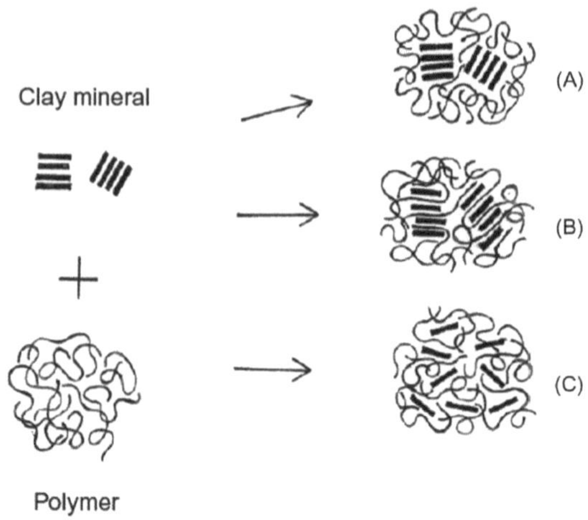

FIGURE 5–4 Different composite structures: (A) separated phase microcomposite; (B) intercalated structure; and (C) exfoliated (delaminated) structure [30].

5.5 Types of Nanocomposites Structures

Types of structures of clay-containing polymer nanocomposites are classified according to the level of intercalation and exfoliation of polymer chains into the silicate galleries. Many parameters including clay nature, organic modifier, types of matrix, and preparation technique affect the intercalation and exfoliation process [1–6,10–16,20–28,33–39,42–54]. Therefore depending on the nature and properties of silicate phase and matrix as well as preparation methodology of nanocomposite, different composite structures could be obtained (Fig. 5–4).

5.5.1 Phase-Separated Microcomposite Structure

When the organic polymer is melt-processed with inorganic clay, such as unmodified silicate, the macromolecules are unable to intercalate within the clay layers and the silicate is dispersed as aggregates or particles with layers stacked together within the polymer matrix. The obtained composite structure is considered as phase-separated. XRD diffractograms of investigated composites show main peaks in the 5–15 angle range of 2θ, which suggests the dispersion of agglomerated microsized particles in the polymer matrix (Fig. 5–4A) [1–6,10–16,20–28,33–39,42–54].

5.5.2 Intercalated Structure

When one or more macrochains are inserted into the gallery and cause an increase of the interlayer spacing, but the periodic array of the clay layer still exists, the intercalated nanocomposite is formed. The presence of macromolecules in the galleries leads to a decrease of electrostatic forces between the layers, but the macrochains are not totally separated. Well-ordered multilayer hybrid morphology, governed by the interactions between the polymer chains and clay layers, is obtained in this configuration (Fig. 5–4B) [1–6,10–16,20–28,33–39,42–55].

5.5.3 Exfoliated (Delaminated) Structure

Delaminated structure is obtained when the insertion of macrochains into the clay galleries results in the separation of the layers into individual layers, which are then dispersed within the polymer matrix. In general, when the polymeric chains cause an increase of interlayer spacing of more than 80–100 Å, the exfoliated structure is obtained (Fig. 5–4C). Due to the well-dispersed state of individual clay layers, high aspect ratio is obtained and lower silicate content is needed for exfoliated nanocomposites. The most significant improvement in polymer properties is obtained for exfoliated structures due to the large surface interactions between the matrix and the filler [20–28,33–39,42–55].

5.6 Functionalized Clays as Fire Retardants in Polymer Nanohybrids

Because of the large use of polymer materials especially in domestic and construction applications, there is a need to reduce their potential for ignition. Conventional flame retardants, such as those with phosphorus, are vastly used to retard the ignition process and to control the burning behavior of polymer-based materials without reducing other properties and overall quality. [56–61]. However, some inorganic flame retardants, for example, aluminum trihydroxide or magnesium dihydroxide, require high level of loading, which causes some undesired effects on the processability and on the quality of the final product.

The fire retardant properties of a range of clays on large scale produced polymers, including PP, polyamide, and PS, have been investigated. Key evidence for the resilience of clay-containing polymer nanocomposites to combustion include [62–64]:

- an increase in the heat distortion temperature;
- an increase in the peak heat release temperature;
- a change in char structure;
- a decrease in char permeability;
- a decrease in the rate of mass loss;
- a tendency to self-extinguish.

MMT has been found to be the most suitable filler as fire retardant for polymer materials, but other clays as well as natural and manmade micas are also used, for example, lithium fluorohectorite or synthetic fluorinated mica. It has been found that at low silicate concentrations the reduction of PHRR is proportional to the fraction of layered silicate used. When the material is heated (before ignition), usually a thin layer of char is formed. The macromolecular char layer often breaks down early in combustion but the nanocomposite char could not fracture. Instead, a tough layer of char—a multilayered silicate structure—is formed and remains stable throughout the whole combustion process. The dimensions of the hybrid sample are largely unaffected by combustion, and the low permeability of the char limited the movement of both gases and liquids through the barrier formed. It is also noted that the rate of mass loss from a polymer–silicate hybrid (when exposed to heat) is significantly reduced compared with the neat polymer. This difference comes into effect after a couple of seconds or even minutes of combustion, when the surface of the silicate–polymer has time to form char on the outer layer of the polymeric material. Heat transfer from an external source promotes thermal decomposition of the organoclay and the matrix. The carbonaceous char formed superficially during combustion is rich in silicates and could be viewed as a sort of ceramic char–layered silicate nanocomposite. Actually, the nanocomposite structure allows the residue formation, which acts as a protective barrier by reducing the heat and mass transfer between the flame and the matrix. That is, the char insulates the underlying matrix from heat and also slows oxygen uptake and the escape of volatile gases produced by polymer degradation. Both these actions interfere with the combustion cycle by reducing the amount of fuel available for combustion and burning. In neat polymer, the thermal degradation continues during the combustion, and all intermediate degradation products volatilize completely. On the other hand, there might be physicochemical adsorption of the intermediate degradation products on filler surfaces in clay-containing polymer nanocomposites. This adsorption may delay volatilization of the degradation products and promote accumulation of the incomplete degradation products on the filler. It is still believed that the excess quaternary ammonium salts used to disperse the silicates may increase the probability of early ignition point [65–76].

It was also observed that when MMT and intumescent fire retardants are combined in a polymer matrix, their synergistic effect is related to their composition ratio. The reason is that ammonia is the main gaseous product, which volatilizes. NH_3 swells phosphor-carbonaceous materials, leading to the formation of the intumescent carbonaceous char residue. Therefore, when the mass of MMT is increased, the negative effect may exceed the positive one of the fire retarding property [62–79].

However, the clay minerals and traditional flame retardants, both halogenated flame retardants [80,81] and phosphorous flame retardants [82,83], may show synergistic effects in the reduction of ignitability of polymer materials. Importantly, the synergistic effect of layered silicates on the flame retardancy of halogenated and phosphorous compounds leads to the reduction of classical FRs mass fraction needed if no nanoadditive is applied. Among different attempts, intercalation of phosphorous flame retardants into the mineral interlayer gaps seems to be a promising approach—the phosphorous flame retardants are shielded by the silicate sheets, which prevent or limit their evaporation during the melt compounding at elevated temperature [84].

Nowadays, it is generally considered that in the polymeric composites nanoadditives enhance the char formation, reinforce its structure, and improve the barrier properties. The char produced insulates the underlying matrix from heat and forms physical barrier, slowing the escape of the volatile gases produced by high-temperature polymer degradation and limiting oxygen access to the burning zone [85–89].

The initial tests on clay-containing polymer nanocomposites show they may have distinct advantages over traditional composites in FR tests. In addition, low concentrations of silicate are necessary in nanocomposites, compared with the amount of additive used in microcomposites, allowing thus easier processing of the former by extrusion, injection molding, or even infusion casting [90,91].

It is well known that the clay-containing polymer nanocomposites are an environmentally friendly alternative to some types of fire retardants, as they contain no additional halogens, phosphates, or aromatics. The silicate remains intact or develops a char barrier at drastic conditions, such as high temperatures [92–94].

Traditional fillers very often severely degrade the physical properties of the matrix, whereby in clay-containing polymer nanocomposites improvements of different physical properties of the macromolecular matrix, including an increase in the dynamic storage modulus and gas permeability, have been observed [94–98].

The most significant advantage offered by clay-containing polymer nanocomposites is the reduction in PHRR. However, compared with the virgin material, the time to ignition (TTI) is usually lowered while the total heat released (THR) remains the same. This means that the nanocomposite ignites faster than the virgin polymer and that, like the neat polymer, the composite is completely combusted [85–89,92–94]. A major limitation of clay-containing polymer nanocomposites is that they only work in the condensed phase and probably do not inhibit the flame propagation in the gas phase [92–94]. Therefore these materials may perform poorly when tested under industrially significant or regulatory fire safety tests. As mentioned previously, the excess of quaternary ammonium salts commonly used to disperse the clays also increases the probability of early ignition [94–98].

For some systems, antagonistic effects were observed with the PHRR and THR increasing for the clay-containing polymer composites with fire retardant system, as compared with the epoxy resin with fire retardant system as a control sample. Morgan [58,62–64] indicated that the observed antagonism could have been due to the lack of uniform silicate dispersion. On the other hand, clay-containing polymer nanocomposites inhibit polymer flow and dripping during flaming combustion, which is a desirable effect for materials used in, for example, the building industry. In general, the synergism reported between clays and conventional fire retardants provides an alternative for improving fire performance without seriously deteriorating appearance and mechanical behavior, and even opens the possibility of formulating self-extinguishing clay-containing nanomaterials.

5.7 Barrier Properties of Functionalized Layered Silicates

One of the most considerable effects of layered silicates on polymer matrix properties is a considerable improvement of barrier properties. Silicate sheets are naturally impermeable, so their role is to increase the barrier properties of macromolecules by creating a maze or tortuous path that retards the diffusion of gas molecules through the matrix. The degree of enhancement in the barrier properties depends on the ability of mineral layers to limit the diffusion of volatiles through the polymer composite. The "tortuous factor" is determined by the ratio of actual walking distance of diffusive molecule to the polymer phase thickness. This factor is affected by the aspect ratio of nanofiller dispersed in the matrix. Increasing the lateral length of the nanofiller sheet as well as increasing the exfoliation or dispersion degree causes larger barrier enhancements in the matrix phase [99–103].

Clays are capable to increase the barrier properties by creating a maze or "tortuous path" that retards the progress of the gas molecules through the matrix (Fig. 5–5). The direct benefit of the formation of such a path is clearly observed in clay-containing PI nanocomposites with substantially improved barrier properties, characterized by a simultaneous decrease in the thermal expansion coefficient. The clay-containing PI nanocomposites with a small fraction of organoclays exhibited reduction in the permeability of gases like O_2, H_2O, He, and CO_2 [6,104–108]. At 2 wt.% nanofiller loading, the permeability coefficient of water vapor was decreased 10 times with synthetic mica relative to neat polymer. By comparing polymer nanomaterials created with layered silicates of various aspect ratios, the permeability was seen to decrease with an aspect ratio increase.

In general, the reduction of permeability arises from the longer diffusive path that the penetrants must overcome in the presence of filler particles. A sheet-like morphology is efficient at maximizing the path length due to the large length-to-width ratio, when compared with other filler shapes such as spheres or cubes [6,104–108].

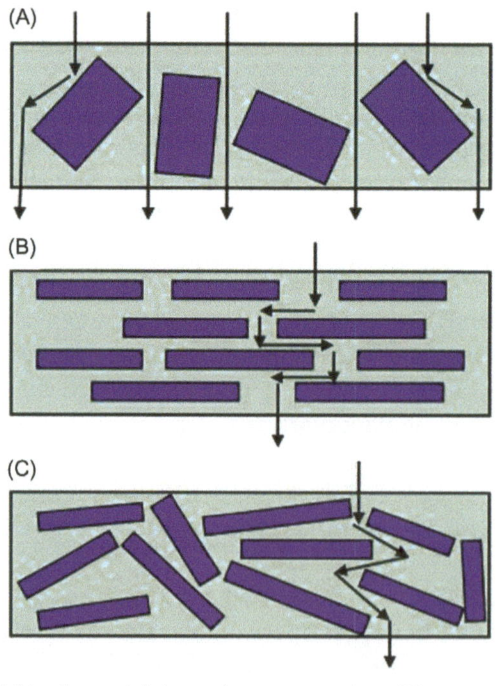

FIGURE 5–5 Tortuous path model in clay-containing polymer composites: (A) conventional composites; (B) intercalated nanocomposites; and (C) exfoliated nanocomposites.

Liu and Wu [109], during their studies of water absorption of PA66 and corresponding clay-containing nanocomposites, have found that with increasing clay content, the water absorption at saturation decreases rapidly from 7.6% for PA66 to 5.2% for the nanocomposite containing 5 wt% of silicate. However, for PA66/clay (10 wt.%) the water absorption was 4.7%. They attributed this reduction to the presence of immobilized and partially immobilized macromolecules in the amorphous phase as the interactions of polymer chains with high surface area clays limit the chain mobility and cause formation of "tightly" and "loosely" bound polymer. Impermeable layered silicate particles lower the overall rate of water transport in the polymer matrix, according to the tortuous path model.

Although a decrease of diffusivity is a well-established result of clay-containing polymer composites formation, contradictory results are reported concerning the saturation uptake values of various gases or solvents. It is worth noticing that in clay-containing polymer nanocomposites the coexistence of phases with different permeabilities could cause complex transport phenomena. The organophilic filler gives rise to superficial adsorption, whereby in a two-phase (crystalline-amorphous) polymer system, the crystalline regions are generally impermeable to penetrant molecules [110–116]. The influence of changes in matrix crystallinity and chain mobility, induced by the presence of the nanofiller, is still the subject of investigations.

5.8 The Influence of Functional Clays on Aging of Polymer Composites

Along with the development of nanotechnology, there has been an increased demand for nanomaterials with higher temperature classification and better thermal performance. Outdoor application stability is a critical issue for the commercial exploitation of the end products. The useful lifetime of a composite material at a given service temperature, is often assessed by determining the time required for a particular property to decrease to the critical value [117,118]. The materials' properties are subject to changes up to the point where they are no longer capable of fulfilling their functions. The comparative prediction of the lifetime of clay-containing polymer composite is often based on accelerated aging tests, which involve combined heat, humidity, UV exposure, etc. to speed up the aging processes. These standard test methods enable determination of the long-term effects of composite materials in a considerably shorter time than natural outdoor exposure tests, such as the Florida test.

However, the behavior of clay-containing polymer nanocomposites under natural and artificial conditions is determined by several factors. Complex polymer products are usually exposed to a variety of weather conditions, such as sunlight, temperature changes, and rain/snow fall. Often the experiments are held in artificial simulated environments to reduce testing time, but one should bear in mind that the results may differ from those in natural conditions due to interactions between atmospheric agents and additives in the stabilized polymer materials [119]. The chemical and physical aging mechanisms of pristine resin may operate simultaneously: The former affects mainly the molecular structure of matrix, while

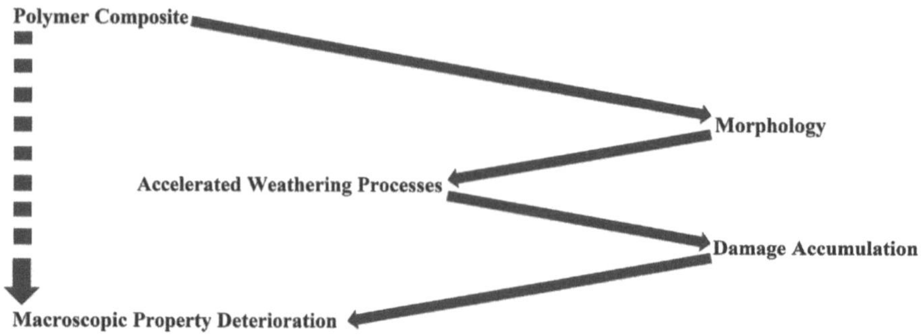

FIGURE 5–6 A scheme for overall task of predicting service lifetime of hybrid materials.

the physical processes are related to the diffusion of the additives toward the surface [120]. The weathering of the clay-containing polymer composites includes photooxidation and cross-linking reactions, which result in significant functionalization of the macromolecular structure. Under environmental conditions, the degradation consists of a stepwise flow out of inorganic components and plasticizer, while under thermal aging conditions, the rearrangement and aggregation of the molecular chains due to annealing are considered as the prevailing phenomena [120–124].

Thermal aging and accelerated weathering of clay-containing polymer nanocomposites toward life predictions usually include tests such as yellowing or discoloration, kinetics of thermal decomposition, mechanical and morphological changes after heat treatment, and determination of volatile products using hyphenated techniques, such as TG-FTIR (Fig. 5–6).

5.8.1 Structural Changes of Polymer/Clay Composites During Thermal Aging

FTIR is a useful method to determine changes in clay-containing polymer composites structure during thermal aging.

Pagacz and Pielichowski, using FTIR, determined the differences in the PVC/MMT nanocomposites after the thermal weathering process [125]. The authors observed progressive weakening of the carbonyl band at 1724 cm^{-1} and double band at 1579 and 1600 cm^{-1}, from conjugated C=C vibrations in aromatic ring of plasticizer, which may suggest emission of the phthalate plasticizer. On the other hand, different species formed during thermal aging could influence the area of carbonyl peak, but there was no dependence on the area of carbonyl peak and time of aging. The most significant changes could be observed in the region of the carbonyl group, and C–C vibrations of benzene ring in the plasticizer molecule. The IR spectra also exhibit broadening of absorbance at 3455 cm^{-1} characteristic of OH groups. In general, the spectra were complex because photooxidation led to several different compounds, such as alcohols, aldehydes, chloroorganic compounds, esters, and ketones, which could give new radicals initiating subsequent reactions [125,126]. These results also suggested that products of thermal degradation could react and form secondary products, as well as favor cross-linking of the polymer chain. The progress of degradation was reflected by the gradual intensity increase of the bands corresponding to unsaturated isolated polyene structures and conjugated double bonds; their formation leads to yellowing of PVC/MMT nanocomposites and worsening of mechanical properties [125,126].

Power spectral density (PSD) analysis is another tool that provides valuable information not only on the height deviation of the roughness profile, but also on its lateral profile. As the authors suggest, the main reason could be that the unfilled rubber vulcanizate undergoes color changes from brown to black upon aging due to thermooxidative reaction, while the nanocomposite changes from light brown to deep brown [119–124]. To study the nature of this aging process, representative samples were analyzed at 100°C and 135°C using FTIR. FTIR spectra analysis confirmed mechanism of degradation of HNBR, which has been discussed in the literature [119–124]. In this mechanism oxygen molecules react with the rubber radicals; hydroperoxides formed could undergo number of reactions leading to higher oxidized moieties, such as alcohols, esters, carboxylic acids, and ketones. As a consequence, the properties of the unfilled elastomer worsen tremendously during aging. The addition of nanoclay protects the sample from oxidative reaction and as a result formation of ester and carbonyl linkages is inhibited, as confirmed by FTIR analysis of filled vulcanizates.

5.8.2 Morphological Developments in Functionalized Clay-Containing Polymer Composites

The heat treatment causes evaporation of volatile compounds and decomposition of less-stable organic additives, such as quaternary onium salts used in the organophilization process, which usually affects the surface microstructure. Study of morphology of PVC/MMT nanocomposites before weathering tests, performed by Pagacz and Pielichowski, showed that the extruded nanocomposites have porous structure, which can influence the thermal and weathering behavior [125]. Chen et al. also reported that pristine PVC has a more homogeneous and smooth surface, while the nanocomposites, especially with OMMT, display cavities of different size [126]. There is a hypothesis that upon accelerated weathering, PVC/MMT composites undergo slow surface erosion, which leads to a mineral-rich surface. In the first stage, plasticizer, as the most volatile additive in a PVC system, is lost by migration and evaporation. The formation of cracks could increase porosity of material and thus open the PVC structure to a more extensive diffusion of oxygen [124–130]. Moreover, Pagacz noted that SEM images of PVC/MMT surfaces after the weathering process show a development of microcracks on the exposed surface of the material. These cracks were visible mainly in composites including organoclay, which suggests that this kind of nanoadditive brings some inhomogeneities into the composite material surface [125]. In addition, it should be pointed out that the presence of water has great influence on the weathering behavior of most polymer matrices, as reported in Refs. [124–130]. It has been found that hydroxyl radicals interact with photodegradation-induced unsaturations in the macrochains, leading to whitening of the surface. This process can compete with yellowing of the polymer matrix due to

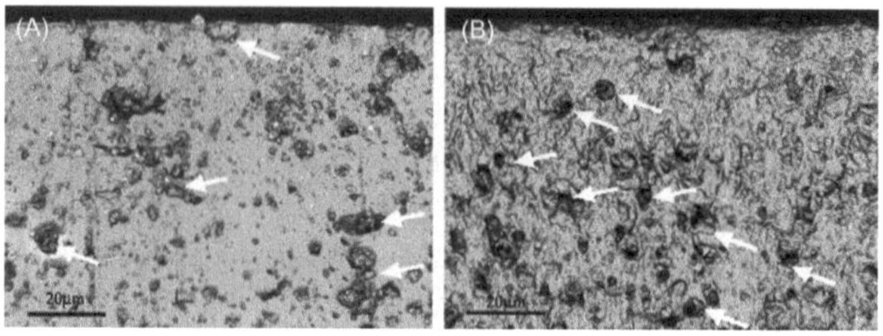

FIGURE 5–7 Laser microscopic image of the fading section around the surface after weathering test: (A) 0 h (the obvious inorganic particles are shown by arrows), and (B) 3000 h (the obvious voids are shown by arrows) [130].

polyenes formation and variability in broad distribution of the measured yellowing index values could be observed. In general, the presence of water favors formation of oxidation products leading to the more pronounced weathering effects (Fig. 5–7).

On the other hand, Choudhury et al. during their research on the effect of organomodified clay on accelerated aging resistance of hydrogenated nitrile rubber nanocomposites and their lifetime prediction found that nanofillers protected the elastomer from surface rupture that took place on oxidation. This marked change in properties upon aging for the pure elastomer was due to cross-linking and degradation of the rubber. Furthermore, changes in the surface morphology of the system on aging reflected a remarkable drop in properties of the pure elastomer. The surfaces of both HNBR and its nanocomposite with layered silicate, subjected to aging, have been studied by scanning electron microscopy and atomic force microscopy. The surface morphology of the unaged sample and its nanocomposite was similar before aging. However, upon aging, a distinct change on surface morphology was observed. The surface of the unfilled elastomer was cracked due to oxidative attack, while the nanoclay protected the rubber surface. There were many cracks as well as bright features probably due to oxidized hard surfaces. However, for the clay-filled elastomer, no significant change in surface morphology was observed. For this reason, to determine the change in surface topography quantitatively, the root mean square (RMS) roughness calculations have been done. The higher value of RMS for rubber composite as compared with pure matrix confirmed the presence of filler particles on the surface. There was a drastic change in roughness taking place upon aging for the pure elastomer, which confirmed the oxidative attack at the surface. For the nanocomposite, only a slight change in roughness was observed on aging. This observation further confirms that clay nanofillers protect the surface of the elastomer matrix from oxidative attack [119–128].

5.9 The Influence of Clay Functionalization on Human Health and the Environment

There is still no full information on the possible effects of low-level exposure of clay nanoadditives on human health. Long-term occupational exposures to clay or organoclay dust may cause structural and functional damage to the lungs. However, available data are inadequate to conclusively establish a dose–response relationship or even a cause-and-effect relationship due to limited information on period and intensity of exposure and to ascertain the role of other factors, such as exposure to tobacco smoke. Long-term exposure to kaolin causes the development of radiologically diagnosed pneumoconiosis in an exposure-related fashion. The composition of the clay (i.e., the quantity and quality of minerals other than kaolinite) is an important determinant of the effects. Bentonite, kaolin, and other types of clays often contain quartz, and exposure to quartz is causally related to silicosis and lung cancer. Statistically significant increases in the incidence of or mortality from chronic bronchitis and pulmonary emphysema have been reported after exposure to quartz [131].

From the limited data available from studies on clay-exposed persons, retained MMT appears to effect only mild nonspecific tissue changes, which are similar to those that have been described in the spectrum of changes of the "small airways mineral dust disease." In some of the studies, radiological abnormalities have also been reported. There exist no reported cases of marked diffuse/nodular pulmonary tissue fibrotic reaction to MMT in the absence of free silica. No quantitative estimates of the potency of clay to cause adverse pulmonary effects can be derived [132]. Long-term exposure to kaolin may lead to a relatively benign pneumoconiosis, known as kaolinosis. Deterioration of lung function has been observed only in cases with prominent radiological alterations. Based on data from china clay workers in the United Kingdom, it could be very roughly estimated that kaolin is at least an order of magnitude less potent than quartz [131–134].

In view of the widespread distribution of clay in nature and its use in an enormous variety of consumer products, the general population's exposure to low concentrations is ubiquitous. There is limited information on occupational exposure to clay dust in mines, processing plants, and user industries. The highest reported values for total dust and respirable dust concentrations were 1430 and 34.9 mg/m^3, although most values were below 10 mg/m^3 for total dust and below 5 mg/m^3 for respirable dust, respectively [131–134].

Kaolin as a natural component of the soil occurs widely in ambient air. Kaolin mining and refining involve considerable exposure, and significant exposure is also expected in polymer composites production. Quantitative information on occupational exposure is available for a few countries and industries only. Respirable dust concentrations in kaolin mining and processing are usually below 5 mg/m^3 [131–135].

The observations of Refs. [133] and [134] showed that clearance of clay stored in foci from the lung is relatively slow. Clay stored in nodules in the lungs that were noted on autopsy stemmed from exposure to dust more than 50 years previous to death at age 79. Three of the studies reported radiological evidence, not supported by autopsy, of significant damage to lungs following exposures to clay or MMT dust of 10–42 years. Dust concentrations were described as high by [135], but were not measured in any of these studies. McNally and Trostler [136] reported significant radiological evidence of lung damage in 28 workers exposed to clay dust for 3.5–16 years, but severity of damage was not correlated with period of exposure or the age of the individual. Twenty-one coworkers with exposures of 1.75–16 years, however, displayed no radiological evidence of damage. The authors ascribed this lack of correlation between damage and period of exposure to the fact that dust concentrations varied widely within the plant and that the men worked at different jobs during the course of their employment and thus were exposed to varying concentrations of dust, which overwhelmed any effects from total period of employment. No attempt was made in any of these studies to approximately reconstruct actual exposures, as has been done, for example, with silica exposure [137]. However, it was also reported that damage to the lungs increased with increasing exposure to clay dust (containing 1% quartz), although radiological data on Bulgarian workers presented in their paper appeared too limited to permit any final conclusions to be drawn. Their data concurred with those presented in Ref. [136] on the absence of radiological evidence of lung damage in many exposed workers. The data also supported the reasonable assumption of greater damage from higher concentrations of dust.

Deposition of radiolabeled fused MMT after inhalation exposure has been studied in mice, rats, dogs, and humans. Deposition in the nasopharynx increases with particle size and is lower in dogs than in rodents. Tracheobronchial deposition was low and independent of animal species and particle size. Pulmonary deposition was considerably higher in dogs than in rodents and decreased with increasing particle size. The removal of particles from the lungs took place by solubilization in situ and by physical clearance. The clearance by mechanical removal was slow, especially in dogs: The halftime was initially 140 days and increased to 6900 days by day 200 postexposure. In humans, there was a rapid initial clearance of 8% and 40% of aluminosilicate particles that were, respectively, 1.9 and 6.1 μm in aerodynamic diameter from the lung region over 6 days. Thereafter, 4% and 11% of the two particle sizes were removed following a halftime of 20 days, and the rest with halftimes of 330 and 420 days. Ultrafine particles (<100 nm) have a high deposition in the nasal area. They could penetrate the alveolar/capillary barrier [129–137].

Toxicity studies on functionalized clay (MMT modified with a quaternary ammonium salt containing methyl *bis*-2-hydroxyethyl group and C14-C18 alkyl side chains—Cloisite 30B) revealed that the genotoxic effect was most likely due to the free quaternary ammonium compounds (QACs) in organofunctionalized Cloisite. In contrast, the in vivo findings showed no induction of DNA strand breaks in liver, kidney, or colon cells from rats given organomodified Cloisite by oral administration, indicating that the test material was not genotoxic in vivo. Tissues of the lumen of the gastrointestinal system were exposed to functionalized clay, but no genotoxic effect was induced. Aluminum was used as tracer element, characteristic of clay particles, but could not be detected in the liver and kidney. This indicated that there was no detectable internal exposure to the clay component of functionalized clay. Furthermore, the negative results of the comet assay in these two organs and the absence of any detectable effect on plasma concentrations of inflammatory cytokines indicated that functionalized clay did not cause any systemic effect to rats at doses up to 1000 mg/kg body weight [138].

Unmodified clay has been reported to cause cytotoxicity in many cell lines. In this sense, Murphy et al. [139] evidenced a decrease in cell viability in HUVEC exposed for 24 h to 0.1 mg/mL clays, although limited results were obtained in the neuronal cell lines. Similarly, CHO cells exposed to1 mg/mL clay for up to 24 h underwent a reduction in cell viability [140]. Baek et al. [141] reported similar results in intestinal cells exposed to the same concentration of clay, showing a concentration- and time-dependent inhibition of cell proliferation after 24–72 h of exposure, as well as cytotoxicity after long-term exposure to clay. It has been also suggested that the organomodifier included in the clay modulates the toxicity of the resulting clay [142]. In this sense, pure clay did not show signs of cytotoxicity in Caco-2 cells, although functionalized clay induced approximately 40% cytotoxicity at the highest concentration tested (226 mg/mL) after 24 h of exposure [142].

In addition, active clay, which is bentonite that is industrially treated with H_2SO_4, produced greater damage in human B lymphoblast cells than the native bentonite [143]. In fact, it has been reported that different modifiers have different effects; hence, clay modified with dimethyl benzyl hydrogenated tallow ammonium (Dellite) showed higher toxicity than those modified with dimethyl dihydrogenated tallow ammonium in five cell lines [144]. This finding was corroborated by analyzing the cytotoxic effects of the modifiers in the same cell lines. Studies evaluating the toxicity of QACs potentially used as clay modifiers are scarce.

Moreover, pure clay showed lower toxicity than the functionalized clay in HepG-2 and Caco-2 cells [145,146]. In addition, the same cell lines showed the different cytotoxic profiles of the three organomodified clay minerals, evidencing that the type of clay mineral, the concentration range, and the origin of the cell line play an important role in the observed toxicity [147,148]. Wagner et al. studied toxicity of unmodified MMT (denoted as UC) and MMT modified with methyl tallow *bis*-2-hydroxyethyl quaternary ammonium (CC), as well as of the ash from incineration at 900°C of UC (UC900) and CC(CC900). Cellular imaging was applied to check if there was a correlation of changes in cell shape with the electric cell-substrate impedance sensing (ECIS) results [149]. It has been found that some changes in cellular shape and size, as well as in cell confluence, occurred (Fig. 5–8).

However, the functionalization of clay was also demonstrated to decrease the toxicity of clay minerals. For instance, the functionalization of clay with oligo(styrene-co-acrylonitrile) resulted in a less toxic material compared with the unmodified clay [150]. In addition, the functionalization of the organoclay could result in a nontoxic material, which is the main objective of the synthesis of these materials intended for human contact. Han et al. [151] reported no significant membrane damage and change in cell viability in four cell types exposed to magnesium and calcium organophyllosilicates.

Currently, there is no clear information on the harmful effects of clay or organoclay present in the polymer nanocomposites on human health. The content of such nanofiller is mostly in the range from 0.1 to 10 wt.%. It is believed that such a small amount of filler relative to the polymer matrix does not adversely affect human health. The dispersed clay is in fact embedded in the resin and its influence on living organisms is then limited. Nevertheless, during mass production of nanocomposites, at the stage of clay functionalization there is a risk of harmful action to the long-term effect of nanofiller. To decrease adverse health effects from occupational exposure to clays, it is recommended to [131]:

- disseminate information on the hazards to exposed workers in an appropriate form;
- enforce and ascertain compliance with the limits by regular exposure monitoring;
- institute appropriate medical monitoring programs to protect populations at risk;

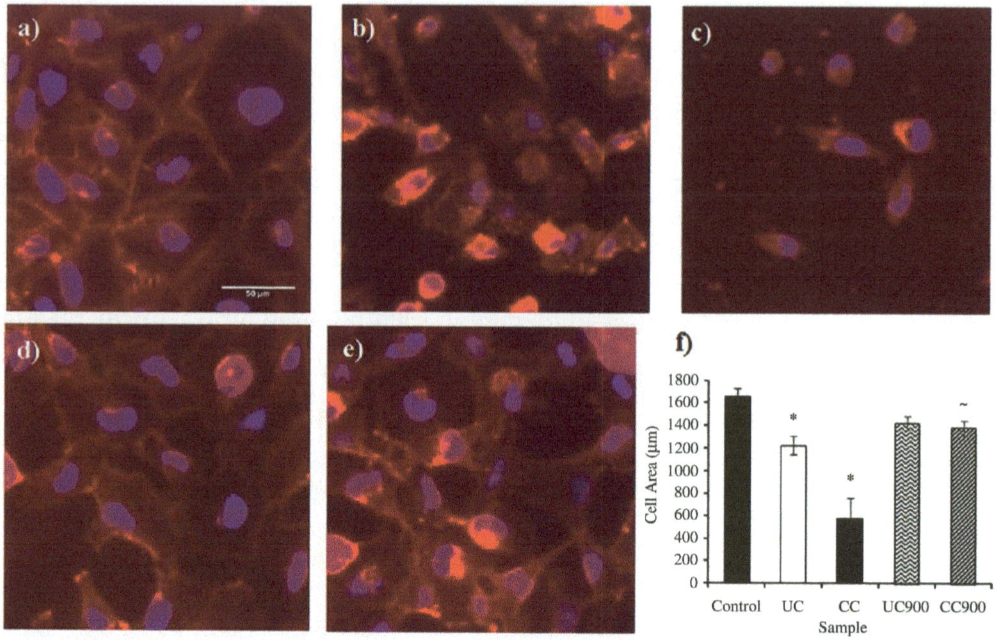

FIGURE 5–8 Fluorescent images of the cell membrane (red) and nucleus (blue) for (A) control cells and cells treated with (B) UC, (C) CC, (D) UC900, and (E) CC900 after 24 h. (F) Cell area (μm) after 24 h of treatment with nanoclays ($n = 3$) [149].

- oblige clay producers to declare the content of hazardous components, such as quartz, of the saleable products;
- prepare guidelines to ensure good workplace practice;
- set limits for occupational exposure to clay materials, taking into consideration the quartz content.

5.10 Concluding Remarks

The current review shows that both unmodified and functionalized clays have still a wide range of applications in composite materials. The functionalization process of layered silicate nanofillers is relatively easy to perform and control; however, for some applications special requirements, for example, ability for high-temperature processing, require special functionalization procedures and/or modifiers. Therefore there is still a search for novel methods of clay-containing composites preparation and functionalization. Although the design and fabrication of this kind of nanocomposites have been successfully developed since the 1990s, the long-term behavior of clay embedded in the polymer matrix, including aging effects, is still the subject of investigation.

Functionalized clay-containing polymer nanocomposites show enhancements in mechanical and barrier properties ("labyrinth effect"), as well as in thermal, optical, and physicochemical properties, compared with pristine polymers or classical microcomposites. However, the relationships between structure, processing, and physicochemical properties still need to be formulated to better answer the question on how the presence of clay influences the nanocomposite characteristics. Better understanding of such relationships shall lead to tailor-made nanostructured polymeric composite materials with superior properties.

References

[1] Hojiyev R, Ulcay Y, Çelik MS. Development of a clay-polymer compatibility approach for nanocomposite applications. Appl Clay Sci 2017;146:548–56.

[2] Rane AV, Kanny K, Abitha VK, Patil SS, Thomas S. Clay–polymer composites: design of clay polymer nanocomposite by mixing. In: Jlassi K, Chehimi MM, Thomas S, editors. Clay-polymer nanocomposites. Elsevier; 2017. p. 113–44.

[3] Zare Y, Fasihi M, Rhee KY. Efficiency of stress transfer between polymer matrix and nanoplatelets in clay/polymer nanocomposites. Appl Clay Sci 2017;143:265–72.

[4] Wilkinson N, Metaxas A, Quinney C, Wickramaratne S, Reineke T, Dutcher CS. pH dependence of bentonite aggregate size and morphology on polymer-clay flocculation.. Colloids Surf A: Physicochem Eng Aspects 2018;537:281–6.

[5] Ayeldeen M, Kitazume M. Using fiber and liquid polymer to improve the behaviour of cement-stabilized soft clay. Geotext Geomemb 2017;45:592–602.

[6] Tan B, Thomas NL. A review of the water barrier properties of polymer/clay and polymer/graphene nanocomposites. J Membr Sci 2016;514:595–612.

[7] Valapa RB, Loganathan S, Pugazhenthi G, Thomas S, Varghese TO. An overview of polymer–clay nanocomposites. In: Jlassi K, Chehimi MM, Thomas S, editors. Clay-polymer nanocomposites. Elsevier; 2017. p. 29–81.

[8] Haase H, Schanz T. Compressibility and saturated hydraulic permeability of clay-polymer composites — experimental and theoretical analysis. Appl Clay Sci 2016;130:62–75.

[9] Saad A, Jlassi K, Omastová M, Chehimi MM. Clay/conductive polymer nanocomposites. In: Jlassi K, Chehimi MM, Thomas S, editors. Clay-polymer nanocomposites. Elsevier; 2017. p. 199–237.

[10] Lopez-Cuesta JM. Flame retardancy properties of clay–polymer nanocomposites. In: Jlassi K, Chehimi MM, Thomas S, editors. Clay-polymer nanocomposites. Elsevier; 2017. p. 443–73.

[11] Jlassi K, Krupa I, Chehimi MM. Overview: clay preparation, properties, modification. In: Jlassi K, Chehimi MM, Thomas S, editors. Clay-polymer nanocomposites. Elsevier; 2017. p. 1–28.

[12] Pongprayoon T, Nuangchamnong R, Yanumet N. Antimicrobial resistance of clay polymer nanocomposites. Appl Clay Sci 2013;86:179–84.

[13] Feng X, Zhang G, Zhuo S, Jiang H, Shi J, Li F, et al. Dual responsive shape memory polymer/clay nanocomposites. Compos Sci Technol 2016;129:53–60.

[14] Kotal M, Bhowmick AK. Polymer nanocomposites from modified clays: recent advances and challenges. Prog Polym Sci 2015;51:127–87.

[15] Kausar A. Physical properties of hybrid polymer/clay composites. In: Thakur VK, Thakur MK, Pappu A, editors. Hybrid polymer composite materials properties and characterisation. Elsevier; 2017. p. 115–32.

[16] Chen RS, Gan SAS. Characterization of recycled thermoplastics-based nanocomposites: polymer-clay compatibility, blending procedure, processing condition, and clay content effects. Compos Part B: Eng 2017;131:91–9.

[17] Bishop MD, Kim K, Palomino AM, Lee JS. Deformation of "tunable" clay–polymer composites. Appl Clay Sci 2014;101:265–71.

[18] Bergaya F, Detellier C, Lambert JF, Lagaly G. Introduction to clay–polymer nanocomposites (CPN). Dev Clay Sci 2013;5:655–77.

[19] Mirzababaei M, Arulrajah A, Ouston M. Polymers for stabilization of soft clay soils. Procedia Eng 2017;189:25–32.

[20] Follain N, Marais S. Gas transport through polymer/clay nanocomposites transport properties of polymeric membranes. Elsevier; 2018. p. 573–614.

[21] Mousa M, Evans ND, Oreffo ROC, Dawson JI. Clay nanoparticles for regenerative medicine and biomaterial design: a review of clay bioactivity. Biomaterials 2018;159:204–14.

[22] Rezaeimalek S, Nasouri A, Huang J, Bin-Shafique S, Gilazghi ST. Comparison of short-term and long-term performances for polymer-stabilized sand and clay. J Traffic Transp Eng 2017;4:145–55.

[23] Momani B, Sen M, Endoh M, Wang X, Koga T, Winter HH. Temperature dependent intercalation and self-exfoliation of clay/polymer nanocomposite. Polymer 2016;93:204–12.

[24] Zare Y. Effects of imperfect interfacial adhesion between polymer and nanoparticles on the tensile modulus of clay/polymer nanocomposites. Appl Clay Sci 2016;129:65–70.

[25] Majka TM, Bogusz K, Pielichowski K. Obniżenie palności kompozytów poliamidowych z montmorylonitem przy zastosowaniu fosforanu wapnia. Przem Chem 2016;95:2030–5.

[26] Majka TM, Leszczyńska A, Kandola BK, Pornwannachai W, Pielichowski K. Modification of organo-montmorillonite with disodium H-phosphonate to develop flame retarded polyamide 6 nanocomposites. Appl Clay Sci 2017;139:28–39.

[27] Majka TM, Leszczyńska A, Pielichowski K. Mechanical properties vs biotic degradation of polyamide/layered silicates nanocomposites. Tech Trans 2017;114:93–111.

[28] Chiu CW, Huang TK, Wang YC, Alamani BG, Lin JJ. Intercalation strategies in clay/polymer hybrids. Prog Polym Sci 2014;39:443–85.

[29] Passador FR, Ruvolo-Filho A, Pessan LA. Nanocomposites ofpolymer matrices and lamellar clays in oliveira oN. In: Ferreira M, Da Róz A, Leite FdL, editors. Nanostructures. Elsevier; 2017. p. 187–207.

[30] Theng BKG. Polymer–clay nanocomposites developments. Clay Sci 2012;4:201–41.

[31] Unuabonah EI, Taubert A. Clay–polymer nanocomposites (CPNs): adsorbents of the future for water treatment. Appl Clay Sci 2014;99:83–92.

[32] Ruiz-Hitzky E, Aranda P, Darder M, Fernandes FM. Fibrous clay mineral–polymer nanocomposites. Dev Clay Sci 2013;5:721–41.

[33] Galimberti M, Cipolletti VR, Coombs M. Applications of clay–polymer nanocomposites. Dev Clay Sci 2013;5:539–86.

[34] Hamidi S, Marandi SM. Clay concrete and effect of clay minerals types on stabilized soft clay soils by epoxy resin. Appl Clay Sci 2018;151:92–101.

[35] Gilazghi ST, Huang J, Rezaeimalek S, Bin-Shafique S. Stabilizing sulfate-rich high plasticity clay with moisture activated polymerization. Eng Geol 2016;211:171–8.

[36] Azzam WR. Behavior of modified clay microstructure using polymer nanocomposites technique. Alex Eng J 2014;53:143–50.

[37] Schoonheydt RA. Reflections on the material science of clay minerals. Appl Clay Sci 2016;131:107–12.

[38] Xia Y, Rubino M, Auras R. Modeling of surfactant release from polymer-clay nanocomposites into ethanol. Polym Test 2016;50:57–63.

[39] Theng BKG. Some practical applications of the clay–polymer interaction. Dev Clay Sci 2012;4:153–99.

[40] Mallakpour S, Rashidimoghadam S. Recent developments in the synthesis of hybrid polymer/clay nanocomposites: properties and applications. In: Thakur VK, Thakur MK, Pappu A, editors. Hybrid polymer composite materials applications. Elsevier; 2017. p. 227–65.

[41] Theng BKG. Polymer behaviour at clay and solid surfaces. Dev Clay Sci 2012;4:47–75.

[42] Huang W. Clay nanopapers. In: Huang H, editor. Nanopapers from nanochemistry and nanomanufacturing to advanced applications a volume in micro andnano technologies. Elsevier; 2018. p. 59–86.

[43] Zhang G, Wu T, Lin W, Tan Y, Chen R, Huang Z, et al. Preparation of polymer/clay nanocomposites via melt intercalation under continuous elongation flow. Compo Sci Technol 2017;145:157–64.

[44] Kredatusová J, Brožek J. Preparation of clay mineral polymer nanocomposites by adsorption in solution. Appl Clay Sci 2012;62–63:94–100.

[45] Zapata PA, Belver C, Quijada R, Aranda P, Ruiz-Hitzky E. Silica/clay organo-heterostructures to promote polyethylene–clay nanocomposites by in situ polymerization. Appli Catal A: Gen 2013;453:142–50.

[46] Ceraulo M, Morreale M, Botta L, Mistretta MC, Scaffaro R. Prediction of the morphology of polymer-clay nanocomposites. Polym Test 2015;41:149–56.

[47] Dhatarwal P, Sengwa RJ, Choudhary S. Effect of intercalated and exfoliated montmorillonite clay on the structural, dielectric and electrical properties of plasticized nanocomposite solid polymer electrolytes. Compos Commun 2017;5:1–7.

[48] Zhou K, Zhang Q, Wang B, Liu J, Wen P, Gui Z, et al. The integrated utilization of typical clays in removal of organic dyes and polymer nanocomposites. J Clean Product 2014;81:281–9.

[49] Liu P. Polymer modified clay minerals: a review. Appl Clay Sci 2007;38:64–76.

[50] Chiu CW, Lin JJ. Self-assembly behavior of polymer-assisted clays. Prog Polym Sci 2012;37:406–44.

[51] Basnayaka L, Subasinghe N, Albijanic B. Influence of clays on fine particle filtration. Appl Clay Sci 2018;156:45–52.

[52] Xu B, Zheng Q, Song Y, Shangguan Y. Calculating barrier properties of polymer/clay nanocomposites: effects of clay layers. Polymer 2006;47:2904–10.

[53] Salam H, Dong Y, Davies I. Development of biobased polymer/clay nanocomposites: a critical review. In: Yu Dong Y, Rehan Umer R, Lau AKT, editors. Fillers and reinforcements for advanced nanocomposites a volume in Woodhead publishing series in composites science and engineering. Elsevier; 2015. p. 101–32.

[54] Faucheu J, Gauthier C, Chazeau L, Cavaillé JY, Mellon V, Pardal F, et al. Properties of polymer/clay interphase in nanoparticles synthesized through in-situ polymerization processes. Polymer 2010;51:4462–71.

[55] Majka TM, Bartyzel O, Raftopoulos KN, Pagacz J, Leszczyńska A, Pielichowski K. Recycling of polypropylene/montmorillonite nanocomposites by pyrolysis. J Anal Appl Pyrolysis 2016;119:1–7.

[56] Costes L, Laoutid F, Brohez S, Dubois P. Bio-based flame retardants: when nature meets fire protection. Mater Sci Eng: R: Rep 2017;117:1–25.

[57] Kiliaris P, Papaspyrides CD. Polymer/layered silicate (clay) nanocomposites: an overview of flame retardancy. Progr Polym Sci 2010;35:902–58.

[58] Levchik SV. Introduction to flame retardancy and polymer flammability. In: Morgan AB, Wilkie CA, editors. Flame retardant polymer nanocomposites. John Wiley & Sons; 2007. p. 1–29.

[59] Yi D, Yang R, Wilkie CA. Full scale nanocomposites: clay in fire retardant and polymer. Polym Degrad Stab 2014;105:31–41.

[60] Gillani QF, Ahmad F, Mutalib MIA, Melor PS, Ullah S, Arogundade A. Effect of dolomite clay on thermal performance and char morphology of expandable graphite based intumescent fire retardant coatings. Procedia Eng 2016;148:146–50.

[61] Ullah S, Ahmad F, Shariff AM, Bustam MA. Synergistic effects of kaolin clay on intumescent fire retardant coating composition for fire protection of structural steel substrate. Polym Degrad Stab 2014;110:91–103.

[62] Hu Y, Yu B, Song L. Novel fire-retardant coatings. In: Wang DY, editor. Novel fire retardant polymers and composite materials. Elsevier; 2017. p. 53–91.

[63] Zia-ul-Mustafa M, Ahmad F, Ullah S, Amir N, Gillani QF. Thermal and pyrolysis analysis of minerals reinforced intumescent fire retardant coating. Prog Org Coat 2017;102:201–16.

[64] Qiu X, Li Z, Li X, Zhang Z. Flame retardant coatings prepared using layer by layer assembly: a review. Chem Eng J 2018;334:108–22.

[65] Choudhury AKR. Flame- and fire-retardant finishes. In: Choudhury AKR, editor. Principles of textile finishing a volume in Woodhead Publishing Series in Textiles. Elsevier; 2017. p. 195–244.

[66] Marosi G, Szolnoki B, Bocz K, Toldy A. Fire-retardant recyclable and biobased polymer composites. In: Wang DY, editor. Novel fire retardant polymers and composite materials. Elsevier; 2017. p. 117–46.

[67] Ray SS. Fire-retardant properties. In: Ray SS, editor. Clay-containing polymer nanocomposites from fundamentals to real applications. Elsevier; 2013. p. 263–71.

[68] Ciesielski M, Burk B, Heinzmann C, Döring M. Fire-retardant high-performance epoxy-based materials. In: Wang DY, editor. Novel fire retardant polymers and composite materials. Elsevier; 2017. p. 3–51.

[69] Bonati A, Merusi F, Bochicchio G, Tessadri B, Polacco G, Filippi S, et al. Effect of nanoclay and conventional flame retardants on asphalt mixtures fire reaction. Constr Build Mater 2013;47:990–1000.

[70] Vahabi H, Lopez-Cuesta JM, Chivas-Joly C. High-performance fire-retardant polyamide materials. In: Wang DY, editor. Novel fire retardant polymers and composite materials. Elsevier; 2017. p. 147–70.

[71] Carosio F, Cuttic F, Medina L, Berglund LA. Clay nanopaper as multifunctional brick and mortar fire protection coating—wood case study. Mater Des 2016;93:357–63.

[72] Hull TR, Witkowski A, Hollingbery L. Fire retardant action of mineral fillers. Polym Degrad Stab 2011;96:1462–9.

[73] Sittisart P, Farid MM. Fire retardants for phase change materials. Appl Energy 2011;88:3140–5.

[74] Han Z, Fina A, Camino G. Organosilicon compounds as polymer fire retardants. In: Papaspyrides CD, Kiliaris P, editors. Polymer green flame retardants. Elsevier; 2014. p. 389–418.

[75] Wang X, Kalali EN, Wan JT, Wang DY. Carbon-family materials for flame retardant polymeric materials. Prog Polym Sci 2017;69:22–46.

[76] Hollingbery LA, Hull TR. The fire retardant behaviour of huntite and hydromagnesite – a review. Polym Degrad Stab 2010;95:2213–25.

[77] Lopez-Cuesta JM. Chapter flame retardancy properties of clay–polymer nanocomposites. In: Jlassi K, Chehimi MM, Thomas S, editors. Clay-polymer nanocomposites. Elsevier; 2017. p. 443–73.

[78] Rothon R, Hornsby P. Fire retardant fillers for polymers in papaspyrides. In: Papaspyrides C, Kiliaris P, editors. Polymer green flame retardants. Elsevier; 2014, pp. 289–321.

[79] Samyn F, Bourbigot S, Jama C, Séverine Bellayer S. Fire retardancy of polymer clay nanocomposites: is there an influence of the nanomorphology? Polym Degrad Stab 2008;93:2019–24.

[80] Wang D, Echols K, Wilkie CA. Cone calorimetric and thermogravimetric analysis evaluation of halogen-containing polymer nanocomposites. Fire Mater 2005;29:283–94.

[81] Weil ED, Hirschler MM, Patel NG, Shaki S. Oxygen index: correlations to other fire tests. Fire Mater 1992;1992(16):159–67.

[82] Zhang S, Horrocks AR. A review of flame retardant polypropylene fibres. Prog Polym Sci 2003;28:1517–38.

[83] Hao X, Gai G, Liu J, Yang Y, Zhang Y, Nan C. Flame retardancy and antidripping effect of OMT/PA nanocomposites. Mater Chem Phys 2006;96:34–41.

[84] Chigwada C, Wilkie CA. Synergy between conventional phosphorus fire retardants and organically-modified clays can lead to fire retardancy of styrenics. Polym Degrad Stab 2003;2003(80):551–7.

[85] Zhang G, Wang J. Preparation of novel flame-retardant organoclay and its application to natural rubber composites. J Phys Chem Solids 2018;115:137–47.

[86] Zotti A, Borriello A, Ricciardi M, Antonucci V, Giordano M, Zarrelli M. Effects of sepiolite clay on degradation and fire behaviour of a bisphenol a-based epoxy. Compos Part B: Eng 2015;73:139–48.

[87] Toldy A, Anna P, Csontos I, Szabó A, Marosi G. Intrinsically flame retardant epoxy resin – fire performance and background – Part I. Polym Degrad Stab 2007;92:2223–30.

[88] Agueda A, Pastor E, Planas E. Different scales for studying the effectiveness of long-term forest fire retardants. Prog Energy Combust Sci 2008;34:782–96.

[89] Cárdenas MA, García-López D, García-Vilchez A, Fernández JF, Merino JC, Pastor JM. Synergy between organo-bentonite and nanofillers for polymer based fire retardant applications. Appl Clay Sci 2009;45:139–46.

[90] Agueda A, Pastor E, Pérez Y, Viegas DX, Planas E. Fire intensity reduction in straw fuel beds treated with a long-term retardant. Fire Saf J 2011;46:41–7.

[91] Rossi S, Bergamo L, Fontanari V. Fire resistance and mechanical properties of enamelled aluminium foam. Mater Des 2017;132:129–37.

[92] Nguyen QT, Ngo TD, Tran P, Mendis P, Bhattacharyya D. Influences of clay and manufacturing on fire resistance of organoclay/thermoset nanocomposites. Composites, Part A: Appl Sci Manuf 2015;74:26–37.

[93] Fina A, Bocchini S, Camino G. Catalytic fire retardant nanocomposites. Polym Degrad Stab 2008;93:1647–55.

[94] Modesti M, Lorenzetti Besco S, Hrelja D, Semenzato S, Bertani R, Michelin RA. Synergism between flame retardant and modified layered silicate on thermal stability and fire behaviour of polyurethane nanocomposite foams. Polym Degrad Stab 2008;93:2166–71.

[95] Misra N, Kumar V, Bahadur J, Bhattacharya S, Mazumder S, Varshney L. Layered silicate-polymer nanocomposite coatings via radiation curing process for flame retardant applications. Prog Org Coat 2014;77:1443–51.

[96] Bodzay B, Bocz K, Bárkai Z, Marosi G. Influence of rheological additives on char formation and fire resistance of intumescent coatings. Polym Degrad Stab 2011;96:355–62.

[97] Jang BN, Costache M, Wilkie CA. The relationship between thermal degradation behavior of polymer and the fire retardancy of polymer/clay nanocomposites. Polymer 2005;46:10678–87.

[98] Kashiwagi T, Harris RH, Zhang X, Briber RM, Cipriano BH, Raghavan SR, et al. Flame retardant mechanism of polyamide 6–clay nanocomposites. Polymer 2004;45:881–91.

[99] Tortora M, Gorrasi G, Vittoria V, Galli G, Ritrovati S, Chiellini E. Structural characterization and transport properties of organically modified montmorillonite/polyurethane nanocomposites. Polymer 2002;43:6147–57.

[100] Ray SS, Yamada K, Okamoto M, Ogami A, Ueda K. New polylactide/layered silicate nanocomposites. 3. High performance biodegradable materials. Chem Mater 2003;15:1456–65.

[101] Lange J, Wyser Y. Recent innovations in barrier technologies for plastic packaging-a review. Packaging Technol Sci 2003;16:149–58.

[102] Giannelis EP. Polymer layered silicate nanocomposites. Adv Mater 1996;8:29–35.

[103] Koh HC, Park JS, Jeong MA, Hwang HY, Hong YT, Ha SY, et al. Prearation and gas permeation properties of biodegradable polymer/layered silicate nanocomposite membranes. Desalination 2008;233:201–9.

[104] Say R, Birlik E, Denizli A, Ersöz A. Removal of heavy metal ions by dithiocarbamate-anchored polymer/organosmectite composites. Appl Clay Sci 2006;31:298–305.

[105] Knowles NC, Jeffries RM, Come B. Benchmarking the thermomechanical behaviour of clays — a progress report — the CEC interclay project. Eng Geol 1996;41:65–71.

[106] Ross B. Release of radioactivity from waste packages. Eng Geol 1989;26:351–72.

[107] Vikas Mittal V. Barrier properties of composite materials. In: Mittal V, editor. Barrier properties of polymer clay nanocomposites. Elsevier; 2009. p. 1–18.

[108] Carrera MC, Erdmann E, Destéfanis HA. Barrier properties and structural study of nanocomposite of HDPE/montmorillonite modified with polyvinylalcohol. J Chem 2013;2013:1–7.

[109] Liu X, Wu Q. Polyamide 66/clay nanocomposites via melt intercalation. Macromol Mater Eng 2002;287:180–6.

[110] Becker O, Varley RJ, Simon GP. Thermal stability and water uptake of high performance epoxy layered silicate nanocomposites. Eur Polym J 2004;40:187–95.

[111] Osman MA, Mittal V, Lusti HR. The aspect ratio and gas permeation in polymer–layered silicate nanocomposites. Macromol Rapid Commun 2004;25:1145–9.

[112] Lu C, Mai YW. Influence of aspect ratio on barrier properties of polymer-clay nanocomposites. Phys Rev Lett 2005;95:088303–1-088303-4.

[113] Dabbaghianamiri M, Das S, Beall GW. Improvement approach for gas barrier behavior of polymer/clay nanocomposite films. Nanomaterials 2017;2:3547–52.

[114] Kalendova A, Merinska D, Jean FG, Slouf M. Polymer/clay nanocomposites and their gas barrier properties. Polym Compos 2013;34:1418–24.

[115] Gaikwad KK, Seonghyuk K. Overview on in polymer-nano clay composite paper coating for packaging application. J Mater Sci Eng 2015;4:1.

[116] Majdzadeh-Ardakani K, Zekriardehani S, Coleman MR, Jabarin Saleh A. A novel approach to improve the barrier properties of PET/clay nanocomposites. Int J Polym Sci 2017;5:1–10.

[117] Garcia D, Black JJ. Fourier transform infrared micro spectroscopy mapping studies of weathered PVC capstock type formulations. II: Outdoor weathering in Pennsylvania. J Vinyl Addit Technol 1999;5:81.

[118] Bockhorn H, Hornung A, Hornung UJ. Mechanisms and kinetics of thermal decomposition of plastics from isothermal and dynamic measurements. J Anal Appl Pyrolysis 1999;50:77.

[119] Saeedi M, Ghasemi I, Karrabi M. Thermal degradation of poly(vinyl chloride): effect of nanoclay and low density polyethylene content. Iranian Polym J 2011;20:423.

[120] Krzymien ME, Day M, Carlsson WDJ. PVC photo-oxidative degradation: identification of volatiles. Macromol Symp 1997;115:27.

[121] Janssen K, Gijsman P, Tummers D. Mechanistic aspects of the stabilization of polyamides by combinations of metal and halogen salts. Polym Degrad Stab 1995;49:127–33.

[122] Zhu J, Uhl FM, Morgan AB, Wilkie CA. Studies on the mechanism by which the formation of nanocomposites enhances thermal stability. Chem Mater 2001;13:4649–54.

[123] Chávez-Montes WM, González-Sánchez G, López-Martínez EI, de Lira-Gómez P, Ballinas-Casarrubias L, Flores-Gallardo S. Effect of artificial weathering on PLA/nanocomposite molecular weight distribution. Polymers 2015;7:760–76.

[124] Matuana L.M., Kamden D.P. Accelerated ultraviolet weathering of PVC/wood-flour composites, Polym Eng Sci 2002;42:1657.

[125] Pagacz J, Pielichowski K. In: Pielichowski K (editor). Proceedings of the 4th Modern Polymeric Materials for Environmental Applications, Teza; 2010.

[126] Chen CC, Teng CC, Yang CH. Preparation and characterization of rigid poly (vinyl chloride)/MMT nanocomposites. J Polym Sci, Part B 2005;43:1465.

[127] Balk€ose D, Egbuchunam TO, Okieimen FEJ. Formulation and properties' evaluation of PVC/(dioctyl phthalate)/(epoxidized rubber seed oil) plastigels. J Vinyl Addit Technol 2008;14:65.

[128] Jonsson S, Ejlertsson J, Svensson BH. Transformation of phthalates in young landfill cells. Waste Manage 2003;23:641.

[129] Real LEP, Ferraria AM, Rego AMB. The influence of weathering conditions on the properties of poly (vinyl chloride) for outdoor applications. An analytical study using surface analysis techniques. Polym Test 2007;26:77.

[130] Ito M, Nagai K. Analysis of degradation mechanism of plasticized PVC under artificial aging conditions. Polym Degrad Stab 2007;92:260.

[131] Pielichowski K, Leszczyńska A, Majka TM. Nanocząstki i nanomateriały - szanse, wyzwania i zagrożenia; Nanocząstki i nanomateriały. Zarząd Główny Polskiego Towarzystwa Toksykologicznego (in polish) 2015;9−32.

[132] Pielichowski K, Michałowski S. toxicity and environmental impact. In: Njuguna J, Pielichowski K, Zhu H, editors. Nanostructured flame retardants: performance. London: Health and environmental safety of nanomaterials Elsevier; 2014. p. 251−73.

[133] Tonning HO. Pneumoconiosis from Fuller's earth. J Ind Hyg Toxicol 1949;31:41−5.

[134] Ungváry G, Timár M, Tátrai E, Bacsy E, Gaal G. Analysis of aluminium silicate storage foci in the lungs. Exp Toxicol Pathol 1983;23:203−14.

[135] Sakula A. Pneumoconiosis due to Fuller's earth. Thorax 1961;1961(16):176−9.

[136] McNally WD, Trostler IS. Severe pneumoconiosis caused by inhalation of Fuller's earth. J Ind Hyg Toxicol 1941;23:118−26.

[137] Rice C, Harris RL, Lumsden JC, Symons MJ. Reconstruction of silica exposure in the North Carolina dusty trades. Am Ind Hyg Assoc J 1984;45:689−95.

[138] Sharma AK, Mortensen A, Schmidt B, Frandsen H, Hadrup N, Larsen EH, et al. In-vivo study of genotoxic and inflammatory effects of the organo-modified Montmorillonite Cloisite®30B. Mutat Res 2014;770:66−71.

[139] Murphy EJ, Roberts E, Horrocks LA. Aluminium silicate toxicity in cell cultures. Neuroscience 1993;55:597−605.

[140] Li PR, Wei JC, Chiu YF, Su ÇHN, Peng FC, Lin JJ. Evaluation on cytotoxicity and genotoxicity of the exfoliated silicate nanoclay. ACS Appl Mater Interfaces 2010;2:1608−13.

[141] Baek M, Lee AJ, Choi SJ. Toxicological effects of a cationic clay, montmorillonite in vitro and in vivo. Mol Cell Toxicol 2012;8:95−101.

[142] Sharma AK, Schmidt B, Frandsen H, Jacobsen NR, Larsen EH, Binderup ML. Genotoxicity of unmodified andorgano-modified montmorillonite. Mutat Res 2010;700:18−25.

[143] Meibian Z, Yezhen L, Xiaoxue L, Qing C, Longxi L, Mingluan X, et al. Studying the cytotoxicity and oxidative stress induced by two kinds of bentonite particles on human B lymphoblast cells in vitro. Chem-Biol Interact 2010;183:390−6.

[144] Janer G, Fernández-Rosas E, MasdelMolino E, González-Gálvez D, Vilar G, López-Iglesias C, et al. In vitro toxicity of functionalised nanoclays in mainly driven by the presence of organic modifiers. Nanotoxicology 2014;8:279−94.

[145] Maisanaba S, Puerto M, Pichardo S, Jordá M, Moreno FJ, Aucejo S, et al. In vitro toxicological assessment of clays for their use in food packaging applications. Food Chem Toxicol 2013;57:266−75.

[146] Maisanaba S, Gutiérrez-Praena D, Pichardo S, Moreno FJ, Jordá M, Cameán AM, et al. Toxic effects of a modified montmorillonite clay on the human in testinal cell line Caco-2. J Appl Toxicol 2014;34:714–25.

[147] Houtman J, Maisanaba S, Puerto M, Gutiérrez-Praena D, Jordá M, Aucejo S, et al. Toxicity assessment of organomodified clays used in food contact materials on human target cell lines. Appl Clay Sci 2014;90:150–8.

[148] Jordá-Beneyto M, Ortuño N, Devis A, Aucejo S, Puerto M, Gutiérrez-Praena D, et al. Use of nanoclay platelets in food packaging materials: technical and cytotoxity approach. Food Addit Contam 2014; A31:354–64.

[149] Wagner A, Eldawud R, White A, Agarwal S, Stueckle TA, Sierros KA, et al. Toxicity evaluations of nanoclays and thermally degraded byproducts through spectroscopical and microscopical approaches. Biochim Biophys Acta, Gen Subj 2017;1861:3406–15.

[150] Liu Q, Liu Y, Xiang S, Mo X, Su S, Zhang J. A poptosisand cytotoxicity of oligo(styrene-co-acrylonitrile)-modified montmorillonite. Appl Clay Sci 2011;51:214–19.

[151] Han HK, Lee YC, Lee MY, Patil AJ, Shin HJ. Magnesium and calcium organophyllosilicates: synthesis and in vitro cytotoxicity study. ACS Appl Mater Interfaces 2011;3:2564–72.

6

Functionalized POSS-Based Hybrid Composites

Hui Zhou, Ming Hui Chua, Jianwei Xu

*INSTITUTE OF MATERIALS RESEARCH AND ENGINEERING, AGENCY FOR SCIENCE, TECHNOLOGY AND RESEARCH (A*STAR), SINGAPORE*

6.1 Introduction

Hybrid composites, in general, are defined as composites that consist of one organic polymer and one or several inorganic additives/fillers at the nanometer or molecular level. Introduction of an inorganic component in a polymer system usually leads to a change in polymer properties including but not limited to mechanical, electronic, thermal, optical properties. To achieve polymeric materials with certain desired properties, inorganic fillers with different structures and geometric shapes (such as fibers, whiskers, platelets, and spheres) are dispersed either physically or chemically in polymeric materials systems at a macro or nanoscale level. This ubiquitous methodology can conveniently combine good processability of thermoplastic polymers together with the advantages of inorganic fillers added, for example, high modulus, oxidation resistance, good thermal stability, etc. Furthermore, once the filler size is reduced to smaller than 100 nm, the resulting nanocomposites may result in drastic alternations in a specific characteristic of hybrid materials, for instance, gas barrier, thermal stability, elasticity, and mechanical properties. The enhanced performance is partially attributed to the changes in surface properties and interfacial interactions between polymer matrix and fillers. Further size reduction of nanosized additive causes substantial increase in surface energy of nanoparticles and subsequent aggregation to discourage the homogeneity of resulted hybrid composites. To solve this issue, the surface of nanoparticles is commonly modified with organic functional groups that are compatible with polymeric materials, then followed by blending, mixing, or grafting with polymer matrix to form hybrid materials.

Polyhedral oligosilsesquioxane (POSS) is a unique 3D nanoparticle and it has been investigated for more than 50 years [1]. The chemical structures of silsesquioxane follow the formula of $(RSiO_{1.5})_n$, in which R indicates the vertex groups, such as hydrogen, halogen, alkyl, alkene, aryl, arylene, etc. [2]. Silsesquioxanes can adopt caged or noncaged structures (Scheme 6–1). The noncaged silsesquioxane would be further classified into (A) partial cage, (B) ladder, and (C) random structures [3]. In contrast, caged silsesquioxanes are called POSS,

Polymer Composites with Functionalized Nanoparticles. DOI: https://doi.org/10.1016/B978-0-12-814064-2.00006-8

which exhibit well-defined and highly symmetric structures with a nanoscopic size of 1–3 nm in diameter including vertex groups [4]. "T" and "Q" are abbreviated from "triplet" and "quartet" from silicon NMR literature, indicating silicon atoms covalently connect to three or four oxygen atoms in POSS structures, respectively [5]. Among them, T_8 cubic POSS with a formula of $R_8Si_8O_{12}$ are the most popular system for academic study and industrial application [6].

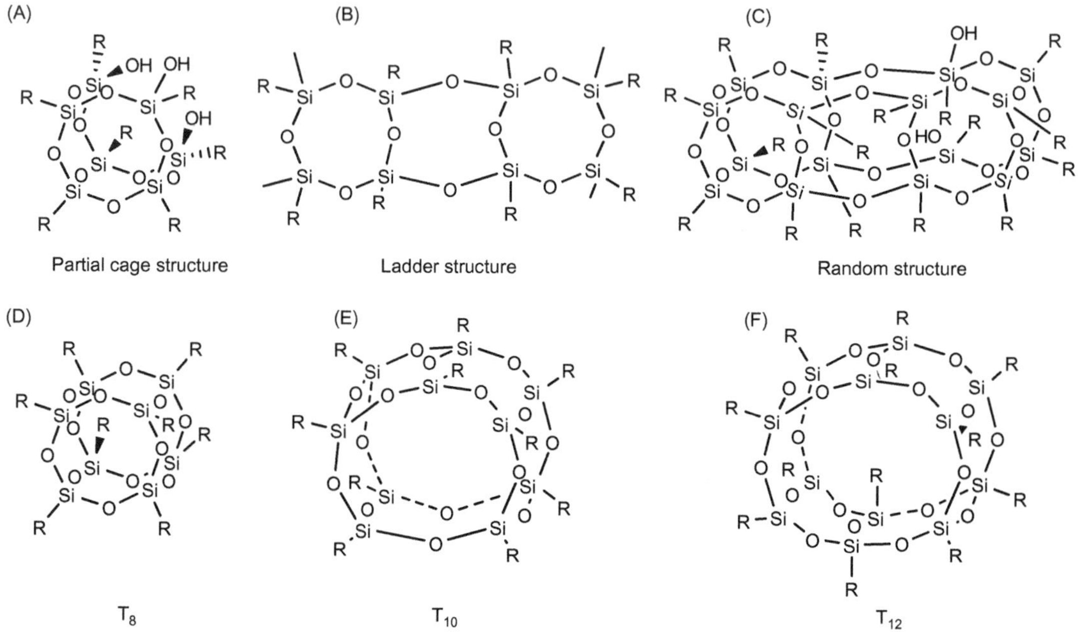

SCHEME 6–1 Chemical structures of silsesquioxanes. (I) Noncaged silsesquioxanes: (A) partial cage structures; (B) ladder; (C) random structure. (II) Caged silsesquioxanes: (D) T_8, (E) T_{10}, (F) T_{12} structures.

POSS can be dispersed in other materials at the molecular level, and thus POSS with nonreactive groups (e.g., isobutyl, cyclopentyl, and phenyl, Scheme 6–2) are always directly selected as nanoscopic fillers [7]. Nevertheless, one or more functional groups (e.g., methacrylate, styrene, norbornene, amine, epoxy, and alcohol) would be introduced into POSS as vertex groups through chemicals methods, allowing them to be covalently attached onto polymeric networks through polymerization or grafting reactions, such as atom transfer radical polymerization (ATRP), reversible addition fragmentation chain transfer polymerization (RAFT), etc. [8]. With these well-developed strategies, numerous POSS-polymer structures have been successfully constructed. Typically, POSS can be incorporated into polymeric materials through four major approaches: (1) POSS as blenders; (2) monofunctional POSS as pendant cages to graft onto the polymer main chain; (3) multifunctional POSS as the initiator centers to form a star-like polymer network; and (4) multifunctional POSS as the initiator center to generate a heavily cross-linked polymer network. The introduction of POSS into polymer chains or polymer networks can efficiently change the molecular interactions and their morphologies, resulting in improved physical properties of hybrid materials, including modulus, strength, glass transition temperature, thermal stability, structural stability, etc. [9,10].

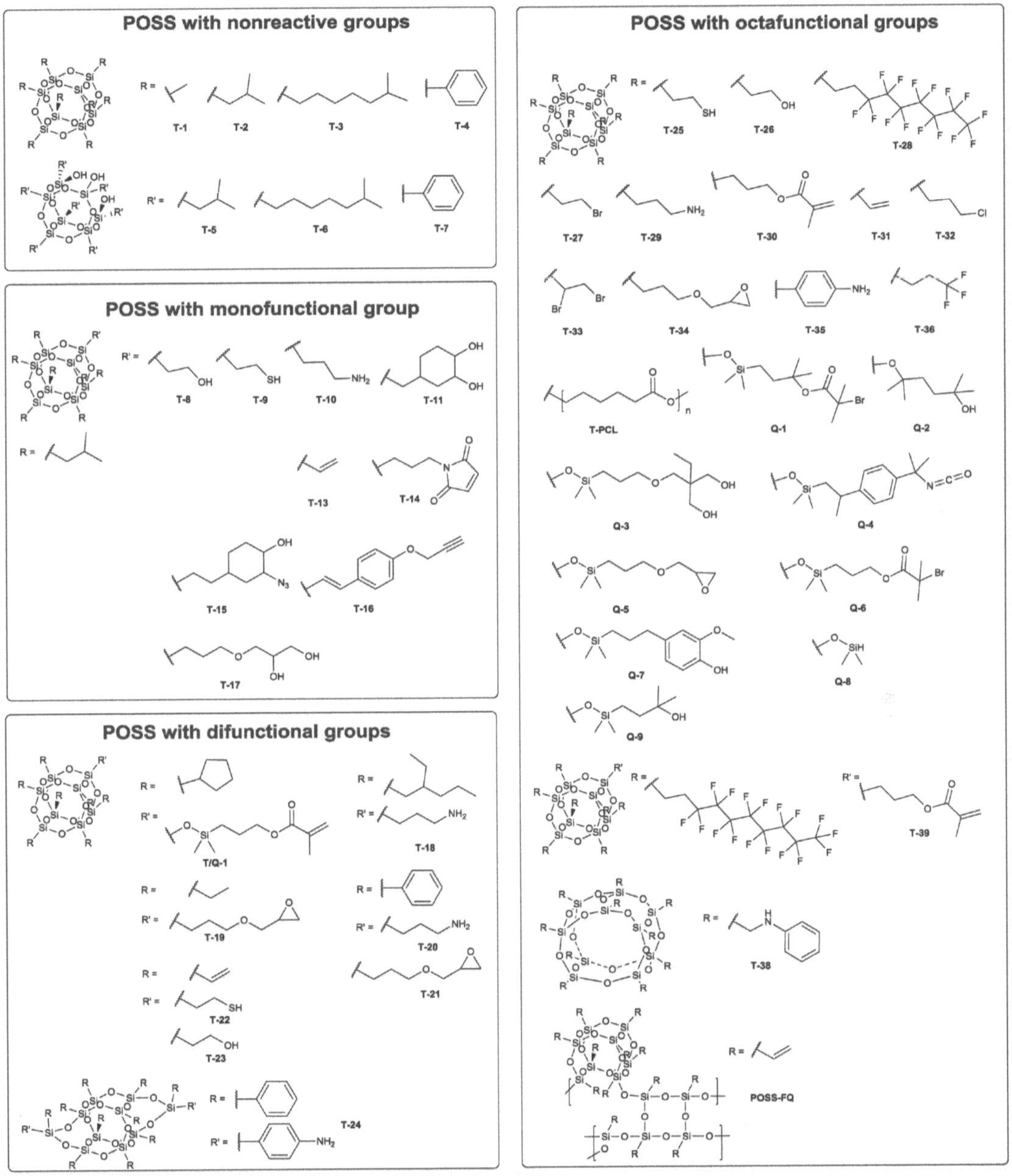

SCHEME 6–2 Summary of POSS structures cited in this chapter.

Recently, the POSS-based hybrid composites have received increasing attention for enhancement of materials performance. As shown in Fig. 6–1, the number of publications on the topic of POSS-based composites has kept increasing continuously since 2001. POSS and its derivatives have shown self-assembly capabilities at the nanoscale. The study of self-assembly of POSS-based hybrid composites at a nanoscale scale has become popular, and our latest reviews have summarized the latest developments on the self-assembly of POSS-based hybrid composites [11a]. In addition, the POSS chemistry [11b–f], applications of POSS in nanocomposites [11g–i], light-emitting diode materials [11j], biomaterials [11l–m], and catalysis systems [11o–r] have been well documented in several review papers and book chapters [11s]. As such, it would be useful to summarize the recent development of POSS-based functional hybrid composites in terms of their performance enhancement, which will directly benefit materials scientists to design and prepare POSS-based hybrid composites with ideal performance.

This chapter summarizes the latest development on POSS-based hybrid composites for performance enhancement of materials, including thermal stability, mechanical strength, electric conductivity, antiatomic oxidation (anti-AO), anti-photodegradation, optical application, crystalline, surface modification, porosity, catalytic activity, biochemistry application, and self-assembly (Fig. 6–2). In addition, this chapter also outlines the latest developments in the application of high-performance POSS composites, such as aerospace craft materials, electronic devices, fluorescence sensors, liquid crystals, coating materials, catalysts, medicine, and cosmetic products. Due to the space limitation, only selected works on POSS-based hybrid composites after 2010 are presented here.

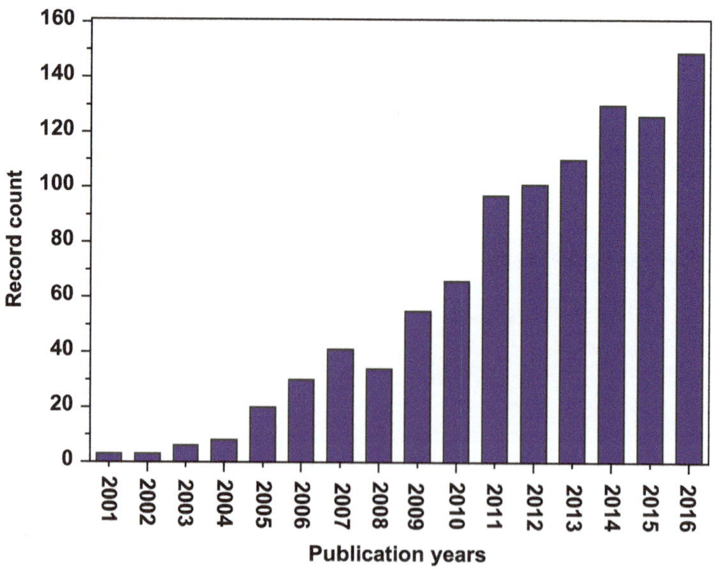

FIGURE 6–1 Publication records of POSS composites from 2001 to 2016. Data are extracted from Web of Science™ using *POSS* and *composites* as the search terms.

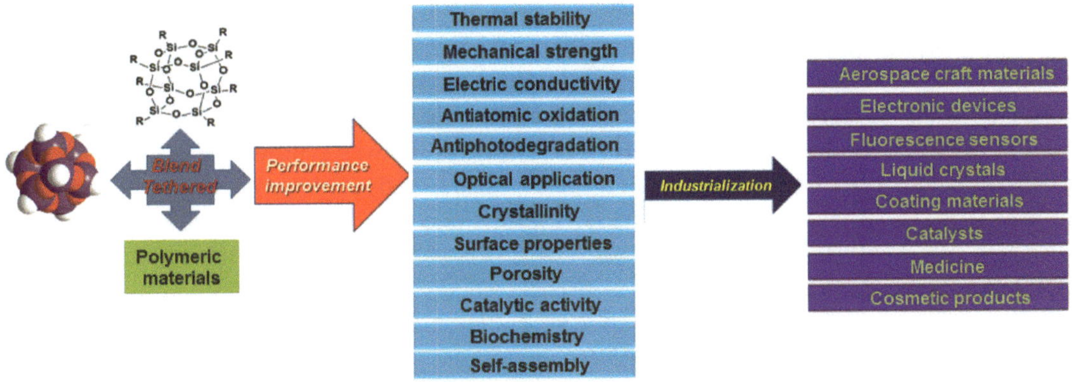

FIGURE 6–2 Summary of functionalized POSS-based hybrid composites and their related functional materials.

6.2 Polyhedral Oligosilsesquioxanes–Based Hybrid Composites for Performance Enhancement of Materials

6.2.1 Thermal Stability

Polymeric materials may undergo physical and chemical changes, such as thermal decomposition and degradation, when they are heated, thus compromising their specific application performance. As such, improved thermal stability of polymeric materials is always desired. Nanoparticle fillers, including POSS, are often incorporated into polymers to enhance thermal stability. In contrast to organic fillers, POSS have the advantage of high-temperature stability. Thus, POSS are extensively integrated into polymeric materials, such as polylactide (PLA), poly(styrene) (PS), polybenzoxazines (PBZ), and graphene oxide (GO), to improve their thermal stability.

Through electrospinning mixture of Ph-POSS and ethylene–propylene–diene monomer (EPDM), composite fibers (EPDM/POSS (**1**)) were prepared by Xue et al. [12], which showed a fibrous morphology with a diameter of the micron scale. By introducing POSS into EPDM, the thermal decomposition temperature was significantly increased while the total heat release of fibers was greatly reduced. A number of amphiphilic hybrid copolymers with thermoresponsive properties were synthesized by Li et al. via ATRP of PEGMA, MA-POSS, and PPGMA [13]. The hybrid copolymers poly(PEGMA–PPGMA–POSSMA) (PEPS) (**2**) formed a core–shell structure with a thick hydration layer in an aqueous media. The micelles of PEPS exhibited better stability than their counterpart due to the incorporation of POSS. In the denaturation process induced by heat, the thermoactive PPG brush in **2** micelles would spontaneously capture and protect the unfolded proteins to avoid aggregation with undesirable proteins, where the efficiencies of protection for lysozyme, lipase, and GFP were 89.3%, 88.7%, and 81.4%, respectively. These high protection efficiencies and good biocompatibility had made micelles **2** to be good artificial chaperones to protect unfolded proteins from toxic aggregation at high temperatures. A hybrid PS composite (POSS/PSS) (**3**) was conveniently prepared by Li et al. from mono-vinyl-substituted POSS and common PS via Friedel–Crafts reaction [14]. In comparison with pristine PS, **3** displayed improved thermal properties, crystallinity, and solubility in organic solvents. In addition, **3** can self-assemble into spheres with tunable diameters in ethanol, which could be easily adjusted by the amount of POSS loading. Huang et al. reported a PBZ matrix (PBZ-POSS) (**4**), which was synthesized via ring opening polymerization (ROP) of allyl-modified benzoxazine and POSS with eight benzoxazine groups (VBa-POSS), which displayed better thermal and mechanical properties than pristine PBZ [15]. Wu et al. [16] prepared another PBZ/POSS (**5**) composite from benzoxazine POSS (OBZ-POSS), which was synthesized via a click reaction of octa-azido functionalized POSS (OVBN3-POSS) and 3,4-dihydro-3-(prop-2-ynyl)-2H-benzoxazine (P-pa). The incorporation of POSS into their PBZs extensively enhanced thermal decomposition temperature of PBZ/POSS hybrid composite by about 50°C. The PBZ was reinforced through incorporation of octakis(dimethylsiloxypropylglycidylether)silsesquioxane (OGPOSS) via ROP of benzoxazine precursors and OGPOSS (up to 10 wt%). The resulting composite PBZ-OGPOSS (**6**) also showed improved thermal stability [17]. Wang et al. reduced functionalize GO to form octa-aminophenyl-POSS (OapPOSS)-rGO (**7**) through direct refluxing GO and OapPOSS without the use of any reducing agents (Fig. 6–3) [18]. Due to the adsorption and barrier effect of **7**, the heat and gas release is inhibited, whereas the formation of graphitized carbons is promoted. In addition, the thermal-oxidative resistance of char layer in **7** is enhanced extensively, leading to its dramatically reduced fire hazards. Bourbigot et al. studied the fire retardancy properties of polyurethane (PU) and POSS-FQ composites. It was found that the peak of heat release rate of the composite was reduced after addition of POSS-EQ. This is due to the formation of an insulating layer at the surface of substrate to inhibit heat and mass transfer and thus to decrease the heat release rate [19]. Michałowski et al. also reported that the flame retardancy property of PU was enhanced by incorporation of octa(3-hydroxy-3-methylbutyldimethylsiloxy) POSS (OCTAPOSS) as a cross-linker [20]. Devaux et al. [21] reported a method to improve the flame retardancy of textile structures through PU/POSS composite **(8)** coating on the surface of polyester or cotton fabrics. Wang et al. prepared biodegradable and biocompatible PEU/POSS composites **(9)** with improved biocompatibility, thermal and hydrolytic stability, stiffness, and strength. With a very low POSS concentration (6 wt%), PEU/POSS composite **9** exhibited significant improvements in mechanical properties and degradation resistance. Composite **9** also showed good support for cell growth without toxicity, indicating that composite **9** has potential for future high-end uses such as biomedical devices, especially at cardiovascular interfaces [22].

6.2.2 Mechanical Strength

Polymeric materials are widely used in engineering because of their light weight but high mechanical strength. For example, polyamide (PA) is one kind of engineering material that is widely used in sportswear, carpets, and automotive applications. POSS has been incorporated into PA to further improve its performance. Gnanasekaran et al. prepared a bionanocomposite membrane from POSS/PA (**10**) with no cracks or voids for gas transport study

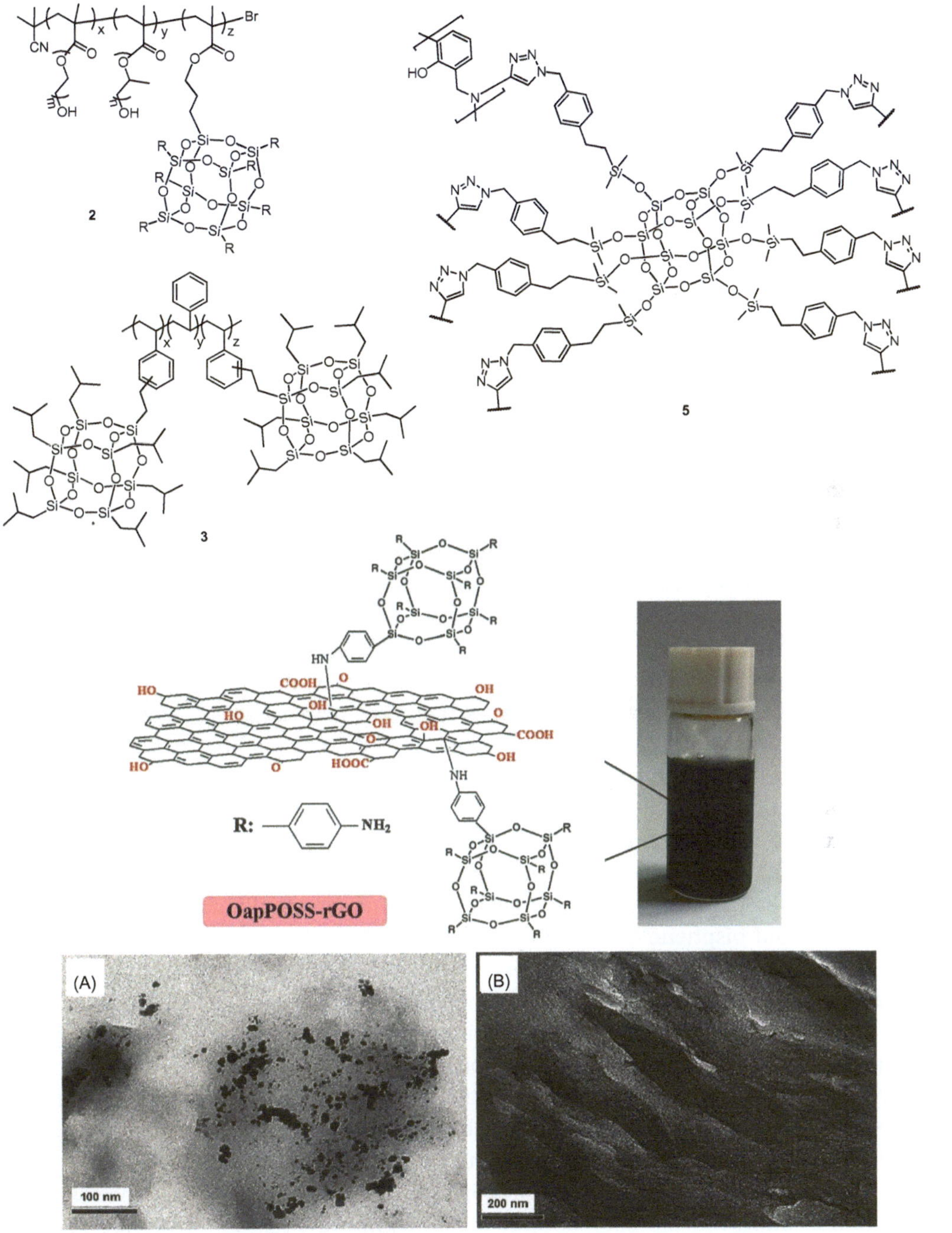

FIGURE 6–3 Chemical structure of PEPS **2**, POSS/PSS **3**, PBZ/POSS **5**, and graphene oxide (OapPOSS-rGO) **7**. Inserted is the photograph of **7** dispersions in THF. TEM images of (A) OapPOSS-rGO and (B) OapPOSS-rGO composites **7**. *Adapted from Wang X, Song L, Yang H, Xing W, Kandola B, Hu Y. J Mater Chem 2012;22:22037–43. Copyright 2012 Royal Society of Chemistry.*

[23]. The physical and gas transport property of the membranes would be improved significantly via adjusting POSS concentration because increasing POSS content will increase T_g, surface roughness and density of membranes, as well as decrease the fractional free volume and surface free energy of membranes. Thus, **10** showed high potential for the application in the food packaging and petrochemical industries. The aromatic PA dendrimers (**11**) were synthesized by Matsumoto et al. with a high yield up to the sixth generation via condensation reaction of octa(*n*-propylamine)-POSS and 3,5-bis(trifluoroacetamido)benzoic acid [24]. The dendrimers showed squashed shapes even after the sixth generation. Its fluorescence emission was enhanced with the increased generation, and the one with trifluoroacetamide terminal emitted at 445 nm under UV excitation. Milliman et al. [25] synthesized nylon 6/POSS (**12**) through melt-blending of aminopropylisobutyl POSS and nylon 6, and studied the effect of processing method on performance of composites . With the same composition, the melt-spun composite fibers showed better modulus and strength than that from injection molded blends, and the best result was obtained when the POSS loading value is 2.5 wt%. The improved properties were mainly attributed to the physical cross-linking in composites. The thermoplastic elastomer hybrids hyperbranched polyurethane (HBP-POSS) (**13)** were prepared by Mahapatra et al. by using $A_2 + B_3$ approach [26]. Hybrid nanomaterials with different properties were easily obtained through changing the feeding ratio of triethanolamine, 4,4′-methylenebis(phenyl isocyanate), POSS-diol, and poly(ε-caprolactone) (PCL)-diol. The nonagglomerated composite can be synthesized by covalent incorporation of POSS into polymeric network (Fig. 6–4A–C). The mechanical properties and thermal stability of composites can be significantly enhanced by the incorporation of POSS (Fig. 6–4D–E). Prządka et al. [27] prepared hybrid elastomers (PU-NCOPOSS (**14**)) using NCOPOSS as the cross-linker. The incorporation of NCOPOSS with 1.2–2.5 wt% loading would improve the mechanical properties of polymers. However, further increase of NCOPOSS concentration caused the reduction of mechanical strength due to an over-cross-linking. Sethuraman et al. studied the effect of NH_2–POSS weight percentages (1%, 3%, and 5%) on the reinforcement of diglycidyl ether bisphenol A (DGEBA) epoxy resin (nBGE-BDGE-POSS (**15**)), where the matrix was further cured by diamines, such as capron-based diamine, octane diol-based ether diamine, and bisphenol-A-based ether diamine [28]. Due to their better performance and improved lifespan, the resulting epoxy matrix can be used for coating, encapsulation, and sealant materials in industry and engineering. Copolymers (nBGE-BDGE-POSS (**16**)) were prepared by Ni et al. from POSS-NH_2 with *n*-butyl glycidyl ether (nBGE) and 1,4-butanediol diglycidyl ether (BDGE) [29]. **16** can be used to improve the toughness and thermal stability of the epoxy matrix by forming an anchor structure in the polymeric networks.

A thermoplastic elastomeric composites based on polystyrene–butadiene–polystyrene (SBS)/POSS (**17**) was fabricated by Bai et al., through a UV-induced thiol–ene click reaction between thiol groups in POSS and double bonds in SBS [30]. The covalent incorporation of POSS into SBS improved the compatibility of the composite, leading to an ordered microphase morphology. The enhanced mechanical properties of SBS may be caused by the morphology changes from SBS's circle inland structure to **17**'s lamella structure. Jiang et al. reinforced the interfacial strength of carbon fiber (CF) with unsaturated polyester resin

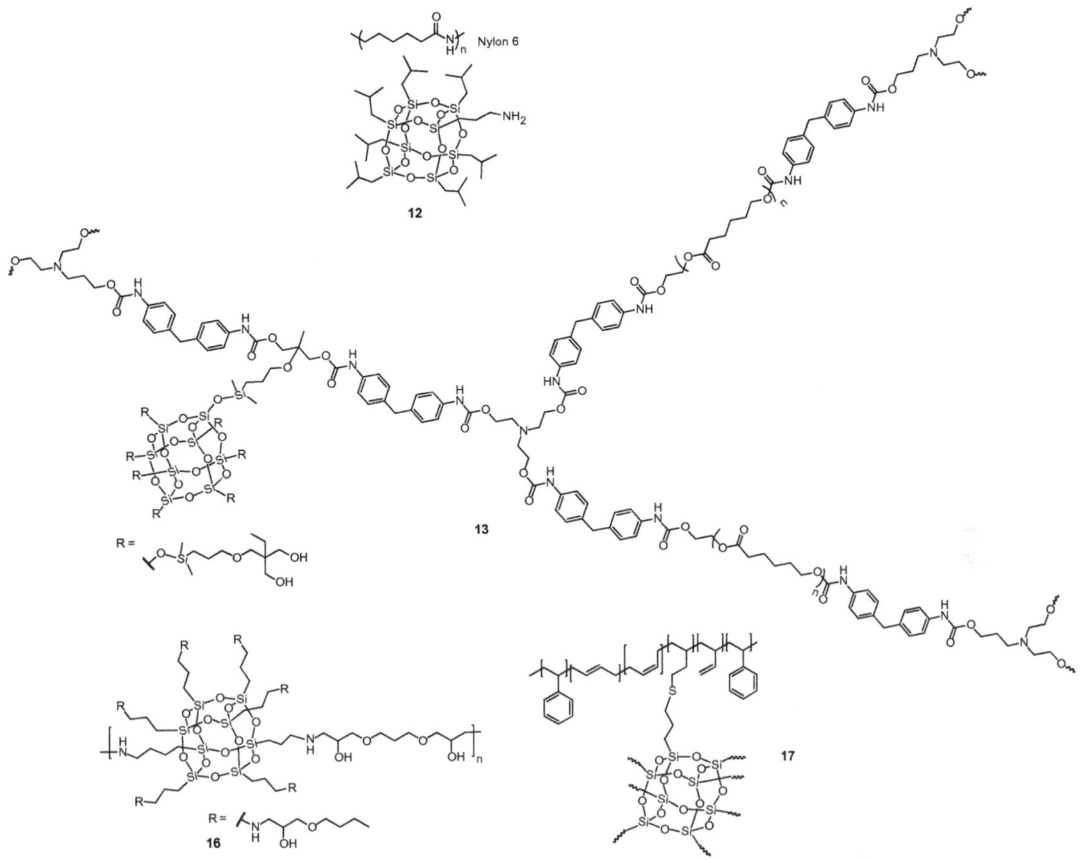

FIGURE 6–4 Chemical structure of nylon 6/POSS **12**, HBP **13**, *n*BGE-BDGE-POSS **16**, and SBS/POSS **17**. (A) Stress–strain curves and (B) moduli of (I) pure HBP (**13**), (II) HBP-POSS-2.5, (III) HBP-POSS-5.0, and (IV) HBP-POSS-7. SEM images of (C) carbon fiber, (D) CF–POSS, and (E) CF–POSS–CNT **19**. *(A–B) Adapted from Mahapatra SS, Yadav SK, Cho JW. React Funct Polym 2012;72:227–32. Copyright 2012 Elsevier Ltd. (C–E) Adapted from Zhao F, Huang Y, Liu L, Bai Y, Xu L. Carbon 2011;49:2624–32. Copyright 2011 Elsevier Ltd.*

(UPR) composites, in which methacryl or methacrylolsobutyl POSS could be uniformly grafted on the surface of CF via a silanization reaction (Fig. 6–5) [31a]. The resulting composites (**18**) exhibited almost similar wettability and roughness of UPR but higher mechanical strength than pristine UPR. Compared with UPR composites reinforced by monomethacrylolsobutyl grafted CF, the UPR composites reinforced by multimethacryl POSS grafted CF showed higher interlaminar shear strength, interfacial shear strength, and impact energy as 67, 93 MPa, and 1.72 J, respectively. Zhao et al. [31b] reported a method to reinforce CF via grafting carbon nanotubes (CNTs) on the surface of CF using octaglycidyldimethylsilyl POSS as a linkage. The interfacial adhesion of resulting CF/POSS/CNT (**19**) was effectively enhanced by increased covalent bonding, mechanical interlocking, improved resin wettability, and increased surface roughness.

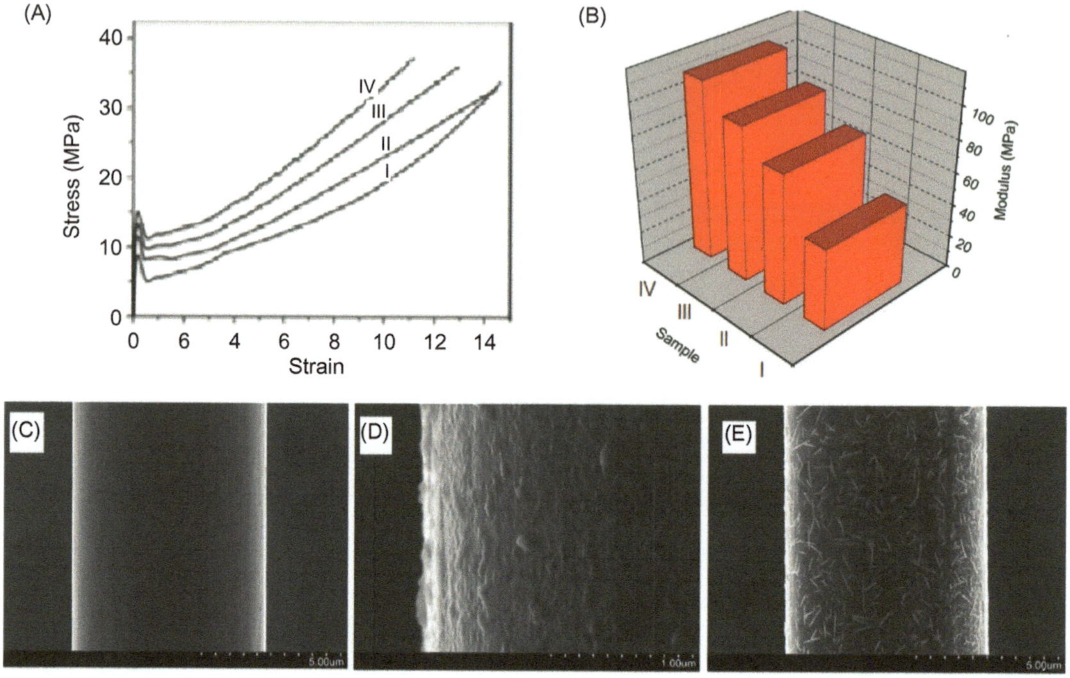

FIGURE 6–4 (*Continued*).

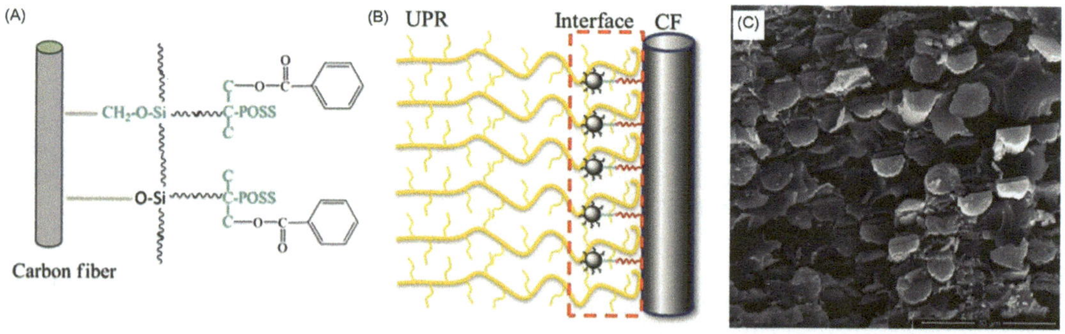

FIGURE 6–5 (A) Chemical structure of CF-UPR **18** and the image of the unsizing CF. (B) 3D AFM images of methacrylolsobutyl POSS **T-12** grafted CF. (C) Morphologies of fractured surface of UPR composites reinforced with **T-12** grafted CF. *Adapted from Jiang D, Xing L, Liu L, Yan X, Guo J, Zhang X, et al. J Mater Chem A 2014;2:18293–303. Copyright 2015 Royal Society of Chemistry.*

6.2.3 Electric Conductivity

Electrolytes are a necessary component in many electronic and energy storage devices, such as solar cells, electrochromic devices, batteries, electrochemical devices, etc. Currently, two major types of electrolytes, liquid and solid electrolytes, have been developed. While liquid electrolytes are generally widely used, they suffer from some drawbacks, such as weak mechanical stability, electrode corrosion, limited operation temperature range, flammability, as well as leakage. To solve these issues, solid electrolyte may be a good alternative to liquid electrolyte. For these purposes, POSS has been incorporated into electrolyte materials to improve their properties. The hybrid branched (BCPs (**20**)) and linear (LCPs)-graft copolymers (LCPs (**21**)) were synthesized by Shim et al. though RAFT polymerization of ethylene glycol dimethacrylate (EGDMA), PEGMA, and MA-POSS, which would be used in high-temperature lithium ion batteries [32]. When the contents of MA-POSS were more than 21 mol%, the structural dimension of BCPs and LCPs could be well maintained even at the working temperature up to 90°C. Under the same condition at 60°C, the BCPs electrolyte showed higher ionic conductivity (1.6×10^{-4} S/cm) than LCPs electrolyte (5.6×10^{-5} S/cm) due to its higher chain mobility. All batteries containing BCPs electrolyte showed good performance without safety problems at 60°C, showing potential for use in high-temperature batteries. Polu et al. investigated the mechanical, thermal, and ionic conductivity properties of solid polymer electrolytes composited by poly(ethylene oxide) (PEO)/POSS-PEG (**22**) and $LiN(-SO_2CF_3)_2$ (LiTFSI) [33]. **22** efficiently inhabited the segmental motion of polymer chains, and improved the ionic conductivity to 5.05×10^{-5} S/cm at 23°C when POSS-PEG loading is 30 wt%. Kim et al. investigated the effects of PEs (**23**) on the performance of composite electrolytes (CE), which is composed of LiTFSI, star-shaped polymer SPP13 (**24**), and poly(ethylene glycol) (PEG)-functionalized POSS derivatives [34]. SPP13 **24** was obtained though polymerization reaction of methacryl-cyclohexyl-POSS and PEG methyl ether methacrylate. The solid CE showed high ionic conductivity of 4.5×10^{-5} S/cm at 30°C, which was almost three times of that of pure SPP13 electrolyte. The solid CE was electrochemically stable without decomposition even when the voltage is up to + 4.2 V. The solid lithium battery using CE electrolyte showed a larger discharge capacity than that using pure SPP13 as electrolyte. Wu et al. synthesized exchange membranes containing $(phenyl)_8$-bisaniline-POSS in the backbone of linear sulfonated polyimides (PI) (SPIs (**25**)) with molar concentrations of 3%, 5%, and 10% [35]. In comparison with other POSS hybrid membranes formed by blending or cross-linking, these membranes showed more uniform distribution of POSS aggregates, and improved performance for proton exchanges (Fig. 6–6). The membranes with 5 and 10 mol% POSS displayed ion conductivity of 0.114 and 0.132 S/cm at 80°C, respectively, which is higher than 0.0998 S/cm for POSS-free SPI membrane. POSS-SPI membranes also displayed improved antioxidative and antihydrolytic capabilities, low swelling ratio with water, and low methanol permeability. Nezakati et al. [36] prepared an electrically conductive composite (POSS–PCL/graphene (**26**)) through introduction of POSS into poly(caprolactone-based urea-urethane) (PCL)/graphene hybrid material. Graphene flakes with 8 nm thickness could be homogeneously dispersed into POSS–PCL hybrid material with different weight concentrations of 0.1%, 2%, 5%, and 10%. The composite with a higher graphene concentration (>5.0 wt%) in POSS–PCL displayed higher conductivity than pristine POSS–PCL (Fig. 6–7).

POSS are also introduced to polymeric materials to enhance their dielectric properties. Sharma et al. [37] reported a material with an ultralow dielectric constant of 1.53 from

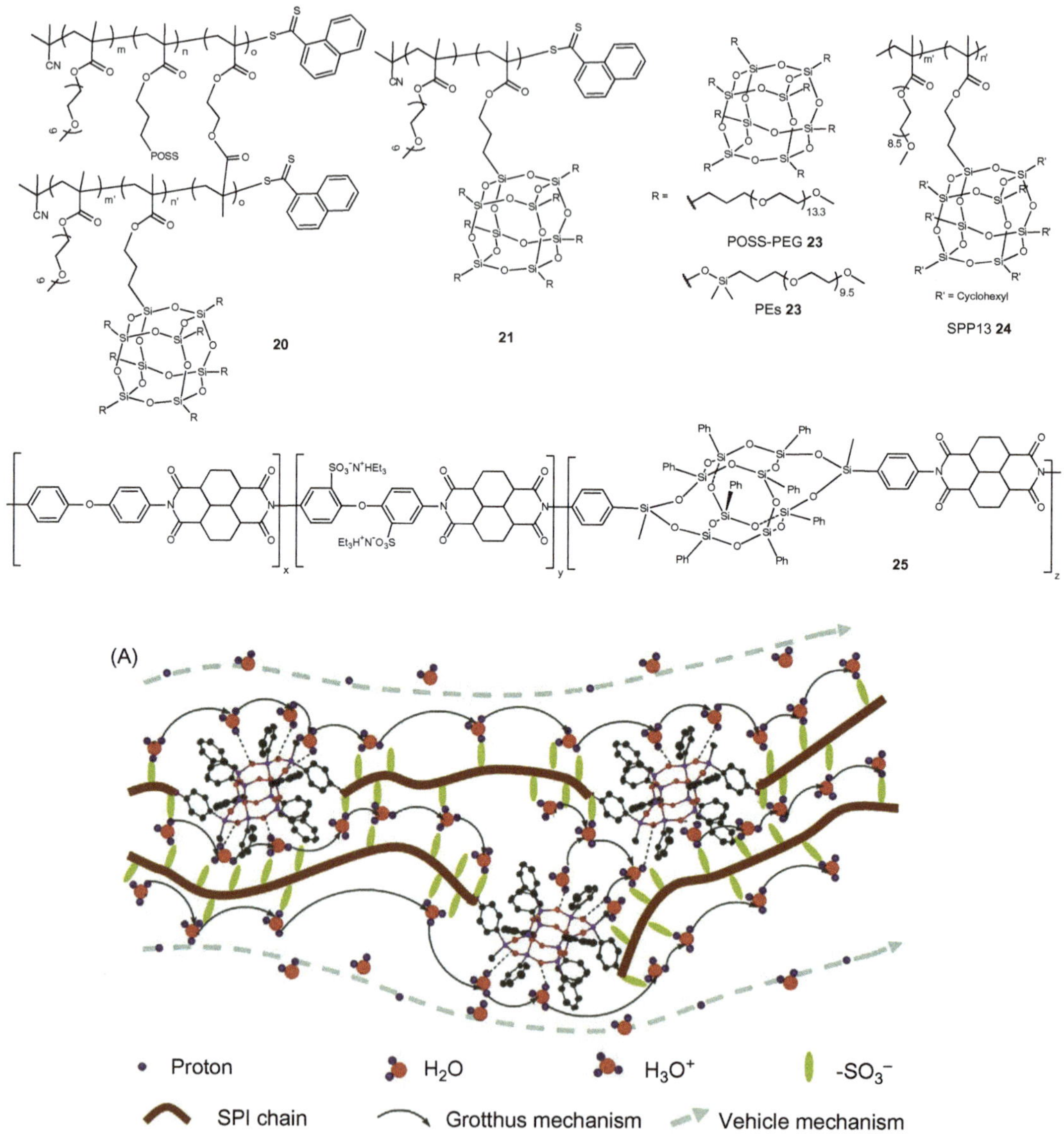

FIGURE 6–6 Chemical structure of BCPs **20**, LCPs **21**, PEO/POSS-PEG **22**, PEs **23**, SPP13 **24**, and SPIs **25**. Proposed mechanisms for proton transportation within SPIs **25** membranes. *Adapted from Wu Z, Zhang S, Li H, Liang Y, Qi Z, Xu Y, et al. J Power Sources 2015;290:42–52. Copyright 2015 Elsevier Ltd.*

fluorodecyl-POSS (FD-POSS) and perfluorocyclopentenyl aryl ether polymers (PFCP). The dielectric constant of POSS-PFCP composites were dramatically reduced with increase of weight percentages of FD-POSS. A material POSS–PU–PBZ **28** with a low *k* value was prepared by Sasi Kumar et al. [38] for microelectronics applications. The *k* value of the material

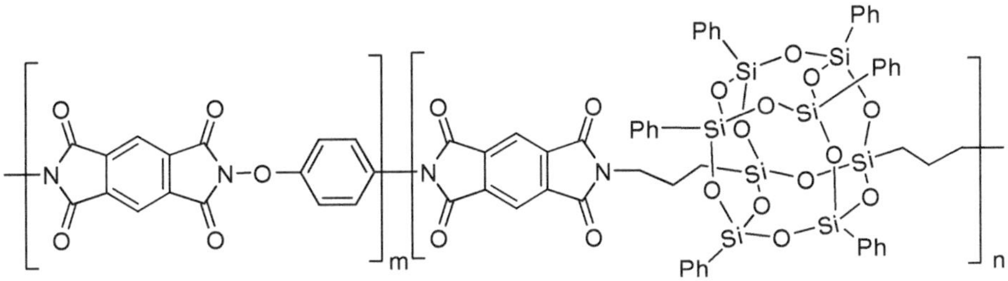

32

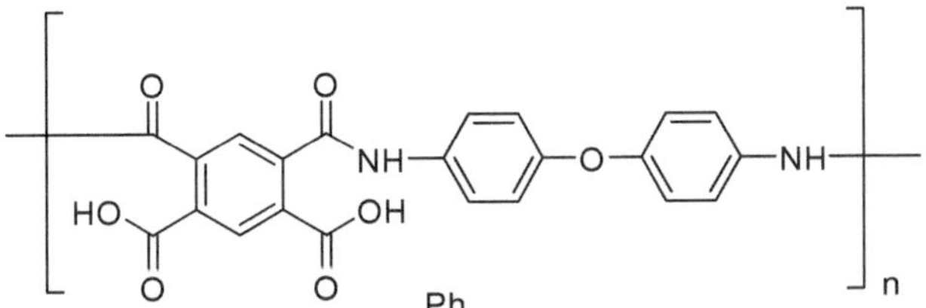

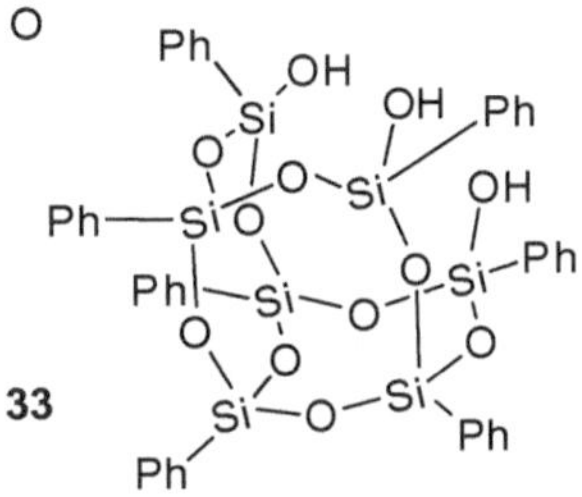

33

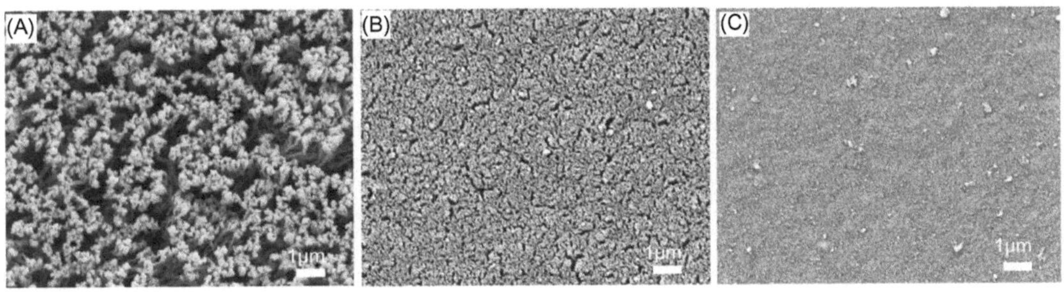

FIGURE 6–7 Chemical structure of POSS-PI **32** and PI/POSS **33**. SEM images of polyimide membranes **32** after exposure to atomic oxygen fluence of 3.87×10^{20} O atoms/cm^2: (A) polyimide, (B) polyimide with 8.8 wt% POSS, (C) polyimide with 21.9 wt% POSS. Note: The scale bar at the bottom of each image indicates 1 μm. *Adapted from Lei X-F, Qiao M-T, Tian L-D, Yao P, Ma Y, Zhang H-P, Zhang Q-Y. Corros Sci 2015;90:223–8. Copyright 2014 Elsevier Ltd.*

was significantly reduced by the introduction of POSS into the matrix. However, the reduction trend of the k value is limited by the POSS–PU–PBZ concentration, and the reverse trend was found at the POSS–PU–PBZ concentration of 30% (k ¼ 1.94) due to POSS agglomeration. PBZ-POSS (**29**) nanocomposites were synthesized via polymerization of methylmethacrylate-POSS and furan-containing benzoxazine by Tseng et al. [39]. The k value of **29** was increased to 1.9 after integration of POSS. Through grafting of 1-allyl-3-methylimidazolium hexafluorophosphate ionic liquid and octa-methylmethacrylate-POSS on the surface of CNT, a composite (rGO-OapPOSS (**30**)) was successfully prepared by Li et al. [40]. Because of the POSS coating layer on the CNT surface, **30** showed high *k* value and low dielectric loss. A hybrid composite Oap-POSS-g-GO (**31**) was prepared by Yu et al. through functionalization graphene with Oap-POSS [41]. **31** showed significantly improved thermal stability than GO (T_d from 75°C to 395°C). Nevertheless, compared with the neat epoxy composites, the *k* and dielectric loss of **31**/epoxy composites (only 0.2 wt% **31**) dropped by 9% and 49%, respectively.

6.2.4 Antiatomic Oxidation

Anti-AO materials are needed to protect spacecraft devices. For example, PI is used as anti-AO material in the external surfaces of the spacecraft, thermal blankets in solar arrays, and space inflatable structures because of its excellent anti-AO performance. To further improve the protective capability, additives such as POSS have been incorporated into PI to enhance their AO resistant properties [11f,57]. POSS-PI matrix would form a SiO_2 layer under AO exposure, which will prevent AO from corroding the inner bulk matrix, leading to increased AO resistance.

Lei et al. [42] prepared POSS-PI (**32**) composites through introduction of diamine POSS to PI backbones via copolymerizing. PI became roughened and showed a "carpet-like" morphology after AO exposure. In contrast, **32** with a POSS weight concentration of 21.9% exhibited less root-mean-square surface roughness, and is likely to form a denser and well-connected passivating layer on the surface after AO exposure, leading to improved AO durability of **32**. **32** with 29.7 wt% POSS exhibited the best AO resistance with a lowest erosion yield of 0.9×10^{-25} cm^3/atom. Therefore, POSS significantly increased AO survivability of PI. These advantages together with good mechanical strength and thermal stability of PI make **32** a potential material for surface protection on spacecraft to avoid AO attack. A PI/POSS (**33**) hybrid nanocomposite was prepared by Duo et al. [43] through polymerization of pyromellitic dianhydride, 4,4′-oxydianiline, and trisilanolphenyl-POSS. Through incorporation of 20 wt% **33** into PI, the erosion yield of the resulting film was reduced from 4.3% cm^3/atom to 1.2×10^{-25} cm^3/atom. To enhance the poly(vinylidene fluoride) (PVDF) stability to AO, Zeng et al. [44] studied the contribution of FP-POSS to PVDF matrix. As the stable core in FP-POSS, the PVDF/FP-POSS matrix (**34**) exhibited enhanced AO durability. This work presented valuable information for materials scientists to design and prepare novel PVDF materials to avoid AO corroding in LEO environment.

6.2.5 Anti-Photodegradation

POSS were integrated into PS to enhance its resistance against photo-induced degradation [45]. With the same condition, POSS-PS (**35**) displayed a much lower concentration of hydroxy and carbonyl groups than pristine PS, indicating its protective role of POSS, and subsequently extending the lifetime of the composites. The partial-caged POSS with triols likely exhibit better protective efficiency than the caged POSS, suggesting the effective role of OH groups in PS stabilization. A POSS reinforced PBZ/EP (**36**) composites were prepared by Selvi et al. through thermocuring matrix of POSS and BZ/DGEBA EP. A series of composites were fabricated through varying weight percentage of POSS to mixture of DGEBA EP and BZ (1:1, w/w). Among them, only PBZ/EP with 5.0% POSS showed significantly enhanced tensile strength [46]. Scanning electron microscopy (SEM) images shown in Fig. 6–8 suggested a passivating silica layer being generated upon UV radiation, which effectively protected the inner composite from further corroding under UV radiation (Fig. 6–8A–C).

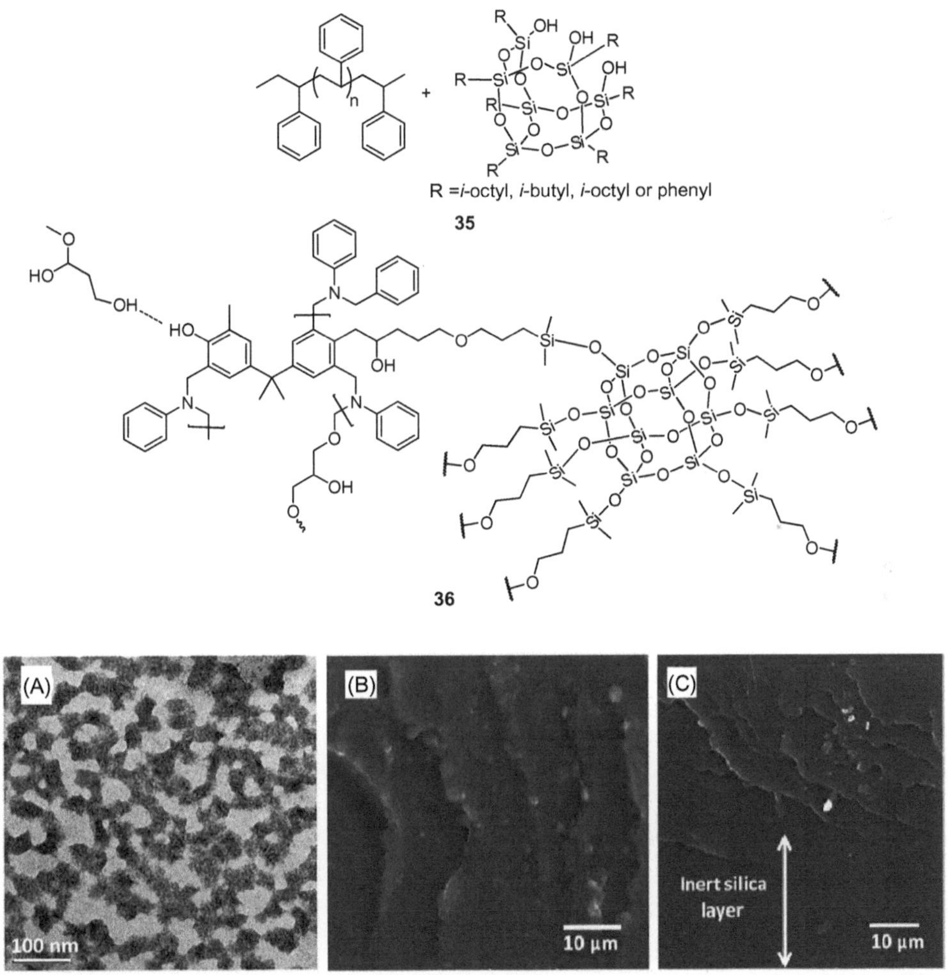

FIGURE 6–8 Chemical structure of POSS-PS **35** and PBZ/EP/POSS nanocomposites **36**. (A) TEM image of **36** with 5.0 wt% POSS. SEM images for before (B) and after (C) UV radiation of **36** with 5.0 wt% POSS. *Adapted from Selvi M, Devaraju S, Vengatesan MR, Go JS, Kumard M, Alagar M. RSC Adv 2014;4:8238–44. Copyright 2014 Royal Society of Chemistry.*

6.2.6 Optical Application

POSS and related compounds are able to adjust the optical transparency of their corresponding polymer hybrid materials. For example, Lungu et al. [47] incorporated POSS into dimethacrylate (UDMA) matrix, and used differential scanning calorimetry (DSC) to study the conversion of methacrylic groups in UDMA-POSS (**37**), which was prepared from urethane UDMA monomer and POSS derivatives, including HISO-POSS, CPENTYLPOSS and MA-POSS. Due to the formation of agglomerates, the transparency of **37** films was decreased significantly. A polymer poly{[9,9-di(hexyl)fluorene]-*alt-co*-[4,7-bis(thiophene-2-yl)-2,1,3-benzothiazole]} (PFDBT-POSS (**38**)) with POSS as pendant groups was synthesized by Liu et al. [48], which would form uniform nanoparticles in an aqueous condition containing matrix of DSPE-PEG2000-Mal and DSPE-PEG2000. **38's** nanoparticles exhibit absorption at 536 nm and a Far-red/NIR emission at 687 nm. The fluorescence quantum yield of **38** increased from 2% to 13.6% via incorporation of POSS onto the side chains, suggesting inhibition of POSS on fluorescence quenching. As shown in Fig. 6–9, the prepared nanoparticle

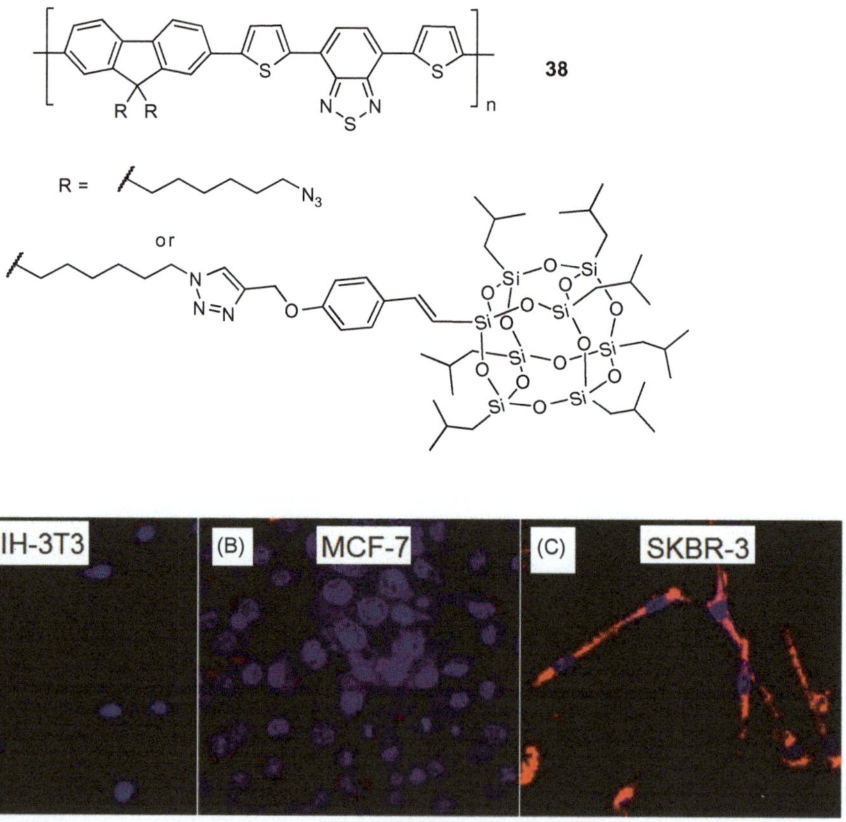

FIGURE 6–9 Chemical structure of PFDBT-POSS **38**. CLSM fluorescence images of cancer cells incubated overnight with 5 nM **38**-Affibody NPs at 37°C: (A) NIH-3T3, (B) MCF-7, and (C) SKBR-3. Scale bar is 30 mm. *Adapted from Liu J, Feng G, Ding D, Liu B. Polym Chem 2013;4:4326–34. Copyright 2013 Royal Society of Chemistry.*

probe PFDBT-POSS-Affibody can be used for cell imaging of HER2 with high photostability and low cytotoxicity, in which MCF-7, NIH-3T3, and SKBR-3 were expressed as cancer cells.

6.2.7 Crystallinity

POSS have been used to develop liquid crystals with various mesophases. A mesogen-jacketed liquid crystalline (LC) material (MJLCP, PNb10POSS (**39**)) was synthesized via ROP of norbornene, where POSS are introduced into polymer chain as pendant groups [49]. As the competitive self-assemblies between POSS and MJLCP groups, **39** exhibited various phase change structures at different temperature, such as the hexagonal columnar phase, and angstrom POSS phase. In addition, the complex phase behavior of **39** is also generated by the POSS effect on the LC behavior of MJLCP. In addition, **39** self-assembled into a hybrid complex at sub-10-nm scale, providing a new method to fabricate nanostructures with a sub-10-nm scale (Fig. 6–10A). PLA-g-MA/POSS-NH_2 (**40**) or PLA-g-MA/POSS-OH (**41**) were

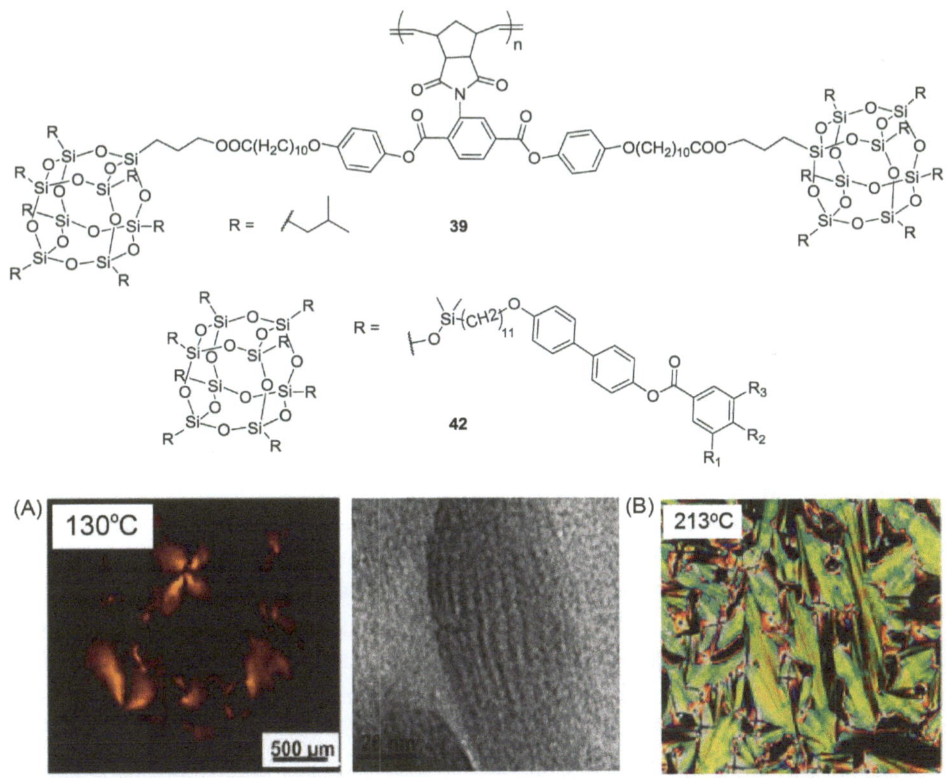

FIGURE 6–10 Chemical structure of PNb10POSS **39** and POSS-LCs **42**. (A) PLM micrographs of **39** at 130°C (left), HR-TEM micrograph of **39** (right). (B) POM image of **43** ($R_1 = R_3 = H$, $2 = -OC_{12}H_{25}$) at 213°C upon cooling. *(A) Adapted from Hou PP, Gu KH, Zhu YF, Zhang ZY, Wang Q, Pan HB, et al. RSC Adv 2015;5:70163–71. Copyright 2015 Royal Society of Chemistry. (B) Adapted from Wang X, Cho CM, Say WY, Tan AYX, He C, Chan HSO, et al. J Mater Chem 2011;21:5248–57. Copyright 2011 Royal Society of Chemistry.*

prepared by Gardella et al. [50] through functionalization polymer matrix with POSS. In this method, 0.7 wt% of POSS-OH would be grafted onto PLA-*g*-MA backbone without any reduction of polymer molecular weight. The crystallinity of PLA was improved by grafting POSS on the polymer backbone due to the effect of crystal nucleation density. Our group also synthesized several LCs (POSS-LCs (**42**)) via a platinum catalyzed hydrosilylation, in which only POSS contained LCs with a single flexible chain as mesogens displayed smectic mesophases (Fig. 6–10B) [51]. The different chains on the POSS core would affect the formation and properties of the mesophase. During the melting process, the steric repulsion induced by excessive chains on POSS could inhabit the reorganization of the flexible region, consequently generating the isotropic phase but not the mesophase.

6.2.8 Surface Properties

Functional surfaces have attracted increasing attention due to their wide applications, including chemical shielding, functional coatings such as superhydrophobic oleophobic coating, stain free textiles, etc. POSS modified with fluorinated segments is a major class of materials with extra low surface energy for functional surface applications. The first covalently bound F-POSS nanocomposites F-POSS/MMA (**43**) was synthesized by Ramirez et al. [52] by RAFT polymerization of MMA and F-POSS macromer, which displayed excellent wetting resistant properties. The cotton fibers coated with **43** were superhydrophobic and oleophobic. A POSS-capped PEO (**44**) nanocomposite was prepared by Wang et al. [53] through introduction of a heptaphenyl-POSS-capped PEO telechelics into epoxy via the Huisgen 1,3-dipolar cycloaddition of α,ω-dialkynyl-terminated PEO and 3-azidopropylheptaphenyl POSS . The integration of **44** significantly reduced the surface free energy and hydrophobicity because of the enrichment of POSS at the surface. The copolymer poly(hydroxylether of bisphenol A) (PH-POSS (**45**)) containing POSS in the backbone was prepared by Wang et al. [54] from bisphenol A and diepoxy. Through changing the ratio of 3,13-diglycidyloxypropyloctaphenyl double decker silsesquioxane to DGEBA, the PH copolymers with different POSS contents were produced. **45** displayed not only improved thermal stability but also increased surface hydrophobicity when compared with pristine PH (Fig. 6–11B). A series of membranes were prepared from Pebax/POSS (**46**) by Le et al. [55], which were used for the recovery of ethanol through pervaporation. The pervaporation performance of the hybrid membranes (Pebax/AL0136) was improved by incorporation of octa(3-hydroxy-3-methylbutyldimethylsiloxy) (POSS-AL0136). The permeation flux and separation factor of **46** for ethanol/water reach the highest values of 183.5 and 4.6 g/m^2h respectively, when the POSS loading value is only 2 wt%. Pan et al. [56] prepared star-shaped copolymers with (*s*-POSS-PMMA-*b*-P(MA-POSS (**47**))) from MMA and MA-POSS initiated by octakis(dibromoethyl) POSS (POSS-$(Br)_{16}$). **47** exhibited enhanced hydrophobic/oleophobic performance and thermal stability. In addition, the surface roughness of films based on **47** was also significantly improved. Thus, these diblock copolymers showed high potential for the solvent dependent coating material. A novel miscible polymer matrix was prepared by Le et al. [57] though blending of POSS-PIs (**48**) and copoly]1,5-naphthalene/3,5-benzoic acid-2,2′-bis(3,4-dicarboxy phenyl)]hexafluoropropanedimide (6FDA-NDA/DABA),

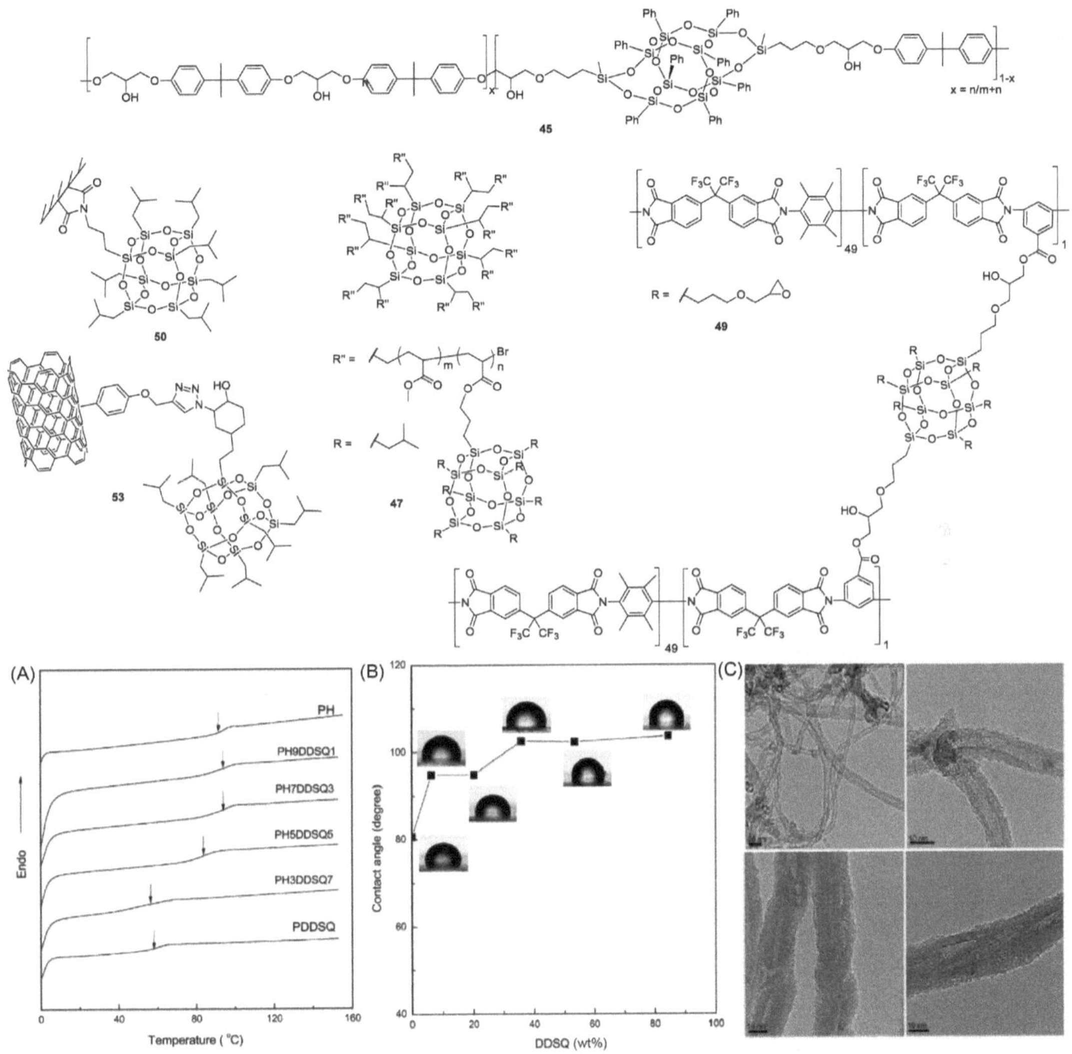

FIGURE 6–11 Chemical structure of PH-POSS **45**, MA-POSS **47**, glycidyl-POSS-PIs **49**, and PSMA/POSS **50**. (A) DSC curves of PH and PH-POSS **45**. (B) Plot of surface water contact angles as a function of the content of POSS in PH-POSS **45**. (C) TEM images of POSS–CNT **53**. *(A–B) Adapted from Wang L, Zhang C, Zheng S. J Mater Chem 2011;21:19344–52. Copyright 2011 Royal Society of Chemistry. (C) Adapted from Yadav SK, Mahapatra SS, Yoo HJ, Cho JW. Nanoscale Res Lett 2011;6:122–7. Copyright 2011 Springer.*

which can be used to fabricate fiber membranes for ethanol dehydration. The membrane selectivity would be effectively improved by the POSS modification because of the antiswelling property of POSS, inhibition of chain mobility, and turntable d-space in polymeric networks. When compared with other membranes, these hollow fibers showed superior fluxes with a factor of 42.0 kg/m^{2}h and good separation for ethanol dehydration with a factor of 4200.

A glycidyl-POSS membrane (glycidyl-POSS-PIs (**49**)) was synthesized by Konietzny et al. [58] from PI and glycidyl-POSS that showed lower fluxes due to heavy cross-linking networks. Monticell et al. [59] modified the polymer films through POSS vaporization and subsequent reaction on the surface. A polymer layer (PSMA/POSS (**50**)) was produced via grafting POSS-NH_2 on poly(styrene-co-maleic anhydride) (PSMA) with controllable layer thickness by varying the amount of POSS added as well as reaction time. The telechelic POSS–PLLA (**51**) and ditelechelic POSS–PCL–POSS (**52**) were prepared by Fernández et al. [60] through click reaction of alkyne contained POSS, azide contained poly(L-lactic acid) (PLLA), and bis-azide contained PCL. **51** and **52** showed increased T_g and retarded crystallization after POSS integration. In addition, surface hydrophobicity and thermal-oxidative stability were obviously improved due to the introduction of POSS. Yadav et al. showed an efficient method to modify CNTs surface by incorporating a variety of nanostructured materials. Among them, POSS–CNT (**53**) was prepared via click reaction of azide modified POSS and alkyne modified CNT [61].

6.2.9 Porosity

Functional porous materials have been well studied for wide applications, including gas separation, gas storage, heterogeneous catalysis, sensing, superhydrophobic coating, etc. POSS has been incorporated into functional polymers as a rigid and dendritic core to construct 3D porous polymeric materials.

Our group developed aggregation-induced emission (AIE)-active copolymers (polyacrylate-POSS (**54**) for detection of explosive vapor [62]. The porous film prepared via electrospinning of **54** showed obvious fluorescence quenching response to the vapor of dinitrotoluene (DNT) and trinitrotoluene (TNT) (Fig. 6–12A). Unlike usual fluorescent conjugated polymers, the fluorescence response of **54** porous film to explosives displayed less dependence on film thickness even as high as 560 nm. This is because the POSS groups facilitate the construction of porous structures, resulting in a high sensitivity to explosive vapor. We also prepared a POSS molecule POSS-T_8A (**55**) containing eight tetraphenylethene (TPE) units through amide linkages, which can form a hydrogen-bonded organic framework (HOF) and only be dissolved in pyridine due to dissociation of hydrogen bonding by pyridine coordination [63]. However, **55** can be well dispersed in other solvents as nanoparticles owing to its stable framework. **55** formed nanoparticles with a diameter of 291.4 nm in DMSO (Fig. 6–12C). **55** exhibited good porosity with Brunauer–Emmett–Teller (BET) surface area of 101.9 m^2/g. The fluorescence of **55** nanoparticles in DMSO could be quenched by Cu^{2+} (Fig. 6–12D) due to the dissociation of HOF structure by Cu^{2+}, which was evidenced by X-ray diffraction experiment. Interestingly, the dissociation of Cu^{2+} with HOF is reversible, and the initial state can be recovered through reaction with CN^- ion, indicating the dynamic nature of **55's** HOF.

6.2.10 Catalytic Activity

A catalytic system was prepared by Gardella et al. [64] through homogenous dispersion of palladium nanoclusters on the nanostructured surface of polymer (amino-POSS/PLLA (**56**) nanofibers, which was formed form POSS-NH_2 and PLLA. The formation of Pd nanoclusters

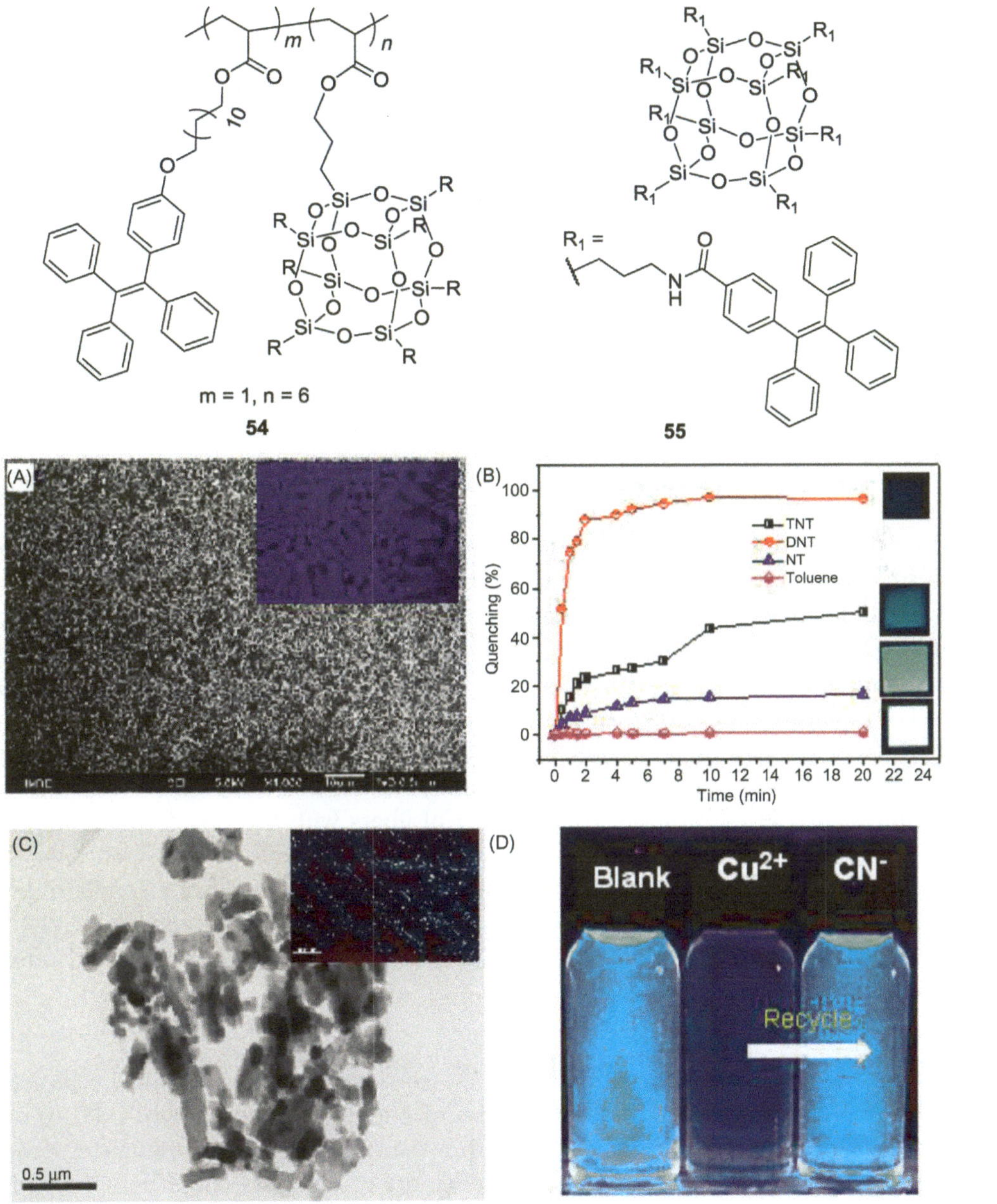

FIGURE 6–12 Chemical structure of polyacrylate-POSS polymer **54** and POSS-T_8A **55**. (A) SEM image of film F4 from **54**. The inserted photo shows the fluorescence image of F4 under UV light (337 nm). (B) Time-dependent fluorescence quenching of F4 upon exposure to analyte vapor. The inserted photos are images of F4 upon exposure to analyte vapor under UV light (365 nm). (C) TEM image of POSS-T_8A **55** particles prepared in the DMSO. (D) Photographs of **55**, POSS-T_8A · Cu^{2+}, and POSS-T_8A · Cu^{2+} with tetrabutylammonium cyanide under UV light (365 nm). *(A–B) Adapted from Zhou H, Ye Q, Neo WT, Song J, Yan H, Zong Y, et al. Chem Commun 2014;50:13785–8. Copyright 2014 Royal Society of Chemistry. (C–D) Adapted from Zhou H, Ye Q, Wu X, Song J, Cho CM, Zong Y, et al. J Mater Chem C 2015;3:11874. Copyright 2014 Royal Society of Chemistry.*

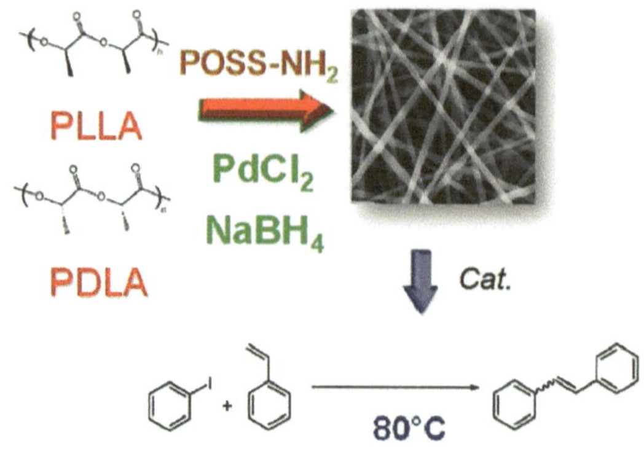

FIGURE 6–13 Preparation of *sc*-PLA/Pd fibers **58** and application in Heck reaction. *Adapted from Monticelli O, Putti M, Gardella L, Cavallo D, Basso A, Prato M, et al. Macromolecules 2014;47:4718–27. Copyright 2014 American Chemical Society.*

would be promoted by the submicrometically dispersed POSS-NH_2 in the PLLA nanofibers via electrospinning, leading to high catalytic activity for stilbene hydrogenation at heterogeneous conditions. Another catalytic system, PSMA/Ti-POSS-NH_2 (**57**), based on metal-containing POSS modified polymer nanofibers, was reported by Cozza et al. [65] through electrospinning, in which the titanium-POSS is attached to PSMA nanofibers via its amino group. The resulting system **57** displayed good catalytic activity for sulforhodamine B degradation of under UV light. In addition, Monticelli et al. [66] prepared functional stereocomplex fibers (Amino-POSS/*sc*-PLA (**58**) by electrospun solution of PLLA, PDLA, and POSS-NH_2. The amino groups in **58** promote coordination of fibers with the metal compound, that is, $PdCl_2$, to achieve homogeneous dispersion of metal nanoclusters on the fiber surface (Fig. 6–13). The application area of this catalytic system was obviously extended due to the improved thermal and chemical resistance of fibers through incorporation of *sc*-PLA. Thus, **58** showed good activity for Heck reaction, and could be easily recovered and reused for multiple catalytic cycles.

6.2.11 Biochemistry Application

Zhang et al. [67] reported the gelation process of *N*-phenylaminomethyl-POSS/PU composites (POSS/PU (**59**) in the curing process and in the stable state after the curing process. It was found that the gelation time of **59** and hence the gel stiffness could be reduced though increasing the curing temperature or POSS concentration in composites. After curing, the critical concentration of POSS in POSS/PU composites was found as 2.5 wt%. The homogeneity of cross-linking POSS/PU composites would be increased via integration of multifunctional POSS into PU. Wang et al. [68] produced an inorganic–organic hybrid hydrogel with tunable properties though a controlled cross-linking of disulfide-linked core/shell precursor from PEG shells and POSS. The hydrogels and aggregates with common compact and loose structures were produced via a thiol–disulfide exchange reaction, which was used as the pH-responsive "on/off" to modify the cross-linking reactions. Consequently, POSS–(SS-PEG)$_8$ (**60**) could be constructed as isolated nanocages with cubic core size of 0.53 nm and eight corner chains size of 5 nm (Fig. 6–14A–C). Unimolecular micelles were reported by Fan et al. [69] from copolymer POSS–(PAA-(PLLA-PEG)$_4$)$_8$ (**61**), where eight linear dendritic-like arms of PAA-(PLLA-PEG)$_4$ are grafted onto a POSS core. Transmission electron microscopy (TEM) images shown in Fig. 6–14D indicated that the micelles exhibit rod-like morphology, whose size can be tuned by varying pH values. This amphiphilic hybrid copolymer showed potential for the controlled drug delivery of doxorubicin. The epoxy-POSS hybrid networks containing tethered POSS or blended POSS were reported by Matějka et al. [70]. For the hybrids with POSS as pendant groups (epoxy-POSS (**62**)), physical cross-linking in the pre–gel stage of the network formation could be conveniently controlled in situ.

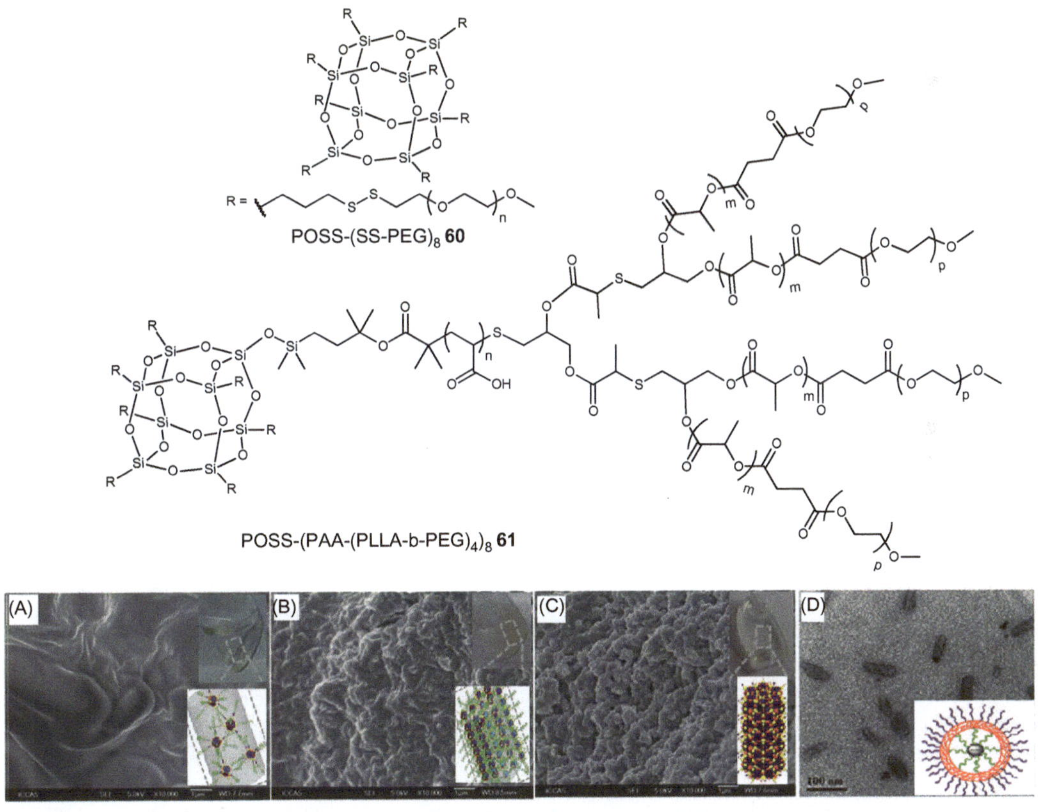

FIGURE 6–14 Chemical structure of POSS–(SS-PEG)$_8$ **60** and POSS–(PAA-(PLLA-b-PEG)$_4$)$_8$ **61**. SEM images of hydrogels or aggregate after basification of **60** solution for (A) 1, (B) 8, (C) 24, and (D) 240 h. (D) TEM images of the micelles formed from **61**. *(A–C) Adapted from Wang X, Li D, Yang F, Shen H, Li Z, Wu D. Polym Chem 2013;4:4596–600. Copyright 2013 Royal Society of Chemistry. (D) Adapted from Mahapatra SS, Yadav SK, Cho JW. React Funct Polym 2012;72:227–32. Copyright 2015 Royal Society of Chemistry.*

6.2.12 Self-Assembly

POSS and its derivatives have received increasing attention in terms of self-assembly research due to their nanosized 3D structure, and convenient modifications via chemistry to produce many types of well-defined nanostructured materials (Table 6–1).

Tan et al. [71] prepared copolymers PLLA-*b*-P(MA-POSS (**63**)) and PDLA-*b*-P(MA-POSS (**64**)) by ATRP of MA-POSS, PLLA, or PDLA, and studied their self-assembly behaviors. In comparison with a blended mixture of PDLA and PLLA, the stereocomplexed nanoparticles with 1:1 weight ratio of **63** and **64** displayed significantly enhanced stability, which could be kept as a stable dispersion for up to 30 days. The enhanced stability is owing to the bulky POSS nanocages, which sterically inhibit their aggregation in solution. Several POSS-helical

Table 6–1 Summary of POSS-Based Composites

POSS	Composite	Code	Incorporation Type	Enhanced Performance	Reference
T-9	EPDM/POSS	**1**	Blended	Thermal stability	[12]
T-12	PEPS	**2**	Tethered	Thermal stability	[13]
T-13	POSS/PSS	**3**	Tethered	Thermal stability	[14]
Q-8	PBZ-POSS	**4**	Tethered	Thermal stability	[15]
Q-8	PBZ/POSS	**5**	Tethered	Thermal stability	[16]
Q-5	PBZ-OGPOSS	**6**	Tethered	Thermal stability	[17]
T-35	OapPOSS-rGO	**7**	Tethered	Thermal stability	[18]
Q-9	PU/POSS	**8**	Blended	Flame retardancy	[21]
POSS-FQ	PEU/POSS	**9**	Blended	Degradation resistance	[22]
T-5	POSS/PA	**10**	Tethered	Mechanical strength	[23]
T-29	PA dendrimers	**11**	Tethered	Mechanical strength	[24]
T-10	Nylon 6/POSS	**12**	Blended	Mechanical strength	[25]
Q-3	HBP-POSS	**13**	Tethered	Mechanical strength	[26]
Q-4	PU-NCOPOSS	**14**	Tethered	Mechanical strength	[27]
T-35	DGEBA epoxy resin-POSS	**15**	Tethered	Mechanical strength	[28]
T-29	*n*BGE-BDGE-POSS	**16**	Tethered	Mechanical strength	[29]
T-4	SBS/POSS	**17**	Tethered	Mechanical strength	[30]
T-12, **T-30**	CF-UPR	**18**	Tethered	Mechanical strength	[31a]
Q-5	CF–POSS–CNT	**19**	Tethered	Mechanical strength	[31b]
T-12	BCPs	**20**	Tethered	Electric conductivity	[32]
T-12	LCPs	**21**	Tethered	Electric conductivity	[32]
T-26	POSS-PEG	**22**	Tethered	Electric conductivity	[32]
Q-6	PEs	**23**	Tethered	Electric conductivity	[33]
Q-8	SPP13	**24**	Tethered	Electric conductivity	[34]
T-22	SPIs	**25**	Tethered	Electric conductivity	[35]
T-PCL	POSS–PCL/graphene	**26**	Blended	Electric conductivity	[36]
T-28	POSS-PFCP	**27**	Blended	Electric conductivity	[37]
Q-7	POSS–PU–PBZ	**28**	Tethered	Electric conductivity	[38]
T-30	PBZ-POSS	**29**	Tethered	Electric conductivity	[39]

(Continued)

Table 6–1 (Continued)

POSS	Composite	Code	Incorporation Type	Enhanced Performance	Reference
T-30	rGo-OapPOSS	**30**	Tethered	Electric conductivity	[40]
T-35	Oap-POSS-g-GO	**31**	Tethered	Electric conductivity	[41]
T-20	POSS-PIs	**32**	Tethered	Antiatomic oxidation	[42]
T-7	PI/POSS	**33**	Tethered	Antiatomic oxidation	[43]
T-36	PVDF/FP-POSS	**34**	Blended	Antiatomic oxidation	[44]
T-2,3,5,6,7	POSS-PS	**35**	Blended	Antiphoto irradiation	[45]
Q-5	BZ/DGEBA EP	**36**	Tethered	Antiphoto irradiation	[46]
T/Q-1, T-30, T-11	UDMA-POSS	**37**	Tethered	Optical application	[47]
T-16	PFDBT-POSS	**38**	Tethered	Optical application	[48]
T-8	PNb10POSS	**39**	Tethered	Liquid crystals	[49]
T-10	PLA-*g*-MA/POSS-NH_2	**40**	Tethered	Liquid crystals	[50]
T-11	PLA-*g*-MA/POSS-OH	**41**	Tethered	Liquid crystals	[51]
Q-8	POSS-LCs	**42**	Tethered	Liquid crystals	[51]
T-37	F-POSS/MMA	**43**	Tethered	Surface modification	[52]
T-27	POSS-capped PEO	**44**	Tethered	Surface modification	[53]
T-31	PH-POSS	**45**	Tethered	Surface modification	[54]
Q-2	Pebax/POSS	**46**	Blended	Surface modification	[55]
T-33	*s*-POSS-PMMA-*b*-P(MA-POSS)	**47**	Tethered	Surface modification	[56]
T-20	POSS-PIs	**48**	Tethered	Surface modification	[57]
T-34	Glycidyl-POSS-PIs	**49**	Tethered	Surface modification	[58]
T-10	PSMA/POSS	**50**	Tethered	Surface modification	[59]
T-12	MA-POSS/PLA	**51**	Tethered	Surface modification	[60]
T-10	POSS–PCL–POSS	**52**	Tethered	Surface modification	[60]
T-15	POSS–CNT	**53**	Tethered	Surface modification	[61]
T-12	Polyacrylate-POSS	**54**	Tethered	Porosity	[62]
T-29	POSS-T_8A	**55**	Tethered	Porosity	[63]
T-10	Amino-POSS/PLLA	**56**	Blended	Catalytic activity	[64]
T-14	PSMA/Ti-POSS-NH_2	**57**	Tethered	Catalytic activity	[65]
T-10	Amino-POSS/*sc*-PLA	**58**	Blended	Catalytic activity	[66]
T-38	POSS/PU	**59**	Tethered	Biochemistry application	[67]
T-25	POSS–$(SS\text{-}PEG)_8$	**60**	Tethered	Biochemistry application	[68]
Q-1	POSS-$(PAA\text{-}(PLLA\text{-}PEG)_4)_8$	**61**	Tethered	Biochemistry application	[69]
T-20, T-10, T-18, T-21, T-19, T-1, T-4, T-2	Epoxy-POSS	**62**	Tethered	Biochemistry application	[70]
T-10	PLLA-*b*-P(MA-POSS)	**63**	Tethered	Self-assembly	[71]
T-10	PDLA-*b*-P(MA-POSS)	**64**	Tethered	Self-assembly	[71]
T-10	POSS-$PBLG_{56}$	**65**	Tethered	Self-assembly	[72]
T-32	POSS–$(P(MEO_2MA\text{-co-}OEGMA))_8$	**66**	Tethered	Self-assembly	[73]
T-Y	tri-PS-APOSS	**67**	Tethered	Self-assembly	[74]
T-y	tri-APOSS	**68**	Tethered	Self-assembly	[74]
T-10, T-X	tetra-POSS	**69**	Tethered	Self-assembly	[75]

polypeptides were prepared by Kuo et al. [72] through ROP of isobutyl-POSS and γ-benzyl-L-glutamate *N*-carboxyanhydride. As shown in Fig. 6–15A, pure POSS-PBLG$_{56}$ copolymer (**65**) formed fibrous structures. Owing to the strong dipolar π-π interactions between poly

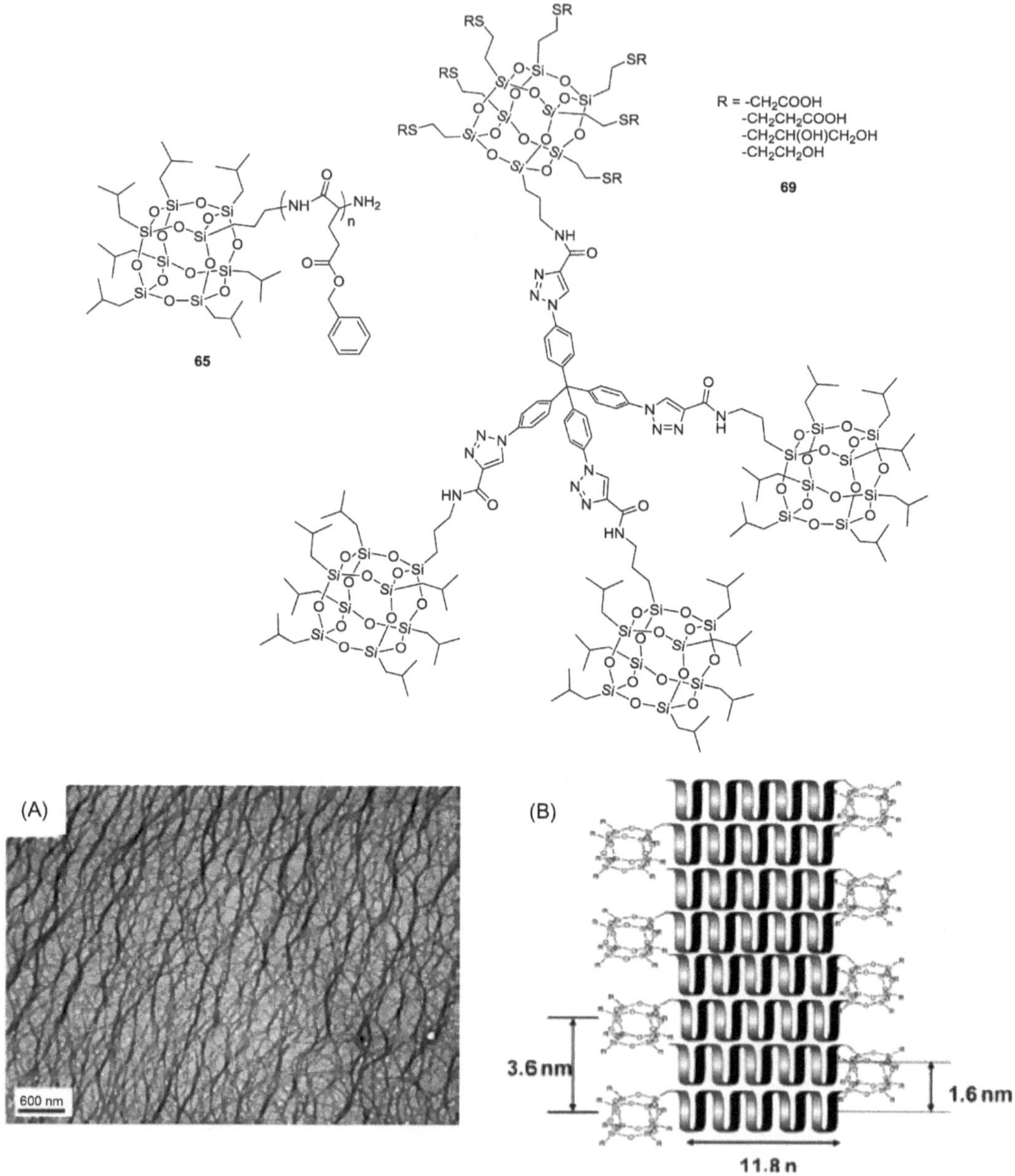

FIGURE 6–15 Chemical structure of POSS-PBLG$_{56}$ **65** and tetra-POSS **69.** (A) TEM images of POSS-PBLG$_{56}$ **65.** (B) Illustration of nanoribbon formed in the network structure of the toluene gel of POSS-PBLG$_{56}$ **63**. *Adapted from Kuo S-W, Lee H-F, Huang W-J, Jeong K-U, Chang F-C. Macromolecules 2009;42:1619–26. Copyright 2009 American Chemical Society.*

Table 6.2 List of abbreviations in this chapter

Abbreviations			
1D	One-dimensional	PBZ	Polybenzoxazines
2D	Two-dimensional	PC	Poly(carbonate)
3D	Three-dimensional	PCL	Poly(ε-caprolactone)
ACPOSS	Acryloxyl-functionalized POSS	PCU	Poly(carbonate-urea)urethane
AFM	Atomic force microscopy	PEG	Poly(ethylene glycol)
AIE	Aggregation-induced emission	PEO	Poly(ethylene oxide)
AL0136	Octa(3-hydroxy-3-methylbutyldimethylsiloxy)	PFCP	Perfluorocyclopentenyl aryl ether polymers
AO	Atomic oxygen	PI	Polyimides
APP	Ammonium polyphosphate	PLA	Polylactide
ATRP	Atom transfer radical polymerization	PLLA	Poly(L-lactic acid)
CE	Composite electrolytes	PMMA	Poly(methyl methacrylate)
CF	Carbon fibers	POM	Poly(oxymethylene)
CNT	Carbon nanotube	POSS	Polyhedral oligomeric silsesquioxane
DGEBA	Diglycidyl ether bisphenol A	PPO	Poly(propylene oxide)
DMS	Degree of micro phase separation	PS	Poly(styrene)
DNT	2,4-Dinitrotoluene	PSMA	Poly(styrene-co-maleic anhydride)
DSC	Differential scanning calorimetry	PSMIPOSS	Poly(styrene-alt-maleimide isobutyl POSS)
EGDMA	Ethylene glycol dimethacrylate	PTMO	Poly(tetramethyleneoxide)
EPDM	Ethylene-propylene-diene monomer	PU	Polyurethane
FD-POSS	Fluorodecyl-POSS	PVDF	Poly(vinylidene fluoride)
FP-POSS	(3,3,3-Trifluoropropyl)8Si_8O_{12}	RAFT	Reversible addition fragmentation chain transfer polymerization
FR/NIR	Far-red/near-infrared	SBS	Polystyrene–butadiene–polystyrene
GO	Graphene oxide	SEM	Scanning electron microscopy
HBP	Hyperbranched polyurethane	SPE	Solid polymer electrolyte
IFSS	Interfacial shear strength	TEM	Transmission electron microscopy
Ma	Molecular weight of the arm	Tg	Glass transition temperatures
OapPOSS	Octa-aminophenyl-POSS	TGA	Thermal gravity analysis
PA	Polyamide	TNT	2,4,6-Trinitrotoluene
PBLG	Poly(γ-benzyl-L-glutamate)	UDMA	Urethane dimethacrylates

(γ-benzyl-L-glutamate) (PBLG) rods and phenyl groups, the $PBLG_{53}$ polymers intertwined to form an irregular fiber aggregation. In contrast, **65** self-assembled antiparallelly to minimize steric hindrance of POSS units, resulting in an ordered microfiber gel. These POSS-$PBLG_{56}$ copolymers formed a smectic ordering in solution, which was evidenced by polarized optical microscope images. The hierarchical self-organization of polymer showed potential for biomimetic application, and thus provided a method to construct materials in a micrometer scale. Several star-shaped hybrid polymers P(MEO_2MA-*co*-OEGMA) with a POSS core (POSS-(P(MEO_2MA-*co*-OEGMA))$_8$ (**66**)) was prepared by Yuan et al. though ATRP [73]. The resulting polymers would form micelles in an aqueous solution because of their amphiphilicity. The responsive temperature of micelles was increased from 29.7°C to

39.1°C by altering the ratio between MEO_2MA and OEGMA, showing potential for applications in smart materials and biomedical fields. Zhang et al. [74] synthesized two triarmed organic–inorganic hybrid materials (tri-PS-APOSS (**67**) and tri-APOSS (**68**)) starting from carboxylic acid–functionalized POSS (APOSS) with/without PS linker. Both **67** and **68** can self-assemble into hollow spherical nanostructures in water/organic mixed solvents though they possess completely different mechanisms and driving forces. The hybrid with PS linkers forms bilayer vesicles similar to surfactants. In contrast, the hybrid without PS linkers assembles into single-layered, vesicle-like "blackberry"-type structure. The trend to aggregate size on change of solvent polarity is different for those two hybrids. It presented a convenient method to control the self-assembly process through minor adjustment of organic–inorganic hybrid structures. Huang et al. [75] studied the assembly behaviors of a series of POSS-based giant molecules (tetra-POSS (**69**)), which were constructed by covalently attaching POSS to the vertex positions of a rigid tetrahedral molecular. The change in symmetry of these giant tetrahedral giant molecules introduced precisely positional interactions, resulting in selectively self-assembled, highly ordered lattice structures, such as Frank–Kasper A15 phase (Table 6–2).

6.3 Conclusions and Perspectives

In this chapter, we summarized the latest development on POSS-based composites according to property enhancement. Introduction of suitable POSS derivatives into the polymer systems leads to improvement in thermal stability, mechanical strength, electric conductivity, AO, and anti-photodegradation. POSS play a paramount role in contributing to the enhancement in other composite properties, for example, surface, porosity, catalytic activity, as well as self-assembly. POSS-based composites also could be used in optical, liquid crystal, and biochemistry applications. Most importantly, a large number of POSS-polymer nanocomposites had been explored to improve the performance of products by many companies, like L'Oréal, Xerox Corporation, Samsung, etc. [76–80]. The materials with POSS-based nanocomposites have shown significantly improved properties for industrial applications, such as photooxidative stability and AO stability, electric conductivity, dielectric properties, high-resolution lithography, biomedicine, and so on. With the advanced developments of organic chemistry, a large number of POSS compounds with a variety of functional groups have been prepared successfully, allowing them to be covalently attached onto polymeric networks through different polymerization or grafting reactions. On the other hand, the modern polymer chemistry has offered many polymerization approaches, including ATRP, RAFT, etc., some of which have shown great potential to precisely introduce POSS into polymeric architectures and therefore tailor make the materials performance for various applications such as biomaterials for tissue engineering, drug delivery, etc. It is expected that POSS-based hybrid composites and their applications will continue to grow, and offer better performance with longer lifetime and more cost-effectiveness in aerospace, electronics, medicine, coatings, and cosmetics applications (Fig. 6–16).

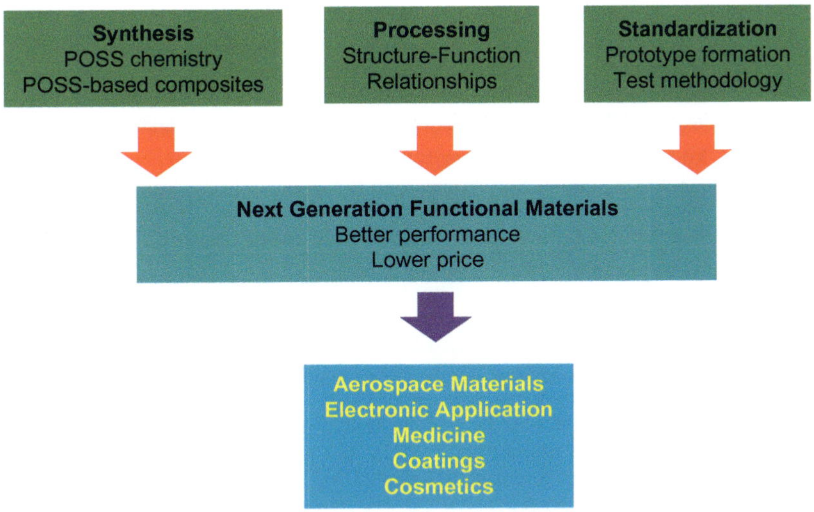

FIGURE 6–16 Next generation POSS-polymer functional composites with better performance and lower price.

Acknowledgment

The authors would like to acknowledge the financial support from the Institute of Materials Research and Engineering (IMRE), Agency for Science, Technology and Research (A*STAR).

References

[1] Scott DW. J Am Chem Soc 1946;68:356–8.

[2] Baney RH, Itoh M, Sakakibara A, Suzuki T. Chem Rev 1995;95:1409–30.

[3] Li G, Wang L, Ni H, et al. J Appl Polym Sci 2013;128:2395–401.

[4] Wu J, Mather PT. Polym Rev 2009;49:25–63.

[5] Hurd CB. J Am Chem Soc 1946;68:364–70.

[6] Kawakami Y, Kakihana Y, Miyazato A, Tateyama S, Hoque MA. Adv Polym Sci 2010;55:1–44.

[7] (a) Vollenberg PHT, Heikens D. Polymer 1989;30 16556-11662.
(b) Petrović ZS, Javni I, Waddon A, Bánhegyi G. J Appl Polym Sci 2000;76:133–51.
(c) Andrady AL, Merkel TC, Toy LG. Macromolecules 2004;37:4329–31.
(d) Cho KS, Hong JI, Chung CI. Polym Eng Sci 2004;44:1702–6.

[8] (a) Lichtenhan JD. Comments Inorg Chem 1995;17:115–30.
(b) Voigt A, Murugavel R, Roesky HW. Organometallics 1996;15:5097–101.
(c) Ellsworth MW, Gin DL. Polym News 1999;24:331–41.
(d) Zhao JQ, Fu F, Liu SM. Polym Polym Compos 2008;16:483–500.
(e) Pittman Jr CU, Li GZ, Ni H. Macromol Symp 2003;196:301–25.

[9] Rubinstein M, Colby RH. Polymer physics. New York: Oxford University Press; 2003.

[10] (a) Bizet S, Galy J, Gerard JF. Polymer 2006;47:8219–27.
(b) Zhang X, Chan ER, Glotzer SC. J Chem Phys 2005;123(18):184718.
(c) Striolo A, McCabe C, Cummings PT. J Chem Phys 2006;125:104904.

(d) Patel RR, Mohanraj R, Pittman CU. J Polym Sci Part B: Polym Phys 2006;44:234–48.
(e) Chan ER, Zhang X, Lee CY, Neurock M, Glotzer SC. Macromolecules 2005;38:6168–80.
(f) Chattopadhyay DK, Webster DC. Prog Polym Sci 2009;34:1068–133.
(g) Ghanbari H, Cousins BG, Seifalian AM. Macromol Rapid Commun 2011;32(14):1032–46.
(h) Kannan RY, Salacinski HJ, Butler PE, Seifalian AM. Acc Chem Res 2005;38(11):879–84.

[11] (a) Ye Q, Zhou H, Xu J. Chem Asian J 2016;11:1322–37.
(b) Baney RH, Itoh M, Sakakibara A, Suzuki T. Chem Rev 1995;95:1409–30.
(c) Laine RM. J Mater Chem 2005;15:3725–44.
(d) Cordes DB, Lickiss PD, Rataboul F. Chem Rev 2010;110:2081–173.
(e) Wang F, Lu X, He C. J Mater Chem 2011;21:2775–82.
(f) Fina A, Monticelli O, Camino G. J Mater Chem 2010;20:9297–305.
(h) Kuo SW, Chang FC. Prog Polym Sci 2011;36:649–1696.
(i) Raftopoulos KN, Pielichowski K. Prog Polym Sci 2016;52:136–87.
(j) Chan KL, Sonar P, Sellinger A. J Mater Chem 2009;19:9103–20.
(k) Kannan RY, Salacinski HJ, Butler PE, Seifalian AM. Acc Chem Res 2005;38(11):879–84.
(l) Wu J, Mather PT. J Macromol Sci Polym Rev 2009;49:25–63.
(m) Tanaka K, Chujo Y. J Mater Chem 2012;22:1733–46.
(n) Tanaka K, Chujo Y. Bull Chem Soc Jpn 2013;86:1231–9.
(o) Abbenhuis HCL. Chem Eur J 2000;6(1):25–32.
(p) Duchateau R. Chem Rev 2002;102(10):3525–42.
(q) Quadrelli EA, Basset JM. Coord Chem Rev 2010;254:707–28.
(r) Kung MC, Riofski MV, Misasghi MN, Kung HH. Chem Commun 2014;50(25):3262–76.
(s) Applications of polyhedral oligomeric silsesquioxane (C. Hartmann-Thompson, editor, Springer, Netherlands, 2011.

[12] Xue M, Zhang X, Ma L, Gu Z, Lin Y, Bao C, Tian X. J Appl Polym Sci 2013;128:2395–401.

[13] Li Z, Tan BH, Jin G, Li K, He C. Polym Chem 2014;5:6740–53.

[14] Li L, Liu H. RSC Adv 2014;4:46710–17.

[15] Huang KW, Kuo SW. Macromol Chem Phys 2010;211:2301–11.

[16] Wu YC, Kuo SW. Polymer 2010;51:3948–55.

[17] Chandramohan A, Devaraju S, Vengatesan MR, Alagar M. J Polym Res 2012;19:9903–12.

[18] Wang X, Song L, Yang H, Xing W, Kandola B, Hu Y. J Mater Chem 2012;22:22037–43.

[19] Bourbigot S, Turf T, Bellayer S, Duquesne S. Polym Degrad Stab 2009;94:1230–7.

[20] Michłowski S, Hebda E, Pielichowski K. J Therm Anal Calorim 2017;130:155–63.

[21] Devaux E, Rochery M, Bourbigot S. Fire Mater 2002;26:149–54.

[22] Wang W, Guo Y, Otaigbe JU. Polymer 2009;50:5749–57.

[23] Gnanasekaran D, Shanavas A, Focke WW, Sadiku R. RSC Adv 2015;5:11272–83.

[24] Matsumoto K, Nishi K, Ando K, Jikei M. Polym Chem 2015;6:4758–65.

[25] Milliman HW, Ishida H, Schiraldi DA. Macromolecules 2012;45:4650–7.

[26] Mahapatra SS, Yadav SK, Cho JW. React Funct Polym 2012;72:227–32.

[27] Prządka D, Jęczalik J, Andrzejewska E, Marciniec B, Dutkiewicz M, Szłapka M. React Funct Polym 2013;73:114–21.

[28] Sethuraman K, Prabunathan P, Alagar M. RSC Adv 2014;4:45433–41.

[29] Ni C, Ni G, Zhang L, Mi J, Yao B, Zhu C. J Colloid Interface Sci 2011;362(1):94–9.

[30] Bai J, Shi Z, Yin J, Tian M. Polym Chem 2014;5:6761–9.

[31] (a) Jiang D, Xing L, Liu L, Yan X, Guo J, Zhang X, et al. J Mater Chem A 2014;2:18293–303.
(b) Zhao F, Huang Y, Liu L, Bai Y, Xu L. Carbon 2011;49:2624–32.

[32] Shim J, Kim DG, Lee JH, Baik JH, Lee JC. Polym Chem 2014;5:3432–42.

[33] Polu AR, Rhee HW. J Ind Eng Chem 2015;31:323–9.

[34] Kim D-G, Shim J, Lee JH, Kwon S-J, Baik J-H, Lee J-C. Polymer 2013;54:5812–20.

[35] Wu Z, Zhang S, Li H, Liang Y, Qi Z, Xu Y, et al. J Power Sources 2015;290:42–52.

[36] Nezakati T, Tan A, Seifalian AM. J Colloid Interface Sci 2014;435:145–55.

[37] Sharma B, Verma R, Baur C, Bykova J, Mabry JM, Smith Jr. DW. J Mater Chem C 2013;1:7222–7.

[38] Sasi kumar R, Alagar M. RSC Adv 2015;5 33008-33008.

[39] Tseng MC, Liu YL. Polymer 2010;51:5567–75.

[40] Li Q, Peng P, Chen G-X, Yoon SW. J Mater Chem C 2014;2:8216–21.

[41] Yu W, Fu J, Dong X, Chen L, Shi L. Compos Sci Technol 2014;92:112–19.

[42] Lei X-F, Qiao M-T, Tian L-D, Yao P, Ma Y, Zhang H-P, Zhang Q-Y. Corros Sci 2015;90:223–38.

[43] Duo S, Ke H, Liu T, Song M, Li M. Nucl Instrum Methods Phys Res Sect B 2013;307:324–7.

[44] Zeng F, Peng C, Liu Y, Qu J. J Phys Chem A 2015;119(30):8359–68.

[45] Tz Dintcheva N, Morici E, Arrigo R, La Mantia FP, Malatesta V, Schwab JJ. Polym Degrad Stab 2012;97:2313–22.

[46] Selvi M, Devaraju S, Vengatesan MR, Go JS, Kumard M, Alagar M. RSC Adv 2014;4:8238–44.

[47] Lungu A, Şulcă NM, Vasile E, Badea N, Pârvu C, Iovu H. J Appl Polym Sci 2011;121:2919–26.

[48] Liu J, Feng G, Ding D, Liu B. Polym Chem 2013;4:4326–34.

[49] Hou PP, Gu KH, Zhu YF, Zhang ZY, Wang Q, Pan HB, et al. RSC Adv 2015;5:70163–71.

[50] Gardella L, Colonna S, Fina A, Monticelli O. Colloid Polym Sci 2014;292:3271–8.

[51] Wang X, Cho CM, Say WY, Tan AYX, He C, Chan HSO, et al. J Mater Chem 2011;21:5248–57.

[52] Ramirez SM, Diaz YJ, Sahagun CM, Duff MW, Lawal OB, Iacono ST, et al. Polym Chem 2013;4:2230–4.

[53] Wang L, Zheng S. Mater Chem Phys 2012;136:744–54.

[54] Wang L, Zhang C, Zheng S. J Mater Chem 2011;21:19344–52.

[55] Le NL, Wang Y, Chung T-S. J Membr Sci 2011;379:174–83.

[56] Pan A, Yang S, He L, Zhao X. RSC Adv 2014;4:27857–66.

[57] Le NL, Chung T-S. J Membr Sci 2014;454:62–73.

[58] Konietzny R, Koschine T, Rätzke K, Staudt C. Sep Purif Technol 2014;123:175–82.

[59] Monticelli O, Fina A, Cozza ES, Prato M, Bruzzo V. J Mater Chem 2011;21:18049–54.

[60] Ferńandez MJ, Ferńandez MD, Cobos M. RSC Adv 2014;4:21435–49.

[61] Yadav SK, Mahapatra SS, Yoo HJ, Cho JW. Nanoscale Res Lett 2011;6:122–7.

[62] Zhou H, Ye Q, Neo WT, Song J, Yan H, Zong Y, et al. Chem Commun 2014;50:13785–8.

[63] Zhou H, Ye Q, Wu X, Song J, Cho CM, Zong Y, et al. J Mater Chem C 2015;3:11874.

[64] Gardella L, Basso A, Prato M, Monticelli O. ACS Appl Mater Interfaces 2013;5(16):7688–92.

[65] Cozza ES, Bruzzo V, Carniato F, Marsano E, Monticelli O. ACS Appl Mater Interfaces 2012;4(2):604–7.

[66] Monticelli O, Putti M, Gardella L, Cavallo D, Basso A, Prato M, et al. Macromolecules 2014;47:4718–27.

[67] Zhang QH, Huang X, Wang XL, Jia XD, Xi K. Polymer 2014;55:1282–91.

[68] Wang X, Li D, Yang F, Shen H, Li Z, Wu D. Polym Chem 2013;4:4596–600.

[69] Fan X, Wang Z, He C. J Mater Chem B 2015;3:4715–22.

[70] Matějka L, Murias P, Pleštil J. Eur Polym J 2012;48:260–74.

[71] Tan BH, Hussain H, Leong YW, Lin TT, Tjiu WW, He CB. Polym Chem 2013;4:1250–9.

[72] Kuo S-W, Lee H-F, Huang W-J, Jeong K-U, Chang F-C. Macromolecules 2009;42:1619–26.

[73] Yuan W, Shen T, Liu X, Ren J. Mater Lett 2013;111:9–12.

[74] Zhang W, Chu Y, Mu G, Eghtesadi SA, Liu Y, Zhou Z, et al. Macromolecules 2017;50:5042–50.

[75] Huang M, Hsu CH, Wang J, Mei S, Dong X, Li Y, et al. Science 2015;348(6233):424–8.

[76] Dimotakis E, Bui HS, Kanji M. EP 2648694A2, 2013, L'Oréal, SA.

[77] (a) Maitra P, Zheng T. US8133478B2, 2012, Avon Products Inc.;(b) P Maitra, T Zheng, US8545823B2, 2013, Avon Products Inc.

[78] Kathan KR, Liddy JA, Adams MJ. US20140332461A1, 2014, Pall Corporation.

[79] (a) DM Pietrantoni, J Wu, JH Herko, L Zhang, L Ma, FJ Lopez, et al., US8025967B1, 2011, Xerox Corporation;(b) DM Pietrantoni, Y Tong, MS Roetker, FJ Lopez, KB Tallman, JH Herko, et al., US8609233B2, 2013, Xerox Corporation;(c) J Wu, L Zhang, L Ma, R Ehmann, US8435632B2, 2013, Xerox Corporation.

[80] Ji SY, Kim SH, Ham SJ. US20140187687, 2014, Samsung Electro-Mechanics Co., Ltd.

7

Polymer Composites with Functionalized Carbon Nanotube and Graphene

Chaohui Wei[1], Bamidele Akinwolemiwa[1], Linpo Yu[2], Di Hu[2], George Z. Chen[2,3]

[1]INTERNATIONAL DOCTORAL INNOVATION CENTRE, FACULTY OF SCIENCE AND ENGINEERING, UNIVERSITY OF NOTTINGHAM NINGBO CHINA, NINGBO, P.R. CHINA [2]DEPARTMENT OF CHEMICAL AND ENVIRONMENTAL ENGINEERING, AND ENERGY ENGINEERING RESEARCH GROUP, FACULTY OF SCIENCE AND ENGINEERING, UNIVERSITY OF NOTTINGHAM NINGBO CHINA, NINGBO, P.R. CHINA [3]DEPARTMENT OF CHEMICAL AND ENVIRONMENTAL ENGINEERING, AND ADVANCED MATERIALS RESEARCH GROUP, FACULTY OF ENGINEERING, UNIVERSITY OF NOTTINGHAM, NOTTINGHAM, UNITED KINGDOM

Abbreviations

BJH	Bulk heterojunction
CFS	Capacitive Faradaic storage
CNTs	Carbon nanotubes
CNM	Carbon nanomaterial
CV	Cyclic voltammogram
CVD	Chemical vapor deposition
DMFC	Direct methanol fuel cell
DSSC	Dye-sensitized solar cell
ECP	Electronically conducting polymer
FS	Faraday storage
GCD	Galvanostatic charge−discharge
GO	Graphene oxide
HA	Hydroxyapatite
J_{sc}	Short circuit current,
LDS	Localized dry spots
NCFS	Noncapacitive Faradaic storage
P3AT	Poly(3-alkylthiophene)
P3OT	Poly(3-octylthio-phene-2,5-diyl)
PAA	Poly(acrylic acid)
PAni	Poly(aniline)
PC	Poly(carbonate)
PC71BM	[6,6]-phenyl C71 butyric acid methyl ester

Polymer Composites with Functionalized Nanoparticles. DOI: https://doi.org/10.1016/B978-0-12-814064-2.00007-X

PE	Poly(ethylene)
PEI	Poly(ethyleneimine)
PEEK	Poly(ether ether ketone) PEMFC proton exchange membrane fuel cells
PEDOT	Poly(3,4-ethylene-dioxythiophene)
PEM	Polymer electrolyte membrane
PESEEK	Poly(ether sulfone ether ketone ketone)
PMMA	Poly(methyl methacrylate)
PLA	Poly(lactic acid)
PP	Poly(propylene)
PPE	Poly(phenyl ether)
PPS	Poly(phenylene sulfide)
PPy	Poly(pyrrole)
PS	Poly(styrene)
PSS	Poly(styrene sulfonates)
PTB7	Poly(thieno[3,4-b] thiophene/benzodithiophene)
PU	poly(urethance)
PVDF	poly(vinylidene fluoride)
PVA	Poly(vinyl alcohol)
PVP	Poly(vinylpyrrolidone)
PW	Poly(oxometalate)
rGO	Reduced graphene oxide
SBS	Styrene–butadiene–styrene
SC	Supercapacitor
SDS	Sodium dodecyl sulfate
SDBS	Sodium dodecyl benzene sulfonate
SPAES	Sulfonated poly(arylene ether sulfone)
XPS	X-ray photoelectron spectroscopy

7.1 Composites of Polymers With Carbon Nanomaterials

The role of polymers and their composites in modern society has been of great importance due to their vast range of applications. Composites of polymers with one or more components usually have improved characteristics due to individual components or a synergy of their combined properties. Due to the wide range of production processes and the possibility of obtaining a large variety of functionalities, carbon nanomaterials (CNMs), particularly carbon nanotubes (CNTs) and graphene, are the most frequently used components in preparing polymer composites. Accordingly, these polymer–CNM composites have widespread uses in electrochemical energy storage and conversion, microelectronics, mechanical, rheological, and tribomechanical applications, etc. Although CNMs exhibit a wide range of properties, their inert surface and ease of entanglement pose huge challenges in their applications, particularly in the preprocessing stages [1,2]. In this vein, various approaches aimed at adjusting the surface chemistry of CNMs are adopted to enhance their dispersion in solvents, particularly water, and also to promote their molecular scale interaction with the polymer matrix. This chapter will discuss the various methods of introducing functional groups into CNMs

(CNTs and graphene) and their resulting properties, coupled with the role they play within the polymer matrix. The functionalization methods briefly covered herein include oxygenation, nitrogenation, sulfonation, and halogenation. The applications of composites of polymers with CNMs will be briefly discussed from the standpoint of their fundamental electrochemical properties. Furthermore, the use of polymer–CNM composites in energy storage and conversion, catalysis, and as structural fillers will be briefly enumerated.

7.1.1 Introduction to the Properties and Applications of Carbon Nanomaterials

Carbonaceous materials could be broadly classified into traditional carbon materials such as carbon blacks and activated carbons, and nanostructured carbons, which include CNTs, graphene, fullerenes, etc. The large specific surface area, high conductivity, high mechanical strength, and good thermal stability of carbonaceous materials are indeed the reasons for their wide-ranging applications.

Since CNTs were discovered in the 1990s, they have been used for a wide variety of applications in nanotechnology [3]. CNTs are cylindrical in shape with rolled up graphite planes with more than one close-end having a hemispherical fullerene structure. The diameter of CNT is usually around 20–30 nm and its length is in the micrometer scale; as such, CNTs are characterized by a large aspect ratio of over 1000. Single-walled carbon nanotubes (SWCNTs) with a single graphene layer rolled up into a cylinder and multiwall carbon nanotubes (MWCNTs) with a number of graphene cylinders rolled together with an interspacing of 0.34 nm (see Fig. 7–1) are the two general types of CNTs. Depending on the atomic

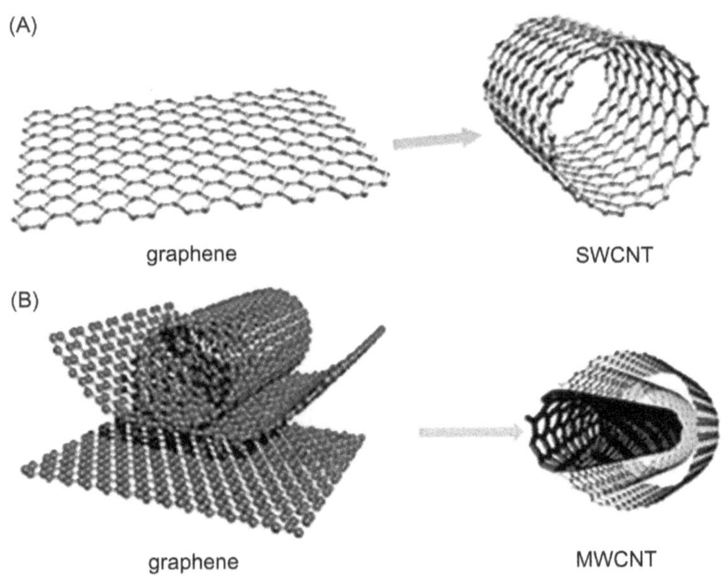

FIGURE 7–1 The structure of graphene and CNTs [3].

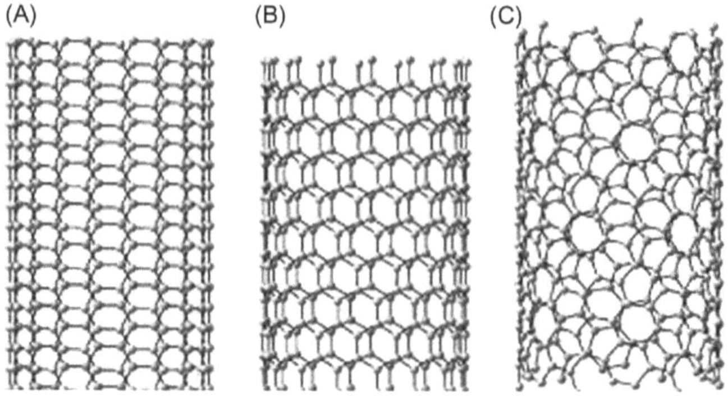

FIGURE 7–2 Molecular models exhibited by SWNTs based on the chirality: (A) armchair, (B) zigzag, and (C) chiral based conformation [4].

Table 7–1 General Properties of Different Carbon Materials [1,7,8]

Property	Graphite	Graphene	Graphene Oxide	SWCNTs	MWCNTs
Electrical conductivity (S/cm)	10^5	10^4–10^6	Low	10^2–10^6	10^3–10^5
Electron mobility [cm^2/(V s)]	2.0×10^4	1.5×10^4–5×10^4	Low	~10^5	10^4–10^5
Thermal conductivity [W/(m K)]	~2000	3000–5000	/	6000	2000
Surface area (m^2/g)	<500	0–3000	/	10–20	200–230
Tensile strength (GPa)	0.01–0.07	130	0.03–0.1	10–52	10–60
Young's modulus (GPa)	0.004–0.03	1000–2000	30	1500–2500	200–950

structure of the carbon bonds, CNTs in general have three types of orientations, which are zigzag, armchair, and chiral arrangements (see Fig. 7–2). Both SWCNTs and MWCNTs could have either metallic or semiconducting properties related to the direction of rolling of the graphite sheets and the tube diameter. Due to the almost one-dimensional structure, electronic transport in metallic CNTs occurs along the tube length and current can be transported without extra heating. CNTs have a high Young's modulus and tensile strength, for instance, an individual SWCNT has Young's modulus of about 37 GPa [5]. Both SWCNTs and MWCNTs can be synthesized by chemical vapor deposition (CVD), laser ablation of carbon or carbon-arc discharge. Generally, the homogeneity of SWCNTs is relatively more difficult to control compared with MWCNTs.

Ever since its isolation and first characterization of its electric field effect, the studies and applications of graphene have attracted numerous attention [6]. Graphene has a two-dimensional (2D) sp^2-bonded carbon atom, which is similar to the molecular structure of CNTs. Owing to its pure sp^2 hybridization network, graphene possesses unique mechanical, thermal, and electrical properties. Graphene has an extremely high elastic modulus of 1 TPa, and impressively high electrical and thermal conductivities. It is also characterized by high intrinsic flexibility, large specific surface area, and a high aspect ratio (see Table 7–1). The thickness of a single sheet of graphene is 0.34 nm, which is considered to be the thinnest 2D material so far. The current development of graphene and its derivatives is generally hampered by the difficulties encountered in the large scale production of high quality sheets. Graphene can be synthesized through mechanical or chemical exfoliation from highly ordered pyrolytic graphite, CVD, and molecular beam epitaxy, etc. [9]. Moreover, one of the most common means of producing graphene is through the chemical or electrochemical reduction of graphene oxide (GO) [10]. GO is an oxidized graphene derivative that disperses easily in water, and as such it serves as a good precursor for the production of graphene from aqueous dispersions. GO can be produced by thermal oxidation of graphite flakes, according to a method first described in [11]. Moreover, all processes currently used in the production of GO are various modifications or improvements to this initial method (i.e., the so-called modified or improved Hummers' method) [12].

7.1.2 Composites of Polymer With Functionalized Carbon Nanomaterials

The large variety, relatively easy synthesis routes, and wide-ranging physical and chemical properties have made polymers one of the most important classes of materials used in various engineering applications. This chapter will selectively review the properties and applications of the composites of some insulating and conducting polymers with CNMs. Most insulating polymers are usually configurational polymers. Notable examples of these configurational polymers are elastomers, thermoplastics, epoxy, block copolymers, etc. Elastomers are highly stretchable and have a large number of cross-linked long chains. Epoxy possesses stiff segments and is also highly cross-linked; consequently it has very high mechanical strength, stiffness, and brittleness. Thermoplastic polymers are flexible and can be shaped under heating, making them relatively easy to fabricate and recycle [2]. Electronically conducting polymers (ECPs) (also termed as intrinsically conducting polymers), which could either be conjugated or nonconjugated, have been widely studied and applied in electrochemical systems (energy storage, energy conversion, catalysis, sensing, etc.) due to their impressive electroactivity. For example, conjugated polymers such as: poly(pyrrole) (PPy), poly(aniline) (PAni), and poly(3,4-ethylene-dioxythiophene) (PEDOT) are widely used in energy storage devices [13]. However, nanostructured ECPs are usually insulating in their dedoped phase and their relatively poor mechanical strength also means that their performance easily deteriorates after a few cycles of insertion/extraction of ions during the process of charging/discharging. Therefore, various porous CNMs are used as conducting and mechanical additives for fabricating electrodes with ECPs [14]. In general, these composites of CNMs and ECPs also possess high interfacial area, increased polymer loading, and controlled interfacial reactions.

When graphene and CNTs are used (as fillers) within the polymer matrix, filler loading is crucial as it determines the ultimate properties of the composites when used for various applications. Moreover, obtaining a homogenous and uniform dispersion of the fillers within the polymer matrix during processing of the composites, is still an important challenge [1].

The ease of entanglement and aggregation of CNTs, coupled with their large aspect ratio, hydrophobicity, and strong van der Waals interaction between tubes, greatly restrict their solubility in most solvents [15]. Graphene also poses similar concerns because strong van der Waals force and $\pi-\pi$ interaction between adjacent sheets also allow graphene sheets to readily restack and agglomerate [16]. Interfacial interaction between CNMs and polymers could in general be hampered if the CNMs are not well dispersed prior to preparing the composites with the polymers.

Surface functionalization of CNMs have been found to be highly effective in resolving the above stated issues of CNM dispersion and in some cases interaction with the polymers. In general, the functional groups on the surfaces of the CNMs determine the wettability, electrokinetic potential, surface charge density as well as the dispersion/aggregative stability [17,18]. These surface groups also act as reaction sites for further modifications by other molecules, and this is the basis of either sidewall or defect functionalization in CNMs. Surface groups also serve as polymerization sites for effective nucleation and polymer formation. Broadly speaking, functionalization of CNTs, which could either be covalent or noncovalent, could occur at end caps and defects or on the sidewalls, or on the basal and edge plane for graphene [17–19]. Covalent functionalization is based on covalent linkage between carbon and other functional groups. Direct covalent functionalization occurs when sp^2 hybridization changes into sp^3 accompanied by a loss of the π-conjugation system of graphene layers. Noncovalent functionalization utilizes surfactants or polymer molecules to wrap individual tubes of CNTs/single layers of graphene, thus preventing them from entangling. Owing to the marked changes in property that surface functionalization creates, the modification of CNMs with a wide variety of dopants, particularly oxygen, sulfur, nitrogen, and halogens, or the use of noncovalent functionalization agents like surfactants or polymers, are important aspects of the technologies that utilize polymer–CNM composites.

7.2 Functionalization of Carbon Nanomaterials and the Effects on Polymer Composites

The physicochemical properties of CNMs on functionalization have been the subject of numerous investigations. This section will summarize the various approaches adopted to functionalize CNMs with oxygen, sulfur, nitrogen, and halogen containing groups. Also, the properties introduced by these functional groups together with their roles in the polymer–CNM composites will be briefly discussed.

7.2.1 Oxygenation of Carbon Nanomaterials

Oxygen modification of CNMs is usually performed by oxidation reactions, which could take place either at edge sites or defects of CNMs. Oxygenated CNMs generally disperse uniformly in a wide variety of solvents, and thus can enhance the interaction between the polymers and CNMs in the composites; such improved interaction can then result in the enhancement of electrical, mechanical, and thermal properties of the composites [13,20].

7.2.1.1 Oxygenation of Carbon Nanomaterials

There are different approaches that can be used to introduce oxygen functionalities on CNMs, which include acid treatment involving the use of HNO_3, H_2SO_4, and H_2O_2 or a mixture of any two [21]. Other methods involve the treatment of the CNMs with plasma [22], ultraviolet radiation, ozone [23], etc.

Strong acid and other oxidizing agents such as HNO_3, H_2O_2, $(NH_4)_2S_2O_8$ [24,25], $HNO_3 + H_2SO_4$ [13,26], $H_2SO_4 + KMnO_4$ [27], $HClO_4$ [28] etc. are used to oxidize the surface of CNTs thereby generating oxygen-containing functional groups such as carboxyl, hydroxyl, epoxy, ketone, etc. Acid treatment is usually accompanied by a wide range of procedures, which include high temperature refluxing, sonication, agitation, high-speed stirring, microwave digestion, or plasma treatment, of the mixture of oxidants and CNTs. Also, these methods of oxidation can remove amorphous carbon and impurities such as the residual metal catalysts contained in the CNTs after the production process. The mixture of HNO_3 and H_2SO_4 at different proportions is the most widely used reagent for the oxidation of CNTs [29]. It is generally accepted that the tips of CNTs are more active than the sidewalls; accordingly, it has been proposed that treatment of CNTs with acid mixtures can first open the tips and introduce oxygen-containing groups on the open ends, coupled with corrosion of the sidewalls. In practice, different carbon oxidizing agents have different oxidation effect on the CNTs. In line with this, the number and type of oxygen-containing groups on the carbon surface would ultimately depend on the type and concentration of the oxidant utilized. Some oxygen groups that can be generated on the carbon surface through oxidative treatment methods are illustrated in Fig. 7–3. X-ray photoelectron spectroscopy (XPS) has been used to demonstrate that HNO_3 and $KMnO_4$ can modify CNTs with more O–C=O bonds (aldehydes, ketones, carboxylic acids, and their derivatives) while moderate oxidants like H_2O_2, oxygen, and $(NH_4)_2S_2O_8$ can only functionalize CNTs into mainly C–O bonds (hydroxyls and/or ethers) as can be seen in Fig. 7–4A and B [25,30]. The cyclic voltammogram (CV) in Fig. 7–4C shows that the surface confined redox reactions of CNTs treated with HNO_3 or $KMnO_4$ are considerably higher than that obtained from CNTs treated with H_2O_2 under the same reaction conditions. This generally implies that HNO_3 or $KMnO_4$ introduces more quinonoid groups to the CNTs than would H_2O_2 or other mild oxidants [31].

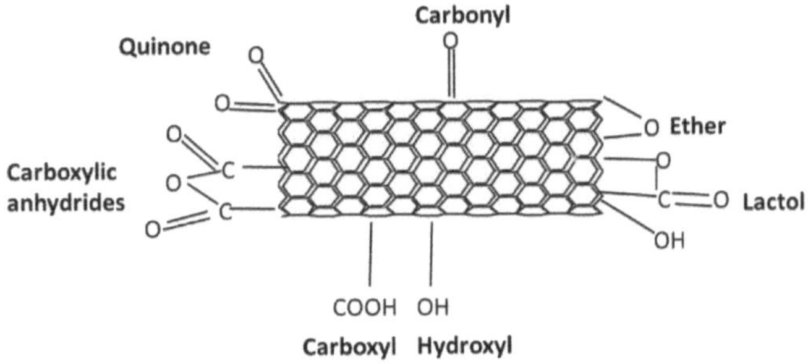

FIGURE 7–3 Different oxygen-containing groups on the surface of CNTs.

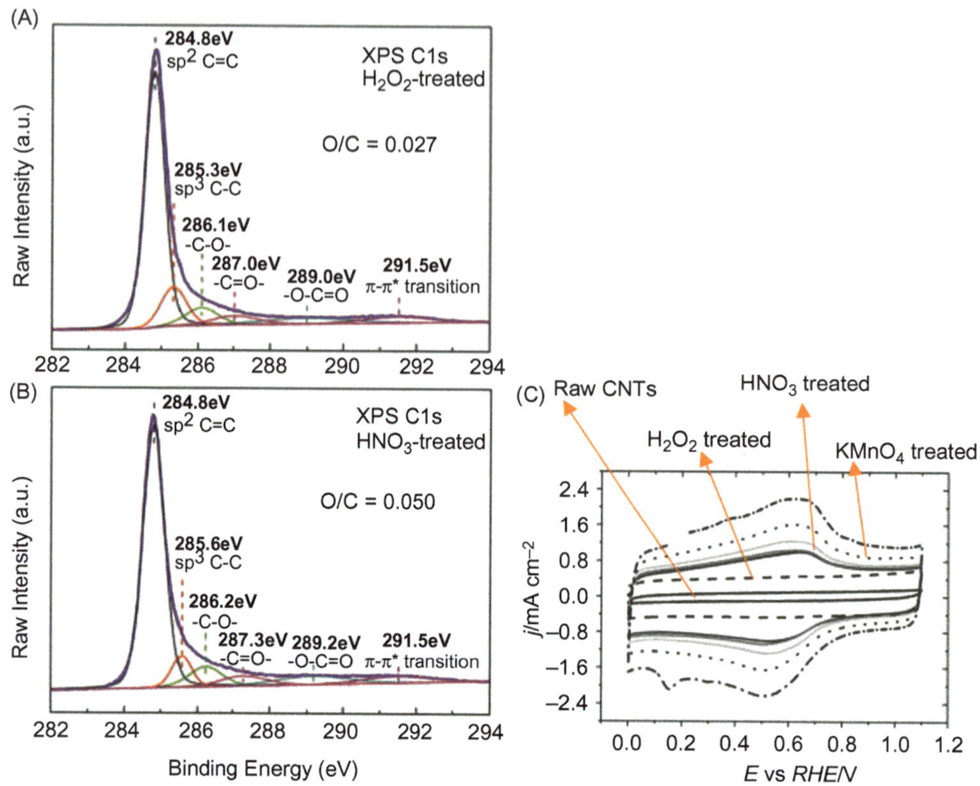

FIGURE 7–4 Deconvolution of C1 in the X-ray spectrograph: (A) H_2O_2 treated, (B) HNO_3 treated [30], and (C) CV of differently treated CNTs in 0.1 M $HClO_4$ [31].

Although direct acid treatment of CNTs would generally improve the dispersion of CNTs in various solvents, this often comes at the expense of reduced crystallinity and electrical conductivity of the CNTs. Alternatively, plasma treatment offers an effective process to chemically modify CNTs and increase its dispersion in various solvents without damaging the structural integrity and bulk properties. In plasma treatment, the excited species and radicals strongly interact with the surface of CNTs; this consequently breaks the π-bonds in C=C, creating active sites for binding of $C-O_x$ groups. It is important to point out that plasma treatment is more advantageous than other methods because its reaction time is short and it is generally nonpolluting. Also, it involves lesser damage to the CNTs, and can introduce a large range of functional groups that can be chosen by adjusting the process conditions and the substance used to generate the plasma [32].

Addition of these polar surface groups will transform the hydrophobic CNTs into hydrophilic materials; this can promote strong interfacial interaction between the functionalized CNTs and polymers, which could aid in the enhancement of the properties required from the composite. For example, this effective dispersion of the CNTs aided by the functional groups can promote the molecular interaction (so-called interfacial conjugation) between the functionalized CNTs when used in composites with conducting polymers such as PPy, PAni, or PEDOT, thus enhancing the conduction and charge transfer efficiency across the composite electrode/electrolyte interface (EEI). Such effective charge transport aided by enhanced molecular interaction between the functionalized-CNT and polymers is also responsible for the improved charge carrier transport (hole and electron) when functionalized CNTs are used to dope P3OT resulting in improved short circuit current [33,34].

Furthermore, the strong adhesion of polymers to functionalized CNTs also enables the effective transfer of the mechanical and thermal properties of the CNTs across the entire network of the polymer–CNM composite. For instance, H bonding between functionalized CNTs and the ester groups of poly(methyl methacrylate) (PMMA) enhances the wetting of the CNTs, which are consequently well wrapped by the polymers; this effectively creates a new interfacial layer for effective stress transfer between the two components of the composites [35].

7.2.1.2 Oxygenation of Graphene

Graphene rarely exists in its isolated form in nature, but substantial amounts of solution-processable functionalized graphene derivatives such as GO can be readily prepared [36]. GO, which is usually produced by different modifications of Hummers' method, is characterized by a large number of oxygen-containing species on both the basal and edge planes. Various attempts at modifying Hummers' method have been adopted in a bid to produce GO with more oxygen groups in a shorter period and with fewer hazards, particularly in relation to the gases generated.

Functional groups on the surface of graphene could improve interfacial interaction between the graphene derivatives and polymers. Although GO is very soluble in aqueous solvent, it has a low conductivity due to the disrupted conjugated networks and a widening of the bandgap induced by localized electrons from the functional groups [33]. However, the electrical conductivity could be regained through reduction and the products are generally called reduced graphene oxides (rGO). Chemical, thermal, electrochemical, hydrothermal, and photothermal methods can be used to reduce GO [34,37]. Chemical reduction is the most commonly adopted procedure, however, the hazardous nature and reduction power of the reductants coupled with the toxicity of gases produced from the reaction are some factors that need to be critically considered in this process. Furthermore, thermal reduction decomposes the oxygen groups of GO, which results in considerable mass loss at the end of the process. Electrochemical reduction is an efficient and environmentally friendly approach, but its major bottleneck remains the issue of scaling-up. The X-ray photoelectron spectrograph in Fig. 7–5 shows that the proportion of sp^2 containing structure of the GO are increased after reduction [38].

Oxidized CNTs and graphene can serve as polymerization dopants and supporting electrolyte to enhance the solution conductivity during electrochemical codeposition of the monomers and the CNMs in the preparation of the polymer–CNM composites. More importantly, GO provides effective nucleation sites during the polymerization process. Generally,

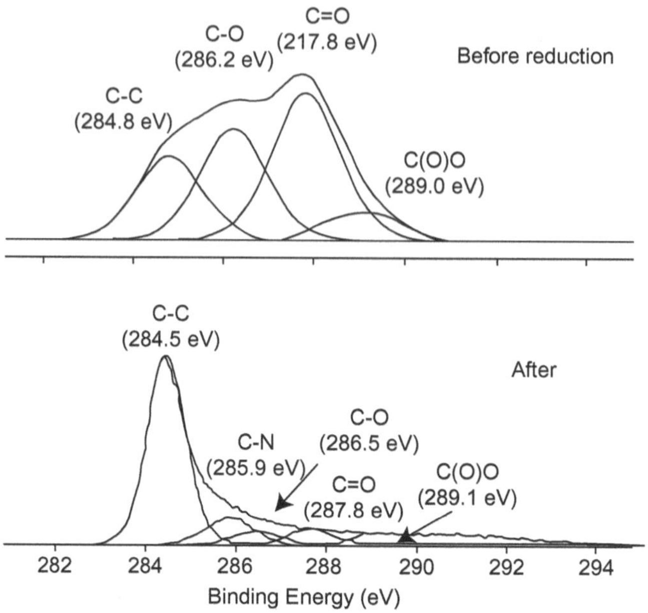

FIGURE 7–5 XPS results of GO (C1s) before and after reduction [38].

the π−π interaction/electrostatic interaction and hydrogen bonding formed between the GO layers and PPy rings enhances the uniform growth of the PPy on the GO implying that the resulting composite would optimally utilize the 3D structure of GO in improving its structural properties. In addition to the porous properties of the GO, which could aid in effective ionic transport, this 3D structure of the composite also serves to accommodate the eventual swelling and shrinkage of the PPy in the course of repeated charge/discharge.

7.2.2 Sulfonation of Carbon Nanomaterials

Sulfonation of CNMs has increased in popularity because sulfur-containing groups are also very effective in increasing the dispersion of CNMs in aqueous solvents and improving the interfacial interaction of the CNM with the polymer matrix. A majority of the current methods used to sulfonate CNMs usually utilize preoxidized CNMs as the starting materials. For example CNTs-$(SO_3H)_n$ was produced by heating a mixture of oxidized CNTs and concentrated sulfuric acid at 300°C for ca. 12−18 hours [39]. It has also been reported that microwave treatment of a mixture of CNTs with sulfuric and nitric acids is an effective and environment-friendly means of modifying CNTs with carboxylate (−OH and −COOH) and sulfonated [(−SO_2 ˙OH)] groups [40]. Aryl diazonium salts are also used as sulfur precursors to sulfonate CNMs [41]. The procedures usually reduces GO with the aid of $NaBH_4$ followed by sulfonation with aryl diazonium salts, which is subsequently reduced by using N_2H_2 [42]. Other reports have also described the use of oleum, sodium nitrate, and sulfanilic acid in the sulfonation of GO [36].

$-SO_3$ groups can create stronger H-bonding moieties compared with $-COOH/OH$, which is more beneficial for interfacial reactions with polymers [43]. The interaction between π and SO_3 bond of sulfonated CNMs with the quinonoid rings of PAni enhances the charge transfer process. In addition, high conductivity of CNMs in the composite create a conducting network for improved electronic conductivity and effective interfacial redox activity [41,44]. S-CNTs or S-graphene in addition to acting as a framework for $-SO_3H$ to enhance proton conductivity in Nafion, also contributes to the improved mechanical strength of the membrane [45–47]. Furthermore, sulfonated CNMs used in polymer electrolyte membrane (PEM) act as a Bronsted state acid, which enhances the proton conductivity and can also reduce the permeation of methanol when the PEM is used in methanol fuel cells. Additionally, inclusion of S-graphene into the matrix of Nafion, has been demonstrated to be capable of reducing the hydrogen crossover, owing to the high surface area of the S-graphene, which increases the path length of the hydrogen transport across the polymer chains. This property is also coupled with the wrinkling of the graphene structure due to the inclusion of the sulfur, and this results in an increase in the surface roughness of the composite, which was shown to be important in increasing the water uptake of the composite membrane [48].

7.2.3 Nitrogenation of Carbon Nanomaterials

The doping of CNMs with heteroatoms, such as N, P, or B has been shown to be effective in altering the properties of CNMs in energy storage and conversion, catalysis, or in sensors. Moreover, nitrogen-containing species are also commonly used heteroatoms; this is mainly due to their abundance, accessibility, and low health hazard. The two main methods used in modifying CNMs with nitrogen species are in situ and posttreatment nitrogenation. CVD is the main in situ N-doping process used [49,50]. Generally, pyridine, melamine, ammonia, acetonitrile, and benzylamine, etc. serve as nitrogen source whilst xylene, ferrocene or hydrocarbons are used as the source of carbon in the CVD process [51–53]. In situ nitrogen doping tends to produce a homogeneously modified material because it is usually carried out during the production of the CNMs. However, a major drawback is that CVD is a relatively expensive process that is difficult to carry out in industrial scale. The defects created on the surface of CNTs can also be used as the site for further functionalization with nitrogen-containing heteroatoms; this can be considered as a posttreatment procedure [54]. Some posttreatment methods that utilize either oxidized or pristine CNMs include the adoption of pyrolysis [55], plasma [56], or hydrazine (N_2H_4) [57]. Different posttreatment methods are selected based on the required N-doping level. In general, it has been suggested that posttreatment nitrogenating procedures usually only modify the surface, with little change to the bulk properties of the CNMs. Current approaches used in doping CNMs with nitrogen can result in three types of nitrogen structure within the carbon lattice; these are graphitic N, pyridinic N, and pyrrolic N. Due to the comparatively higher electronegativity of N (3.04) to C (2.55), N-doping or P-doping effects can be introduced on the surface depending on the location in which the N atoms are situated on the carbon lattice [58].

N-doping will add extra electron to the graphitic structure without deteriorating the original properties and also enhance the surface energy, thus improving the adhesive properties of the CNM surface [59]. In addition, N-doping sites on the wall of CNTs increase the wettability of the CNMs and might also contribute to nucleation. Also, nitrogen sites can act as electron donors or acceptors [60,61]. It has been shown that the introduction of N-CNTs into the matrix of the poly(thieno[3,4-b] thiophene/benzodithiophene)/[6,6]-phenyl C71 butyric acid methyl ester (PTB7/PC71BM) copolymer can enhance the dissociation of excitons across the collector–active material interface. Accordingly, bulk heterojunction (BHJ) solar cells using this composite material will exhibit improved light absorption and charge dissociation within the active layers coupled with enhanced charge collection efficiency [62].

7.2.4 Halogenation of Carbon Nanomaterials

Owing to the high electronegativity and reactivity of halogens, their incorporation into the lattice of CNMs would result in significant change in the electronic properties and distortion of the structures of the carbon.

Graphene fluorination can be achieved through reacting it with F_2/XeF_2, and, also through the liquid or mechanical exfoliation of graphite–fluorite [63,64]. F-graphene exhibits comparable mechanical strength to pristine graphene, coupled with a widened bandgap (>3 eV), which makes it the thinnest wide-bandgap semiconductor [65]. Different halonium salts have also been used to functionalize CNTs with F, Cl, I or Br atoms [66]. It is also important to point out that unlike most functionalization methods, fluorination increases the hydrophobicity of graphene [67,68]; furthermore in comparison with fluorinated carbons, other halogen atoms have a lower binding energy. For example, the C–Cl bond, which is less stable than the C–F bond, would decompose at low temperatures, thus precluding the former for use in composites that are prepared at relatively high temperatures. It is also important to note that fluorinated CNTs, which disperse effectively in poly(propylene) (PP) matrix, have been shown to improve the rheological and electrical properties of the composites [69].

The improved conductivity of polymer and iodine-doped CNT composites was attributed to the increased conductivity of the doped CNT and the tendency for the CNT to generate smaller polymer barriers within the matrix [70]. It has also been demonstrated that chlorinated phenyl isocyanate functionalized rGO, used to dope the cyanoethyl pullulan polymer, aids in improving the dielectric properties of the composites. This increase in dielectric performance was attributed to the effects of the polarizable C–Cl bonds, and increased interfacial polarization of the composite [71].

7.2.5 Noncovalent Functionalization

Although covalent functionalization is effective in increasing the solubility of CNMs in various solvents, which consequently aids their dispersion in the polymer matrix, in some cases, it could affect the structural integrity of the CNMs. Conversely, noncovalent functionalization could promote the molecular interconnection between the CNM and the polymer without forming any chemical bonds, and thus, could have minimal impact on the structural properties of the CNMs.

Surfactants are generally adopted to aid the effective and homogeneous dispersion of CNTs in a number of media relevant in the preparation of polymer–CNM composites. In water, the hydrophilic head of the surfactant molecule adheres to the surface of CNTs leaving the hydrophobic tail to interact with water molecules, as illustrated in Fig. 7–6. The application of surfactants is usually accompanied by sonication, which can aid in the unbundling of interwoven CNTs or the exfoliation of stacked graphene platelets into individual tubes or sheets respectively. Ionic surfactants such as sodium dodecyl benzenesulfonate, sodium dodecyl sulfate, localized dry spots, etc. and nonionic surfactants, for example, Tween-20, Tween-60, Tween-80, and Triton X-100, have been shown to have the ability to effectively disperse CNTs and graphene in different solvents [73,74]. It has also been demonstrated that CNTs with longer tubes have higher conductivity and lower percolation threshold when used as fillers in the preparation of polystyrene composite [75], thus making the surfactant assisted route a preferred method of preparing polystyrene–CNT composites. In addition to surfactants, polymers such as polyvinylpyrrolidone, poly(3-alkylthiophene), and polymeric maleimide can adsorb or wrap around CNTs, which can aid in their unbundling and easy dispersal in different polymer matrix [76,77]. In line with this, conductive polymeric materials such as poly(styrene sulfonates) (PSS):PEDOT have been used to replace conventional ionic surfactants to aid the dispersion of SWCNTs in the matrix of poly(styrene) (PS), and the resulting composite displayed a markedly low percolation threshold [78].

The limited effects on the structural properties of the CNMs treated through noncovalent functionalization, and their effectiveness in improving their dispersibility within the polymer matrix, is indeed the major importance of this method. For instance, EpPy-16, which is

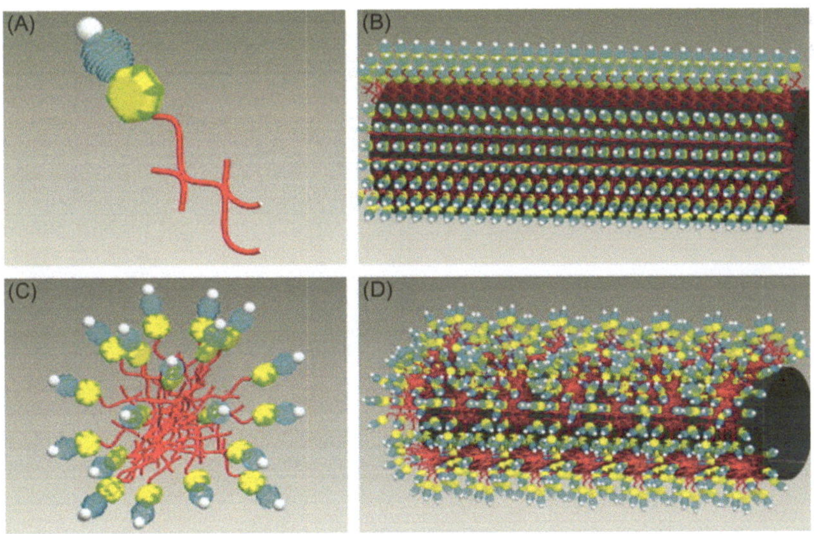

FIGURE 7–6 Schematics of CNT functionalization using noncovalent methods by surfactant triton X-100 [72].

comprised of two epoxide-containing pyrene derivatives, and which adsorbed on the surface of CNTs, have been shown to be good dispersants of the CNTs in the epoxy matrix, and this was attributed to the strong intermolecular π-stacking interaction that the EpPy-16 induces. Accordingly, such strong interconnection was shown to boost the efficient strength transfer from the CNT to the epoxy, and since this interaction also prevents the CNTs from sliding under external tension and load, the resulting composite displayed improved shear modulus [79]. Similarly, for graphene/epoxy composite, the addition of poly(glycidyl methacrylate) (Py-PGMA) can significantly assist in the effective dispersal of graphene, which subsequently aids in improved interaction with the epoxy resin to achieve a large contact area, which promotes heat flow and phonon diffusion and thus enables the epoxy–graphene composite to display highly improved thermal conductivity [80].

The chemical functionalization methods discussed in the preceding sections can be broadly classified into two categories: that is, direct and defect functionalization processes. Direct functionalization is usually adopted for the oxygenation of CNM while other functionalization methods such as sulfonation, nitrogenation, and halogenation can be carried out through both the direct and defect functionalization processes. Generally, most defect modifications of CNMs usually employ oxidized CNMs as precursors. This is illustrated in Fig. 7–7 [1]. Defect functionalization utilizes the structural deformities generated by oxygen groups on the surface of CNMs (sidewalls and open ends) as the site for the addition of other functional groups, which are usually difficult to introduce through direct functionalization. This method significantly increases the variety and properties of surface groups that can be introduced on the

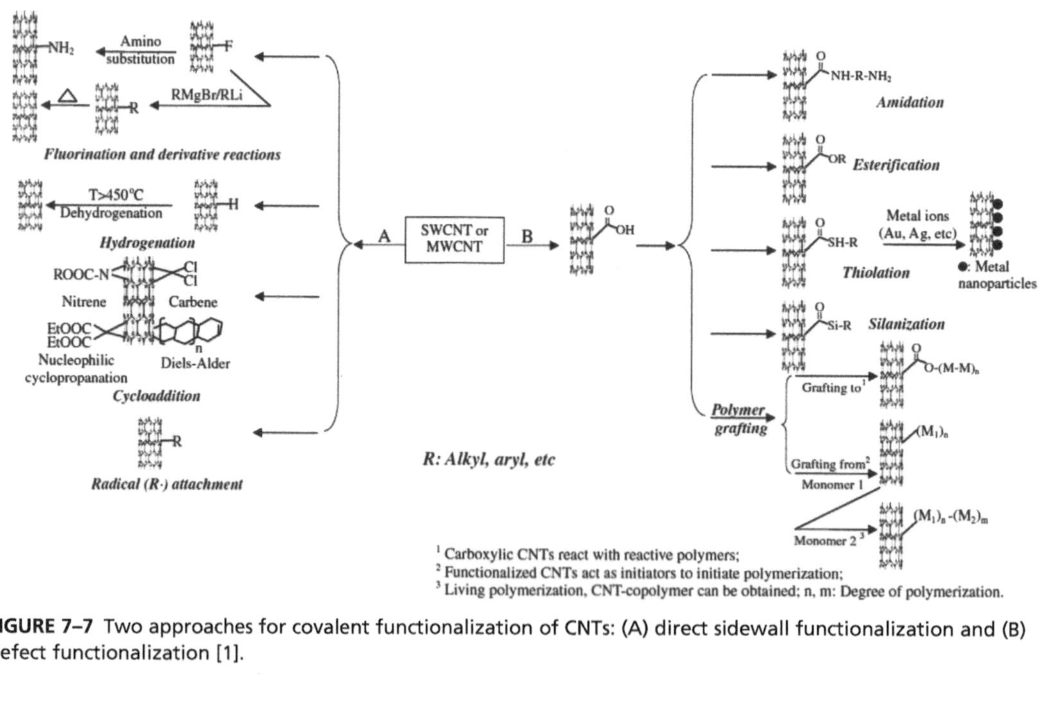

FIGURE 7–7 Two approaches for covalent functionalization of CNTs: (A) direct sidewall functionalization and (B) defect functionalization [1].

CNMs. For instance, some reports have demonstrated the addition of highly active isocyanate groups on the surface of oxidized CNTs; this introduction of isocyanate groups is particularly relevant as it aids the effective reaction of the CNTs with sources of other functional groups such as amines, carboxylic acid, ethanol, chloroform, etc., which could make these materials useful for doping a wide variety of polymers for various applications [81]. For example, amine treated isocyanated CNT, was utilized to dope polyurethane (PUE), and it was shown that the existence of isocyanate groups contributes to the enhancement of the tensile strength and scratch resistance of the prepared PUE–CNT composite [82]. It has also been shown that the introduction of isocyanate groups in GO reduces its hydrophilicity, which subsequently aids their exfoliation into graphene nanoplatelets in polar solvents [83]; such nanosheets can then be applied to dope polymers [84].

Generally, the length of individual tubes of CNTs in the composite is in the range of 0.5–1.0 μm. Below this critical value, CNTs cannot transfer their stiffness or mechanical properties to the polymer matrix [1]. Accordingly, different functionalization methods could be selected to suit the desired property of the CNM used in the preparation of the required composite. Although achieving a homogeneous dispersion of the CNMs in the polymer matrix is of utmost importance, it is also worth noting that the interaction between the polymer and the CNM in the composite is extremely important because even well-dispersed CNTs cannot provide effective reinforcement in the absence of sufficient intramatrix interaction. Thus, for structural enhancements of polymers using CNMs as fillers, chemical bonding and hence strong interaction of CNMs with the polymer matrix can be enhanced with increased reactive functional groups and availability of free-bonding sites. Functional groups can also be useful in enhancing percolation.

7.2.6 Preparation of Polymer–Carbon Nanomaterial Composites

There are a wide variety of procedures that can be used to prepare composites of polymers and functionalized (or pristine) CNMs. The key to process selection would be to achieve effective dispersal of the CNMs in the polymer matrix coupled with good adhesion and/or interaction between the components of the composites. The surface groups introduced through the functionalization process also inform the choice of the method used to prepare the polymer–CNM composite. This is because surface groups in general influence the wettability of the CNMs, which in turn affects their dispersion in different solvents. This is of practical importance because the preliminary stage of the preparation of most polymer–CNM composites usually involve the dispersion of the CNMs, for example in aqueous solvents. Currently, some of the main methods used in the preparation of composites of polymers and CNMs are solution mixing, melt processing, in situ polymerization, latex technology, chemical grafting, and layer-by-layer assembly (LBLA). During these processes, high power dispersion methods such as ultrasonication, high-speed shearing, etc. are generally adopted to aid in the effective dispersion of the CNMs in the polymer matrix, resulting in composites with highly improved interfacial interaction between the components [1,2].

The most common among the aforementioned methods is solution mixing and it is usually adopted in the preparation of films. The typical procedure for solution mixing consists of three steps. Firstly, the CNMs are dispersed in an adequate solvent that can also disperse the polymers. Effective dispersion of the CNMs could then be achieved through sonication, agitation, or mechanical mixing. Secondly, the polymers are added into the dispersion at room or elevated temperature. Finally, the desired composite material is obtained by precipitating or casting. Some polymers that can be used to prepare composites through the solution-mixing method include PMMA [85], poly(vinyl alcohol) (PVA) [86], and PS [87] etc. In enhancing the electrical conductivity of polymers through the use of CNMs (discussed in Section 7.3.1), the solution-mixing method is usually effective in the preparation of composites with a low electrical percolation threshold, moreover, a suitable solvent has to be selected. Furthermore, during the removal of the solvent, which is usually done through evaporation, reaggregation of the CNMs could occur. Accordingly, techniques such as spin casting or electrospinning can then be utilized to produce solution-mixed composites with minimal aggregation of the CNMs.

In melt-blending, the CNMs are directly dispersed into a melt of the polymer under high temperature, and effective dispersion is usually achieved through the application of shear forces, for example with the use of screw-extrusion mechanisms [88]. Although this approach avoids the challenge of selecting an adequate solvent as is the case in solution mixing, it generally possesses a comparatively lower dispersion of the CNM in the polymer matrix and the materials are usually subjected to high stress; furthermore, the viscosity of the polymers could aggravate the issue of agglomeration/stacking of the CNMs. Consequently, this process is used for the introduction of CNMs as low content fillers in thermoplastics such as PP [89], PS [90], poly(carbonate) (PC) [91], etc.

In situ polymerization is essentially a chemical method for preparing polymer–CNM composites. In this process, monomers are used in place of polymers as the starting material. CNMs are usually dispersed in the monomers with or without solvents, followed by the addition of polymerization reagents (e.g., oxidants) or the application of electric current or voltage, to initiate the reaction. The principal advantage of this approach is that the growth of the polymers on the surface of the CNMs can lead to the formation of covalent bonds between the polymers and the CNMs in the composite. In electrochemical codeposition, the treated CNTs (or CNMs in general), which usually comprise of oxygen surface groups, can also serve as supporting electrolyte, and the mass of the final polymer–CNT composite can be easily modulated by varying the amount of charge applied [92]. This method is usually preferred when the polymers are insoluble or unstable at elevated temperatures; also, in situ polymerization is utilized when the desired loading of the CNMs is high. Fig. 7–8 illustrates the procedures that can be adopted in the preparation of polymer–CNT composites through electrochemical codeposition. Generally, composites of CNMs and conducting polymers such as PPy, PAni, PEDOT [13], or nonconducting polymers, for example PS [94] and epoxy [95], can be prepared through in situ polymerization.

Latex technique is particularly useful for polymers that can emulsify, and it involves the dispersion and mixing together of surfactant-treated CNMs in a polymer emulsion.

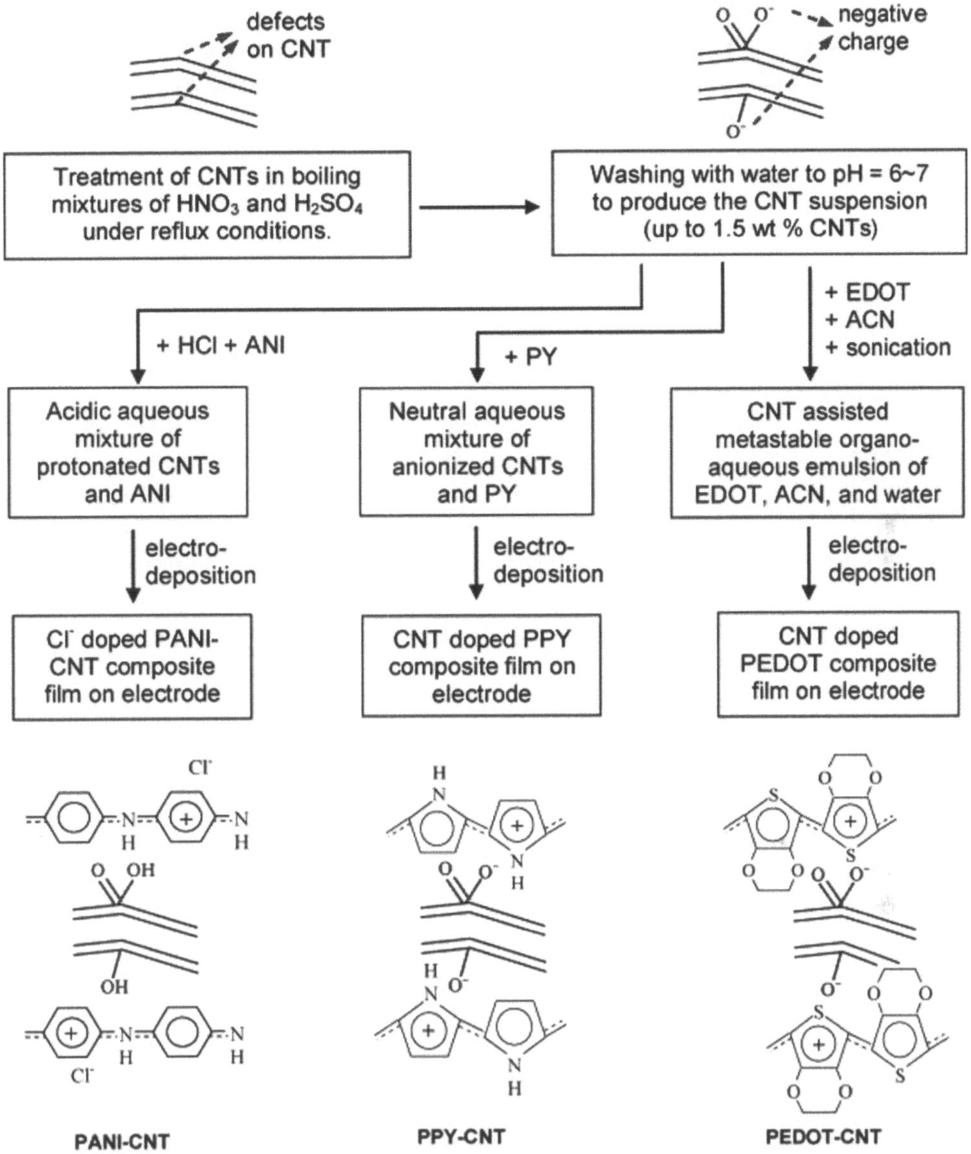

FIGURE 7–8 Preparation procedures of polymer–CNT composite through electrochemical codeposition [93].

Subsequently, freeze-drying or melt-processing could be utilized to obtain the final composites; since water is usually the solvent used in this case, the process is relatively safe and environmentally friendly. Additionally, this procedure is easy to scale-up for the production of large quantities of polymer–CNM composites. Also, the procedures involved in the latex technique can be used to promote effective interaction of the CNM in the polymer matrix, as has been demonstrated for a PP–rGO composite [96,97].

LBLA is a procedure used to prepare ultrastrong, robust, and thin composite films [98]. Some techniques used to carry out LBLA includes immersive, spin, spray, electromagnetic, and fluidic methods, and these procedures are adopted based on the desired morphology of the end product. This process is usually carried out by soaking a positively charged substrate in a polyanion solution, which is then transferred to a polycation solution that strongly adheres to the surface of the polyanion film on the substrate. This procedure is then repeated a number of times depending on the required thickness. Compared with other polymerization methods, LBLA can be used to produce ultrathin composite films owing to the molecular-scale adhesion of the layers. LBLA method can also achieve molecular precision in the control of the distribution and amount of CNM in the polymer matrix. The final morphology of the composites can also be altered by the deposition mode, solvent removal procedures, or the application of shear force [99,100].

It is also important to point out that the aforementioned methods of preparing polymer–CNM composites could be integrated depending on the required properties of the composites. For instance, solution-mixing and melt-processing were applied together to prepare PP–CNM [101] composites whilst latex technology and LBLA have been adopted to produce PS–CNM composites [102]. Table 7–2 gives a summarized comparison of various methods for preparing polymer–CNM composites that were discussed in this section.

7.3 Properties and Applications of Polymer Composites With Carbon Nanomaterials

Due to their low cost, ease of processing, and versatile thermal, mechanical, electronic, and optoelectronic properties, polymers have grown to become one of the most important classes of materials in modern society. In a bid to improve their properties or to enhance their effectiveness for specific applications, the preparation of composites of polymers with CNMs has received widespread applications. The equally versatile properties of CNTs and graphene presented in the preceding sections have made them the two major classes of CNMs that are at the cutting edge of industrial and research applications of polymer–CNM composites. As has been described already, functionalization of these CNMs has been demonstrated to be an important first step in obtaining high-performance polymer–CNM composite. This is because the functional groups introduced usually promote the interaction between the polymer–matrix and the CNMs (CNTs and graphene) at the molecular scale. For example, –COO functional groups introduced during the acid treatment of CNTs allow the CNTs to act as negatively charged species, which aids the electrochemical codeposition of CNTs with ECPs by serving as coions in the deposition process. Additionally, these negatively charged CNTs also improve the charge storage characteristics of the produced ECP–CNTs composite [92]. Polymer reinforcement particularly for structural and/or thermal applications is also another broad area of polymer–CNM application [103].

In this section, the properties of functionalized CNMs and their composites with polymers will be discussed from the standpoint of their general applications. Particular attention will

Table 7–2 Comparison of Different Procedures for the Preparation of Polymer–Carbon Nonmaterial Composite

Methods	Polymers	Treatment of CNMs	Solvent Needed	Advantages	Disadvantages
Solution mixing	PMMA, PVA, PS, PU, PAni etc.	Predispersion	Yes	Easy to process and scale-up	Low content of CNM loading and limited range of solvents
Melt processing	PP, PS, PC, PLA PET, etc.	No	No	Easy to process and scale-up	Low content of CNM loading, low uniformity in dispersion of CNMs and influence of high stress
In situ polymerization	PPy, PAni, PEDOT, PS, epoxy, etc.	Predispersion	Case-specific	Uniform dispersion of CNMs, strong interaction between polymer and CNM, high loading of CNM	Need filtration to obtain the final product, and the electrochemical method is difficult to scale-up.
Latex technology	Epoxy, PP, PS, rubber, etc.	Surfactant aided	predispersion	Yes (usually water)	Easy to process and scale up, environmentally friendly
Polymers have to form emulsion					
LBLA	PW, PEI, etc.	Predispersion	Yes	Precision control at molecular scale	Difficult to scale-up

be paid to the electrical and electrochemical properties of polymer–CNM composites, and their use in energy storage and conversion, microelectronics, optoelectronics, etc. Also, the use of CNMs to enhance the mechanical/rheological and thermal properties of polymer composites will be briefly discussed.

7.3.1 Electrical Properties and Applications of Polymer–Carbon Nanomaterial Composites

The electrical properties of polymers range from insulators to conductors. Conducting polymers are a diverse class of electroactive materials that display different charge transfer mechanisms that have been exploited in energy storage, energy conversion, sensing technologies, etc. (see Section 7.3.2). Generally, CNMs could be used to increase the electrical conductivity of all classes of polymers. In this section, the use of CNMs to enhance the electrical properties of nonconducting and conducting polymers will be discussed from the standpoint of their general applications.

The close to metallic conductivity of CNMs has been exploited to enhance the conductivity of nonconducting polymers. In general, various descriptions of the percolation theory have been proposed to describe the mechanism of conductivity enhancement of CNM (conducting materials) and nonconducting polymers. Notably, percolation-dominated electrical conductivity has been linked to the aspect ratio, nanostructure of the conducting filler material, the filling ratio, type of surface functional groups on the CNM, etc. [104–106]. Fundamentally, the percolation theory is based on the notion that carbonaceous materials (at a given weight percentage known as the percolation threshold) will form a continuous conducting network in-between the polymer matrix, and through mechanisms such as the tunneling effect, the nonconducting polymer can then transition from an insulator to a conductor [107,108].

Traditional carbonaceous materials (e.g., carbon black, soot, graphite, etc.) usually have a high threshold to induce percolation owing to their high density and low mechanical strength which could lead to unstable conductive networks. However, due to their high mechanical strength and relatively low density, pristine or functionalized CNTs together with graphene and its derivatives have been shown to provide a lower threshold for percolation, and as such have been used to improve the conductivity of a wide variety of nonconducting polymers [104]. For instance, enhanced conductivity of polystyrene achieved through the incorporation of either CNTs or graphene sheets has been linked to an electric field induced dynamic percolation. It was shown that graphene sheets readily form conductive networks than CNTs in this polymer matrix, and this was partly attributed to the geometry of the nanoparticles [109].

Generally, the type and presence of functional groups on the CNM, and the processing method of the nonconducting polymer–CNM composite, are important determinants of the percolation threshold and the sheet conductivity. It has been proposed that a percolation threshold of approximately 0.1% can be achieved from any nonconducting polymer–CNT composite that is well processed [110]. For example, significant increase in the conductivity achieved with a percolation threshold of ca. 0.1% has been reported when completely exfoliated graphene sheets were incorporated into the polymer matrix of polystyrene [111]. Understanding and correlating the role of the geometry, morphology, and surface groups of CNMs in enhancing the electrical, rheological, mechanical, and thermal properties of composites of nonconducting polymers and CNMs is the basis for their applications in construction, aviation, electronics, defense, etc.

7.3.2 Electrochemical Properties and Applications of Polymer–Carbon Nanomaterial Composites

The use of polymer–CNM composites for energy storage and conversion, sensors, microelectronics, catalysis, etc. has been increasing rapidly. This is because the composites usually have improved properties that either of the components might not possess individually. For example, in energy storage, composites of acid-treated CNTs and PPy, in addition to having an improved electrochemical performance over the PPy material, also increase the ease of processing the composites into electrode materials. The latter characteristic is due to the improved mechanical property provided by the CNT. Indeed this improvement in the rheological properties of some polymer–CNTs composites is the reason why industrially important production processes such as ink-jet printing can be used to produce improved polymer–CNTs electrodes for energy storage (for example in supercapacitors) or for conductive networks in microelectronics [112]. This section will discuss the electrochemical properties of polymer–CNM composites in relation to their applications in energy storage and conversion, microelectronics, etc. In this section, except where stated otherwise, an electrode is in the solid phase (responsible for conduction of holes or electrons), whilst the electrolyte is in the liquid phase, and conducts only ions.

Fundamentally, pristine CNMs would store charge via the electrochemical double layer (EDL) mechanism. This is generally attributed to the electrostatic separation of electrons (on the electrode) and counterions (from the electrolyte) at the EEI within the matrix of the CNM. In this regard, the EDL mechanism is generally described to be storage under ideal polarization, that is, without the transfer of charge across the EEI. For this reason, the EDL mechanism has always been linked to the surface area of the CNM, which is predominantly due to micropores, which can for example accommodate hydrated ions. Consequently, CNTs which are predominantly mesoporous might not necessarily display high-EDL capacitances (when compared with activated carbons for example). However this does not hinder their use for EDL storage applications. This is because mesopores have been shown to serve as conduits for large ions, implying that CNTs can be used to improve the ionic conduction of predominantly microporous activated carbon used for EDL storage, and have even been shown to significantly contribute to the EDL capacitance of some polymer–CNT composites. On the other hand, graphene-based materials, which can be processed to display relatively high-EDL capacitances, can be utilized directly as EDL electrodes, or physically mixed with polymers to improve the EDL capacitance. In general, polymer–CNM composites with CNTs or graphene and some of its derivatives would usually have substantial increase in EDL capacitance. Still it should be pointed out that the other class of charge storage mechanism, that is, Faradaic storage (FS), which is characterized by the exchange of charge across the EEI, is an important aspect of the use of polymer–CNM composites. This is because FS mechanisms are generally capable of storing more charge and this is also the main storage mechanism displayed by conducting polymers. Accordingly, FS mechanisms are at the forefront of the design of novel polymer composite electrodes for next generation supercapacitor, battery, or hybrid devices.

FS mechanisms involve the transfer of charge across the EEI. There are generally two kinds of FS mechanisms, that is, noncapacitive Faradaic storage (NCFS) and capacitive Faradaic storage (CFS). NCFS has been described as due to the transfer of localized valence electrons across a single energy band corresponding to the redox potential of the material and leading to peak shaped cyclic voltammograms. Such storage mechanisms can in general be explained with the aid of the Nernst equation, and could be broadly referred to as Nernstian mechanisms. For example, oxygen groups on the surface of functionalized CNTs display surface confined Faradaic mechanisms, which could be described as the redox reactions of fused quinonoid groups [37,63]. Generally, heteroatom doped carbon material, for example redox reactions between pyridine and pyridinic groups, are also sources of increased Faradaic charge storage [51,113,114].

CFS, which is the technologically relevant pseudocapacitve mechanism displayed by ECPs such as PPy, PAni, and PEDOT and some transition metal oxides such as MnO_2 RuO_2, etc., have been described as due to the transfer of delocalized valence electrons. In these materials, which are mostly semiconducting, due to electron delocalization, valence electrons can transfer over different energy gaps corresponding to a wide range of electrode potential leading to the passage of a capacitive current. Accordingly, the electrochemical properties of materials displaying CFS are similar to those of EDL capacitance, that is, rectangular CVs and triangular galvanostatic charge–discharge plots. Moreover, pseudocapacitive charge storage is characterized by much greater capacitance values in the order of 200–1000 F/g or greater, as opposed to a maximum of about 300 F/g obtainable from EDL capacitance displayed by advanced porous carbon electrodes such as activated carbon [115]. Recently, the theoretical proof of pseudocapacitance in semiconducting materials such as ECPs and transition metal oxides was described through first principle treatments of model graphene structures. These structures include *fully oxidized graphene*, which is an insulator, and *partially oxidized graphene*, which possesses semiconducting properties and displays EDL capacitance with significant pseudocapacitance. The total removal of the oxygen groups on the partially oxidized graphene results in the formation of *pristine graphene*, which has metallic conductivities, and displays only EDL capacitance. Herein, transfers of delocalized electrons were shown to be the origin of the pseudocapacitive properties displayed by the partially reduced GOs. This non-Nernstian but capacitive Faradaic charge storage mechanism can thus be used to describe the charge storage of ECPs, since the delocalization system of the valence electrons in these ECPs might be linked to their discontinuous conjugated-π bonds [37].

It is also important to point out that these so-called charge storage mechanisms are in reality redox (or charge) transfer mechanisms of materials in general, and are broadly referred to as storage mechanisms when the purpose of use of the material is as electrodes for energy storage or conversion. Accordingly, their fundamental understanding and exploitation are at the basis of applications where charge transfer across the EEI is required, and these include but are not limited to photovoltaics, electrocatalysis, electrochromism, etc. In general, all redox polymers can display either or both the CF and NCF mechanisms. In line with this, the two broad classes of electroactive polymers are ECPs, and the general class of so-called nonconjugated redox polymers (NCRP). ECPs store charge across their conjugated backbones, through the movement of electrons, similar to the case of metals (or in most cases as semiconductors). On the other hand, NCRPs are referred to as localized-state conductors, because electrons usually hop along redox groups that are adjacent to each other, and the molecular chains connecting redox sites usually experience little displacements on electron hopping. Within an appropriate applied electrode potential, films of ECPs such as PAni, PPy, or PEDOT would store charge via the pseudocapacitve mechanism, and this property can be described using the concept of the delocalization of valence electrons. On the other hand, the localized properties of redox polymers, for example, polyvinyl-ferrocene

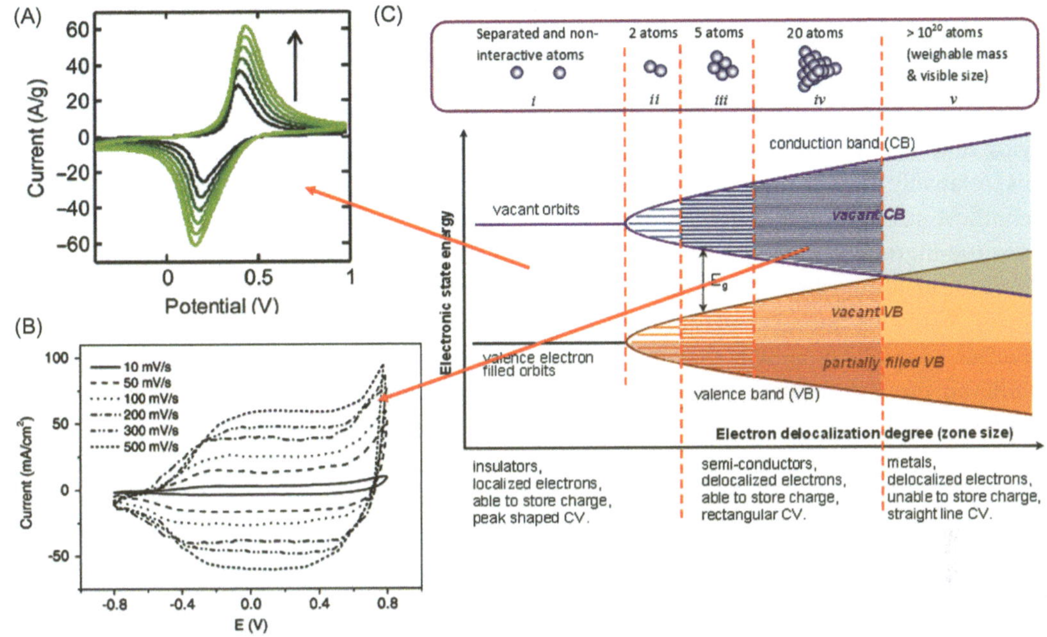

FIGURE 7–9 (A) Noncapacitive (Nernstian) mechanism displayed by the nonconjugated polymer of PVFc [116]. (B) Capacitive Faradaic mechanism exhibited by a film of the ECP (i.e. PPy) [118]. (C) Schematic illustration of the band model describing localized and delocalized valence electrons and their relationship to charge storage mechanisms [14].

(PVFc), can store charge via the Nernstian mechanism [116]. Also a ferrocenyl bithiophenic conducting polymer has been described to display a mixture of pseudocapacitive and Nernstian charge storage mechanism within certain potential range [117]. Fig. 7–9 shows the cyclic voltammograms of polymer materials displaying noncapacitive Faradaic and capacitive Faradaic charge storage, illustrated with the aid of the band model.

In light of these general electrochemical properties of conducting polymers, the charge storage properties of these materials would be dependent on the porosity of the electrode materials prepared, nature of the ions in the electrolyte (for example hydrated crystal radius of ions in aqueous electrolytes), conductivity of the different phases of these materials on charging/discharging, etc. It is the understanding of these material properties that can then inform the use of functionalized CNMs in the preparation of composites to enhance or introduce properties that make them effective for the desired application.

7.3.2.1 Composites of Electronically Conducting Polymers and Carbon Nanomaterials for Energy Storage Applications

The electrochemical properties of the composites of ECPs and acid-treated CNTs have been demonstrated to be very important for improved interfacial charge transfer. The role of the negatively charged (usually acid treated) partially oxidized CNTs in the storage properties of composite electrodes are not completely understood. Moreover, it has been proposed that negatively charged CNT particles increase the ionic conductivity of the electrolyte solution during the process of electrochemical codeposition of the polymer–CNM composite from a mixture of the monomer CNM [13,119,120].

Electrochemical codeposition has been used to produce PPy–CNT composite films of varying and uniform thicknesses. PPy, which is a versatile electrode material especially due to its relatively high capacitance per unit volume, is unfortunately characterized by low mechanical strength, which leads to rapid deterioration in its electrochemical performance during the ingression/deingression of ions when charging/discharging [121]. It is along these lines that the composites of PPy and CNT have found far-reaching applications as pseudocapacitive electrodes due to structural advantages that can be obtained from the CNT.

Earlier works have shown that the negatively charged CNT nanoparticles also contribute considerably to the charge storage mechanism of the composite. This conclusion was drawn because on reducing the PPy–CNT composite, the expected polarization due to the insulating properties of the reduced PPy was significantly decreased owing to the presence of the entrapped particles of acid-treated CNTs, which act as large anions [92]. Accordingly, it was shown that the oxidation of the electrodeposited PPy–CNT is accompanied by the desorption of small cations such as K^+, which greatly improves the electrochemical charge/discharge properties of these composites, through a reduction in the electrode polarization, and an avoidance of the insertion of large anions, which might increase electrode degradation [92,122].

Together with the increase in the specific capacitance on the addition of partially oxidized CNTs to the polymer matrix, it has also been shown that ion and conductive paths are introduced by the CNT in the polymer matrix [119]. As seen from Fig. 7–10A1 electrodeposition of PPy on acid-treated CNT can lead to increased interfacial conjugation between the polymer film and the functionalized CNT. This improved molecular interaction between the polymer and the CNT, coupled with enhanced ion transport by virtue of the porous characteristics of the CNT, is the reason for the enhanced electrochemical performance as demonstrated by the increased capacitance of the PPy–CNT composite when compared with the PPy electrode (see Fig. 7–10A2).

A critical consideration of Fig. 7–10B–D can be used to describe the general need for preparing composites of polymers with aligned (Fig. 7–10B) and unaligned (Fig. 7–10C) CNTs. In general, when the polymers are coated at the tube ends, such that there is effective molecular interaction between the tubes and the polymers, electronic transport between the polymer film and the tubes is greatly improved. Also, due to the porous nature of the tubes, there would be enhanced ion transport and the tubes could aid in reducing the effects of internal stress due to ion movement.

Another important feature of such composites is the good electronic interaction between the substrate (or current collector) and the tubes, which creates an increase in conductivity of the entire composite, and thus improves the charge efficiency of the charging/discharging process. Broadly speaking, the conceptual diagrams in Fig. 7–10B–D may be generalized to include other forms of CNMs such as graphene. This is because, apart from the aforementioned properties, the morphological properties of these CNMs could also aid in the

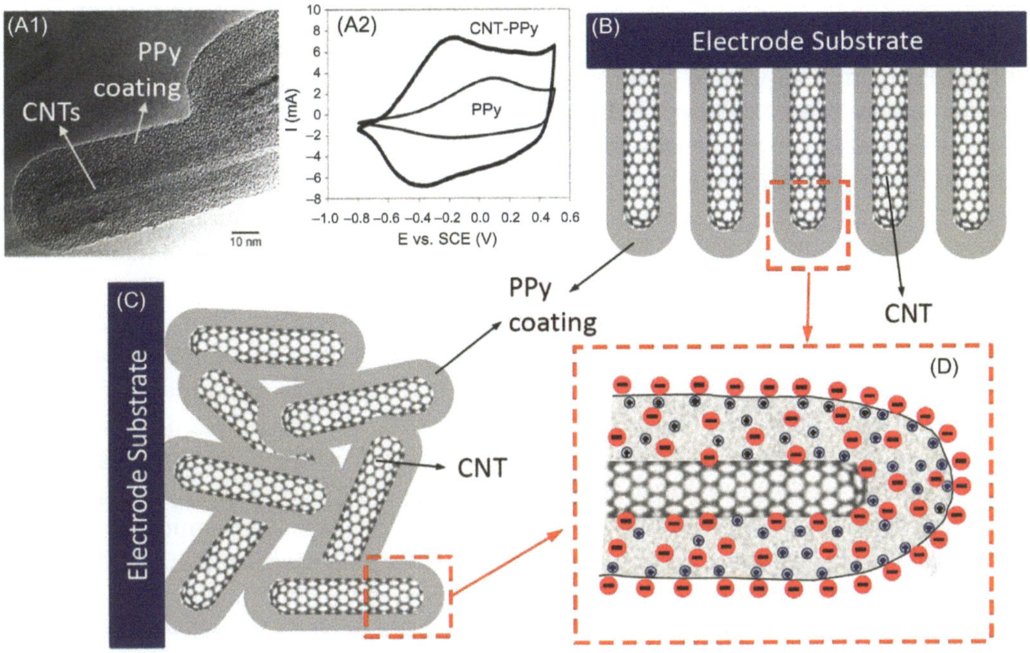

FIGURE 7–10 (A1) SEM micrograph of coelectrodeposited PPy and acid-treated CNTs (A2) CVs of the CNT–PPy composite in (A1) compared with electrodeposited PPy [123]. Schematic diagrams of polymer coated (B) aligned CNTs and (C) unaligned CNTs on an electrode substrate, and (D) the CNT end with a polymer coating in which the positive signs represent the cationic polymer chains and the negative signs are the movable charge balancing anions [124].

nucleation and/or the formation of the polymer film, that is, acting as a site for effective film formation. Accordingly, optimizing this molecular interaction between the polymers and the CNMs, is the basis for various methods of synthesizing these composites and even the functionalization of the CNMs.

It is also important to point out that ions from surfactants have been used to act as negative charge for the stabilization of emulsions of polymer–CNMs used for electropolymerization. For example, sodium dodecyl benzene sulfonate acting as supporting electrolyte has been adopted for the electro codeposition of PPy on rGO. In this case, a thin layer of the rGO was precasted on the substrate and the surface groups of the rGO then provides nucleation sites for the codeposition of the PPy–rGO film. In general, although thicker films resulted in an increase in specific capacity, the PPy–rGO composite did not increase in conductivity, implying that the presence of the rGO does not necessarily increase the conductivity of the composite, but rather aids in improving the mechanical strength and also increasing the amount of loaded polymer [125]. Other reports have demonstrated that the role of rGO in the PPy–rGO composite is mainly to improve the capacitive performance through the increase in ion access for pseudocapacitive charging [126]. The coelectrodeposition of PAni–CNM composites is also another widespread method of producing high-performance electrodes. It has been earlier demonstrated that due to the porous nature of PAni, the acid-treated CNTs serve mainly to enhance the structure of the composite, because there was no substantial difference in the measured capacitance of the PAni film and the PAni–CNT composite due to the inherent porosity of the PAni fibrils. Moreover, mechanical stress resulting from cyclic charging–discharging of the electrode was shown to be reduced in the PAni–CNT composite electrode [119]. The structural characteristics of beads of rGO have made it possible to prepare an electrode material that sandwiches the rGO beads in between PAni films through an electrodeposition process. Such structures were able to load over 90% of PAni films on the composite of PAni and rGO-beads. However, from an electrode performance point of view, high PAni loading did not necessarily result in increased electrode capacitance. This is because the maximum electrode capacitance was obtained from the PAni–rGO composite with 63% of loaded PAni, compared with one with over 90% of loaded polymer. In such composites the mechanical properties of the CNM are mainly exploited, as they displayed relatively long cycle stability [127]. A recent approach to attaining high electrode capacitances using PAni–CNM composites as binder-free electrodes (which is the general target of all electrodeposition processes) has been demonstrated by electrodepositing PAni films on a novel rGO–CNT material supported on nickel. Since it has been shown that introducing CNTs into rGO as spacers improves the electrical conductivity, the produced PAni–rGO–CNT electrodes displayed improved electrochemical performance. Although in this work comparisons of the PAni–rGO–CNT and PAni electrodes were not shown, the good cycle stability of the PAni–rGO–CNT electrode suggests that the rGO–CNT support acts to improve both the mechanical and electrode/electrolyte interfacial properties as claimed [128]. PEDOT is also another well-described ECP for energy storage purposes. In comparison with PPy and PAni, PEDOT is an electron-rich n-doped polymer, with a relatively wide potential window albeit with a lower specific capacitance, mostly due to its high molecular weight. Moreover, its high conductivity and good interfacial properties have been exploited through the production of various PEDOT–CNM composites. Earlier reports have critically demonstrated the means by which an acetonitrile/water mixture can serve as a relatively stable emulsion for the stabilization of the acid-treated CNT and 3,4-ethylene-dioxythiophene (EDOT) monomer prior to the coelectrodeposition of the PEDOT–CNT film. This is because the insoluble EDOT monomer can be dissolved in acetonitrile, whilst the acid-treated CNT can be dissolved in water, leading to a stable composite precursor emulsion [129]. Moreover, the broad range of chemical methods used in preparing PEDOT–CNM composites implies that issues with the insolubility of the EDOT monomer might not necessarily hamper their widespread production [130]. For example, an oxidative chemical deposition process was adopted in producing uniform coats of PEDOT on aligned-CNT (ACNT). In general, the use of ACNT, which provided improved ion transport properties to the composite, was able to display over 30% increase in the specific capacitance of the PEDOT–ACNT composite [131]. Another recent approach at producing PEDOT–CNT composites has also exploited a templated method wherein varying thickness of the PEDOT film was grown on ACNT. It was shown that improved ion transport from the

ACNT can lead to thicker coats of PEDOT resulting in increased specific capacitance [132]. This method basically demonstrates that using ACNTs or CNMs with uniform morphology could help mitigate the problem of reduced performance when the thickness of polymer electrode films are too large.

7.3.2.2 Composites of Polymers and Carbon Nanomaterials for Other Electrochemical Applications

Energy conversion systems such as fuel cells, solar photovoltaics, and other charge transfer applications such as electrochromism and catalysis have also utilized composites of polymers and CNMs. In addition to its good interfacial properties, which have been exploited for use in energy storage, the low bandgap properties of PEDOT have also made it useful for solar photovoltaics and other optoelectrical applications. Notably, the commercially available polymer composite of PEDOT:PSS has been established as a versatile material in photovoltaics. For example, it has been shown that the polymer–CNM composite PSS:PEDOT–CNT, which was produced through a surfactant assisted electropolymerization route, performed as a highly effective counterelectrode for dye-sensitized solar cells (DSSCs). This property was linked to the improved conductivity on the addition of CNT to the polymer matrix, which also displayed an improved catalysis towards I_3^-/I^- redox reaction [133]. Indeed this improved conductivity on the addition of CNTs to PSS:PEDOT has rapidly increased the use of these composites for DSSCs in general [134]. The role of CNMs in improving the performance of polymers for photovoltaic applications has been described mainly from the standpoint of the improvement of electron–hole transport due to the increased interfacial activity of the polymer–CNM composites [135]. For example, interlayer deposition of graphene oxide between the matrix of the photoactive polymers poly(3-hexylthiophene) (P3HT) and phenyl-C61-buyric acid methyl ester reduced the rate of electron–hole recombination and leakage current, which ultimately increased the solar conversion efficiency [136]. Also, the introduction of polymer junctions on the addition of CNMs has been shown to increase the open circuit voltage (OCV) of organic polymer solar cells [137].

In a bid to produce improved transparent counter electrodes for photovoltaic applications, the interfacial activity of CNMs especially as it relates to their donor/acceptor interfacial surface area and electronic conduction has made them useful as additives to conventional polymer blends. Notably, GO has been used as an effective electron acceptor in combination with P3HT) in a BHJ solar cell. Herein it was postulated that the increase in current efficiency of the BHJ with configuration Indium tin oxide (ITO) | PEDOT:PSS | P3HT-GO | Al is due to the enhancement in charge transfer by the GO [138]. The ability for the bandgap of graphene and some of its derivatives to be tuned is the basis for the application of GO with tunable bandgaps as electron acceptors in BHJ. For example, through a pulsed laser approach, GO-ethylene-dinitro-benzoyl (EDNB) was synthesized. This pulsed laser GO-EDNB nanomaterial displayed effective bandgap tuning coupled with surface functionalization, which aided its dispersion in the polymer matrix. Essentially, when this pulsed laser GO-EDNB nanomaterial was used to prepare a composite with poly[*N* -9′-heptadecanyl-2,7-carbazole-alt-5,5-(4′,7′-di-2-thienyl-2′,1′,3′-benzothiadiazole)] as electron donor, a high OCV of 1.17 V was obtained [139]. GO inserted as hole transport layer for the BHJ with configuration ITO | GO | P3HT | [6,6]-phenyl-C61-butyric acid methyl ester (PCBM) | Al, also displayed a substantial increase in short circuit current density and fill factor. This essentially showed that by virtue of their good interfacial activity, optimized graphene derivatives could be used as effective hole transport layers in BHJ solar cells [140].

The improved interfacial properties that can be obtained from polymer–CNM composites have also been reported to enhance the performance of membranes for various classes of fuel cells owing to the catalytic properties of their surfaces, particularly with regard to active sites for proton transport [141–143] or effective oxygen reduction catalysis [144]. Also, the incorporation of CNMs into membranes for direct methanol fuel cells have been shown to reduce the permeability of methanol into the membrane, whilst also increasing the proton conductivity [45,145].

7.3.3 Mechanical and Structural Enhancements in Polymer–Carbon Nanomaterial Composites

Due to their high mechanical strength and thermal stability, polymers are usually used as structural materials for a diverse range of applications. It has been shown that strong interaction between functionalized CNMs and the polymer matrix can enhance load transfer to the fillers within the composite. Along these lines incorporating CNMs into polymers for structural enhancements is an important means of designing robust polymers for diverse applications. For example, the tensile strength of PS for structural applications has been improved through the preparation of its composites with fluorinated SWCNT; this increase in mechanical strength was attributed to the formation of strong covalent bonds between SWCNT and the polymer matrix [146]. For efficient design of composites, it is imperative that the optimal loading of the fillers is well dispersed in the polymer matrix. For example, GO has been used as reinforcement fillers for PP, wherein it was shown that plasticizing effects set in when the filler loading exceeds a certain amount [147]. Broadly speaking, effective load transfer can be achieved through optimal processing methods carried out with CNMs with adequate morphologies and functional groups [148]. Table 7–3 provides a brief summary of the improvements that can be obtained when CNMs with different functional groups are used as fillers in polymer composites.

7.4 Concluding Remarks

Composites of polymers with CNMs (CNTs or graphene) have been broadly reviewed from the standpoint of their properties, preparation, and general applications. It has been highlighted that achieving effective dispersion of the CNM in the polymer matrix is vital for most applications.

Accordingly, various methods of introducing functional groups into CNMs to aid their dispersion in, and interaction with, the polymer matrix have been summarized. Whilst covalent functionalization would modify the CNMs with oxygen, nitrogen, phosphorus, and halogen

Table 7–3 Summary of Property Improvements From Addition of CNMs Into Polymers

Methods	CNMs	Electrochemical Energy Storage	Energy Conversion	Thermal and Mechanical Properties
Oxygenation	CNTs	PPy (increased capacitance and cycling stability) PAni (improved cycling stability) [13,153] PTh (increased capacitance and cycling stability) [155]	P3OT (increased J_{SC} and V_{OC} in a BHJ solar cell) [149,150] P3OT (enhanced photoresponse of transparent electrode) [154]	PMMA PEEK (improved mechanical strength) [35,151] Epoxy (improved thermal conductivity) [152]
	Graphene	PPy, PAni, PEDOT (improved interfacial activity and stability) [156,157]	Nafion (improved proton conductivity of PEMFC) [46] PEDOT:PSS (increased thermoelectric conversion efficiency) [158] PPy/PEDOT (improved catalysis of I^-/I_3^- reaction in a DSSC) [159]	PVA (increased tensile strength) [160] PS (improved electrical conductivity) [111] PU (increased electrical conductivity and thermal stability) [84] PP (improved thermal oxidative stability) [147]
Sulfonation	CNTs	PAni (improved interfacial interaction) [41]	Nafion (improve proton conductivity) [47]	PS (increased tensile strength) [161] PESEEK (increase in Young's modulus) [162]
	Graphene	PPy (improved conductivity) [44]	Nafion (reduced methanol uptake and enhanced proton conductivity in a DMFC) [45], SPEN (reduced methanol permeability, increased proton conductivity in a DMFC) [145]	PVA (increase of Young's modulus) [43]
Nitrogenation	CNMs	PAni (improved charge capacity) [61]	PTB7 and PC71BM (enhanced charge separation and transport in a BHJ solar cell) [62]	PS, PBB/PLA Epoxy (increased tensile strength and Young's modulus) [148], PS (improved tensile strength) [53]
Halogenation	CNTs	PPEs (conductive and stable electrodes) [70]	PP (enhanced conductivity for electrostatic discharge applications) [69]	Epoxy (increase of Young's modulus) [67] PP (improved tensile strength) [146]
Noncovalent methods	CNMs	PEDOT (increased capacitance and cycle stability) [163]	PEDOT:PSS (increased conversion efficiency in a BHJ solar cell) [164]	PVA (reduced thermal degradation rate) [165], Epoxy (enhanced thermal conductivity) [80]

containing heteroatoms, noncovalent functionalization is used to physically disentangle CNTs and unstack graphene sheets with the assistance of ionic, nonionic, or polymeric surfactants.

The challenges in the functionalization of CNMs which is usually carried out through the more effective chemical route, is in the process conditions, which in most cases involve hazardous reagents and effluents that are bad for the environment. Furthermore, various methods of preparing polymer–CNM composites also face challenges of scale-up and/or efficiency, a case in point being the electrochemical codeposition of polymer and CNM composites.

In addition to the preparation processes, it is also important to point out that fundamental understanding of the mechanisms of interaction between the CNMs and the polymers in the composites is also very important. This is because adequate understanding of the roles of the morphology, functional groups, and physicochemical properties of the CNMs is necessary if resource-effective and high-performance composites are to be prepared. Accordingly, future research efforts should be geared towards integrating the understanding obtained from fundamental theoretical analysis with experimental findings.

An important example of this theoretical and simulation based approach is the fundamental study of the charge storage properties of polymers with the aid of DFT methods to gain insight into their charge transfer mechanisms to understand the extent to which CNMs can be used to further enhance their electrochemical properties for a wide range of applications particularly in energy storage and conversion. Advanced imaging analysis has and would continue to play a pivotal role in understanding the structural properties of polymer–CNM composites. Essentially efforts in the development of novel polymer–CNM composites should be geared towards the aforementioned and other related areas, to ensure that materials of the future would not only display impressive properties, but would also be prepared through processes that are sustainable.

References

[1] Ma P-C, Siddiqui NA, Marom G, Kim J-K. Dispersion and functionalization of carbon nanotubes for polymer-based nanocomposites: a review. Composites Part A: Appl Sci Manuf 2010;41(10):1345–67.

[2] Hu K, Kulkarni DD, Choi I, Tsukruk VV. Graphene-polymer nanocomposites for structural and functional applications. Prog Polym Sci 2014;39(11):1934–72.

[3] Ando Y, Iijima S. Preparation of carbon nanotubes by arc-discharge evaporation. Jpn J Appl Phys Part 2 Lett 1993;32:L107-L; Vidu R, Rahman M, Mahmoudi M, Enachescu M, Poteca TD, Opris I. Nanostructures: a platform for brain repair and augmentation. Front Syst Neurosci 2014;8:91.

[4] Popov VN. Carbon nanotubes: properties and application. Mater Sci Eng: R: Rep 2004;43(3):61–102.

[5] Kreupl F, Graham AP, Duesberg GS, Steinhögl W, Liebau M, Unger E, et al. Carbon nanotubes in interconnect applications. Microelectronic Eng 2002;64(1):399–408.

[6] Novoselov KS, Geim AK, Morozov SV, Jiang D, Zhang Y, Dubonos SV, et al. Electric field effect in atomically thin carbon films. Science 2004;306(5696):666–9.

[7] Xie X-L, Mai Y-W, Zhou X-P. Dispersion and alignment of carbon nanotubes in polymer matrix: a review. Mater Sci Eng: R: Rep 2005;49(4):89–112.

[8] Chrissafis K, Bikiaris D. Can nanoparticles really enhance thermal stability of polymers? Part I: an overview on thermal decomposition of addition polymers. Thermochim Acta 2011;523(1):1–24.

[9] Bonaccorso F, Lombardo A, Hasan T, Sun Z, Colombo L, Ferrari AC. Production and processing of graphene and 2d crystals. Mater Today 2012;15(12):564–89.

[10] Kulkarni DD, Kim S, Chyasnavichyus M, Hu K, Fedorov AG, Tsukruk VV. Chemical reduction of individual graphene oxide sheets as revealed by electrostatic force microscopy. J Am Chem Soc 2014;136 (18):6546–9.

[11] Hummers Jr WS, Offeman RE. Preparation of graphitic oxide. J Am Chem Soc 1958;80(6):1339.

[12] Chen J, Yao B, Li C, Shi G. An improved Hummers method for ecofriendly synthesis of graphene oxide. Carbon 2013;64:225–9.

[13] Peng C, Zhang S, Jewell D, Chen GZ. Carbon nanotube and conducting polymer composites for supercapacitors. Prog Nat Sci 2008;18(7):777–88.

[14] Chen GZ. Supercapacitor and supercapattery as emerging electrochemical energy stores. Int Mater Rev 2017;62(4):173–202.

[15] Yue L, Pircheraghi G, Monemian SA, Manas-Zloczower I. Epoxy composites with carbon nanotubes and graphene nanoplatelets—dispersion and synergy effects. Carbon 2014;78(Suppl. C):268–78.

[16] Yang S-Y, Lin W-N, Huang Y-L, Tien H-W, Wang J-Y, Ma C-CM, et al. Synergetic effects of graphene platelets and carbon nanotubes on the mechanical and thermal properties of epoxy composites. Carbon 2011;49(3):793–803.

[17] Banerjee S, Hemraj-Benny T, Wong SS. Covalent surface chemistry of single-walled carbon nanotubes. Adv Mater 2005;17(1):17–29.

[18] Georgakilas V, Otyepka M, Bourlinos AB, Chandra V, Kim N, Kemp KC, et al. Functionalization of graphene: covalent and non-covalent approaches, derivatives and applications. Chem Rev 2012;112(11):6156–214.

[19] Dyke CA, Tour JM. Covalent functionalization of single-walled carbon nanotubes for materials applications. J Phys Chem A 2004;108(51):11151–9.

[20] Ramanathan T, Abdala AA, Stankovich S, Dikin DA, Herrera-Alonso M, Piner RD, et al. Functionalized graphene sheets for polymer nanocomposites. Nat Nanotechnol 2008;3:327.

[21] Datsyuk V, Kalyva M, Papagelis K, Parthenios J, Tasis D, Siokou A, et al. Chemical oxidation of multiwalled carbon nanotubes. Carbon 2008;46(6):833–40.

[22] Liu X, Xu F, Zhang K, Wei B, Gao Z, Qiu Y. Characterization of enhanced interfacial bonding between epoxy and plasma functionalized carbon nanotube films. Compos. Sci Technol 2017;145:114–21.

[23] Petersen EJ, Lam T, Gorham JM, Scott KC, Long CJ, Stanley D, et al. Methods to assess the impact of UV irradiation on the surface chemistry and structure of multiwall carbon nanotube epoxy nanocomposites. Carbon 2014;69:194–205.

[24] Muñoz-Vélez M, Valadez-González A, Herrera-Franco P. Effect of fiber surface treatment on the incorporation of carbon nanotubes and on the micromechanical properties of a single-carbon fiber-epoxy matrix composite. Express Polym Lett 2017;11:9.

[25] Wepasnick KA, Smith BA, Schrote KE, Wilson HK, Diegelmann SR, Fairbrother DH. Surface and structural characterization of multi-walled carbon nanotubes following different oxidative treatments. Carbon 2011;49(1):24–36.

[26] Yoonessi M, Lebrón-Colón M, Scheiman D, Meador MA. Carbon nanotube epoxy nanocomposites: the effects of interfacial modifications on the dynamic mechanical properties of the nanocomposites. ACS Appl Mater Interfaces 2014;6(19):16621–30.

[27] Dutko O, Plachá D, Mikeska M, Martynková GS, Wróbel P, Bachmatiuk A, et al. Comparison of selected oxidative methods for carbon nanotubes: structure and functionalization study. J Nanosci Nanotechnol 2016;16(8):7822–5.

[28] Vanyorek L, Meszaros R, Barany S. Surface and electrosurface characterization of surface-oxidized multi-walled N-doped carbon nanotubes. Colloids Surf. A: Physicochem Eng Asp 2014;448:140–6.

[29] Gómez S, Rendtorff NM, Aglietti EF, Sakka Y, Suárez G. Surface modification of multiwall carbon nanotubes by sulfonitric treatment. Appl Surf Sci 2016;379:264–9.

[30] Li M, Wu Y, Zhao F, Wei Y, Wang J, Jiang K, et al. Cycle and rate performance of chemically modified super-aligned carbon nanotube electrodes for lithium ion batteries. Carbon 2014;69:444–51.

[31] Steimecke M, Ruemmler S, Bron M. The effect of rapid functionalization on the structural and electrochemical properties of high-purity carbon nanotubes. Electrochim. Acta 2015;163:1–8.

[32] Chen C, Liang B, Ogino A, Wang X, Nagatsu M. Oxygen functionalization of multiwall carbon nanotubes by microwave-excited surface-wave plasma treatment. J Phys Chem C 2009;113(18):7659–65.

[33] Pei S, Cheng H-M. The reduction of graphene oxide. Carbon 2012;50(9):3210–28.

[34] Dreyer DR, Park S, Bielawski CW, Ruoff RS. The chemistry of graphene oxide. Chem Soc Rev 2010;39 (1):228–40.

[35] Weng B, Xu F, Salinas A, Lozano K. Mass production of carbon nanotube reinforced poly(methyl methacrylate) nonwoven nanofiber mats. Carbon 2014;75(Suppl. C):217–26.

[36] Xu Y, Wang Y, Liang J, Huang Y, Ma Y, Wan X, et al. A hybrid material of graphene and poly (3,4-ethyldioxythiophene) with high conductivity, flexibility, and transparency. Nano Res 2009;2(4):343–8.

[37] Li J, O'Shea J, Hou X, Chen GZ. Faradaic processes beyond Nernst's law: density functional theory assisted modelling of partial electron delocalization and pseudocapacitance in graphene oxides. Chem Commun 2017;53(75):10414–17.

[38] Stankovich S, Dikin DA, Piner RD, Kohlhaas KA, Kleinhammes A, Jia Y, et al. Synthesis of graphene-based nanosheets via chemical reduction of exfoliated graphite oxide. Carbon 2007;45 (7):1558–65.

[39] Wei Z, Wan M, Lin T, Dai L. Polyaniline nanotubes doped with sulfonated carbon nanotubes made via a self-assembly process. Adv Mater 2003;15(2):136–9.

[40] Wang Y, Iqbal Z, Mitra S. Rapidly functionalized, water-dispersed carbon nanotubes at high concentration. J Am Chem Soc 2006;128(1):95–9.

[41] Zhu Z-Z, Wang G-C, Sun M-Q, Li X-W, Li C-Z. Fabrication and electrochemical characterization of polyaniline nanorods modified with sulfonated carbon nanotubes for supercapacitor applications. Electrochim Acta 2011;56(3):1366–72.

[42] Si Y, Samulski ET. Synthesis of water soluble graphene. Nano Lett 2008;8(6):1679–82.

[43] Layek RK, Samanta S, Nandi AK. The physical properties of sulfonated graphene/poly(vinyl alcohol) composites. Carbon 2012;50(3):815–27.

[44] Liu A, Li C, Bai H, Shi G. Electrochemical deposition of polypyrrole/sulfonated graphene composite films. J Phys Chem C 2010;114(51):22783–9.

[45] Chien H-C, Tsai L-D, Huang C-P, Kang C-Y, Lin J-N, Chang F-C. Sulfonated graphene oxide/nafion composite membranes for high-performance direct methanol fuel cells. Int J Hydrogen Energy 2013;38 (31):13792–801.

[46] Lee DC, Yang HN, Park SH, Kim WJ. Nafion/graphene oxide composite membranes for low humidifying polymer electrolyte membrane fuel cell. J Membr Sci 2014;452(Suppl. C):20–8.

[47] Kannan R, Parthasarathy M, Maraveedu SU, Kurungot S, Pillai VK. Domain size manipulation of perflouorinated polymer electrolytes by sulfonic acid-functionalized MWCNTs to enhance fuel cell performance. Langmuir 2009;25(14):8299–305.

[48] Sahu AK, Ketpang K, Shanmugam S, Kwon O, Lee S, Kim H. Sulfonated graphene–nafion composite membranes for polymer electrolyte fuel cells operating under reduced relative humidity. J Phys Chem C 2016;120(29):15855–66.

[49] Arjmand M, Chizari K, Krause B, Poetschke P, Sundararaj U. Effect of synthesis catalyst on structure of nitrogen-doped carbon nanotubes and electrical conductivity and electromagnetic interference shielding of their polymeric nanocomposites. Carbon 2016;98:358–72.

[50] Wang X, Sun G, Routh P, Kim D-H, Huang W, Chen P. Heteroatom-doped graphene materials: syntheses, properties and applications. Chem Soc Rev 2014;43(20):7067–98.

[51] Zhang LL, Zhao X, Ji H, Stoller MD, Lai L, Murali S, et al. Nitrogen doping of graphene and its effect on quantum capacitance, and a new insight on the enhanced capacitance of N-doped carbon. Energy Environ Sci 2012;5(11):9618–25.

[52] Panchakarla L, Govindaraj A, Rao C. Boron-and nitrogen-doped carbon nanotubes and graphene. Inorg Chim Acta 2010;363(15):4163–74.

[53] Fragneaud B, Masenelli-Varlot K, Gonzalez-Montiel A, Terrones M, Cavaillé J-Y. Efficient coating of N-doped carbon nanotubes with polystyrene using atomic transfer radical polymerization. Chem Phys Lett 2006;419(4):567–73.

[54] Khazaee M, Xia W, Lackner G, Mendes RG, Rümmeli M, Muhler M, et al. Dispersibility of vapor phase oxygen and nitrogen functionalized multi-walled carbon nanotubes in various organic solvents. Sci Rep 2016;6.

[55] Xu G, Ding B, Nie P, Shen L, Wang J, Zhang X. Porous nitrogen-doped carbon nanotubes derived from tubular polypyrrole for energy-storage applications. Chem—A Eur J 2013;19(37):12306–12.

[56] Nolan H, Mendoza-Sanchez B, Kumar NA, McEvoy N, O'Brien S, Nicolosi V, et al. Nitrogen-doped reduced graphene oxide electrodes for electrochemical supercapacitors. Phys Chem Chem Phys 2014;16 (6):2280–4.

[57] Wang H, Maiyalagan T, Wang X. Review on recent progress in nitrogen-doped graphene: synthesis, characterization, and its potential applications. ACS Catal 2012;2(5):781–94.

[58] Schiros T, Nordlund D, Pálová L, Prezzi D, Zhao L, Kim KS, et al. Connecting dopant bond type with electronic structure in N-doped graphene. Nano Lett 2012;12(8):4025–31.

[59] Wood KN, O'Hayre R, Pylypenko S. Recent progress on nitrogen/carbon structures designed for use in energy and sustainability applications. Energy Environ Sci 2014;7(4):1212–49.

[60] Lai L, Wang L, Yang H, Sahoo NG, Tam QX, Liu J, et al. Tuning graphene surface chemistry to prepare graphene/polypyrrole supercapacitors with improved performance. Nano Energy 2012;1(5):723–31.

[61] Haq AU, Lim J, Yun JM, Lee WJ, Han TH, Kim SO. Direct growth of polyaniline chains from N-doped sites of carbon nanotubes. Small 2013;9(22):3829–33.

[62] Lu L, Xu T, Chen W, Lee JM, Luo Z, Jung IH, et al. The role of N-doped multiwall carbon nanotubes in achieving highly efficient polymer bulk heterojunction solar cells. Nano Lett 2013;13(6):2365–9.

[63] Cheng X, Long J, Wu R, Huang L, Tan L, Chen L, et al. Fluorinated reduced graphene oxide as an efficient hole-transport layer for efficient and stable polymer solar cells. ACS Omega 2017;2(5):2010–16.

[64] Chronopoulos DD, Bakandritsos A, Pykal M, Zbořil R, Otyepka M. Chemistry, properties, and applications of fluorographene. Appl Mater Today 2017;9:60–70.

[65] Nair RR, Ren W, Jalil R, Riaz I, Kravets VG, Britnell L, et al. Fluorographene: a two-dimensional counterpart of Teflon. Small 2010;6(24):2877–84.

[66] He M, Swager TM. Covalent functionalization of carbon nanomaterials with iodonium salts. Chem Mater 2016;28(23):8542–9.

[67] Zhu J, Kim J, Peng H, Margrave JL, Khabashesku VN, Barrera EV. Improving the dispersion and integration of single-walled carbon nanotubes in epoxy composites through functionalization. Nano Lett 2003;3 (8):1107–13.

[68] Shen B, Chen J, Yan X, Xue Q. Synthesis of fluorine-doped multi-layered graphene sheets by arc-discharge. RSC Adv 2012;2(17):6761–4.

[69] Lee J-I, Yang S-B, Jung H-T. Carbon nanotubes – polypropylene nanocomposites for electrostatic discharge applications. Macromolecules 2009;42(21):8328–34.

[70] Sankapal BR, Setyowati K, Chen J, Liu H. Electrical properties of air-stable, iodine-doped carbon-nanotube–polymer composites. Appl Phys Lett 2007;91(17):173103.

[71] Kim JY, Lee WH, Suk JW, Potts JR, Chou H, Kholmanov IN, et al. Chlorination of reduced graphene oxide enhances the dielectric constant of reduced graphene oxide/polymer composites. Adv Mater 2013;25(16):2308–13.

[72] Geng Y, Liu MY, Li J, Shi XM, Kim JK. Effects of surfactant treatment on mechanical and electrical properties of CNT/epoxy nanocomposites. Composites Part A: Appl Sci Manufactr 2008;39(12):1876–83.

[73] Sun Z, Nicolosi V, Rickard D, Bergin SD, Aherne D, Coleman JN. Quantitative evaluation of surfactant-stabilized single-walled carbon nanotubes: dispersion quality and its correlation with zeta potential. J Phys Chem C 2008;112(29):10692–9.

[74] Smith RJ, Lotya M, Coleman JN. The importance of repulsive potential barriers for the dispersion of graphene using surfactants. New J Phys 2010;12(12):125008.

[75] Grossiord N, Loos J, Van Laake L, Maugey M, Zakri C, Koning CE, et al. High-conductivity polymer nanocomposites obtained by tailoring the characteristics of carbon nanotube fillers. Adv Funct Mater 2008;18(20):3226–34.

[76] Hsiao A-E, Tsai S-Y, Hsu M-W, Chang S-J. Decoration of multi-walled carbon nanotubes by polymer wrapping and its application in MWCNT/polyethylene composites. Nanoscale Res Lett 2012;7(1):240.

[77] Morishita T, Matsushita M, Katagiri Y, Fukumori K. Noncovalent functionalization of carbon nanotubes with maleimide polymers applicable to high-melting polymer-based composites. Carbon 2010;48 (8):2308–16.

[78] Hermant MC, Klumperman B, Kyrylyuk AV, van der Schoot P, Koning CE. Lowering the percolation threshold of single-walled carbon nanotubes using polystyrene/poly (3,4-ethylenedioxythiophene): poly (styrene sulfonate) blends. Soft Matter 2009;5(4):878–85.

[79] Zhao S, Song Z, Cui J, Li C, Yan Y. Improving dispersion and integration of single-walled carbon nanotubes in epoxy composites by using a reactive noncovalent dispersant. J Polym Sci Part A: Polym Chem 2012;50(21):4548–56.

[80] Teng C-C, Ma C-CM, Lu C-H, Yang S-Y, Lee S-H, Hsiao M-C, et al. Thermal conductivity and structure of non-covalent functionalized graphene/epoxy composites. Carbon 2011;49(15):5107–16.

[81] Zhao C, Ji L, Liu H, Hu G, Zhang S, Yang M, et al. Functionalized carbon nanotubes containing isocyanate groups. J Solid State Chem 2004;177(12):4394–8.

[82] Lopes MC, Ribeiro H, Santos MCG, Seara LM, Ferreira FLQ, Lavall RL, et al. High performance polyurethane composites with isocyanate-functionalized carbon nanotubes: improvements in tear strength and scratch hardness. J Appl Polym Sci 2017;134:2.

[83] Stankovich S, Piner RD, Nguyen ST, Ruoff RS. Synthesis and exfoliation of isocyanate-treated graphene oxide nanoplatelets. Carbon 2006;44(15):3342–7.

[84] Wang X, Hu Y, Song L, Yang H, Xing W, Lu H. In situ polymerization of graphene nanosheets and polyurethane with enhanced mechanical and thermal properties. J Mater Chem 2011;21(12):4222–7.

[85] Ryu SH, Cho H-B, Moon JW, Kwon Y-T, Eom NSA, Lee S, et al. Highly conductive polymethly (methacrylate)/multi-wall carbon nanotube composites by modeling a three-dimensional percolated microstructure. Composites Part A: Appl Sci Manufactr 2016;91:133–9.

[86] Chen Y, Zhang L, Zhan H, Wang JN. New processing method to fabricate high-performance carbon-nanotube/polyvinyl alcohol composite films. Carbon 2016;110:490–6.

[87] Ding P, Zhang J, Song N, Tang S, Liu Y, Shi L. Anisotropic thermal conductive properties of hot-pressed polystyrene/graphene composites in the through-plane and in-plane directions. Compos Sci Technol 2015;109:25–31.

[88] Shah R, Kausar A, Muhammad B, Shah S. Progression from graphene and graphene oxide to high performance polymer-based nanocomposite: a review. Polym Plastics Technol Eng 2015;54(2):173−83.

[89] Yuan B, Bao C, Song L, Hong N, Liew KM, Hu Y. Preparation of functionalized graphene oxide/polypropylene nanocomposite with significantly improved thermal stability and studies on the crystallization behavior and mechanical properties. Chem Eng J 2014;237(Suppl. C):411−20.

[90] Han Y, Wang T, Gao X, Li T, Zhang Q. Preparation of thermally reduced graphene oxide and the influence of its reduction temperature on the thermal, mechanical, flame retardant performances of PS nanocomposites. Composites Part A: Appl Sci Manufactr 2016;84:336−43.

[91] Choi EY, Choi LW, Kim C. Noncovalent functionalization of multi-walled carbon nanotubes with hydroxyl group-containing pyrene derivatives for their composites with polycarbonate. Carbon 2015;95:91−9.

[92] Chen GZ, Shaffer MS, Coleby D, Dixon G, Zhou W, Fray DJ, et al. Carbon nanotube and polypyrrole composites: coating and doping. Adv Mater 2000;12(7):522−6.

[93] Peng C, Jin J, Chen GZ. A comparative study on electrochemical codeposition and capacitance of composite films of conducting polymers and carbon nanotubes. Electrochim Acta 2007;53(2):525−37.

[94] Khan MU, Reddy KR, Snguanwongchai T, Haque E, Gomes VG. Polymer brush synthesis on surface modified carbon nanotubes via in situ emulsion polymerization. Colloid Polym Sci 2016;294 (10):1599−610.

[95] Cha J, Jin S, Shim JH, Park CS, Ryu HJ, Hong SH. Functionalization of carbon nanotubes for fabrication of CNT/epoxy nanocomposites. Mater Des 2016;95:1−8.

[96] Kaseem M, Hamad K, Ko YG. Fabrication and materials properties of polystyrene/carbon nanotube (PS/CNT) composites: a review. Eur Polym J 2016;79:36−62.

[97] Aguilar-Bolados H, Lopez-Manchado MA, Brasero J, Avilés F, Yazdani-Pedram M. Effect of the morphology of thermally reduced graphite oxide on the mechanical and electrical properties of natural rubber nanocomposites. Compos Part B: Eng 2016;87:350−6.

[98] Li H, Pang S, Wu S, Feng X, Müllen K, Bubeck C. Layer-by-layer assembly and UV photoreduction of graphene−polyoxometalate composite films for electronics. J Am Chem Soc 2011;133(24):9423−9.

[99] Richardson JJ, Björnmalm M, Caruso F. Technology-driven layer-by-layer assembly of nanofilms. Science 2015;348(6233):aaa2491.

[100] Lee SW, Kim B-S, Chen S, Shao-Horn Y, Hammond PT. Layer-by-layer assembly of all carbon nanotube ultrathin films for electrochemical applications. J Am Chem Soc 2009;131(2):671−9.

[101] Song P, Liu L, Fu S, Yu Y, Jin C, Wu Q, et al. Striking multiple synergies created by combining reduced graphene oxides and carbon nanotubes for polymer nanocomposites. Nanotechnology 2013;24 (12):125704.

[102] Fan W, Zhang C, Tjiu WW, Liu T. Fabrication of electrically conductive graphene/polystyrene composites via a combination of latex and layer-by-layer assembly approaches. J Mater Res 2013;28(4):611−19.

[103] Wang M, Pramoda K, Goh SH. Enhancement of the mechanical properties of poly (styrene-coacrylonitrile) with poly (methyl methacrylate)-grafted multiwalled carbon nanotubes. Polymer 2005;46(25):11510−16.

[104] Zhang H-B, Zheng W-G, Yan Q, Jiang Z-G, Yu Z-Z. The effect of surface chemistry of graphene on rheological and electrical properties of polymethylmethacrylate composites. Carbon 2012;50(14):5117−25.

[105] Coelho PHDSL, Marchesin MS, Morales AR, Bartoli JR. Electrical percolation, morphological and dispersion properties of MWCNT/PMMA nanocomposites. Mater Res 2014;17:127−32.

[106] Du F, Scogna RC, Zhou W, Brand S, Fischer JE, Winey KI. Nanotube networks in polymer nanocomposites: rheology and electrical conductivity. Macromolecules 2004;37(24):9048−55.

[107] Rahaman M, Aldalbahi A, Govindasami P, Khanam NP, Bhandari S, Feng P, et al. A new insight in determining the percolation threshold of electrical conductivity for extrinsically conducting polymer composites through different sigmoidal models. Polymers 2017;9(10):527.

[108] Grunlan JC, Mehrabi AR, Bannon MV, Bahr JL. Water-based single-walled-nanotube-filled polymer composite with an exceptionally low percolation threshold. Adv Mater 2004;16(2):150–3.

[109] Pang H, Chen C, Zhang Y-C, Ren P-G, Yan D-X, Li Z-M. The effect of electric field, annealing temperature and filler loading on the percolation threshold of polystyrene containing carbon nanotubes and graphene nanosheets. Carbon 2011;49(6):1980–8.

[110] Bauhofer W, Kovacs JZ. A review and analysis of electrical percolation in carbon nanotube polymer composites. Compos Sci Technol 2009;69(10):1486–98.

[111] Stankovich S, Dikin DA, Dommett GH, Kohlhaas KM, Zimney EJ, Stach EA, et al. Graphene-based composite materials. Nature 2006;442(7100):282–6.

[112] Song J-W, Kim J, Yoon Y-H, Choi B-S, Kim J-H, Han C-S. Inkjet printing of single-walled carbon nanotubes and electrical characterization of the line pattern. Nanotechnology 2008;19(9):095702.

[113] Lin T, Chen I-W, Liu F, Yang C, Bi H, Xu F, et al. Nitrogen-doped mesoporous carbon of extraordinary capacitance for electrochemical energy storage. Science 2015;350(6267):1508–13.

[114] Lota G, Grzyb B, Machnikowska H, Machnikowski J, Frackowiak E. Effect of nitrogen in carbon electrode on the supercapacitor performance. Chem Phys Lett 2005;404(1):53–8.

[115] Conway BE. Electrochemical supercapacitors: scientific fundamentals and technological applications. Springer Science & Business Media; 2013.

[116] Mao X, Yan EH, Rutledge GC, Hatton TA. Enhanced redox transformation efficiency in unconjugated electroactive polymer/carbon nanotube hybrids. Chem Mater 2016;28(2):543–8.

[117] Peng C, Zhou X, Chen GZ, Moggia F, Fages F, Brisset H, et al. Internally referenced analysis of charge-transfer reactions in a new ferrocenyl bithiophenic conducting polymer through cyclic voltammetry. Chem Commun 2008;48:6606–8.

[118] Fan L-Z, Maier J. High-performance polypyrrole electrode materials for redox supercapacitors. Electrochem Commun 2006;8(6):937–40.

[119] Li J, Ma PC, Chow WS, To CK, Tang BZ, Kim JK. Correlations between percolation threshold, dispersion state, and aspect ratio of carbon nanotubes. Adv Funct Mater 2007;17(16):3207–15.

[120] Hu D, Peng C, Chen GZ. Electrodeposition of nonconducting polymers: roles of carbon nanotubes in the process and products. ACS Nano 2010;4(7):4274–82.

[121] Snook GA, Kao P, Best AS. Conducting-polymer-based supercapacitor devices and electrodes. J Power Sources 2011;196(1):1–12.

[122] Snook GA, Chen GZ, Fray DJ, Hughes M, Shaffer M. Studies of deposition of and charge storage in polypyrrole–chloride and polypyrrole–carbon nanotube composites with an electrochemical quartz crystal microbalance. J Electroanal Chem 2004;568:135–42.

[123] Hughes M, Chen GZ, Shaffer MS, Fray DJ, Windle AH. Electrochemical capacitance of a nanoporous composite of carbon nanotubes and polypyrrole. Chem Mater 2002;14(4):1610–13.

[124] Chae JH, Zhou X, Chen GZ. From electrochemical capacitors to supercapatteries. Green 2012;2(1):41–54.

[125] Chen J, Wang Y, Cao J, Liu Y, Zhou Y, Ouyang J-H, et al. Facile co-electrodeposition method for high-performance supercapacitor based on reduced graphene oxide/polypyrrole composite film. ACS Appl Mater Interfaces 2017;9(23):19831–42.

[126] Zhu C, Zhai J, Wen D, Dong S. Graphene oxide/polypyrrole nanocomposites: one-step electrochemical doping, coating and synergistic effect for energy storage. J Mater Chem 2012;22(13):6300–6.

[127] Ouyang A, Cao A, Hu S, Li Y, Xu R, Wei J, et al. Polymer-coated graphene aerogel beads and supercapacitor application. ACS Appl Mater Interfaces 2016;8(17):11179–87.

[128] Xiong C, Li T, Zhu Y, Zhao T, Dang A, Li H, et al. Two-step approach of fabrication of interconnected nanoporous 3D reduced graphene oxide-carbon nanotube-polyaniline hybrid as a binder-free supercapacitor electrode. J Alloys Comp 2017;695:1248–59.

[129] Peng C, Snook GA, Fray DJ, Shaffer MS, Chen GZ. Carbon nanotube stabilised emulsions for electrochemical synthesis of porous nanocomposite coatings of poly [3, 4-ethylene-dioxythiophene]. Chem Commun 2006;44:4629–31.

[130] Bai X, Hu X, Zhou S, Yan J, Sun C, Chen P, et al. In situ polymerization and characterization of grafted poly (3, 4-ethylenedioxythiophene)/multiwalled carbon nanotubes composite with high electrochemical performances. Electrochim Acta 2013;87:394–400.

[131] Ghaffari M, Kosolwattana S, Zhou Y, Lachman N, Lin M, Bhattacharya D, et al. Hybrid supercapacitor materials from poly (3, 4-ethylenedioxythiophene) conformally coated aligned carbon nanotubes. Electrochim Acta 2013;112:522–8.

[132] Lachman N, Xu H, Zhou Y, Ghaffari M, Lin M, Bhattacharyya D, et al. Tailoring thickness of conformal conducting polymer decorated aligned carbon nanotube electrodes for energy storage. Adv Mater Interfaces 2014;1:7.

[133] Rhee Y, Ko M, Jin H, Jin J-H, Min NK. Photovoltaic performance of multi-wall carbon nanotube/PEDOT: PSS composite on the counter electrode of a dye-sensitized solar cell. Jpn J Appl Phys 2014;53(8S3) 08NJ2.

[134] Shin H-J, Jeon SS, Im SS. CNT/PEDOT core/shell nanostructures as a counter electrode for dye-sensitized solar cells. Synth Metals 2011;161(13-14):1284–8.

[135] Kymakis E, Amaratunga G. Single-wall carbon nanotube/conjugated polymer photovoltaic devices. Appl Phys Lett 2002;80(1):112–14.

[136] Liu C, Yu Z, Neff D, Zhamu A, Jang BZ. Graphene-based supercapacitor with an ultrahigh energy density. Nano Lett 2010;10(12):4863–8.

[137] Keru G, Ndungu PG, Nyamori VO. A review on carbon nanotube/polymer composites for organic solar cells. Int J Energy Res 2014;38(13):1635–53.

[138] Hill CM, Zhu Y, Pan S. Fluorescence and electroluminescence quenching evidence of interfacial charge transfer in poly (3-hexylthiophene): graphene oxide bulk heterojunction photovoltaic devices. ACS Nano 2011;5(2):942–51.

[139] Stylianakis MM, Sygletou M, Savva K, Kakavelakis G, Kymakis E, Stratakis E. Photochemical synthesis of solution-processable graphene derivatives with tunable bandgaps for organic solar cells. Adv Optical Mater 2015;3(5):658–66.

[140] Li S-S, Tu K-H, Lin C-C, Chen C-W, Chhowalla M. Solution-processable graphene oxide as an efficient hole transport layer in polymer solar cells. ACS Nano 2010;4(6):3169–74.

[141] Alvarez A, Guzman C, Rivas S, Godinez L, Sacca A, Carbone A, et al. Composites membranes based on nafion and PAMAM dendrimers for PEMFC applications. Int J Hydrogen Energy 2014;39(29):16686–93.

[142] Zhao Q, An QF, Ji Y, Qian J, Gao C. Polyelectrolyte complex membranes for pervaporation, nanofiltration and fuel cell applications. J Membr Ssci 2011;379(1–2):19–45.

[143] Son H, Cho M, Nam J-D, Cho S, Chung C, Choi H-G, et al. Depression of methanol-crossover using multilayer proton conducting membranes prepared by layer-by-layer deposition onto a porous polyethylene film. J Power Sources 2006;163(1):66–70.

[144] Zhou X, Tang S, Yin Y, Sun S, Qiao J. Hierarchical porous N-doped graphene foams with superior oxygen reduction reactivity for polymer electrolyte membrane fuel cells. Appl Energy 2016;175:459–67.

[145] Cheng T, Feng MN, Huang YM, Liu XB. SGO/SPEN-based highly selective polymer electrolyte membranes for direct methanol fuel cells. Ionics 2017;23(8):2143–52.

[146] McIntosh D, Khabashesku VN, Barrera EV. Nanocomposite fiber systems processed from fluorinated single-walled carbon nanotubes and a polypropylene matrix. Chem Mater 2006;18(19):4561–9.

[147] Song P, Cao Z, Cai Y, Zhao L, Fang Z, Fu S. Fabrication of exfoliated graphene-based polypropylene nanocomposites with enhanced mechanical and thermal properties. Polymer 2011;52(18):4001–10.

[148] Gojny FH, Wichmann MHG, Köpke U, Fiedler B, Schulte K. Carbon nanotube-reinforced epoxy-composites: enhanced stiffness and fracture toughness at low nanotube content. Compos Sci Technol 2004;64(15):2363–71.

[149] Landi BJ, Raffaelle RP, Castro SL, Bailey SG. Single-wall carbon nanotube-polymer solar cells. Prog Photovoltaics: Res Appl 2005;13(2):165–72.

[150] Somani SP, Somani PR, Umeno M. Carbon nanotube incorporation: a new route to improve the performance of organic–inorganic heterojunction solar cells. Diamond Relat Mater 2008;17(4):585–8.

[151] Rong C, Ma G, Zhang S, Song L, Chen Z, Wang G, et al. Effect of carbon nanotubes on the mechanical properties and crystallization behavior of poly(ether ether ketone). Compos Sci Technol 2010;70(2):380–6.

[152] Yang S-Y, Ma C-CM, Teng C-C, Huang Y-W, Liao S-H, Huang Y-L, et al. Effect of functionalized carbon nanotubes on the thermal conductivity of epoxy composites. Carbon 2010;48(3):592–603.

[153] Peng C, Zhou X, Zhang S, Chen G. Electrochemical considerations in supercapacitors with nanocomposites. ECS Trans 2011;33(27):107–16.

[154] Kymakis E, Servati P, Tzanetakis P, Koudoumas E, Kornilios N, Rompogiannakis I, et al. Effective mobility and photocurrent in carbon nanotube–polymer composite photovoltaic cells. Nanotechnology 2007;18(43):435702.

[155] Fu C, Zhou H, Liu R, Huang Z, Chen J, Kuang Y. Supercapacitor based on electropolymerized polythiophene and multi-walled carbon nanotubes composites. Mater Chem Phys 2012;132(2):596–600.

[156] Hassan M, Reddy KR, Haque E, Faisal SN, Ghasemi S, Minett AI, et al. Hierarchical assembly of graphene/polyaniline nanostructures to synthesize free-standing supercapacitor electrode. Compos Sci Technol 2014;98:1–8.

[157] Cao J, Wang Y, Chen J, Li X, Walsh FC, Ouyang J-H, et al. Three-dimensional graphene oxide/polypyrrole composite electrodes fabricated by one-step electrodeposition for high performance supercapacitors. J Mat Chem A 2015;3(27):14445–57.

[158] Yoo D, Kim J, Kim JH. Direct synthesis of highly conductive poly (3, 4-ethylenedioxythiophene): poly (4-styrenesulfonate)(PEDOT: PSS)/graphene composites and their applications in energy harvesting systems. Nano Res 2014;7(5):717.

[159] Sekkarapatti Ramasamy M, Nikolakapoulou A, Raptis D, Dracopoulos V, Paterakis G, Lianos P. Reduced graphene oxide/Polypyrrole/PEDOT composite films as efficient Pt-free counter electrode for dye-sensitized solar cells. Electrochim Acta 2015;173(Suppl. C):276–81.

[160] Zhao X, Zhang Q, Chen D, Lu P. Enhanced mechanical properties of graphene-based poly (vinyl alcohol) composites. Macromolecules 2010;43(5):2357–63.

[161] Ying D, Hong HP, Guiver M, Welsh JS. Reinforced films based on cross-linked water-soluble sulfonated carbon nanotubes with sulfonated polystyrene. J Nanosci Nanotechnol 2009;9(9):5150–6.

[162] Zhou W, Xiao J, Chen Y, Zeng R, Xiao S, Nie H, et al. Sulfonated carbon nanotubes/sulfonated poly (ether sulfone ether ketone ketone) composites for polymer electrolyte membranes. Polym Adv Technol 2011;22(12):1747–52.

[163] Antiohos D, Folkes G, Sherrell P, Ashraf S, Wallace GG, Aitchison P, et al. Compositional effects of PEDOT-PSS/single walled carbon nanotube films on supercapacitor device performance. J Mater Chem 2011;21(40):15987–94.

[164] Hong W, Xu Y, Lu G, Li C, Shi G. Transparent graphene/PEDOT–PSS composite films as counter electrodes of dye-sensitized solar cells. Electrochem Commun 2008;10(10):1555–8.

[165] Konidari M, Soulas D, Papadokostaki K, Sanopoulou M. Study of the effect of modified and pristine carbon nanotubes on the properties of poly (vinyl alcohol) nanocomposite films. J Appl Polym Sci 2012;125(S1).

8

Polymer Composites With Metal Nanoparticles: Synthesis, Properties, and Applications

Laura Tamayo[1], Humberto Palza[2], Julián Bejarano[2,3], Paula A. Zapata[4]

[1]DEPARTAMENTO DE QUÍMICA, FACULTAD DE CIENCIAS, UNIVERSIDAD DE CHILE, LAS PALMERAS, CASILLA, SANTIAGO, CHILE [2]DEPARTAMENTO DE INGENIERÍA QUÍMICA Y BIOTECNOLOGÍA, FACULTAD DE CIENCIAS FÍSICAS Y MATEMÁTICAS, UNIVERSIDAD DE CHILE, LAS PALMERAS, CASILLA, SANTIAGO, CHILE [3]DEPARTAMENTO DE QUÍMICA FARMACOLÓGICA Y TOXICOLÓGICA, FACULTAD DE CIENCIAS QUÍMICAS Y FARMACÉUTICAS, UNIVERSIDAD DE CHILE, SERGIO LIVINGSTONE, INDEPENDENCIA, SANTIAGO, CHILE [4]GRUPO POLÍMEROS, FACULTAD DE QUÍMICA Y BIOLOGÍA, UNIVERSIDAD DE SANTIAGO DE CHILE, USACH, CASILLA 40, CORREO 33, SANTIAGO, CHILE

8.1 Introduction

Gold and silver nanoparticles present a broad range of properties, allowing the development of novel technologies from sensors to biomedical devices. One of the best routes to take advantage of these properties is by mixing these nanoparticles with polymers to obtain nanocomposites having functionalities associated with the metal nanoparticles together with the main characteristics of the polymer matrix. These nanocomposite materials have markedly improved properties [1]. Moreover, these improvements are achieved at low nanoparticle concentrations (1−10 wt.%). In this context, the nanocomposites retain the low density and processability of the matrix, allowing the design of specific products [2]. Polymers can also be used in hybrid metal materials for stabilization of the nanoparticles and adding surface functionalization. Nowadays, polymer/metal nanocomposites are recognized as having unique properties such as antimicrobial behavior, catalytic activity in diverse types of reactions, and conductivity properties, therefore they may be applied in water treatment, the textile industry, food packaging, sensors, and medical devices. In the first section of this chapter the main synthesis route of the nanoparticles together with their functionalization is described. Then the main properties of the polymer/AuNPs or AgNPs nanocomposites and their applications are explained.

Polymer Composites with Functionalized Nanoparticles. DOI: https://doi.org/10.1016/B978-0-12-814064-2.00008-1

8.2 Synthesis of Metal Nanoparticles

There are several synthesis routes for obtaining metal nanoparticles that can be classified as chemical, physical, and biological; these are briefly described in the following section.

8.2.1 Chemical Methods

The most widely used techniques in this kind of method are chemical reduction in aqueous media, electrochemical reduction, solvothermal synthesis, hydrothermal synthesis, sonochemistry synthesis, and chemically assisted radiation.

Synthesis by chemical reduction consists of the reduction of a dissolved metal salt (metal precursor) by means of a reducing agent. The most commonly used reducing agents for synthesis of silver and gold nanoparticles are sodium borohydride, sodium citrate, citric acid, ascorbic acid, carbohydrates (glucose, maltose, and starch), and hydrazine and hydroxylamine compounds. To control the size of the nanoparticles and also to avoid their agglomeration, the presence of a stabilizer or capping agent is necessary. The stabilizing agent is often used as functionalizer to control the reactivity of the nanoparticle. The chemical characteristics of the functionalizer will directly influence the final properties of the nanoparticles, and the properties of the resulting polymeric composite material. The functionalizers can be biomolecules, organic and inorganic molecules. Among the most used biomolecules are peptides, enzymes, DNA, and proteins [3–5] In general, the organic molecules used for functionalization must contain reactive functional groups that have a strong interaction with the surface of the nanoparticle through physisorption, chemisorption, or a covalent bond. The rest of the structure should present good compatibility with the surrounding media (organic solvent, water, polymer matrix, etc.). Brust–Schiffrin method is a typical preparation for AuNPs that has been widely used in biomedical applications [6]. Another classical method of reduction is that published by Brust–Schiffrin, where the aqueous chloroaurate ions ($AuCl_4^-$) are transferred to an organic phase (toluene) using a phase-transfer agent (tetraoctylammonium bromide) and then reduced by sodium borohydride ($NaBH_4$) in the presence of a stabilizer (dodecanethiol), which generates AuNPs of 1.5–5 nm diameter. This is a simple technique for the direct synthesis of functionalized metal clusters on the surface, allowing to control their growth determined by the surface coverage, however, it is only possible to produce hydrophobic nanoparticles [7].

In the case of electrochemical synthesis, the reduction process occurs through the application of an electrochemical potential. Zou et al. have electrodeposited AuNPs and graphene-type flower structures on glassy carbon electrodes (Au/f-GE/GCE) (Fig. 8–1), obtaining an excellent response for the determination of nitrates, with high reproducibility and stability [8].

In the solvothermal or hydrothermal synthesis (when the solvent is water), the metallic precursor is dissolved in the solvent in a closed container, which is heated above its boiling point, generating a pressure greater than atmospheric pressure. In general, this kind of

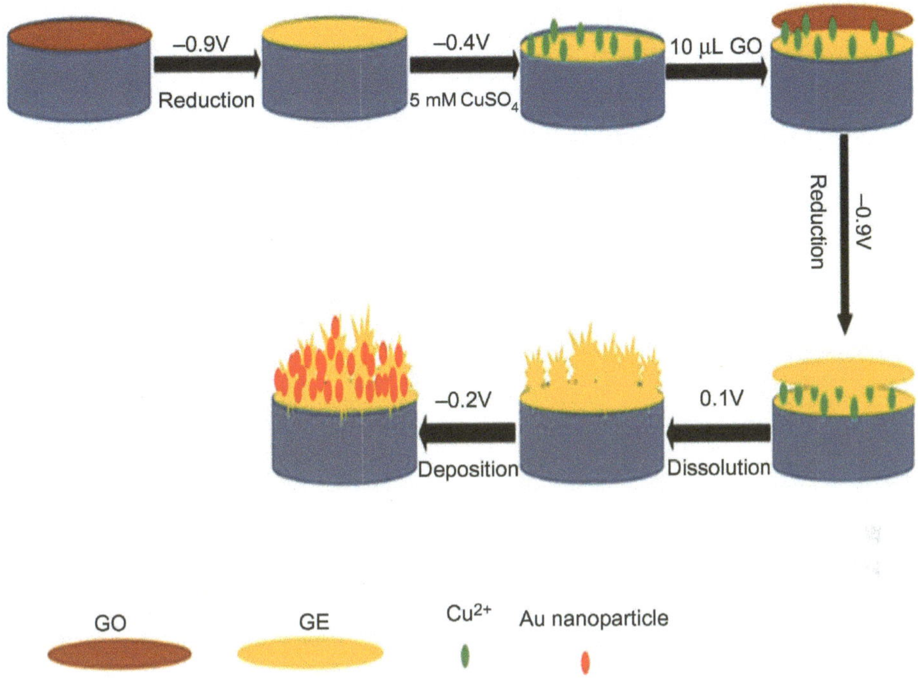

FIGURE 8–1 Schematic illustration for fabricating graphene-type flower (f-GE) structures on glassy carbon electrodes (GCE), 3D Au/f-GE/GCE [8]. *Adapted from Zou CE, Yang B, Bin D, Wang J., Li S, Yang P, et al. Electrochemical synthesis of gold nanoparticles decorated flower-like graphene for high sensitivity detection of nitrite. J Colloid Interface Sci 2017;488:135–41.*

synthesis requires long reaction times. However, controlling some factors such as temperature, reaction time, solvent effects, ligand effects, facet-specific capping agent effects, and reductant effects, it is possible to obtain different well-defined morphologies as shown in Fig. 8–2 [9]. Sonochemistry synthesis otherwise is based on acoustic cavitation, formation, growth, and the implosive collapse of bubbles in a liquid, promoting the reduction of metal ions. This method presents some advantages as the reaction times and the energy required are considerably lower [10].

8.2.2 Physical Methods

AgNPs and AuNPs can be produced by evaporation and condensation methods such as that generated in a tubular furnace at atmospheric pressure, where the sample is vaporized in a carrier gas. However, for the synthesis of AgNPs or AuNPs, long heating times are required to achieve thermal stability and therefore high energy.

These metal nanoparticles have also been prepared using laser ablation of metal plates in an acetone–water mixture, followed by laser irradiation of the colloidal solution. In general, with this method it is possible to obtain nanoparticles completely free of contaminants and

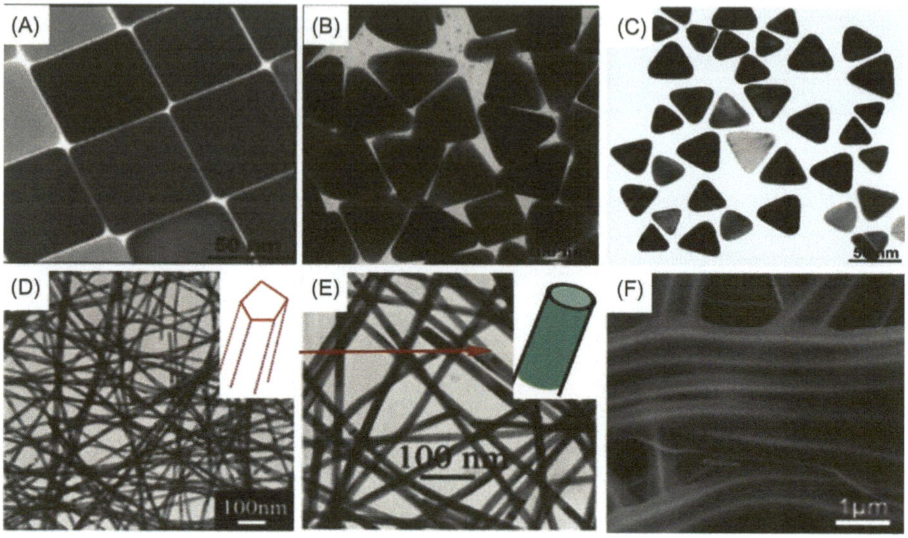

FIGURE 8–2 Electron microscopy images of different morphology Ag nanocrystals by solvothermal methods: (A–E) TEM images of Ag of (A) nanocubes, (B) triangles, (C) triangular plates, (D) pentagonal nanocables, (E) cylindrical nanocables. The inset shows the corresponding structure model, respectively. (F) SEM images of Ag/cross-linked poly(vinyl alcohol) coaxial nanocables [9]. *Adapted from Lai J, Niu W, Luque R, Xu G. Solvothermal synthesis of metal nanocrystals and their applications. Nano Today 2015;10(2):240–67.*

with good size reproducibility. The solvents reported in this method are water, organic solvents, ionic liquids, and supercritical fluids [11–14]. Some studies have shown that after the radiation process the nanoparticles change their spherical morphology to the formation of cubes with edge lengths of 50–200 nm, due to a reassembly process that occurs in the ripening of NPs prepared using chemical reduction methods (Fig. 8–3). Although the synthesis of nanoparticles by this method does not require stabilizers, some studies have used cyclodextrins [15], proteins [16], and oligonucleotides [17] to improve the polydispersity of AuNPs.

8.2.3 Biological Methods

The biological methods are based on the reduction of the metallic precursor and stabilization of the nanoparticle with molecules produced by living organisms such as bacteria, fungi, yeasts, plants, and algae.

Roots of ginseng plants (Panax ginseng) have been used for the synthesis of AgNPs and AuNPs. The AgNPs are obtained after 2 h and the AuNPs after 5 min of reaction, yielding sizes of 100 nm and 10–40 nm, respectively [18] Other syntheses involve the use of *Bacillus licheniformis*, a probiotic used in the industry for the production of the bacitracin antibiotic. The synthesis of AgNPs using this bacterium is based on the incubation of *B. licheniformis* in the presence of 1 mM silver nitrate, yielding sizes between 19 and 63 nm [19].

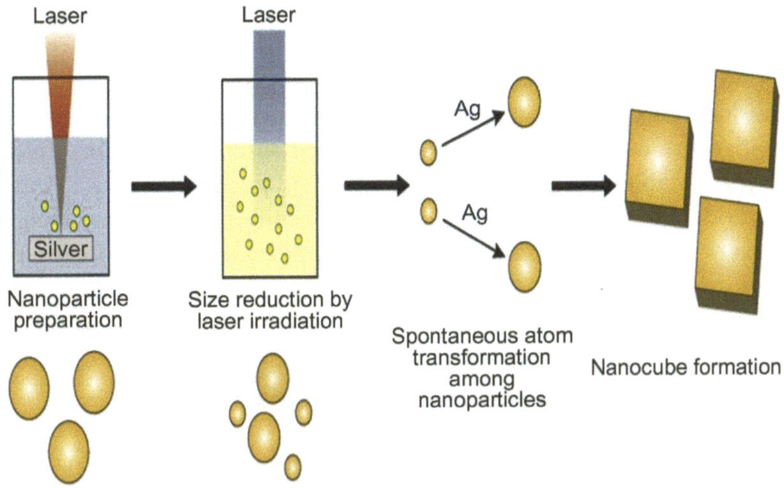

FIGURE 8–3 Preparation of silver nanoparticles with spherical morphology by laser ablation, (red beam) followed by laser irradiation (blue beam), and subsequent transformation to nanocubes [11]. *Adapted from Tsuji T, Kikuchi M, Kagawa T, Adachi H, Tsuji M. Morphological changes from spherical silver nanoparticles to cubes after laser irradiation in acetone–water solutions via spontaneous atom transportation process. Colloids Surf A Physicochem Eng Asp 2017;529:33–7.*

The biosynthesis of silver and AuNPs from *Aspergillus terreus*, an isolated fungus from soils of southern India, had also been reported. The nanoparticles obtained showed high stability, with sizes between 8 and 20 nm for silver and 10 and 50 nm for AuNPs with spherical morphology [20]. The synthesis of nanoparticles using algae has also shown highly stable nanoparticles. The synthesis of AuNPs from *Sargassum tenerrimum* and *Turbinaria conoides*, two species of brown algae, capable of reducing Au(III) ions to AuNPs between 5 and 57 nm has been studied. The AuNPs presented different catalytic activity, depending on the alga species used for their synthesis [21]. The synthesis of AuNPs from yeast is less widely used, but it has also shown excellent results. For example, the use of *Magnusiomyces ingens* to reduce $HAuCl_4$ generated nanoparticles of different morphology including spheres, triangles, and hexagons [22]. The use of biological methods allows good control of the size of the nanoparticles, but it is difficult to obtain large amounts of them. In this sense, chemical methods make it possible to obtain nanoparticles on a large scale.

8.3 Functionalization of Silver and Gold Nanoparticles

The development of functionalized nanoparticles has had great impact in terms of its potential application in biomedical, electronic, food packaging, and optical materials. The surface chemistry of functionalized nanoparticles has a direct relationship with characteristics such as stability, monodispersity, and polar/apolar nature, among others. When nanoparticles are incorporated into polymer matrices, the most likely phenomenon to occur is aggregation, and in this sense it is ideal to have their functionalization improve the compatibility with the matrix and therefore its spatial distribution.

The nanoparticle can be functionalized with polymer chains by self-assembly, controlling the separation distance between them. In general, self-assembly requires favorable enough interactions to form stable structures at equilibrium or near equilibrium, but avoiding the trapping of materials in unstructured local energy minima. The intermolecular interaction includes hydrogen bonds, metal coordination, electrostatic attraction/repulsion, dipolar interactions, van der Waals forces, and hydrophobic interactions. Through the technique of self-assembly it is possible to obtain highly ordered nano-, micro-, and macrostructures.

8.3.1 Functionalization of Gold Nanoparticles

A clear example of functionalization of AuNPs to facilitate their incorporation into a polymer matrix was reported by Cobierre et al. [23] by covalently binding an alkanethiol to the surface of AuNPs, producing a high degree of dispersion in a polystyrene (PS) polymer matrix.

Templeton [24] and Brust [25] have focused on obtaining monolayer-functionalized nanoparticles. This is a versatile route involving a two-step process, whereby the nanoparticles protected with the alkanethiolate monolayer are initially synthesized, followed by a partial or complete exchange of the monolayer with functionalized ligands (p-mercaptophenol derivatized) at a later stage. Fig. 8–4 illustrates this approach with an outline of the creation of AuNPs functionalized through synthesis in the solution phase [26] followed by the place-exchange technique.

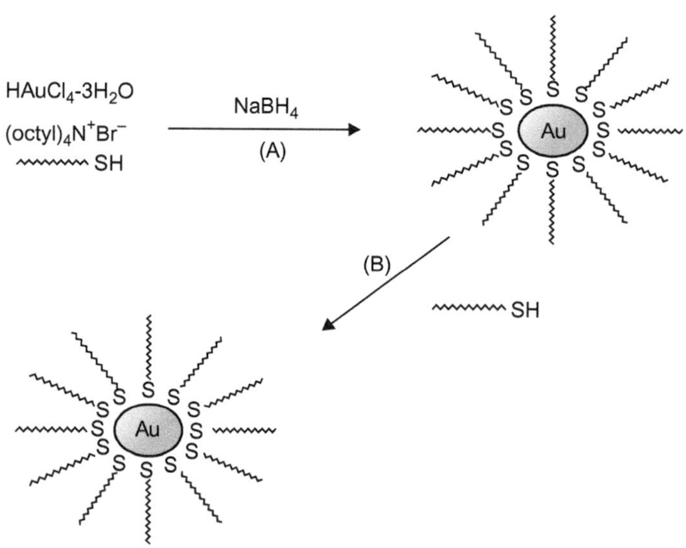

FIGURE 8–4 (A) The Brust–Schiffrin solution-phase gold nanoparticle synthesis, followed by (B) the place-exchange process developed by Murray. *Adapted from Copyright ACS. Shenhar R, Norsten TB, Rotello VM. Polymer-mediated nanoparticle assembly: structural control and applications. Adv Mater 2005;17(6):657–69 [26].*

The use of polymers rather as entities used to space and regulate the distance between particles could be advantageous, as it can generate simpler building blocks. Some works used poly(amidoamine) dendrimers (PAMAM) to assemble AuNPs and control the separation distance between them [27].

8.3.2 Functionalization of Silver Nanoparticles

Some works have shown the functionalization of AgNPs with biological molecules to improve their antibacterial response. In general, the functionalization of AgNPs with biomolecules can be achieved through chemisorption phenomena, for example with thiol groups or amines, electrostatic interactions, covalent binding with functional groups, or noncovalent affinities (Pearson's theory). Morales-Avila et al. [28] have functionalized AgNPs with ubiquicidin 29-21 (UBI), a known peptide with antibacterial activity. The results obtained showed that the effect of functionalization with peptides generates a synergistic phenomenon in the antibacterial property of the nanoparticles.

Taheri et al. [29] have developed AgNPs functionalized with a monolayer of 2-mercaptosuccinic acid (Fig. 8–5), obtaining excellent antibacterial efficacy against *Staphylococcus epidermidis* (*S. epidermidis*), *Staphylococcus aureus* MRSA [*S. aureus* (MRSA)], and *Pseudomonas aeruginosa* (*P. aeruginosa*).

Tamayo et al. have functionalized AgNPs with oleic acid for subsequent incorporation into polyethylene (PE) polymer matrices (Fig. 8–6A) [30] The oleic acid promotes the homogeneous dispersion of the nanoparticles in the PE matrix (Fig. 8–6B).

A different and effective form of functionalization is observed in the work of Fontana et al. [31], who studied the functionalization of AgNPs in interconnected systems, using

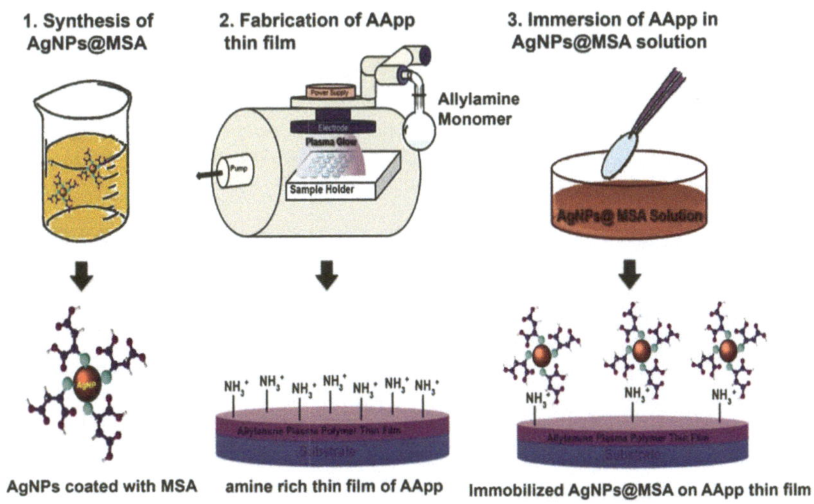

FIGURE 8–5 Schematic representation of the strategy for the generation of antibacterial coatings. *Adapted from Copyright Elsevier. Taheri S, Cavallaro A, Christo SN, Smith LE, Majewski P, Barton M, et al. Substrate independent silver nanoparticle based antibacterial coatings. Biomaterials 2014;35(16):4601–9 [29].*

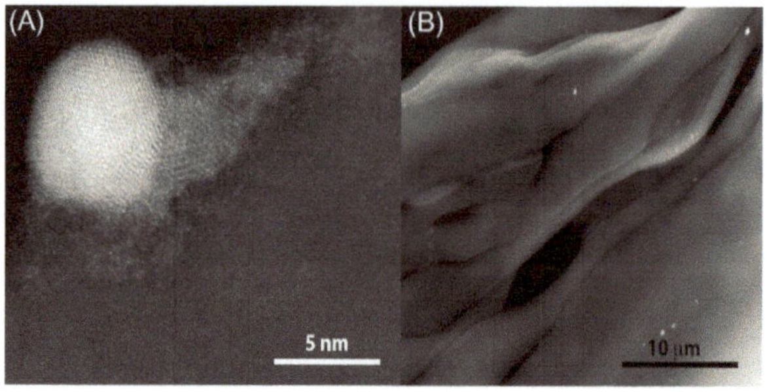

FIGURE 8–6 (A) Silver nanoparticles functionalized with oleic acid. (B) Functionalized silver nanoparticles homogeneously distributed in a polyethylene matrix.

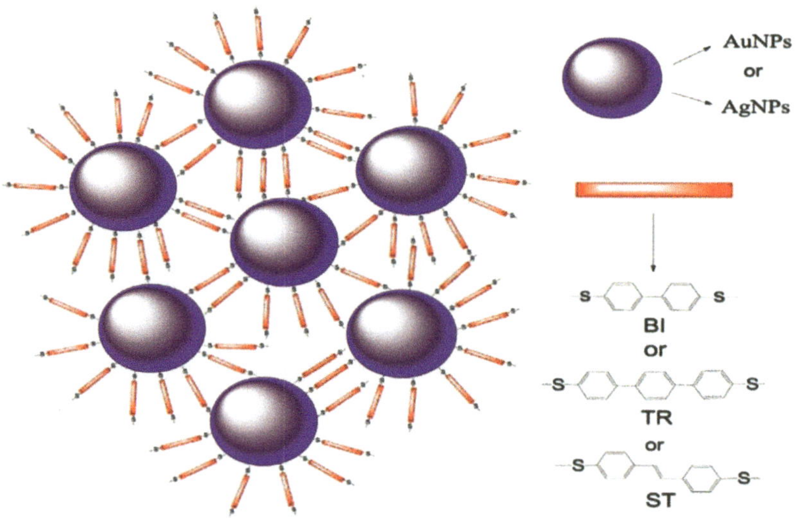

FIGURE 8–7 Silver nanoparticles functionalized by bifunctional dithiols. *Adapted from Copyright Elsevier ACS. Fontana L, Bassetti M, Battocchio C, Venditti I, Fratoddi I. Synthesis of gold and silver nanoparticles functionalized with organic dithiols. Colloids Surf A Physicochem Eng Asp 2017;532:282–9 [31].*

organic dithiols, as shown in Fig. 8–7. The functionalization of AgNPs with bifunctional thiols of different lengths allows modulating their optical properties.

8.4 Methods of Preparation of Metal/Polymer Nanocomposites

Several methods have been used to prepare metal–polymer nanocomposites. For all of them, ex situ and in situ methods are the main routes of preparation.

8.4.1 In Situ Method

Nanocomposites prepared by this method consist of polymerizing monomers in the presence of the nanoparticles, which must be dispersed in the monomeric solution or solvent before polymerization. Another approach is by the metal ion reduction into the polymer matrix or the metal ion reduction and polymerization simultaneously. In situ ion reduction into the polymer matrix can improve nanoparticle stabilization and consequently control nanoparticle size. The high surface energy of the nanoparticles promotes thermodynamic stabilization through the interaction of the polymer chains with nanoparticle surfaces, thus controlling nanoparticle growth and agglomeration [32]. When the metal ions are reduced in the presence of the polymer, the latter can present different forms, the most studied of which have been the continuous films, porous matrices, spheres, and polymeric fibers. Zhang et al. [33] have achieved excellent results in obtaining AuNPs on bacterial cellulose (BC) fibers. The reduction of $AuHCl_4$ using polyethyleneimine (PEI) on the BC fibers has led to obtaining fibers decorated with AuNPs (Fig. 8–8).

Li et al. prepared a nanocomposite based on AuNPs and poly(4-vinylpyridine) (Au at PVP) through surface-initiated atom-transfer radical polymerization (SI-ATRP) at ambient conditions (Fig. 8–9). The Au at PVP nanocomposites as a smart supporter can entrap

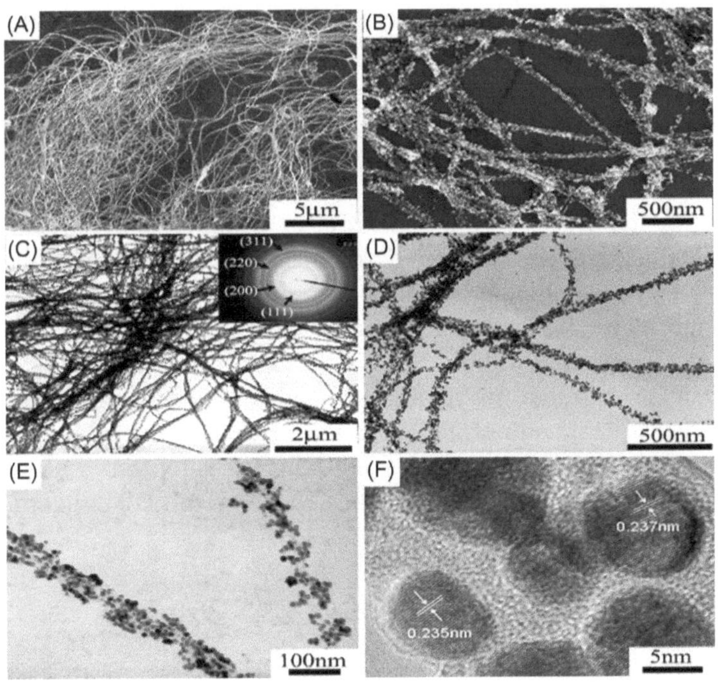

FIGURE 8–8 (A,B) SEM, (C–E) TEM, and (F) HRTEM images of Au–BC nanocomposites obtained after 1 h of reaction at pH 8.2. The inset shows the related ED pattern. *Adapted from [PEI] = 1 g L 1 [$HAuCl_4$] = 1 mM. Copyright Wiley. Zhang T, Wang W, Zhang D, Zhang X, Ma Y, Zhou Y, et al. Biotemplated synthesis of gold nanoparticle–bacteria cellulose nanofiber nanocomposites and their application in biosensing. Adv Funct Mater 2010;20(7):1152–60 [33].*

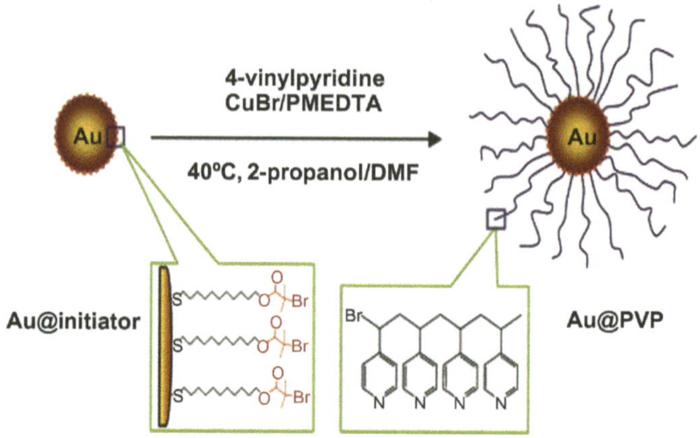

FIGURE 8–9 Schematic fabrication of Au at PVP nanoparticles by SI-ATRP. *Adapted from Copyright ACS. Li D, He Q, Cui Y, Li J. Fabrication of pH-responsive nanocomposites of gold nanoparticles/poly (4-vinylpyridine). Chem Mater 2007;19(3):412–17 [34].*

transition metal ions by their efficient coordinating segments, and subsequently the metal ions can be reduced in situ to construct novel pH-stimuli bimetallic catalysts [34].

In another approach, the reduction of the metal ion and the polymerization of a monomer can be carried out simultaneously under UV radiation conditions or microwave assistance [35]. Yagci et al. reported a simultaneous UV-induced radical polymerization of a typical acrylic resin and gold nanoparticle formation by the reduction of $HAuCl_4$ in the presence of 1-[4-(2-hydroxyethoxy)phenyl]-2-hydroxy-2-methyl-1-propane-1-one (Irgacure 2959) as photoinitiator. Thus, AuNPs are formed in situ during the polymer network formation [36]. A similar approach was reported by Sangermano et al. using 3,4-epoxycyclohexylmethyl-3′,4′-epoxycyclohexanecarboxylate (EEC) to serve as the cationically polymerizable monomer, and 2,2-dimethoxy-2-phenylacetophenone (DMPA) as a good source of electron donor radicals by irradiation. This technique allowed obtaining a homogeneous distribution of the nanoparticles within the polymer network without any macroscopic agglomeration [37].

There are several studies that use in situ polymerization to obtain silver nanocomposites based on polyacrylates (PAs) and polyaniline (PANI). However, even when a good dispersion of the nanoparticles in the polymer is achieved, polymerization times are long and require temperature [38,39]. In this sense, polymerizations assisted with UV radiation are more effective, requiring shorter reaction times.

8.4.2 Ex Situ Method

This method is based on synthesizing metal nanoparticles and polymer separately. Subsequently the incorporation of the nanoparticles into the polymer can be obtained by two main techniques: melt compounding or solution blending. Yeo et al. prepared nanocomposites based on polypropylene (PP) and AgNPs by means of the coextruded melt state technique. Master batches were made up of mixtures of PP chips and nanosized silver powder. Scanning electron microscope images obtained showed that the silver powder consists of uniform spherical nanoparticles, but some particles are aggregated by the interaction force of nanosilver powder [40].

Other nanocomposites have been prepared by other techniques such as deposition of nanoparticles on the polymer's surface, by spraying or dip coating, but these are less used. In general, using ex situ methods, the spatial distribution of the nanoparticles in the polymer is not optimal. For instance, a spray coating technique can be used to deposit uniform nanoparticles or obtain patterned nanoparticles on different surfaces. Madaria et al. demonstrated that the application of a simple spray coating technique allowed obtaining large scale, highly uniform, and conductive silver nanowire films on arbitrary substrates [41].

8.5 Properties of Metal/Polymer Nanocomposites

The properties of polymer/AuNPs or AgNPs nanocomposites are explained below. With respect to these properties, the nanocomposites may be applied in water treatment, textile industry, food packaging, sensor, and medical devices.

8.5.1 Antibacterial Properties of the Silver and Gold Nanocomposites

Polymer nanocomposites including AgNPs and AuNPs are recognized as having long term antimicrobial activity against different bacteria and fungi, in addition to low toxicity to human cells and tissues, so nanoparticles have been used as antibacterial materials in different polymer matrices [42]. Antimicrobial materials play an important role in the academic and technological fields due to applications in food packaging, textile industry, and medicinal devices such as catheters, cannulas, and others [43].

8.5.1.1 Gold Nanoparticles

The antimicrobial activity of AuNPs has been demonstrated [44–46] although their bacterial inhibition mechanisms are still unclear. Many reports present that AuNPs could damage the bacterial wall, leading to leakage of cytoplasmic content as the cause of the bacterial cell death [44,46,47]. Another hypothesis considering the mechanism of AuNPs' biocidal activity focuses on reactive oxygen [reactive oxygen species (ROS), e.g., singlet oxygen] as a potential cause of bacterial cell damage and death [48]. On the other hand, Cui et al. [49] stated a molecular mechanism for the antibacterial ability of the AuNPs without including the participation of ROS formation (Fig. 8–10). Another antibacterial mechanism based on classical optical theories was proposed by Li et al. [50] from the viewpoint of the localized surface plasmon resonance (LSPR) of the AuNPs. Studies have demonstrated that the antibacterial activity shown for AuNPs is lower than that of other conventional antibacterial nanoparticles like AgNPs, mainly due to the higher chemical stability [48].

AuNPs have been combined with polymers to obtain antibacterial nanocomposites or hybrid nanoparticles. For instance, Chamundeeswari et al. [51] prepared hybrid

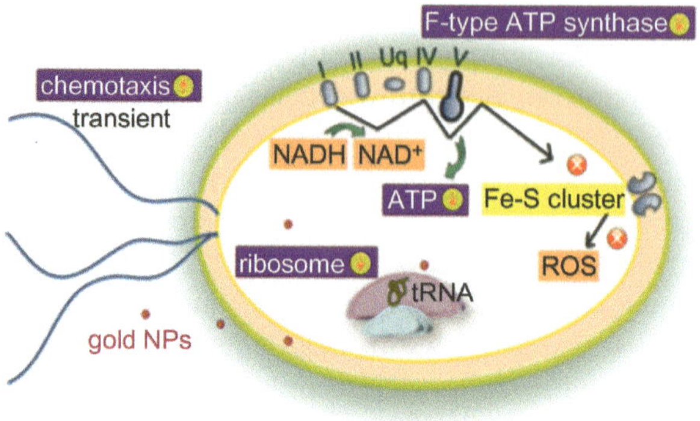

FIGURE 8–10 Schematic diagram of the bactericidal mechanism of AuNPs on *E. coli*. AuNPs induce the down-regulation of the oxidative phosphorylation pathway (F-type ATP synthase and ATP level) and ribosome pathways, and the transient up-regulation of chemotaxis. AuNPs do not induce the change of ROS-related processes. *Adapted from Copyright Elsevier. Cui Y, Zhao Y Tian Y, Zhang W, Lü X, Jiang X. The molecular mechanism of action of bactericidal gold nanoparticles on* Escherichia coli. *Biomaterials 2012;33(7):2327–33 [49].*

nanoparticles composed of chitosan-capped AuNPs coupled with ampicillin. The results revealed the antimicrobial efficacy of chitosan-AuNP-Amp showing a twofold increase in activity compared with that of free ampicillin. Barkoula et al. [52] investigated the antibacterial activity of six common polymers (polyvinyl alcohol (PVA), hydroxypropylmethylcellulose, polycaprolactone, sodium alginate (SA), and chitosan) before and after the formation of nanocomposites with AuNPs by a solvent casting method [52]. All the AuNPs-loaded polymer films showed significantly greater antibacterial activity than neat polymer films. Nanocomposites based on PANI with AuNPs were obtained by Boomi et al. [53]. These nanocomposites exhibited antibacterial activity against Gram-positive as well as Gram-negative bacterial strains [53] AuNPs were also incorporated into polyurethane (PU), polycaprolactam, polycarbonate (PC), and polymethylmethacrylate (PMMA), showing for all nanocomposites a lower formation of *Eschericha coli* (*E. coli*) biofilms [54]. Taking advantage of this antibiofilm capacity, Russo et al. [55] developed a PMMA-based bone cement loaded with AuNPs, which showed the reduction in thickness of the biofilm and also a decrease of the ratio between live and dead bacteria per unit area.

8.5.1.2 Silver Nanocomposites

Silver is known for antimicrobial properties and has been used for years in the medical field. The antimicrobial mechanism from silver is only partially understood. It is widely known that the main antibacterial effect of Ag nanoparticles or Ag-based nanocomposites is due to their partial oxidation and release of Ag^+ ions [56]. The silver ions avidly bind to proteins and nucleic acids, causing structural changes in the bacterial cell walls, membranes, and nucleic acids that affect viability [57,58]. Another mechanism suggests that due to their small size, AgNPs penetrate into the bacterial cell and cause directly changes in the deoxyribonucleic acid (DNA) of the cell [59]. Marambio-Jones and Hoek postulated that the toxicity of

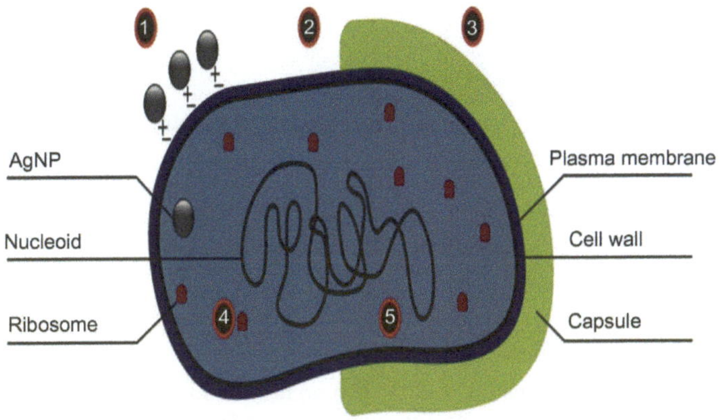

FIGURE 8–11 Proposed mechanism of the antibacterial action of silver nanoparticles: 1—electrostatic attraction; 2—production of free radicals, changes in permeability, disturbance of respiration, leakage of intracellular content; 3—modulation of phosphotyrosine profiles of proteins, involved in the cell cycle progression and in the synthesis of capsular polysaccharides; 4—interaction with SH-groups, inhibition of protein synthesis and function; 5—interaction with phosphorus-containing molecules (DNA). *Adapted from Copyright Elsevier. Durán N, Durán M, de Jesus MB, Seabra AB, Fávaro WJ, Nakazato G. Silver nanoparticles: a new view on mechanistic aspects on antimicrobial activity. Nanomedicine: Nanotec Biol Med 2016;12(3):789–99 [61].*

silver could also be due to ROS generation by AgNPs and silver ions [60]. Fig. 8–11 shows the different possible mechanisms of AgNPs against bacteria [61].

To obtain new antimicrobial materials, silver particles have been incorporated in different polymers like PU, polyester (PES) [62], polyamide (PA) [63], PP [64], PA [65], PS [66], and PE [30,67] for antimicrobial applications. Recently, poly(3-hydroxybutyrate) (PHB) nanocomposites with biosynthesized AgNPs (c. 76–95 nm) with antimicrobial properties against *Salmonella enterica* and *Listeria monocytogenes* was obtained [68].

Different authors have explained that the effectiveness of the bacterial properties of these composites are related silver ion release. To have an efficient and "permanent" activity, a stable and prolonged release of silver ions at a suitable concentration is crucial; a too slow or too fast release of Ag^+ ions would be inappropriate for the main applications such as intelligent food packaging, and in the medicine field as catheters and implants. Metallic silver is relatively unreactive, but in aqueous environments silver ions are released, and the antimicrobial activity depends on the intracellular accumulation of low concentrations of silver ions [69].

Damm et al. [70] and Kumar et al. [43] prepared nanocomposites from PA/AgNPs by the melt process with efficacy against *E. coli* due to the high release of Ag^+. Other authors studied the influence of the matrix on the biocidal properties, finding that the release of Ag^+ ions in PP was much lower than from PA/Ag nanocomposites, reflected in decreased antimicrobial properties. This behavior was due to the poor water affinity of the PP matrix compared with the PA matrix [71] Zapata et al. [72] and Tamayo et al. [30] reported that the antimicrobial properties of PE/AgNPs nanocomposites against *E. coli* and *L. monocytogenes* were dependent on the silver amount in the AgNPs and the release of silver ions from the

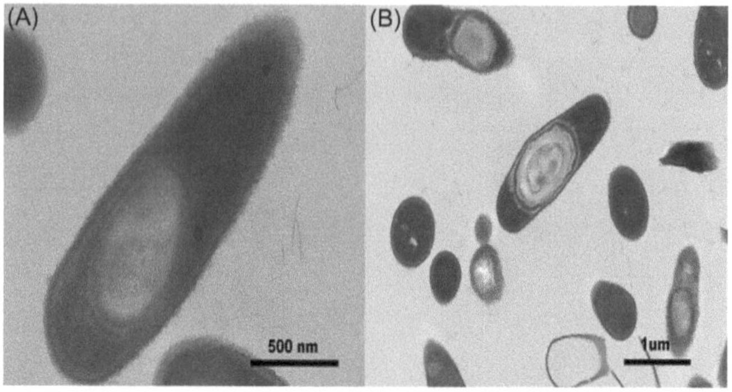

FIGURE 8–12 *L. monocytogenes* (ISP 6508) incubated for 16 h on the nanocomposite surface: (A, B) PE-AgNp nanocomposite. *Adapted from Copyright Elsevier. Tamayo LA, Zapata PA, Vejar ND, Azócar MI, Gulppi MA, Zhou X, et al. Release of silver and copper nanoparticles from polyethylene nanocomposites and their penetration into* Listeria monocytogenes. *Mater Sci Eng C 2014;40:24–31 [30].*

nanocomposites used. The interaction between nanoparticles from nanocomposites and *L. monocytogenes* bacteria was examined by Transmission electron microscopy (TEM) as displayed in Fig. 8–12. Resulting morphological changes involve separation of the cytoplasmic membrane from the cell wall, which is known to be an effect of plasmolysis.

The influence of the morphology (silver nanofibers; Ag-Nfbs) against nanospheres (AgNPs) into PE was also studied [67] and it was found that PE/AgNPs nanocomposites shown greater biocidal activity than PE/Ag-Nfbs [72].

Ag/TiO_2 has attracted attention because TiO_2 is a promising material for various applications, including antimicrobial activity. Colloidal TiO_2 nanoparticles/tubes were synthesized using PVA as a capping agent. Ag nanoparticles were photocatalytically deposited on all the synthesized samples under UV–vis light from silver nitrate solution. The synthesized metal–metal oxide polymer nanocomposites are shown to be very efficient toward destroying *E. coli* Gram-negative bacteria [73].

The combination of AgNPs with water soluble biopolymers will produce new polymer/AgNPs nanocomposites with antimicrobials properties. Based on this, various natural polymers such as gum acacia, glucose, starch, gelatin, SA, and carboxymethylcellulose have been reported [74–76].

8.5.2 Catalytic Activity of the Gold and Silver Nanocomposites

The catalytic characteristics of Au and Ag metals can be explained by the extent of their d-band vacancy as Au and Ag having it completely occupied [77]. Unlike Au and Ag can shed electrons, thereby creating d-band holes and becoming catalytically active.

8.5.2.1 Gold Nanocomposites

Au-based catalysts are very good catalysts in the oxidation of carbon monoxide (CO) in automotive exhaust systems [78]. Moreover, AuNPs supported on different substrates have been demonstrated to have catalytic activity in diverse types of reactions, such as carbon monoxide oxidation, Suzuki couplings, reduction of nitro derivatives, and reduction of organic dyes [79].

Another type of Au nanocomposites are the bimetallic Ag/Au particles, which have a catalytic performance 25-fold higher than that of the monometallic Au nanoparticles. This Au-based bimetallic nanocomposite was embedded in poly(N-isopropylacrylamide)/polyethyleneimine (PNIPAm/PEI) core–shell microgels, and showed controllable catalytic activities [80].

In the literature there are many studies related to different configurations of nanocomposites or hybrid nanoparticles for the Au-based catalyst. Some of them arechitosan/Au/CNT films [81], Au/poly (amidoamine) (PAMAM), and Au/poly (propyleneimine) (PPI) dendrimers [82], Au/alginate [83,84], among others [85].

8.5.2.2 Silver Nanocomposites

Recent research has shown that AgNPs composites have excellent catalytic activity [86]. Supported metal nanoparticles show catalytic activity towards various organic reactions such as coupling, cycloaddition, and reduction of dyes and of nitrophenols. Sulfonated polystyrene/chitosan (SPS/CS)/AgNPs, montmorillonite/Ag, and PVA/AgNPs [87] presented activity to catalyze the reduction of methylene blue (MB) [88], and alginate/AgNPS was evaluated for the reduction of p-nitrophenol to p-aminophenol by NaBH4 [89].

Polymer microspheres have become one of the most promising substrate candidates for AgNPs due to their porous character, which provides better accessibility to reactants and larger specific surface area. AgNPs are loaded on raspberry-like multihollow polymer microspheres through seeded swelling polymerization and subsequent in situ chemical reduction. The excellent catalytic performance in the degradation of MB of the prepared polymer/Ag nanocomposite after 10 catalytic cycles demonstrates good stability and durability of AgNPs [90].

8.5.3 Electronic Properties of the Silver and Gold Nanocomposites

Many polymers are electrically insulating, but the conductive metal nanoparticles provide an acceptable conductivity in polymer/metal nanocomposites. The need for flexible, lightweight, electrically conductive devices led to the generation of smart materials called conductive polymer nanocomposites (CPCs). The electrical conductivity can be obtained based on the formation of charge transfer complexes inside the polymer chain network, due to the addition of metal nanoparticles [56]. The conductive nanocomposites can be used in various applications such as rechargeable batteries, electrodes, shields, sensors, etc.

8.5.3.1 Gold/Polymer Nanocomposites

Some work on the preparation of gold dispersed into conducting polymer matrices has been carried out, such as polythiophene [91,92], polypyrrole (PPy) [93], PANI [94,95], and poly (3,4-ethylenedioxythiophene) polystyrene sulfonate (PEDOT:PSS) [96]. Elhalawany et al. [97] showed the preparation of highly conductive PANI/Au composites. TEM analysis showed that the Au nanoparticles take triangle and tetrapod structures that gave rise to a very high conductivity of the prepared composites due to fast electron transport and efficient charge transfer.

High aspect ratio gold nanowires (Au-NWs) were uniformly dispersed into a poly(vinylidene difluoride–trifluoroethylene) matrix. The bulk electric conductivity of nanocomposite films at room temperature obeys a percolation behavior at a low threshold of 2.2 vol%, and this was confirmed using the surface resistivity value. An electric conductivity of 100 S m^{-1} is achieved for a 3 vol% filler content [98].

AuNPs have been incorporated into cellulose for increased electric conductivity [99]. Synthesized gold-coated BC-PANI composites for electroconductive applications were obtained by Faria-Tischer et al. [100], who got a large increase in the electric conductivity of (BC) PANI composites. They observed a more defined redox wave shape in the presence of AuNPs, which is an indication of better diffusional processes through the solid material.

8.5.3.2 Silver/Polymer Nanocomposites

Silver (Ag) has higher electrical (6.3×10^7 S m^{-1}) and thermal (429 W $m^{-1}K^{-1}$) conductivities than other noble metals, and it also has the ability to enhance electrical and optical properties of the polymer composites [101,102]. Conducting polymer nanocomposites such as PANI, PPy, and polythiophene have received a great interest in the conducting polymer nanocomposites [103–105].

On the other hand, AgNPs and nanotubes have also been incorporated into PVA. Increasing the concentration of the AgNPs embedded in the composites obtained enhanced the conductivity of PVA [106–108].

The use of the AgNWs with conducting polymer matrices to produce nanocomposite materials has recently received much attention, especially in organic electronics [102,109]. Examples of conductive polymer nanocomposite with AgNWs are (PEDOT:PSS) [110,111], PC [112], and PA [113] or other polymers like polyetheretherketone (PEEK), PET, PDMS, PVA, PU, PVDF, and PLA [114–117]. Supercapacitor devices utilizing AgNWs have been reported, with AgNWs used in conjunction with PEDOT:PSS [118] and PPy [119]. Fig. 8–13 shows the synthesis of the AgNWs and the preparation of AgNWs/PPy nanocomposites with capacitor properties by in situ polymerization.

8.5.4 Water Treatment Using Gold and Silver Polymer Nanocomposites

Numerous studies have shown that nanomaterials have vast capability and potential in water treatment. Au and Ag nanoparticles have been extensively used for the detection of trace levels of organic contaminants or to adsorb the contaminants to inactivate bacteria [120–122]. There are still inherent disadvantages of the direct use of free nanoparticles in water and wastewater treatment process. For instance, it is desirable to develop a material or a device that could minimize the release or mobilization of the nanomaterials while maintaining their high reactivity [123].

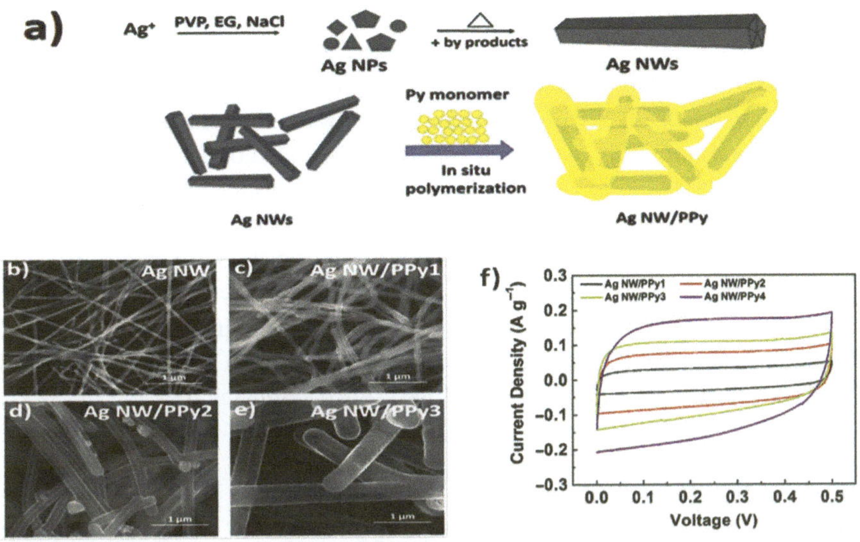

FIGURE 8–13 (A) Schematic illustration of core–shell AgNW/PPy nanocomposite formation. High-resolution SEM images of (B) AgNW, and (C)–(E) AgNW/PPy nanocomposite electrodes with different PPy thicknesses. (F) CV results of the fabricated AgNW/PPy nanocomposite supercapacitors in 1 M KOH electrolyte at a scan rate of 1 mV s^{-1}. *Adapted from Copyright Elsevier. Yuksel R, Alpugan E, Unalan H. Coaxial silver nanowire/polypyrrole nanocomposite supercapacitors. Organ Electron 2018;52:272–80 [119].*

8.5.4.1 Nanocomposites Using Gold Nanoparticles for Water Treatment

Bovine serum albumin (BSA)-capped gold NPs (c. 12 nm) in agarose gel membrane (Au NPs/BSA/AGM) has been developed for removing Hg^{2+} or Pb^{2+} in saline solution [124–126]. On the other hand, Wei et al. [127] prepared gold nanoparticle/bacterial cellulose (AuNP/BC) nanocomposites by citrate-mediated in situ reduction of Au^{3+} in the presence of a BC hydrogel. They found that sorption not only occurred on the top surface of the AuNP/BC nanocomposite, but also within the nanocomposite matrix. In addition, agitation can intensify the surface-enhanced Raman spectroscopy (SERS) response fourfold.

8.5.4.2 Nanocomposites Using Silver Nanoparticles for Water Treatment

Nanocomposites of polyurethane foam with AgNPs were obtained for use as water filters. The filter was tested against *E. coli* at a flow rate of 0.5 L min^{-1}. The *E. coli* in the effluent was below the detection limit [128,129]. The biocidal effect in the coal with AgNPs filter was also studied against *E. coli*, *Saccharomyces cerevisiae*, and *Pichia pastoris*, and strong lethal activity was observed within a few seconds [130] There is a serious problem that commercially available water filters contain silver, which has raised the risk of silver ingestion. Rosa et al. [131] studied silver leaching from coal samples into water and evaluated colloidal silver and silver nanoparticle bioaccessibility. Compared with AgNPs, the colloidal form releases more silver from water leaching and makes it more bioaccessible [132].

Spherical beads of PVA–SA membrane were prepared by the solution casting method to biosynthesize AgNPs [133]. The nanocomposite was effective against different microorganisms such as *Enterobacter* sp., *E. coli, Candida albicans, S. aureus,* and *Klebsiella pneumoniae.* The formation of beads can offer many advantages that include easy removal of the AgNPs after the treatment process, reuse of the immobilized nanosilver for repeating the water treatment processes, and strong binding with the nanoparticles that inhibits their leakage to the treated water samples. Therefore, these nanocomposites can be used for the decontamination of drinking water to free it from antibiotic resistant microbes.

Djerahov et al. [134] prepared a stable CS-AgNPs colloid by dispersing the AgNPs sol in chitosan solution, and further used it to obtain a cast film with high stability under storage and good mechanical strength for handling. It showed efficient separation and extraction of Al(III), Mn(II), Cd(II), Cu(II), Co(II), Fe(III), Ni(II), Pb(II), and Zn(II).

8.5.5 Food Packaging Applications

The main function of packaging is to protect food from the external environment, avoiding, for instance, its deterioration by the action of microorganisms [135]. Active food packaging systems are supposed to perform a role different from providing an inert barrier to external conditions, as shown in Fig. 8–14. In the active package active agents are added directly to the packaging material. This agent can release active biocidal substances to improve the quality of the food, extend its shelf life, and prevent or delay its spoilage [136,137]. AgNPs have been used as active agents due to their properties [138,139]. There are no reports of the incorporation of AuNPs for use as food packaging.

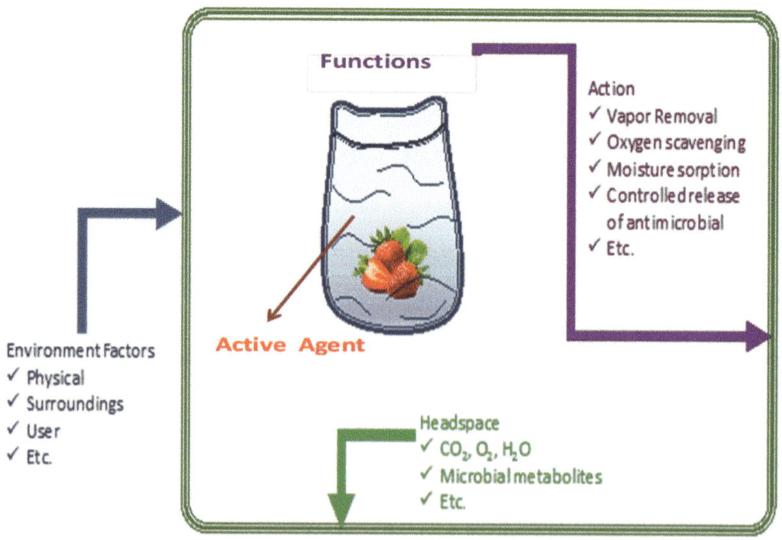

FIGURE 8–14 Active packaging functions.

Antimicrobial nanocomposite systems can include either nanosized antimicrobial agents or conventional biocidal agents together with a nanoparticle such as clay. AgNPs or metallic silver may be readily incorporated into thermoplastic polymeric materials. The interfacial properties of these PE/Clay/AgNPs composites improve with the introduction of different types of compatibilizers. The obtained nanocomposites showed enhanced barrier properties and antimicrobial properties against *E. coli* bacteria [140].

AgNPs have been incorporated into polyvinylpyrrolidone, PE, or polyvinyl chloride (PVC) to preserve and increase the shelf life of asparagus [141], Chinese jujube [142], orange juice [143], apple slices [144], barberry [145], strawberries [146], minced beef [147], and chicken breast [148].

On the other hand, silver-doped zeolites have also been incorporated in polymers for use in food packaging [149] as PE [150] and polylactide [151]. Silver-doped zeolites showed antimicrobial activity against *E. coli* and *S. aureus*.

A green synthetic approach for preparing antimicrobial AgNPs has been developed by using carbohydrates such as agar, sucrose, soluble waxy starch, and cellulose [135,152,153]. Absorbent pads are made of cellulose material with AgNPs, and it is widely used in fresh fruits (kiwi), melon juice, and beef [154]. *Pseudomonas* sp. and *Enterobacteriaceae* counts were below those of the control samples [155].

Chitosan with AgNPs has been investigated as packaging for its strong antimicrobial function [156,157]. Chitosan/AgNPs were also incorporated in rice starch. These silver-loaded chitosan matrices were effective in the inhibition of the growth of Gram + and Gram- microorganisms [158]. Fayas et al. [159] biosynthesized AgNPs and they was incorporated into alginate film by the casting method. The nanocomposites show effective antimicrobial properties against *E. coli* and *S. aureus*. The effect of the nanocomposite film was also studied in carrots and pears, increasing the shelf life of the vegetables.

Gammariello et al. [160] showed the effectiveness of agar/AgNPs montmorillonite to prolong the shelf life of Fior di Latte cheese. Their results indicated that the shelf life of cheese stored in the traditional packaging was three days, while the shelf life of the Ag-MMT coated cheese was extended to more than five days.

A nanostructured starch-based film containing clay and AgNPs has been proposed [161]. The presence of nanoparticles was shown to improve mechanical and gas barrier properties with excellent antimicrobial results.

8.5.6 Textile Industry

Textiles suffer the attack of different microorganisms; therefore antimicrobial agents are used to prevent undesirable effects such as degradation of color, staining, and deterioration of fibers; production of unpleasant odors; and increased potential health risks [162]. To remove these disadvantages, modification of textiles by producing polymeric nanocomposites and also surface modification of textiles with metals have been developed. A rapid global increase in the production of textile materials for medical, healthcare, hygiene, and sports applications renewed the interest for antimicrobial finishing in the last decade [163].

The major research on the functionalization of textile materials with AgNPs is focused on cotton (Co) fibers. The nanocomposite cotton/AgNPs presented antimicrobial properties against bacteria like *S. aureus, K. pneumoniae,* and *Streptococcus faecalis* [164–166]. AgNPs have been deposited into cotton fabrics, and they have also been incorporated into different polymers as polypyrrole [167], polyvinylidene fluoride (PVDF) [168] in the presence/absence of an acrylate binder [169], poly(ethylenimine) (PEI)/fluorinated decyl polyhedral oligomeric silsesquioxane (F-POSS) [170], and chitosan(CS)/3-glycidyloxypropyltrimethoxysilane (GPTMS) [171]. The polymer/AgNPs/cotton nanocomposites exhibited excellent antibacterial activity against *E. coli* and *S. aureus* after washing cycles.

Plasma treatment can be applied for the activation of the Co fiber surface, facilitating the deposition of AgNPs and induced increase of antibacterial properties [172,173]. Gittard et al. proposed the application of supercritical CO_2 [174]. Impregnated Co fabrics exhibited antifungal activity against *C. albicans.* It was shown that Ag precursors dissolved in supercritical CO_2 are able to diffuse into the Co fabric and react with Co. Impregnated Co fabrics exhibited antifungal activity against *C. albicans.*

PES [175], PA [62], polyethylene terephthalate (PET) [176], and PP [177] are the most widely used polymers in the textile field, with the incorporation of AgNPs. The interaction between inorganic particles and polymer is poor, therefore the textile surfaces had been activated by appropriate chemical or physicochemical treatment [178,179].

The AgNPs have been modified with ZnO [180] or TiO_2 [181–184,178] to increase photocatalytic and therefore antimicrobial properties of the textile materials. The AgNPs have also been modified with silica for application in textile materials [185]. Functionalization of textile materials using fabric dyes modified with silver or ZnO nanoparticles was reported by Pulit-Prociak et al. [180] and Mahltig et al. [186]. These composites had high impact on the viability and growth ability of *E. coli, Micrococcus luteus,* and *S. cerevisiae.* Such modified materials may be used to produce fabrics with antimicrobial properties that also effectively block harmful UV radiation. There are no reports of the incorporation of AuNPs for application in the textile field.

8.5.7 Polymer Composites With Metal Nanoparticles for Sensors

Metals, in particular gold and silver, have been extensively used to prepare nanomaterials for a broad range of sensors, since the binding event between the recognition element and the analyte can alter the physicochemical properties of the transducer metal nanoparticles [187,188]. These interactions can generate a detectable response signal associated with plasmon-resonance absorption, conductivity, redox behavior, etc. AuNPs are further highlighted for biosensors as they allow the immobilization of enzymes by amine groups and cysteine residues from the macromolecules that are known to bind strongly with the gold colloids maintaining their enzymatic and electrochemical activity for a considerably long time [189].

The fabrication of sensors based on metal nanoparticles embedded in polymeric matrices is of recent technological interest due to their improved behavior. One of the most extensively studied polymer composite sensors is based on electrodes with deposited conductive polymers, where metal nanoparticles act as "electron antennae" efficiently channeling electrons between the electrode and the electroactive species. The nanoparticles therefore promote better electron transfer between the electrode surface and the electrolyte [190]. It further avoids enzyme leakage in biosensors, while it allows rapid diffusion of substrates and products [191]. For instance, in dopamine sensors, the oxidation current is improved by absorption of this molecule on the metal surface [192]. The deposition of Au nanoclusters on nanoporous conductive polymers formed a composite nanoelectrode array sensor with increased electronic conductivity and mass transportation rate associated with the dopamine electrochemical reactions [193]. By this route, electrochemical sensors for glucose and H_2O_2 are further produced [191]. Examples of conductive polymers electropolymerized with AuNPs for glucose detection are polypyrrole, PANI, and (3-mercaptopropyl)-trimethoxysilane [188]. Graphene nanosheets can improve the behavior of these sensors, as reported for poly(diallyldimethyl ammonium chloride), functionalized for self-assembly of graphene/AuNPs heterostructure, enhancing H_2O_2 sensing [194]. The electrodeposition approach can be extended to nonconductive polymers such as chitosan, producing biosensors by adding specific enzymes active for glucose detection [195].

Another natural polymer biosensor was prepared using polyhydroxyalkanoate with embedded AuNPs cast on an indium–tin oxide glass plate for the analysis of artemisinin [189]. This system increased the differential pulse voltammetric peak current for artemisinin, producing greater sensitivities. Electrochemical sensors based on polymers and AgNPs have been recently studied mainly by adding carbon nanostructures. For instance, AgNPs /carbon nanotubes/chitosan films were used as the immobilization matrix for the enzymes to make a sensitive glucose biosensor [196]. DNA sensors based on AgNPs /poly(trans-3-(3-pyridyl) acrylic acid)/multiwalled carbon nanotube-modified glassy carbon electrode was reported using an electropolymerization method [197]. A similar DNA sensor can also be obtained using polydopamine and graphene compounds [198]. By adding the glucose oxidase enzyme in this system, a glucose biosensor was produced using a gold electrode [199].

Another polymer/metal nanocomposite sensor is based on the plasmonic response of metal nanoparticles. For instance, a 20-nm-thick swellable poly-2-vinylpyridine brush layer with AuNPs and supported on a gold substrate can be used as a highly sensitive pH-responsive nanosensor with short response times. This polymer is a stimuli-responsive system facilitating efficient transduction mechanisms for use in sensor applications [200]. A 50-nm shift in the plasmon-resonance position was observed as a result of changing pH caused by shrinking of the brush thickness from 22 to 7 nm [201]. A similar system was used to prepare layer-by-layer ultrathin pH-responsive plasmonic membranes of cross-linked poly-methacrylic acid hydrogels having gold nanorods [202]. The deswelling of the hydrogel film under a pH change from 8 to 5 causes a 21-nm blue shift of the longitudinal plasmon peak due to the increased side-by-side interactions of adjacent gold nanorods. By this approach polymer/silver nanoparticle sensors can be produced such as poly(vinyl alcohol)-capped AgNPs used for colorimetric hydrogen peroxide sensors [203]. AgNPs have the catalytic ability to decompose hydrogen peroxide that induces the degradation of AgNPs, changing the plasmon response of the system [204]. The same compounds can be used for the production of a universal thin film sensor for Hg ions in aqueous medium associated with a blue shift. This process occurred together with the decrease in the surface plasmon-resonance absorption upon interaction with mercury, making the sensing highly selective. Guar gum polysaccharide having AgNPs can be used for detection of aqueous ammonia based also on the surface plasmon resonance of the metal nanoparticles [205].

These metal nanoparticles can also be used to produce conductive polymer composite sensors. Using a layer-by-layer method, the production of vapor-sensitive thin-film resistors comprising AuNPs and different types of organic dendrimers such as polyphenylene, poly(propyleneimine), and poly(amidoamine) was reported [206]. In this case, AuNPs were utilized to provide electric conductivity to the composites while the dendrimers served to cross-link the nanoparticles and to provide sites for the selective sorption of analyte molecules.

8.5.8 Silver and Gold Nanocomposites for Medical Application

Due to these relevant properties, AuNPs and AgNPs have been extensively used in medical devices, drug delivery application, gene regulation, imaging (as contrast agents), antiinflammatory, and anticancer treatments (photodiagnostic and photothermal therapy) [207–209].

8.5.8.1 Gold Nanocomposites for Medical Devices

AuNPs have unique electrochemical and optical properties that have stimulated their application in the interfacial recognition of biological events with signal transduction and the design of biosensing devices exhibiting novel functions. The excellent biocompatibility, conductivity, catalytic properties, high surface-to-volume ratio, and high density of AuNPs facilitate their extensive application in the construction of optical, electrochemical, and piezoelectric biosensors with enhanced analytical performance with respect to other biosensor designs [210]. Similarly, and exploiting the physical properties of AuNP-based biosensors, they can be used to detect with greater efficiency, sensitivity, and rapidity, toxic ions, drugs, glucose, oligonucleotides, virus, biomarkers for cancer, other relevant diseases (e.g., α-fetoprotein, immunoglobulins, etc.) [211,212]. Nascimento et al. [213] developed a dengue genosensor using a positively charged gold nanoparticle–polyaniline hybrid composite (AuPANI) immobilized on a gold electrode with negatively charged serotype-specific primers (ST) as shown in Fig. 8–15 [213].

Other types of medical devices using AuNPs have been developed. Perez-Lopez et al. [214] prepared solid-supported gold catalysts by in situ generation of AuNPs in a polyethylene glycol (PEG)-grafted low cross-linked PS matrix. This study demonstrated that AuNPs in a small medical device can boost the effectiveness of current chemotherapies used to treat lung cancer cells without damaging healthy tissue [214]. In another approach, Alsaiari et al. [215] designed and made a smart membrane coating composed of colloidal lysozyme-templated gold nanoclusters as nanofillers into a poly(ethylene oxide)/poly(butylene terephthalate) amphiphilic polymer matrix. This mixed-matrix coating can successfully sense

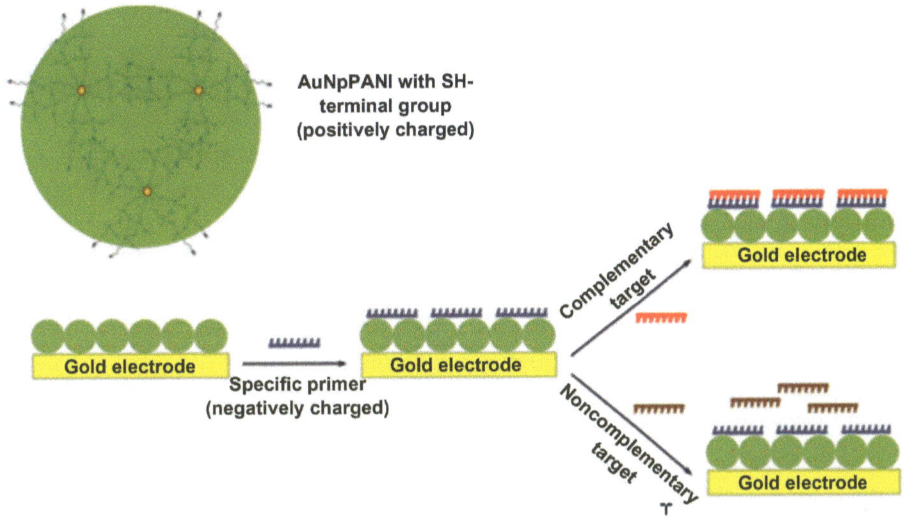

FIGURE 8–15 Schematic representation of the AuNpPANI nanocomposite and the biosensor construction. *Adapted from Copyright Elsevier. Nascimento HPO, Oliveira MDL, de Melo CP, Silva GJL, Cordeiro MT, Andrade CAS. An impedimetric biosensor for detection of dengue serotype at picomolar concentration based on gold nanoparticles-polyaniline hybrid composites. Colloids Surf B Biointerfaces 2011;86(2):414–19 [213].*

and inhibit bacterial contamination via a controlled release mechanism that is only triggered by bacteria [215]. The nanosystem was coated on a radiographic dental imaging device (photostimulable phosphor plate) that is prone to oral bacteria contamination. The antibiotic kanamycin was controllably released to inhibit bacterial growth instantly [215]. On the other hand, Tamai et al. [216] used two radiopaque gold markers on stents to facilitate the identification of both ends of the poly-L-lactic acid coronary stents in humans.

AuNPs may be used as tissue engineering and regenerative medicine (TERM), a multidisciplinary field that has been promising to develop viable therapeutic alternatives to organ and tissue transplants [217,218]. Shevach et al. [219] developed a hybrid patch based on natural extracellular matrix scaffold (3D porous templates to guide the cell growth) and AuNPs for cardiac tissue engineering. The researchers showed that the AuNPs were able to increase the conductivity of the natural material significantly without affecting the mechanical properties relevant for cardiac tissue engineering [219]. A thermosensitive conductive hydrogel based on CS with AuNPs was prepared to provide electrical cues. CS-AuNPs scaffolds supported viability, metabolism, migration, proliferation, and differentiation of mesenchymal stem cells along with the development of uniform cellular constructs for myocardial regeneration [220].

Polymer/AuNPs tissue engineering has been prepared to improve osteoblastic induction, skin care and regeneration, and musculoskeletal construct [221–223]. Akturk et al. [224] described the enhancement of degradation behavior and mechanical properties, the excellent in vitro biocompatibility, and stimulation of wound healing in collagen sponges with AuNPs as potent skin substitutes.

8.5.8.2 Silver Nanocomposites for Medical Devices

AgNPs have become of intense interest in biomedical applications because of their antibacterial, antifungal, antiviral, and antiinflammatory activity. AgNPs have been widely used for diagnosis [225], treatment [226], drug delivery [227], medical devices [228], coatings [229], wound dressings [226], and medical textiles [230]. Most of the applications of AgNPs and nanocomposites for medical devices have been used because of their antibacterial properties. The main purpose is to avoid infections during clinical procedures or implantation.

One of the most popular applications of AgNPs in medical devices is for coating medical catheters, including central venous catheters and neurosurgical catheters [231]. AgNPs can decrease the incidence of bacterial infection and complications after surgery, therefore they have been widely accepted for use in medical catheters. Examples of commercially available catheters containing AgNPs are Silverline (Spiegelberg GmbH & Co., KG, Germany) and ON-Q Silver Soaker (I-Flow Corporation, United States) used to prevent catheter-centered infections [231]. Most commonly, polyurethanes, already established as plastic catheters, have been modified with AgNPs [232].

Andara et al. [233] developed a new nanocomposite with AgNPs and diamond-like carbon as a surface coating for heart valves and stents. They found that the surface of the nanocomposite showed antithrombogenic and antibacterial properties. Ghanbari et al. [234] and Fu et al. [235] also constructed antibacterial multilayer films containing AgNPs, and investigated their antibacterial, mechanical, and hemodynamic properties in vitro for use in cardiovascular implant coating.

AgNPs have also been used in dental instruments and bandages. Resin composites with antibacterial activity may be useful to decrease the frequency of secondary caries around restorations [236]. Magalhães et al. [237] showed that incorporating AgNPs in light-activated endodontic cement provided a significantly enhanced antibactericidal activity against *Streptococcus milleri, S. aureus,* and *Enterococcus faecalis.*

AgNPs can be used for biodiagnosis of the detection of serum p53 protein as a biomarker of head and neck squamous cell carcinoma [238,239] as well as to detect the interaction between amyloid β-derived diffusible ligands (ADDL) [240] and the antiADDL antibody, which are related to the development of Alzheimer's disease. AgNPs have also been used in nanocomposites for tissue engineering taking advantage of their antibacterial capacity. Alginate/hydroxyapatite/AgNPs composites have been used for bone regeneration scaffolds [241,242]. Recently, the same author developed an injectable paste for bone regeneration composed of nanocomposite microbeads of alginate, chitlac, hydroxyapatite, and AgNPs [243]. These microbeads displayed bactericidal effects against *S. aureus, P. aeruginosa,* and *S. epidermidis,* and did not exert any cytotoxic effect in osteoblast-like cells.

In addition to the antimicrobial activity provided by AgNPs, there may be additional benefits to the wound healing cascade [244]. The sequence of events that initiate reepithelialization and subsequently control the process is mediated by matrix metalloproteinases [245,246]. Bhowmick and Koul [247] developed an elastic, nonadhesive, and antimicrobial hydrogel PVA scaffold loaded with AgNPs using the freeze–thaw method. The scaffold was able to maintain a moist wound environment for 96 h, while exhibiting a sustained release of AgNPs [247].

Several polymer matrices have been studied with the incorporation of AgNPs for applications as implants and drug delivery systems, such as nanocomposites based on gelatin [248], polyurethane [249], agarose-chitosan [250], and poly (e-caprolactone) [251].

8.6 Conclusions

By embedding gold and AgNPs into polymer matrices a novel family of multifunctional materials can be developed taking advantage of both the intrinsic properties of the metal at the nanoscale and the good behavior of organic macromolecules. These nanocomposites allow the technological application of unique properties of metals emerging at the nanoscale as the polymer composite can be produced at the macroscale, and even processed by the standard equipment of the plastic industry such as extrusion. The silver and AuNPs can improve some intrinsic property of the polymer matrix, for instance the electric conductivity in conductive polymers, or even add new functionalities such as antimicrobial capacity and bioactivity in medical applications. It is noteworthy that the processing method, morphology, size, and surface characteristics of the nanoparticles are key aspects to consider in the preparation of the polymer nanocomposites and the final properties that are desired. For instance, optical properties and bioactivity depend on morphology and surface charge, respectively. The effect between polymer and metal nanoparticles opens the development of novel applications such as water treatment, active packaging, textiles, sensors, and medical devices, among other important technologies.

Acknowledgments

P.A. Zapata acknowledges the financial support of FONDECYT Regular N° 1170226.

L. Tamayo is grateful for funding from FONDECYT Iniciación grant 11160230 and PAI-CONICYT 79170015.

References

[1] Rong M, Zhang M, Pan S, Friedrich K. Interfacial effects in polypropylene-silica nanocomposites. J App Polym Sci 2004;92:1771–81.

[2] Kontuo E, Niaounakis M. Thermo-mechanical properties of LLDPE/SiO_2 nanocomposites. Polymer 2006;47:1267–80.

[3] Kontuo X, Rosi NL, Wang Y, Huo F, Mirkin CA. Asymmetric functionalization of gold nanoparticles with oligonucleotides. J Am Chem Soc 2006;128(29):9286–7.

[4] Niemeyer CM, Ceyhan B, Noyong M, Simon U. Bifunctional DNA–gold nanoparticle conjugates as building blocks for the self-assembly of cross-linked particle layers. Biochem Biophys Res Commun 2003;311 (4):995–9.

[5] Pruneanu S, Coroş M, Pogacean F. Bio-functionalized metallic nanoparticles with applications in medicine. Handbook Nanoparticle. 2016. p. 803–17.

[6] Brust M, Walker M, Bethell D, Schiffrin DJ, Whyman R. Synthesis of thiol-derivatised gold nanoparticles in a two-phase liquid–liquid system. J Chem Soc Chem Commun 1994;7:801–2.

[7] Bahadur KCR, Thapa B, Bhattarai N. Gold nanoparticle-based gene delivery: promises and challenges. Nanotechnol Rev 2014;3:269.

[8] Zou CE, Yang B, Bin D, Wang J, Li S, Yang P, et al. Electrochemical synthesis of gold nanoparticles decorated flower-like graphene for high sensitivity detection of nitrite. J Colloid Interface Sci 2017;488:135–41.

[9] Lai J, Niu W, Luque R, Xu G. Solvothermal synthesis of metal nanocrystals and their applications. Nano Today 2015;10(2):240–67.

[10] Xu H, Zeiger BW, Suslick KS. Sonochemical synthesis of nanomaterials. Chem Soc Rev 2013;42(7):2555–67.

[11] Tsuji T, Kikuchi M, Kagawa T, Adachi H, Tsuji M. Morphological changes from spherical silver nanoparticles to cubes after laser irradiation in acetone–water solutions via spontaneous atom transportation process. Colloids Surf A Physicochem Eng Asp 2017;529:33–7.

[12] Amendola V, Polizzi S, Meneghetti M. Laser ablation synthesis of gold nanoparticles in organic solvents. J Phys Chem B 2016;110(14):7232–7.

[13] Wender H, Andreazza ML, Correia RR, Teixeira SR, Dupont J. Synthesis of gold nanoparticles by laser ablation of an Au foil inside and outside ionic liquids. Nanoscale 2011;3(3):1240–5.

[14] Saitow KI, Okamoto Y, Yano YF. Fractal of gold nanoparticles controlled by ambient dielectricity: synthesis by laser ablation as a function of permittivity. J Phys Chem C 2012;116(32):17252–8.

[15] Kabashin AV, Meunier M, Kingston C, Luong JH. Fabrication and characterization of gold nanoparticles by femtosecond laser ablation in an aqueous solution of cyclodextrins. J Phys Chem B 2003;107(19):4527–31.

[16] Petersen S, Barchanski A, Taylor U, Klein S, Rath D, Barcikowski S. Penetratin-conjugated gold nanoparticles – design of cell-penetrating nanomarkers by femtosecond laser ablation. J Phys Chem C 2010;115(12):5152–9.

[17] Petersen S, Barcikowski S. In situ bioconjugation: single step approach to tailored nanoparticle-bioconjugates by ultrashort pulsed laser ablation. Adv Funct Mater 2009;19(8):1167–72.

[18] Ingh P, Kim YJ, Wang C, Mathiyalagan R, Yang DC. The development of a green approach for the biosynthesis of silver and gold nanoparticles by using Panax ginseng root extract, and their biological applications. Artif Cells Nanomed Biotechnol 2016;44(4):1150–7.

[19] Shanthi S, Jayaseelan BD, Velusamy P, Vijayakumar S, Chih CT, Vaseeharan B. Biosynthesis of silver nanoparticles using a probiotic Bacillus licheniformis Dahb1 and their antibiofilm activity and toxicity effects in Ceriodaphnia cornuta. Microb Pathog 2016;93:70–7.

[20] Balakumaran MD, Ramachandran R, Balashanmugam P, Mukeshkumar DJ, Kalaichelvan PT. Mycosynthesis of silver and gold nanoparticles: optimization, characterization and antimicrobial activity against human pathogens. Microbiol Res 2016;182:8–20.

[21] Ramakrishna M, Babu DR, Gengan RM, Chandra S, Rao GN. Green synthesis of gold nanoparticles using marine algae and evaluation of their catalytic activity. J Nanostruct Chem 2016;6(1):1–13.

[22] Zhang X, Qu Y, Shen W, Wang J, Li H, Zhang Z, et al. Biogenic synthesis of gold nanoparticles by yeast *Magnusiomyces ingens* LH-F1 for catalytic reduction of nitrophenols. Colloids Surf A Physicochem Eng Asp 2016;497:280–5.

[23] Corbierre MK, Cameron NS, Sutton M, Mochrie SG, Lurio LB, Rühm A, et al. Polymer-stabilized gold nanoparticles and their incorporation into polymer matrices. J Am Chem Soc 2001;123(42):10411–12.

[24] Templeton AC, Wuelfing WP, Murray RW. Monolayer-protected cluster molecules. Acc Chem Res 2000;33(1):27–36.

[25] Brust M, Fink J, Bethell D, Schiffrin DJ, Kiely C. Synthesis and reactions of functionalized gold nanoparticles. J Chem Soc Chem Commun 1995;16:1655–6.

[26] Shenhar R, Norsten TB, Rotello VM. Polymer-mediated nanoparticle assembly: structural control and applications. Adv Mater 2005;17(6):657–69.

[27] Frankamp BL, Boal AK, Rotello VM. Controlled interparticle spacing through self-assembly of Au nanoparticles and poly (amidoamine) dendrimers. J Am Chem Soc 2002;124(51):15146–7.

[28] Morales-Avila E, Ferro-Flores G, Ocampo-García BE, López-Téllez G, López-Ortega J, Rogel-Ayala DG, et al. Antibacterial efficacy of gold and silver nanoparticles functionalized with the ubiquicidin (29–41) antimicrobial peptide. J Nanomater 2017; ID 5831959.

[29] Taheri S, Cavallaro A, Christo SN, Smith LE, Majewski P, Barton M, et al. Substrate independent silver nanoparticle based antibacterial coatings. Biomaterials 2014;35(16):4601–9.

[30] Tamayo LA, Zapata PA, Vejar ND, Azócar MI, Gulppi MA, Zhou X, et al. Release of silver and copper nanoparticles from polyethylene nanocomposites and their penetration into *Listeria monocytogenes*. Mater Sci Eng C 2014;40:24–31.

[31] Fontana L, Bassetti M, Battocchio C, Venditti I, Fratoddi I. Synthesis of gold and silver nanoparticles functionalized with organic dithiols. Colloids Surf A Physicochem Eng Asp 2017;532:282–9.

[32] Kumar AP, Depan D, Tomer NS, Singh RP. Nanoscale particles for polymer degradation and stabilization—trends and future perspectives. Prog Polym Sci 2009;34(6):479–515.

[33] Zhang T, Wang W, Zhang D, Zhang X, Ma Y, Zhou Y, et al. Biotemplated synthesis of gold nanoparticle–bacteria cellulose nanofiber nanocomposites and their application in biosensing. Adv Funct Mater 2010;20(7):1152–60.

[34] Li D, He Q, Cui Y, Li J. Fabrication of pH-responsive nanocomposites of gold nanoparticles/poly (4-vinylpyridine). Chem Mater 2007;19(3):412–17.

[35] Breitwieser D, Moghaddam MM, Spirk S, Baghbanzadeh M, Pivec T, Fasl H, et al. In situ preparation of silver nanocomposites on cellulosic fibers–microwave vs. conventional heating. Carbohydr Polym 2013;94(1):677–86.

[36] Yagci Y, Sangermano M, Rizza G. In situ synthesis of gold-cross-linked poly (ethylene glycol) nanocomposites by photoinduced electron transfer and free radical polymerization processes. Chem Commun 2008;24:2771–3.

[37] Sangermano M, Yagci Y, Rizza G. In situ synthesis of silver – epoxy nanocomposites by photoinduced electron transfer and cationic polymerization processes. Macromolecules 2007;40(25):8827–9.

[38] Tamboli MS, Kulkarni MV, Patil RH, Gade WN, Navale SC, Kale BB. Nanowires of silver–polyaniline nanocomposite synthesized via in situ polymerization and its novel functionality as an antibacterial agent. Colloids Surf B Biointerfaces 2012;92:35–41.

[39] Vodnik VV, Božanić DK, Džunuzović E, Vuković J, Nedeljković JM. Thermal and optical properties of silver–poly (methylmethacrylate) nanocomposites prepared by in-situ radical polymerization. Eur Polym J 2010;46(2):137–44.

[40] Yeo SY, Lee HJ, Jeong SH. Preparation of nanocomposite fibers for permanent antibacterial effect. J Mater Sci 2003;38(10):2143–7.

[41] Madaria AR, Kumar A, Zhou C. Large scale, highly conductive and patterned transparent films of silver nanowires on arbitrary substrates and their application in touch screens. Nanotechnology 2011;22 (24):245201.

[42] de Azeredo, Henriette MC. Nanocomposites for food packaging applications. Food Res Int 2009;42:1240–53.

[43] Kumar R, Münstedt H. Silver ion release antimicrobial polyamide/silver composites. Biomaterials 2005;26:2081–8.

[44] Zhou Y, Kong Y, Kundu S, Cirillo JD, Liang H. Antibacterial activities of gold and silver nanoparticles against *Escherichia coli* and bacillus Calmette-Guérin. J Nanobiotechnology 2012;10(1):19.

[45] Pissuwan D, Cortie CH, Valenzuela SM, Cortie MB. Functionalised gold nanoparticles for controlling pathogenic bacteria. Trends Biotechnol 2010;28(4):207–13.

[46] Badwaik VD, Vangala LM, Pender DS, Willis CB, Aguilar ZP, Gonzalez MS, et al. Size-dependent antimicrobial properties of sugar-encapsulated gold nanoparticles synthesized by a green method. Nanoscale Res Lett 2012;7(1):623.

[47] Rai A, Prabhune A, Perry CC. Antibiotic mediated synthesis of gold nanoparticles with potent antimicrobial activity and their application in antimicrobial coatings. J Mater Chem 2010;20 (32):6789–98.

[48] Zhang W, Li Y, Niu J, Chen Y. Photogeneration of reactive oxygen species on uncoated silver, gold, nickel, and silicon nanoparticles and their antibacterial effects. Langmuir 2013;29(15):4647–51.

[49] Cui Y, Zhao Y, Tian Y, Zhang W, Lü X, Jiang X. The molecular mechanism of action of bactericidal gold nanoparticles on *Escherichia coli*. Biomaterials 2012;33(7):2327–33.

[50] Li J, Zhou H, Qian S, Liu Z, Feng J, Jin P, et al. Plasmonic gold nanoparticles modified titania nanotubes for antibacterial application. Appl Phys Lett 2014;104:261110.

[51] Chamundeeswari M, Sobhana SSL, Jacob JP, Kumar MG, Devi MP, Sastry TP, et al. Preparation, characterization and evaluation of a biopolymeric gold nanocomposite with antimicrobial activity. Biotechnol Appl Biochem 2010;55(1):29–35.

[52] Barkoula NM, Alcock B, Cabrera NO, Peijs T. Fatigue properties of highly oriented polypropylene tapes and all-polypropylene composites. Polym Polym Compos 2008;16(2):101–13.

[53] Boomi P, Prabu HG. Synthesis, characterization and antibacterial analysis of polyaniline/Au-Pd nanocomposite. Colloids Surf A Physicochem Eng Asp 2013;429:51–9.

[54] Sawant SN, Selvaraj V, Prabhawathi V, Doble M. Antibiofilm properties of silver and gold incorporated PU, PCLm, PC and PMMA nanocomposites under two shear conditions. PLoS One 2013;8(5):1–9.

[55] Russo T, Gloria A, De Santis R, D'Amora U, Balato G, Vollaro A, et al. Preliminary focus on the mechanical and antibacterial activity of a PMMA-based bone cement loaded with gold nanoparticles. Bioact Mater 2017;2(3):156–61.

[56] Zare Y, Shabani I. Polymer/metal nanocomposites for biomedical applications. Mat Sci Eng C 2016;60:195–203.

[57] Hoon S, Hwang Hwan Y, Sung Chul Y. Antibacterial properties of padded PP/PE nonwovens incorporating nano-sized silver colloids. J Mat Sci 2005;40:5413–18.

[58] Shearer AE, Paik JS, Hoover DG, Haynie SL, Kelly MJ. Potential of an antibacterial ultraviolet-irradiated nylon film. Biotechnol Bioeng 2000;67:141–6.

[59] Haide A, Haider S, Kang I, Kumar A, Kummara MR, Kamal T, et al. A novel use of cellulose based filter paper containing silver nanoparticles for its potential application as wound dressing agent. Int J Biolog Macrom 2018;108:455–61.

[60] Marambio-Jones C, Hoek EMV. A review of the antibacterial effects of silver nanomaterials and potential implications for human health and the environment. J Nanopart Res 2010;12:1531–51.

[61] Durán N, Durán M, de Jesus MB, Seabra AB, Fávaro WJ, Nakazato G. Silver nanoparticles: a new view on mechanistic aspects on antimicrobial activity. Nanomedicine: Nanotec, Biol Med 2016;12 (3):789–99.

[62] Ilic V, Saponjic Z, Vodnik V, Molina R, Dimitrijevic S, Jovancic P, et al. Antifungal efficiency of corona pretreated polyester and polyamide fabrics loaded with Ag nanoparticles. J Mat Sci 2009;44:3983–90.

[63] Damm C, Münstedt H, Rösch A. Long-term antimicrobial polyamide6/silver-nanocomposites. J Mat Sci 2007;42:6067–73.

[64] Sang YY, Sung HG. Preparation and characterization of polypropylene/silver nanocomposite fibers. Polym Int 2003;52:1053–7.

[65] Falleta E, Bonini M, Fratini E, Lo Nostro A, Pesavento G, Becheri A, et al. Cluster of poly(acrilates) and silver nanoparticles: structure and applications for antimicrobial fabrics. J Phys Chem C 2008;112:11758–66.

[66] Youssef AM, Abdel-Aziz MS. Preparation of polystyrene nanocomposites based on silver nanoparticles using marine bacterium for packaging. Polym Plast Technol Eng 2013;52:607–13.

[67] Zapata PA, Tamayo LA, Azócar MI, Rabagliati FM, Paez M. Polyethylene/silver-nano fi ber composites: a material for antibacterial films. Mat Sci Eng C 2016;69:1282–9.

[68] Castro-Mayorga L, Freitas F, Reis M, Prieto A, Lagaron JM Biosynthesis of silver nanoparticles and poly-hydroxyburyrate nanocomposites of interest in antimicrobial applications, Int J Biol Macromol <https://doi.org/10.1016/j.ijbiomac.2017.12.007>.

[69] Pinto RJB, Marques P, Neto CP, Trindade T, Daina S, Sadocco P. Antibacterial activity of nanocomposites of silver and bacterial or vegetable cellulosic fibers. Acta Biomat 2009;5:2279–89.

[70] Damm C, Münstedt H, Rösch A. The antimicrobial efficacy ofpolyamide 6/silver-nano- and microcomposites. Mater Chem Phys 2008;108:61–6.

[71] Radheshkumar C, Munstedt H. Antimicrobial polymers from polypropylene/silver composites-Ag release measurement by anode stripping voltammetry. React Funct Polym 2006;66:780–8.

[72] Zapata PA, Tamayo L, Páez M, Cerda E, Azócar I, Rabagliati FM. Nanocomposites based on polyethylne and nanosilver particles produced by metallocenic "in situ" polymerization: synthesis, characterization and antimicrobial behavior. Eur Polym J 2011;47:1541–9.

[73] Guin D, Manorama SV, Latha JNL, Sinh S. Photoreduction of silver on bare and colloidal TiO_2 nanoparticles/nanotubes: synthesis, characterization, and tested for antibacterial outcome. J Phys Chem C 2007;111:13393–7.

[74] Kong H, Jang J. Antibacterial properties of novel poly (methyl methacrylate) nanofiber containing silver nanoparticles. Langmuir 2008;24(5):2051–6.

[75] Thomas V, Namdeo M, Murali Y, Mohan S, Bajpai M. Review on polymer, hydrogel and microgel metal nanocomposites: a facile nanotechnological approach. J Macromol Sci Part A: Pure Appl Chem. 2007;45 (1):107–19.

[76] Shao W, Liu X, Min H, Dong G, Feng Q, Zuo S. Preparation, characterization, and antibacterial activity of silver nanoparticle-decorated graphene oxide nanocomposite. ACS Appl Mater Interf 2015;7:6966–73.

[77] Alshammari A, Narayana V. Supported Gold Nanoparticles as Promising Catalysts, Catalytic Application of Nano-Gold Catalysts N. Kumar. Intech Open 2016; <https://doi.org/10.5772/64394>. Available from: <https://www.intechopen.com/books/catalytic-application-of-nano-gold-catalysts/supported-gold-nanoparticles-as-promising-catalysts>.

[78] Hvolbæk B, Janssens TVW, Clausen BS, Falsig H, Christensen CH, Nørskov JK. Catalytic activity of Au nanoparticles. Nano Today 2007;2(4):14–18.

[79] Ramirez O, Bonardd S, Saldías C, Radic D, Leiva Á. Biobased chitosan nanocomposite films containing gold nanoparticles: obtainment, characterization, and catalytic activity assessment. ACS Appl Mater Interfaces 2017;9(19):16561–70.

[80] Tan NPB, Lee CH, Li P. Green synthesis of smart metal/polymer nanocomposite particles and their tuneable catalytic activities. Polymers (Basel) 2016;8(105):1–14.

[81] Bibi S, Price GJ, Yasin T, Nawaz M. Eco-friendly synthesis and catalytic application of chitosan/gold/carbon nanotube nanocomposite films. RSC Adv 2016;6(65):60180–6.

[82] Esumi K, Houdatsu H, Yoshimura T. Comparison of gold-PAMAM nanocomposites for and dendrimer action antioxidant. J Japan Soc Colour Mater 2004;77(11):487–92.

[83] Li Y, Li G, Li W, Yang F, Liu H. Greenly Synthesized gold–alginate nanocomposites catalyst for reducing decoloration of azo-dyes. Nano 2015;10(8):1550108.

[84] Jia L, Zhou T, Xu J, Li X, Dong K, Huang J, et al. The enhanced catalytic activities of asymmetric Au-Ni nanoparticle decorated halloysite-based nanocomposite for the degradation of organic dyes. Nanoscale Res Lett 2016;11(71):1–7.

[85] Cho HR, Lee JH. Surface-selective deposition of a reusable gold catalyst on a thin film of metal hydroxide nanocrystals. Eur J Inorg Chem 2015;2015(4):640–5.

[86] Jana S, Ghosh S, Nath S, Pande S, Praharaj S, Panigrahi S, et al. Show more synthesis of silver nano shell-coated cationic polystyrene beads: a solid phase catalyst for the reduction of 4-nitrophenol. App Cat A: Gen 2006;313:41–8.

[87] Saghita P, Sarada K, Muraleedharan K. One-pot synthesis of poly vinyl alcohol (PVA) supported silver nanoparticles and its efficiency in catalytic reduction of methylene blue. Trans Nonferrous Met Soc China 2016;26:2693–700.

[88] Zha T, Song L, Chen P, Nie W, Zhou Y. Nonsolvent/solvent-induced phaseseparation to multi-porous sulfonated polystyrene/chitosan/silver particlesand their application in adsorbing chromium ion (III) and reduction of methylene blue. Colloids Surf A Physicochem Eng Asp 2015;481:423–30.

[89] You J, Zhao C, Cao J, Zhou J, Zhang L. Fabrication of high-density silver nanoparticles on the surface of alginate microspheres for application in catalytic reaction. J Mater Chem A 2014;2:8491–9.

[90] Tian Q, Yu X, Zhang L, Yu D. Monodisperse raspberry-like multihollow polymer/Ag nanocomposite microspheres for rapid catalytic degradation of methylene blue. J Coll Interf Sci 2017;491:294–304.

[91] Zhai L, McCoullough RD. Regioregular polythiophene/gold nanoparticle hybrid materials. J Mat Chem 2004;14:141–3.

[92] Paruzel B, Pavlova E, Pfleger J, Šlouf M, Halašová K, Mikmeková S. Poly(3-hexylthiophene)/gold nanoparticle nanocomposites: relationship between morphology and electrical conductivity. Chem Papers 2017;71:401–8.

[93] Sun L, Shi Y, Li B, Chu L, He Z, Li J. Synthesis and characterization of polypyrrole/Au nanocomposites by microemulsion polymerization. Colloids Surf A Physicochem Eng Asp 2012;397:8–11.

[94] Afzal A, Akhta MJ, Nadeem M, Hassan MM. Investigation of structural and electrical properties of polyaniline/gold nanocomposites. J. Phys Chem C 2009;113(40):17560–5.

[95] Pillalamarri SK, Blum FD, Tokuhiro AT, Bertino MF. One-pot synthesis of polyaniline – metal nanocomposites. Chem Mat 2005;17:5941–4.

[96] Shin NR, Choi SH, Kim JY. Highly conductive PEDOT: PSS electrode films hybridized with gold-nanoparticle-doped-carbon nanotubes. Synthetic Met 2014;192:23–8.

[97] Elhalawany N, Elmelegy H, Nayfeh F. Synthesis, characterization and electrical properties of highly conductive polyaniline/gold and/or platinum nanocomposites. Synthetic Met 2015;205:145–52.

[98] Lonjon A, Demont L, Dantras P, Lacabanne E. Structural and electrical properties of gold nanowires/P(VDF-TrFE) nanocomposites. J Phys D: App Phys 2010;43(34):1–7.

[99] Liu Z, Li M, Turyanska L, Makarovsky O, Patanè A, Wu W, et al. Self-assembly of electrically conducting biopolymer thin films by cellulose regeneration in gold nanoparticle aqueous dispersions. Chem Mat 2010;22(8):2675–80.

[100] Faria-Tischer CA, Costa I, Tozetti LH, Dall'Antonia M. Structure and effects of gold nanoparticles in bacterial cellulose–polyaniline conductive membranes. RSC Adv 2016;12:9571–80.

[101] Araki T, Jiu J, Nogi M, Koga H, Nagao S, Sugahara T, et al. Low haze transparent electrodes and highly conducting air dried films with ultra-long silver nanowires synthesized by one-step polyol method. Nano Res 2014;7:236–45.

[102] Abbasi NM, Yu H, Wang L, ul-Abdin. Preparation of silver nanowires and their application in conducting polymer nanocomposites. Mat Chem Phys 2015;166:1–15.

[103] Afzal AB, Akhtar M, Nadeem M, Ahmad M, Hassan M, Yasin T, et al. Structural and electrical properties of polyaniline/silver nanocomposites. J Phys D Appl Phys 2009;42:015411.

[104] Gupta K, Jana PC, Meikapa AK. Optical and electrical transport properties of polyaniline–silver nanocomposite. Synthetic Met 2010;160:1566–73.

[105] Khalil A, Hassan M, Ward A. Novel nanofibrillated cellulose/polyvinylpyrrolidone/silver nanoparticles films with electrical conductivity properties. Carbohydr Polym 2017;157:503–11.

[106] Mahendia S, Tomar AK, Kumar S. Electrical conductivity and dielectric spectroscopic studies of PVA–Ag nanocomposite films. J. Alloys Comp 2010;508(2):406–11.

[107] Yu DG, Lin WC, Lin CH, Chang LM, Yang MC. An in situ reduction method for preparing silver/poly (vinyl alcohol) nanocomposite as surface-enhanced Raman scattering (SERS)-active substrates. Mater Chem Phys 2007;101(1):93–8.

[108] Mukherjee PS, Das AK, Dutta B, Meikap AK. Role of silver nanotube on conductivity, dielectric permittivity and current voltage characteristics of polyvinyl alcohol-silver nanocomposite film. J Phys Chem Solids 2017;111:266–73.

[109] Zhang Z, Wang H, Li S, Li L, Li D. Transparent and flexible cellulose nanofibers/silver nanowires/acrylic resin composite electrode. Compos Part A Appl Sci Manuf 2015;76:309–15.

[110] Park Y, Bormann L, Müller-Meskamp L, Vandewal K. Efficient flexible organic photovoltaics using silver nanowires and polymer based transparent electrodes. Org Electron 2016;36:68–72.

[111] Lee J, Lee P, Lee HB, Hong S, Lee I, Yeo J, et al. Room-temperature nanosoldering of a very long metal nanowire network by conducting-polymer-assisted joining for a flexible touch-panel application. Adv Funct Mat 2013;23:4171–6.

[112] Kazeminezhad I, Barnes A, Holbrey J, Seddon K, Schwarzacher W. Templated electrodeposition of silver nanowires in a nanoporous polycarbonate membrane from a nonaqueous ionic liquid electrolyte. Appl Phys A 2007;86:373–5.

[113] Hu W, Niu X, Li L, Yun S, Yu Z, Pei Q. Intrinsically stretchable transparent electrodes based on silver-nanowireecrosslinked-polyacrylate composites. Nanotechnology 2012;23:344002.

[114] Xu F, Zhu Y. Highly conductive and stretchable silver nanowire conductors. Adv Mater 2012;24:5117–22.

[115] Fallahi H, Azizi H, Ghasemi I, Karrabi M. Preparation and properties of electrically conductive, flexible and transparent silver nanowire/poly (lactic acid) nanocomposites. Org Electron 2017;44:74–84.

[116] Rivière L, Lonjon A, Dantras E, Lacabanne C, Olivier P, Gleizes NR. Silver fillers aspect ratio influence on electrical and thermal conductivity in PEEK/Ag nanocomposites. Eur Polym J 2016;85:115–25.

[117] Zhi J, Zhao W, Liu XY, Chen AR, Liu ZQ, Huang FQ. Highly conductive ordered mesoporous carbon based electrodes decorated by 3D graphene and 1D silver nanowire for flexible supercapacitor. Adv Funct Mater 2014;24:2013–19.

[118] Yu ZA, Li C, Abbitt D, Thomas J. Flexible sandwich-like Ag-nanowire/PEDOT: PSS-nanopillar/MnO2 high performance supercapacitors. J Mater Chem A 2014;2:10923–9.

[119] Yuksel R, Alpugan E, Unalan H. Coaxial silver nanowire/polypyrrole nanocomposite supercapacitors. Organ Electron 2018;52:272–80.

[120] Mulvihill M, Tao A, Benjauthrit K, Arnold J, Yang P. Surface-enhanced Raman spectroscopy for trace arsenic detection in contaminated water. Angew Chem Int Ed 2008;47:6456–60.

[121] Sajanlal PR, Pradeep T. Electric-field-assisted growth of highly uniform and oriented gold nanotriangles on conducting glass substrates. Adv Mat 2008;20(5):980–3.

[122] Liu M, Wang Z, Zong S, Chen H, Zhu D, Wu L, et al. SERS detection and removal of mercury(II)/silver(I) using oligonucleotide-functionalized core/shell magnetic silica sphere@Au nanoparticles. ACS Appl Mater Interface 2014;6:7371–9.

[123] Zhang Y, Wu B, Xu H, Liu H, Wang M, He Y, et al. Nanomaterials-enabled water and wastewater treatment. Nano Impact 2016;3–4:22–39.

[124] Lee YF, Nan FH, Chen MJ, Wu HY, Ho CW, Chen YY, et al. Detection and removal of mercury and lead ions by using gold nanoparticle based gel membrane. Anal Meth 2012;4:1709–17.

[125] Lu F, Astruc D. Nanomaterials for removal of toxic elements from water. Coord Chem Rev 2018;356:147–64.

[126] Sumesh E, Bootharaju MS, Pradeep AT. A practical silver nanoparticle-based adsorbent for the removal of Hg2 + from water. J Hazard Mater 2011;189(1–2):450–7.

[127] Wei X, Bhojappa S, Lin LS, Viadero RC. Performance of nano-magnetite for removal of selenium from aqueous solutions. Environ Eng Sci 2012;29(6):526–32.

[128] Sharma V, Yngard RA, Lin Y. Silver nanoparticles: green synthesis and their antimicrobial activities. Adv Coll Interface Sci 2009;145:83–96.

[129] Jain P, Pradeep T. Potential of silver nanoparticle-coated polyurethane foam as an antibacterial water filter. Biotechnol Bioeng 2005;90(1):59–63.

[130] Le Pape H, Solano-Serena F, Contini P, Devillers C, Maftah A, Leprat P. Involvement of reactive oxygen species in the bactericidal activity of activated carbon fibre supporting silver: bactericidal activity of ACF (Ag) mediated by ROS. J Inorg Biochem 2004;98(6):1054–60.

[131] Rosa LR, Rosa RD, Silva da Veiga MAM. Colloidal silver and silver nanoparticles bioaccessibility in drinking water filters. J Environ Chem Eng 2016;4:3451–8.

[132] Waisberg M, Black WD, Waisberg CM, Hale B. The effect of pH, time and dietary source of cadmium on the bioaccessibility and adsorption of cadmiumto/from lettuce (Lactuca sativa L. cv Ostinata). Food Chem Toxicol 2014;42:835–42.

[133] Abu-Saied MA, Taha TH, El-Deeb NM, Hafez EE. Polyvinyl alcohol/sodium alginate integrated silver nanoparticles as probable solution for decontamination of microbes contaminated water. Int J Biol Macromol 2017;. Available from: https://doi.org/10.1016/j.ijbiomac.2017.10.047.

[134] Djerahov L, Vasileva P, Karadjova I, Kurakalva RM, Aradh KK. Chitosan film loaded with silver nanoparticles-sorbent for solid phase extraction of Al(III), Cd(II), Cu(II), Co(II), Fe(III), Ni(II), Pb(II) and Zn(II). Carbohydr Polym 2016;147:45–52.

[135] Rhim JW, Park HM, Ha CS. Bio-nanocomposites for food packaging applications. Prog Polym Sci 2013;38:1629–52.

[136] López C, Sánchez C, Battle R, Nerín C. Development of flexible antimicrobial films using essential oils as active agents. J Agr Food Chem 2007;55:8814–24.

[137] Zapata PA, Palza H. Polyethylene-based biocomposites and bionanocomposites. In: Visakh PM, Luftl S, editors. Polyethylene-based bio-and nanocomposite for packaging applications. Wiley; 2016. <https://doi.org/10.1002/9781119038467.ch10>.

[138] Martinez-Abad A, Lagaron JM, Ocio MJ. Development and characterization of silver-based antimicrobial ethylene-vinyl alcohol copolymer (EVOH) films for food-packaging applications. J Agr Food Chem 2012;60:5350–9.

[139] Carbone M, Donia D, Sabbatella G, Antiochia R. Silver nanoparticles in polymeric matrices for fresh food packaging. J King Saud Univ – Sci 2016;28:273–9.

[140] Ibarra-Alonso MC, Sánchez-Valdes S, Ramírez-Vargas E, Fernandez-Tavizón S, Romero-Garcia J, Ledezma-Perez AS, et al. Preparation and characterization of polyethylene/clay/silver nanocomposites using functionalized polyethylenes as an adhesion promoter. J Adhes Sci Technol 2015;29:1911–23.

[141] An J, Zhang M, Wang S, Tang J. Physical, chemical and microbiological changes in stored green asparagus spears as affected by coating of silver nanoparticles-PVP. LWT – Food Sci Technol 2008;41:1000–7.

[142] Li H, Li F, Wang L, Sheng J, Xin X, Zhao L, et al. Effect of nano-packing on preservation quality of Chinese jujube(Ziziphus jujube Mill. var. inermis (Bunge) Rehd). Food Chem 2009;114:547–52.

[143] Emamifar A, Kadivar M, Shahedi M, Soleimanian-Zad S. Effect of nanocomposite packaging containing Ag and ZnO on inactivation of Lactobacillus plantarum in orange juice. Food Control 2011;22:408–13.

[144] Zhou L, Lu S, He G, He Q, Shi B. Effect of PE/Ag_2O nano-packaging on the quality of apple slices. J Food Quality 2011;34:171–6.

[145] Motlagh NV, Mosavian MTH, Mortazavi SA. Effect of polyethylene packaging modified with silver particles on themicrobial, sensory and appearance of dried barberry. Packaging Techn Sci 2012;26:39–49.

[146] Yang FM, Li HM, Li F, Xin ZH, Zhao LY, Zheng YH, et al. Effect of nano-packing on preservation quality of fresh strawberry (Fragaria ananassa Duch. Cv Fengxiang) during storage at 4ºC. J Food Chem 2010;75:C236–40.

[147] Mahdi SS, Vadood R, Nourdahr R. Study on the antimicrobial effect of nanosilver tray packaging of minced beef at refrigerator temperature. Global Vet 2012;9:284–9.

[148] Panea B, Ripoll G, González J, Fernández-Cuello A, Albertí P. Effect of nanocomposite packaging containing different proportions of ZnO and Ag on chicken breast meat quality. J Food Eng 2014;123:104–12.

[149] Llorens A, Lloret E, Picouet PA, Trbojevich R, Fernandez A. Metallic-based micro and nanocomposites in food contact materials and active food packaging. Trends Food Sci Techn 2012;24:19–29.

[150] Kamisoglu K, Aksoy EA, Hasirci N, Bac N. Preparation and characterization of antibacterial zeolite-polyurethane composites. J Appl Polym Sci 2008;110:2854–61.

[151] Fernandez A, Soriano E, Hernandez-Munoz P, Gavara R. Migration of antimicrobial silver from composites of polylactide with silver zeolites. J Food Sci 2010;75:186–93.

[152] Rhim JW, Wang LF, Lee Y, Hong SI. Preparation and characterization of bio-nanocomposite films of agar and silver nanoparticles: laser ablation method. Carbohydr Polym 2014;103:456–65.

[153] De Moura MR, Mattoso LH, Zucolotto V. Development of cellulose-based bactericidal nanocomposites containing silver nanoparticles and their use as active food packaging. J Food Eng 2012;109:520–4.

[154] Lloret E, Picouet P, Fernandez A. Matrix effects on the antimicrobial capacity of silver based nanocomposite absorbing materials. LWT Food Sci Technol 2012;49:333–8.

[155] Fernandez A, Picouet P, Lloret E. Cellulose-silver nanoparticle hybrid materials to control spoilage-related microflora inabsorbent pads located in trays of fresh-cut melon. Int J Food Microbiol 2010;142:222–8.

[156] Wei D, Sun W, Quian W, Ye Y, Ma X. The synthesis of chitosane-based silver nanoparticle and their antibacterial activity. Carbohydr Res 2009;344:2375–82.

[157] Sanpui P, Murugadoss A, Prasad PVD, Sankar Ghosh S, Chattopadhyay A. The antibacterial properties of a novel chitosan–Ag-nanoparticle composite. Int J Food Microbiol 2008;124:142–6.

[158] Yoksan R, Chirachanchai S. Silver nanoparticles loaded chitosan starch based films: fabrication and evaluation of tensile, barrier and antimicrobial properties. Mat Sci Eng 2010;30:891–7.

[159] Fayaz AM, Balaji K, Girilal M, Kalaichelvant PT, Venkatesan R. Mycobased synthesis of silver nanoparticles and their incorporation into sodium alginate films for vegetable and fruit preservation. J Agr Food Chem 2009;57(14):6246–52.

[160] Gammariello D, Conte A, Buonocore GG, Del Nobile MA. Bio-based nanocomposite coating to preserve quality of Fior di latte cheese. J Dairy Sci 2011;94:5298–304.

[161] Abreu AS, Oliveira M, de Sa A, Rodrigues RM, Cerqueira MA, Vicente AA, et al. Antimicrobial nanostructured starch based films for packaging. Carbohydr Polym 2015;129:127–34.

[162] Dastjerdi R, Montazer M. A review on the application of inorganic nano-structured materials in the modification of textiles: focus on anti-microbial properties. Colloids Surf B Biointerfaces 2010;79:5–18.

[163] Radetić M. Functionalization of textile materials with silver nanoparticles. J Mat Sci 2013;48:95–107.

[164] Chen CY, Chiang CL. Preparation of cotton fibers with antibacterial silver nanoparticles. Mater Lett 2008;62:3607–9.

[165] El-Rafie MH, Mohamed AA, Shaheen TI, Hebeish A. Antimicrobial effect of silver nanoparticle produced by fungal process on cotton fabrics. Carbohydr Polym 2010;80:779–82.

[166] Ilić V, Šaponjić Z, Vodnik V, Potkonjak B, Jovančić P, Nedeljković J, et al. The influence of silver content on antimicrobial activity and color of cotton fabrics functionalized with Ag nanoparticles. Carbohydr Polym 2009;78:564–9.

[167] Firoz Babu K, Dhandapani P, Maruthamuthu S, Anbu Kulandainathan M. One pot synthesis of polypyrrole silver nanocomposite on cotton fabrics for multifunctional property. Carbohydr Polym 2012;90:1557–63.

[168] Sathishkumar M, Sneha K, Yun YS. Immobilization of silver nanoparticles synthesized using Curcuma longa tuber powder and extract on cotton cloth for bactericidal activity. Bioresour Technol 2010;101:7958–65.

[169] Hebeish A, El-Naggar M, Fouda MM, Ramadan M, Al-Deyab SS, El-Rafie M. Highly effective antibacterial textiles containing green synthesized silver nanoparticles. Carbohydr Polym 2011;86:936–40.

[170] Wu M, Ma B, Pan T, Chen S, Su J. Silver-nanoparticle-colored cotton fabrics with tunable colors and durable antibacterial and self-healing. Superhydrophobic Prop 2016;26:569–76.

[171] Buşilă M, Muşat V, Textor T, Mahltig B. Synthesis and characterization of antimicrobial textile finishing based on Ag: ZnO nanoparticles/chitosan biocomposites. RSC Adv 2015;5(28):21562–71.

[172] Gorjanc M, Bukošek V, Gorenšek M, Vesel A. The influence of water vapor plasma treatment on specific properties of bleached and mercerized cotton fabric. Text Res J 2010;80:557–67.

[173] Gorjanc M, Bukošek V, Gorenšek M, Mozetič M. CF4 plasma and silver functionalized cotton. Text Res J 2010;80:2204–13.

[174] Gittard SD, Hojo D, Hyde GK, Scarel G, Narayan RJ, Parsons GN. Antifungal textile s formed using silver deposition in supercritical carbon dioxide. J Mater Eng Perform 2010;19:368–73.

[175] Perelshtein I, Applerot G, Perkas N, Guibert G, Mikhailov S, Gedanken A. Sonochemical coating of silver nanoparticles on textile fabrics (nylon, polyester and cotton) and their antibacterial activity. Nanotechnology 2008;18:19(24):245705.

[176] Dastjerdi R, Montazer M, Shahsavan S. A new method to stabilize nanoparticles on textile surfaces. Colloids Surf A Physicochem Eng Asp 2009;345:202–10.

[177] Dastjerdi R, Mojtahedi MRM, Shoshtari AK. Investigating the production and properties of Ag/TiO_2/PP antibacterial nanocomposite filament yarns. J Tex Inst 2010;101:204–13.

[178] Radetic M. Functionalization of textile materials with TiO_2 nanoparticles. J Photochem Photob C: Photochem Rev 2013;16:62–76.

[179] Dastjerdi MRM, Mojtahedi AM, Shoshtari M. Processing, properties of nanocomposite filament yarns with various filler concentrations from two different modification methods. Macromol Symp 2008;274 (1):154–65.

[180] Pulit-Prociak J, Chwastowski J, Kucharski A, Banach M. Functionalization of textiles with silver and zinc oxide nanoparticles. Appl Surf Sci 2016;385:543–53.

[181] Jiang S, Newton E, Yuen CWM, Kan CW. Application of chemical silver on polyester and cotton blended fabric. Textile Res J 2007;77(2):85–91.

[182] Dastjerdi R, Montazer M. A review on the application of inorganic nano-structured materials in the modification of textiles: focus on anti-microbial properties. Colloids Surf B Biointerfaces 2010;79:5–18.

[183] Gorensek M, Recelj P. Nanosilver functionalized cotton fabric. Textile Res J 2007;77(3):138–41.

[184] Montazer M, Behzadnia A, Moghadam MB. Superior self-cleaning features on wool fabric using TiO_2/Ag nanocomposite optimized by response surface methodology. J Appl Polym Sci 2012;125(S2):E356–63.

[185] Panwar K, Jassal M, Agrawal A. In situ synthesis of Ag–SiO_2 Janus particles with epoxy functionality for textile applications. Particuology 2005;19:107–12.

[186] Ibănescu M, Musat V, Textor T, Badilita V, Mahltig B. Photocatalytic and antimicrobial Ag/ZnO nanocomposites for functionalization of textile fabric. J Alloys Comp 2014;610:244–9.

[187] Rajesh, Ahuja T, Kumar D. Recent progress in the development of nano-structured conducting polymers/nanocomposites for sensor applications. Sens Actuators B 2009;136:275–86.

[188] Saha K, Agasti S, Kim C, Li X, Rotello VM. Gold nanoparticles in chemical and biological sensing. Chem Rev 2012;112:2739–79.

[189] Phukon P, Radhapyari K, Konwar BK, Khan R. Natural polyhydroxyalkanoate – gold nanocomposite based biosensor for detection of antimalarial drug artemisinin. Mater Sci Eng C 2014;37:314–20.

[190] Atta NF, Galal A, El-Ads EH. Gold nanoparticles-coated poly(3,4-ethylene-dioxythiophene) for the selective determination of sub-nano concentrations of dopamine in presence of sodium dodecyl sulfate. Electrochim Acta 2012;69:102–11.

[191] Pingarrón JM, Yáñez-Sedeño P, González-Cortés A. Gold nanoparticle-based electrochemical biosensors. Electrochim Acta 2008;53:5848–66.

[192] Kumar SS, Mathiyarasu J, Phani KL. Exploration of synergism between a polymer matrix and gold nanoparticles for selective determination of dopamine. J Electroanal Chem 2005;578(8):95–103.

[193] Li J, Lin X. Simultaneous determination of dopamine and serotonin on gold nanocluster/over oxidized-polypyrrole composite modified glassy carbon electrode. Sens Actuators B 2007;124:486–93.

[194] Fang Y, Guo S, Zhu C, Zhai Y, Wang E. Self-assembly of cationic polyelectrolyte-functionalized graphene nanosheets and gold nanoparticles: a two-dimensional heterostructure for hydrogen peroxide sensing. Langmuir 2010;26(13):11277–82.

[195] Luo XL, Xu JJ, Du Y, Chen HY. A glucose biosensor based on chitosan–glucose oxidase–gold nanoparticles biocomposite formed by one-step electrodeposition. Anal Biochem 2004;334:284–9.

[196] Lin J, He C, Zhao Y, Zhang S. One-step synthesis of silver nanoparticles/carbon nanotubes/chitosan film and its application in glucose biosensor. Sens Actuators B 2009;137:768–73.

[197] Zhang Y, Zhang K, Ma H. Electrochemical DNA biosensor based on silver nanoparticles/poly(3-(3-pyridyl) acrylic acid)/carbon nanotubes modified electrode. Anal Biochem 2009;387:13–19.

[198] Huang KJ, Liu YJ, Wang HB, Wang YY. A sensitive electrochemical DNA biosensor based on silver nanoparticles-polydopamine@graphene composite. Electrochim Acta 2014;118:130–7.

[199] Ruan C, Shi W, Jiang H, Suna Y, Liu X, Zhang X, et al. One-pot preparation of glucose biosensor based on polydopamine–graphene composite film modified enzyme electrode. Sens Actuators B 2013;177:826–32.

[200] Stuart MAC, Huck WTS, Genzer J, Müller M, Ober C, Stamm M, et al. Emerging applications of stimuli-responsive polymer materials. Nat Mater 2010;9:101–13.

[201] Tokareva I, Minko S, Fendler JH, Hutter E. Nanosensors based on responsive polymer brushes and gold nanoparticle enhanced transmission surface plasmon resonance spectroscopy. J Am Chem Soc 2004;126:15950–1.

[202] Kozlovskaya V, Kharlampieva E, Khanal BP, Manna P, Zubarev ER, Tsukruk VV. Ultrathin layer-by-layer hydrogels with incorporated gold nanorods as pH-sensitive optical materials. Chem Mater 2008;20(24):7474–85.

[203] Filippo E, Serra A, Manno D. Poly(vinyl alcohol) capped silver nanoparticles as localized surface plasmon resonance-based hydrogen peroxide sensor. Sens Actuators B 2009;138:625–30.

[204] Ramesh GV, Radhakrishnan TP. A Universal sensor for mercury (Hg, HgI, HgII) based on silver nanoparticle-embedded polymer thin film. ACS Appl Mater Interfaces 2011;3(4):988–94.

[205] Pandey S, Goswami GK, Nanda KK. Green synthesis of biopolymer–silver nanoparticle nanocomposite: an optical sensor for ammonia detection. Int J Biol Macromol 2012;51:583–9.

[206] Krasteva N, Besnard I, Guse B, Bauer RE, Mullen K, Yasuda A, et al. Self-assembled gold nanoparticle/dendrimer composite films for vapor sensing applications. Nano Lett 2002;2:551–5.

[207] Jain PK, ElSayed IH, El-Sayed MA. Au nanoparticles target cancer. Nano Today 2007;2(1):18–29.

[208] Ghosh P, Han G, De M, Kim CK, Rotello VM. Gold nanoparticles in delivery applications. Adv Drug Deliv Rev 2008;60(11):1307–15.

[209] Boisselier E, Astruc D. Gold nanoparticles in nanomedicine: preparations, imaging, diagnostics, therapies and toxicity. Chem Soc Rev 2009;38(6):1759.

[210] Zhang Y, Chu W, Foroushani AD, Wang H, Li D, Liu J, et al. New gold nanostructures for sensor applications: a review. Materials (Basel) 2014;7(7):5169–201.

[211] Ahmed SR, Kim J, Tran VT, Suzuki T, Neethirajan S. In situ self-assembly of gold nanoparticles on hydrophilic and hydrophobic substrates for influenza virus-sensing platform. Sci Rep 2017;7:44495.

[212] Barash O, Peled N, Tisch U, Bunn PA, Hirsch FR, Haick H. Classification of lung cancer histology by gold nanoparticle sensors. Nanomed Nanotechnol Biol Med 2012;8(5):580–9.

[213] Nascimento HPO, Oliveira MDL, de Melo CP, Silva GJL, Cordeiro MT, Andrade CAS. An impedimetric biosensor for detection of dengue serotype at picomolar concentration based on gold nanoparticles-polyaniline hybrid composites. Colloids Surf B Biointerfaces 2011;86(2):414–19.

[214] Pérez-López AM, Rubio-Ruiz B, Sebastián V, Hamilton L, Adam C, Bray TL, et al. Gold-triggered uncaging chemistry in living systems. Angew Chem Int Ed 2017;56(41):12548–52.

[215] Alsaiari SK, Hammami MA, Croissant JG, Omar HW, Neelakanda P, Yapici T, et al. Colloidal gold nanoclusters spiked silica fillers in mixed matrix coatings: simultaneous detection and inhibition of healthcare-associated infections. Adv Healthc Mater 2017;6(6):1–9.

[216] Tamai H, Igaki K, Kyo E, Kosuga K, Kawashima A, Matsui S, et al. Initial and 6-month results of biodegradable poly-l-lactic acid coronary stents in humans. Circulation 2000;102(4):399–404.

[217] Vial S, Reis RL, Oliveira JM. Recent advances using gold nanoparticles as a promising multimodal tool for tissue engineering and regenerative medicine. Curr Opin Solid State Mater Sci 2017;21(2):92–112.

[218] Baranes K, Shevach M, Shefi O, Dvir T. Gold nanoparticle-decorated scaffolds promote neuronal differentiation and maturation. Nano Lett 2016;16(5):2916–20.

[219] Shevach M, Fleischer S, Shapira A, Dvir T. Gold nanoparticle-decellularized matrix hybrids for cardiac tissue engineering. Nano Lett 2014;14:5792–6.

[220] Baei P, Jalili-firoozinezhad S, Rajabi-zeleti S, Tafazzoli-shadpour M. Electrically conductive gold nanoparticle-chitosan thermosensitive hydrogels for cardiac tissue engineering. Mater Sci Eng C 2016;63:131–41.

[221] Ribeiro M, Ferraz MP, Monteiro FJ, Fernandes MH, Beppu MM, Mantione D, et al. Antibacterial silk fibroin/nanohydroxyapatite hydrogels with silver and gold nanoparticles for bone regeneration. Nanomed Nanotechnol Biol Med 2017;13(1):231–9.

[222] BarathManiKanth S, Kalishwaralal K, Sriram M, Pandian S, Youn H, Eom S, et al. Anti-oxidant effect of gold nanoparticles restrains hyperglycemic conditions in diabetic mice. J Nanobiotechnol 2010;8(1):16.

[223] Smith SE, Snider CL, Gilley DR, Grant DN, Sherman SL, Ulery BD, et al. Homogenized porcine extracellular matrix derived injectable tissue construct with gold nanoparticles for musculoskeletal tissue engineering applications. J Biomater Nanobiotechnol 2017;8(2):125–43.

[224] Akturk O, Kismet K, Yasti AC, Kuru S, Duymus ME, Kaya F, et al. Collagen/gold nanoparticle nanocomposites: a potential skin wound healing biomaterial. J Biomater Appl 2016;31(2):283–301.

[225] Uchihara T. Silver diagnosis in neuropathology: principles. Acta Neuropathol. 2007;483–99.

[226] Sibbald RG, Contreras-Ruiz J, Coutts P, Fierheller M, Rothman A, Woo K. Bacteriology, inflammation, and healing: a study of nanocrystalline silver dressings in chronic venous leg ulcers. Adv Skin Wound Care 2007;20(10):549–58.

[227] Skirtach AG, Muñoz Javier A, Kreft O, Köhler K, Piera Alberola A, Möhwald H, et al. Laser-induced release of encapsulated materials inside living cells. Angew Chem Int Ed 2006;45 (28):4612–17.

[228] Galiano K, Pleifer C, Engelhardt K, Brossner G, Lackner P, Huck C, et al. Silver segregation and bacterial growth of intraventricular catheters impregnated with silver nanoparticles in cerebrospinal fluid drainages. Neurol Res 2008;30(3):285–7.

[229] Moore K. A new silver dressing for wounds with delayed healing. Wounds UK 2006;2(2):70–8.

[230] Vigneshwaran N, Kathe AA, Varadarajan PV, Nachane RP, Balasubramanya RH. Functional finishing of cotton fabrics using silver nanoparticles. J Nanosci Nanotechnol 2007;7(6):1893–7.

[231] Chaloupka K, Malam Y, Seifalian AM. Nanosilver as a new generation of nanoproduct in biomedical applications. Trends Biotechnol 2010;28(11):580–8.

[232] Lackner P, Beer R, Broessner G, Helbok R, Galiano K, Pleifer C, et al. Efficacy of silver nanoparticles-impregnated external ventricular drain catheters in patients with acute occlusive hydrocephalus. Neurocrit Care 2008;8(3):360–5.

[233] Andara M, Agarwal A, Scholvin D, Gerhardt RA, Doraiswamy A, Jin C, et al. Hemocompatibility of diamondlike carbon-metal composite thin films. Diam Relat Mater 2006;15(11–12 SPEC. ISS.):1941–8.

[234] Ghanbari H, Viatge H, Kidane AG, Burriesci G, Tavakoli M, Seifalian AM. Polymeric heart valves: new materials, emerging hopes. Trends Biotechnol 2009;27(6):359–67.

[235] Fu J, Ji J, Fan D, Shen J. Construction of antibacterial multilayer films containing nanosilver via layer-by-layer assembly of heparin and chitosan-silver ions complex. J Biomed Mater Res Part A 2006;79 (3):665–74.

[236] Yoshida K, Tanagawa M, Matsumoto S, Yamada T, Atsuta M. Antibacterial activity of resin composites with silver-containing materials. Eur J Oral Sci 1999;107(4):290–6.

[237] Magalhães APR, Santos LB, Lopes LG, Estrela CRDA, Estrela C, et al. Nanosilver application in dental cements. ISRN Nanotechnol 2012;2012:1–6 ID 365438, 6.

[238] Zhou W, Ma Y, Yang H, Ding Y, Luo X. A label-free biosensor based on silver nanoparticles array for clinical detection of serum p53 in head and neck squamous cell carcinoma. Int J Nanomedicine 2011;6:381–6.

[239] Haes AJ, Van Duyne RP. A nanoscale optical biosensor: sensitivity and selectivity of an approach based on the localized surface plasmon resonance spectroscopy of triangular silver nanoparticles. J Am Chem Soc 2002;124(35):10596–604.

[240] Haes AJ, Hall WP, Chang L, Klein WL, Van Duyne RP. A localized surface plasmon resonance biosensor: first steps toward an assay for Alzheimer's disease. Nano Lett 2004;4(6):1029–34.

[241] Marsich E, Bellomo F, Turco G, Travan A, Donati I, Paoletti S. Nano-composite scaffolds for bone tissue engineering containing silver nanoparticles: preparation, characterization and biological properties. J Mater Sci Mater Med 2013;24(7):1799–807.

[242] González-Sánchez MI, Perni S, Tommasi G, Morris NG, Hawkins K, López-Cabarcos E, et al. Silver nanoparticle based antibacterial methacrylate hydrogels potential for bone graft applications. Mater Sci Eng C 2015;50:332–40.

[243] Porrelli D, Travan A, Turco G, Crosera M, Borgogna M, Donati I, et al. Antibacterial-nanocomposite bone filler based on silver nanoparticles and polysaccharides. J Tissue Eng Regen Med 2017;. Available from: https://doi.org/10.1002/term.2365.

[244] Lu S, Gao W, Gu HY. Construction, application and biosafety of silver nanocrystalline chitosan wound dressing. Burns 2008;34(5):623–8.

[245] Widgerow AD. Nanocrystalline silver, gelatinases and the clinical implications. Burns 2010;36 (7):965–74.

[246] Liu X, Lee PY, Ho CM, Lui VCH, Chen Y, Che CM, et al. Silver nanoparticles mediate differential responses in keratinocytes and fibroblasts during skin wound healing. Chem Med Chem 2010;5 (3):468–75.

[247] Bhowmick S, Koul V. Assessment of PVA/silver nanocomposite hydrogel patch as antimicrobial dressing scaffold: synthesis, characterization and biological evaluation. Mater Sci Eng C 2016;59:109–1019.

[248] García-Astrain C, Chen C, Burón M, Palomares T, Eceiza A, Fruk L, et al. Biocompatible hydrogel nanocomposite with covalently embedded silver nanoparticles. Biomacromolecules 2015;16(4):1301–10.

[249] Hung HS, Hsu SH. Biological performances of poly(ether)urethane-silver nanocomposites. Nanotechnology 2007;18:47.

[250] Kumar N, Desagani D, Chandran G, Nath N, Karthikeyan G, Waigaonkar S, et al. Biocompatible agarose-chitosan coated silver nanoparticle composite for soft tissue engineering applications. Artif Cells Nanomed Biotechnol 2018;46(3):637–49.

[251] Madhavan RV, Rosemary MJ, Nandkumar MA, Krishnan KV, Krishnan LK. Silver nanoparticle impregnated poly (ε-caprolactone) scaffolds: optimization of antimicrobial and noncytotoxic concentrations. Tissue Eng Part A 2011;17(3–4):439–49.

9

Polymer Nanocomposites With Decorated Metal Oxides

Serap Hayat Soytaş[1], Oğuzhan Oğuz[2], Yusuf Ziya Menceloğlu[1,2]

[1]SABANCI UNIVERSITY NANOTECHNOLOGY RESEARCH AND APPLICATION CENTER (SUNUM), TUZLA, ISTANBUL, TURKEY [2]FACULTY OF ENGINEERING AND NATURAL SCIENCES, MATERIALS SCIENCE AND NANO ENGINEERING, SABANCI UNIVERSITY, TUZLA, ISTANBUL, TURKEY

9.1 Introduction

Polymer nanocomposites have been at the forefront research field in the polymer community for the last few decades as they find applications in areas from tires to microelectronic devices, from water filtration membranes to lithium-ion batteries, from piezoelectric sensors to biomedical devices, and many other important areas. Metal oxide nanocomposites, in particular, have received considerable attention as the metal oxides offer unique features in the area of photocatalysis, drug delivery, insulation, magnetism, flame retardants, and UV absorbents. The nanocomposites combine the functionalities of polymer matrices, which include easy processability, light weight, flexibility, and cost-effectiveness, with the specific features of the metal oxides. Interfacial interactions between the surface of the metal oxides and the polymer matrix play an important role to achieve the desired mechanical, optical, electrical, and thermal properties. Since the nanostructures tend to form aggregates because of their high specific surface area and volume effects, it is difficult to achieve a nanocomposite with well-dispersed nanostructures. Agglomeration and insufficient dispersion of the nanostructures not only are ineffective to realize the desired properties of the nanocomposites but also may degrade the inherent properties of the polymers. This problem could be addressed by changing the surface chemistry of the nanostructures. Surfaces of the metal oxide nanostructures can be modified via a reaction with a coupling agent, growing or attaching polymer chains and exchanging the existing capping agents with others. The most commonly utilized chemical agents are silane coupling agents, which are used to create organic surface groups such as amines ($-NH_2$) [1], vinyls ($-CH=CH2$) [2], acryloxies ($-O-C(H \text{ or } CH_3)=CH_2$) [3], epoxies ($-CHOCH_2$) [4], and thiols ($-SH$) [5]. These surface functional groups can provide specific interactions with the polymer matrix and, in addition, can be used for further functionalization. Development of controlled polymerization techniques, such as atom transfer radical polymerization (ATRP) [6], reversible addition – fragmentation chain transfer (RAFT) [6], and nitroxide-mediated polymerization [7], presented new prospects for synthetic

Polymer Composites with Functionalized Nanoparticles. DOI: https://doi.org/10.1016/B978-0-12-814064-2.00009-3

chemists to functionalize the metal oxide nanoparticles with various polymer chains. Furthermore, ligand exchange strategy provides an opportunity to add specifically designed species on the surface of metal oxide nanoparticles by replacing the original ligands.

The modified metal oxide nanoparticles can be incorporated into the polymer matrices via various strategies including solution and melt blending, sol–gel processing, and growth of polymers in the presence of nanomaterials as well as in situ formation of metal oxides in the polymer matrix.

This chapter provides an overview of the synthesis and surface functionalization of various metal oxide nanostructures and the strategies for their incorporation into the polymer matrices. Furthermore, the properties and potential applications of the nanocomposites with decorated metal oxides are discussed.

9.2 Synthesis of Metal Oxide Nanoparticles

Synthesis of metal oxide nanoparticles may be categorized into bottom-up and top-down approaches. The top-down approach refers to a fabrication technique that starts with large objects and uses removing methods such as mechanical, chemical, and electrochemical processes to shrink them down to the nanoscale [8–12]. On the other hand, the bottom-up approach uses the methods where atomic or molecular dimensional components self-assemble together to form larger organized particles [13–18]. Although the mostly used top-down approaches such as ultrasonication and ball milling are cost-effective and scalable and form nanoparticles down to a few nanometers, they generally generate irregular particles and introduce defects. Moreover, the ball milling process is prone to introduce impurities. Therefore, the bottom-up approach has been recognized as the most promising method to produce metal oxide nanoparticles. It also provides possibilities to manipulate the properties such as size, shape, surface area, and surface functionalization of the final nanoparticles. The bottom-up approaches may be categorized based on the precursors, either liquid or gaseous, utilized to synthesize the metal oxides. The widely used wet chemical methods utilizing the liquid precursors are sol–gel process, microemulsion technique, precipitation/coprecipitation method, and solvothermal/hydrothermal technique. Chemical vapor deposition (CVD) and the pulse laser deposition (PLD) are the most common physical bottom-up approaches that use gaseous precursors. Both approaches will be discussed in this section with some examples.

9.2.1 Wet Chemical Methods

9.2.1.1 Sol–Gel Processing

Sol–gel process has been intensely investigated for the synthesis of metal oxide nanostructures since the first introduction of the concept of silica gel by Ebelmen [19] and Graham [20] in the mid-19th century. Since then, SiO_2 nanoparticles are the mostly synthesized nanoparticles using sol–gel method. The most commonly utilized method for the synthesis of silica nanoparticles is the Stöber method, which was first reported in 1968 [21]. Although it is challenging to fabricate silica particles with diameters less than 100 nm, most of the nano-sized monodispersed nanoparticles are synthesized by the modified version of this original idea. The sol–gel process involves the formation of two consecutive phases of a "sol," which is a colloidal suspension of solid particles in a liquid, and a "gel," which is an inorganic integrated network formed by polycondensation of the sol [22]. The precursors in sol–gel reactions are metal alkoxides, which are hydrolyzed in either acidic or alkali conditions and followed by the condensation of the hydrolysis product forming a polymeric network surrounded with solvent molecules. Acidic and basic catalysis of silicon alkoxides are depicted in Scheme 9–1. The properties of the final product are profoundly affected by the rate of hydrolysis and condensation reactions, which depend on factors such as the nature of the alkoxy group, electronegativity of the metal atom, the structure of the metal alkoxides, and the solvent used [23]. For example, particles with smaller sizes can be obtained with slower rate of hydrolysis [24].

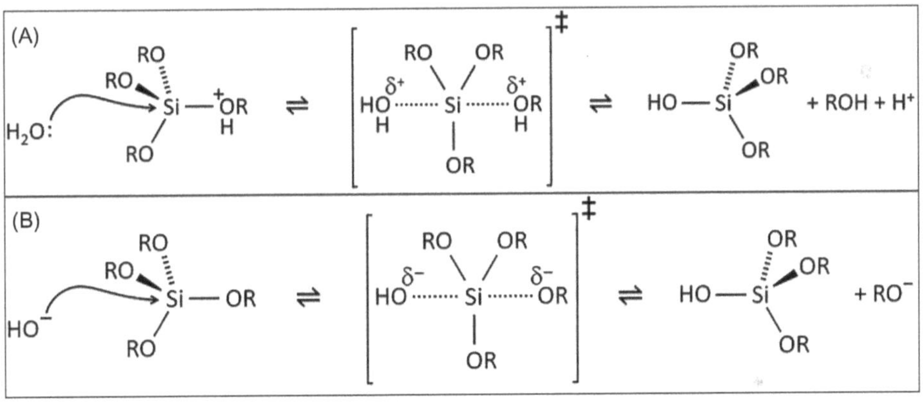

SCHEME 9–1 (A) Acid catalyzed and (B) based catalyzed hydrolysis of silicon alkoxide.

The particle size of nanoparticles synthesized by sol–gel process is controlled by concentration of reactants, concentration of catalyst, temperature, mixing mode, rate of hydrolysis, and condensation reaction and drying conditions [22,24]. The particle growth in the sol–gel process follows nucleation and aggregation [25], which are affected by the rate of hydrolysis and condensation reactions. Unstable, negatively charged particles with a diameter of less than 10 nm in the initial solution tend to reduce the total surface area by forming aggregates. In that stage, competition between nucleation and aggregation, which would affect the final particle size, depends on reactants concentrations [26]. Greasley et al. [27] showed that increasing the concentration of tetraethoxysilane (TEOS) from 0.045 to 0.17 M resulted in an increase in the particle size (Fig. 9–1A) as well as an increase in total yield. Furthermore, increasing ammonium hydroxide (the catalyst) concentration also caused an increase in particle size. They also investigated the effect of water content as well as amount of ammonium hydroxide on both particle size and particle size distribution. High ammonium hydroxide concentration resulted in not only increased particle size but also widened particle size

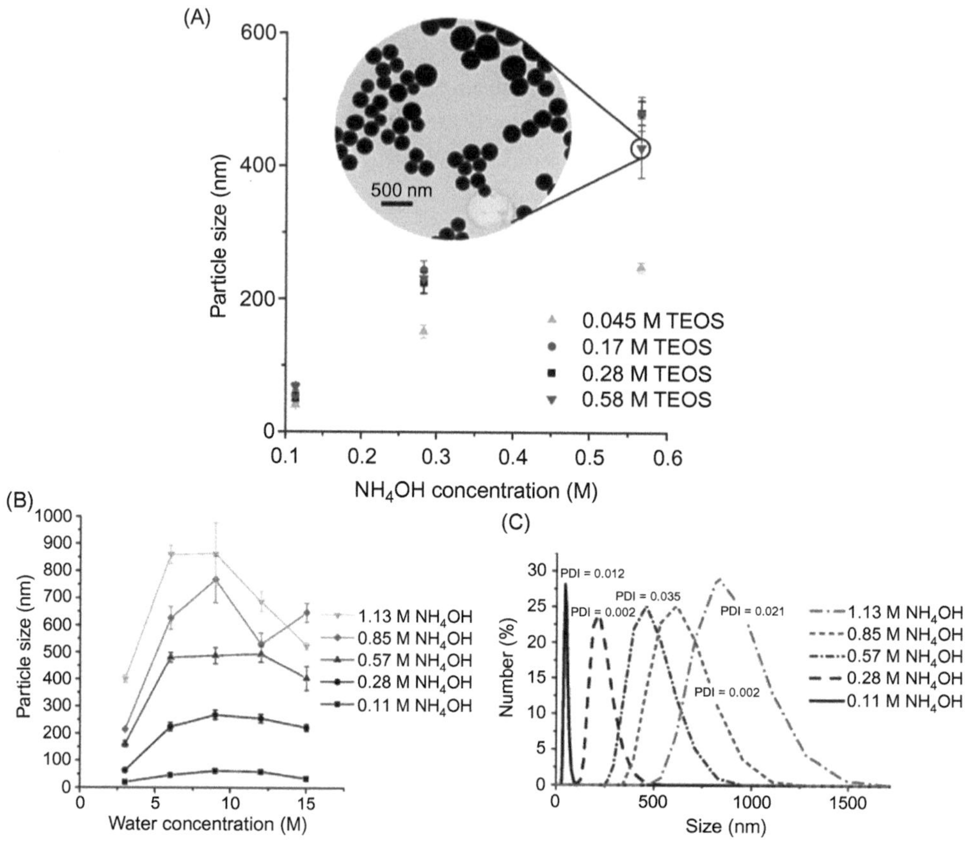

FIGURE 9–1 Effect of (A) TEOS and (B) water and ammonium hydroxide concentration on the mean particle size, and (C) ammonium hydroxide concentration on particle size distribution of silica particles. *Reprinted with permission from Greasley SL, Page SJ, Sirovica S, Chen S, Martin RA, Riveiro A, et al. Controlling particle size in the Stöber process and incorporation of calcium. J Colloid Interface Sci 2016;469(1):218. Copyright © (2016) Elsevier.*

distribution (Fig. 9–1B and C). For the same ammonium hydroxide concentration, particle size increases up to 9 M water concentration then start decreasing with further dilution (Fig. 9–1B).

Rahman et al. [28] reported an optimized technique that utilizes low frequency ultrasound during reactant mixing to obtain a highly dispersed, homogeneous, and stable silica nanoparticles with an average diameter of 7.1 nm with a high yield of 75%.

Menceloglu et al. [29] investigated the effect of different solvents, namely methanol, ethanol, n-propanol, isopropanol, and n-butanol, in the ammonia-catalyzed synthesis of SiO_2 nanoparticles. They reported that particle size along with the initial hydrolysis rate increased with the molecular weight of the primary alcohols while secondary alcohols yielding SiO_2 nanoparticles with larger particles size; in comparison. SiO_2 nanoparticles with the smallest particle size were obtained using methanol yielding the highest hydrolysis rate.

FIGURE 9–2 Effect of drying process on the morphology of the dried colloidal silica nanoparticles. *Reprinted from Rahman IA, Padavettan V. Synthesis of silica nanoparticles by sol–gel: size-dependent properties, surface modification, and applications in silica-polymer nanocomposites—a review. J Nanomater 2012;5:1–15.*

Although the desired particle size and morphology of metal oxide particles are obtained by the application of optimum reaction conditions, the particles may agglomerate during drying if it is not controlled [30,31]. The presence of water during the drying process may result in formation of agglomerates while well-dispersed nanoparticles can be obtained with a controlled process such as freeze-drying [32]. The effect of drying conditions on the morphology of silica can be seen in Fig. 9–2.

Rahman et al. [31] investigated the effect of three different techniques, namely, alcohol dehydration, freeze-drying, and oven drying, on the dispersion and agglomeration of silica nanoparticles. The closely packed particles forming dense islands of aggregates with voids and microcracks on the surface were observed when the particles were dried with conventional oven drying. Since water was forming bridges between the nanoparticles, constant removal of water with the heating process caused a sudden increase in the concentration of the sol getting particles close to each other. Then, at a critical concentration, the particles collide forming aggregates under the influence of attractive force such as van der Waals and hydrogen bonding of silanol groups. This showed that the presence of water during drying process encourages the agglomeration process. They were able to remove water effectively by alcohol dehydration process and obtain monodispersed, isolated nanoparticles with the average diameter of 6.7 nm with a consecutive oven drying. Similarly, elimination of water through solid–gas transformation using freeze-drying technique provided nanoparticles of 6.9 nm in diameter with narrow size distribution.

9.2.1.2 Coprecipitation Method

Precipitation/coprecipitation is a standard synthesis technique to prepare pure/multinary metal oxide nanoparticles. In a typical coprecipitation reaction, a salt precursor, generally a nitrate, chloride, or oxychloride, is dissolved in aqueous solution and then the corresponding hydroxides are precipitated by the addition of a base such as ammonium hydroxide or sodium hydroxide. After washing the produced ammonium or sodium salt, the hydroxides are calcined resulting in metal oxide powders. If more than one precursor salt is used in the starting solution, multinary metal oxides can be obtained by coprecipitation of the corresponding hydroxides [33]. The control of particle size and distribution is difficult if the rate of precipitation is not slow (controlled); however, it was reported that nanosized metal oxide powder is synthesized with a modified coprecipitation method by keeping the pH level constant [34].

Coprecipitation is the most commonly used method for the synthesis of magnetic nanoparticles. Pure iron oxides and ferrites, which are iron oxides chemically combined with one or more additional metals including Mn, Co, Cu, Mg, Zn, and Ni, are conveniently synthesized from aqueous solutions of metal salts by the addition of a base under oxygen-free atmosphere at either room or elevated temperature [35]. The following reaction takes place to form the magnetic nanoparticles in the basic aqueous solution:

$$M^{2+} + 2\,Fe^{3+} + 8OH^{-} \rightarrow MFe_2O_4 + 4H_2O$$

where M can be Fe^{2+}, Mn^{2+}, Co^{2+}, Cu^{2+}, Mg^{2+}, Zn^{2+}, and Ni^{2+}. The type of the salt used, the reaction temperature and pH value, the ratio of M^{2+}/Fe^{3+}, the ionic strength of the solution affects the size, shape, and composition of magnetic nanoparticles [36–38].

Iida et al. [36] investigated the effect of iron salts, namely sulfates and chlorides, and the ratio of ferrous ions to ferric ions on the size and magnetic properties of Fe_3O_4. The formation of larger particles ($\sim$37 nm) were observed with the use of ferrous salts only compared with the particles with the size of approximately 9 nm synthesized with both ferrous and ferric salts. The size of the Fe_3O_4 magnetic nanoparticles was mainly varied with the valence of the ions instead of the type of the counter ion of the iron salt used in the synthesis. The saturation magnetization values of the magnetic nanoparticles from ferrous sulfate and chloride were 86.6 and 81.0 emu/g, respectively, while that of nanoparticles synthesized with both ferrous and ferric sulfate and chloride, 46.7 and 55.4 emu/g, respectively. This reduction in saturation magnetization was explained as an effect of thermal fluctuation that rises near the surface of the synthesized magnetic nanoparticles or due to the much finer size of the particles.

Pure single-phase multiferroic material $BiFeO_3$ was synthesized using coprecipitation method avoiding formation of impurities by controlling parameters based on thermodynamic laws [34]. The main impurities in this system are Bi_2O_3, which occurs due to the evaporation of Bi component of the salts with low decomposition temperature, and $Bi_2Fe_4O_9$, which is a thermodynamically stable substitution of $BiFeO_3$ [39]. The yield of pure $BiFeO_3$ powders improved with decreasing pH level from 11 to 9.3 as characterized with X-ray diffraction (XRD) (Fig. 9–3A). The amount of impurities was further reduced with the use of excess Bi at pH 9.3 as seen in Fig. 9–3B. The pure $BiFeO_3$ was obtained with the excess Bi of 3%. Bi content of smaller or larger than 3% showed the presence of impurities.

Ternary transition metal oxide, $CuFeMnO_4$, with magnetic and catalytic properties was synthesized using coprecipitation method and consecutive calcination at 500–900°C [40]. The stoichiometric amount of nitrates, $Cu(NO_3)_2 \cdot 3H_2O$, $Fe(NO_3)_3 \cdot 9H_2O$, $Mn(NO_3)_2$, were used as the precursor salt, and were coprecipitated with NaOH at the pH level of 11 $\pm$ 0.5.

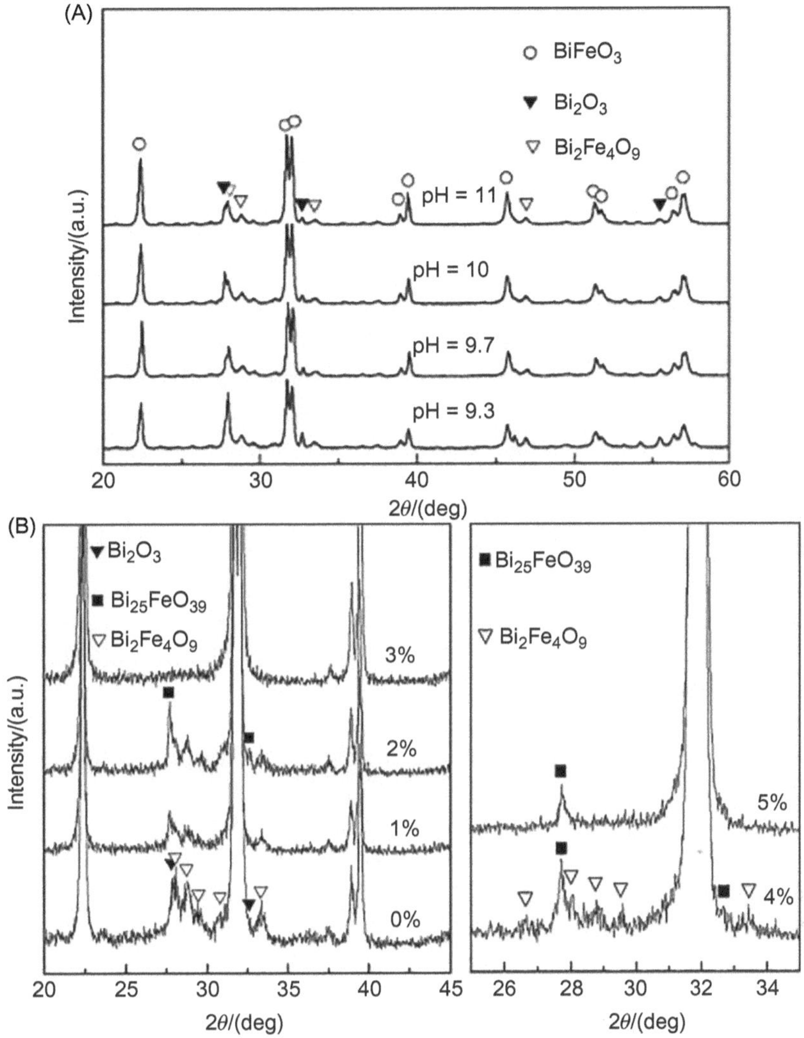

FIGURE 9–3 XRD patterns of (A) $BiFeO_3$ prepared at different pH levels and (B) with different amounts of excess Bi at pH 9.3. *Reprinted with permission from Ke H, Wang W, Wang Y, Xu J, Jia D, Lu Z, Zhou Y. Factors controlling pure-phase multiferroic* $BiFeO_3$ *powders synthesized by chemical coprecipitation. J Alloys Compd 2011; 509(5):2193. Copyright © (2011) Elsevier.*

The dried nanopowders calcined at 500°C, 700°C, and 900°C. Although the precipitate obtained from three precursors is composed of amorphous hydroxides, the formation of crystal phases was observed with the increase of calcination temperature. While the sample calcined at 500°C contained multiple phases such as CuO, Fe_3O_4, and Mn_3O_4, single-phase crystalline $CuFeMnO_4$ formed after calcination at 700°C. XRD pattern of the sample calcined at 900°C showed the strongest and sharpest peak corresponding to the spinel structure of

$CuFeMnO_4$ although the peaks of CuO were also observed, showing the segregation of CuO at high temperatures.

Coprecipitation is also an excellent tool for the synthesis of highly dispersed, well-intermixed, and uniform noble metal supported metal oxides. $Cu/ZnOAl_2O_3$ catalyst [41–43], $Ni/MgAlO_x$ catalyst [44], $Pd/MgO/MgAl_2O_4$ catalyst [45], and $Pd_2Ga/MgO/MgGa_2O_4$ intermetallic catalyst [46] were successfully synthesized using the coprecipitation method.

9.2.1.3 Microemulsion Technique

Water-in-oil microemulsion is a widely used, powerful technique to synthesize uniform size metal oxide nanoparticles. Danielsson and Lindman [47] described the microemulsion as "a system of water, oil and an amphiphile which is a single optically isotropic and thermodynamically stable liquid solution." In a typical water-in-oil microemulsion system, the water nanodroplets, which are thermodynamically stabilized by the amphiphile, called surfactant, are suspended in the bulk oil phase and act as a nanoreactor for nanoparticle synthesis. The surfactant in this system reduces the interfacial tension between the water and oil phases by specific interactions of its polar or ionic part and its nonpolar hydrophobic part with each phase, respectively, so that a transparent solution is established [48]. Different types of surfactants containing branched or straight hydrocarbon chain of 8–18 carbon atoms and hydrophilic part are used in water-in-oil microemulsions. Some examples of the surfactants can be seen in Fig. 9–4.

The size of the nanoparticles synthesized via microemulsion method can be controlled by the size of water nanoreactors, which can be tuned by the water-to-surfactant molar ratio. Size of the water nanoreactors, the micelles, has a linear dependence on the water-to-surfactant molar ratio, W_0, proposed by Pileni [50] as expressed with the following equation:

$$R_w = \frac{3 \cdot V_{AQ}}{A_h} \times W_0 = k \times W_0$$

where V_{AQ} is water molecule volume and A_h is area of the surfactant polar group. It was also reported that it is necessary to use very low metal precursor concentrations (10^{-3}–10^{-4} M) to minimize the size of the particles in dispersed phase of the microemulsion [51]. Although the calculated and experimentally measured radii of the nanoparticles agree at low precursor concentration up to 0.08 wt.%, radius differs from the calculated value for the precursor concentration of 0.16 wt.%, indicating the significant effect of precursor concentration on the microemulsion droplet size.

The complex effect of precursor concentration was shown in literature in the synthesis of molybdenum oxide nanoparticles using water-in-oil microemulsion method from the water solution of $(NH0)_6Mo_7O_{24}$/dioctyl sodium sulfosuccinate (AOT)/pentadecane (PD) [52,53]. The minimum particle size was obtained at W_0 ~2.5 and at the lowest studied concentration of the precursor in the microemulsion, which is $C_{Mo} = 160$ ppm.

Other factors affecting the particle size are pH of the water phase and the temperature of the thermal treatment. Since the pH affects the state of the precursor in microemulsion

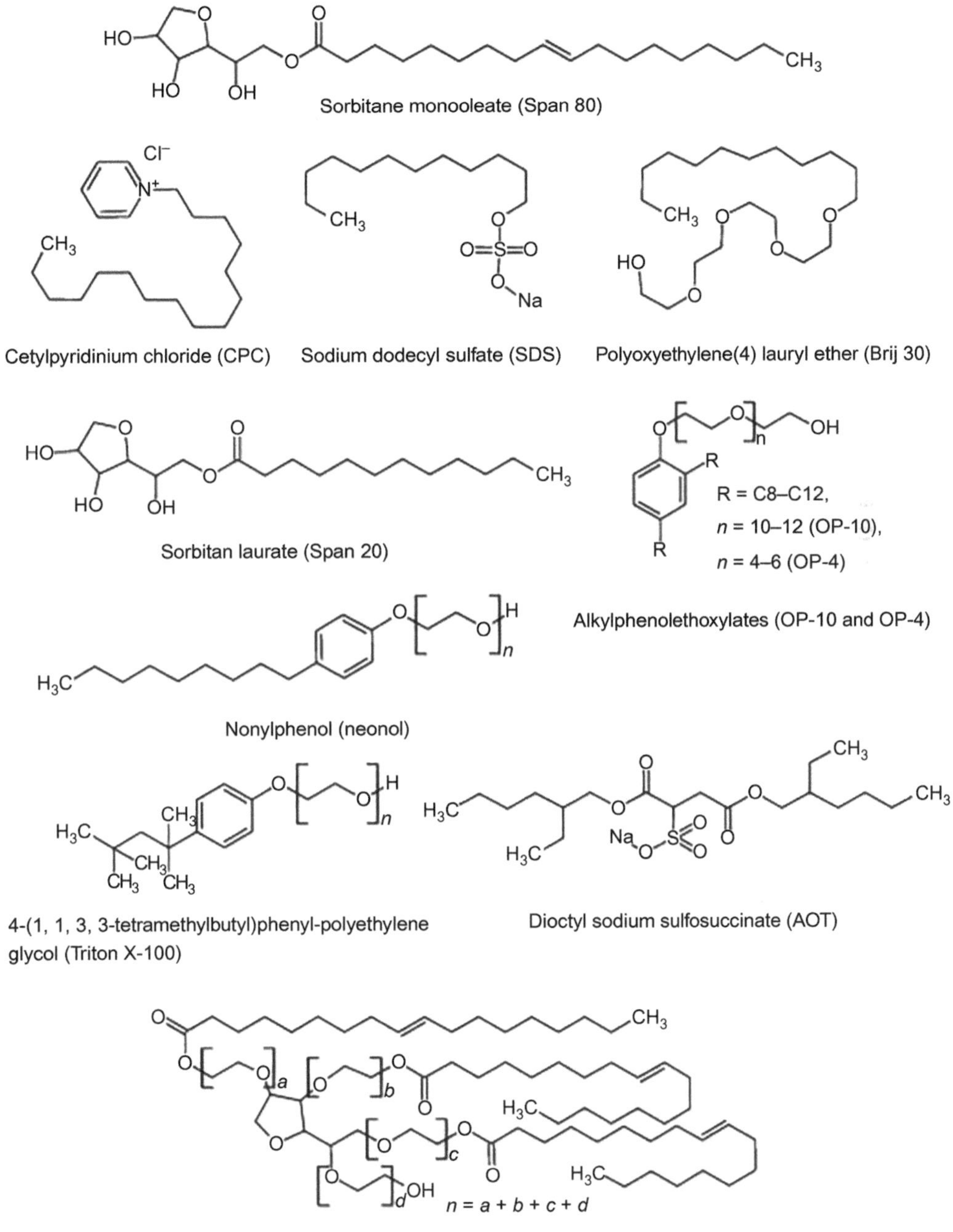

FIGURE 9–4 Structural formulas and names of some surfactants. *Reprinted with permission from Khadzhiev SN, Kadiev KM, Yampolskaya GP, Kadieva MK. Trends in the synthesis of metal oxide nanoparticles through reverse microemulsions in hydrocarbon media. Adv Colloid Interface Sci 2013;197–8:134. Copyright © (2013) Elsevier.*

droplet, size of the particles forming is affected by the pH of the water phase. The size of Al_2O_3 particles synthesized by the thermal decomposition of microemulsion water solution of $Al(NO_3)_3$/AOT/PD increased from 129.5 to 174.8 with the increase of pH values from 3.5 to 9 while further increase in pH level caused a reduction of particle size, forming particles with the radius of 41 and 38 nm at pH = 10 and 12, respectively (Table 9–1) [49].

Temperature of thermal treatment as well as the duration of the treatment also affects the size of metal oxide nanoparticle synthesized in microemulsion. The size of the MoO_3 and Al_2O_3 nanoparticles tended to decrease with increase of the temperature within the range of 150–250°C forming the minimum sized particles at ~240°C with the duration of 60 minutes [53]. Further exposure to the high temperature is followed with the formation of aggregates.

9.2.1.4 Solvothermal/Hydrothermal Technique

Solvothermal method is one of the oldest techniques for the synthesis of metal oxides. The first application of this method can be dated back to 1845 when Schafhautl performed the first hydrothermal synthesis, which is a special type of solvothermal synthesis where the solvent is water, and obtained quartz microcrystals [54]. In a typical solvothermal process, a solvent, water in the case of hydrothermal synthesis, facilitates the interaction between reactants under moderate to high pressure (typically 1–10,000 atm) and temperature (typically 100–1000°C).

Generally, a suitable capping agent or surfactant is utilized to provide stabilization against agglomeration and to help dissolution of the particles in different solvents. However, synthesis of highly crystalline Fe_3O_4 magnetic nanoparticles via hydrothermal process without any surfactant was reported [55]. Fe_3O_4 nanoparticles with the size of 40 nm synthesized at 140°C for 6 hours exhibited a saturation magnetization of 85.8 emu/g, which is close to that of the corresponding bulk Fe_3O_4 (92 emu/g). Zheng et al. [56] also synthesized α-Fe_2O_3 nanoparticles in an organic solvent (*N,N*-dimethyl formamide) without any surfactant. They showed

Table 9–1 Mean Radius (Rh) and Polydispersity Indices (PI) of Alumina Nanoparticles Synthesized From the Microemulsion of $Al(NO_3)_3$/AOT/PD in Terms of pH of Water Phase

Reverse ME Composition			Characteristics of Dispersion		
C_{Al} (ppm)	W_0	pH	T (°C)	R_h (nm)	PI[a]
353	7.3	3.5	207	129.5	7.340
		4		162.0	1.039
		9		174.8	1.088
		10		41.0	6.913
		12		38.0	7.296

[a]According to DLS, polydispersity index <0.1 and >0.3 correspond to monodisperse and polydisperse system, respectively.
Reprinted with permission from Khadzhiev SN, Kadiev KM, Yampolskaya GP, Kadieva MK. Trends in the synthesis of metal oxide nanoparticles through reverse microemulsions in hydrocarbon media. Adv Colloid Interface Sci 2013;140:197–98:140.

the shape evolution of α-Fe_2O_3 nanoparticles with the reaction time at 180°C (Fig. 9–5). The shape evolution was observed in the following sequence: First, the formation of primary particles (5–8 nm), next, the formation of secondary particle (15–30 nm) to minimize the surface energy, then, aggregation of the secondary particles forming flower-like structures (40–50 nm), and finally, the formation of polyhedral nanoparticles through Ostwald ripening keeping the size the same as the size of the particle in the previous stage.

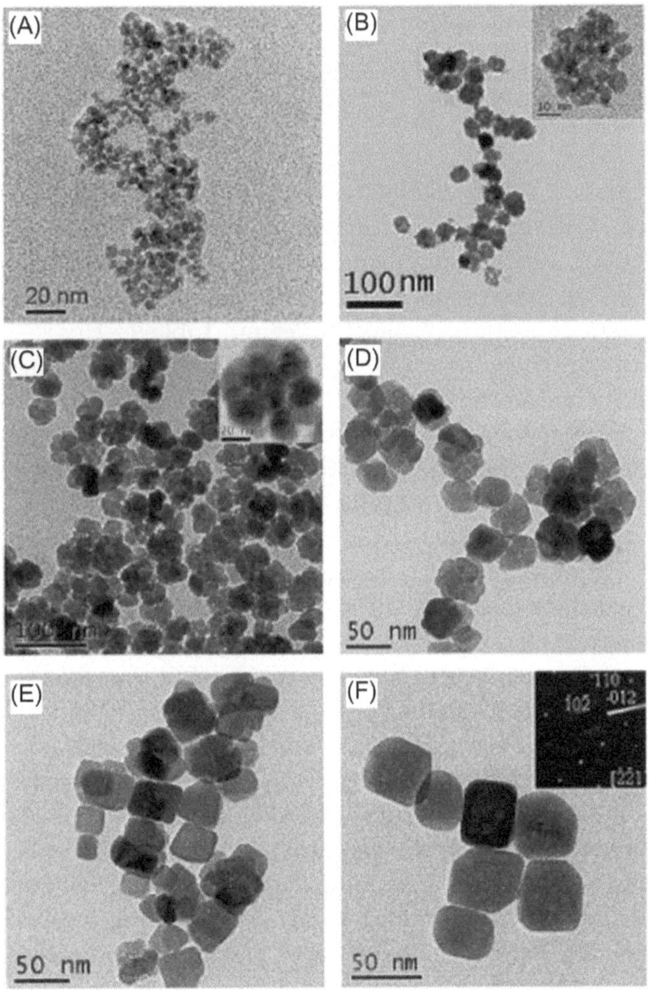

FIGURE 9–5 TEM images of α-Fe_2O_3 nanoparticles prepared at 180°C for different reaction periods: (A) 1 h, (B) 2 h, (C) 3 h, (D) 5 h, (E) 10 h, and (F) 30 h. *Reprinted with permission from Zheng Y, Cheng Y, Wang Y, Bao F. Synthesis and shape evolution of a-Fe_2O_3 nanophase through two-step oriented aggregation in solvothermal system. J Crystal Growth 20015;284(1–2):223. Copyright © Elsevier.*

Table 9–2 Size, Shape, and Magnetic Parameters of α-Fe_2O_3 Nanoparticle

Sample	Shape	Particle Size (TEM) (nm)	Coercivity H_c (Oe)	Remanent Magnetization M_r (emu/g)	Magnetic Properties
α-Fe_2O_3	Rod-like	215 ± 20[a] 28 ± 10	46.94	0.002754	Weak ferromagnetism
α-Fe_2O_3/SDS	Rhombohedral	~400	99.21	0.001657	Weak ferromagnetism
α-Fe_2O_3/DBS	Rhombohedral	~330	530.3	0.01612	Weak ferromagnetism
α-Fe_2O_3/HPC	Rhombohedral	~290	156.9	0.002554	Weak ferromagnetism
α-Fe_2O_3/CTAB	Rhombohedral	~230	77.75	0.001043	Weak ferromagnetism

[a]Major axis: 215 ± 20 nm; minor axis: 28 ± 10.

On the other hand, α-Fe_2O_3 nanoparticles were also synthesized via hydrothermal method using several surfactants, such as sodium dodecylsulfonate (SDS), sodium dodecylebenzene sulfonate (DBS), cetyltrimethyl ammonium bromide (CTAB), and hexadecyipyridinium chloride (HPC) [57]. Size, shape, and magnetic properties of the α-Fe_2O_3 nanoparticle could be controlled with different surfactants as seen in Table 9–2.

Titirici et al. [58] reported synthesis of Fe_2O_3, NiO, Co_3O_4, CeO_2, MgO, and CuO hollow spheres via one-pot hydrothermal approach. They first formed carbon spheres with metal ions incorporated into their shell via hydrothermal route at 180°C in the presence of a carbon source and metal salts and then, hollow metal oxide spheres were obtained after removal of carbon via calcination at 550°C for 4 hours.

Synthesis of titanium dioxide (TiO_2), which is an important photocatalyst, is one of the mostly investigated hydrothermal/solvothermal processes. Several factors such as temperature, pH, pressure, reaction duration, stirring, solvent type, and presence of mineralizers were studied by numerous researchers [59–65]. It has been shown that the pH, temperature, and presence of mineralizers influence not only the size but also the morphology (spheroid or rod-like) and phase (rutile or anatase) of the crystallites [59]. Aruna et al. [61] showed effect of pH and stirring on the particle size, which decreases with stirring and at low pH values. Reaction time and duration affected the size and the morphology of the TiO_2 anatase nanocrytals, which underwent several stages including blocky-pinacoidal, elongated rod-like, wedge-shaped morphologies and finally reaching a bipyramidal shape [64]. Kandiel et al. [65] showed the effect of use of urea on the morphology of the nanoparticles, being small anatase nanoparticles at low concentration of urea and turning into brookite rods at high concentrations of urea. They proposed the routes shown Fig. 9–6 based on their experimental observations.

Recently, Zhang et al. [62] showed the synthesis of template free TiO_2 hollow nanoparticles via hydrothermal process followed by a critical liquid-phase deposition (LDP). They obtained TiO_2 nanotubes via hydrothermal reaction of TiO_2 particles under alkaline conditions. Then, calcination of TiO_2 nanotubes at 600°C causes a complete loss of tubular

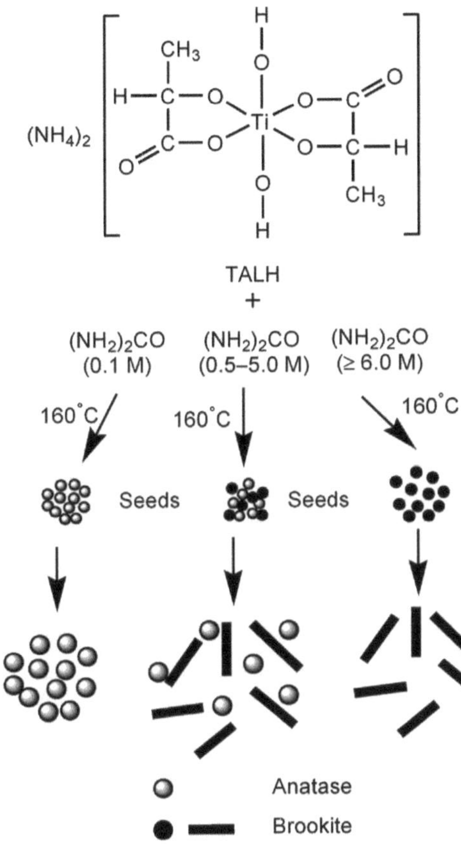

FIGURE 9–6 Proposed routes for the formation of anatase nanoparticles and brookite rods. *Reprinted with permission from Kandiel TA, Feldhoff A, Robben L, Dillert R, Bahnemann DW. Tailored titanium dioxide nanomaterials: anatase nanoparticles and brookite nanorods as highly active photocatalysts. Chem Mater 2010; 22:2054. Copyright © (2010) American Chemical Society.*

structure forming TiO_2 nanosheets. When TiO_2 nanotubes were modified by LDP followed by calcination at 600°C, hollow TiO_2 nanoparticles with an average inner diameter of about 3–5 nm were obtained.

9.2.2 Physical Methods—Gas-to-Solid Transformation

9.2.2.1 Chemical Vapor Deposition

CVD is a process of the deposition of solid materials from a volatile precursor that can be a gaseous, liquid, or solid material. Although reproducible, ultrapure, and uniform nanoparticles can be achieved with the CVD technique, it is a complex system that has several independent parameters such as system pressure, flow rate, and partial pressures, which require close monitoring and control. All these parameters require a strict optimization to obtain a

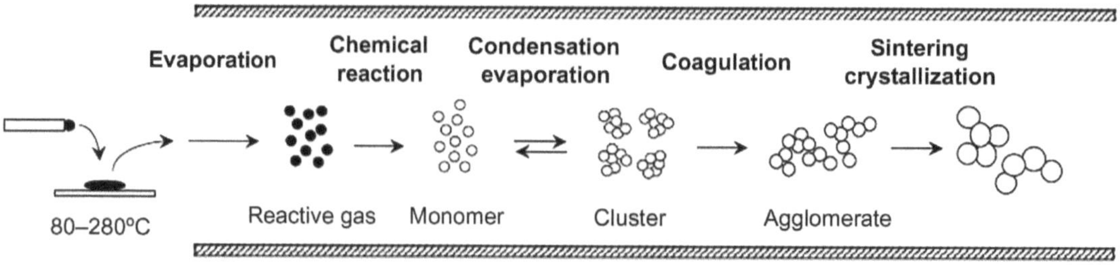

FIGURE 9–7 A typical particle formation in a CVD system. *Reprinted with permission from Nakaso K, Han B, Ahn KH, Choi M, Okuyama K. Synthesis of nonagglomerated nanoparticles by an electrospray assisted chemical vapor deposition (ES-CVD) method. J Aerosol Sci 2001;34(7):876 [66]. Copyright © (2003) Elsevier.*

product with desired properties. In addition, the precursors and the other chemicals used in this process are generally toxic, inflammable, explosive, or corrosive in nature. A typical CVD system, as shown in Fig. 9–7, contains three main components: A precursor supply system to generate precursors and deliver them to the reactor with the help of a carrier gas; a CVD reactor where the deposition reaction is initiated with the help of an external energy that comes from a heat, plasma, or light supply; and an exhaust system to remove the reaction byproducts.

There are different types of CVD processes that are named based on the source of external energy or the source of the precursor. Thermally activated or pyrolytic CVD processes that are performed at normal to ultralow pressures use thermal energy through resistance heating, radio frequency (RF) heating, or infrared heating. High-quality epitaxial thin film growth, for example, is generally performed by high-temperature CVD method [67,68]. Plasma-assisted CVD, which mostly uses a RF field to create plasma discharge, and photo CVD, which uses lasers to enhance the reaction rates, are the chemical deposition techniques performed at lower temperatures and comparably at lower pressures. Synthesis of iron oxide nanoparticles [69–71], ZnO nanowires, nanoribbons and needle-like rods [72–74], and gold nanoparticle decorated Al_2O_3, SiO_2, and TiO_2 for the oxidation of CO and of H_2 [75] were a few examples of metal oxide nanostructures synthesized with CVD processes.

9.2.2.2 Pulsed Laser Deposition

PLD is a relatively recent technique providing a simple, versatile, and contamination-free deposition of nanoparticles and thin films of various materials especially suited for oxide growth [76]. In PLD, material vaporized from the target using high-power pulsed laser beam is deposited on a substrate located opposite to the target. The laser beam enters a high vacuum chamber, in which the target and the substrate are located, through a window and is focused on the target of the desired composition. Depending on the background gases, for example, O_2 or Ar, used during the process, composition of the deposited materials is controlled. A schematic diagram of typical PLD setup is shown in Fig. 9–8.

PLD is a much faster technique compared with other deposition methods with pulse durations ranging from nanoseconds at pressures in the order of 100 Pa [78] to femtoseconds

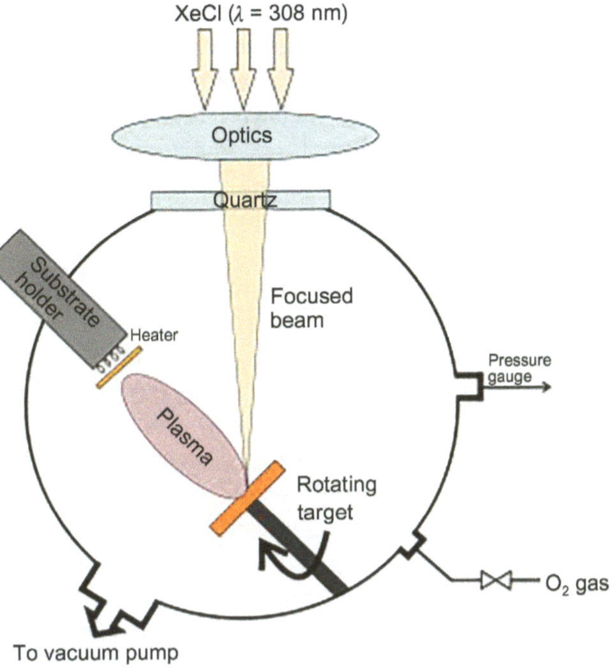

FIGURE 9–8 A schematic diagram of typical PLD configuration. *Reprinted from Huotari J, Kekkonen V, Puustinen J, Liimatainen J, Lappalainen J. Pulsed laser deposition for improved metal-oxide gas sensing layers. Procedia Eng 2016;168:1067 [77].*

in a vacuum [79] and to picoseconds in a liquid [80]. Vanadium dioxide (VO_2) nanoparticles were synthesized using PLD process by increasing the annealing time [81]. First, VO_2 film with the thickness of 100 nm was grown on a silicon substrate and then the film crystallized into nanoparticles as the annealing time increases in 250 mTorr oxygen at 450°C.

Blakemore et al. [82] reported a successful synthesis of surfactant-free, unsupported, size-controlled (<5 nm) cobalt oxide (Co_3O_4) nanoparticles, which are highly active catalysts for oxygen evolution in basic water, using PLD method in liquids. Other examples of metal oxide nanostructures synthesized with PLD technique are tungsten trioxide (WO_3) and zinc oxide (ZnO)-modified stannic oxide (SnO_2) [83,84], titanium dioxide (TiO_2) [85], zinc oxide (ZnO) [86,87], monoclinic gallium oxide (β-Ga_2O_3) [78], nickel oxide (NiO) [88], and calcium-iron complex oxides ($CaFe_2O_4$) [89].

9.3 Surface Modification of Metal Oxide Nanoparticles

9.3.1 Chemical Treatments

The chemical treatment is one of the most applied methods for surface functionalization of nanoparticles to improve their dispersion stability in liquid media and compatibility in any polymer matrix. Silane coupling is the most studied chemical treatment since the publication of Plueddemann in 1962, studying more than hundred different organosilanes as coupling agents in glass-reinforced polyester and epoxy composites. Organofunctional silanes are hybrids that consist of a hydrolyzable group such as methoxy, ethoxy, or acetoxy, which would react with metal oxide surface, and an organofunctional group such as amino, epoxy, or acryloxy, which would interact with the polymer matrix. In a typical reaction of silane coupling agent with metal oxide surface, alkoxy groups of the organosilane would hydrolyze and condense forming a covalent bond with the hydroxyl groups on the surface of the metal oxides.

Ma et al. [90] modified the surface of TiO_2 nanoparticles with polar *N*-(2-aminoethyl) 3-aminopropyl-trimethoxysilane (AEAPS) and found that low-density polyethylene (LDPE) nanocomposites with modified nanoparticles shows improved dielectric breakdown strength and space charge distribution compared with the nanocomposites with pristine nanoparticles. Qi et al. [91] modified the TiO_2 nanoparticles to obtain hydrophobic surfaces with 3-methacryloxypropyl-trimethoxysilane (MPS) for application in composited cool materials with high solar reflectance. Ju et al. [92] modified the MgO nanoparticles with three different silane coupling agents, that is, AEAPS, *n*-dodecyltrimethoxysilane (DTES), and vinyltris(2-methoxyethoxy)silane (VMES), and investigated the electrical insulating properties of LDPE nanocomposites containing modified MgO nanoparticles. Zhao et al. [3] investigated the effect of three silane coupling agents, that is, 3-glycidyloxypropyltrimethoxysilane, 3-aminopropyltriethoxysilane (APTES), and 3-methacryloxypropyl-trimethoxysilane, on the polyimide (PI)/SiO_2 interface using molecular dynamic simulations. A variety of other surface modifiers utilizing physical interactions, such as hydrogen bonding, have also been utilized for the functionalization of the metal oxide nanoparticles. Some examples of these surface modifiers include bovine serum albumin [93], ethylenediaminetetraacetic acid (EDTA) [94], vitamin B [95,96], 3,4,5,6-tetrabromo-*N*-(4-hydroxy-phenyl)-phthalamic acid [97], and *s*-triazine [98].

9.3.2 Polymer Grafting

Polymer grafting is another method to change the surface functionality of metal oxide nanoparticles, in which a synthetic polymer covalently bonds to the surface of the inorganic nanoparticles forming structures called organic–inorganic nanocomposites. There are two approaches to attach polymer chains onto the surface of metal oxide particles. The first method is called "grafting to," which is a process of attaching preformed polymer chains containing reactive groups onto the surface of the nanoparticles. On the other hand, in the "grafting from" protocol, first initiator-terminated monolayer is adsorbed on the surface of the nanoparticles followed by a subsequent polymer chain growth via several surface-initiated polymerization methods such as ATRP, RAFT, and anionic, cationic, or ring-opening polymerizations.

In the "grafting to" method, there is an activation barrier of attaching the polymer chains onto the nanoparticles surface, which appears once the tethered chains start to overlap. Therefore, the number of polymer chains that can be attached to the nanoparticle surfaces is limited in this approach [99]. Also, the total number of chains tethered to the surface, called grafting density, decreases with the polymer chain length, presumably due to the steric hindrance exerted by previously grafted chains. However, the length of the polymer chains is better controlled and characterized in the "grafting to" method since they are synthesized independently before the grafting process [100]. On the other hand, the higher number of polymer chains can successfully be tethered to the nanoparticle surface via "grafting from" protocol. The grafting density can be controlled by the coverage of the adsorbed initiating sides. Therefore, the "grafting from" method is more preferred in context of compatibilization metal oxide nanoparticles in a polymer matrix [101]. However, Zhao et al. [102] recently showed that the nanoparticles with a physical adsorbed poly(styrene-*b*-2-vinylpyridine) copolymer (principally a "grafting to" procedure) showed enhanced compatibility with polystyrene (PS) matrix compared with nanoparticles that are functionalized with the same copolymer at a comparable grafting density via "grafting from" strategy using RAFT polymerization. They claimed that the "grafting to" process provides a more uniform coverage on the surface of the nanoparticles than the "grafting from" one for a given grafting density since polymer chains grow on randomly located sites so that the final surface coverage is anisotropic and asymmetric, especially at low grafting density and long grafts while the polymer chain attachment continues until all sites are occupied in the case of the "grafting to" process. More recently, Asai et al. [103] reported a model to calculate the graft distribution of nanoparticle surface and concluded that "grafting to" process provides more homogeneous distribution of grafts on the nanoparticle surface than the "grafting from" counterpart for a given grafting density.

9.3.3 Ligand Exchange

Metal oxide nanostructures can be synthesized in the presence of capping agents or ligands to establish a colloidal stability; however, those surface functional groups may not be suitable for the application of the synthesized metal oxides. Therefore, as-synthesized metal oxides must undergo surface modification via a ligand exchange reaction. Ligands are defined depending on the number of electrons that the neutral ligand contributes to the metal oxide nanostructure-ligand bond as L-type (two electrons), X-type (one electron), and Z-type (zero electron). L-type ligands are Lewis bases, X-type ligands are single-electron donors in their neutral form, and Z-type ligands are Lewis acids [104]. Ligand-nanocrystal bond classification and exchange reactions can be seen in Fig. 9–9.

Lattuada and Hatton [105] reported the synthesis of oleic acid (OA) stabilized magnetic Fe_2O_4 nanoparticles and subsequent replacement of the oleic groups via ligand exchange reaction with various reactive ligands, which are ricinoleic acid, galataric acid, 2-bromo-2-methyl propionic acid, citric acid, and succinic anhydride. Those reactive ligands were used for further functionalization of the nanoparticles with variety of water-soluble polymers and polyelectrolytes. Dong et al. [106] reported synthesis of functional nanocrystals, including iron oxide (Fe_2O_4) nanoparticles and titanium dioxide (TiO_2) nanorods, and applied ligand exchange strategy for sequential surface engineering of those colloidal nanocrystals. First,

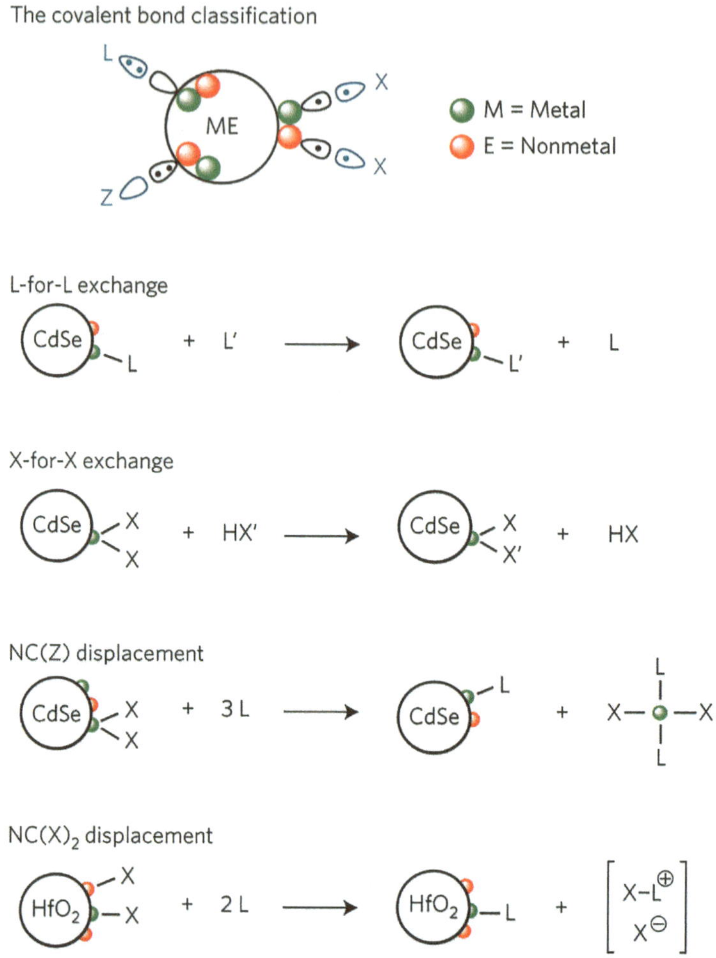

FIGURE 9–9 Ligand bond classification and exchange reactions. *Reprinted with permission from De Roo J, Van Driessche I, Martins JC, Hens Z. Colloidal metal oxide nanocrystal catalysis by sustained chemically driven ligand displacement. Nat Mater 2016;15(5):517–21. Copyright © (2016) Springer Nature.*

they used nitrosonium tetrafluoroborate ($NOBF_4$) to replace the original organic ligands. Then, variety of other capping agents, that is, OA, oleylamine (OAm), tetradecylphosphonic acid (TDPA), hexylamine (HAm), and poly(vinyl pyrrolidone) (PVP), were exchanged with $NOBF_4$ as seen in Fig. 9–10.

9.4 Polymer–Metal Oxide Nanocomposite Materials

Polymer nanocomposites are of interest to the scientific and industrial community because of their high performance and unique properties. Various types of inorganic nanoparticles

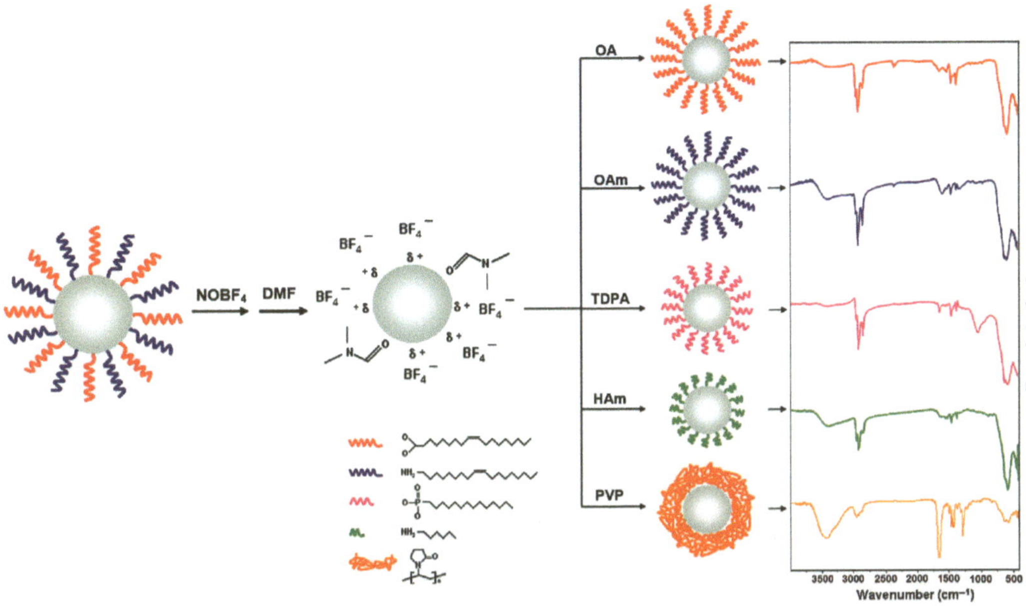

FIGURE 9–10 Sequential ligand exchange process of Fe_3O_4 nanoparticles by various capping agents. *Adapted with permission from Dong A, Ye X, Chen J, Kang Y, Gordon T, Kikkawa JM, et al. A generalized ligand-exchange strategy enabling sequential surface functionalization of colloidal nanocrystals. J Am Chem Soc 2011;133 (4):998–1006. Copyright © (2011) American Chemical Society.*

have been used in polymer–inorganic nanocomposites to realize desired mechanical, thermal, optical, and electrical properties. Incorporation of those nanoparticles in a polymer matrix is a critical step to achieve the best performance for the desired application. The interactions between the particle surface and the polymer matrix play an important role for the miscibility of nanoparticles and polymer matrix. In this section, methods used for polymer–inorganic metal oxide nanoparticle nanocomposite preparation and their applications will be discussed.

9.4.1 Methods for Preparation of Nanocomposites

9.4.1.1 Blending

The most conventional and industrially viable process for the preparation of nanocomposites is direct blending nanoparticles into the polymer matrix. Since the nanoparticles have a high tendency to agglomerate, realizing a successful dispersion of the nanoparticles in the polymer matrix is the major concern in the blending process. The surface functionalization and decoration with various organic molecules or polymers, as described in the previous sections, help to accomplish the favored interaction between the two components. There are two main practices in blending: Melt and solution blending. Melt blending is a high-temperature process requiring high-shear mixers or extruders and temperatures above the melting point or the glass transition point of the polymer. It is a cost-effective, environmentally friendly, and industrially preferred process since it does not require the use of any organic solvent. High temperatures and long processing times are generally required to ensure homogeneous mixing; however, degradation of the polymer or the immobilization of the organic or polymeric groups from the surface of the functionalized nanoparticles may arise from the high mixing temperature and/or long processing time of the melt blending procedure. On the other hand, solution blending has the advantage of being a low temperature and a low viscosity process. The specific interactions between the functionalized nanoparticles and the polymer matrix can readily be attained in solution blending. The major disadvantages of this method are the requirement of the large amount of the solvent and the cost and environmental issues associated with the use and recovery of the solvent.

9.4.1.2 In Situ Nanocomposite Formation

In situ incorporation of metal oxide nanoparticles in a polymer matrix can be achieve via two different approaches. One is the dispersing nanoparticles in a monomer solution prior to the polymerization and subsequent growth of polymer chains around the nanoparticles, which could already be functionalized or could be linked to the growing polymer chains with special interactions. Resulting polymer–inorganic composite can satisfy the desired interactions at the interface assuming a good dispersion of the nanoparticles in the initial solution. The second one is the in situ formation of nanoparticles in a polymer matrix or in the presence of monomer. The interfacial interactions between the inorganic nanoparticles and the polymer can be manipulated at the molecular level so that the agglomeration of nanoparticles is prevented. The reduction of metal salts in the presence of the polymer can realize the immobilization of the nanoparticles in a polymer matrix. Various reduction methods such as chemical, electrochemical, microwave assisted, or radiation-induced reduction can be applied for the nanoparticle formation. In this process, interaction between the metal ion and the polymer plays an important role in the final properties of the nanoparticles such as size, size distribution, and their dispersion in the matrix polymer. Menceloglu et al. [107–110] utilized electrostatic interactions between functional monomers, such as vinyl pyrrolidone, acrylic acid, and 2-acrylamido-2-methylpropane sulfonic acid, copolymerized with acrylonitrile (AN) and precursor metal salts to control the size and the distribution of the resulting metal and metal oxide nanoparticles.

Recently, a new reactor granule technology was developed for the in situ formation of highly dispersed metal oxide nanoparticles [111]. Polypropylene (PP)/TiO_2 and Al_2O_3 nanocomposites were fabricated by impregnating the PP granules with corresponding alkoxides and then conversion of the alkoxide to oxides by either melt mixing or melt mixing after exposure to water vapor.

Wan et al. [112–114] synthesized polyaniline (PANI)/TiO_2, PANI/Fe_3O_4, and PANI/γ-Fe_3O_4 nanocomposites via polymerization of β-naphthalene sulfonic acid (β-NSA) and anilinium in the presence of metal oxide nanoparticles and nanotubes. β-NSA and anilinium were self-assembled forming a micelle having the nanoparticle as the "core." Polymerization was induced by addition of ammonium persulfate resulting in formation of 1D PANI/metal oxide nanocomposites. Similarly, Lu et al. [115] reported the synthesis of PANI/Fe_3O_4 nanocomposites by ultrasound-assisted polymerization of the aqueous solutions of aniline, ammonium peroxydisulfate, and phosphoric acid (H_3PO_4) in the presence of aniline functionalized Fe_3O_4 nanoparticles.

9.4.1.3 Sol–Gel Processing

Sol–gel synthesis of inorganic metal oxide nanoparticles can be performed in the presence of a polymer or in parallel with the formation of the polymer. The nanoparticle formation proceeds via hydrolysis and condensation of an organometallic precursor such as TEOS, aluminum isopropoxide, or tetrabutyl titanate in the presence of an acidic or basic catalyst as in conventional sol–gel reaction and the polymer acts as a stabilizer for the forming nanoparticles. Therefore, homogeneous distribution of the nanoparticles in the polymer matrix is satisfied. The miscibility of the oxide nanoparticles and the polymer matrix can also be enhanced via covalent incorporation of an organometallic group into the polymer chains [116,117]. The resulting polymers are then hydrolyzed and condensed to produce polymer–inorganic hybrid materials.

TiO_2/PI hybrid films have been fabricated via sol–gel reaction of tetraethyl orthotitanate [$Ti(OEt)_4$] [118]. First polyamic acid (PAA) was synthesized at room temperature and then Ti $(OEt)_4$ was added along acetylacetone to prevent early gelation forming the precursor sol. After casting films using the PAA/$Ti(OEt)_4$ sol precursor, PI/TiO_2 hybrid films were formed by heat treatment at 300°C.

9.4.2 Properties of Polymer–Metal Oxide Nanocomposites

The main reason to incorporate inorganic metal oxide nanoparticles into polymer matrices is to produce a material with improved mechanical, optical, electrical, or thermal properties desired for its applications.

9.4.2.1 Mechanical Properties

The mechanical properties such as tensile strength, Young's modulus, hardness, toughness, and flexural modulus can be modified by the incorporation of nanoparticles in the polymer matrices. The well-dispersed metal oxide nanoparticles promote an effective stress transfer from polymer matrix to the nanoparticles leading to a higher strength; however, nonhomogeneous distribution of the filler nanoparticles produces the formation of more defects in the structure leading to weakening of the mechanical properties.

Unsaturated polyester nanocomposites were reinforced with natural fiber, jute, along with Al_2O_3 and ZrO_2 nanoparticles, which are modified with silanes, with different filler loadings [119]. Incorporation of metal oxides significantly improved the mechanical properties of the composite material as seen in Fig. 9–11A. Microhardness of the composites also improved with the addition of fillers up to 20%. Addition of modified metal oxide particles

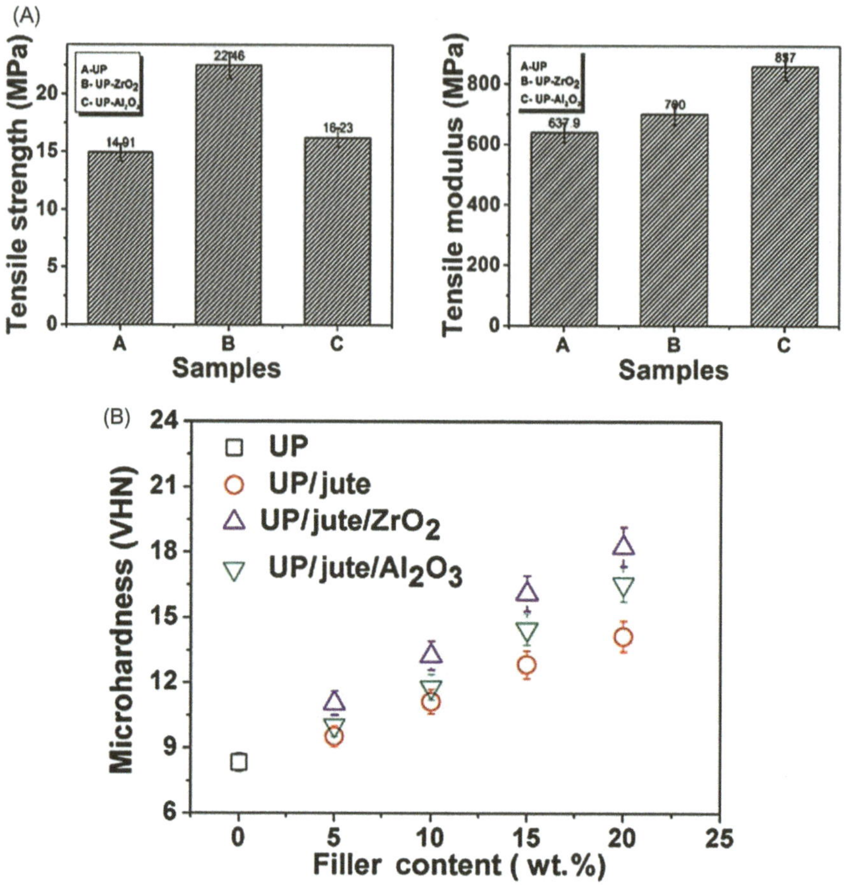

FIGURE 9–11 (A) Tensile strength, modulus and (B) microhardness of unsaturated polyester/modified metal oxide nanocomposites. *Reprinted with permission from Adhikari J, Biswas B, Chabri S, Bandyapadhyay NR, Sawai P, Mitra BC, et al. Effect of functionalized metal oxides addition on the mechanical, thermal and swelling behaviour of polyester/jute composites, Eng Sci Technol Int J 2017;20:763. Copyright © (2017) Elsevier.*

showed greater enhancement in microhardness compared with that of the only jute fiber–added composites (Fig. 9–11B).

Menceloglu et al. [120] investigated the effect of filler content on the structure–property behavior of SiO_2 supported polyethylene oxide based polyurethaneurea (PU) nanocomposites. PU/SiO_2 nanocomposites exhibited improved elastic modulus and tensile strength as well as increased soft segment glass transition temperature and melting temperatures with increasing amount of silica filler content.

Schechtel et al. [121] fabricated 4-allylcatechol functionalized TiO_2/thiolene polymer nanocomposites as a potential candidate for dental restorative materials. They observed that effective longitudinal modulus of the composite films increases from 6.2 to 37.5 GPa with an increase in the weight fraction of modified TiO_2 nanoparticles from 0% to 90%.

Synergetic effect of carbon nanofiber (CNF) and Fe_2O_3 in improving mechanical properties of styrene–butadiene rubber (SBR) nanocomposites was recently reported [122]. While both the addition of Fe_2O_3 and CNF improved the mechanical properties of the nanocomposites, addition of Fe_2O_3-decorated CNF into the SBR showed even greater increase in that of the nanocomposite materials. Tensile strength of Fe_2O_3-CNF containing nanocomposite was increased by 13% and 5% relative to that of control and CNF embedded nanocomposites, respectively. Elongation at break of the same nanocomposites was higher (710%) compared with those with only CNF (490%).

9.4.2.2 Optical Properties

Metal oxide nanoparticles provide specific optical properties such as UV absorption, specific color absorption in the visible region, dichroism, and photoluminescence. Metal oxide–polymer nanocomposites combine these specific features with the processability of a polymer and find a wide range of applications in the fields of solar cells, light-emitting diodes, transistors, optoelectronic packaging, and coatings.

Hajibeygi et al. [123] recently reported azo-dye surface modified ZnO nanoparticles reinforced polyamide (PA) self-colored nanocomposites. PA/modified ZnO nanocomposites showed an intensity reduction and blue shift in UV–vis spectroscopy and also a violet emission and an increased intensity in photoluminescence spectroscopy, as seen in Fig. 9–12.

Tungsten trioxide (WO_3) is known for its photochromic properties when exposed to UV irradiation. Sangpraserdsuk et al. [124] investigated optical and photochromic properties of both unmodified and modified tungsten trioxide–reinforced polycarbonate (PC) composites.

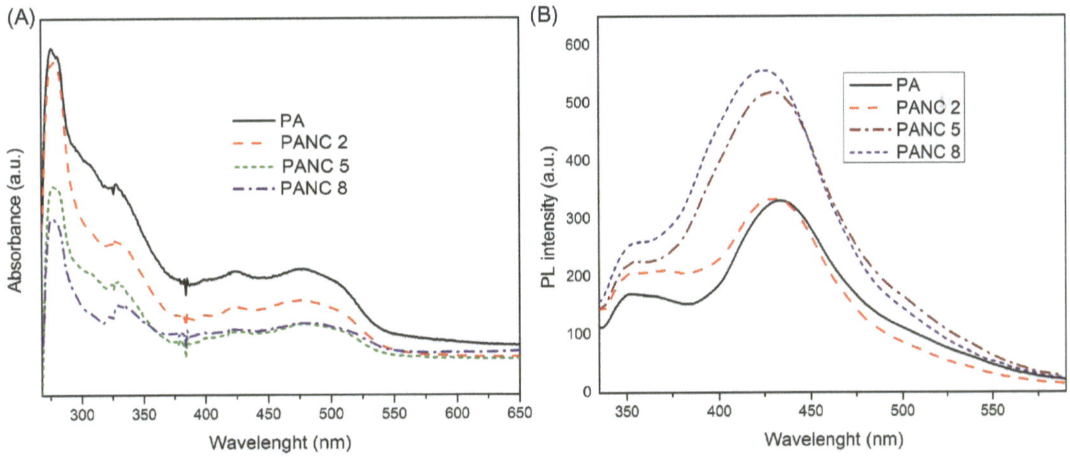

FIGURE 9–12 (A) UV–vis spectra and (B) photoluminescence spectra of PA and PA/modified ZnO nanocomposites (PANC) with different amounts of ZnO. *Reprinted with permission from Hajibeygi M, Shabanian M, Omidi-Ghallemohamadi M, Khonakdar HA. Optical, thermal and combustion properties of self-colored polyamide nanocomposites reinforced with azo dye surface modified ZnO nanoparticles. Appl Surf Sci 2017;416:634–5. Copyright © (2017) Elsevier.*

Tungsten trioxide particles were functionalized with γ-glycidoxypropyltrimethoxy silane (GPTMS) through silane coupling reaction. Spectral behavior of the PC changed by the addition of tungsten trioxide filler regardless of added amount from being a neat infrared (NIR) reflective material to NIR absorbing material after exposure to sunlight. NIR absorbance of the composite material was further increased when silane-functional tungsten trioxide was used. The photochromic properties of the PC composite films were also improved by the use of silane-functional tungsten trioxide compared with the pristine tungsten trioxide with the same amounts. The silane-modified tungsten trioxide provided higher color contrast as well as reduced bleaching time than its neat counterpart.

9.4.2.3 Electrical Properties

Metal oxide–polymer nanocomposites with homogeneously dispersed fillers of the size less than 100 nm provide significant improvement in electrical properties such as dielectric strength, electrical conductivity, and voltage endurance. The polymer/nanoparticle surface area per unit volume of the nanocomposite determined by the particle size affects the level of improvement in all these properties.

The development of ultrahigh electric insulator materials for high-voltage direct-current cables is important to raise the upper voltage limit of intercontinental electric power transmissions. Several researchers investigated the polyethylene/metal oxide nanocomposites and the parameters, such as interfacial area and chemistry, and particle aggregation affecting the final properties. Pourrahimi et al. [125] modified the surface of zinc oxide (ZnO) nanoparticles with alkyl-containing silanes with alkyl groups being methyl (C1), octyl (C8), and octadecyl (C18) and investigated the effect of the surface modification, amount, and size of the nanoparticles on electrical insulation properties of polyethylene nanocomposites. Charging current was lower for all nanocomposite systems than for the unfilled polyethylene. The conductivity decreased with increasing amount of modified ZnO nanoparticles; the highest investigated amount of 3% being the most effective in reducing conductivity especially in the case of the nanocomposite with C8-coated ZnO (Fig. 9–13). They also showed that the smallest interparticle distance was observed for the nanocomposite containing 3 wt.% of C8-coated ZnO nanoparticles, which also showed the lowest conductivity (2.4×10^{-16} S/m) (Fig. 9–13C).

The same group also reported nanocomposites of LDPE with moisture-resistant magnesium oxide (MgO) nanoparticles modified with silane coating [126]. These nanocomposites exhibited high interfacial adhesion and well-dispersed nanoparticles compared with the nanocomposites with unmodified MgO as well as a significant reduction in the electrical conductivity. Moreover, hydrocarbon functional silsesquioxane coated MgO nanoparticles caused three orders of magnitude on average reduction in the volume conductivity (7×10^{-16} S/m) in LDPE nanocomposites compared with the pristine LDPE (2×10^{-13} S/m) [127]. Uniformly dispersed surface functional MgO nanoparticles also delayed the onset of thermal degradation of the nanocomposites by 100°C.

Sampreeth et al. [128] synthesized PANI/cerium-doped titanium oxide (Ce-TiO_2) nanocomposites via one-step in situ polymerization technique and evaluated the effect of

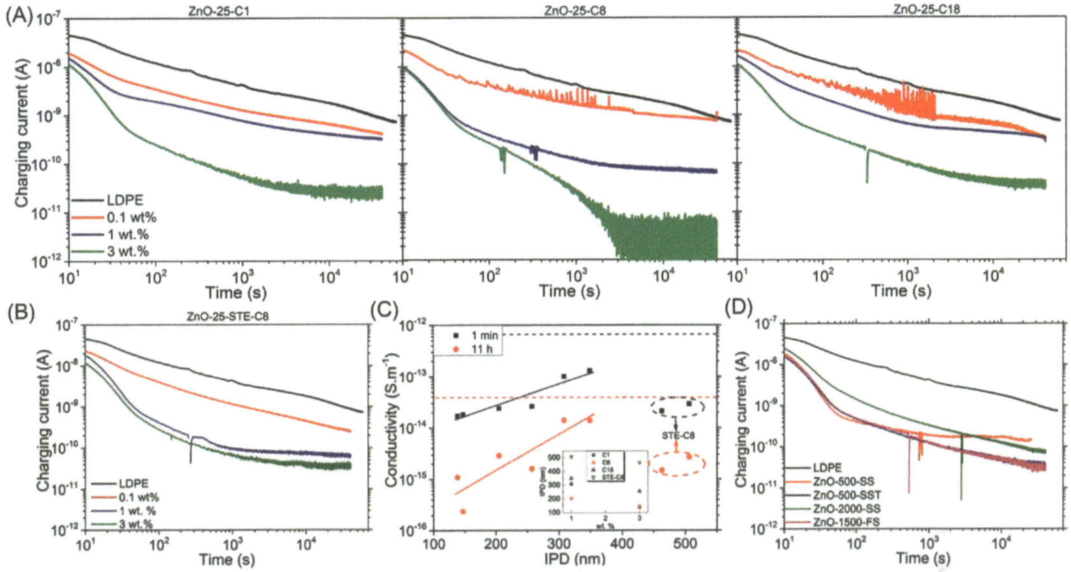

FIGURE 9–13 (A and B) Charging current of pristine LDPE and its nanocomposites based on different weight fractions of ZnO nanoparticles with different surface coatings at 2.6 kV ($E = 32.5$ kV/mm) and 60°C. (C) DC conductivity plotted as a function of interparticle distance for the composites based on 25 nm ZnO nanoparticles with different surface functionalities. The black and red dashed lines correspond to the conductivity values of unfilled LDPE at 1 min and 11 h, respectively. The inset of the graph shows the interparticle distance versus the weight fraction of the nanoparticles. (D) Charging current of composites based on 3 wt.% of submicron- and micrometer-sized ZnO particles. *Reprinted with permission from Pourrahimi AM, Hoang TA, Liu D, Pallon LKH, Gubanski S, Olsson RT, et al. Highly efficient interfaces in nanocomposites based on polyethylene and ZnO nano/hierarchical particles: a novel approach toward ultralow electrical conductivity insulations. Adv Mater 2016;28:8654. Copyright © (2016) John Wiley & Sons.*

nanoparticle content on electrical properties. All PANI/Ce-TiO_2 nanocomposites containing different amount of filler exhibited higher AC conductivity, higher dielectric constant, and higher DC conductivity compared with the pristine PANI. The nanocomposites containing 7% Ce-doped TiO_2 showed the most significant improvement among all.

9.4.2.4 Thermal Properties

The properties such as thermal stability, flame resistance, flame retardancy, and thermal conductivity of a polymer can be enhanced by the addition of metal oxide nanoparticles. The thermal performance of the composite materials, besides the inherent properties of the polymer and the nanoparticles, is influenced by factors such as nanoparticle agglomeration, particle size, functional groups on the particle surface, nanoparticle networking in the polymer matrix, and polymer/nanoparticle system surface energy.

Polylactic acid (PLA)/MgO nanocomposites were synthesized by in situ melt condensation of L-lactic acid and surface-hydroxylated MgO nanoparticles [129]. PLA was grafted onto the MgO nanoparticles through the surface hydroxyl groups and interfacial interaction between PLA and MgO was further improved with H-bonding between the ester groups of PLA and the remaining hydroxyl groups on the MgO. Thermal stability of PLA/MgO nanocomposites was significantly improved compared with that of pure PLA.

Kang et al. [130] investigated the effect of interfacial interaction between the polymer and filler surface in epoxy/silica nanocomposites. They functionalized the surface of SiO_2 nanocomposites with epoxide (S-epoxide), amine (S-amine), and isocyanate (S-NCO) and removed the surface silanol groups with calcination. While the nanocomposites with calcination and S-NCO showed weak interface and aggregation of the nanoparticles, aggregation of the fillers were highly suppressed using S-epoxide and S-amine nanoparticles, which provides high interfaces in the nanocomposites. The coefficient of thermal expansion (CTE) of the nanocomposites was reduced with increasing amount of the fillers regardless of the surface functionality although nanocomposites with high interface provided a reduction in CTE to a greater extent.

PS-encapsulated SiO_2 were used as filler in poly(ethylene terephthalate) (PET) nanocomposites and increased the crystallization temperature of PET nanocomposites compared with both unmodified and modified SiO_2 [131].

Friederich et al. [132] investigated fire retardant properties of poly(methyl methacrylate) (PMMA)/metal oxide nanocomposite. They added three different metal oxides, titanium dioxide (TiO_2), alumina (Al_2O_3), and boehmite (AlOOH), into PMMA matrix by melt blending. Time-to-ignition (TTI), total heat release (THR), and peak heat release rate (pHRR) were calculated based on the cone calorimetry measurements for the nanocomposites containing 0%, 5%, 10%, and 15% nanofiller. TTI increased up to 22% by the addition of 15% filler. The peak heat release rate was reduced 32% by the addition of 15% boehmite and 46% by the addition of the same amount of alumina and titanium dioxide. Moreover, alumina was the most effective filler in the amount of 15% for the THR rate by showing a reduction of 28% (Fig. 9–14).

9.4.3 Applications of Polymer–Metal Oxide Nanocomposites

Decorated metal oxide nanostructures provide improved interfacial interaction between the nanofillers and polymer matrix and introduce specific features to their nanocomposites resulting in superior mechanical, optical, electrical, and thermal properties. These materials with improved properties find application in a variety of fields such as mechanics, electronics, sensors, optics, energy, and biomedicine. Table 9–3 summarizes potential applications of decorated metal oxide–polymer nanocomposites.

9.5 Conclusion and Outlook

Polymer nanocomposites containing functional metal oxide nanomaterials are an important subdivision of polymer-based nanocomposites. Metal oxide nanoparticles not only provide reinforcement and improvement of mechanical properties of polymers to which they are added but also add unique functionalities they possess, such as antibacterial, UV absorbance,

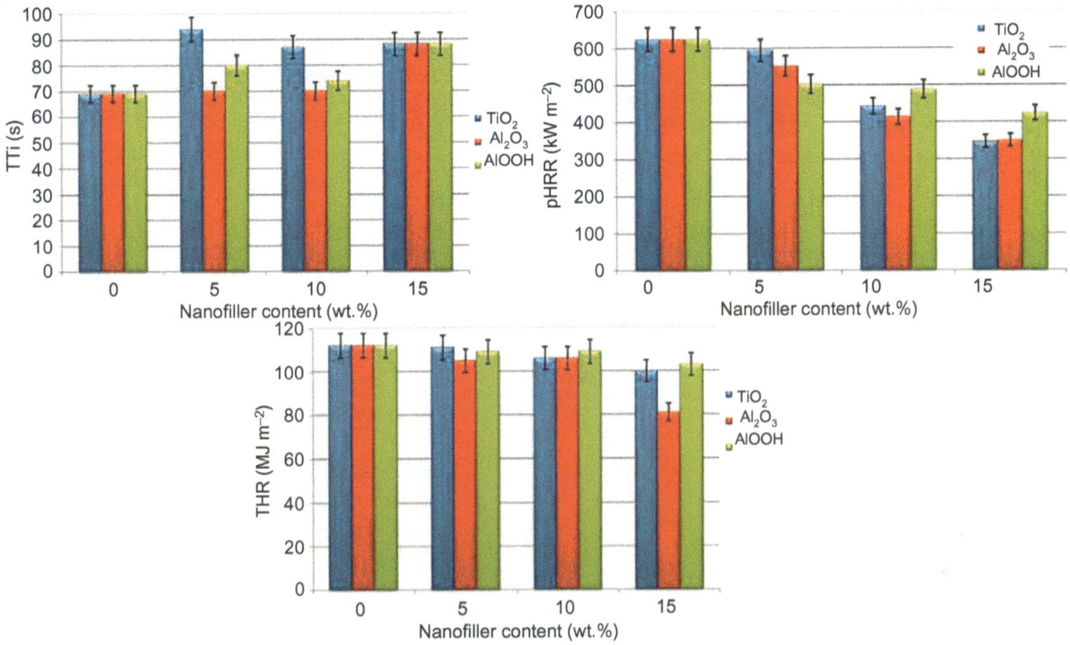

FIGURE 9–14 TTI, pHRR, and THR of PMMA nanocomposites with TiO_2, Al_2O_3, and AlOOH nanoparticles. *Adapted with permission from Friederich B, Laachachi A, Ferriol M, Ruch D, Cochez M, Toniazzo VR. Tentative links between thermal diffusivity and fire-retardant properties in poly(methyl methacrylate)–metal oxide nanocomposites. Polym Degrad Stab 2010;95:1188–9. Copyright © (2010) Elsevier.*

Table 9–3 Potential Applications of Decorated Metal Oxide/Polymer Nanocomposites

Metal Oxide	Surface Function	Polymer Matrix	Application	Reference
SiO_2	Sulfonated-HPA	Nafion	DMFC membrane	[133]
	MPS	PS/PET	Engineering plastics reinforcement	[131]
	Epoxy ring, amine, isocyanate	Epoxy	Coatings, molding compounds	[130]
	Amino, epoxy and alkyl silanes	Epoxy	Adhesives, electronic packaging	[134]
	TMSPMA	PMMA	Bioactive bone substitute, dental applications	[135]
	Amino	PEG	Bioimaging	[136]
	ATRP initiator (isobutyryl bromide)	HPBBEA-g-PDMAEMA	Biomedicine	[137]
	APTES	LBL-PEG	Controlled drug release	[138]

(Continued)

Table 9–3 (Continued)

Metal Oxide	Surface Function	Polymer Matrix	Application	Reference
TiO_2	MPS	AN-St-A terpolymer	Solar reflectance, environmental applications	[91]
	AEAPS	LDPE	Electrical insulator	[90]
	GPTES	PTAI	Sensors	[4]
	4-allylcatechol	Thiolene	Dental restorative materials	[121]
	γ-MPS	High-density polyethylene	Bone repair	[139]
	DMPA	PU	Self-cleaning and antibacterial coatings	[140]
	Phenyl dichlorophosphate	Epoxy resin	Fire retardant nanocomposites	[141]
ZnO	EDTA	PVC	Biomedical devices	[94]
	PVA	PCL	Tissue engineering, food packaging	[142]
	Monoethanolamine	PMMA-PS	UV shielding coating	[143]
	4-[(2-Hydroxy-1-naphthyl)azo] benzenesulfonic acid	PA	Optical applications	[123]
	Alkyl silanes	LDPE	Electrical insulator	[125]
	VTES, PTES	PS	Antistatic materials	[2]
	MPTMS	PS	UV shielding in coatings, films, cosmetics	[144]
	Bromo-acetamide (ATRP initiator)	PHEA	UV absorber	[145]
	Alkyl ammonium salt	SEBS and SEBS-MA	HV insulation	[146]
MgO	AEAPS, DTES, VMES	LDPE	Electrical insulator	[92]
	OTMS	LDPE	Electrical insulator	[126]
	Hydroxyl groups	PLA	Biomedical materials, textiles, packaging materials	[129]
Fe_3O_4	$NOBF_4$	POA, PVP	Magnetic applications	[106]
	MPTES	PMMA	Magnetic and electronic applications	[147]
	Oleic acid, oleylamine	PEG, PLA, PHEMA, PNIPAm	Biomedical and separation applications	[105]
	Oleic acid	SEBS	MW communication devices	[148]
	Oleic acid	PDMS	Electromagnetic circuits	[149]
Al_2O_3	Silane	Unsat. PE	Packaging applications	[119]
	P(St-MA)	PC	Optical applications	[150]
	APDMS	PMMA	Dental and high impact glasses	[151]
	PDMS, PET, Terathane	PMMA	Flame retardant materials	[152]
ZrO_2	Citric acid and ascorbic acid	PVC	Engineering plastics	[153]
	Silane	Unsat. PE	Packaging applications	[119]
	GPTMS	Cyanate ester/ azomethine	High dielectric applications, consumer electronics	[154]
WO_3	GPTES	PC	Photochromic glass, building applications	[124]
SnO_2	Sulfonates	Nafion	PEMFC membrane	[155]
Clay	16-BAC	ABS	3D printing material	[156]
	CTAB	ABS	Engineering thermoplastics	[157]
	4-Acetylbiphenyl ammonium salt	PS, high impact PS, ABS	Transportation, construction, and electrical applications	[158]

photoluminescence, photocatalytic, magnetic, thermal, and flame resistance. Compatibility and dispersion stability of the metal oxide nanomaterials with the polymer matrices have a significant effect to achieve the desired properties. Improved interfacial interactions between the oxide nanofillers and the polymer matrices can be achieved with the surface modification of the metal oxides. Various strategies, such as chemical coupling, polymer grafting, and ligand exchange, can be applied for the functionalization. Development of novel strategies for the functionalization and diversification of the functional groups would open new horizons for the applications of the nanocomposites. We have to emphasize that new synthetic and processing strategies for nanocomposite production remain to be exploited by synthetic chemists and polymer scientists.

References

[1] Mallakpour S, Barati A. Efficient preparation of hybrid nanocomposite coatings based on poly(vinyl alcohol) and silane coupling agent modified TiO_2 nanoparticles. Prog Org Coat 2011;71(4):391–8.

[2] Ma C-CM, Chen Y-J, Kuan H-C. Polystyrene nanocomposite materials—preparation, mechanical, electrical and thermal properties, and morphology. J Appl Polym Sci 2006;100(1):508–15.

[3] Zhao Y, Qi X, Ma J, Song L, Yang Y, Yang Q. Interface of polyimide-silica grafted with different silane coupling agents: molecular dynamic simulation. J Appl Polym Sci 2018;135(4):45725.

[4] Ghaemy M, Qasemi S, Ghassemi K, Bazzar M. Nanostructured composites of poly(triazole-amide-imide)s and reactive titanium oxide by epoxide functionalization: thermal, mechanical, photophysical and metal ions adsorption properties. J Polym Res 2013;20(10).

[5] Robertson CG, Lin CJ, Rackaitis M, Roland CM. Influence of particle size and polymer – filler coupling on viscoelastic glass transition of particle-reinforced polymers. Macromolecules 2008;41:2727–31.

[6] Li C, Benicewicz BC. Synthesis of well-defined polymer brushes grafted onto silica nanoparticles via surface reversible addition – fragmentation chain transfer polymerization. Macromolecules 2005;38:5929–36.

[7] Bartholome C, Beyou E, Bourgeat-Lami E, Chaumont P, Zydowicz N. Nitroxide-mediated polymerizations from silica nanoparticle surfaces: "graft from" polymerization of styrene using a triethoxysilyl-terminated alkoxyamine initiator. Macromolecules 2003;36:7946–52.

[8] Giri PK, Bhattacharyya S, Singh DK, Kesavamoorthy R, Panigrahi BK, Nair KGM. Correlation between microstructure and optical properties of ZnO nanoparticles synthesized by ball milling. J Appl Phys 2007;102(9):093515.

[9] Sharma S, Pande SS, Swaminathan P. Top-down synthesis of zinc oxide based inks for inkjet printing. RSC Adv 2017;7(63):39411–19.

[10] Priyadarshana G, Kottegoda N, Senaratne A, De Alwis A, Karunaratne V. Synthesis of magnetite nanoparticles by top-down approach from a high purity ore. J Nanomater 2015;2015:1–8.

[11] Chen JS, Zhu T, Yang XH, Yang HG, Lou XW. Top-down fabrication of r-Fe_2O_3 single-crystal nanodiscs and microparticles with tunable porosity for largely improved lithium storage properties. J Am Chem Soc 2010;132(38):13162–4.

[12] Salah N, Habib SS, Khan ZH, Memic A, Azam A, Alarfaj E, et al. High-energy ball milling technique for ZnO nanoparticles as antibacterial material. Int J Nanomed 2011;6:863–9.

[13] Cunningham A, Bürgi T. Bottom-up organisation of metallic nanoparticles. In: Rockstuhl C, Scharf T, editors. Amorphous nanophotonics. Berlin, Heidelberg: Springer; 2013. p. 1–37.

[14] Sui R, Charpentier P. Synthesis of metal oxide nanostructures by direct sol–gel chemistry in supercritical fluids. Chem Rev 2012;112(6):3057–82.

[15] Thiagarajan S, Sanmugam A, Vikraman D. Facile methodology of sol–gel synthesis for metal oxide nanostructures. In: Chandra U, editor. Recent applications in sol–gel synthesis. Croatia: InTechOpen; 2017.

[16] Deshmukh R, Niederberger M. Mechanistic aspects in the formation, growth and surface functionalization of metal oxide nanoparticles in organic solvents. Chem Eur J 2017;23(36):8542–70.

[17] Therese GHA, Kamath PV. Electrochemical synthesis of metal oxides and hydroxides. Chem Mater 2000;12:1195–204.

[18] Suh SK, Yuet K, Hwang DK, Bong KW, Doyle PS, Hatton TA. Synthesis of nonspherical superparamagnetic particles: in situ coprecipitation of magnetic nanoparticles in microgels prepared by stop-flow lithography. J Am Chem Soc 2012;134:7337–43.

[19] Ebelmen M. Recherches sur les combinaisons des acides borique et silicique avec les éthers. Ann Chim Phys 1846;16:129–66.

[20] Graham T. On the properties of silicic acid and other analogous colloidal substances. J Chem Soc 1864;17:318–27.

[21] StöBer W, Fink A, Bohn E. Controlled growth of monodisperse silica spheres in the micron size range. J Colloid Interface Sci 1968;26(1):62–9.

[22] Hench LL, West JK. The sol–gel process. Chem Rev 1990;90(1):33–74.

[23] Lofgreen JE, Ozin GA. Controlling morphology and porosity to improve performance of molecularly imprinted sol–gel silica. Chem Soc Rev 2014;43:911–33.

[24] Brinker CJ, Scherer GW. Sol–gel science: the physics and chemistry of sol–gel processing. Boston, MA: Academic Press; 1990.

[25] Vanblaaderen A, Vangeest J, Vrij A. Monodisperse colloidal silica spheres from tetraalkoxysilanes: particle formation and growth-mechanism. J Colloid Interface Sci 1992;154:481–501.

[26] Lee KT, Sathyagal AN, Mccormick AV. A closer look at an aggregation model of the Stober process. Colloids Surf A Physicochem Eng Asp 1998;144:115–25.

[27] Greasley SL, Page SJ, Sirovica S, Chen S, Martin RA, Riveiro A, et al. Controlling particle size in the Stöber process and incorporation of calcium. J Colloid Interface Sci 2016;469(1):213–23.

[28] Rahman IA, Vejayakumaran P, Sipaut CS, Ismail J, Abubakar M, Adnan R, et al. An optimized sol–gel synthesis of stable primary equivalent silica particles. Colloids Surf A Physicochem Eng Asp 2007;294 (1–3):102–10.

[29] Malay O, Yilgor I, Menceloglu YZ. Effects of solvent on TEOS hydrolysis kinetics and silica particle size under basic conditions. J Solgel Sci Technol 2013;67(2):351–61.

[30] Jafarzadeh M, Rahman IA, Sipaut CS. Synthesis of silica nanoparticles by modified sol–gel process: the effect of mixing modes of the reactants and drying techniques. J Solgel Sci Technol 2009;50(3):328–36.

[31] Rahman IA, Vejayakumaran P, Sipaut CS, Ismail J, Chee CK. Effect of the drying techniques on the morphology of silica nanoparticles synthesized via sol–gel process. Ceram Int 2008;34(8):2059–66.

[32] Rahman IA, Padavettan V. Synthesis of silica nanoparticles by sol–gel: size-dependent properties, surface modification, and applications in silica-polymer nanocomposites—a review. J Nanomater 2012;2012:1–15.

[33] Behrens M. Coprecipitation: an excellent tool for the synthesis of supported metal catalysts–from the understanding of the well known recipes to new materials. Catal Today 2015;246:46–54.

[34] Ke H, Wang W, Wang Y, Xu J, Jia D, Lu Z, et al. Factors controlling pure-phase multiferroic $BiFeO_3$ powders synthesized by chemical coprecipitation. J Alloys Compd 2011;509(5):2192–7.

[35] Kalia S, Kango S, Kumar A, Haldorai Y, Kumari B, Kumar R. Magnetic polymer nanocomposites for environmental and biomedical applications. Colloid Polym Sci 2014;292(9):2025–52.

[36] Iida H, Takayanagi K, Nakanishi T, Osaka T. Synthesis of Fe_3O_4 nanoparticles with various sizes and magnetic properties by controlled hydrolysis. J Colloid Interface Sci 2007;314(1):274–80.

[37] Lee J, Isobe T, Senna M. Magnetic properties of ultrafine magnetite particles and their slurries prepared via in-situ precipitation. Colloids Surf A Physicochem Eng Asp 1996;109:121–7.

[38] Kim JH, Kim SM, Kim YI. Properties of magnetic nanoparticles prepared by coprecipitation. J Nanosci Nanotechnol 2014;14:8739–44.

[39] Muneeswaran M, Jegatheesan P, Giridharan NV. Synthesis of nanosized $BiFeO_3$ powders by coprecipitation method. J Exp Nanosci 2013;8(3):341–6.

[40] Ni H, Gao Z, Li X, Xiao Y, Wang Y, Zhang Y. Synthesis and characterization of $CuFeMnO_4$ prepared by coprecipitation method. J Mater Sci 2018;53(5):3581–9.

[41] Behrens M, Kasatkin I, Kühl S, Weinberg G. Phase-pure Cu, Zn, Al hydrotalcite-like materials as precursors for copper rich $Cu/ZnO/Al_2O_3$ catalysts. Chem Mater 2010;22(2):386–97.

[42] Behrens M, Kissner S, Girsgdies F, Kasatkin I, Hermerschmidt F, Mette K, et al. Knowledge-based development of a nitrate-free synthesis route for Cu/ZnO methanol synthesis catalysts via formate precursors. Chem Commun 2011;47(6):1701–3.

[43] Kühl S, Tarasov A, Zander S, Kasatkin I, Behrens M. Cu-based catalyst resulting from a Cu, Zn, Al hydrotalcite-like compound: a microstructural, thermoanalytical, and in situ XAS study. Chem Eur J 2014;20(13):3782–92.

[44] Mette K, Kühl S, Düdder H, Kähler K, Tarasov A, Muhler M, et al. Stable performance of Ni catalysts in the dry reforming of methane at high temperatures for the efficient conversion of CO_2 into syngas. ChemCatChem 2014;6(1):100–4.

[45] Ota A, Kunkes EL, Kröhnert J, Schmal M, Behrens M. Particle size effect in methane activation over supported palladium nanoparticles. Appl Catal A 2013;452:203–13.

[46] Ota A, Kröhnert J, Weinberg G, Kasatkin I, Kunkes EL, Ferri D, et al. Dynamic surface processes of nanostructured Pd_2Ga catalysts derived from hydrotalcite-like precursors. ACS Catal 2014;4(6):2048–59.

[47] Danielsson I, Lindman B. The definition of microemulsion. Colloids Surf 1981;3(4):391–2.

[48] Tadros TF. Applied surfactants: principles and applications. Weinheim: Wiley-VCH Verlag GmbH & Co. KGaA; 2005.

[49] Khadzhiev SN, Kadiev KM, Yampolskaya GP, Kadieva MK. Trends in the synthesis of metal oxide nanoparticles through reverse microemulsions in hydrocarbon media. Adv Colloid Interface Sci 2013;197–198:132–45.

[50] Pileni MP. Reverse micelles as microreactors. J Phys Chem 1993;97(27):6961–73.

[51] Chen JP, Lee KM, Sorensen GM, Klabunde KJ, Hadjipanayis GC. Magnetic properties of microemulsion synthesized cobalt fine particles. J Appl Phys 1994;75(10):5876–8.

[52] Kadieva MK, Khadzhiev SN, Kadiev KM, Gyul'maliev AM, Yakovenko TV. The formation of nanosized molybdenum oxide particles in a hydrocarbon medium. Petroleum Chemistry 2011;51(1):16–23.

[53] Kadieva MK, Khadzhiev SN, Kadiev KM, Yakovenko TV. Synthesis of nanosized particles of components of petroleum refining catalysts in a hydrocarbon medium from inverted microemulsions. Petrol Chem 2011;51(6):426–34.

[54] Byrappa K, Yoshimura M. Handbook of hydrothermal technology a technology for crystal growth and materials processing. In: Mcguire GE, Rossnagel SM, Bunshah RF, editors. Materials science and process technology series, ed.. New York: Noyes Publications/William Andrew Publishing, LLC; 2001.

[55] Wang J, Sun J, Sun Q, Chen Q. One-step hydrothermal process to prepare highly crystalline Fe_3O_4 nanoparticles with improved magnetic properties. Mater Res Bull 2003;38(7):1113–18.

[56] Zheng Y, Cheng Y, Wang Y, Bao F. Synthesis and shape evolution of a-Fe_2O_3 nanophase through two-step oriented aggregation in solvothermal system. J Cryst Growth 2005;284(1–2):221–5.

[57] Jing Z, Wu S. Synthesis and characterization of monodisperse hematite nanoparticles modified by surfactants via hydrothermal approach. Mater Lett 2004;58(27–28):3637–40.

[58] Titirici M-M, Antonietti M, Thomas A. A generalized synthesis of metal oxide hollow spheres using a hydrothermal approach. Chem Mater 2006;18(16):3808–12.

[59] Cheng H, Ma J, Zhao Z, Qi L. Hydrothermal preparation of uniform nanosize rutile and anatase particles. Chem Mater 1995;7:663–71.

[60] Komarneni S, Rajha RK, Katsuki H. Microwave-hydrothermal processing of titanium dioxide. Mater Chem Phys 1999;61:50–4.

[61] Aruna ST, Tirosh S, Zaban A. Nanosize rutile titania particle synthesis via a hydrothermal method without mineralizers. J Mater Chem 2000;10:2388–91.

[62] Zhang Y, Hu H, Chang M, Wei H, Chen D, Zhang M, et al. Template-free scalable synthesis of TiO_2 hollow nanoparticles for excellent photoelectrochemical application. J Mater Sci 2018;53:2102–14.

[63] Yin J, Gao W, Yan S, Man J, Cui J, Hao H, et al. Facile synthesis of CTAB-SM-TiO_2 nanophotocatalyst and its superior photocatalytic performance. Mater Sci Semicond Process 2018;74:284–91.

[64] Horvat B, Recnik A, Drazic G. The growth of anatase bipyramidal crystals during hydrothermal synthesis. J Cryst Growth 2012;347(1):19–24.

[65] Kandiel TA, Feldhoff A, Robben L, Dillert R, Bahnemann DW. Tailored titanium dioxide nanomaterials: anatase nanoparticles and brookite nanorods as highly active photocatalysts. Chem Mater 2010;22:2050–60.

[66] Nakaso K, Han B, Ahn KH, Choi M, Okuyama K. Synthesis of non-agglomerated nanoparticles by an electrospray assisted chemical vapor deposition (ES-CVD) method. J Aerosol Sci 2003;34(7):869–81.

[67] Ivanov PG, Watts SM, Lind DM. Epitaxial growth of CrO_2 thin films by chemical-vapor deposition from a Cr8O21 precursor. J Appl Phys 2001;89(2):1035–40.

[68] Lu P, He S, Lia FX, Jia QX. Epitaxial growth of RuO_2 thin films by metal-organic chemical vapor deposition. Thin Solid Films 1999;340:140–4.

[69] Martelli S, Mancini A, Giorgi R, Alexandrescu R, Cojocaru S, Crunteanu A, et al. Production of iron-oxide nanoparticles by laser-induced pyrolysis of gaseous precursors. Appl Surf Sci 2000;154–155:353–9.

[70] Morjan I, Alexandrescu R, Soare I, Dumitrache F, Sandu I, Voicu I, et al. Nanoscale powders of different iron oxide phases prepared by continuous laser irradiation of iron pentacarbonyl-containing gas precursors. Mater Sci Eng C, 2003;23(1–2):211–16.

[71] Bomatí-Miguel O, Mazeina L, Navrotsky A, Veintemillas-Verdaguer S. Calorimetric study of maghemite nanoparticles synthesized by laser-induced pyrolysis. Chem Mater 2008;20(2):591–8.

[72] Yao BD, Chan YF, Wang N. Formation of ZnO nanostructures by a simple way of thermal evaporation. Appl Phys Lett 2002;81(4):757–9.

[73] Wu J-J, Liu S-C. Low-temperature growth of well-aligned ZnO nanorods by chemical vapor deposition. Adv Mater 2002;14(3):215–18.

[74] Umar A, Kim SH, Lee Y-S, Nahm KS, Hahn YB. Catalyst-free large-quantity synthesis of ZnO nanorods by a vapor–solid growth mechanism: structural and optical properties. J Cryst Growth 2005;282:131–6.

[75] Okumura M, Nakamura S, Tsubota S, Nakamura T, Azuma M, Haruta M. Chemical vapor deposition of gold on Al_2O_3, SiO_2, and TiO_2 for the oxidation of CO and of H_2. Catal Lett 1998;51:53–8.

[76] Dwivedi A. Recent advances in pulsed laser ablated plasma plumes: a review. Surf Rev Lett 2007;14 (1):57–69.

[77] Huotari J, Kekkonen V, Puustinen J, Liimatainen J, Lappalainen J. Pulsed laser deposition for improved metal-oxide gas sensing layers. Procedia Eng 2016;168:1066–9.

[78] Lam M, Hong MH, Yuan S, Chong TC. Growth of β-Ga_2O_3 nanoparticles by pulsed laser ablation technique. Appl Phys A 2004;79(8):2099−102.

[79] Teghil R, D'alessio L, Bonis AD, Ferro D, Galasso A, Lanza G, et al. Ultra-short pulse laser ablation of $Al_{70}Cu_{20}Fe_{10}$ alloy: nanoparticles generation and thin films deposition. Thin Solid Films 2009;517(6):1880−6.

[80] Bärsch N, Jakobi J, Weiler S, Barcikowski S. Pure colloidal metal and ceramic nanoparticles from high-power picosecond laser ablation in water and acetone. Nanotechnology 2009;20(44):445603−12.

[81] Suh JY, Lopez R, Feldman LC, Haglund Jr RF. Semiconductor to metal phase transition in the nucleation and growth of VO_2 nanoparticles and thin films. J Appl Phys 2004;96(2):1209−13.

[82] Blakemore JD, Gray HB, Winkler JR, MüLler AM. CO_3O_4 nanoparticle water-oxidation catalysts made by pulsed-laser ablation in liquids. ACS Catal 2013;3:2497−500.

[83] Huotari J, Kekkonen V, Haapalainen T, Leidinger M, Sauerwald T, Puustinen J, et al. Pulsed laser deposition of metal oxide nanostructures for highly sensitive gas sensor applications. Sens Actuator B-Chem 2016;236:978−87.

[84] Huotari J, Lappalainen J, Puustinen J, Baur T, Alépée C, Haapalainen T, et al. Pulsed laser deposition of metal oxide nanoparticles, agglomerates, and nanotrees for chemical sensors. Procedia Eng 2015;120:1158−61.

[85] Rella R, Spadavecchia J, Manera MG, Capone S, Taurino A, Martino M, et al. Acetone and ethanol solid-state gas sensors based on TiO_2 nanoparticles thin film deposited by matrix assisted pulsed laser evaporation. Sens Actuator B-Chem 2007;127(2):426−31.

[86] Chau JLH, Yang M-C, Nakamura T, Sato S, Yang C-C, Cheng C-W. Fabrication of ZnO thin films by femtosecond pulsed laser deposition. Opt Laser Technol 2010;42(8):1337−9.

[87] Kawakami M, Hartanto AB, Nakata Y, Okada T. Synthesis of ZnO nanorods by nanoparticle assisted pulsed-laser deposition. Jpn J Appl Phys 2003;42:L33−5.

[88] Kaneko S, Ito T, Hirabayashi Y, Ozawa T, Okuda T, Motoizumi Y, et al. Optimizing coverage of metal oxide nanoparticle prepared by pulsed laser deposition on nonenzymatic glucose detection. Talanta 2011;84(2):579−82.

[89] Sasaki T, Terauchi S, Koshizaki N, Umehara H. The preparation of iron complex oxide nanoparticles by pulsed-laser ablation. Appl Surf Sci 1998;127-129:398−402.

[90] Ma D, Hugener TA, Siegel RW, Christerson A, Mårtensson E, Önneby C, et al. Influence of nanoparticle surface modification on the electrical behaviour of polyethylene nanocomposites. Nanotechnology 2005;16(6):724−31.

[91] Qi Y, Xiang B, Tan W, Zhang J. Hydrophobic surface modification of TiO_2 nanoparticles for production of acrylonitrile-styrene-acrylate terpolymer/TiO_2 composited cool materials. Appl Surf Sci 2017;419:213−23.

[92] Ju S, Zhang H, Chen M, Zhang C, Chen X, Zhang Z. Improved electrical insulating properties of LDPE based nanocomposite: effect of surface modification of magnesia nanoparticles. Compos Part A Appl Sci Manuf 2014;66:183−92.

[93] Mallakpour S, Darvishzadeh M. Nanocomposite materials based on poly(vinyl chloride) and bovine serum albumin modified ZnO through ultrasonic irradiation as a green technique: optical, thermal, mechanical and morphological properties. Ultrason Sonochem 2018;41:85−99.

[94] Mallakpour S, Javadpour M. Antimicrobial, mechanical, optical and thermal properties of PVC/ZnO-EDTA nanocomposite films. Polym Adv Technol 2017;28(3):393−403.

[95] Mallakpour S, Adnany Sadaty M. Preparation and characterization of nanocomposites based on poly(vinyl alcohol) and vitamin B1-modified TiO_2 and evaluation of the optical, mechanical, and thermal properties. Colloid Polym Sci 2016;294(12):2099−107.

[96] Mallakpour S, Khani Z. Surface modified SiO_2 nanoparticles by thiamine and ultrasonication synthesis of PCL/SiO_2-VB1 NCs: morphology, thermal, mechanical and bioactivity investigations. Ultrason Sonochem 2018;41:527–37.

[97] Mallakpour S, Behranvand V. Surface treatment of nano ZnO using 3,4,5,6-tetrabromo-*n*-(4-hydroxy-phenyl)-phthalamic acid as novel coupling agent for the preparation of poly(amide–imide)/ZnO nanocomposites. Colloid Polym Sci 2014;292(9):2275–83.

[98] Dinari M, Haghighi A. Ultrasound-assisted synthesis of nanocomposites based on aromatic polyamide and modified ZnO nanoparticle for removal of toxic Cr(VI) from water. Ultrason Sonochem 2018;41:75–84.

[99] Minko S. Grafting on solid surfaces: "grafting to" and "grafting from" methods. In: Stamm M, editor. Polymer surfaces and interfaces: characterization, modification and applications. Berlin, Heidelberg: Springer; 2008.

[100] Hu L, Percheron AL, Chaumont D, Brachais C-HLN. Microwave synthesis of core-shell structured biocompatible magnetic nanohybrids in aqueous medium. In: Chandra U, editor. Microwave heating. Croatia: InTech; 2011.

[101] Francis R, Joy N, Aparna EP, Vijayan R. Polymer grafted inorganic nanoparticles, preparation, properties, and applications: a review. Polym Rev 2014;54(2):268–347.

[102] Zhao D, Di Nicola M, Khani MM, Jestin J, Benicewicz BC, Kumar SK. Role of block copolymer adsorption versus bimodal grafting on nanoparticle self-assembly in polymer nanocomposites. Soft Matter 2016;12(34):7241–7.

[103] Asai M, Zhao D, Kumar SK. Role of grafting mechanism on the polymer coverage and self-assembly of hairy nanoparticles. ACS Nano 2017;11(7):7028–35.

[104] De Roo J, Van Driessche I, Martins JC, Hens Z. Colloidal metal oxide nanocrystal catalysis by sustained chemically driven ligand displacement. Nat Mater 2016;15(5):517–21.

[105] Lattuada M, Hatton TA. Functionalization of monodisperse magnetic nanoparticles. Langmuir 2007;23:2158–68.

[106] Dong A, Ye X, Chen J, Kang Y, Gordon T, Kikkawa JM, et al. A generalized ligand-exchange strategy enabling sequential surface functionalization of colloidal nanocrystals. J Am Chem Soc 2011;133 (4):998–1006.

[107] Birkan B, Menceloglu YZ, Gulgun MA. Carbon nanofibers containing catalyst nanoparticles. Turkey: Sabanci University, TUBITAK; 2008. p. 1–26.

[108] Demir MM, Gulgun MA, Menceloglu YZ, Erman B, Abramchuk SS, Makhaeva EE, et al. Palladium nanoparticles by electrospinning from poly(acrylonitrile-coacrylic acid)-$PdCl_2$ solutions. Relations between preparation conditions, particle size, and catalytic activity,. Macromolecules 2004;37:1787–92.

[109] ShakooriOskooie M, Menceloglu YZ, Unal S, Hayat Soytaş S. Rapid microwave-assisted synthesis of platinum catalyst nanoparticles immobilized in electrospun carbon nanofibers. ACS Appl Mater Interfaces 2018; [Under review].

[110] S. Hayat Soytaş, Unal S, Cebeci FC, Menceloglu YZ. Self-standing CNF/SnOx composite anodes for lithium-ion batteries with exceptional cycle-life. Electrochim Acta 2018; [Under review].

[111] Maira B, Chammingkwan P, Terano M, Taniike T. Reactor granule technology for fabrication of functionally advantageous polypropylene nanocomposites with oxide nanoparticles. Compos Sci Technol 2017;144:151–9.

[112] Zhang Z, Wan M. Nanostructures of polyaniline composites containing nano-magnet. Synth Met 2003;132:205–12.

[113] Zhang Z, Wan M, Wei Y. Electromagnetic functionalized polyaniline nanostructures. Nanotechnology 2005;16:2827–32.

[114] Zhang L, Wan M. Polyaniline/TiO_2 composite nanotubes. J Phys Chem B 2003;107:6748–53.

[115] Lu X, Mao H, Chao D, Zhang W, Wei Y. Ultrasonic synthesis of polyaniline nanotubes containing Fe_3O_4 nanoparticles. J Solid State Chem 2006;179:2609–15.

[116] Hsiue G-H, Kuo W-J, Huang Y-P, Jeng R-J. Microstructural and morphological characteristics of PS–SiO_2 nanocomposites. Polymer 2000;41:2813–25.

[117] Crivello JV, Song KY, Ghoshal R. Synthesis and photoinitiated cationic polymerization of organic – inorganic hybrid resins. Chem Mater 2001;13:1932–42.

[118] Chiang P-C, Whang W-T, Tsai M-H, Wu S-C. Physical and mechanical properties of polyimide/titania hybrid films. Thin Solid Films 2004;447–448:359–64.

[119] Adhikari J, Biswas B, Chabri S, Bandyapadhyay NR, Sawai P, Mitra BC, et al. Effect of functionalized metal oxides addition on the mechanical, thermal and swelling behaviour of polyester/jute composites. Eng Sci Technol Int J 2017;20:760–74.

[120] Oguz O, Simsek E, Soz CK, Heinz OK, Yilgor E, Yilgor I, et al. Effect of filler content on the structure–property behavior of poly(ethylene oxide) based polyurethaneurea–silica nanocomposites. Polym Eng Sci 2018;58:1097–107.

[121] Schechtel E, Yan Y, Xu X, Cang Y, Tremel W, Wang Z, et al. Elastic modulus and thermal conductivity of thiolene/TiO_2 nanocomposites. J Phys Chem 2017;121:25568–75.

[122] Song SH. Synergistic effect of carbon nanofiber decorated with iron oxide in enhancing properties of styrene butadiene rubber nanocomposites. J Appl Polym Sci 2017;134:45376–84.

[123] Hajibeygi M, Shabanian M, Omidi-Ghallemohamadi M, Khonakdar HA. Optical, thermal and combustion properties of self-colored polyamide nanocomposites reinforced with azo dye surface modified ZnO nanoparticles. Appl Surf Sci 2017;416:628–38.

[124] Sangpraserdsuk T, Phiriyawirut M, Ngaotrakanwiwat P, Wootthikanokkhan J. Mechanical, optical, and photochromic properties of polycarbonate composites reinforced with nano-tungsten trioxide particles. J Reinf Plast Compos 2017;36(16):1168–82.

[125] Pourrahimi AM, Hoang TA, Liu D, Pallon LKH, Gubanski S, Olsson RT, et al. Highly efficient interfaces in nanocomposites based on polyethylene and ZnO nano/hierarchical particles: a novel approach toward ultralow electrical conductivity insulations. Adv Mater 2016;28:8651–7.

[126] Pourrahimi AM, Pallon LKH, Liu D, Hoang TA, Gubanski S, Hedenqvist MS, et al. Polyethylene nanocomposites for the next generation of ultralow-transmission-loss HVDC cables: insulation containing moisture-resistant MgO nanoparticles. ACS Appl Mater Interfaces 2016;8:14824–35.

[127] Pallon LKH, Hoang AT, Pourrahimi AM, Hedenqvist MS, Nilsson F, Gubanski S, et al. The impact of MgO nanoparticle interface in ultra-insulating polyethylene nanocomposites for high voltage DC cables. J Mater Chem A 2016;4:8590–601.

[128] Sampreeth T, Al-Maghrabi MA, Bahuleyan BK, Ramesan MT. Synthesis, characterization, thermal properties, conductivity and sensor application study of polyaniline/cerium-doped titanium dioxide nanocomposites. J Mater Sci 2018;53:591–603.

[129] Li Y, Sun XS. Preparation and characterization of polymer-inorganic nanocomposites by in situ melt polycondensation of L-lactic acid and surface-hydroxylated MgO. Biomacromolecules 2010;11:1847–55.

[130] Kang S, Hong SI, Choe CR, Park M, Rim S, Kim J. Preparation and characterization of epoxy composites filled with functionalized nanosilica particles obtained via sol–gel process. Polymer 2001;42(3):879–87.

[131] Wu T, Ke Y. Melting, crystallization and optical behaviors of poly (ethylene terephthalate)-silica/polystyrene nanocomposite films. Thin Solid Films 2007;515(13):5220–6.

[132] Friederich B, Laachachi A, Ferriol M, Ruch D, Cochez M, Toniazzo VR. Tentative links between thermal diffusivity and fire-retardant properties in poly(methyl methacrylate)-metal oxide nanocomposites. Polym Degrad Stab 2010;95:1183–93.

[133] Hyun-Jongkim Y-G, Shul H. , Sulfonic-functionalized heteropolyacid–silica nanoparticles for high temperature operation of a direct methanol fuel cell. J Power Sources 2006;158(1):137–42.

[134] Guo Q, Zhu P, Li G, Wen J, Wang T, Lu D, et al. Study on the effects of interfacial interaction on the rheological and thermal performance of silica nanoparticles reinforced epoxy nanocomposites. Compos Part B: Eng 2017;116:388–97.

[135] Rhee S-H, Choi J-Y. Preparation of a bioactive poly(methyl methacrylate)–silica nanocomposite. J Am Ceram Soc 2002;85(5):1318–20.

[136] Maurice V, Slostowski C, Herlin-Boime N, Carrot G. Polymer-grafted silicon nanoparticles obtained either via peptide bonding or click chemistry. Macromol Chem Phys 2012;213(23):2498–503.

[137] Li X, Hong C-Y, Pan C-Y. Preparation and characterization of hyperbranched polymer grafted mesoporous silica nanoparticles via self-condensing atom transfer radical vinyl polymerization. Polymer 2010;51(1):92–9.

[138] Du P, Zhao X, Zeng J, Guo J, Liu P. Layer-by-layer engineering fluorescent polyelectrolyte coated mesoporous silica nanoparticles as pH-sensitive nanocarriers for controlled release. Appl Surf Sci 2015;345:90–8.

[139] Hashimoto M, Takadama H, Mizuno M, Kokubo T. Enhancement of mechanical strength of TiO_2/high-density polyethylene composites for bone repair with silane-coupling treatment. Mater Res Bull 2006;41(3):515–24.

[140] Charpentier PA, Burgess K, Wang L, Chowdhury RR, Lotus AF, Moula G. Nano-TiO_2/polyurethane composites for antibacterial and self-cleaning coatings. Nanotechnology 2012;23(42):425606.

[141] Wu Y, Song L, Hu Y. Fabrication and characterization of TiO_2 nanotube–epoxy nanocomposites. Ind Eng Chem Res 2011;50(21):11988–95.

[142] Mallakpour S, Nouruzi N. Effect of modified ZnO nanoparticles with biosafe molecule on the morphology and physiochemical properties of novel polycaprolactone nanocomposites. Polymer 2016;89:94–101.

[143] Zeng X-F, Kong X-R, Ge J-L, Liu H-T, Gao C, Shen Z-G, et al. Effective solution mixing method to fabricate highly transparent and optical functional organic – inorganic nanocomposite film. Ind Eng Chem Res 2011;50(6):3253–8.

[144] Tang E, Liu H, Sun L, Zheng E, Cheng G. Fabrication of zinc oxide/poly(styrene) grafted nanocomposite latex and its dispersion. Eur Polym J 2007;43(10):4210–18.

[145] Liu P, Wang T. Poly(hydroethyl acrylate) grafted from ZnO nanoparticles via surface-initiated atom transfer radical polymerization. Curr Appl Phys 2008;8(1):66–70.

[146] Helal E, David E, Fréchette M, Demarquette NR. Thermoplastic elastomer nanocomposites with controlled nanoparticles dispersion for HV insulation systems: correlation between rheological, thermal, electrical and dielectric properties. Eur Polym J 2017;94:68–86.

[147] Bach LG, Islam MR, Kim JT, Seo S, Lim KT. Encapsulation of Fe_3O_4 magnetic nanoparticles with poly (methyl methacrylate) via surface functionalized thiol-lactam initiated radical polymerization. Appl Surf Sci 2012;258(7):2959–66.

[148] Yang T-I, Brown RNC, Kempel LC, Kofinas P. Magneto-dielectric properties of polymer–nanocomposites. J Magn Magn Mater 2008;320(21):2714–20.

[149] Pisanello F, De Paolis R, Lorenzo D, Nitti S, Monti G, Fragouli D, et al. Radiofrequency characterization of polydimethylsiloxane–iron oxide based nanocomposites. Microelectron Eng 2013;111:46–51.

[150] Chandra A, Turng L-S, Gopalan P, Rowell RM, Gong S. Study of utilizing thin polymer surface coating on the nanoparticles for melt compounding of polycarbonate/alumina nanocomposites and their optical properties. Compos Sci Technol 2008;68(3-4):768–76.

[151] Siegel RW, Chang SK, Ash BJ, Stone J, Ajayan PM, Doremus RW, et al. Mechanical behavior of polymer and ceramic matrix nanocomposites. Scripta Mater 2001;44:2061–4.

[152] Cinausero N, Azema N, Cochez M, Ferriol M, Essahli M, Ganachaud F, et al. Influence of the surface modification of alumina nanoparticles on the thermal stability and fire reaction of PMMA composites. Polym Adv Technol 2008;19(6):701–9.

[153] Mallakpour S, Ezhieh AN. A simple and environmentally friendly method for surface modification of ZrO_2 nanoparticles by biosafe citric acid as well as ascorbic acid (vitamin C) and its application for the preparation of poly(vinyl chloride) nanocomposite films. Polym Compos 2017;38(8):1756–65.

[154] Ariraman M, Kumar RS, Alagar M. Design of cyanate ester/azomethine/ZrO_2 nanocomposites high-k dielectric materials by single step sol–gel approach. J Appl Polym Sci 2014;131(24):41097.

[155] Scipioni R, Gazzoli D, Teocoli F, Palumbo O, Paolone A, Ibris N, et al. Preparation and characterization of nanocomposite polymer membranes containing functionalized SnO_2 additives. Membranes (Basel) 2014;4(1):123–42.

[156] Weng Z, Wang J, Senthil T, Wu L. Mechanical and thermal properties of ABS/montmorillonite nanocomposites for fused deposition modeling 3D printing. Mater Design 2016;102:276–83.

[157] Wang S, Hu Y, Song L, Wang Z, Chen Z, Fan W. Preparation and thermal properties of ABS/montmorillonite nanocomposite. Polym Degrad Stab 2002;77:423–6.

[158] Chigwada G, Wang D, Jiang DD, Wilkie CA. Styrenic nanocomposites prepared using a novel biphenyl-containing modified clay. Polym Degrad Stab 2006;91(4):755–62.

Further Reading

Plueddemann EP, Clark HA, Nelson LE, Hoffman KR. Silane coupling agents for reinforced plastics. Mod Plast 1962;39:135–87.

Laurent S, Forge D, Port M, Roch A, Robic C, Elst LV, et al. Magnetic iron oxide nanoparticles: synthesis, stabilization, vectorization, physicochemical characterizations, and biological applications. Chem Rev 2008;108:2064–110.

10

Optical Properties of Polymer Nanocomposites With Functionalized Nanoparticles

Andrey L. Stepanov[1,2]

[1]KAZAN PHYSICAL-TECHNICAL INSTITUTE, RUSSIAN ACADEMY OF SCIENCES, KAZAN, RUSSIAN FEDERATION [2]KAZAN NATIONAL RESEARCH TECHNOLOGICAL UNIVERSITY, KAZAN, RUSSIAN FEDERATION

10.1 Introduction

The task of designing new polymer-based composite materials containing metal nanoparticles (MNPs) is of current interest particularly in the fabrication of nonlinear optical switches, magnetooptic data storages, directional connectors, etc. The nonlinear optical properties of these composites stem from the dependence of their refractive index on incident light intensity. This effect is associated with MNPs, which exhibit a high nonlinear third-order susceptibility when exposed to ultrashort (ps or fs) laser pulses [1]. Light-induced electron excitation in MNPs (the so-called surface plasma resonance (SPR)) [2], which shows up most vividly in the range of linear absorption, gives rise to nonlinear optical effects in the same spectral range. In practice, the SPR effect may be enhanced by raising the nanoparticle concentration in the host matrix, that is, by increasing the volume fraction of the metal phase (filling factor). Systems with a higher filling factor offer a higher nonlinear third-order susceptibility [1].

MNPs may be embedded in a polymer matrix in a variety of ways. These are such techniques as chemical synthesis in an organic solvent [2], vacuum deposition on viscous polymers [3], plasma polymerization combined with metal evaporation [4], etc. However, they all suffer from disadvantages, such as a low filling factor or a large distribution in size and shape of nanoparticles, which offsets the good optical properties of composites. One promising method is ion implantation [5], which provides controllable synthesis of MNPs at various depths under the surface and unlimitedly high impurity doses. Despite the intensive study of MNP synthesis by ion implantation in dielectrics, such as nonorganic glasses and crystals, which was started in 1973 by Davenas et al. with Na and K ions [6] and in 1975 Arnold with Au ions [7], the formation of nanoparticles in organic matrices was realized only at the beginning of the 1980s by Koon et al. in their experiments on implantation of Fe ions into

Polymer Composites with Functionalized Nanoparticles. DOI: https://doi.org/10.1016/B978-0-12-814064-2.00010-X

polymers [8]. By ion implantation, it is possible to produce almost any metal–dielectric (specifically, metal–polymer) composite.

Noble metals exhibit the most pronounced SPR effect and, hence, the highest nonlinearity of the MNP optical properties in dielectrics. Quite recently such materials were fabricated by Ag-ion implantation into epoxy resins [9], polymethylmethacrylate (PMMA) [10], and polyethylene terephthalate [11]. The aim of this review chapter is to present the SPR-related linear optical properties of MNPs introduced into a polymer matrix by implantation and vapor deposition.

10.2 Synthesis of Metal Nanoparticles in Polymethylmethacrylate by Ion Implantation

10.2.1 Experimental Procedure

As substrates, 1.2-mm-thick PMMA plates, which are optically transparent in a wide spectral range (400–1000 nm), were used. Substrates were implanted by 30 keV Ag^+ ions with doses in the range from $3.1 \cdot 10^{15}$ to $7.5 \cdot 10^{16}$ ion/cm^2 at ion current density of 4 μA/cm^2 in a vacuum of 10^{-6} Torr using an ILU-3 ion implanter. In a control experiment, Xe-ion implantation into PMMA at the same conditions was performed. Spectra of optical density (OD) were measured from 300 to 900 nm by a spectrophotometer Hitachi 330. All spectra were recorded in a standard differential mode to normalize substrate effects. The samples obtained were examined with transmission electron microscopy (TEM).

Optical spectra of spherical MNPs embedded in various dielectric media were simulated in terms of the Mie electromagnetic theory [12], which allows one to estimate the extinction cross section σ_{ext} for a light wave incident on a particle. This value is related to the intensity loss ΔI_{ext} of an incident light beam I_0 passing through a transparent particle-containing dielectric medium due to absorption σ_{abs} and elastic scattering σ_{sca}, where $\sigma_{ext} = \sigma_{abs} + \sigma_{sca}$. Following the Lambert–Beer law

$$\Delta I_{ext} = I_0(1 - e^{-\#\sigma_{ext}\, h}) \qquad (10-1)$$

where h is the thickness of the optical layer and # is the density of nanoparticles in a sample. The extinction cross section is connected to the extinction constant γ as $\gamma = \#\sigma_{ext}$.

Experimental spectral dependencies of OD are given by

$$\text{OD} = -\lg(I/I_0) = \gamma \log(e)h \qquad (10-2)$$

hence, for samples with electromagnetically noninteracting nanoparticles, it possible to put OD $\sim \sigma_{ext}$. Therefore, experimental OD spectra are compared with modeled spectral dependences that are expressed through σ_{ext} calculated by the Mie theory.

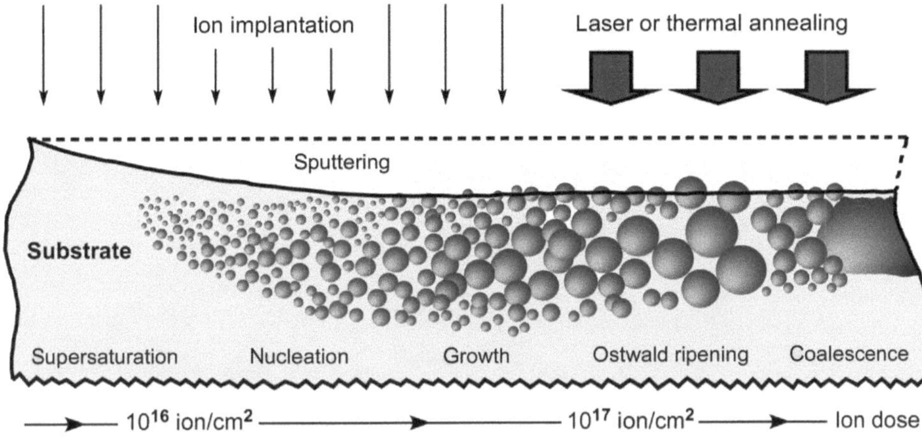

FIGURE 10–1 Basic physical stages of nanoparticle synthesis by ion implantation in dependence on ion dose.

10.2.2 Ion Synthesis of Metal Nanoparticles in Dielectric Matrix

Ion implantation is an effective tool for introducing single impurities into the surface layer to a depth of several micrometers [5]. The surface modification of the material depends on its properties, as well as on ion implantation parameters (ion type and energy, ion current density, target temperature, etc.). A critical implantation parameter is ion dose F_0, which determines the implant amount. Depending on the modification of an insulating target (polymers, inorganic materials, ionic crystals, minerals, etc.), ion implantation may be conventionally divided into low-dose and high-dose implantation (Fig. 10–1). In the former case ($F_0 \leq 5.0 \cdot 10^{14}$ ion/cm^2), the stopped ions are dispersed (isolated from one another) in the insulating matrix. The energy of ions implanted is transferred to the matrix through the excitation of electronic shells (ionization) and nuclear collisions. This causes radiation-induced defects, which, in turn, may reversibly or irreversibly modify the material structure [5]. Various types of polymer structure damage have been observed [13]: Breaking of covalent bonds in macromolecules, generation of free radicals, cross linkage, oxidation and carbonization of irradiated layers, formation of new chemical bonds between atoms of the insulator or between ions implanted, etc. In addition, ion implantation may be accompanied by the intense sputtering of the surface exposed or, sometimes, by the swelling of the polymer.

High-dose implantation may also be divided into dose (or time) stages (Fig. 10–1). At F_0 between 10^{15} and 10^{17} ion/cm^2, the equilibrium solubility of metallic implants in insulators (in particular, polymers) is usually exceeded, causing the nucleation and growth of MN (Fig. 10–2). The dose threshold value depends on the type of the insulator and implant. For 25 keV silver ions implanted into $LiNbO_3$, the threshold dose was found to be $F_0 \approx 5.0 \cdot 10^{15}$ ion/cm^2 [14], but for 30 keV silver ions implanted into epoxy resin, $F_0 \approx 10^{16}$ ion/cm^2 [10].

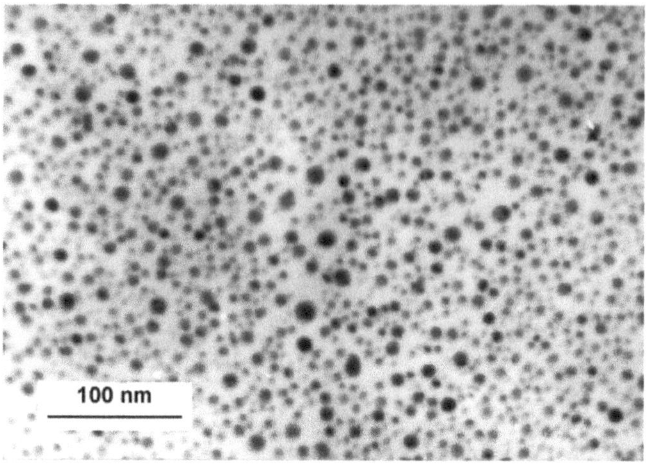

FIGURE 10–2 Micrograph of silver nanoparticles produced by Ag^+ implantation into PMMA at a dose of $5.0 \cdot 10^{16}$ ion/cm^2.

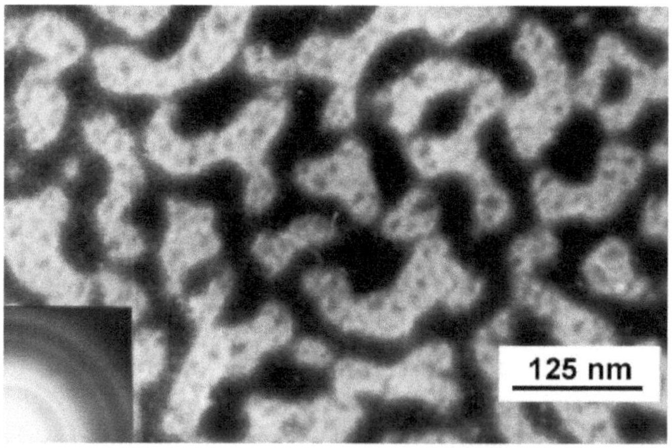

FIGURE 10–3 Micrograph of cobalt labyrinth structures produced by Co^+ implantation into viscous epoxy resins at a dose of $2.5 \cdot 10^{17}$ ion/cm^2.

At the next stage of high-dose implantation, starting from $F_0 \geq 10^{17}$ ion/cm^2, the existing MNPs coalesce to form MN aggregates or quasi-continuous films in the surface layer (Fig. 10–1). For example, the irradiation of epoxy resin by 40 keV cobalt ions at higher-than-threshold doses favors the formation of thin labyrinth structures (Fig. 10–3) [15]. The MN distribution established in the insulator after coalescence or Ostwald ripening may be disturbed by postimplantation thermal or laser annealing.

In such situation, composites where MNPs are disperse and isolated from each another, that is, synthesized at ion doses of 10^{15}–10^{17} ion/cm^2 in PMMA, were studied. In a case of implantation by heavy Ag^+ ions at relatively low energy 30 keV, nuclear collisions prevail during interaction of accelerated ions with atoms of the insulator. They displace atoms in the polymer matrix and break some of the chemical bonds of the substrate. Along with this, target atoms effectively lose electrons and the implanted Ag^+ ions deionize with the formation of neutral silver atoms (Ag^0). In principle, Ag atoms may combine with arising organic radicals and polymer ions or take part in the oxidation reaction. However, because of a great difference in Gibbs free energy between Ag atoms and atoms of PMMA elements, Ag–Ag bonding is energetically more favorable.

The formation of MNPs proceeds in several stages: The accumulation of and subsequent supersaturation by Ag^0 atoms in a local surface region of the polymer, the formation of nuclei consisting of several atoms, and the growth of silver particles from the nuclei. Assuming that the nanoparticles nucleate and grow via the successive attachment of silver atoms (which are neutralized by embedded Ag^+ ions), one may conclude that this process is governed simultaneously by the diffusion coefficient and local concentration of silver atoms, that is, depends on the matrix temperature. In this work, ion implantation was performed under identical conditions; specifically, the polymer during irradiation was kept at room temperature.

As follows from electron microscopy data, silver ion implantation under the experimental conditions considered results in the formation of silver nanoparticles in PMMA. For example, the cross-sectional micrograph in Fig. 10–1 and plane view in Fig. 10–1 (a dose of $5.0 \cdot 10^{16}$ ion/cm^2) show dark spherical nanoparticles against the bright field (polymer). The irradiation of PMMA by xenon ions did not result in such patterns. Microdiffraction patterns demonstrate that the nanoparticles have the fcc structure of metallic silver. The patterns consist of thin rings (corresponding to polycrystalline nanoparticles) imposed on wide diffuse faint rings from the amorphous polymer matrix. By comparing the experimental diffraction patterns with standard ASTM data, it is possible to conclude that implantation does not form any chemical compounds involving silver ions.

From Rutherford backscattering spectrometry (RBS) spectra (Fig. 10–4), it is seen that the silver implantation layer (i.e., the depth where the nanoparticles are located) is almost independent of the ion dose. The dose dependence is observed only for the silver peak RBS intensities with the widths and positions of the lines remaining unchanged. It is known that the implantation depth of an ion depends largely on its energy (accelerating voltage) [5] provided that the chemical constitution of the surface irradiated does not change dramatically. The similarity of the RBS spectra shown in Fig. 10–4 implies that the arising MNPs do not restrict the penetration depth of silver ions at the higher dose. Thus, in PMMA, the implantation dose, being responsible for the amount of the implant, influences directly the MN size but does not affect the implant distribution profile at present conditions. As was noted previously, the particles nucleate at a dose of $\sim 10^{16}$ ion/cm^2 (low-dose implantation). For silver in PMMA, this dose provides MN with a size of about 2 nm. However, at a dose of $5.0 \cdot 10^{16}$ ion/cm^2 (Fig. 10–2), the particles grow to ~ 10 nm.

Experimental optical absorption spectra for PMMA irradiated by xenon and silver ions at various doses are shown in Fig. 10–5. It is evident that the xenon irradiation of PMMA does

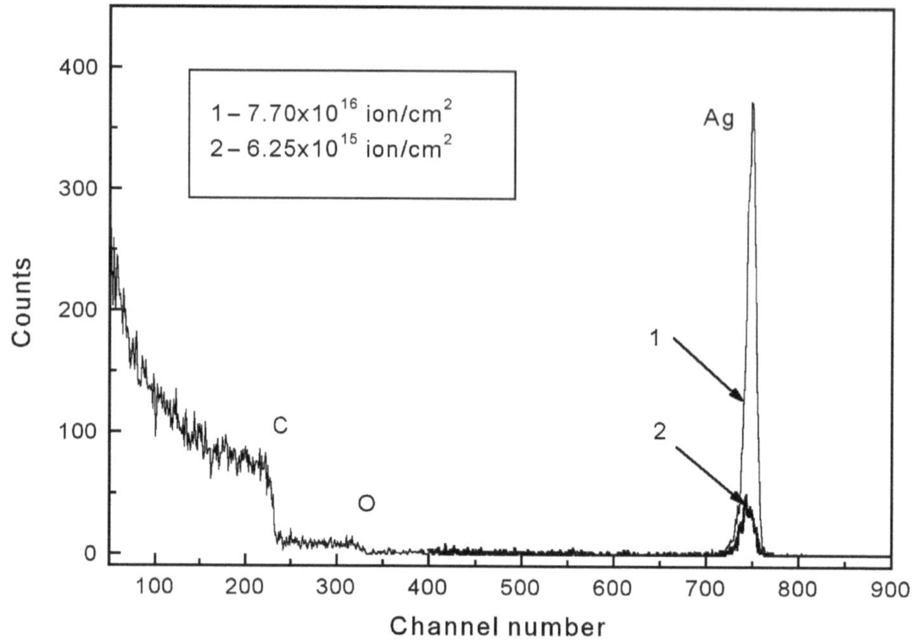

FIGURE 10–4 RBS spectra from PMMA irradiated by silver ions for doses of (1) $7.7 \cdot 10^{16}$ and (2) $6.25 \cdot 10^{15}$ ion/cm^2.

not produce nanoparticles, as also follows from the micrographs (Fig. 10–2). In Fig. 10–5A, as the xenon ion dose increases, the absorption of the polymer in the visible (especially in the close-to-ultraviolet (UV)) range also increases monotonically. This indicates the presence of radiation-induced structure defects in the PMMA. The absence of absorption bands in these spectral curves is noteworthy. The implantation of silver not only generates radiation-induced defects but also causes the nucleation and growth of MNPs. Therefore, along with the absorption intensity variation as in Fig. 10–5A, an absorption band associated with silver nanoparticles is observed (Fig. 10–5B). For the lowest dose, the peak of this band is near 420 nm and shifts to longer waves (up to 600 nm) with dose, with the band broadening significantly. The peak of this band is not high, although it is related to the SPR effect in the silver nanoparticles. The low intensity of SPR absorption is not typical of silver nanoparticles in PMMA and cannot be explained by the polymer environment of the particles. When silver particles were synthesized in PMMA by the convection technique [16], the SPR intensity was very high, unlike our experiment. Nor can the weak SPR absorption be explained by any features of the implantation process. For comparison, Fig. 10–5 shows the OD spectrum for inorganic silica glass (SiO_2) irradiated by silver ions under the same conditions used in this work for the ion synthesis (silica glass has the refractive index $n \approx 1.5$ close to that of PMMA) [10]. It is seen that the absorption of Ag nanoparticles in the glass (Fig. 10–5B) is much more intense (even in view of the background absorption due to matrix structure imperfections) than the absorption of the particles in the polymer. Note that the particle size

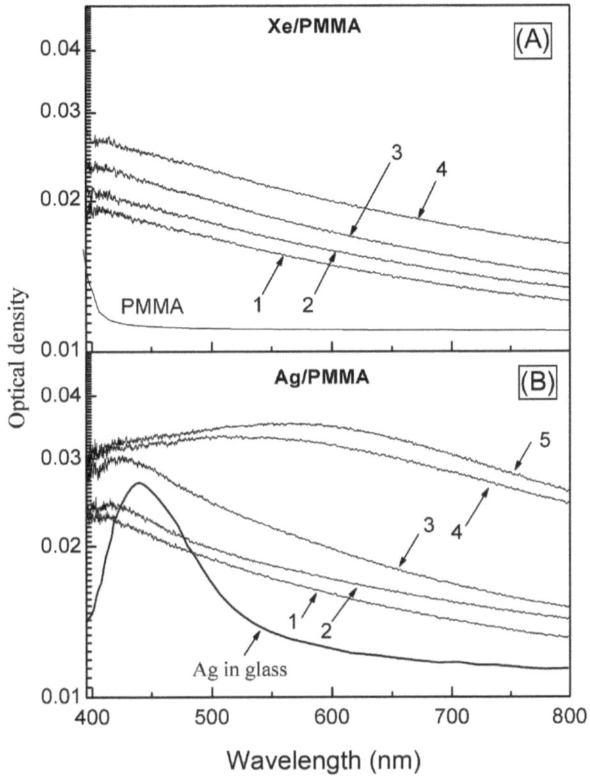

FIGURE 10–5 Experimental optical density spectra from PMMA irradiated by (A) xenon and (B) silver ions for doses of (1) $0.3 \cdot 10^{16}$, (2) $0.6 \cdot 10^{16}$, (3) $2.5 \cdot 10^{16}$, (4) $5.0 \cdot 10^{16}$, and (5) $7.5 \cdot 10^{16}$ ion/cm^2. The spectrum taken from silica glass irradiated by silver ions ($5.0 \cdot 10^{16}$ ion/cm^2) [10] is shown for comparison.

distributions in the glass and PMMA are nearly the same. In the following, the optical properties of the Ag–PMMA composite are simulated and various effects that may clarify the SPR absorption of Ag nanoparticles synthesized in PMMA by ion implantation are discussed.

10.2.3 Extinction of Implanted Silver Nanoparticles in Dielectric Matrix

The attenuation (extinction) of an optical wave propagating in a medium with MN depends on the amount of the SPR effect and the light scattering efficiency. The wavelength of optical radiation, the particle size, and the properties of the environment are governing factors in this process. Within the framework of classical electrodynamics (the Maxwell equations), the problem of interaction between a plane electromagnetic wave and a single spherical particle was exactly solved in terms of optical constants of the interacting objects by Mie [12,17]. According to the Mie theory, the extinction and scattering cross sections are expressed as an infinite sum of spherically symmetric partial electric and magnetic waves that generate fields similar to those generated by the particle when it is viewed as an excited multipole. The extinction cross section is generally given by

$$\sigma_{\text{ext}} = \frac{2\pi}{|\mathbf{k}|}\sum_{L=1}^{\infty}(2L+1)Re(a_L + b_L), \quad (10-3)$$

where **k** is the wavenumber and L is the order of spherical multipole excitation in the particle.

The case $L = 1$ corresponds to a dipole; $L = 2$, to a quadrupole; etc. The Mie coefficients a_L and b_L are expressed through the Riccati–Bessel cylindrical functions ψ_L and η_L of variables mx or x, where $mx = \varepsilon_{Ag}/\varepsilon_{MMA}$ is the ratio of the optical constants of the particles and polymer environment and is $x = |\mathbf{k}| R$ is the dimensional parameter, where R is the radius of the particle. These coefficients are given by [12,17]

$$a_L = \frac{m\Psi_L(mx)\Psi'_L(x) - \Psi'_L(mx)m\Psi_L(x)}{m\Psi_L(mx)\eta'_L - \Psi'_L(mx)\eta_L(x)} \quad (10-4)$$

$$b_L = \frac{\Psi_L(mx)\Psi'_L(x) - m\Psi'_L(mx)m\Psi_L(x)}{\Psi_L(mx)\eta'_L - m\Psi'_L(mx)\eta_L(x)} \quad (10-5)$$

In the general case, the Mie electromagnetic theory imposes no limitations on the wavelength of optical radiation. Therefore, the operation on the optical constants of the particles and matrix results in extinction spectra, the so-called Mie resonance bands [2], which agree well with experimental data. However, the Mie theory, which relies on the spectral dependence of the optical constants, does not allow one to penetrate deep into the physics of Mie optical peaks exhibited by the particles. Yet, independent investigations [2] into the behavior of silver nanoparticles showed that Mie resonances are due to the SPR effect, so that analytical Mie spectra may be compared with experimental data.

Simulated extinction spectra for Ag nanoparticles embedded in a polymer matrix for comparison with experimental data are shown in Fig. 10–5. In theoretical calculations, the complex value of the optical constant ε_{Ag} in the visible range [18] was used, which was obtained by measurements on a set of fine silver particles. Such an approach [18] takes into account limitations imposed on the electron free path in particles of different size and electron scattering at the particle–insulator interface [2] and thus yields a more exact value of ε_g than the procedure of correcting optical constants for bulk silver. The complex value of ε_{MMA} for the polymer matrix was found elsewhere [19]. The extinction was calculated for particles of size between 1 and 10 nm (according to the MN sizes in Fig. 10–2).

At the early stage of simulation, consider the simple case where Ag nanoparticles are incorporated into the PMMA matrix. Associated extinction spectra for different metal particle sizes are shown in Fig. 10–6. These spectra feature a wide band, which covers the entire spectral range. In the given range of particle sizes, the position of the SPR absorption maximum (near 440 nm) is almost independent of the particle size. However, the extinction band intensity grows while the band itself somewhat narrows with increasing particle size.

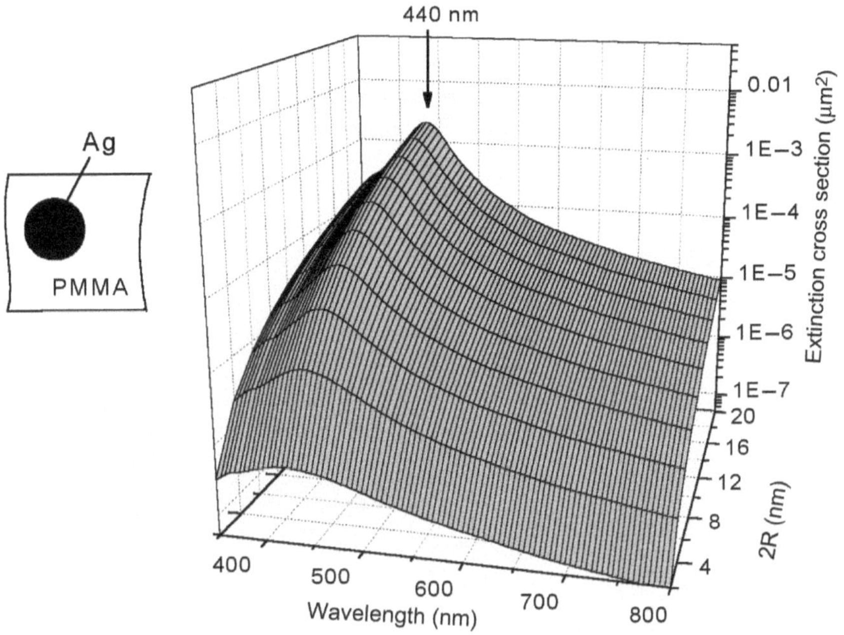

FIGURE 10–6 Modeled optical extinction spectra for silver nanoparticles embedded in PMMA versus particle size.

Comparing the analytical and experimental spectra, it is seen that, to the greatest extent, Fig. 10−6 refers to the situation where PMMA is irradiated by silver ions with doses between $0.33 \cdot 10^{16}$ and $2.5 \cdot 10^{16}$ ion/cm^2 (Fig. 10−6B, curves 1−3). This dose range corresponds to the early stage of MNPs nucleation and growth in the OD spectral band with a maximum between 420 and 440 nm. Thus, one may conclude that ion implantation in this dose range results in the formation of Ag nanoparticles, as also revealed microscopically. It may be supposed that radiation-induced defects in the PMMA have an insignificant effect on the MNPs' optical properties in this case. However, at higher implantation doses, the recorded OD spectra and the analytical spectra shown in Fig. 10−6 diverge; hence, the structure of the metal−polymer composite should be considered.

To explain the experimental dependences corresponding to high-dose silver implantation into PMMA, we will first elucidate the difference between implantation into polymers and inorganic insulators (silicate glasses, single crystals, minerals, etc.). The most important distinction is that, as the absorbed dose grows, so does the number of dangling chemical bonds along the track of an ion. Because of this, gaseous hydrogen, low-molecular hydrocarbons (e.g., acetylene), CO, and CO_2 evolve from the matrix [13]. In particular, ion-irradiated PMMA loses HCOOCH3 methoxy groups [20]. The evolution of several organic fractions leads to the accumulation of carbon in the polymer layer irradiated, and radiation-induced chemical processes may cause chain linking. Eventually, an amorphous hydrogenated carbon layer is produced. Polymer carbonization starts with the formation of polycyclic

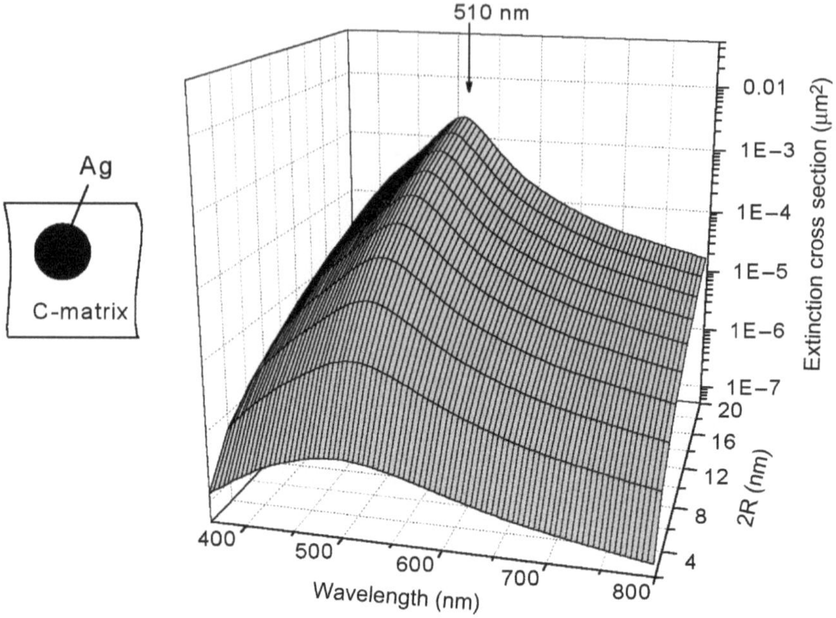

FIGURE 10–7 Modeled optical extinction spectra for silver nanoparticles embedded in the C-matrix versus particle size.

compounds (in essence, primary carboniferous clusters) and, at higher doses, ends up with the formation of the well-developed carbonized phase via carbon cluster linking.

In view of the specific phase structure of the polymer irradiated, it is of interest to analyze the optical properties (extinction) of Ag nanoparticles embedded in the amorphous carbon matrix (C-matrix). For this system, the extinction cross-sectional spectra versus particle size dependence (Fig. 10–7) was simulated in the same way as for the MNPs–PMMA system, that is, by using complex optical constants ε_C for amorphous carbon, which were taken from [21]. As before (Fig. 10–6), throughout the particle size interval, the extinction spectra exhibit a single broad band, which covers the visible range, with a peak at longer waves (510 nm). The calculated long-wave shift of the peak, which is observed upon changing the matrix, may be assigned to a longer wave OD band in the experimental spectra for the PMMA, which arises when the Ag-ion dose exceeds $2.5 \cdot 10^{16}$ ion/cm^2 (Fig. 10–5B; curves 3 and 4). It seems that this spectral shift may be associated with the fact that the pure polymeric environment of the Ag nanoparticles turns into the amorphous carbon as the implantation dose rises. The broader extinction bands in the C-matrix (Fig. 10–7) compared with the PMMA (Fig. 10–6) also count in favor of this supposition, since the broadening of the extinction bands is observed in the experiments as well (Fig. 10–5B). In a number of experiments, however, the carbonization of the polymer surface layer depended on the type of the polymer and ion, as well as on the process parameters, and completed at doses of (0.5–5.0) 10^{16} ion/cm^2 but the entire material was not carbonized. The carbon clusters may reach several tens of nanometers in size [22]. Thus, the assumption that the polymer irradiated is completely carbonized, which was used in the simulation (Fig. 10–7), does not meet the real situation when the process lasts for a long implantation time. In the following, the variation of the extinction spectra with amount of carbon in the PMMA layer is analyzed in terms of a model that considers the optical properties of Ag MNPs covered by the amorphous carbon shell.

Extinction spectra for nanoparticles represented as a silver core covered by a carbon shell in an insulating matrix (PMMA) will be analyzed in terms of the Mie relationships for shelled cores [23]. Here, an additional interface for which electrodynamic boundary conditions must be set up arises. Plasmon–polariton modes may be excited in both the core and the shell. These modes, interacting through the inner interface, are responsible for the resulting extinction spectrum. The Mie coefficients a_L and b_L for a homogenous sphere are replaced by associated expressions for a single spherical core with one shell: [24,25].

$$a_L = -\frac{m\Psi_L(mx)[\Psi'_L(x) + T_L\vartheta'_L(x)] - \Psi'_L(mx)\Psi_L(x)[\Psi_L(x) + T_L\vartheta_L(x)]}{m\xi_L(mx)[\Psi'_L(x) + T_L\vartheta'_L(x)] - \xi'_L(mx)[\Psi_L(x) + T_L\vartheta_L(x)]}, \quad (10-6)$$

$$b_L = -\frac{\Psi_L(mx)[\Psi'_L(x) + S_L\vartheta'_L(x)] - m\Psi'_L(mx)\Psi_L(x)[\Psi_L(x) + S_L\vartheta_L(x)]}{\xi_L(mx)[\Psi'_L(x) + S_L\vartheta'_L(x)] - m\xi'_L(mx)[\Psi_L(x) + S_L\vartheta_L(x)]}, \quad (10-7)$$

where the functions T_L and S_L are given by

$$T_L = -\frac{m\Psi_L(mx)[\Psi'_L(x) + \vartheta'_L(x)] - \Psi'_L(mx)\Psi_L(x)[\Psi_L(x) + \vartheta_L(x)]}{m\vartheta_L(mx)[\Psi'_L(x) + \vartheta'_L(x)] - \vartheta'_L(mx)[\Psi_L(x) + \vartheta_L(x)]}, \quad (10-8)$$

$$S_L = -\frac{\Psi_L(mx)[\Psi'_L(x) + \vartheta'_L(x)] - m\Psi'_L(mx)\Psi_L(x)[\Psi_L(x) + \vartheta_L(x)]}{\vartheta_L(mx)[\Psi'_L(x) + \vartheta'_L(x)] - m\vartheta'_L(mx)[\Psi_L(x) + \vartheta_L(x)]}. \quad (10-9)$$

Optical extinction spectra for a Ag nanoparticle with a fixed size of the core (4 nm) and a varying thickness of the carbon shell (from 0 to 5 nm) are shown in Fig. 10–8. The maximum of the SPR bands of the particles is seen to shift from 410 nm (uncovered particle, Fig. 10–7) to $\sim$510 nm. Simultaneously, the SPR band intensity decreases, while the UV absorption increases, so that the absorption intensity at 300 nm and a shell thickness of 5 nm exceeds the SPR absorption of the particles. Both effects (namely, the shift of the SPR band to longer waves and the increased absorption in the near UV) agree qualitatively with the variation of the experimental OD spectra (Fig. 10–5B) when the implantation dose exceeds 2.5 10^{16} ion/cm^2. Thus, our assumption that the increase in the carbonized phase fraction with implantation dose and the variation of the OD spectra (Fig. 10–5B) go in parallel is sustained by the simulation of the optical extinction for complex particles (sheathed cores, Fig. 10–8).

In spite of the fact that the model dependences on the carbon shell thickness and the experimental dose dependences agree qualitatively, discrepancies still exist, particularly in the position of the long-wave maximum in the OD spectra and in the breadths of the simulated and experimental spectra. Possible reasons for such quantitative discrepancies are discussed in the following section.

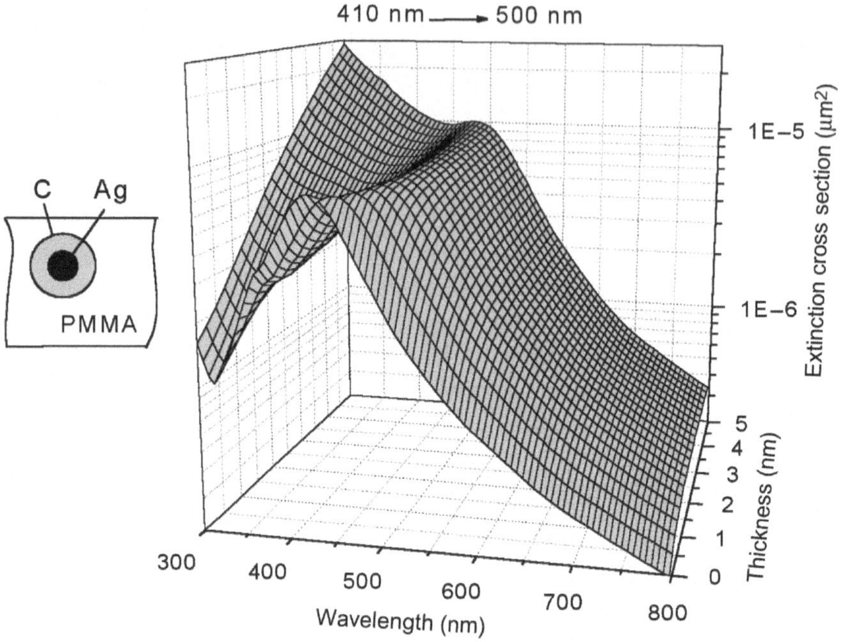

FIGURE 10–8 Modeled optical extinction spectra for 4 nm silver nanoparticles with the carbon shell that are placed in the PMMA matrix versus shell thickness.

10.2.4 Interface Effects in The Silver Core–Carbon Shell Structure on Optical Extinction

Interest in carbon-based composites with MNPs goes back a long way. Examples are the studies of electric and optical properties of layers with copper [26] or silver [27,28] nanoparticles, etc. It was found in optical absorption experiments that copper and silver nanoparticles [27,28] dispersed in carbon matrices exhibit a weak SPR effect as in the present work (Figs. 10–5B, 10–6, and 10–7).

When analyzing the optical properties of nanoparticles embedded in a medium, one should take into account effects arising at the particle–matrix interface, such as the static and dynamic redistributions of charges between electronic states in the particles and the environment in view of their chemical constitution [29].

Consider first the charge static redistribution. When an atom is deposited (adsorbed) on the MN surface, the energy levels of this atom ε_a change their positions compared with those in the free state [29,30] (Fig. 10–9). When the number of the adsorbed matrix atoms becomes significant, their contact generates a wide distribution of density of states. Most frequently, the adsorbed atoms are separated from surface atoms of the metal by a tunnel barrier. The gap between the energy positions ε_a of the adsorbed atoms and the Fermi level ε_F of the particles depends on the type of the adsorbate (Fig. 10–9). In addition, the overlap

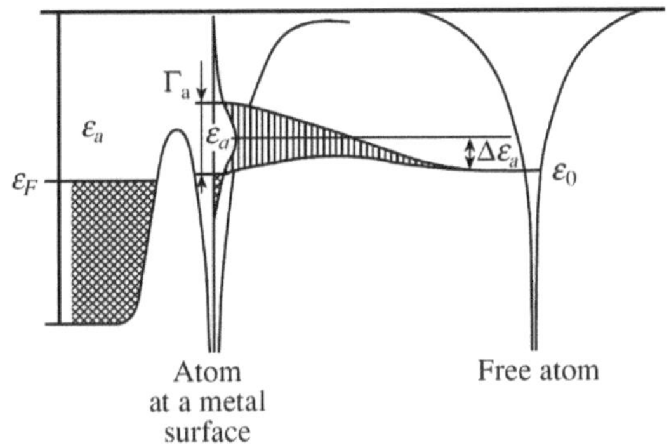

FIGURE 10–9 Electron energy levels in an atom adsorbed on a metal surface [29]. A free atom (to the right) reaches the surface (to the left). Γ_a is the spread of energy levels ε_a. Electron levels in the conduction band of the metal are occupied up to the Fermi level ε_F.

between the energy positions of the matrix atoms and the energy positions of the silver surface atoms depends on the rate with which the electrons tunnel through the barrier. Accordingly, the conduction electron density in the particles embedded will change compared with that in the particles placed in a vacuum (without adsorbates): It decreases if the electrons tunnel toward the adsorbed atoms or increases when the electrons tunnel in the reverse direction. Eventually, equilibrium between the particle and the matrix sets in; that is, a constant electrical charge (Coulomb barrier) forms at the nanoparticle surface.

Such a charge static redistribution due to the deposition of an adsorbate on the particle surface and the respective change in the electron concentration in the MN were also observed in the SPR absorption spectra [2,29]. In metals (silver, sodium, aluminum, etc.), where free conduction electrons dominate, the SPR spectral maximum $h\omega_{max}$ depends on the concentrations of electrons, *N*, and nanoparticles as

$$h\omega_{max} \approx [N/(\varepsilon_0 m_{eff})]^{1/2}[2\varepsilon_m + 1 + \chi_1^{inter}]^{1/2}, \qquad (10-10)$$

where ε_m is the permittivity of the matrix and specifies the contribution of the real part of the susceptibility of interband optical transitions in a metal, and m_{eff} is the effective mass of an electron.

It was shown [31] that the incorporation of Ag nanoparticles into the carbon matrix of C_{60} fullerene (or the deposition of carbon on the nanoparticle surface) reduces the concentration of 5*sp* electrons in the particle roughly by 20%, since they are trapped by matrix molecules. According to Eq. (10–10), the decrease in *N* is bound to shift the MNP extinction spectrum toward longer waves, as also demonstrated by comparing the experimental spectra of the particles in free space (without an adsorbate) with those of the particles in the C_{60} matrix [31]. Samples studied in [31] were similar to those obtained by ion implantation in our work (a carbonized layer near silver particles implanted into the polymer). Thus, the shift of the SPR extinction band into the longer wave range with implantation dose in this experiment

(Fig. 10–5) may be explained by the formation of a carbon sheath around silver nanoparticles. This sheath traps conduction electrons of the particles. The simulation (Fig. 10–8) also demonstrates the shift of the SPR maximum. However, the effect of charge dynamic redistribution is disregarded in the Mie theory. Therefore, the long-wave shift of the SPR band due to the charge static redistribution at the particle–matrix interface is an additional reason why the experimental spectra are observed at longer waves than the model ones (Figs. 10–7 and 10–8).

Along with the charge static redistribution at the interface, the charge at the same interface may also change dynamically, that is, with a high rate [29]. After the static state of the charge has been established and the Fermi level at the interface has been stabilized, the MNPs electrons optically excited above the Fermi level (hot electrons) may tunnel (by fluctuations) to the matrix over or through the static barrier (Fig. 10–9). Levels occupied by the electrons in the intermediate (between the particle and the matrix) state depend on the chemical constitution of the materials. Within a residence lifetime, the electrons may tunnel again from the acceptor levels of the matrix to the particle and this process may occur over and over.

The charge dynamic variation in time at the particle–matrix interface causes the electron concentration in the particle to fluctuate. Fluctuation influences directly the SPR relaxation. The lifetime of excited conduction electrons in the particle defines the SPR spectral width. Here, the contribution from electron scattering by the interface (because of restrictions imposed on the electron free path [2]) adds up with the charge dynamic variation at the interface. Thus, the temporal capture of conduction electrons from the particle broadens the SPR-related extinction spectra. This was demonstrated with a set of silver nanoparticles embedded in the C_{60} matrix [31]. Silver nanoparticles in the carbon matrix exhibit the much broader SPR band than in free space. We may therefore suppose that, as the dose rises, the charge dynamic redistribution may broaden the SPR spectra of silver nanoparticles synthesized by ion implantation in PMMA. This is because implantation carbonizes the irradiated layer with increasing absorbed dose and raises the amount of acceptor levels on the MNP surface, which change the relaxation time of electrons excited. Since the classical Mie theory disregards the charge dynamic redistribution, the model spectra (Fig. 10–8) must be narrower than the experimental spectra, which is the case.

10.3 Fabrication of Viscous Polymer Composites With Metal Nanoparticles

One method of polymer—MNPs composite material production that is effective and rather widely applied is based on the deposition of a thermally evaporated metal (tin, selenium, indium, gold, copper, and silver) onto the surface of a molten polymer, that is, an organic material having a temperature above glass transition temperature T_g [32–40]. The morphology of the metal–polymer transition layer (interface) was found to undergo substantial changes depending on the type of polymer, its temperature, the deposition rate, the type of metal, and the subsequent annealing of the composite material. This method can be used for various polymers, such as styrynhexomethacrylate ($T_g = 60°C$, $T_{exp} = 60–120°C$) [33–36], polymethylmethacrylate ($T_g = 106°C$, $T_{exp} = 150°C$) [37], polystyrene ($T_g = 93°C$, $T_{exp} = 200°C$) [38], and polycarbonate ($T_g = 235°C$, $T_{exp} = 250°C$) [39], where T_{exp} is temperature of experimental deposition of metal. This method uses the saturation of the surface layer of a polymer by metallic atoms to form a supersaturated solid solution of impurity atoms for nucleation and growth of MNPs. The degree of filling of the polymer volume with a metal is rather high.

As noted previously [32–40], the thermal evaporation technique used for MNPs fabrication is characterized by heating of a polymer to a temperature $T > T_g$ to transform organic substrates into a liquid or softened state. This procedure requires rather high temperatures, often more than 200°C. Unfortunately, the rate of metal diffusion into the bulk of a polymer at such high temperatures is very high; as a result, it is almost impossible to control the size of MNPs in a polymer. Therefore, the reproducibility of the composite material parameters is quite low.

A new method, which is also based on the deposition of a thermally evaporated metal but onto a polymer (epoxy resin) chemically diluted by a solvent at room temperature rather than onto a very heated organic matrix, is proposed and described here. As a main result, it is possible to achieve the relaxation viscous fluid state (molecular mobility) of a polymer matrix at lower temperatures (below $T_g = 120°C$ for epoxy resin). By changing of the epoxy hardener concentration everyone can control the degree of dynamic viscosity of epoxy resin. After the finishing of metal deposition, epoxy resin transforms into a solid state in a certain time due to polymerization. Thus, the metal–polymer composite layers can be formed by thermal evaporation of silver onto the surface of viscous fluid epoxy resin and such composites with MNPs are also interesting for an application in optoelectronics for a control of light signal transferring along a polymer waveguide.

10.3.1 Formation of Silver Nanoparticles During Deposition on Viscous Epoxy Resin

As was mentioned, silver nanoparticles can be synthesized in a viscous fluid epoxy matrix by thermal evaporation of the metal in vacuum onto softened (below the glass transition temperature) high molecular organic substrates at room temperature [3]. For experimental demonstration of such possibility for MNPs fabrication in polymers, the amount of silver deposited onto the surface of epoxy polymer was chosen to be $4.8 \cdot 10^{-6}$ g/cm^2, the deposition rate was $4.4 \cdot 10^{14}$ atoms/cm^2, and the deposition time was 60 seconds. At the chosen silver deposition rate, the partial vapor pressure in a vacuum chamber was so low that metal particles could form when silver atoms drift from an evaporator to a substrate. As a viscous fluid polymer matrix, liquid epoxy resin in a mixture with dibutylphthalate and polyethylenepolyamine was used. The chosen chemical formula of polymer provided the optimum conditions of forming an epoxyamine polymer from the standpoint of the cure rate. The lifetime of the viscous fluid polymer matrix was about 2 hours, and, then, its physicomechanical properties were affected

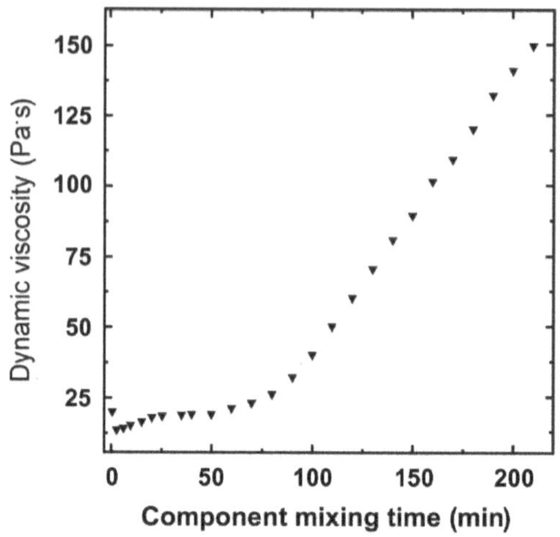

FIGURE 10–10 Epoxy dynamic viscosity dependence on component mixing time.

by the formation of a solid spatial network polymer structure. The dynamic viscosity of the polymer matrix during the deposition of silver was changed from 20 to 120 Pa · s. In an additional specific experiment, samples were prepared at the deposition times varying from 5 to 80 seconds for processing onto epoxy resin with a viscosity of 20 Pa · s.

The dependence of the dynamic viscosity μ of the epoxy polymer on the cure time t (Fig. 10–10) was measured with a capillary viscosimeter according to a standard procedure [15]. The presented curve $\mu(t)$ is typical for the various epoxies and is consistent with the epoxy curing kinetics. For the time range $t < 110$ minutes, the low viscosity of the polymer is explained by the prevailing interdiffusion processes of the epoxy molecules, and their concentration is averaged in macrovolume. The subsequent composite curing followed by the polymer viscosity increases due to the chemical cross-linking between the molecules and the rigid polymer network formation. It is known that the polymer network nucleation centers are "cybotaxes"—the associates comprising some epoxy molecules with a "kinetically gainful" orientation. Before the molecules will cross-link, the cybotaxes have no distinct shape or interface. The situation, however, changes when the polymer network is formed. Many solid microblocks appear in the liquid composite, thus providing a high degree of volume nonuniformity of physical properties in the polymer material. Upon growth of the spatial network, the viscous flow polymer transforms into a glassy one.

Epoxy resin layers of ~10 μm thickness were deposited by centrifuging onto thin glass substrates. Silver was deposited in a VUP-5 vacuum device. The residual vacuum during silver deposition was 10^{-3} Pa. After silver deposition, the samples were held in a vacuum for 3 hours until complete curing.

When silver is deposited by thermal vacuum evaporation onto epoxy resin substrates having different viscosities, metal atoms can diffuse into the bulk of the substrate to various depths depending on its viscosity. Such silver deposition causes the formation of silver nanoparticles in the near surface volume of the polymer. The formation of MNPs in all samples is directly supported by the appearance of a characteristic selective absorption band induced by the surface plasmon resonance in silver particles in optical spectral range [2].

The micrographs in Fig. 10–11 show examples of disperse MN-containing structures studied by TEM normal to the sample surface (left) and the cross sections of these structures (right) [3]. It can conventionally distinguish three types of metallic structures in the near surface volume of the polymer depending on its initial viscosity. For example, in the polymer viscosity range from 20 to 30 Pa · s, a near surface layer 110–120 nm thick (Fig. 10–11A and B) that contains ultrafine (1.5–2 nm) silver particles is detected. However, such small particles cannot be resolved by TEM because of an insufficient resolution of electron microscope. The MNP sizes in this sample correspond to the minimum critical size (1.5–2 nm) at which plasmon optical absorption can manifest itself in MNPs [2].

As the viscosity of epoxy resin increases from 30 to 90 Pa · s, the following bilayer structure forms (Fig. 10–2C and D): A surface monolayer of spherical silver nanoparticles with an average diameter of 6–8 nm and a layer 35–60 nm thick in the bulk of the polymer next to the first layer, which contains ultrafine (1.5–2 nm) MNPs. Similar sample characterized by a layer with ultrafine MNPs was described by the modeled optical spectra of a composite layered nanomaterial with MNPs that was evidence of plasmon absorption of MN in such composites. Note that a bilayer structure is also fabricated when a thermally evaporated metal is deposited onto the surface of a molten (heated) polymer [33–37].

It was observed that the average MN size and the distribution of the MN sizes in the near surface layer of epoxy increase with the polymer viscosity. As an example, Fig. 10–12 presents the histograms of the size distributions of silver nanoparticles for samples with an epoxy resin viscosity of 36 (Fig. 10–12A) and 48 Pa · s (Fig. 10–12B). For the same composites, Fig. 10–13 shows optical absorption spectra with the typical selective bands of MN plasmon absorption. The maximum absorption of a sample with a viscosity of 48 Pa · s is located at a longer wavelength compared with the more viscous sample. This fact also indicates the formation of bigger MNPs as the epoxy resin viscosity increases [12,17].

When silver is deposited onto epoxy resin with a viscosity of 60–100 Pa · s, a disperse metallic thin film consisting of spherical silver nanoparticles forms on the polymer surface (Fig. 10–2E and F). The formation of MNPs in the bulk of the polymer as the viscosity exceeds 100 Pa · s is not detected. At a higher viscosity of epoxy resin (>120 Pa · s), a thin island of silver film forms on the polymer surface (Fig. 10–2G and H), as in the case of traditional deposition of a metal onto the surface of solid materials [41]. Thus, three types of thin films, which have different structures, can form on the surface of a polymer depending on the ratio of the volume diffusion and surface diffusion of deposited silver atoms.

The formation of a composite layer with ultrafine MNPs when silver is deposited onto the surface of a polymer with a viscosity of 20–30 Pa · s at room temperature is most interesting because the condition of the polymer is close to being almost liquid. Therefore, this viscosity range of the polymer substrate was analyzed in more detail. To study the MNP growth kinetics in a viscous fluid polymer, a series of experiments on the deposition of silver onto liquid epoxy resin (20 Pa · s) at a different deposition time from 5 to 80 seconds were performed.

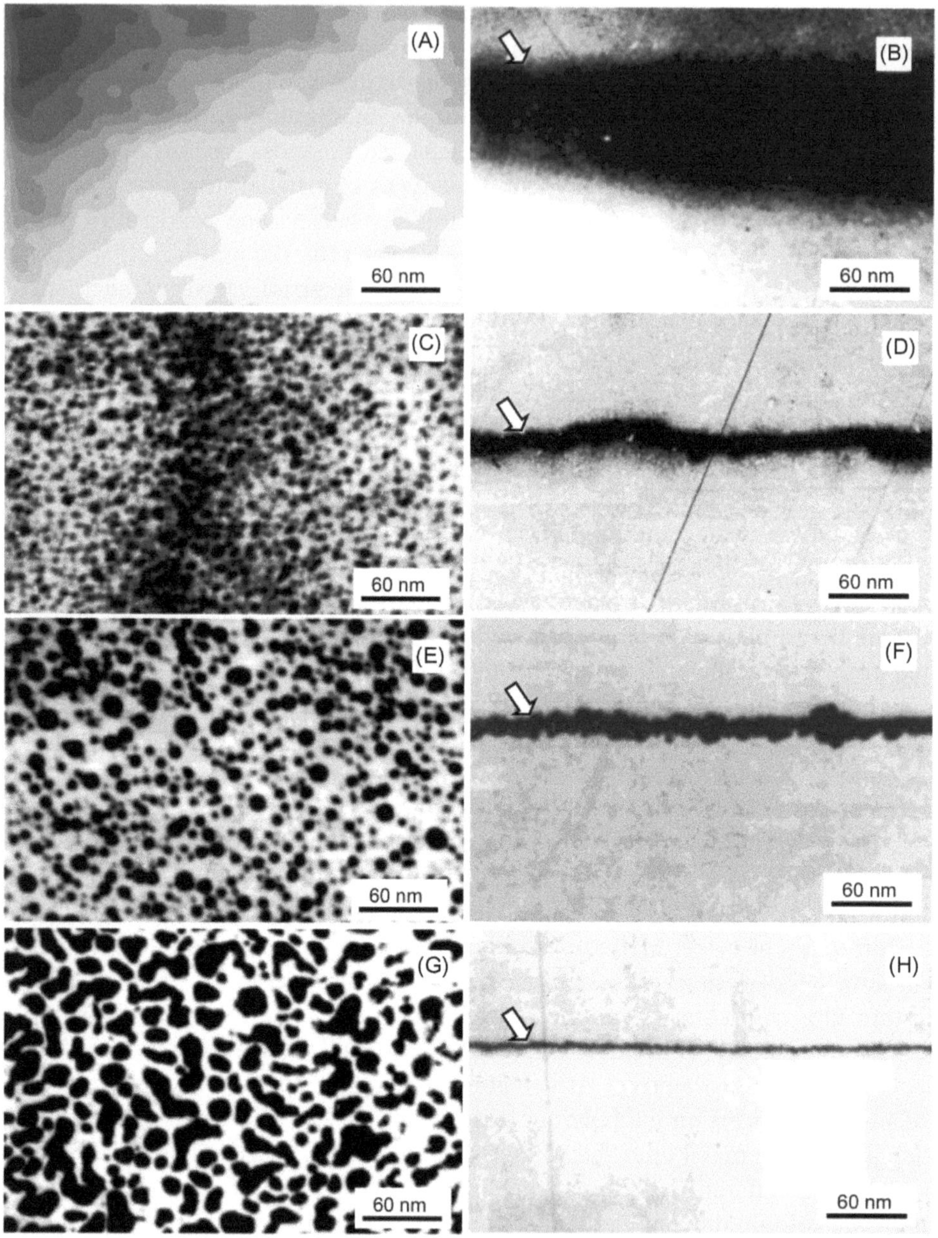

FIGURE 10–11 Micrographs of the silver disperse films synthesized by vacuum deposition onto liquid epoxy resin as a function of its viscosity. (A, C, E, G) Observation is normal to the sample surface and (B, D, F, H) cross sections of samples. The epoxy resin viscosity during deposition was (A, B) 20, (C, D) 36, (E, F) 86, and (H) 120 Pa · s. Arrows indicate the position of the polymer surface.

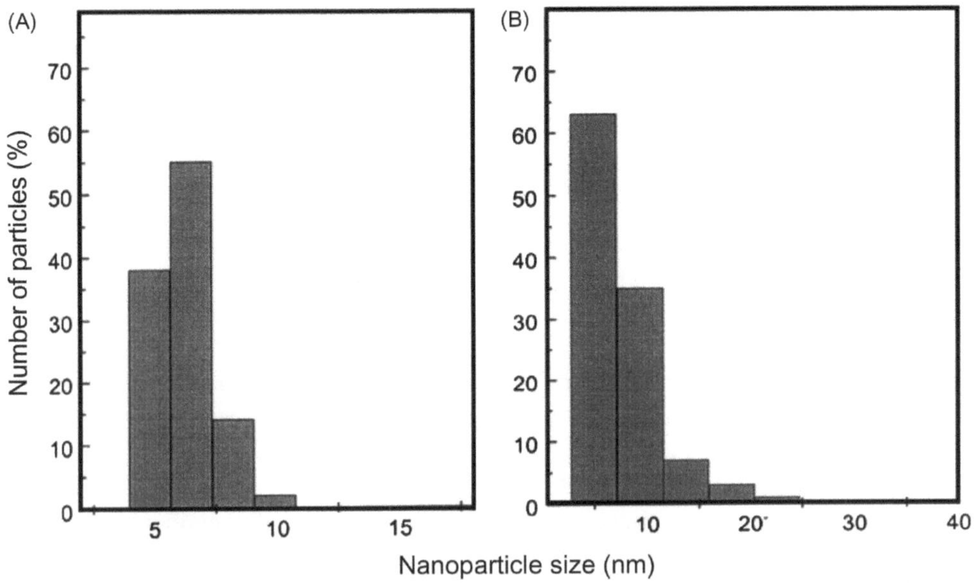

FIGURE 10–12 Histogram of the size distribution of silver nanoparticles for deposition onto epoxy resin with a different viscosity of (A) 36 and (B) 48 Pa · s.

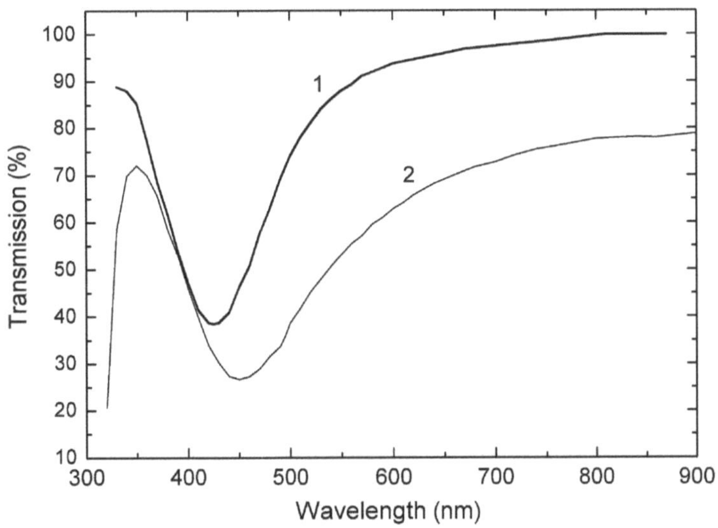

FIGURE 10–13 Optical transmission spectra for the samples prepared when silver was deposited onto epoxy resin with a different viscosity of (1) 36 and (2) 48 Pa · s.

In these samples, the silver particle size was estimated only optically because low TEM resolution. Since the MNP size ($\sim 1.5-2$ nm) is significantly smaller than the electron free path length in a bulk silver, the average nanoparticle size in these samples can be estimated from optical plasmon absorption band width $\Delta W_{1/2}$ using the equation [2]:

$$d = 2V_F / \Delta W_{1/2}, \qquad (10-11)$$

where d is the nanoparticle diameter and $V_F = 1.4 \cdot 10^{15}$ nm/s is the electron velocity at the Fermi surface.

Using the values of d, the optical absorption spectra of the composite materials in terms of the Bruggeman effective medium were simulated as described in [42]. Based on the results of model simulations, the silver nanoparticle with size d density per area unit N in each sample was estimated. Fig. 10–14 presents the dependence of N on time t of silver deposition onto liquid epoxy resin with a viscosity of 20 Pa · s.

As follows from this dependence, MNPs near the polymer surface have nucleated in a time of ~ 17 seconds after the beginning of deposition. Also, the concentration of silver atoms near the polymer surface at which optically detectable MNPs form can be estimated. During the diffusion of silver atoms, their concentration C_{Ag} at distance h from the polymer surface can be described as

$$C_{Ag} = \frac{2G}{D}\left[\left(\frac{Dt}{\pi}\right)^{1/2} e^{\frac{-2h}{4Dt}} - \frac{h}{2} erfc\left(\frac{h}{\sqrt{Dt}}\right)\right], \qquad (10-12)$$

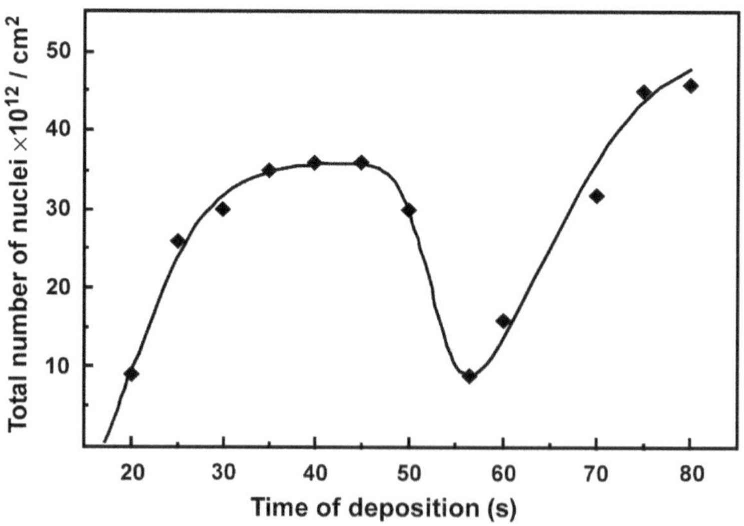

FIGURE 10–14 The number of ultrafine silver nanoparticles per unit area N when the metal is deposited onto epoxy resin with a viscosity of 20 Pa · s as a function of the deposition time (points). Simulation data is the solid line.

where G is the silver atomic flux onto the polymer surface and D is the diffusion coefficient [43].

The diffusion coefficient of silver atoms in epoxy resin can be estimated using the Stokes–Einstein formula [44]

$$D = kT/3\pi\phi\mu \qquad (10-13)$$

where μ is the dynamic viscosity of the polymer, T is the temperature of polymer, φ is the silver atom diameter, and k is the Boltzmann constant. According to this estimation, the volume diffusion coefficient of silver atoms is $D = 8 \cdot 10^{-10}$ cm^2/s. Then, the value of C_{Ag} after 17 seconds deposition in the near surface region with a thickness on the order of the silver atom diameter is $\sim 7 \cdot 10^{20}$ atoms/cm^3, which indicates a critical supersaturation of the near surface layer in the polymer. Within this time, silver nanoparticles with an average size $d = 1.6$ nm are fabricated, as follows from the optical data. In the next 25 seconds of deposition, the depth of MNPs in the polymer reaches 50–55 nm due to silver diffusion, the number of particles increases, and their average diameter increases monotonically by 0.1–0.2 nm due to the entrapment of silver atoms by the nanoparticles formed in the previous stage (Fig. 10–5). After the next 10–15 seconds, the concentration of MNPs decreases quickly and their average size increases until 2.5 nm.

The Ostwald ripening can explain the decrease in the number of particles that accompanies their growing, when the smallest particles dissolve and the metal atoms that become free join bigger MNPs [45]. In this case, the average interparticle distance increases noticeably, and MNPs can renucleate and grow in the space that is free of nanoparticles in the substrate if deposition continues [41]. The observed increase in the number of silver particles in viscous fluid epoxy resin after 57 seconds from the beginning of deposition corresponds to MNP renucleation. Since the number of particles increases in a 50-nm-thick layer, the average MNP size at this time decreases to 1.4–1.6 nm. However, longer deposition leads to an increase in both the MNP concentration and average size.

As follows from TEM study, there is a certain depth profile of the size distribution of nanoparticles. The biggest silver nanoparticles ($d \sim 4$–5 nm) are located near the surface layer of the polymer, and the smaller the MN size are placed deeper from the surface. This behavior is clear and related to the depth profile of the impurity concentration. At higher concentration, there is higher nanoparticle growth rate. Obviously, an increase in the epoxy viscosity at the same silver deposition rate should decrease the time it takes for a layer consisting of large MNPs to form. The silver diffusion into the bulk of the polymer decreases; the thickness of the layer with ultrafine particles decreases; and, at a certain viscosity, only a disperse surface metallic film will be formed.

Discussions on the formation mechanism of a composite layer with MNPs in the near surface region of a liquid polymer ($T > T_g$) are related to the immersion of MNPs in this layer [32–40]. According to [36], two forces act on the MNPs located on the surface of a liquid polymer provided wetting is complete or partial. The van der Waals force tends to immerse MNPs in the polymer volume, whereas the entropy force (which is caused by the compression of polymer chains near particles) tends to push them to the polymer surface. The result of these forces forms a potential well in the dependence of the free energy of particles on the immersion depth. Due to this minimum, particles or their ensembles are stabilized under the surface of a viscous fluid polymer at a depth smaller than the MNP diameter. The model calculations [40] demonstrate that, as the particle size decreases, the potential well depth decreases rapidly: In particular, for gold nanoparticles smaller than 50 nm, the potential well is absent. This means that the immersion mechanism does not operate for sufficiently small MNPs. This conclusion is also true of our work, since the silver nanoparticle size did not exceed 5 nm under our experimental conditions.

When describing the results of our experiments on the thermal deposition of silver onto epoxy resin having temperature $T < T_g$ due to chemical dilution, another method should be noted that can decrease the glass transition temperature of the surface of a polymer as compared with its bulk T_g. As follows from the results of [40], when the surface of polystyrene is subjected to UV radiation during the deposition of gold nanoparticles onto it from a solution, the nanoparticles are also immersed in a thin near surface layer of the polymer, which indicates a decrease in the glass transition temperature of the polymer surface. However, it is obvious that this technique differs substantially from the technological approach observed here.

10.3.2 Synthesis of Silver Nanoparticles in Viscous Epoxy Resin by Ion Implantation

Ion implantation is a promising method for the formation of MNPs in a polymer matrix, since it can perform a controlled synthesis of MNPs at various depths under the surface of an irradiated matrix at an unlimited content of an implanted impurity. Koon et al. [8] were the first to carry out studies in the field of ion synthesis of MNPs in polymers. The number of works dealing with the ion synthesis of MNPs in polymers is now continuously growing.

The following new approach has recently been discussed: It consists of the ion synthesis of MNPs in organic matrices that are in a viscous fluid relaxed state during implantation in contrast to the traditional irradiation of a glassy polymer [46]. The relaxation state of viscous fluid silicone polymer implanted with Fe ions was shown to affect the nucleation and growth of iron and cobalt nanoparticles. The size, shape, and crystal structure of the synthesized MNPs were found to be controlled by the diffusion mobility of metal atoms and small clusters. Various layered structures of ferromagnetic MNPs were grown. The authors of the work [46] assumed that, when magnetic (iron, cobalt, etc.) ions are implanted into polymers, the dipole–dipole interaction between MNPs affects the synthesis of MNPs, in particular, the formation of particles of various shapes (from a spherical to an acicular shape).

As shown previously [47] the effect of the relaxation state in an irradiated viscous fluid polymer on the ion synthesis of MNPs can be also used with epoxy resin. To exclude the effect of the magnetic dipole interaction described previously, diamagnetic silver ions were implanted into viscous fluid epoxy resin and fabricated metal nanostructures were studied for the first time [9]. For that glassy and viscous fluid epoxy resin was irradiated by 30 keV

$Ag+$ ions at a current density of $4\,\mu A/cm^2$ with doses in the range of $2.2 \cdot 10^{16}$–$7.5 \cdot 10^{16}$ ion/cm^2 at room temperature. For comparative experiments, epoxy resin was also implanted by 40 keV Ar^+ ions at a current density of $4\,\mu A/cm^2$ in the ion beam at doses of $3.2 \cdot 10^{16}$ and $1.0 \cdot 10^{17}$ ion/cm^2. The average calculated penetration depth of silver ions into epoxy resin was 30 nm, and this depth for argon ions was ~50 nm. As a viscous fluid polymer matrix, a liquid epoxy resin mixed with dibutylphthalate and polyethylenepolyamine was used. The lifetime of the viscous fluid polymer matrix was about 2 hours; then, its physicomechanical properties were affected by the formation of a spatial network structure in the polymer. The viscosity of the epoxy composite was 30 Pa · s at the beginning of implantation process and increased by about 12 Pa · s upon irradiation. Thus, the ion implantation time did not exceed the composite lifetime.

Epoxy layers ~10 μm thick were deposited onto thin glass substrates by spin coating. With allowance for radiation heating, the epoxy substrate temperature was about 60°C, which is several tens of degrees Celsius lower than the upper boundary of the range of allowable nondestructive heating of this polymer. Once irradiation is finished, the epoxy composite transforms into a glassy state according to the cure (polymerization) kinetics of this compound (Fig. 10–10).

Fig. 10–15 shows electron micrographs of the silver ion implanted layers in epoxy resin that is in a glassy or viscous fluid state during irradiation. Implantation is seen to result in the formation of silver nanoparticles in a polymer matrix in both types of samples. Note that MNPs visible in TEM form in the viscous fluid sample at a lower dose ($2.2 \cdot 10^{16}$ ion/cm^2) as compared with the glassy polymer (~$5.2 \cdot 10^{16}$ ion/cm^2). This circumstance indicates more effective MN nucleation in the viscous fluid material. Moreover, the MNPs formed in the viscous fluid medium at all ion doses are spherical, whereas the nanoparticles formed in the glassy matrix have an irregular droplike shape.

The electron diffraction patterns of all polymer samples with ion-synthesized MNPs consist of narrow polycrystalline rings corresponding to the face-centered cubic lattice of metallic silver superimposed on the weak broad diffuse rings of the polymer matrix. It should also be noted that the intensities of the polycrystalline rings in the glassy polymer are significantly weaker than those in the electron diffraction pattern taken from the irradiated viscous fluid polymer, which indicates a higher content of the metallic phase in the latter case. We did not detect any chemical compounds with silver ions in the irradiated matrices.

The estimates of the size parameters of the synthesized MNPs, such as mean particle size d_{mean}, particle density per unit area $N_{implant}$, and average cubic deviation Δimplant of the particle sizes from d_{mean}, are given in Table 10–1. As is seen from Table 10–1 and Fig. 10–15, d_{mean} and $\Delta_{implant}$ increase monotonically and $N_{implant}$ decreases as the dose of silver ions implanted into epoxy resin in various relaxation states increases. In the viscous fluid medium, the values of d_{mean} and $N_{implant}$ are higher and the scatter of the sizes $\Delta_{implant}$ of ion-synthesized MNPs is lower as compared with the glassy polymer at the same doses. Therefore, it can be concluded that the number of implanted silver ions that form MNPs in the viscous fluid matrix is larger; hence, the corresponding filling factor $f_{implant}$ should be higher.

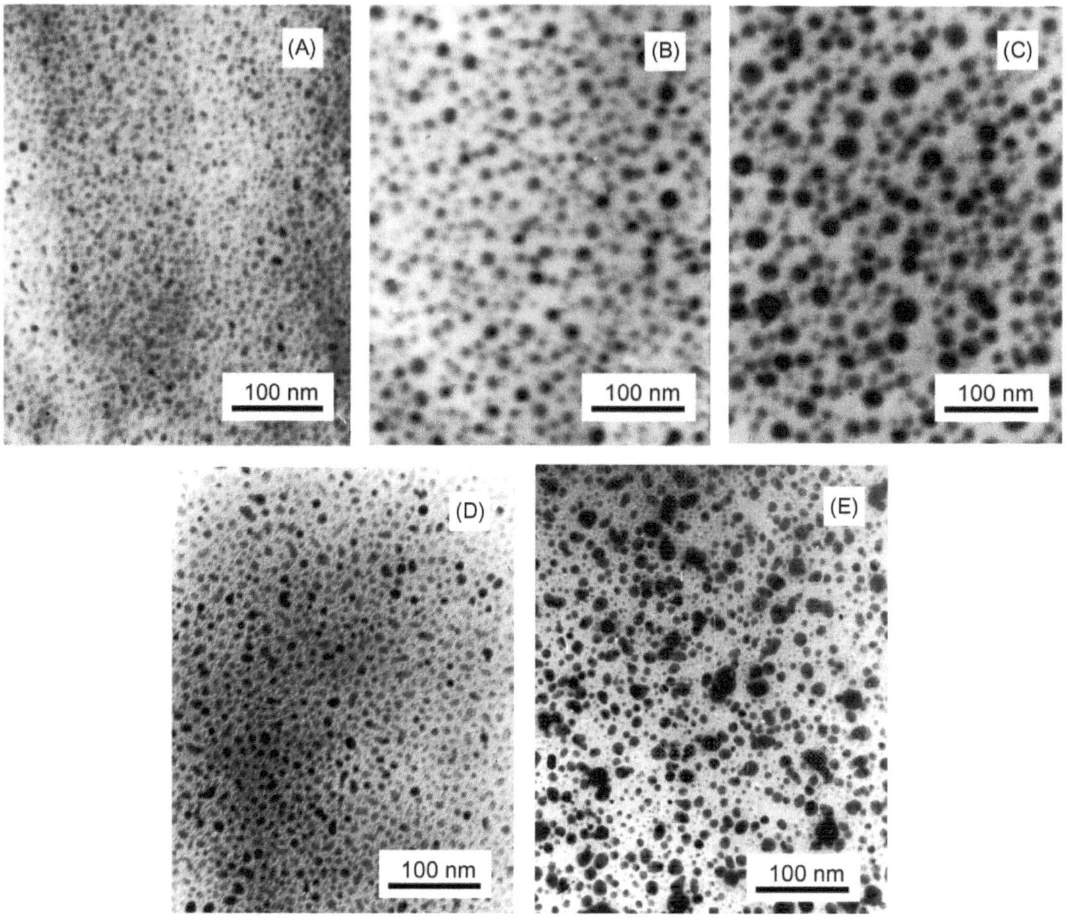

FIGURE 10–15 Micrographs of the epoxy layers with MNPs synthesized by the implantation of 30 keV silver ions at a current density of 4 μA/cm^2 in the ion beam into (A, B, C) viscous fluid and (D, E) glassy polymer matrices at a dose of (A) $2.2 \cdot 10^{16}$, (B) $5.2 \cdot 10^{16}$, (C) $7.5 \cdot 10^{16}$, (D) $5.2 \cdot 10^{16}$, and (E) $7.5 \cdot 10^{16}$ ion/cm^2.

Table 10–1 Granulometric Characteristics of the Silver Nanoparticles Ion-Synthesized in Epoxy Resin

Sample	State of Polymer	Dose 10^{16} (ion/cm^2)	d_{mean} (nm)	$\Delta_{implant}$ (nm)	$N_{implant} \times 10^{11}$ (cm^{-2})	$f_{implant}$
1	Viscous	2.2	5.0	2.0	3.90	0.12
2	Viscous	5.2	16.3	6.0	2.10	0.35
3	Viscous	7.5	22	11.1	1.38	0.49
5	Glassy	5.2	8.3	10.2	1.20	0.25
6	Glassy	7.5	13.8	14.2	0.90	0.35

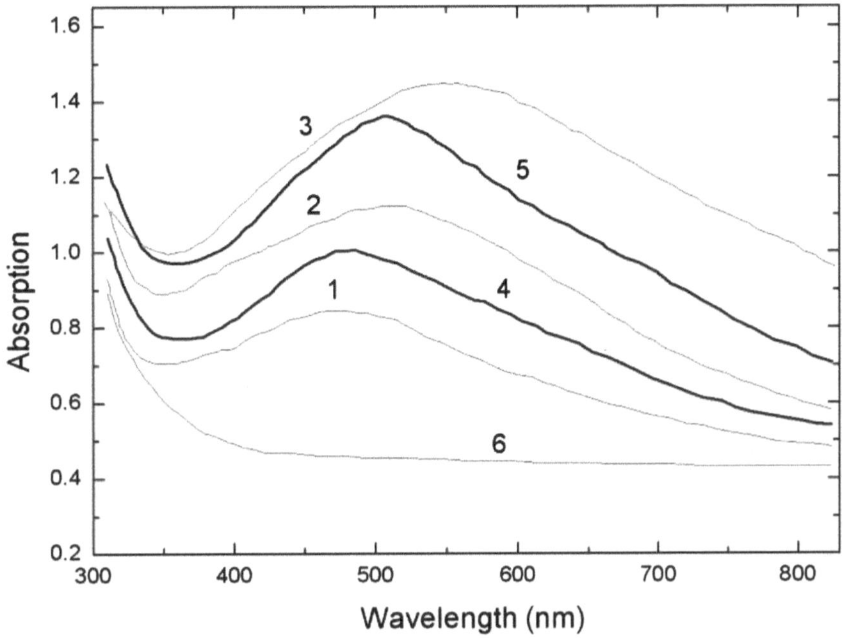

FIGURE 10–16 Optical absorption spectra [ln(1/*A*), where *A* is the transmission] of the epoxy layers with MNPs synthesized by the implantation of 30 keV silver ions at a current density of 4 $\mu A/cm^2$ in the ion beam into (1–3) viscous fluid and (4, 5) glassy polymer matrices at a dose of (1) $2.2 \cdot 10^{16}$, (2) $5.2 \cdot 10^{16}$, (3) $7.5 \cdot 10^{16}$, (4) $5.2 \cdot 10^{16}$, and (5) $7.5 \cdot 10^{16}$ ion/cm^2. (6) Absorption spectrum of the glassy epoxy resin after its irradiation by argon ions at a dose of $3.2 \cdot 10^{16}$ ion/cm^2.

To estimate the factor of filling of an epoxy matrix with metallic silver upon ion implantation, we use the optical absorption spectra of the synthesized composite materials (Fig. 10–16).

As is seen from Fig. 10–16, the optical spectra of all epoxy resin samples implanted by silver ions have a broad selective band, which is caused by the SPR absorption of MNPs [2]. This band is absent in the absorption spectrum of the epoxy resin irradiated by argon ions (Fig. 10–16, curve 6). All spectra exhibit intense absorption dominating in the UV region, which is caused by radiation defects and structural distortions in the polymer [5]. At a low dose ($2.2 \cdot 10^{16}$ ion/cm^2), the maximum of the SPR absorption of the viscous fluid sample is located near 490 nm (Fig. 10–16, curve 1). As the implantation dose increases, the SPR band absorption intensity increases monotonically and its maximum shifts toward long wavelengths, up to 570 nm (Fig. 10–16, curves 1–3). The changes in the optical spectra of the glassy epoxy resin with increasing dose are similar (Fig. 10–16, curves 4 and 5). However, the longer wavelength shift in the absorption maximum (to 520 nm) and its intensity are significantly smaller than for the viscous fluid polymer at the same implantation doses (Fig. 10–16, curves 3 and 5). The more intense SPR absorption bands of the ion-synthesized silver nanoparticles in the viscous fluid epoxy resin also indicate more effective MN nucleation and growth during implantation as compared with the glassy polymer. As follows from the Maxwell–Garnet theory, the position of the maximum in the absorption band of MNPs depends on the filling of the composite material with a metal [2,48]. The position of the maximum in the SPR absorption of MNPs at a longer wavelength corresponds to a higher value of $f_{implant}$. It is assumed the absence of a strong electromagnetic interaction between MNPs (whose size is smaller than the light wavelength) in the samples under study, use the dielectric constants of silver [18] and the polymer matrix [21] at the wavelengths of the maxima in the SPR absorption of MNPs, and determine the corresponding values of $f_{implant}$ from the well-known expression [2,7]

$$\varepsilon_1^{Ag} = \frac{(2 + f_{implant})}{(1 - f_{implant})} n^2_{polymer} \qquad (10-14)$$

where $\varepsilon_1{}^{Ag}$ is the real part of the dielectric constant of silver nanoparticles and $n_{polymer}$ is the refractive index of the dielectric matrix with MNPs.

The values of $f_{implant}$ calculated for the samples under study are given in the table. It is seen that, at the same implantation doses, $f_{implant}$ is significantly higher in the viscous fluid epoxy layers than in the glassy polymer.

As noted previously, ion implantation in materials leads to radiation defects, such as the rupture of the covalent bonds of macromolecules, the formation of free radicals, cross bonding, oxidation, and carbonization [45]. The formation of such defects in polymers increases the absorbability in the optical region and, especially, in the UV region.

As follows from Fig. 10–8, the UV absorption intensity and, hence, the number of radiation defects in the glassy polymer are larger than in the viscous fluid epoxy resin at the same implantation doses. The smaller number of structural defects in the implanted viscous fluid medium manifests itself in the higher optical transparency of these samples almost over the entire visible region (see Fig. 10–17).

The transmission spectra of the glassy and viscous fluid polymers irradiated by argon ions, which can only cause structural defects in the irradiated matrix, under the same conditions demonstrate higher transmission in the viscous fluid epoxy resin. Obviously, the high mobility of macromolecules in the viscous fluid matrix favors rapid restoration of broken bonds in the polymer and its molecular structure during irradiation. It was assumed that the efficiency of carbonization in the viscous fluid medium during ion implantation is lower than that in the glassy polymer, since it is carbon aggregates that cause significant absorption in the longer wavelength spectral region of the irradiated polymer.

To understand the formation of MNPs in this experiment, the process of polymerization of epoxy resin should be taken into account. As was discussed previously, during glass transition, the epoxy composite passes through several stages from the liquid state of the initial mixture to a fully solid polymer. The viscosity of epoxy resin changes in time according to these stages. At a polymer viscosity of 30 Pa · s, epoxy molecular groups begin to open and the epoxy system can be considered as a low viscosity homogeneous molecular liquid. No polymer blocks and supramolecular structure form in the polymer.

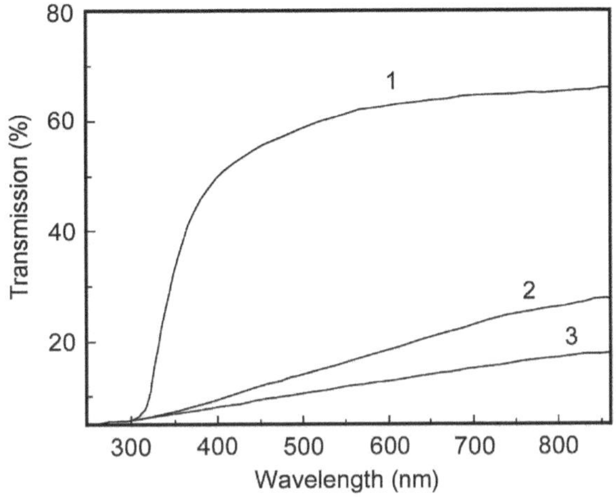

FIGURE 10–17 Transmission spectra of (1) unirradiated glassy epoxy resin and (2, 3) the same polymer implanted by 40 keV argon ions at a current density of 4 μA/cm^2 in the ion beam at a dose of 10^{17} ion/cm^2. (2) Viscous fluid and (3) glassy states of the epoxy resin during irradiation.

Thus, ion implantation in an epoxy polymer at a viscosity of 30 Pa · s (homogeneous liquid medium) leads to the formation of a supersaturated solution of silver atoms in the irradiated organic layer. Then, in the absence of internal defects in the matrix (which are inherent in solid materials), silver nanoparticles can homogeneously nucleate in the liquid medium similarly to the condensation of metallic drops in an inert gas atmosphere [45]. The further growth of MNPs is supported by the diffusion of impurity atoms toward their nucleation centers. Such conditions provide the formation of spherical particles and a rather uniform size distribution of MNPs (see Fig. 10–15).

In contrast, the presence of its own and additional radiation defects in glassy epoxy resin stimulates the heterogeneous nucleation of MNPs [39]. For this mechanism, the formation of MNPs and their size distribution depend on the system of defects and are also caused by the low mobility of silver atoms in the glassy polymer [45]. As a result, the MNPs acquire an asymmetrical droplike shape and the size distribution of the MNP is rather high (see Table 10–1).

Droplike ion-synthesized nanoparticles were also observed earlier, for example, in glassy polymethylmethacrylate implanted by iron ions [19]. However, it should be noted that the silver nanoparticles ion-synthesized in glassy polymers sometimes have a spherical shape [9]. These polymers, which have lower softening and glass transition temperatures as compared with epoxy resin, are likely to be heated under an ion beam and to transform into a liquidlike state. As a result, impurity atoms acquire a high mobility, which leads to the formation of spherical MNPs.

During ion implantation in a viscous fluid medium at a relatively low dose ($\leq 10^{16}$ ion/cm^2), the high mobility of silver atoms can lead to the smearing of the depth profile of impurity atoms. According to Eq. (10–3), the volume diffusion coefficient of silver atoms is five to six orders of magnitude higher than D_{Ag} in glassy polymers [19]. It is obvious that, at the early stage of implantation, a very high value of D_{Ag} can substantially decrease the local concentration of silver atoms in the near surface layer of the polymer and, thus, can decrease the efficiency of MN nucleation. However, at a slightly lower value of D_{Ag}, the mobility of silver atoms increases the formation probability of nucleation centers, which serve as sinks for implanted impurity atoms. As is seen from the results obtained (Fig. 10–15), MNPs nucleate in viscous fluid epoxy resin with a viscosity of 30 Pa · s at a lower implantation dose than in the glassy polymer. Thus, at the chosen viscosity of epoxy resin, the mobility of silver atoms leads to effective MN nucleation rather than to a decrease in their local concentration in the implanted layer. It is also obvious that the growth rate of MNPs is higher in the viscous fluid polymer, which manifests itself in the formation of coarse particles (see Table 10–1).

The high mobility of silver atoms in the viscous fluid medium results in high efficiency of their accumulation on formed MNP nucleation centers due to an increase in the getter sphere size R_{gett}, which is proportional to D_{Ag} [49].

$$R_{gett} \propto \sqrt{D_{Ag} \cdot t_{growth}} \qquad (10-15)$$

where t_{growth} is the MNP growth time. If t_{growth} is assumed to be the same in all irradiated samples, R_{gett} in the viscous fluid epoxy resin is three orders of magnitude higher than in the glassy polymer. According to Ref. [49], the higher the value of R_{gett}, the larger the MNP size, which was observed in our experiment (see Fig. 10–15 and Table 10–1).

Thus, a new method for synthesizing a disperse film that consists of silver nanoparticles on the surface and in the volume of a viscous fluid polymer (epoxy resin) kept at room temperature when the metal deposited by thermal vacuum evaporation is proposed. The structure and surface morphology of metal–polymer layers as a function of the viscosity of the polymer substrate during metal deposition is analyzed. We considered the dynamics of MN formation depending on the time of metal deposition onto the surface of a viscous fluid epoxy resin is considered and the formation mechanism of a nanostructured metal–polymer layer is discussed. In addition, it is shown that ion implantation in an epoxy resin that is in a viscous fluid state during irradiation can be used to synthesize nanoparticles of precious metals. Due to the viscous fluid state of the substrate, the diffusion coefficient of an implanted impurity without a special purpose increase in the polymer temperature can be increased. As a result, the nucleation and growth of nanoparticles are stimulated at low energy ion implantation doses and a high factor of filling of the polymer with the metal can be reached. With a viscous fluid epoxy resin, the formation of irreversible structural radiation defects in the irradiated organic material as compared with the implanted glassy polymer can be reduced and, thus, the uniformity of the size distribution of ion-synthesized MNPs can be increased.

10.4 Conclusion

A new method for synthesizing of thin layers that contain silver nanoparticles on the surface and in the volume of a viscous fluid polymer (epoxy resin) kept at room temperature when the metal deposited by thermal vacuum evaporation is proposed. The structure and surface morphology of metal–polymer layers as a function of the viscosity of the polymer substrate during metal deposition is analyzed. The dynamics of MNP formation depending on the time of metal deposition onto the surface of a viscous fluid epoxy resin is considered and the formation mechanism of a nanostructured metal–polymer layer is discussed. In addition, it is shown that ion implantation into a PMMA and epoxy resin that is in a viscous fluid or solid state during irradiation could be used to synthesize MNPs. Based on the Mie theory, optical extinction spectra for metal particles in the polymer and carbon matrices are simulated and compared with optical spectra for complex silver core–carbon shell nanoparticles. The formation of silver nanoparticles in polymer carbonized by ion irradiation is observed. Due to the viscous fluid state of the substrate, the diffusion coefficient of an implanted impurity without a special purpose increase in the polymer temperature could be increased. As a result, the nucleation and growth of nanoparticles are stimulated at low energy ion implantation doses and a high factor of filling of the polymer with the metal can be reached. With a viscous fluid epoxy resin, the formation of irreversible structural radiation defects in the irradiated organic material as compared with the implanted glassy polymer could be reduced and, thus, the uniformity of the size distribution of ion-synthesized MNPs will be increased.

Acknowledgments

I wish to thank my partners and coauthors V.I. Nuzhdin, V.F. Valeev, and Yu.N. Osin for fruitful discussions. Also, I am grateful to the Alexander von Humboldt Foundation and the DAAD in Germany, Austrian Scientific Foundation in the frame of the Lisa Meitner Fellowship and the Royal Society in the United Kingdom for financial support. This work was also supported by the Russian Foundation for Basic Research No. 17-08-00850.

References

[1] Flytzanis C, Hache F, Klein MC, Ricard D, Rousignol P. Nonlinear optics in composite materials. Amsterdam: Elsevier Science; 1991.

[2] Kreibig U, Vollmer M. Optical properties of metal clusters. Berlin: Springer; 1995.

[3] Abdullin SN, Stepanov AL, Osin YN, Khaibullin IB. Kinetics of silver nanoparticle formation in a viscous-flow polymer. Surf Sci 1998;395:L242–5.

[4] Heilmann A. Polymer films with embedded metal nanoparticles. Berlin: Springer; 2003.

[5] Townsend PT, Chandler PJ, Zhang L. Optical effects of ion implantation. Cambridge: Cambridge University Press; 1994.

[6] Davenas J, Perez A, Thevenard P, Dupuy CHS. Correlation between absorption bands and implanted alkali ions in LiF. Phys Stat Sol A 1973;19:679–86.

[7] Arnold GW. Near-surface nuleation and crystallization of an ion-implanted lithia-alumina-silica glass. J Appl Phys 1975;46:4466–73.

[8] Koon NC, Weber D, Pehrsson P, Sindler AI. Magnetic properties of iron implanted polymers and graphite. Mater Res Soc Symp Proc 1984;27:445–9.

[9] Stepanov AL, Abdullin SN, Khaibullin RI, Valeev VF, Osin YuN, Bazarov VV, et al. Ion synthesis of colloidal silver nanoclusters in the organic substrate. Mater Res Soc Symp Proc 1995;392:267–72.

[10] Stepanov AL, Abdullin SN, Petukhov VY, Osin YN, Khaibullin RI, Khaibullin IB. Formation of metal-polymer composites by ion implantation. Phil Mag B 2000;80:23–8.

[11] Wu Y, Zhang T, Andong L, Zhou G. The nano-structure and properties of Ag-implanted PET. Surf Coat Techol 2002;157:262–6.

[12] Mie G. Beiträge zu Optik trüber Medien, speziell kolloidaler Metallösungen. Ann Phys 1908;25:377–445.

[13] Sviridov DV. Chemical aspects of implantation of high-energy ions into polymeric materials. Russian Chem Rev 2002;71:315–27.

[14] Deying S, Saito Y, Suganomata S. Optical properties of $LiNbO_3$ implanted with Ag + ions. Jpn J Appl Phys 1994;33:L966–9.

[15] Abdullin SN, Stepanov AL, Osin YN, Khaibullin RI, Khaibullin IB. Synthesis of metallic dispersion and continuous films in the viscous polymer by implantation of cobalt ions. Surf Coat Technol 1998;106:214–19.

[16] Scheunemann W, Jäger H. Über die temperaturabhängigkeit optischer einenschaften lkeiner silberteilchen. Z Physik 1973;265:441–54.

[17] Bohren CF, Huffman DR. Absorption and scattering of light by small particles. New York: Wiley; 1983.

[18] Quinten M. Optical constants of gold and silver clusters in the spectral range between 1.5 and 4.5 eV. Z Phys B 1996;101:211–17.

[19] Zener C. Theory of growth of spherical precipitates from solid solutions. J Appl Phys 1949;20:950–4.

[20] Pignatato B, Fragala ME, Puglisi O. AFM and XPS study of ion bombarded polymethylmethacrylate. Nucl Instr Meth Phys Res 1997;131:141–8.

[21] Palik ED. Handbook of optical constants of solids. London: Academic; 1997.

[22] Rao GR, Wang ZL, Lee EH. Microstructural effects on surface mechanical properties of ion-implanted polymers. J Mater Res 1993;8:927–33.

[23] Aden AL. Scattering of electromagnetic waves from two concentric spheres. J Appl Phys 1951;22:1242–6.

[24] Sinzig J, Radtke U, Quinten M, Kreibig U. Binary clusters: homogeneous alloys and nucleus-shell structures. Z Phys D 1993;26:242–5.

[25] Sinzig J, Quinten M. Scattering and absorption by spherical multilayer particles. Appl Phys A 1994;58:157–62.

[26] Ivanov-Omskii VI, Tolmatchev AV, Yastrebov SG. Optical absorption of amorphous carbon doped with copper. Phil Mag B 1996;13:715–22.

[27] Biederman H, Chmel Z, Fejfar A, Misina M, Pesicka J. Temperature induced structural rearrangements of Ag/a-C:H composite films and their dc electrical conduction. Vacuum 1990;40:377–80.

[28] Stenzel O, Kupfer H, Pfeifer T, Lebedev A, Schulze S. Probing ultra-thin amorphous carbon films by means of nanometric silver islands. Opt Mater 2000;15:159–265.

[29] Kreibig U. Handbook of optical properties. In: Hummel RE, Wissmann P, editors. Optics of small particles, interfaces and surface, Vol. 2. London: CRC; 1997.

[30] Holzl J, Schulte F, Wagner H. Solid surface physics. Berlin: Springer; 1979.

[31] Kreibig U, Gartz M, Hilger A. Mie resonances: sensors for physical and chemical cluster interface properties. Ber Bunsenges Phys Chem 1997;101:1593–602.

[32] Goffe L. Photographic migration imaging—new concept in photography. W Photogr Sci Eng 1971;15:304–8.

[33] Kovacs GJ, Vincett PS. Formation and thermodynamic stability of a novel class of useful materials: close-packed monolayers of submicron monodisperse spheres just below a polymer surface. J Colloid Interface Sci 1982;90:335–51.

[34] Kovacs GJ, Vincett PS. Subsurface particulate film formation in softenable substrates: present status and possible new applications. Thin Solid Films 1983;100:341–53.

[35] Kovacs GJ, Vincett PS, Tremblay C, Pundsack AL. Vacuum deposition onto softenable substrates: formation of novel subsurface structures. Thin Solid Films 1983;101:21–40.

[36] Kovacs GJ, Vincett PS. Subsurface particle monolayer and film formation in softenable substrates: techniques and thermodynamics criteria. Thin Solid Films 1984;111:65–81.

[37] Pattabi M, Sastry MS, Sivaramakrishnan V. Silver-island films deposited on a substrate above its softening temperature. Phys Rev B 1989;39:9959–65.

[38] Payne RS, Swann A, Mills PJ. Thermal evaporation of tin on to softened polystyrene substrates. J Mater Sci 1990;25:3133–8.

[39] Bechtolsheim CV, Zaporojtchenko V, Faupel F. Interface structure and formation between gold and trimethylcyclohexane polycarbonate. J Mater Res 1999;14:3538–43.

[40] Rudoy VM, Ershov BG, Sukhov NL, Dementeva OV, Zaitseva AV, Seliverstov AF, et al. Colloid J 2002;64:755–8.

[41] Maissel LI, Gland R. Handbook of thin film technology. New York: McGraw-Hill; 1972.

[42] Bruggeman DAG. Berechnung verschiedener physikalischer konstanten von heterogenen Substanzen. Ann. Phys. B (Leipzig) 1935;24:636–64.

[43] Robertson D, Pundsack AL. Subsurface particle growth kinetics in physical vapour deposition. J Appl Phys 1981;52:455–62.

[44] Reif F. Fundamentals of statistical and thermal physics. New York: McGraw-Hill; 1965.

[45] Zettlemoyer AC. Nucleation. New York: Marcel Dekker; 1969.

[46] Khaibullin RI, Osin YN, Stepanov AL, Khaibullin IB. Ion synthesis of Fe and Ag granular films in viscous and solid state polymers. Vacuum 1998;51:289–94.

[47] Stepanov AL, Khaibullin RI, Abdullin SN, Osin YN, Valeev VF, Khaibullin IB. Study of thin cobalt films produced in polymer by ion implantation. Proc Inst Phys Conf Ser 1995;147:357–61.

[48] Maxwell Garnet JC. Colors in metal glasses and in metallic films. Philos Trans R Soc Lond 1904;203:385–420.

[49] Maurer RD. Nucleation and growth in a photosensitive glass. J Appl Phys 1958;29:1–7.

Further Reading

Johnson PB, Christy RW. Optical constants of transition metals. Phys Rev B 1974;9:5056–70.

11

Organic Conjugated Polymer-Based Functional Nanohybrids: Synthesis Methods, Mechanisms and Its Applications in Electrochemical Energy Storage Supercapacitors and Solar Cells

Kakarla Raghava Reddy[1], B. Hemavathi[2], Geetha R. Balakrishna[2], A.V. Raghu[3], S. Naveen[3], M.V. Shankar[4]

[1]THE SCHOOL OF CHEMICAL AND BIOMOLECULAR ENGINEERING, THE UNIVERSITY OF SYDNEY, SYDNEY, NSW, AUSTRALIA [2]CENTRE FOR NANO AND MATERIAL SCIENCES, JAIN UNIVERSITY, JAIN GLOBAL CAMPUS, BANGALORE, KARNATAKA, INDIA [3]DEPARTMENT OF BASIC SCIENCES, SCHOOL OF ENGINEERING AND TECHNOLOGY, JAIN UNIVERSITY, BANGALORE, KARNATAKA, INDIA [4]DEPARTMENT OF MATERIALS SCIENCE & NANOTECHNOLOGY, YOGI VEMANA UNIVERSITY, KADAPA, ANDHRA PRADESH, INDIA

11.1 Introduction

During the past few decades, electronic technology has experienced tremendous growth especially due to the advent of advanced and next generation nanoelectronic devices for various potential applications. Development of alternative materials in place of silicon semiconductors in alternative energy conversion/storage resources is crucial, leading to an interdisciplinary field called molecular electronics, comprising organic materials such as organic conducting polymers, proteins and pigments, etc. The emergence of organic conducting polymers with the discovery of iodine doped *trans*-polyacetylene, $(CH)_x$, by Shirakawa et al. made organic conducting polymers an indispensable material among their counterparts due to their versatility and exclusive properties [1]. Since then, much research has been focused on developing new organic conducting polymers such as polyaniline (PANI), polyanisidine, poly(o-toluidine), polythiophene, PEDOT, polypyrrole (PPy), poly(*p*-phenylenediamine), polyfuran, poly(vinyl carbzazole), polyaminophenol, poly(p-phenylene), their derivatives and copolymers for various electronic applications.

Polymer Composites with Functionalized Nanoparticles. DOI: https://doi.org/10.1016/B978-0-12-814064-2.00011-1

PANI is an organic conducting polymeric material that captured research interest after the discovery of conjugated polyacetylene. PANI has advantages such as ease of synthesis process, low-price, and excellent polymerization yield [2]. It has been researched due to its high electrical conductivity, and reversible (acid and base) doping and dedoping routes recognized with polymer chain containing nitrogen atom [3]. Graphene (GN) is an one-atom-thick sp^2-bonded ultrathin and transparent nanosheet that has been comprehensively researched and used in sensors [4], fuel cells [5], capacitors [6], etc. 3D structured GN nanosheets and their chemically functionalized GN materials have fascinated scientists and engineers due to their large l/d ratio, and excellent electrical and mechanical characteristics [7].

The introduction of GN nanosheets into polymeric matrices to generate new type hybrid nanocomposite materials [8,9] and a good dispersion of GN sheets in the polymeric matrix provides hybrid composites with improved properties of mechanical strength, thermal stability, and electrical conductivity. The prevalent improvement in the properties is because of the large aspect and surface-to-volume ratios of GN sheets in the polymer composite. Also, the nanoscale GN fillers can provide an excellent electron transfer pathway as it forms active nucleation sites for PANI [9]. This kind of conducting polymer nanohybrid has been developed into sensors, supercapacitors, and various other nanoelectronics applications.

11.2 Organic Conjugated Polymers

In recent years, the development of a variety of intrinsically conducting polymers (ICPs) has become one of the most promising fields in polymer chemistry, nanotechnology, and material science. Organic conjugated polymers find tremendous applications in different fields, especially in the electronics industry.

11.2.1 Types of Organic Conjugated Polymers

Organic conjugated polymers can be predominantly classified into three categories: (1) intrinsically/inherently electrically conjugated polymers (ICPs), (2) organic conducting polymeric hybrid composites, and (3) ionically conducting polymers [1]. A brief description of the conducting polymers is presented below due to the fact that the study deals with the synthesis of multifunctional nanocomposites comprising of organic conjugated/conducting polymers.

11.2.1.1 Intrinsically Electrically Conjugated Polymers

The organic conjugated polymers that consist of an electrical or electronic characteristic of a metal in conjunction with the mechanical properties, like processability, etc., are called "intrinsically conducting polymers" (ICPs) and otherwise known as "synthetic metal" [10]. MacDirmid et al. [11] revealed that organic conducting polymers could be doped to obtain conducting polymers. Polyacetylene was doped with bromine and iodine [11,12].

The typical examples of the intrinsically or inherently organic conjugated polymers are PANI, polyacetylene, polytoluidine, PPy, polyanisidine, polythiophene, PEDOT, poly(phenylene), poly(diphenylamine) (PDPA), poly(phenylene vinylene), their derivatives and copolymers, etc., Scheme 11−1 shows some of the intrinsically organic conjugated polymers.

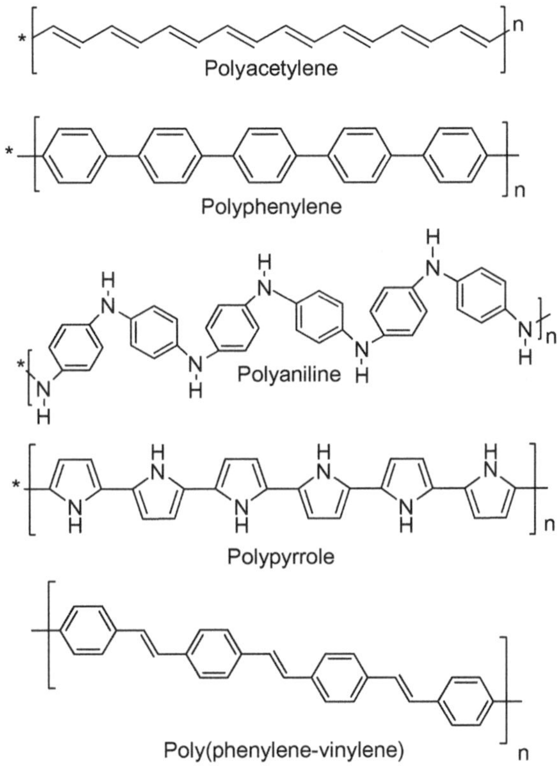

SCHEME 11–1 Examples of intrinsically/inherently electrically organic conjugated polymers.

The special electronic characteristics of the organic conjugated polymers are obtained from the existence of delocalized π−electrons. The atom extends over a long distance of the conjugated polymeric chain when it contains planar molecular structure in the backbone. The essential properties of the delocalized π−electrons system, which differentiate a typical conjugated polymer from a conventional polymer with σ-bonds, are (1) the small electronic (π) band gap (~1−3.5 eV), (2) ease of doping by oxidation or reduction, (3) net charge mobilities in the backbone of polymer, and (4) existence of quasiparticles [13].

The electrical and optical characteristics of ICPs depend on the electronic properties and basically on the chemical behavior of repeating unit. The general requirement of the structure for these polymers was well recognized many years ago. The electronic conducting properties of organic polymer are proportional to both the density and drift mobility of carriers. For the ICPs to have high electrical conductivity, it is required to have significant improvement in the carrier mobility and the density of carriers [13].

11.2.1.2 Organic Conjugated Polymer Hybrid Composites

A conducting polymer should be processable for successful application. However, the conducting polymers synthesized via the chemical route as well as electrochemical route are insoluble and infusible powders or films that have poor mechanical properties. Studies have been attempted to resolve this issue by making a blend of conducting polymers with the conventional polymers or thermoplastics or inorganic metal particles/carbon materials.

The mixture of a polymer and another component (polymer, metal particles, etc.) can be homogeneous or heterogeneous. These are blends or composites. The conducting polymeric composites are synthesized by mixing organic polymers with the electrically conducting nanoparticles such as nanocarbon black, metallic fine power or flakes, or nanofibers. It was found that the electrical conductivity of these hybrid composite nanomaterials is dependent on the content of the nanofiller, its size, and morphological structures. The percolation threshold is relatively high ($>$25 V %).

The blends and composites have been synthesized by using different techniques including mechanical blending, dispersion swelling emulsion techniques, and diffusion. The chemical methods of preparation offer large-scale production of the conductive polymer-based composites. However, in electrochemical methods, the films of conductive polymers are electrochemically deposited on the surface of metallic or the semiconducting electrodes and the thickness of films could be controlled by electrochemical parameters. A blend or composite can be prepared by the mechanical mixture of a conductive and insulating polymer (engineering polymers) using a counter ion rotating mixer or a double screw extruder. In this case, the conductivity of the composite strongly depends on the miscibility of the components and the rheological properties. Another method involves codissolution of the blend components in the common organic solvents followed by the vaporization of the organic solvent. Here, the phase separation may occur during the organic solvent vaporization because of different solubility of the polymer in the common solvent. The miscibility is an important factor in the polymer solutions. Interfacial polymerization is another approach to produce the organic conducting polymers with excellent properties.

11.2.1.3 Ionically Organic Conducting Polymers

Ion conducting polymers comprise of ionic salts dispersed in the heteropolymers such as polyethylene oxide. Ionic conducting polymers are divided into two categories. The classification is based on the temperature at which the polymeric material shows appreciable conductivity relative to its glass transition temperature Tg. The first group shows appreciable ionic conductivity at temperatures above Tg of the respective polymer. The second group has ionic conductivity at temperatures (typically, room temperature) below Tg of the polymer. Nafion and blend of polyvinyl alcohol (PVA) with phosphoric acid (H_3PO_4) are examples of polymers in this group.

11.3 Synthesis of Electrically Conducting Organic Conjugated Polymers

Synthetic strategies are useful in forming conjugated structures in polymers [14]. Organic conjugated polymers can be fabricated through various polymerization methodologies such as the chemical oxidation, electropolymerization, photochemical method, metathesis method, emulsion (mini and micro) method, inclusion technique, solid-state method, and plasma polymerization. Chemical polymerization is the most preferred route for the fabrication of organic conducting polymers in bulk form.

11.3.1 Commonly Used Organic Conjugated Polymers

PANI and PPy polymers are important polymers because of their chemical and environmental stability in air in comparison with other organic conducting polymers. The discussion on PANI is presented below as the present study deals with the PANI-based nanocomposites.

11.3.1.1 Polyaniline

PANI polymer is a model of phenylene derivative–based polymeric material that consists of a chemically flexible amino group in the polymeric chain, which is bound on any side of the phenylene ring. The transferring of protons (protonation), deprotonation and different physicochemical characteristics of PANI is due to presence of –NH– group in the polymer. There are four states, leucomeraldine base (LEB), emeraldine (EB), pernigraniline (PNB), and emeraldine salt, shown in Scheme 11–2.

11.3.1.2 Advantages of Polyaniline

Among all the organic conjugated polymers, PANI has an exceptional representation, because of the following: (1) inexpensive aniline monomer; (2) easy synthesis; (3) environmental stability; (4) simple doping by protonic acids; (5) polymerization progress has excellent yield; (6) obtained salt is highly stable and shows good electrical conductivity; (7) when the salt reacts with a base like NaOH or KOH solution, PANI salt changes into PANI base; (8) PANI is the only conjugated polymeric material whose electrical conductivity and electronic structure can reversibly be optimized by both reactions of chemical oxidation and protonation; (9) PANI shows an excellent stability and the electrochemical characteristics; and (10) a conducting polymer with nitrogen atoms occupies the bridging to the benzenoid and quinoid rings in its backbone structure and plays an important role in π-bond formation and electrical intrachain conduction.

PANI has different characteristics from other organic conjugated polymers in three aspects. Firstly, conjugation bands in PANI are not symmetric and hence the Fermi level is not in the center of valence and the conjugation bands are entirely asymmetric in nature. Secondly, nitrogen atoms and carbon chemical rings are involved in the conjugation of monomer units. This type of conjugation is entirely different from PPy, as the heteromolecules in PPy do not donate to the formation of the π-bonds. Thirdly, the electronic structure

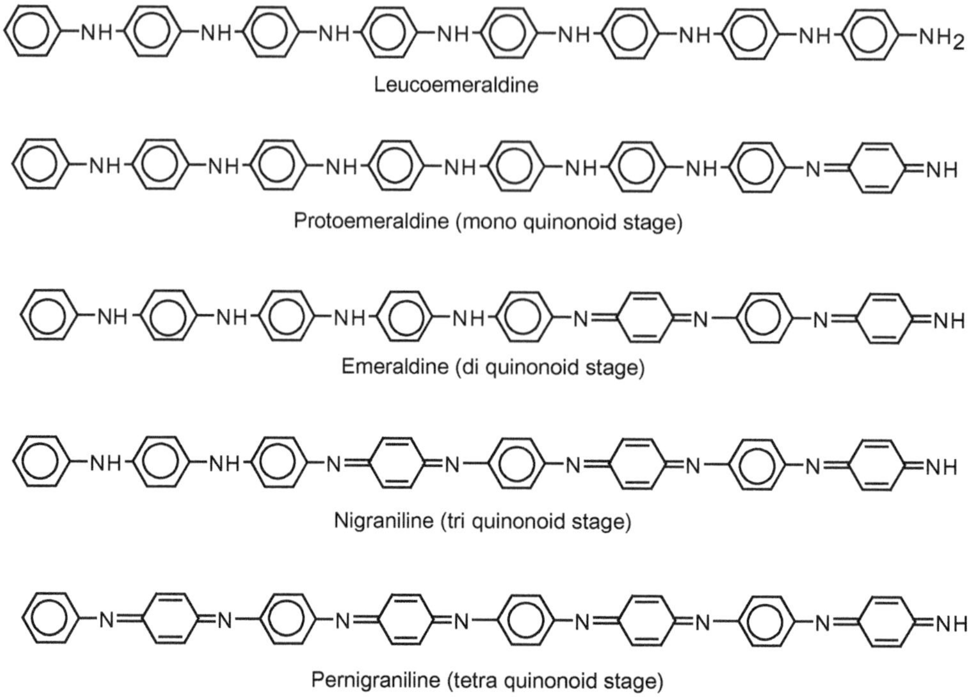

SCHEME 11–2 Structures of aniline octamers.

of the PANI can be changed via variation in the number of protons or electrons in the repeat units.

11.3.1.3 Synthesis Strategy of Polyaniline

The most favorable technique for the preparation of PANI is chemical oxidative polymerization, which requires both strong acids (e.g., hydrochloric or sulfuric acid) as the medium and ammoniumpersulfate as an oxidizing agent (e.g., ammonium persulfate) [15,16]. Since the chemical polymerization is a two-electron exchange process, a mole of persulfate is required for the oxidation of aniline monomer molecules. However, for practical purposes, a small quantity of oxidant is used to avoid excessive oxidation leading to degradation of PANI. PANI may exist in various oxidation states depending on the conditions of polymerization. The principal role of the oxidizing agent is to take away an electron from the nitrogen atom of $-NH_2$ group in aniline. The polymerization reaction is summarized as follows:

$$\text{Ar-NH}_2 \xrightarrow{\text{Ar-NH}_2} \text{Ar-NH-Ar-NH}_2 \xrightarrow[2e,-2H]{\text{Ar-NH}_2}$$
$$\text{Ar-NH-Ar-NH-Ar-NH}_2 \xrightarrow[\text{coupling}]{\text{Further oxidative}} \text{Polyaniline}$$

The factors affecting the polymerization process are the pH of the solution, types of acid used, concentration of aniline, salvation and electronegativity of a conjugated base connected with strong acids like HCl, H_2SO_4 or HNO_3, concentration of the oxidant, temperature of polymerization, etc.

The electropolymerization of aniline monomer is generally conducted in aqueous protonic acid by applying potentials/current at the working electrode (WE). The electrochemical synthesis is thus attained by galvanostatic method, where the stable current is in the range of 1 to 10 mA, or potentiostatic method, with potential in the range of -0.2 V to 0.9 V [15,16].

11.3.1.4 Polymerization Mechanism of Aniline Monomer

Chemical oxidative polymerization of aniline monomer has generally been conducted in acidic medium [15,16]. The acids as medium can the dissolve the aniline monomer molecules at the initial stages of polymerization and formation of PANI salt after polymerization. The mechanism of polymerization of aniline can be explained as follows. The aniline monomer is converted into an anilinium ion in the presence of acid solution as a medium, and the chemical reaction results in the generation of high-yield PANI with protonation and partial oxidation. Upon oxidation, aniline radical cation is formed. The reaction of *N*- and *ortho*-radical cations with successive rearomatization of the dication results in the formation of *p*-amino diphenylamine (PADPA). Further, the chemical oxidation of the diradical dications results in formation of pernigraniline PANI salt with completely oxidized state. While "head-to-tail" coupling is predominant, a minor proportion of coupling via *ortho* position may occur, leading to the deficiency of conjugation in the final polymerized product of PANI [16].

On complete consumption of oxidizing agent, the unreacted aniline monomer molecules decrease to the pernigraniline form to generate the green emeraldine salt of PANI. A visible color change can be noticed during polymerization. The formation of PADPA can be visually seen by the appearance of a pink color and the formation of protonated pernigraniline can be seen in the color of the product. The emeraldine salt (green color product) precipitates out from the medium. The emeraldine base form of PANI can be converted into nonconducting polymer as the fully oxidized form of pernigraniline ($n = 0$) or the reduced form of leucoemeraldine ($n = 1$) at proper experimental reaction conditions. Imine sites in the base form of polymer are more susceptible for protonation in acidic conditions, resulting in the generation of emeraldine salt (green color). The electrical conducting properties of PANI are influenced by the rate of oxidation and protonation. Also, other factors such as chemical structure and morphology of the polymer obtained from different parameters in the polymerization process also affect the conducting properties of PANI [15,16].

11.3.1.5 Inorganic and Organic Dopants for Polyaniline

PANI can be easily doped by nonredox doping method. PANI can be doped using two very different chemical reactions, that is, (1) doping with protonic acids (nonredox), and (2) doping by oxidation. The doping of the nonconductive form of an emeralide base with the protonic acids, leads to the fully protonated form of the imine nitrogen atoms, and results in the completely protonated form of PANI as an emeraldine salt [15,16].

The doping of polymer by an oxidation process (p-doping) results in the formation of LEB of PANI. This participates in the oxidation process of σ or π-system rather than just the п-system of the PANI, as is normally the case of p-doping system. There are plenty of reports available on the doping of PANI by dopants. Mineral acids, high molecular weight long-chain organic sulfonic acids, phosphoric acids and their esters, etc. have been used as dopants for PANI. Doping of PANI by hydrochloric acid (HCl) has been extensively studied. Sulfuric acid (H_2SO_4), perchloric acid ($HClO_4$), phosphoric acid (H_3PO_4), etc., have also used as dopants for PANI. The organic acidic compounds with low molecular weights such as acetic acid, acrylic acid, propionic acid, formic acid, etc. have been tried as organic dopants for the fabrication of conducting PANI. High molecular weight long-chain organic sulfonic acidic compounds such as camphorsulfonic acid, methanesulfonic acid, p-toluenesulfonic acid, polystyrenesulfonic acid (PSSA), dodecylbenzenesulfonic acid (DBSA), etc., were effectively employed for doping of PANI. Other than the abovementioned acid dopants, esters of phosphoric acid as well as phthalic acid have also been tried as nonredox dopant ions.

It is known that the dopant ion can alter the properties of the conducting PANI. It can affect properties like conductivity, crystallinity, solubility, and hence, processability. Among these, the organic sulfonic acids have been found to impart many interesting properties like high solubility and thermal stability to PANI. The sulfonic acids like DBSA, PSSA, etc., which have long-chain alkyl groups, can plasticize the PANI formed by increasing the interchain separation, thereby facilitating the solvent molecules penetrating into the PANI lattice. Also, the alkyl chain can act as an effective solvent compatible group to improve the solubility of the polymer. These groups can also act as compatibilizers as well as plasticizers when blended with other polymers. The efficiency of some of these dopants to provide better properties and processability to PANI /its composites was examined in detail. The doping process in conducting polymers can be carried out by various routes such as electrochemical polymerization, solution polymerization, self-assembly, ion exchange, and irradiation.

11.3.2 Polyaniline-Based Nanomaterials

The nanoscale size materials have superior physical and chemical properties. The self-assembly of the nanostructures into the ordered form is necessary to use them for practical multifunctional applications. The advanced nanostructured polymeric materials and nanodevices with enhanced physicochemical properties can be produced by a coupling of nanounits and new chemical methodologies to strongly assemble them. The polymeric nanomaterials can form with various size and morphological structures such as 0D spherical nanoparticles, hollow nanoparticles, 1D nanorods, nanofibers, core–shell or core–sheath nanostructures, nanotubes, nanoarrays, hybrid composite nanostructures, other hierarchical structures, etc., and these nanostructured polymers can provide a variety of benefits compared with bulk materials due to the tunable surface multifunctionalities of nanoscale polymers [12]. PANI-based nanomaterials have been potentially applied in various fields such as energy storage systems, sensing devices (chemical, vapor, gas, and biological molecules), anticorrosion protection, energy conversion, fabrication of conductive and magnetic composites using PANI as nanofiller, etc.

11.3.2.1 Fabrication Methods of Polyaniline-Based Nanomaterials

Various physical and chemical synthesis methodologies have been used for the synthesis of PANI materials with different nanostructures, and their hybrids of organic–organic, and inorganic–organic polymer composites. Among the various methodologies, the template technique was found to be an efficient and effective tool to synthesize the nanomaterials of PANI [17]. The template route requires the incorporation of guest molecules like organic or inorganic additives in the empty spaces of the host material. Such empty spaces serve as the template to generate products with desired nanodimensions, orientation, and morphological nanostructures. Generally, the template technique is generally categorized into two processes, that is, the soft template, and the hard template. In the hard template method, the materials like anodic porous Al_2O_3 membrane, zeolite, and track etched polycarbonate are used for fabrication of products with controlled properties. Liquid crystals, organic surfactants (anionic, cationic, and nonionic), and cyclodextrin, etc. are used for the fabrication of nanomaterials through the soft template route. In addition to these methods, the template-free method has been employed to fabricate the organic conjugated polymers in the nanoscale [17].

11.3.2.2 Soft Template Methods

Various organic surfactant molecules with amphiphilic characteristics are widely utilized as the soft templates in the reaction system for the formation of micelles that act as nanoreactors [18]. The emulsions consist of a mixture of surfactants, oil, and water in the reaction system in which oil and aqueous systems are separated by the amphiphile monolayers [19]. During the micro or miniemulsion polymerization process, the organic monomer molecules are placed in the following positions: (1) fine droplets of monomer molecules, (2) swollen micelles with inactive monomer molecules, (3) micelles that are active and become monomer/swollen polymeric particles where polymerization reaction takes place, and (4) monomer molecules in an aqueous system. The conventional polymerization process is different from an emulsion polymerization in its characteristics, namely (1) the transparent reaction system, and (2) absence of monomer droplets and micelles in the reaction system. The microemulsion polymerization process allows the fabrication of narrow size distributed very small latex nanoparticles with the dimensions in the range of 10–100 nm, whereas the conventional polymerization allows the preparation of products with larger size and without uniform morphology. This is due to the formation of micelles in the emulsion polymerization in the form of various shapes that act as nanoreactors for the fabrication of PANI with various morphologies such as spheres, thin layer, lamelle, tubular, core–shell, hollow, rods, etc. [20].

11.3.2.3 Hard Template Methods

As mentioned above, in the hard template process, the common templates include the porous structured inorganic alumina membranes and organic polycarbonate membranes with pore sizes of 10–100 μm. Especially, one dimensional nanostructures such as nanofibers, nanorods, nanotubes, and nanochains of organic conducting polymers are synthesized using the hard template methods [21]. In this method, the monomer molecules are completely filled in the pores of the hard template, which creates self-assembled morphological structures. As the polymerization reaction proceeds upon dropwise addition of initiator solution, the organic monomer molecules present in the pores of the templates start polymerizing and grow on the porous-structured walls of inorganic or organic membranes, resulting in the formation of different lengths of 1D structured polymers. While short polymerization results in a nanotubular conducting polymer with a thin wall size, and long polymerization time gives a nanotube with a higher wall thickness [22].

11.3.2.4 Template-Free Methods

In comparison with the hard and soft template methods, the template-free methods provide a rational and functional approach to generate large scale highly uniform one dimensional nanofiber morphology with a product of high purity. The template-free approach can be used in various chemical polymerization techniques such as solution, bulk, radiolysis/γ-irradiation, interfacial (aqueous–organic), dispersion, and suspension [19]. PANI micro and nanorods were fabricated via electrochemical polymerization by embedding iridium and platinum needles into PANI microrods. The micro and nanotubular structured PANI was fabricated through in situ chemical polymerization of aniline in organic solvents containing different organic dopants such as 10-camphorsulfonic acid, β-naphtalenesulfonic acid, and azobenzene sulfonic acid. PANI nanofibers were fabricated by a gamma ray–assisted radiolytic route that involves irradiating the aqueous solution containing the reaction mixture of aniline monomer molecules, HCl as inorganic dopant, and ammonium persulfate as an initiator [23]. Likewise, inorganic silica oxide–conducting polymer composite was prepared as core–shell nanoparticles through this gamma-irradiation method that initiated an in situ chemical oxidative polymerization on the surface of aniline monomer molecules adsorbed SiO_2 particle. Another physical technique, the electrospinning process, has been widely used as a simple, cost-effective, highly efficient, and multifaceted technique for the fabrication of high-quality 1D long nanofibers with uniform thickness [24]. Some other techniques like sonication and microwave irradiation have been tried for fabricating nanofibers of conducting polymers.

11.3.3 Nanostructured Hybrid Organic Conjugated Polymers

The polymer nanocomposite materials can be broadly classified into three categories such as (1) fibrous, (2) laminar, and (3) particulate. The nanocomposite materials with amorphous or crystalline form of 1D, 2D, and 3D structures can be fabricated by an encapsulation of micro or nanosized particles into organic polymer matrix, or inorganic metal/metal oxides or ceramic material, resulting in changing the physicochemical properties of materials. It was clearly demonstrated in the literature that the organic–organic, inorganic–organic type conducting polymer-based hybrid nanocomposite materials possess superior thermal, mechanical, electrical, and electrochemical properties to those of their separate components [24].

11.3.3.1 Organic Conjugated Polymer-Based Hybrid Structured Nanocomposites Through Guest and Host Interactions

A variety of organic conjugated polymer nanocomposites (inorganic–organic, and organic–organic) can be fabricated through guest/host interactions with organic and inorganic counterparts [23]. In inorganic–organic nanocomposites, organic polymers act as the host and inorganic guest materials act as the guest. In another type of organic–inorganic-based nanocomposites, the organic conjugated polymer acts as guest and is embedded into the host's inorganic porous layers. Both types of composites can form by strong covalent chemical bonds or physical blending. The numerous organic conjugated polymer-based hybrid composites were fabricated using vast combination of various organic conjugated polymers (e.g., PANI, PTh, PPy, poly(*o*-toluidine), poly(anisidine), PPV, their derivatives and copolymers), and inorganic metal/metal oxide materials with different sizes and shapes (e.g., Cu, Au, Ag, Pt, Pd, Ru, CdS, MnS, SiO_2, Fe_2O_3, Fe_3O_4, CuO, CeO_2, MnO_2, TiO_2, ZnO, ZrO_2, $BaTiO_3$, RuO_2, their alloys, etc.) and carbon nanostructured materials (e.g., carbon black, carbon nanotubes (CNTs), 3D GN nanosheets, graphite oxide, porous carbons, carbon fibers, and carbon quantum dots), and their multifunctional properties and different applications have been widely investigated [25,26].

11.3.3.2 Organic Conjugated Polymer-Based Hybrids prepared by Core–Shell Strategy

The encapsulation of inorganic nanoparticles in the matrix of organic conjugated polymers has been used as popular technique for the preparation of nanocomposites [23]. Various nanometallic and nanometal oxide particles were embedded into electrically conductive organic polymers. These nanocomposites have different physical and chemical properties than the pristine polymer and inorganic metallic materials. The core–shell type inorganic metal oxide–polymer nanocomposites are prepared by polymerization of organic monomer molecules on the magnetite nanoparticles surface by using an oxidant, result in formation of uniform and core–shell type superparamagnetic Fe_3O_4/polymer composite [26].

11.3.4 Different Strategies to Synthesis of Conducting Polyaniline-Based Hybrid Nanostructured Composites

Incorporation of inorganic materials into PANI have resulted composite with better mechanical, thermal, electrical, and magnetic, catalytic, and optical properties [27]. One can distinguish these composites based on incorporation of inorganic systems in PANI. The interactions between inorganic materials and PANI in the composite matrix are through any one of strong chemical bonds (coordination, ionic, and covalent links), weak bonds (hydrogen bonds, hydrophilic/hydrophobic interactions, van der Waals), or without the chemical interactions. The ultimate properties of hybrid polymer composites are based on surface area, size, dispersion rate of nanofillers in polymer matrix, concentration of filler morphology, and polymers in the composites.

Several strategies were established to synthesize nanohybrids of metallic nanoparticles and PANI. Xu et al. [28] prepared core–shell type nanobelt/nanoribbon shape of metal oxide (MoO_3, VO_2, SnO_2)-PANI nanocomposites via in situ polymerization. In this method, aniline was dissolved in aqueous solution containing metal oxide nanopowder under ultrasonication and then the reaction mixture was transferred to the Teflon autoclave followed by the dropwise addition of APS solution as the oxidant. The resultant solution was sealed and kept at 140°C for 4 hours for the formation of the composite. Han et al. [29] synthesized PANI/NiO composites with morphology of rectangular tube by using sodium dodecylbenzenesulfonate as surfactant and dopant. Pan et al. [30] reported preparation of the PANI/MnO_2 nanotubes via template method. Likewise, nanostructured composites consisting of PANI and different metal/metal oxides have been synthesized using (1) in situ, (2) ex situ, (3) electrochemical polymerization, (4) soapless emulsion polymerization, (5) ring-opening polymerization, (6) atom-transfer radical polymerization, etc. [31–33].

11.3.5 Potential Applications of Conducting Polyaniline-Based Hybrid Nanostructured Composites

Conducting polymer-based nanocomposites are known to have tremendous applications in various fields [31]. Metal nanoparticles coated with PANI find applications in electrochromic devices, secondary batteries, solar cells, and sensors for gases, depending on the properties of nature of metal nanoparticles in the composite. For example, a TiO_2-SiO_2/PANI hybrid composite showed better photocatalytic activity for phtodegradation of methyl orange in the presence of sunlight, in comparison with pristine TiO_2 [34]. PANI/SWNTs hybrids with the low content of high aspect ratio of CNTs were fabricated for printable conductive inks for potential applications in organic nanoelectronics [35]. PANI/V_2O_5 nanohybrids were fabricated and utilized as cathode electrode materials for Li-ion secondary rechargeable batteries [36].

PANI/TiO_2 hybrids were synthesized with high dielectric constant by in situ chemical polymerization approach and used in electronic circuits as capacitors [34]. PANI/Au composite finds use of dopamine sensor [37]. Likewise, many hybrid nanocomposites have been synthesized with diverse inorganic nanoparticles like CeO_2, SiO_2, clay, CNT, Au, Pd, Ag, etc. [37–43]. These composites find applications in chemical and biological sensors, optically transparent conducting materials, substitute for carbon nanomaterials, electrochromic devices, actuators, field emission display devices, supercapacitors, photovoltaic cells, transistor and data storage, surface protection, etc. [44–46].

11.3.5.1 Electrochemical Energy Storage Devices Based on Polyaniline Nanohybrids

Supercapacitors are electrochemical capacitors with specific energy comparable with batteries and power density comparable with conventional capacitors, and hence have both high specific energy and power densities. The supercapacitors can be classified into two types on the basis of energy storage mechanism: pseudocapacitors and electric double layer capacitors (EDLCs) [47,48]. The pseudocapacitors are capacitors that function based on the fast and reversible redox reactions. The Faradaic redox reactions will take place at the surface and at the bulk of the electrode, and it is basically consists of a conducting polymer and/or carbon or inorganic oxides. In EDLCs, the charge accumulates in an electric double layer that is created at the interface between an electrode and an electrolyte, and carbon nanostructured materials are used as the electrode materials [6,49].

The electrochemical behaviors of the active material are typically analyzed with three-electrode configuration consisting of an electrolyte solution of organic and aqueous mixture, counterelectrode (CE), WE, and reference electrode (RE). The electrochemical parameters such as the electrochemical specific capacitance, cycle life, and the charge–discharge rate are measured with the galvanostatic charge–discharge (GCD), cyclic voltammetry (CV) and electrochemical impedance spectroscopy (EIS) experiments. From the constant current charge–discharge curve of GCD, specific capacitance can be determined using the following equation.

$$C = I\Delta t\ /m\Delta V$$

Where, C is the specific capacitance of the electrode, I is the current, Δt is the discharge time, ΔV is the potential window, and m is the mass of active materials. The cycle life of the active material is ascertained with the symmetry of the charge–discharge CV curve for multiple cycles. From CV curves, the specific capacitance is calculated using the following equation:

$$C = A/(vfm)$$

Where, A is an integral area of cyclic voltammogram, f is the scan rate, v is a voltage window, and m is the mass of the active electrode material.

PANI is considered the highest potential material due to its high electrochemical capacitive properties and an ease of synthesis [50]. PANI is a well-known pseudocapacitor, and exhibits an exceptional doping and dedoping chemistry [51,52]. PANI forms in three oxidation states, that is, (1) leucoemeraldine, (2) emeraldine, and (3) nigraniline. The undoped PANI is in an insulating/very low conducting base form and the completely doped form of PANI has high electrical conductivity. The polyemeraldine salt is known to possess very good electrical conductivity due to protonation of its imine nitrogens. The specific capacitance of a dedoped PANI is much higher when compared with the doped PANI [53] because, in doped PANI, the conveniently bonded chain reduces the PANI capacity. However, dedoping will create accessibility of the electroactive material and also improve morphology [54,55]. It was observed that the better morphology increases the electrochemical accessibility due to higher surface area, which significantly improves the supercapacitor performance [56]. PANI has a relatively weak cycling life and a low capacitance at high charge/discharge rate, which restricts its practical applications. The capability of the material will be improved with nanoscale binding technique, where PANI-based hybrids make it a better electrode material [57]. PANI with coral-like nanostructures are synthesized through the chemical oxidative polymerization of aniline with sunset yellow (SY) as organic dopant and surfactant, and it was utilized as an efficient supercapacitor electrode material that showed an electrochemical specific capacitance of 467 F/g [58].

A lot of research is going on to functionalize nanoscale carbons, inorganic metal, and metal oxides with organic conducting polymers to prepare nanocomposites [21,59,60] to reinforce the mechanical, electrical, and electrochemical properties [61], which increases the total capacitance and cycle life [62]. GN nanosheets with superior stability and surface area can be combined with PANI to obtain excellent pseudocapacitance nanocomposites with synergistic properties. GN has required functional groups that will interact with PANI and thus improve the specific capacitance and the stability [63,64].

PANI/carbon binary composites possess advantages such as excellent electronic conductivity, good mechanical strength, large specific surface area, and good cyclic stability. PANI with carbon nanocomposites were extensively studied with a blend of CNTs, GN, GO, rGO, etc. PANI/GN nanohybrids are synthesized by various methods by various research groups to tailor the property of the electrodes. The synthetic methods include chemical precipitation method [65], electrochemical method [66], etc. The synthesis method is very crucial for the performance and the flexibility of the electrode. Coating of PANI over CNTs forms a flexible 1D sheet, forming covalent or van der Waals bonds, and showing an excellent specific capacitance of 976.5 F/g at a discharge rate of 0.4 A/g. CNT with 2D GN sheets, will not only increase the accessible area, but also facilitate the ion diffusion, leading to improved electrical conductivity and reduced intrinsic iR-drop [67]. A binary composite of tube-like PANI and GO composites showed a specific capacitance of 437.8 F/g at 4 A/g with the capacitance retention of 90% of value after 1000 cycles, indicating good cycle stability [68]. Zhu et al. fabricated a 3D N-doped G/PANI composite foam with porous structure to allow the penetration and diffusion of electrolyte. A specific capacitance of 528 F/g at 0.1 A/g with good cyclic stability of 95.9% capacitance retention was achieved after 5000 long charge−discharge cycles [69]. A 3D cross-linked PANI with nitrogen-doped porous structured carbon composite facilitates more contact area between an electrode material and an electrolyte and hence more active sites for the charge storage of electrode. A specific capacity of 755 F/g at 1 A/g with excellent cycling stability in 1 M H_2SO_4, was obtained. Fig. 11−1 shows the schematic diagram of the cross-linked fiber and the pathway for electron transfer [70]. Lee et al. fabricated a $Mn_3(PO_4)_2$ firmly anchored on branched structures of PANI (PANI-M90) synthesized via the sonochemical method for the enhanced performance of the electrode (Fig. 11−2). Hence, the morphology of the composite is crucial for the quick and easy transfer of charges [71]. Another 3D solvent-free PANI/RGO aerogel synthesized via self-assembled hydrothermal method was employed as an electrochemical supercapacitor electrode material that displayed a specific capacitance of 480 F/g at 1 A/g with very good cyclability [72].

A 3D architecture of the PANI/GN nanohybrid has grabbed attention as its surface area is electrochemically accessible and a 3D porous nanocomposite is formed by depositing PANI

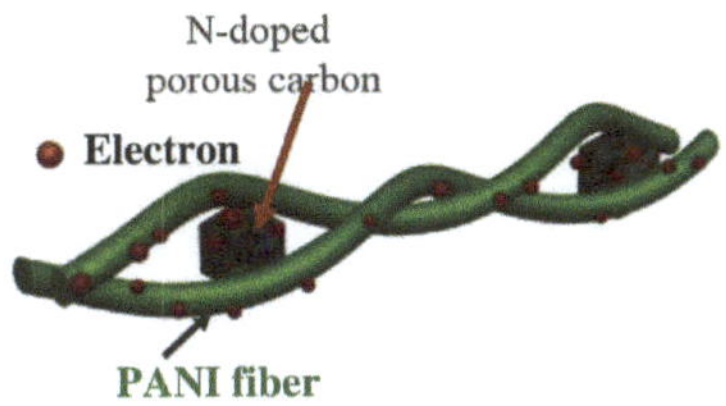

FIGURE 11–1 Schematic representation of PANI and N-doped porous carbon composite. *Reproduced from Guo S, Shen H, Tie Z, Zhu S, Shi P, Fan J, et al. J Power Sources 2017;359:285–294 with permission of Elsevier.*

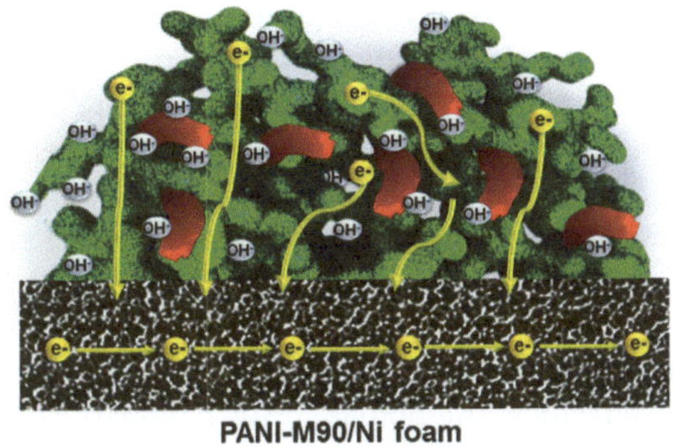

FIGURE 11–2 Schematic illustration of surface reaction at PANI-M90 electrode. *Reproduced from Lee CC, Omar FS, Numan A, Duraisamy N, Ramesh K, Ramesh S. J Solid State Electrochem 2017;21:3205–3213 with permission of Springer.*

over a 3D scaffold [73,74]. A freestanding 3D structure is more advantageous compared with others, however, synthesis method is crucial for the application. Still, the specific capacitance of nanocarbon is not exciting and hence, enhancement on the specific capacitance with binary nanocomposites of PANI/metal oxides is well researched.

A binary composite of PANI with metal oxides has been studied extensively due to its inherent pseudocapacitance. Various transition metal oxides such as nanostructured MnO_2 [75], MoO_3 [76], TiO_2 [77], and RuO_2 [78] have also been investigated with very good electroactive properties. Among the metal oxides, MnO_2 is highly researched due to easy cycle stability, cost, favorable high surface area, and very good charge storage characteristics [79–81].

Zhang et al. synthesized PANI/MnO_2 nanocomposites with encapsulated MnO_2 by exchange chemical reaction of PANI with MnO_2 in organic solvent of N-methyl-2-pyrrolidone. The nanohybrids exhibited a synergistic effect of PANI and MnO_2 and showed a specific capacitance of 330 F/g. It also retained 94% of the initial value even after 1000 cycles [75]. Another method used soaking 1D nanofibers of PANI in a $KMnO_4$ to get PANI/MnO_2 composite with uniform loading and the hybrid electrode showed the specific capacitance of 383 F/g [82]. Another core–shell composite of MnO_2 coated with PANI exhibited good specific density of 565 F/g with 77% retainment of the initial value [83]. A binary composite of manganese phosphate [$Mn_3(PO_4)_2$] with PANI synthesized via the sonochemical method showed a specific capacity of 347 C/g at 1 A/g [71]. A freestanding PANI/CNTs –decorated activated carbon fiber felt as electrode shows a real capacitance of 5611 mF/cm^2 [84]. A skeleton/skin structured RGO/CNTs with PANI composite fiber electrode as tested with all solid-state symmetric supercapacitor showed a specific capacitance of 193.1 F/cm^3 at 1 A/cm^3, with retention of 92.60% after 2000 cycles [85].

The application of metal oxide–based nanostructured composites is very encouraging, but they pose serious limitation in terms of stability in acidic electrolyte. Many synthetic protocols have been reported wherein the metal oxide does not come into direct contact with the electrolyte, thus improving the stability of the electrode. Even though binary composites of PANI have been researched extensively, the performance of the electrode could not reach its full capacity. This led to the development of the ternary blend of PANI/carbon/metal oxide nanocomposites with much higher conductivities, specific capacitance, and cycling stabilities. To obtain the synergistic behavior of the all the components of the composites, it is very important to optimize the interaction between the components in the nanostructure. A ternary composite electrode, PANI/MoO_3/GN nanoplates were synthesized with a simple synthetic procedure using both MoO_3 and GNP in the matrix and in situ polymerization of aniline [86].

A pseudocapacitance electrode of MWCNT/MnO_2/PANi nanocomposite was again synthesized by in situ polymerization method [87]. Ag nanoparticles dispersed on PANI/GN nanocomposite synthesized via in situ polymerization exhibited high specific capacitance of 591 F/g while retaining 96% of the initial value [88]. The SWCNTs/PANI@rGO ternary nanohybrid electrode was fabricated via chemical reduction method and as-prepared hybrid nanocables showed a specific capacitance of 845 F/g, with good cycling stability and retaining about 96% of its initial capacitance after 1000 cycles [89]. Another ternary hybrid, 3D CuO/PANI@rGO electrode by an in situ chemical polymerization method with spontaneous oxidization of Cu nanowire was synthesized and used as a supercapacitor electrode. A maximum specific capacitance of 634.4 F/g at 1.0 A/g and with very high energy-density value of 126.8 Wh/kg was obtained [90]. $NiCo_2S_4$@PANI core–shell hybrid nanowires grown on the surface of carbon fibers with distinctive features have enhanced electrochemical specific performance for supercapacitors. A specific capacitance of 1823 F/g at 2 mA/cm^2 and an excellent cycling stability was obtained for the $NiCo_2S_4$@PANI/CF core/shell electrode. The interesting morphology of the composite is shown in Fig. 11–3 [91]. $NiCo_2O_4$/PANI@CNTs hybrid composites having a thin layer of $NiCo_2O_4$ on the surface of PANI embedded CNTs (Fig. 11–4) showed an excellent specific capacitance of 2250 F/g at a scan rate of 5 mV/s [92]. The enhancement of the properties of the nanocomposites can be attributed to the interaction between the components (π-π, n-π) leading to a better electrode material for the supercapacitor application (Table 11–1).

11.3.5.2 Energy Conversion Systems Based on Polyaniline Nanohybrids

Solar energy conversion is a cheap and green option to produce energy, and organic conjugated polymers have garnered much attention as material for solar energy conversion due to

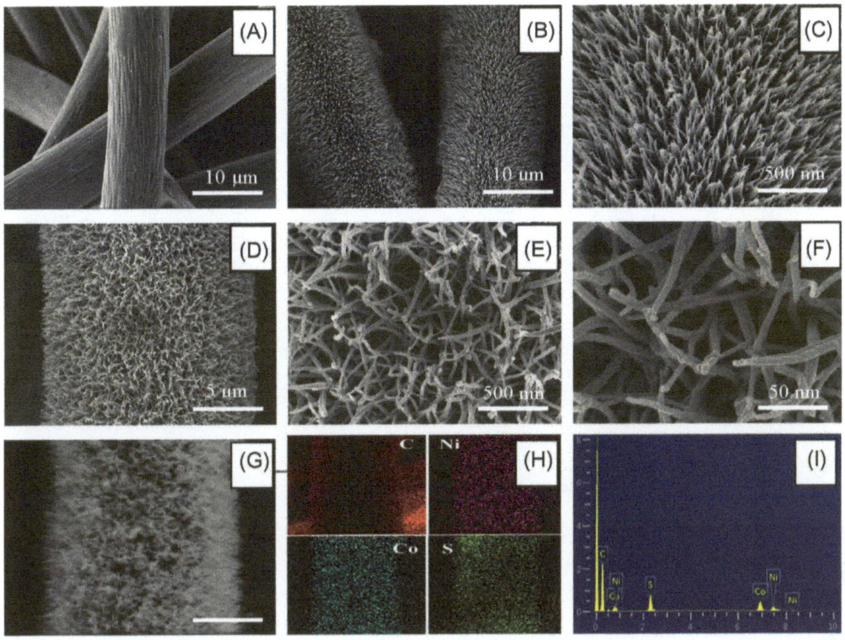

FIGURE 11–3 (A) SEM image of bare carbon fiber; (B) low-magnification and (C) high-magnification SEM images $NiCo_2S_4$/carbon fiber composites; (D) low-magnification, and (E, F) high-magnification SEM images $NiCo_2S_4$/PANI@carbon fiber composites; (G, H, I) EDS spectra element mappings. *Reproduced from Liu X, Wu Z, Yin Y. Chem Eng J 2017;323:330–339 with permission of Elsevier.*

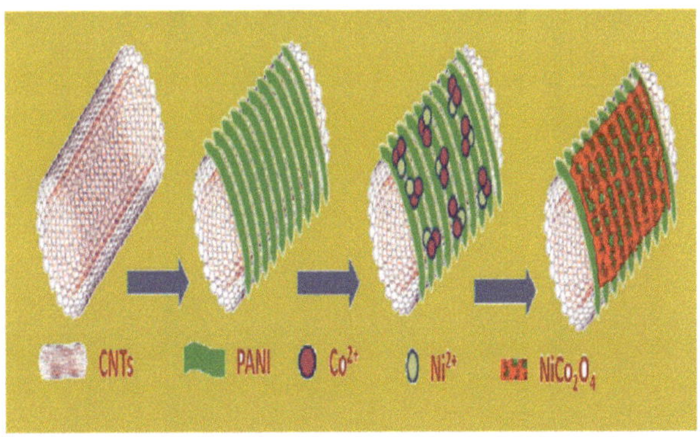

FIGURE 11–4 Schematic of synthesis of $NiCo_2O_4$/PANI@CNTs hybrid composites. *Reproduced from Chaudhary G, Sharma AK, Bhardwaj P, Kant K, Kaushal I, Mishra AK. J Energy Chem 2017;26:175–181 with permission of Elsevier.*

their low cost and ease of processability. Polymer solar cells (PSCs) and dye-sensitized solar cells (DSSCs) are the most studied photovoltaic devices in recent years. The conventional silicon solar cell requires an expensive fabrication process, whereas, the organic solar cell (OSC) is very cheap as it can be processed with doctor blade or spin coating techniques.

Table 11–1 The Electrochemical Performance of Various Polyaniline-Based Hybrid Supercapacitors

Electrode	Electrolyte Solution	Specific Capacitance(F/g)	Energy Density (Specific Energy) Wh/kg	Power Density (Specific Power) kW/kg	Capacity Retention (%)	Reference
PANI	IM H_2SO_4	503	96.23	8.88	85% at 10,000 cycles	[93]
PANI	IM H_2SO_4	532			–	[94]
PANI	0.IM Na_2SO_4	330			94% at 1000 cycles	[75]
PANI/SY	IM H_2SO_4	467			–	[58]
PANI	Nafion	269			65% at 10,000 cycles	[95]
PANI	1 M camphorsulfonic acid	33			–	[96]
PANI	Gel polymer electrolyte	672		7.5	–	[97]
PANI/C	0.5 M H_2SO_4	1486			10% at 1000 cycles	[98]
PANI/C	1 M H_2SO_4	439	60.83	0.5	–	[99]
PANI/CNT	PVA/H_3PO_4	440	7.03	5.2	93% at 1000 cycles	[100]
PANI/CNT	1 M H_2SO_4	322	22	0.083	–	[101]
PANI/MWCNT	IM $NaNO_3$	328			94% at 1000 cycles	[102]
PANI/MWCNT	0.1 M H_2SO_4	560			–	[103]
PANI/GN	2 M H_2SO_4	480	8.3		>70% at 1000 cycles	[104]
N-GE/PANI	1 M H_2SO_4	528			95.9% at 5000 cycles	[69]
PANI/N-doped porous carbon	1 M H_2SO_4	755			–	[70]
PANI/GO	1 M H_2SO_4	662	64.5	1.16	–	[105]
PANI/GO	1 M KCl	437			90% at 2000 cycles	[68]
PANI/MnO_2	0.1 M Na_2SO_4	715	200		96.5% at 5000 cycles	[106]
PANI/MnO_2	1 M H_2SO_4	626	17.8	0.6	–	[107]
PANI/MnO_2	1 M Na_2SO_4	498			95% at 5000 cycles	[108]
PANI/MnO_2/TiN	0.5 M Na_2SO_4	674			–	[109]
PANI/TiO_2/GO	1 M H_2SO_4	430			–	[110]
MoS_2/PANI/G	1 M Na_2SO_4	618			78% at 2000 cycles	[111]
SSCNTs/PANI/RGO	3 M NaOH	845			96% at 1000 cycles	[89]
CuO/PANI/rGO	1 M Na_2SO_4	634.4	126.8	114.2	97.4% at 10,000 cycles	[90]
CNT/Fe_3O_4/PANI	H_2SO_4-PVA	201	28	5.3	3.6% at 10,000 cycles	[112]
$NiCo_2S_4$PANI/CF	6 M KOH	1823	64.92	0.276	86.2% at 5000 cycles	[91]
$NiCo_2O_4$/PANI@CNTs	H_2SO_4 and HNO_3	2250			–	[92]

Also, the electronic characteristics of the polymers can be tuned as per requirement. In an organic photovoltaic cell, the incident photon generates an exciton and this exciton is dissociated into an electron and hole at the donor/acceptor interface. These generated free charges move under the applied electric field to generate electricity. These polymers are especially used as photoactive materials, which generate an exciton with the incident photon. PANI has been used as a photoactive material as it possess interesting electronic properties. A p–n junction layer is created with PANI as p-type material. An OSC typically consists of two electrodes sandwiching a hybrid of two materials with the properties of donor and acceptor. The p–n junction consists of polymer–fullerene or polymer–polymer bulk heterojunctions. The current–voltage (I–V) measurement of the solar cell devices was studied to interpret the performance of the device. Other parameters studied during device measurement are open-circuit voltage (Voc), short-circuit current density (Jsc), fill factor (FF), and efficiency (η).

PANI doped with formic acid was used to fabricate PANI/n-silicon heterojunction by spin coating of the PANI polymer over n-type silicon substrates. The p–n junction diode displayed rectifying ratio of about 100. The Jsc obtained was 45 $\mu A/cm^2$ with Voc = 400 mV, and $\eta = 0.3\%$ [113]. Nanocrystalline TiO_2/PANI interface p–n junction solar cell created with spin coating showed a Jsc of 45.8 $\mu A/cm^2$ and Voc of 0.387 V [114]. H_2SO_4 doped PANI as spin coated over n-MoSe2 crystal to fabricate the p–n junction solar cell and the efficiency, η and the FF obtained are 1% and 0.26 respectively [115].

A hybrid composite of PANI with cadmium sulfide nanoparticles (CdS NPs) are synthesized by in situ chemical oxidative polymerization of aniline monomers in an acetic acid medium over CdS NPs to get nanotubular PANI coated with CdS nanoparticles. The hybrid DSSC fabricated with PANI/CdS composite showed pen circuit voltage of 0.73 V, and a power conversion efficiency of 3.96% [116]. Photovoltaic properties DSSCs incorporated with GN oxide nanosheet/PANI/ethylene oxide, nanohybrid gel electrolyte showed an improved PV property as GOS-PANI nanohybrids facilitate extended electron transfer for the electrochemical reduction of I^{3-}. The highest photoenergy conversion efficiency was 5.63% [117]. PANI/ZnS hybrid composites synthesized by polymerizing aniline monomers in the presence of acetic acid over ZnS nanoparticles were used in fabrication of DSSC, and power conversion efficiency of $\eta = 3.38\%$ was obtained [118].

11.4 Conclusion

Organic conducting polymers have attracted considerable attention as molecular electronics due to their innumerable advantages compared with the inorganic materials. The ability of the polymer material to be tuned to the required application, ease of processability, flexibility, and interesting morphology are a few of the advantages. PANI has been a potential candidate for various applications such as supercapacitor electrodes and energy conversion.

The success of the PANI-based hybrid composites for application depends on the architecture, morphology, and the interaction of the components present in the composite to get synergistic properties. PANI-based supercapacitor electrodes are extensively discussed here with both binary and ternary composites of PANI with carbon and/or metal oxides. PANI, being a pseudocapacitor, can interact with the other components to further improve performance. A 3D ternary composite of PANI with carbon and metal oxide has shown exceptional supercapacitance and durability for practical application. However, still the practical application of the PANI-based composite will critically depend on the architecture and the interaction between the components of the nanocomposite.

PANI-based photoactive material for solar energy conversion has also been discussed here and even though the performance of the device is not very encouraging, its improvement is still very promising.

References

[1] Shirakawa H, Louis EJ, MacDiarmid AG, Chiang CK, Heeger AJ. J Chem Soc, Chem Commun 1977;578–80. Available from: https://doi.org/10.1039/C39770000578.

[2] Wang H, Hao Q, Yang X, Lu L, Wang X. ACS App Mater Interfaces 2010;2:821–8.

[3] Bogdanović U, Pašti I, Ćirić-Marjanović G, Mitrić M, Ahrenkiel SP, Vodnik V. ACS App Mater Interfaces 2015;7:28393–403.

[4] Yang YJ. Sens Actuators B: Chem 2015;221:750–9.

[5] Liu M, Zhang R, Chen W. Chem Rev 2014;114:5117–60.

[6] Stoller MD, Park S, Zhu Y, An J, Ruoff RS. Nano Lett 2008;8:3498–502.

[7] Wei D, Liu Y, Wang Y, Zhang H, Huang L, Yu G. Nano Lett 2009;9:1752–8.

[8] Stankovich S, Dikin DA, Dommett GHB, Kohlhaas KM, Zimney EJ, Stach EA, et al. Nature 2006;442:282.

[9] Yan J, Wei T, Shao B, Fan Z, Qian W, Zhang M, et al. Carbon 2010;48:487–93.

[10] MacDiarmid AG. Angew Chem, Int Ed. Engl 2001;40:2581–90.

[11] Chiang CK, Fincher CR, Park YW, Heeger AJ, Shirakawa H, Louis EJ, et al. Phys Rev Lett 1977;39:1098–101.

[12] Shirakawa H. Angew Chem, Int Ed. Engl 2001;40:2574–80.

[13] Salaneck WR, Brédas JL. Solid State Commun 1994;92:31–6.

[14] Kumar D, Sharma RC. Eur Polym J 1998;34:1053–60.

[15] Eftekhari A, Jafarkhani P. Polym J 2006;38:651.

[16] Kuramoto N, Takahashi Y. React Funct Polym 1998;37:33–8.

[17] Huang J, Virji S, Weiller BH, Kaner RB. J Am Chem Soc 2003;125:314–15.

[18] Qiu H, Wan M. J Polym Sci, Part A Polym Chem 2001;39:3485–97.

[19] Eicher T. J für Praktische Chemie/Chemiker-Zeitung 1998;340:487–8.

[20] Jang J, Bae J. Angew Chem, Int Ed. Engl 2004;43:3803–6.

[21] Kumari P, Khawas K, Nandy S, Kuila BK. Electrochim Acta 2016;190:596–604.

[22] Duan X, Huang Y, Cui Y, Wang J, Lieber CM. Nature 2001;409:66.

[23] Liu J, Wan M. J Mater Chem 2001;11:404–7.

[24] Pillalamarri SK, Blum FD, Tokuhiro AT, Story JG, Bertino MF. Chem Mater 2005;17:227–9.

[25] Gurunathan K, Murugan AV, Marimuthu R, Mulik UP, Amalnerkar DP. Mater Chem Phys 1999;61:173–91.

[26] Novák P, Müller K, Santhanam KSV, Haas O. Chem Rev 1997;97:207–82.

[27] Li L, Jiang J, Xu F. Eur Polym J 2006;42:2221–7.

[28] Xu J, Li X, Liu J, Wang X, Peng Q, Li Y. J Polym Sci A Polym Chem 2005;43:2892–900.

[29] Han J, Song G, Guo R. J Polym Sci A Polym Chem 2006;44:4229–34.

[30] Pan LJ, Pu L, Shi Y, Song SY, Xu Z, Zhang R, et al. Adv Mater 2007;19:461–4.

[31] Odian G. Principles of Polymerization. John Wiley & Sons, Inc; 2004. p. 372–463. Available from: https://doi.org/10.1002/047147875X.ch5.

[32] Sih BC, Wolf MO. Chem Commun 2005;3375–84. Available from: https://doi.org/10.1039/B501448D.

[33] Mallick K, Witcomb MJ, Scurrell MS. Eur Polym J 2006;42:670–5.

[34] Li J, Zhu L, Wu Y, Harima Y, Zhang A, Tang H. Polymer 2006;47:7361–7.

[35] Blanchet GB, Fincher CR, Gao F. Appl Phys Lett 2003;82:1290–2.

[36] Huguenin F, Ferreira M, Zucolotto V, Nart FC, Torresi RM, Oliveira ON. Chem Mater 2004;16:2293–9.

[37] Feng X, Mao C, Yang G, Hou W, Zhu JJ. Langmuir 2006;22:4384.

[38] Ashis D, Sukanta D, Amitabha D, De SK. Nanotechnology 2004;15:1277.

[39] He Y. Mater Chem Phys 2005;92:134–7.

[40] Mo Z, Zuo D, Chen H, Sun Y, Zhang P. Eur Polym J 2007;43:300–6.

[41] Oliveira MM, Castro EG, Canestraro CD, Zanchet D, Ugarte D, Roman LS, et al. J Phys Chem B 2006;110:17063–9.

[42] Peng Z, Guo L, Zhang Z, Tesche B, Wilke T, Ogermann D, et al. Langmuir 2006;22:10915–18.

[43] Ginic-Markovic M, Matisons JG, Cervini R, Simon GP, Fredericks PM. Chem Mater 2006;18:6258–65.

[44] Pron A, Rannou P. Prog Polym Sci 2002;27:135–90.

[45] Argun AA, Aubert P-H, Thompson BC, Schwendeman I, Gaupp CL, Hwang J, et al. Chem Mater 2004;16:4401–12.

[46] Coakley KM, McGehee MD. Chem Mater 2004;16:4533–42.

[47] Simon P, Gogotsi Y. Nat Mater 2008;7:845.

[48] Zhang LL, Zhao XS. Chem Soc Rev 2009;38:2520–31.

[49] Frackowiak E. Phys Chem Chem Phy 2007;9:1774–85.

[50] Palaniappan S, Devi SL. J Appl Polym Sci 2008;107:1887–92.

[51] Yang P, Ding Y, Lin Z, Chen Z, Li Y, Qiang P, et al. Nano Lett 2014;14:731–6.

[52] Yuan L, Lu X-H, Xiao X, Zhai T, Dai J, Zhang F, et al. ACS Nano 2012;6:656–61.

[53] Ansari SA, Parveen N, Han TH, Ansari MO, Cho MH. Phys Chem Chem Phys 2016;18:9053–60.

[54] Wu W, Pan D, Li Y, Zhao G, Jing L, Chen S. Electrochim Acta 2015;152:126–34.

[55] Patil DS, Shaikh JS, Dalavi DS, Karanjkar MM, Devan RS, Ma YR, et al. J Electrochem Soc 2011;158: A653–7.

[56] Sharma V, Sahoo A, Sharma Y, Mohanty P. RSC Adv 2015;5:45749–54.

[57] Fan LZ, Hu YS, Maier J, Adelhelm P, Smarsly B, Antonietti M. Adv Funct Mater 2007;17:3083–7.

[58] Shi M, Zhang Y, Bai M, Li B. Synth Met 2017;233:74–8.

[59] Liao QY, Li SY, Cui H, Wang C. J Mater Chem A 2016;4:8830–6.

[60] Sahoo S, Zhang S, Shim J-J. Electrochim Acta 2016;216:386–96.

[61] Sun J, Huang Y, Fu C, Wang Z, Huang Y, Zhu M, et al. Nano Energy 2016;27:230–7.

[62] Fan W, Miao Y-E, Zhang L, Huang Y, Liu T. RSC Adv 2015;5:31064–73.

[63] Zhu Z-z, Wang G-c, Sun M-q, Li X-w, Li C-z. Electrochim Acta 2011;56:1366–72.

[64] Zhang K, Hu H, Yao W, Ye C. J Mater Chem A 2015;3:617–23.

[65] Eftekhari A. J Appl Polym Sci 2006;102:6060–3.

[66] Eftekhari A, Jafarkhani P. J Electroanal Chem 2014;717-718:110–18.

[67] Wu Q, Xu Y, Yao Z, Liu A, Shi G. ACS Nano 2010;4:1963–70.

[68] Cheng X-W, Zhao D-L, Wu L-L, Ding Z-W, Hu T, Meng S. Electrochim Acta 2018;265:348–54.

[69] Zhu J, Kong L, Shen X, Chen Q, Ji Z, Wang J, et al. Appl Surf Sci 2018;428:348–55.

[70] Guo S, Shen H, Tie Z, Zhu S, Shi P, Fan J, et al. J Power Sources 2017;359:285–94.

[71] Lee CC, Omar FS, Numan A, Duraisamy N, Ramesh K, Ramesh S. J Solid State Electrochem 2017;21:3205–13.

[72] Gao Z, Yang J, Huang J, Xiong C, Yang Q. Nanoscale 2017;9:17710–16.

[73] Kulkarni SB, Patil UM, Shackery I, Sohn JS, Lee S, Park B, et al. J Mater Chem A 2014;2:4989–98.

[74] Zhou Q, Li Y, Huang L, Li C, Shi G. J Mater Chem A 2014;2:17489–94.

[75] Zhang X, Ji L, Zhang S, Yang W. J Power Sources 2007;173:1017–23.

[76] Kumar V, Lee PS. J Phys Chem C 2015;119:9041–9.

[77] Zhang P, Liu Z, Liu Y, Fan H, Jiao Y, Chen B. Electrochim Acta 2015;184:1–7.

[78] Zhao D, Guo X, Gao Y, Gao F. ACS Appl Mater Interfaces 2012;4:5583–9.

[79] Yu L, Gan M, Ma L, Huang H, Hu H, Li Y, et al. Synth Met 2014;198:167–74.

[80] Li Q, Liu J, Zou J, Chunder A, Chen Y, Zhai L. J Power Sources 2011;196:565–72.

[81] Han G, Liu Y, Zhang L, Kan E, Zhang S, Tang J, et al. Sci Rep 2014;4:4824.

[82] Jiang H, Ma J, Li C. J Mater Chem 2012;22:16939–42.

[83] Sun X, Gan M, Ma L, Wang H, Zhou T, Wang S, et al. Electrochim Acta 2015;180:977–82.

[84] Y Li and C Chen, *Polyaniline/carbon nanotubes-decorated activated carbon fiber felt as high-performance, free-standing and flexible supercapacitor electrodes*, 2017.

[85] Liu D, Du P, Wei W, Wang H, Wang Q, Liu P. J Colloid Interface Sci 2018;513:295–303.

[86] Das AK, Karan SK, Khatua BB. Electrochim Acta 2015;180:1–15.

[87] Sk MM, Yue CY, Jena RK. Synth Met 2015;208:2–12.

[88] Shabani Shayeh J, Ehsani A, Ganjali MR, Norouzi P, Jaleh B. Appl Surf Sci 2015;353:594–9.

[89] Tang L, Yang Z, Duan F, Chen M. J Mater Sci Mater Electron 2017;28:15804–18.

[90] Zhu S, Wu M, Ge M-H, Zhang H, Li S-K, Li C-H. J Power Sources 2016;306:593–601.

[91] Liu X, Wu Z, Yin Y. Chem Eng J 2017;323:330–9.

[92] Chaudhary G, Sharma AK, Bhardwaj P, Kant K, Kaushal I, Mishra AK. J Energy Chem 2017;26:175–81.

[93] Dhawale DS, Vinu A, Lokhande CD. Electrochim Acta 2011;56:9482–7.

[94] Khdary NH, Abdesalam ME, Enany G El. J Electrochem Soc 2014;161:G63–8.

[95] Kim BC, Kwon JS, Ko JM, Park JH, Too CO, Wallace GG. Synth Met 2010;160:94–8.

[96] Zhang X, Goux WJ, Manohar SK. J Amer Chem Soc 2004;126:4502–3.

[97] Prasad KR, Munichandraiah N. Electrochem Solid-State Lett 2002;5:A271–4.

[98] Bavio MA, Acosta GG, Kessler T. Int J Hydrogen Energy 2014;39:8582–9.

[99] Xinping H, Bo G, Guibao W, Jiatong W, Chun Z. Electrochim Acta 2013;111:210–15.

[100] Otrokhov G, Pankratov D, Shumakovich G, Khlupova M, Zeifman Y, Vasil'eva I, et al. Electrochim Acta 2014;123:151–7.

[101] Mi H, Zhang X, An S, Ye X, Yang S. Electrochem Commun 2007;9:2859–62.

[102] Dong B, He B-L, Xu C-L, Li H-L. Mater Sci Eng B 2007;143:7–13.

[103] Zhou Y, Qin Z-Y, Li L, Zhang Y, Wei Y-L, Wang L-F, et al. Electrochim Acta 2010;55:3904–8.

[104] Liu X, Shang P, Zhang Y, Wang X, Fan Z, Wang B, et al. J Mater Chem A 2014;2:15273–8.

[105] Khalid M, Tumelero MA, Zoldan VC, Pla Cid CC, Franceschini DF, Timm RA, et al. RSC Adv 2014;4:34168–78.

[106] Prasad KR, Miura N. Electrochem Solid-State Lett 2004;7:A425–8.

[107] Jaidev RI, Jafri AK, Mishra, Ramaprabhu S. J Mater Chem 2011;21:17601–5.

[108] Sumboja A, Foo CY, Yan J, Yan C, Gupta RK, Lee PS. J Mater Chem 2012;22:23921–8.

[109] Xia C, Xie Y, Du H, Wang W. J Nanoparticle Res 2015;17:30.

[110] Su H, Wang T, Zhang S, Song J, Mao C, Niu H, et al. Solid State Sci 2012;14:677–81.

[111] Sha C, Lu B, Mao H, Cheng J, Pan X, Lu J, et al. Carbon 2016;99:26–34.

[112] Li J, Lu W, Yan Y, Chou T-W. J Mater Chem A 2017;5:11271–7.

[113] Zaidan KM, Hussein HF, Talib RA, Hassan AK. Energy Procedia 2011;6:85–91.

[114] Liu Z, Zhou J, Xue H, Shen L, Zang H, Chen W. Synth Met 2006;156:721–3.

[115] Patel A, Pataniya P, Patel KD, Solanki GK, Pathak VM. AIP Conf Proc 2017;1837:040047.

[116] Shit A, Chal P, Nandi AK. PhysChem Chem Phys 2016;18:31732–43.

[117] Hsu Y-C, Tseng L-C, Lee R-H. J Polym Sci B: Polym Phys 2014;52:321–32.

[118] Shit A, Chatterjee S, Nandi AK. Phys Chem Chem Phys 2014;16:20079–88.

12

Medical Applications of Polymer/Functionalized Nanoparticle Systems

Aswathy Ravindran Girija

FUTURE INDUSTRIES INSTITUTE, UNIVERSITY OF SOUTH AUSTRALIA, MAWSON LAKES CAMPUS, MAWSON LAKES, SA, AUSTRALIA

12.1 Introduction

Nanotechnology is an emerging technology based on the design, characterization, fabrication, and application of structures and devices by controlling their size and shape at nanometer scale. Nanotechnology has garnered significant achievements in various fields including materials, engineering, environment, energy, and the biomedical sector. Nanotechnology-based medical advancements have gained immense attention in recent times. The interaction of nanoscale materials with biological molecules widens research and development in the nanomedicine field. The prospective medical applications of nanotechnology are mostly in the diagnostics, development of novel drug delivery systems, and fabrication of superior prosthetics. Nanotechnology offers the flexibility for designing nanomaterials with desired size, shape, and biocompatibility, which are extremely important for biomedical applications including drug delivery and imaging applications. Application of nanotechnology in biomedical field, broadly mentioned as nanomedicine, includes development of efficient nanocarriers (polymeric nanoparticles, liposomes, dendrimers, inorganic nanoparticles, etc.), precise delivery of therapeutics (passive and active delivery of drugs), improving different treatment strategies with nanomaterials (radiotherapy, phototherapy, photodynamic therapy, immunotherapy, gene therapy, magnetic hyperthermia, etc.) and developing diagnostic modalities for rapid and precise detection of diseases [1–5]. A few of the adverse effects of conventional drug delivery include nonspecificity, nonefficacy, ineffective formulation (high dose), and unwanted side effects. Nanotechnology-based drug delivery systems with improved specificity can often reduce the adverse effects of conventional drug delivery systems. The efficient targeting, superior safety profile, controlled drug release profile, and enhanced bioavailability are few of the key advantages of nanoparticle-based drug delivery systems. Recent research is focused on engineering and development of

Polymer Composites with Functionalized Nanoparticles. DOI: https://doi.org/10.1016/B978-0-12-814064-2.00012-3

multifunctional nanosystems to function as effective theragnostic (therapeutic and diagnostic) tools against disorders including cancer, neurodegenerative disorders, or infectious diseases [6–15].

Among various types of nanoparticles, polymeric nanoparticles have garnered considerable attention owing to their versatile nature including their physical and chemical properties, biocompatibility, biodegradability, and ease of functionalization [5–8,16–19]. The physicochemical and pharmacological properties of encapsulated drugs in nanoparticles can be enhanced by appropriate design of drug delivery systems, by carriers including polymer- or lipid-based nanoparticles [6]. Polymeric nanoparticles are excellent nanocarriers of drugs that are loaded either through encapsulation or covalent bonding and are tuned to respond to stimuli for the controlled release of loaded therapeutic contents. The surface functionalization strategies offer prospects for decorating polymeric nanoparticle surface with targeting ligands, which are recognized by receptors expressed in large amounts in cancer cells [16,18]. Nanocarriers are often designed to maskbody's natural immune system and to improve biodistribution of encapsulated or loaded drugs. In this chapter different strategies of functionalization, bioconjugation approaches of polymeric nanoparticles, drug delivery systems, and imaging applications are discussed.

12.2 Functionalization and Bioconjugation of Nanoparticles

Although nanoparticles offer immense potential and promise for biomedical applications, numerous concerns still exist that must be properly addressed. The major concerns are the stability and safety of nanoparticles for their application in the biomedical field. Most of the biological processes happen in an aqueous environment. Therefore stability of nanoparticles in water is a major factor to tap the benefits of nanoparticles. The large surface to area ratio, use of heavy metals, chemical reactivity, and ability to generate reactive oxygen species are factors that must be addressed while considering the safety to humans and the environment [20]. Unmodified or bare nanomaterials can induce cytotoxicity, gene toxicity, and some materials are harmful to the development of embryos. Therefore it is crucial to modify nanomaterials before using them directly for biomedical applications. Surface functionalization strategies are often used to overcome issues associated with stability and safety. Effective surface functionalization prevents nanomaterials from agglomeration, nonspecific interactions, prevents the leaching of toxic core material, and allows suitable surface functionality for conjugation of biomolecules for targeting applications [21]. In general surface functionalization strategies impart (1) nanoparticles with enhanced water solubility, (2) reduced toxicity associated with nanoparticles in certain cases, (3) additional functionalities for promoting the conjugation of biomolecules for targeted drug delivery applications, and (4) stabilization of the physical properties of synthesized nanomaterials (Fig. 12–1). Some of the major strategies for surface functionalization of nanoparticles based on the effect on parameters relevant to stability, nonspecific binding, and biocompatibility include (1) ligand exchange, (2) ligand modification, and (3) ligand addition (encapsulation), as depicted in Fig. 12–2.

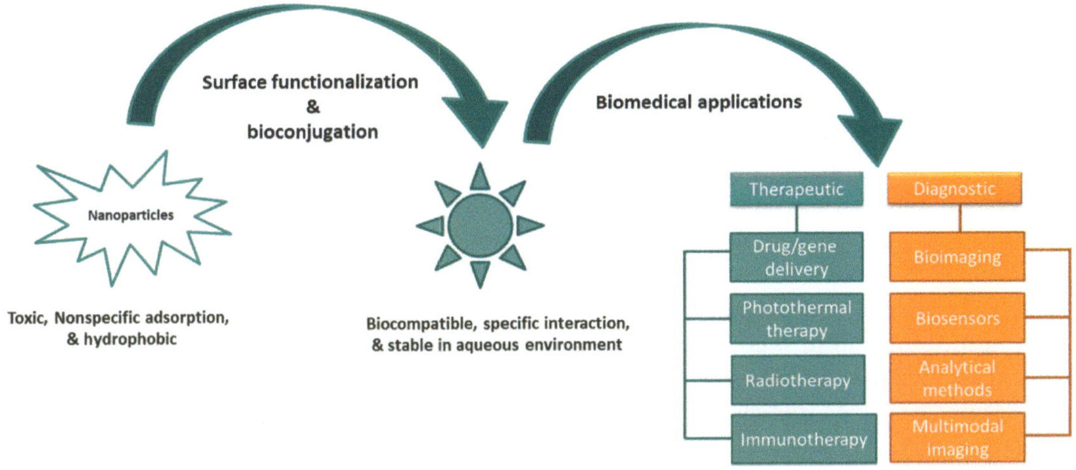

FIGURE 12–1 Significance of surface functionalization and bioconjugation strategies for biomedical applications.

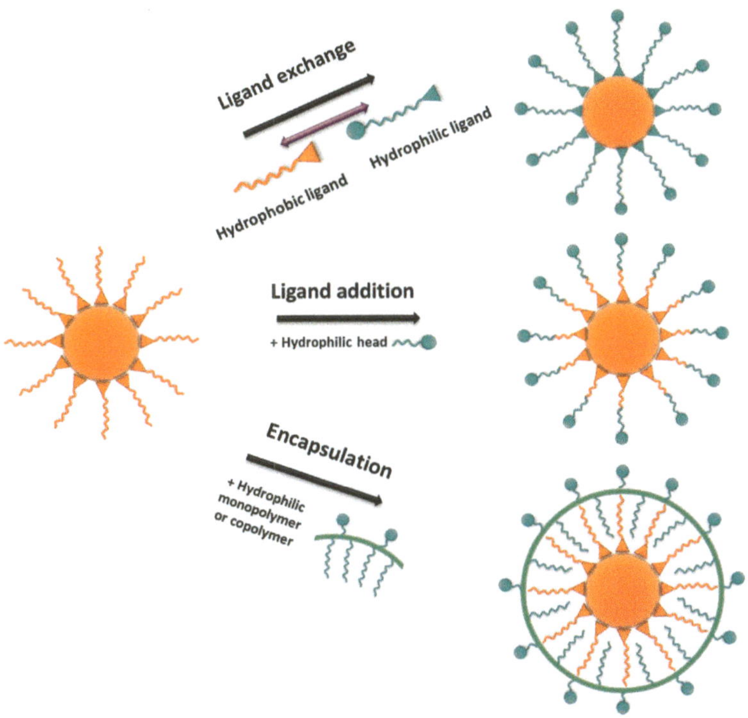

FIGURE 12–2 Schematic of commonly employed surface functionalization strategies.

12.2.1 Ligand Exchange

Ligand exchange facilitates the modification of the surfaces of nanoparticles by making them water soluble for medical and biological applications. It involves exchange of initial hydrophobic ligands with other strong hydrophilic ligands, which permits the transfer of nanoparticles from organic phase to aqueous phase. A wide range of hydrophilic ligands have been reported to exchange the ligand on the surface of nanomaterial to impart aqueous solubility. Some of them are small molecules, poly(ethylene glycol) (PEG) derivatives, and biological molecules.

Small molecules with high affinity functional head group, including thiol, carboxyl, amine, phosphine groups, are often employed for generating water-soluble nanoparticles. In some instances the colloid stability reduces owing to the desorption of ligands from NPs [22]. Bidentate ligands (dihygrolipoic acid and dithiocarbamate ligands) that can increase the number of anchor points on the nanoparticle surface are often used to increase the colloidal stability of nanoparticles [23–25]. Other commonly used ligands include alkyl thiol terminated ligand molecules, tetraalkylammonium salts such as tetraoctylammonium bromide, hexadecyltrimethylammonium bromide oligomeric phosphine (oxide) ligands, amphiphilic species such as 2,3-dimercaptosuccinic acid, cyclodextrin, etc. [26–30]. As small ligands depend on electrostatic interaction for the stabilization of nanoparticles, they tend to agglomerate owing to the change in pH or salt concentration, which is considered as the disadvantage of the ligand exchange process.

Several polymeric ligands have been reported to impart nanoparticles with good stability and water solubility and to overcome poor colloidal stability imparted by small ligands. For biomedical applications, PEG is the most commonly used ligand for the stabilization of nanoparticles. PEG utilizes its ether group in the PEG chains for establishing hydrogen bonding and steric stabilization for imparting aqueous solubility thus enabling its use over a wide range of pH and high salt concentration. Several studies suggest that PEG modified nanomaterials are biocompatible and exhibit longer circulation time in vivo [31,32].

Biomolecules including sugars (glucose, sucrose), peptides (RGD, TAT), nucleic acids (aptamers), and their derivatives are also used as exchange ligands [33,34]. The advantage of using biomolecules as ligand exchange is that they serve both functionalities including stabilization and biofunctionalization, so that modified nanoparticles can function as targeted nanoparticles without prior modification.

12.2.2 Ligand Modification

Ligand modification is another strategy of phase transfer where existing ligand stabilizing the nanoparticle in its original phase is rendered hydrophilic or hydrophobic to transfer and stabilize it in second phase. It can be attained by addition of a hydrophilic agent that interacts with hydrophobic ligand (cyclodextrin complex formation with oleic acid) or modifying a compound that could change the polarity (e.g., oxidation of oleic acid to form carboxyl group or vicinal diol), or covalent attachment with amphiphilic agents (e.g., conjugation with

V-shaped ligand) [28,35,36]. Although this strategy is efficient they are restricted to systems that maintain colloidal stability of nanoparticles during the reaction and are compatible with each other.

12.2.3 Ligand Addition

Another strategy of phase transfer is ligand addition, which involves addition of a molecular layer on the external surface of particles, which adsorbs on the original ligand molecules without removal of the same and changes the surface properties of the nanoparticles accordingly. The approach is attained by the addition of inorganic materials as a layer and encapsulation of the intercalating hydrophobic polymers layer. Amphiphilic molecules with a hydrophobic and a hydrophilic part with aliphatic chains and a polar charged end group are often used as phase-transfer agents.

Silica coating is one of the most commonly used approaches for surface coating of nanoparticles. In addition to water solubility silica coating offers superior colloidal stability, reduced nonspecific interaction, and good biocompatibility [37]. In this process, a layer of silanes was bound to nanoparticle surface and using same silane layer, inorganic silica shell (a polymeric, cross-linked) was deposited, which could be further used to add different chemical functionality. Different types of nanoparticles including Au, semiconductor quantum dots, magnetic nanoparticles, and upconversion nanoparticles of various size and shape have been successfully coated with silica layer. However in some instances, the physical properties of nanoparticles may be affected by the increase in hydrodynamic size of the NPs. In the case of silica coated magnetic nanoparticles, reduction of saturation magnetization has been reported [38].

Polymer encapsulation involves coating of amphiphilic polymers over hydrophobic ligand capped nanoparticles through hydrophobic interactions. The hydrophobic part of the polymer intercalates with the ligand present on the surface of the nanoparticle and renders the hydrophilic part to the solution. The functional group (carboxyl, amine groups) present in amphiphilic molecule allows bioconjugation of nanoparticles with targeting ligands. Phospholipids, amphiphilic coblock polymers [poly(acrylic acid) (PAA)], poly(maleic anhydride alt-1-tetradecene), and PEGylated polymer are reported to make hydrophilic nanoparticles [39]. The major advantage of polymer coating is that, with an ideal amphiphilic polymer, a range of nanoparticles (Au nanoparticles, semiconductor quantum dots, magnetic nanoparticles, etc.) can be made easily hydrophilic with enhanced colloidal stability over a range of pH and high salt concentrations.

12.2.4 Bioconjugation

It is understood from the previous section that nanoparticles after surface functionalization enhance the hydrophilicity, biocompatibility, and colloidal stability. However for biomedical applications they are often tailored with appropriate bioconjugation with specific ligands of interest (e.g., antibodies, peptides, proteins, aptamers, nucleic acid, etc.). There are several strategies for the conjugation of biomolecules and they depend on collective factors including size, shape, surface charge, intrinsic property of the nanomaterial, functional groups available on the surface of nanoparticles, and the nature, type, size, and chemical composition of the biomolecule for bioconjugation. In addition to the abovementioned factors, there are several other aspects that affect bioconjugation. Depending on application, the degree of attachment of biomolecule to nanoparticle varies. Therefore controlled valency needs to be attained. In some instances overconjugation of biomolecules may affect binding capability and impair the function of the system. Some binding strategies can result in heterogeneous attachment of molecules thus damaging the function of the final product. For a few biomedical imaging applications with quantum dots and upconversion nanoparticles, it is essential to maintain a distance from the biomolecule. For drug delivery applications, easily breakable bonds, for example pH or acid labile bonds, are superior for enhancing the release of drugs under stimuli-responsive (acidic conditions) release. However, extreme care should be taken during bioconjugation strategies for not significantly altering the properties of both biomolecule and nanoparticle, which might result in loss of activity of biomolecule or loss of intrinsic property of nanomaterial. Some of the commonly employed methods based on physical interactions and covalent chemistry are discussed.

Physical interactions include electrostatic, hydrophilic/hydrophobic, and affinity interactions. Among the techniques in physical interactions, most commonly used strategy is electrostatic interaction, which relies on ionic attraction between opposite charges. The conjugation of negatively charged plasmid DNA or nucleic acid to cationic nanoparticles is a well-documented example [40–42]. Hydrophobic–hydrophilic interactions for bioconjugation with nanoparticles are often used for adsorbing water-insoluble drugs where nanoparticles are designed with hydrophobic layers that adsorb the drugs, and released intracellularly when hydrophobic coating degenerates [43]. Affinity interactions utilize the affinity to ligands where the surface of nanoparticles is modified with streptavidin, which specifically binds biotinylated moieties. Affinity interactions are very effective, stable, and are the strongest among all noncovalent interactions [44].

Covalent linkages are stable and strong bonds formed between functional groups (amino, carboxylic acid, and thiol) present on surface of nanoparticles and biomolecules either directly or by coupling strategies depending on the functional group. Some of them include direct conjugation, carbodiimide coupling reaction, and click chemistry. The direct conjugation approach utilizes the functional groups present on the nanoparticle surfaces for the formation of bonds with the biomolecule. Functionalization offers various functional groups, such as amine, sulfiydryl, and aldehyde to the surface of the nanoparticle, facilitating conjugation with biomolecules including small peptides, antibodies, dyes, etc. In carbodiimide coupling reaction 1-ethyl-3-dimethylaminopropyl carbodiimide is employed as a coupling agent [45] for coupling of small biomolecules with amine groups, small peptides, proteins, amine-modified nucleic acids, and aptamers to the carboxylic-functionalized nanoparticles. This approach is also used for the coupling between amine-functionalized nanoparticles and carboxylic groups of biomolecules. The coupling efficiency is improved by n-hydroxysuccinimide (NHS) or sulfo NHS, which function as stabilizing agents for the formation of a succinimide ester intermediate. Click chemistry is another approach for direct conjugation of biomolecules, in which an efficient and rapid reaction that requires mild reaction conditions results in biocompatible linkages at neutral pH [46]. The reaction comprises of coupling of an alkyne to an azide, releasing a 1,2,3-triazole ring under catalysis by Cu(I) resulting in a stable rigid bond that aids in bioconjugation.

12.3 Medical Applications

Polymeric nanoparticles offer exciting potential in medicine with remarkable progress that could revolutionize the treatment and diagnosis of disease in near future. However, the sensitivity and efficacy of current clinical diagnosis and treatment practice are major drawbacks. Polymeric nanoparticles–based therapeutic and treatment strategies could significantly address sensitivity and efficacy by avoiding unwanted side effects by offering precise or targeted delivery of therapeutic and diagnostic agents to the affected or diseased site, permitting extreme accuracy and effectiveness. Among different medical applications of polymeric nanoparticles, much research has been done in the area of drug delivery. In the next section, drug delivery and imaging applications of polymeric nanoparticles are discussed.

12.3.1 Drug Delivery

Drug delivery is one of the most promising medical applications of nanoparticles. It refers to synthetic approaches and formulation strategies for delivering therapeutic agents with desired therapeutic effects to the human body. There are several disadvantages to the currently employed conventional drug delivery system. These include high dose of drug, low bioavailability, uncontrollable drug release, nonspecific interaction of drug with normal healthy tissues leading to side effects, nonspecific biodistribution, etc. Drug delivery mediated by nanoparticles overcomes many of these disadvantages associated with conventional drug delivery. Large surface to volume ratio and the prospect of encapsulating lipophilic drugs and the design of the same to various routes of administration are the major advantages of drug delivery mediated with nanoparticles. Effective control over dosage, slow and sustained release of therapeutic agents, and frequency of administration can be controlled by a nanotechnology-based drug delivery approach. The research in advanced nano drug delivery strategies has opened new avenues for the precise and smart drug delivery to target tissues/organs. Owing to the fascinating and intrinsic nanoscale properties, various nanomaterials are being exploited for smart drug delivery. However, nanotechnology-based nano drug carriers should address the stability of nanoformulation in physiological conditions, superior drug loading capability, and safety of nanoparticles in terms of biocompatibility and biodegradability.

In recent decades, polymeric nanoparticles or nanoparticles modified with polymer have been extensively used as nanocarriers for drug delivery owing to their properties such as superior biocompatibility, easy biodegradability, ease of preparation, and remarkable biomimetic potential. Polymer-based nanoparticles offer stability to various forms of active therapeutic drugs and play a substantial role in smart drug delivery with enhanced efficacy owing to the precise delivery of therapeutic agents to target sites to impart their action. The control over the size of nanoparticles, surface properties, solubility, and sustained release of drug at the target site are some of the ideal requirements for successful design of a drug delivery nanosystem. However, smart drug delivery systems respond to various stimuli that determine controlled prescheduled targeted release of the drug from smart nanocarriers thus enhancing their potential in a therapeutic regime. Polymeric nanoparticles are ideal nanocarriers for hydrophobic drugs that enhance the bioavailability of the same through their smart release properties. In addition they are effective in the stabilization and protection of therapeutic payload (proteins, peptides, or DNA molecules) and have demonstrated their potential in delivering the payload [7,47–50]. Most of the targeted nanoformulations enhance systemic circulation and diminish any undesirable cellular uptake, and inhibit the denaturation of active therapeutic cargo. Owing to their beneficial properties in the drug delivery regime several polymer-based systems have been approved by the US Food and Drug Administration (FDA) for human consumption.

In a smart drug delivery system, premature release of encapsulated/loaded/adsorbed drugs is prevented to a great extent before reaching target tissues or organs and then released at proper rates at the target site upon active or passive targeting strategy. These smart or stimuli-responsive nanoformulations respond to either endogenous or exogenous stimuli (Fig. 12–3). Some of the endogenous triggers that enhance the drug release include pH, hormones, enzymes, glucose, or redox gradients [51,52]. Exogenous triggers are external factors such as light, temperature, electric pulse, magnetic field, ultrasound (US), high energy radiation, etc. A brief discussion on the rationales of some commonly employed stimuli-responsive nanoformulations is provided below.

Among different types of stimuli-responsive drug release, pH is the most commonly studied trigger for drug release from nanoparticles [53–56]. The pH in the body varies depending on different tissues/organs. Blood and normal healthy tissues have pH of 7.4; the gastrointestinal (GI) tract expresses different pH at different regions; cellular compartments have

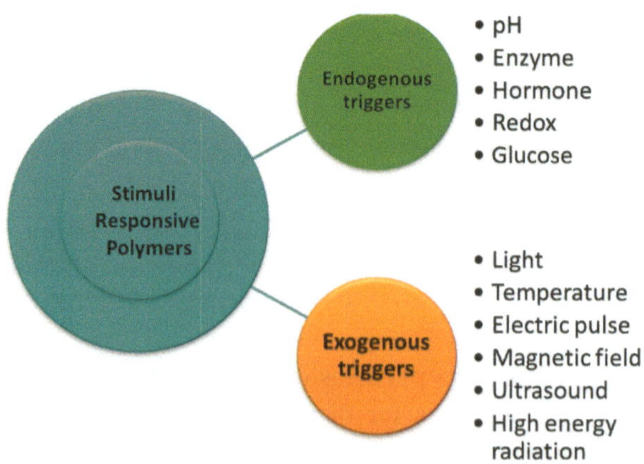

FIGURE 12–3 Endogenous or exogenous stimuli-responsive polymers.

different pH, for example, lysosomes (pH 4.5–5.0) and endosomes (pH 5.0–6.5); and disease microenvironments such as inflammatory, ischemic, and tumor tissues vary in pH as well. Based on pH variation, nanoparticles are designed to release drugs in pH sensitive tissue such as cancer cells or in endosomes or lysosomes [57,58]. Nanoparticles, especially pH responsive nanodrug carriers, take advantage of this reduced pH maintained by tumor cells. Therefore, difference in pH in various tissues is one of the significant design rationales for advancement in advanced smart drug delivery systems.

A commonly used approach to attain organ-specific drug release is the formulation of nanoparticles that demonstrate a pH-dependent swelling. Acrylic-based polymers, especially poly(methacrylic acid) (PMAA), are widely studied. Owing to the protonation of carboxyl groups, nanoparticles maintain a hydrophobic and collapsed state in the acidic pH of the stomach. As pH increases, after gastric passage, nanoparticles swell because of hydrogen bond breakage and carboxyl ionization [59]. Derivatives of PMAA such as PMAA-PEG diblock copolymers, polyethylacrylate (PMAA-PEA), and polymethacrylate (PMAA-PMA) display pH-dependent dissolution that responds to intestinal regions with varying pH [60].

Tumor tissues and cells have been shown to exhibit acidic microenvironment with a pH of 6.5–6.9 [61]. Owing to a high rate of metabolism, extracellular pH values in tumor tissue are acidic (below 7.0) [62,63]. Polymer-based NPs designed for pH-dependent drug release modify their physicochemical properties, such as swelling and enhanced solubility based on varying pH levels. In several studies pH-labile cross-linkers have been used to conjugate drug molecules and polymer chains. When the particles contact with desired pH, the cross-linkers destabilize and polymeric nanoparticles undergo swelling, thus releasing drugs from the nanoparticles [64–72]. Some commonly used pH-labile cross-linkers include ester, orthoester, hydrazone, imine, vinyl ether, carboxy dimethylmaleic anhydride, etc. In addition, other approaches including charge reversal of nanoparticles from neutral/negative to positive in tumor tissue are used to enhance retention of nanoparticles in tumors [68].

Temperature is another appropriate factor for controlled drug release, and temperature in some pathophysiological conditions, such as inflammation and tumors, is slightly higher than normal tissues [73–76]. Smart drug delivery utilizes this slight temperature difference to trigger the release of drugs from nanocarriers to the affected tissue [77,78]. Polymeric thermoresponsive nanocarriers offer an excellent mode for delivering drugs in a controlled manner to the target site. Thermoresponsive polymeric nanocarriers retain their contents at physiological temperature of 37°C, and release them when temperature is increased (40–45°C). Some of the commonly studied thermoresponsive polymers are poly(*N*-isopropyl acrylamide) [P(NIPAM)], poly(*N,N*-diethylacrylamide), poly(*N*-vinylalkylamide), poly(*N*-vinylcaprolactam), pluronics (Pluronic F-127), tetronics, PLGA–PEG–PLGA triblock copolymers, phosphazene derivatives, and chitosan, where most of them are designed for three dimensional hydrogel networks by the cross-linking of hydrophilic polymers [79–85]. Thermogels from PEG/polyester copolymers such as poly(D,L-lactic acid-co-glycolic acid) (PLGA), poly(″-caprolactone) (PCL), poly(″-caprolactone-co-D,L-lactic acid) (PCLA), and poly (″-caprolactone-co-D,L-lactic acid) (PCGA) also exhibit thermoresponsive behavior [86–89]. Some of the exciting properties of thermoresponsive polymers include phase transition, enhancing drug solubility, and controlled temperature responsive drug release [90]. P(NIPAM), one of the widely used thermoresponsive polymers, is a material with a lower critical solution temperature that exists as an aqueous gel network at room temperature and is delivered as an injectable implant to the body. However, owing to the cytotoxicity contributed by the quaternary ammonium in the chemical structure, nonbiodegradability, and ability to elicit immune response, use of poly NIPAAM is limited for drug delivery.

Field-responsive polymers are those that respond to external fields such as light, electric, electromagnetic, magnetic, and US. Fast response time and controlled drug release rate by controlling signal are the major advantages of field-responsive polymer. Light-responsive polymer systems have attractive feature that respond to external light, triggering drug release at the desired target from the nanoparticles. Light-sensitive carriers perform drug release by on/off mechanism owing to phase transition in response to exposing nanoparticles to light, so that they may either open or close upon light irradiation [52]. These polymers are either UV-sensitive, visible-sensitive, or IR-sensitive systems based on the wavelength of light that elicits phase transition of polymer. IR- and visible light–sensitive polymers are preferred over UV-sensitive polymers owing to safety [79,91]. Photosensitive materials such as chromophores (e.g., trisodium salt of copper chlorophyllin) are incorporated to generate visible light–sensitive hydrogels. In contrast, owing to high infrared light absorbency of water, IR light elicits response in hydrogels without the presence of chromophores [92,93]. However light penetration depth, slow response toward the stimulus, and varying response owing to the leaching of unbound chromophores are the limitations associated with light-responsive systems [94]. External magnets function as stimuli to release drugs from drug and magnetic nanoparticle encapsulated polymeric materials. Magnetic stimuli offer an efficient and enhanced drug release under programmable exposure of external magnetic field [95–97]. The inherent magnetic properties of magnetic nanoparticles are exploited to acquire the benefits. The magnetic nanoparticles in polymeric nanoformulation offer added benefit in that they serve as excellent contrast agents for imaging applications. When MNPs are encapsulated in nanocarriers such as micelles, liposomes, or solid nanoparticles, they perform as multifunctional theragnostic formulations extending their application both in diagnostic and therapeutic regimes [98–107].

Another example for field-responsive release is based on electric fields. Some polymers composed of a large number of ionizable groups modify their physical properties upon change in electric current and transfer electric energy into mechanical energy and are used for controlled drug delivery. The chemical property also offers them pH responsive behavior. The electric current leads to change in pH that interferes with hydrogen bonding in the polymer chains, resulting in drug release. The selection of electric current that results in drug release without stimulating or disturbing the nerve endings in neighboring tissue is the major challenge in electroresponsive drug release [108,109]. Chitosan, hyaluronic acid, and alginate are some of the naturally occurring electroresponsive materials and allyl amine, acrylonitrile, vinyl alcohol, vinylacrylic acid, and methacrylic acid are commonly used synthetic electroresponsive polymers. US has been extensively used in advanced medical procedures including diagnosis and therapy owing to its tissue penetration property and safety.

Ultrasonic sensitive nanocarriers for triggered drug release can be controlled at preferred sites by tuning the US frequency, time of exposure, and number of cycles [110]. US-responsive polymers are classified into biodegradable polymers such as polylactides, polyglycolides, and nonbiodegradable polymers such as ethylene vinyl acetate, poly(lactide-co-glycolide), etc. These nanocarriers also function as theragnostic agents thus functioning both in the diagnostic and/or therapeutic applications.

12.3.2 Imaging

Among different types of nanoparticles, polymeric nanoparticles, or polymer functionalized nanoparticles are highly promising owing to their stability, structural versatility, and ease of incorporation of multiple functionalities. Polymer-based nanoparticles have the capability to incorporate multimodal functionalities through highly selective organic reactions that do not impede existing biochemical processes. Chemical reactions such as click chemistry, thiol–yne and thiol–ene chemistry, and regioselective polyester–drug conjugation chemistry are utilized to conjugate drug molecules, organic dyes, or other contrast yielding nanoparticles and offer versatility to the polymeric nanoformulations [111]. Some of the benefits of these nanomaterials for imaging applications are their extremely small (nano) size, which aids their enhanced uptake in target cells, accumulation of contrast agents at the targeted site, multimodal imaging potential, multiplexing capability, and theragnostic potential, which incorporates both therapeutic and diagnostic capability. Several polymeric nanoformulations have demonstrated excellent performance in both preclinical and clinical studies and owing to their safety and performance several polymeric nanoparticles have been approved by the FDA [112]. In general polymer-based nanoparticles for imaging purposes are developed either by covalent conjugation or physical encapsulations of imaging moieties to polymeric matrix (Fig. 12–4A,B). Based on the choice of contrast agents, these materials can function as MR, X-ray, or optical contrast agents for imaging applications (Fig. 12–4C).

12.3.2.1 Magnetic Resonance Imaging

Magnetic resonance imaging (MRI) is one of the most widely used, noninvasive, and safe diagnostic techniques for detection of various abnormalities associated with the brain, central nervous system, and various types of tumors. In addition, MRI offers a viable and safe mode for vulnerable patients with complications associated with radioisotopes or ionizing radiation.

MRI offers three dimensional images with good spatial resolution of tissue samples and facilitates detection or differentiation of diseased lesions from normal tissues. Depending on longitudinal (T1) and transverse (T2) relaxation processes of MR contrast agents, they are categorized into T1 contrast agents (shortened T1 relaxation time with bright images,) and T2 (shortened T2 relaxation time with dark images; negative contrast agents) [11,113]. Commonly used T1 contrast agents are gadolinium-based and some of the commercially available, clinically approved Gd-based contrast agents are ProHance, Magnevist, and Doterm [114]. T2 contrast agents include super paramagnetic materials such as iron oxide

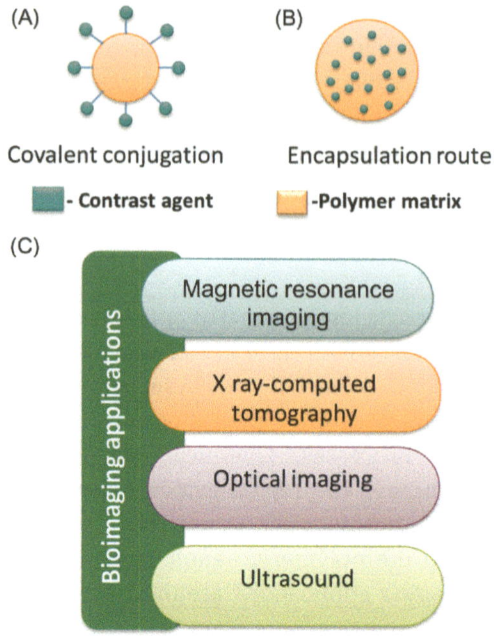

FIGURE 12–4 General route of synthesis of polymer-based nanoparticles for imaging purposes. (A) Covalent conjugation and (B) encapsulation method. (C) Bioimaging applications of polymeric nanoparticles based on the contrast agents.

and ultrasmall paramagnetic iron oxide materials, which yield dark T2-weighted images. In addition to the T1 and T2 agents, T1-T2 dual contrast agents are also used to gain extra advantages and have been explored for disease diagnosis [114,115]. The development of highly sensitive MR agents integrated into polymeric nanoparticles have gained immense attention owing to the precise detection regions of interest, enhanced retention time, and reduced toxicity offered by polymeric nanoparticles [116].

Gd-based chelates such as Gd-DOTA, Gd-DTPA, and Gd-DTPA-BMA are most commonly studied T1 contrast agents. However free Gd ions that are leached from the complex have been found to induce nephrotoxicity at higher concentration. On the other hand, for T2 agents their solubility and magnetic saturation resulting in signal loss are the major concerns. Therefore to overcome the toxicity issues, enhancing solubility and circulation time, avoid magnetic saturation, and attaining target specificity, biocompatible polymeric nanoparticles are generally used as coating material through covalent conjugation or physical encapsulation/electrostatic attraction [113,115,117,118].

Polymeric nanoconstructs integrated with magnetic nanoparticles are often loaded with drug molecules to function as theragnostic agents. In several studies nanoparticles are wrapped with PEG for enhancing the circulation time and for attaching targeting moieties to attain site specific targeting. Polylactic acid-polyethylene glycol-poly-L-Lys was conjugated with anti-VEGF and Gd-DTPA and the nanoconstruct presented significant signal enhancing ability and enhanced accumulation in mice with tumor overexpressing VEGF [119]. Other examples include application of amphiphilic copolymers of polyhydroxyethy-laspartamide with Gd, terpolymer system composed of starch with PMAA, and polysorbate 80 along with Gd, Gd-conjugated dendrimer nanoclusters with polyamidoamine (PAMAM), and various other covalently integrated Gd-based biocompatible polymeric nanoparticles have been studied for their imaging applications [120–125]. A few examples of Gd encapsulated polymeric nanostructures include fatty acid coated gadolinium diethylenetriamine pentaacetic acid (GA-DTPA-FA), Gd-DTPA conjugated PAMAM dendrimers into polymersomes, and Gd-DTPA encapsulated into PLGA particles have been successfully used as contrast agents for MRI [126–129].

Several biocompatible polymers are used as surface coating material for SPIONs [11,130,131]. This includes folate targeted PEG-PCL coated SPIONs [FA-PEG-PCL-SPION], self-assembly of polymeric micelles of SPIONs, and drug loaded acid labile poly(aspartate)-g-poly (ethyleneglycol)-dodecylaminehydrazone-(adriamycin-levulinic acid) (PASP-g-PEG-DDA-Hyd-ADR-LEV), amphiphilic hydrotropic block copolymer, PEG-block-poly(2-[4-vinylbenzyloxy]-*N*, *N*-diethylnicotinamide), (PEG-b-P(VBODENA)) encapsulating SPIONs and drug, nanogel-caged polymersomes, PLGA-SPIONs, etc. [132–136]. Magnetic nanoparticles function as dual representative for both MRI (imaging) and magnetic hyperthermia (therapeutic) applications. Researchers demonstrated encapsulation of SPIONs and drug (curcumin) in PLGA and decorated with cancer specific aptamer. The nanoconstruct was employed for multimodal imaging for MRI, photoacoustic imaging (arising from SPIONs) and for optical imaging (arising from curcumin) signifying the multimodal imaging capability of SPION and drug encapsulated polymeric nanoformulation. The therapeutic potential mediated by drug mediated cytotoxicity, magnetic hyperthermia, and photothermal ablation of cancer cells were also investigated. These studies demonstrate the superior effectiveness of magnetic polymeric nanoparticles as promising theragnostic nanomaterials.

12.3.2.2 Computed Tomography

X-ray computed tomography (CT) is one of the most widely used, noninvasive clinical imaging modalities for the detection of various disorders, including tumors, with superior spatial resolution. CT imaging modality uses specific X-ray instrument setup for analysis and quantifies X-ray absorption with materials of high atomic number (Z) to enhance the sensitivity of the CT image. Extensive research in CT imaging has led to technological advances resulting in the development of dual energy and multislice CT for enabling better differentiation of diseased tissue from normal tissue. Nontoxic elements with a high Z-number and exceptional X-ray attenuation features such as iodinated molecules, barium sulfate, bismuth, and gold containing molecules and nanomaterials, composite ceramics with iron oxide and lanthanide are being used as contrast agents [137–142].

Although iodinated molecules are widely used as X-ray contrast enhancers, their rapid renal clearance, allergic reactions in patients, and the necessity to perform immediately after administration are some of the major challenges associated with X-ray CT. Polymeric nanoparticles encapsulating X-ray contrast enhancers can overcome some of these major drawbacks associated with iodinated contrast agents [143,144]. Most of the CT molecular contrast agents are integrated either by covalent conjugation or encapsulation into nanoparticles of various types such as liposomes, lipoproteins, emulsions, and polymeric nanoparticles [145–151]. Some of the reported polymeric nanoconstructs in which the contrast agents are attached to polymeric nanocarriers are polyiohexol (hexamethylene diisocyanate-mPEG-polylactide) integrated with iohexol, MAOETIB (2-hydroxyethyl methacrylate), and MAOETIB-GMA (2-hydroxyethyl methacrylate, glycidyl methacrylate) linked to 2,3,5-triiodobenzoic acid, etc. [150–152]. In some studies the contrast agents are encapsulated into polymeric nanoparticles. Encapsulation of iodinated oils into PBD-PEO-iodine [poly(butadiene)-b-poly(ethyleneglycol)], encapsulation of iodostearic acid, ethyl diiodo stearate into cROMP [poly(styrene)-b-(polyacrylic acid)], encapsulation of gold nanoparticles into PLGA are some of the reported polymeric nanoconstructs studied for X-ray–CT [141,153,154]. The current progress in the integration of different contrast into polymeric nanoparticles gives them much modal imaging potential.

12.3.2.3 Optical Imaging

There are several advantages for optical imaging over traditional imaging techniques such as MRI, PET, CT, etc. The advantages include high spatial resolution, noninvasive real time monitoring, and lower cost than other contrast agents. Optical imaging probes integrated with polymeric nanoparticles offer an ideal nanoplatform for diagnostics and supports in therapeutic applications. Among the contrast agents for optical imaging, near-infrared imaging probes (NIRF) are promising contrast agents for in vivo imaging owing to minimal absorption of light by tissue [155,156]. The wavelength ranging from 700 to 900 nm is known as the "diagnostic window" and NIRF dyes are active in this wavelength range. Commonly studied NIRF dyes include molecules of squaraine, cyanine, phthalocyanine, and porphyrin derivatives. In addition to NIRF dyes there are several nanoparticles, including semiconductor quantum dots, upconversion nanoparticles, Au nanoparticles, that are active in the NIR region with excellent optical properties and are being studied for fluorescence-guided surgery, thus permitting deep tissue imaging, which is highly beneficial for clinical diagnostics [157,158]. Although inorganic materials offer excellent fluorescent properties for noninvasive deep tissue imaging, the inherent toxicity limits their direct application in humans. Owing to the presence of heavy metals in these nanomaterials they exert toxicity in biological systems. Polymeric nanoparticles mask the toxicity and permit their application in in vivo imaging. In the case of NIRF dyes, polymeric nanomaterials are used to overcome stability issues associated with their solubility and inability to localize in target tissue, and to extend their half-life and avert leaching of dyes for in vivo application. Most of the dyes for bioimaging applications are either covalently attached or encapsulated into the polymeric matrix.

Several NIRF dyes are chemically attached to polymeric matrix to enhance the stability owing to their chemical versatile properties. NIRF dyes in the cyanin family (Cy; Cy5, Cy5.5, and Cy7), owing to their reactive amino group, are often exploited to attach to the polymeric matrix. Cy5-PLA, Cy5.5-PEI, Cy5.5-PEG, Cy5.5-glycol-chitosan, and Cy7-polymethylmethacrylate copolymer are some widely studied dye–polymer nanoformulations where the dye complex was covalently attached to the polymer [159–165]. Another class of NIRF dyes are squaraine based, where squaraine polymeric conjugates, including squaraine catenane endoperoxide conjugated polystyrene nanoparticles, *N*-propanesulfonate-benzothiazolium squaraine, and *N*-propanesulfonate-benzoindolium squaraine conjugated to bovine serum albumin nanoparticles, have been studied for optical imaging [166–168]. In addition to incorporation of fluorescent dyes, targeting ligands are also attached to nanoparticles for the precise delivery of targeted nanoparticles to tissue of interest, which is highly advantageous in the development of nanoformulations against cancer, biodistribution studies, and whole body imaging.

Indocyanine green (ICG) is a widely used NIRF dye that is used for deep tissue imaging. In addition to NIR imaging potential, ICG exhibits excellent photothermal properties that are used for photothermal ablation studies. However the dye suffers a few drawbacks including short plasma half-life of about 3 minutes, poor aqueous stability, and rapid clearance from the body. To overcome these disadvantages, ICG has been encapsulated within biocompatible polymeric nanoparticles to preserve its intrinsic fluorescence for noninvasive in vivo optical imaging [153,169–171]. ICG has been successfully encapsulated into various types of polymers including dextran, PLGA, poly(allylamine hydrochloride), PLCL-Poloxomer, and Eudragit for imaging and photothermal ablation studies in cancer models [172–180]. The dye encapsulated polymeric nanoparticles demonstrated superior retention efficiency in tumor sites, reduced fluorescence quenching, and lower clearance rate in in vivo studies [181]. Porphyrin and phthalocyanine derivatives including meso-tetra(4-hydroxyphenyl)porphyrin, zinc(II) phthalocyanine are another category of NIRF dyes that are successfully encapsulated into polymeric nanoparticles [182,183]. Currently development of multimodal imaging incorporating one or more imaging modalities compensates for the drawback of individual modality and offers a precise imaging nanoplatform.

12.4 Conclusion and Future Outlook

Polymeric nanomaterials with their unique physicochemical and biological properties are a promising tool for the advancement of drug delivery and diagnostic applications. The tunable surface properties of polymeric nanoparticles offer versatile scaffolds for various biomedical applications. From the synthetic perspective, surface functionalities should be endowed with properties including stability in complex biological media, biocompatibility, and specificity for active targeting for precise drug delivery. The proper understanding of the mechanism of functionalized polymeric nanoparticles and biointerface phenomena offers superior features for development of advanced functional nanomaterials. The comprehensive study on cellular uptake mechanism and nanoparticle structure–activity interaction to envisage their effects in a biological environment would be effective in developing efficient therapeutic and theragnostic capabilities. Apart from the potential advantages, the safety issues concerning the nanomaterials must be addressed effectively before shifting to nano-based biomedical applications. In addition to the medical applications in drug delivery, improvement in diagnostics and development of implants, research efforts in stem cell research, regenerative medicine, and nutraceuticals are some of the prime research sectors that can be improved by nanotechnology-based innovations.

References

[1] Erb SE. Nanotechnology in drug delivery. Drug Deliv Syst 2009;24:63–70.

[2] Farokhzad OC, Langer R. Impact of nanotechnology on drug delivery. ACS Nano 2009;3(1):16–20.

[3] Riehemann K, Schneider SW, Luger TA, et al. Nanomedicine-challenge and perspectives. Angew Chem Int Ed 2009;48(5):872–97.

[4] Sanhai WR, Sakamoto JH, Canady R, et al. Seven challenges for nanomedicine. Nat Nanotechnol 2008;3 (5):242–4.

[5] Brigger I, Dubernet C, Couvreur P. Nanoparticles in cancer therapy and diagnosis. Adv Drug Deliv Rev 2002;54(5):631–51.

[6] Sinha R, Kim GJ, Nie S, et al. Nanotechnology in cancer therapeutics: bioconjugated nanoparticles for drug delivery. Mol Cancer Ther 2006;5(8):1909–17.

[7] Panyam J, Labhasetwar V. Biodegradable nanoparticles for drug and gene delivery to cells and tissue. Adv Drug Deliv Rev 2003;55(3):329–47 7 = 51-removed.

[8] Hans ML, Lowman AM. Biodegradable nanoparticles for drug delivery and targeting. Curr Opin Solid State Mater Sci 2002;6(4):319–27.

[9] Laurent S, Forge D, Port M, et al. Magnetic iron oxide nanoparticles: synthesis, stabilization, vectorization, physicochemical characterizations, and biological applications. Chem Rev 2008;108(6):2064–110.

[10] Kelkar SS, Reineke TM. Theranostics: combining imaging and therapy. Bioconjug Chem 2011;22 (10):1879–903.

[11] Lee N, Hyeon T. Designed synthesis of uniformly sized iron oxide nanoparticles for efficient magnetic resonance imaging contrast agents. Chem Soc Rev 2012;41(7):2575–89 11 = 114 removed.

[12] Zhang F, Lees E, Amin F, et al. Polymer-coated nanoparticles: a universal tool for biolabelling experiments. Small 2011;7(22):3113–27.

[13] Cabral H, Nishiyama N, Kataoka K. Supramolecular nanodevices: from design validation to theranostic nanomedicine. Acc Chem Res 2011;44(10):999–1008.

[14] Mout R, Moyano DF, Rana S, et al. Surface functionalization of nanoparticles for nanomedicine. Chem Soc Rev 2012;41(7):2539–44.

[15] Khandare J, Calderon M, Dagia NM, et al. Multifunctional dendritic polymers in nanomedicine: opportunities and challenges. Chem Soc Rev 2012;41(7):2824–48.

[16] Kamaly N, Xiao Z, Valencia PM, et al. Targeted polymeric therapeutic nanoparticles: design, development and clinical translation. Chem Soc Rev 2012;41(7):2971–3010.

[17] Elsabahy M, Wooley KL. Design of polymeric nanoparticles for biomedical delivery applications. Chem Soc Rev 2012;41(7):2545–61.

[18] Lee DE, Koo H, Sun IC, et al. Multifunctional nanoparticles for multimodal imaging and theragnosis. Chem Soc Rev 2012;41(7):2656–72.

[19] Mizrahy S, Peer D. Polysaccharides as building blocks for nanotherapeutics. Chem Soc Rev 2012;41 (7):2623–40.

[20] Nel A, Xia T, Madler L, et al. Toxic potential of materials at the nanolevel. Science 2006;311 (5761):622–7.

[21] Kirchner C, Liedl T, Kudera S, et al. Cytotoxicity of colloidal CdSe and CdSe/ZnS nanoparticles. Nano Lett 2005;5(2):331–8.

[22] Aldana J, Wang YA, Peng X. Photochemical instability of CdSe nanocrystals coated by hydrophilic thiols. J Am Chem Soc 2001;123(36):8844–50.

[23] Dubois F, Mahler B, Dubertret B, et al. A versatile strategy for quantum dot ligand exchange. J Am Chem Soc 2006;129(3):482–3.

[24] Goldman ER, Balighia ED, Mattoussi H, et al. Avidin: a natural bridge for quantum dot-antibody conjugates. J Am Chem Soc 2002;124(22):6378–82.

[25] Mattoussi H, Mauro JM, Goldman ER, et al. Self-assembly of CdS-ZnS quantum dot bioconjugates using an engineered recombinant protein. J Am Chem Soc 2000;122(49):12142–50.

[26] Pazos-Pérez N, Gao Y, Hilgendorff M, et al. Magnetic-noble metal nanocomposites with morphology-dependent optical response. Chem Mater 2007;19(18):4415–22.

[27] Kim J, Kim HS, Lee N, et al. Multifunctional uniform nanoparticles composed of a magnetite nanocrystal core and a mesoporous silica shell for magnetic resonance and fluorescence imaging and for drug delivery. Angew Chem Int Ed Engl 2008;47(44):8438–41.

[28] Wang Y, Wong JF, Teng X, et al. "Pulling" nanoparticles into water: phase transfer of oleic acid stabilized monodisperse nanoparticles into aqueous solutions of α-cyclodextrin. Nano Lett 2003;3(11):1555–9.

[29] Huh YM, Jun YW, Song HT, et al. In vivo magnetic resonance detection of cancer by using multifunctional magnetic nanocrystals. J Am Chem Soc 2005;127(35):12387–91.

[30] Kim S, Bawendi MG. Oligomeric ligands for luminescent and stable nanocrystal quantum dots. J Am Chem Soc 2003;125(48):14652–3.

[31] Lipka J, Semmler-Behnke M, Sperling RA, et al. Biodistribution of PEG-modified gold nanoparticles following intratracheal instillation and intravenous injection. Biomaterials 2010;31(25):6574–81.

[32] Ballou B, Lagerholm BC, Ernst LA, et al. Noninvasive imaging of quantum dots in mice. Bioconj Chem 2003;15(1):79–86.

[33] Pinaud F, King D, Moore HP, et al. Bioactivation and cell targeting of semiconductor CdSe/ZnS nanocrystals with phytochelatin-related peptides. J Am Chem Soc 2004;126(19):6115–23.

[34] Schofield CL, Haines AH, Field RA, et al. Silver and gold glyconanoparticles for colorimetric bioassays. Langmuir 2006;22(15):6707–11.

[35] Zubarev ER, Xu J, Sayyad A, et al. Amphiphilic gold nanoparticles with V-shaped arms. J Am Chem Soc 2006;128(15):4958–9.

[36] Wang M, Peng ML, Cheng W, et al. A novel approach for transferring oleic acid capped iron oxide nanoparticles to water phase. J Nanosci Nanotechnol 2011;11(4):3688–91.

[37] Giaume D, Poggi MI, Casanova D, et al. Organic functionalization of luminescent oxide nanoparticles toward their application as biological probes. Langmuir 2008;24(19):11018–26.

[38] Salgueiriño-Maceira V, Correa-Duarte MA, Spasova M, et al. Composite silica spheres with magnetic and luminescent functionalities. Adv Funct Mater 2006;16(4):509–14.

[39] Pellegrino T, Manna L, Kudera S, et al. Hydrophobic nanocrystals coated with an amphiphilic polymer shell: a general route to water soluble nanocrystals. Nano Lett 2004;4(4):703–7.

[40] Sapsford KE, Algar WR, Berti L, et al. Functionalizing nanoparticles with biological molecules: developing chemistries that facilitate nanotechnology. Chem Rev 2013;113(3):1904–2074.

[41] McBain SC, Yiu HHP, El Haj A, et al. Polyethyleneimine functionalized iron oxide nanoparticles as agents for DNA delivery and transfection. J Mater Chem 2007;17:2561–5.

[42] Park IK, Ng CP, Wang J, et al. Determination of nanoparticle vehicle unpackaging by MR imaging of a T2 magnetic relaxation switch. Biomaterials 2008;29(6):724–32.

[43] Liu Y, Zhang B, Yan B. Enabling anticancer therapeutics by nanoparticle carriers: the delivery of paclitaxel. Int J Mol Sci 2011;12(7):4395–413.

[44] Gunn J, Wallen H, Veiseh O, et al. A multimodal targeting nanoparticle for selectively labeling T cells. Small 2008;4(6):712–15.

[45] Chastellian M, Petri A, Hofman H. Particle size investigations of a multistep synthesis of PVA coated superparamagnetic nanoparticles. J Colloid Interface Sci 2004;278(2):353–60.

[46] Kolb HC, Finn MG, Sharpless KB. Click chemistry: diverse chemical function from a few good reactions. Angew Chem Int Ed Engl 2001;40(11):2004–21.

[47] Cohen H, Levy RJ, Gao J, et al. Sustained delivery and expression of DNA encapsulated in polymeric nanoparticles. Gene Ther 2000;22:71896–905.

[48] Soppimath KS, Aminabhavi TM, Kulkarni AR, et al. Biodegradable polymeric nanoparticles as drug delivery devices. J Control Release 2001;70(1–2):1–20.

[49] Kawashima Y. Nanoparticulate systems for improved drug delivery. Adv Drug Deliv Rev 2001;47 (1):1–2.

[50] Cui Z, Mumper RJ. Plasmid DNA-entrapped nanoparticles engineered from microemulsion precursors: in vitro and in vivo evaluation. Bioconj Chem 2002;13(6):1319–27.

[51] Kelley EG, Albert JN, Sullivan MO, et al. Stimuli-responsive copolymer solution and surface assemblies for biomedical applications. Chem Soc Rev 2013;42(17):7057–71.

[52] Mura S, Nicolas J, Couvreur P. Stimuli-responsive nanocarriers for drug delivery. Nat Mater 2013;12:991–1003 52 = 91 and91 removed.

[53] Gao W, Chan JM, Farokhzad OC. pH-responsive nanoparticles for drug delivery. Mol Pharm 2010;7 (6):1913–20.

[54] Ganesh VA, Baji A, Ramakrishna S. Smart functional polymers—a new route towards creating a sustainable environment. RSC Adv 2014;4(95):53352–64.

[55] Liu J, Huang Y, Kumar A, et al. pH-Sensitive nano-systems for drug delivery in cancer therapy. Biotechnol Adv 2014;32(4):693–710.

[56] Yu P, Yu H, Guo C, et al. Reversal of doxorubicin resistance in breast cancer by mitochondria-targeted pH-responsive micelles. Acta Biomater 2015;14:115–24.

[57] Du JZ, Tang YQ, Lewis AL, et al. pH-Sensitive vesicles based on a biocompatible zwitterionic diblock copolymer. J Am Chem Soc 2005;127(51):17982–3.

[58] Chen W, Meng FH, Li F, Ji SJ, Zhong ZY. pH-responsive biodegradable micelles based on acid-labile polycarbonate hydrophobe: synthesis and triggered drug release. Biomacromolecules 2009;10 (7):1727–35.

[59] Colombo P, Sonvico F, Colombo G, et al. Novel platforms for oral drug delivery. Pharm Res 2009;26 (3):601–11.

[60] Dai S, Tam KC, Jenkins RD. Aggregation behavior of methacrylic acid/ethyl acrylate copolymer in dilute solutions. Eur Polym J 2000;36(12):2671–7.

[61] Dai J, Nagai T, Wang X, et al. pH-sensitive nanoparticles for improving the oral bioavailability of cyclosporine A. Int J Pharm 2004;280(1–2):229–40.

[62] Lin Y, Sonaje K, Lin KM, et al. Multi-ion-crosslinked nanoparticles with pH-responsive characteristics for oral delivery of protein drugs. J Control Release 2008;132(2):141–9.

[63] Lin Y, Chang C, Wu Y, et al. Development of pH-responsive chitosan/heparin nanoparticles for stomach-specific anti-Helicobacter pylori therapy. Biomaterials 2009;30(19):3332–42.

[64] Sengupta S, Eavarone D, Capila I, et al. Temporal targeting of tumour cells and neovasculature with a nanoscale delivery system. Nature 2005;436(7050):568–72.

[65] Rozema DB, Lewis DL, Wakefield DH, et al. Dynamic polyconjugates for targeted in vivo delivery of siRNA to hepatocytes. Proc Natl Acad Sci USA 2007;104(32):12982–7.

[66] Xu Z, Gu W, Chen L, et al. A smart nanoassembly consisting of acid-labile vinyl ether peg-dope and protamine for gene delivery: preparation and in vitro transfection. Biomacromolecules 2008;9 (11):3119–26.

[67] Wang B, Xu C, Xie J, et al. pH controlled release of chromone from chromone-Fe3O4 nanoparticles. J Am Chem Soc 2008;130(44):14436–7.

[68] Mok H, Park JW, Park TG. Enhanced intracellular delivery of quantum dot and adenovirus nanoparticles triggered by acidic pH via surface charge reversal. Bioconjugate Chem 2008;19(4):797–801.

[69] Griset AP, Walpole J, Liu R, et al. Expansile nanoparticles: synthesis, characterization, and in vivo efficacy of an acid-responsive polymeric drug delivery system. J Am Chem Soc 2009;131(7):2469–71.

[70] Tong R, Cheng J. Ring-opening polymerization-mediated controlled formulation of polylactide-drug nanoparticles. J Am Chem Soc 2009;131(13):4744–54.

[71] Aryal S, Hu CJ, Zhang L. Polymer-cisplatin conjugate nanoparticles for acid-responsive drug delivery. ACS Nano 2010;4(1):251–8.

[72] Liu R, Zhang Y, Zhao X, et al. pH-responsive nanogated ensemble based on gold-capped mesoporous silica through an acid-labile acetal linker. J Am Chem Soc 2010;132(5):1500–1.

[73] Danhier F, Feron O, Préat V. To exploit the tumor microenvironment: passive and active tumor targeting of nanocarriers for anti-cancer drug delivery. J Control Release 2010;148(2):135–46.

[74] Shi Y, van Steenbergen MJ, Teunissen EA, et al. Π–Π stacking increases the stability and loading capacity of thermosensitive polymeric micelles for chemotherapeutic drugs. Biomacromolecules 2013;14 (6):1826–37.

[75] Shi Y, van den Dungen ET, Klumperman B, et al. Reversible addition–fragmentation chain transfer synthesis of a micelle-forming, structure reversible thermosensitive diblock copolymer based on the N-(2-Hydroxy propyl) methacrylamide backbone. ACS Macro Lett 2013;2(5):403–8.

[76] Shi Y, Cardoso RM, Van Nostrum CF, et al. Anthracene functionalized thermosensitive and UV-crosslinkable polymeric micelles. Polym Chem 2015;6:2048–53.

[77] Zhao Y, Fan X, Liu D, et al. PEGylated thermo-sensitive poly (amidoamine) dendritic drug delivery systems. Int J Pharm 2011;409(1–2):229–36.

[78] Adelsberger J, Kulkarni A, Jain A, Wang W, Bivigou-Koumba AM, Busch P, et al. Thermoresponsive PS-b-PNIPAM-b-PS micelles: aggregation behavior, segmental dynamics, and thermal response. Macromolecules 2010;43(5):2490–501.

[79] Qiu Y, Park K. Environment-sensitive hydrogels for drug delivery. Adv Drug Deliv Rev 2001;53 (3):321–39.

[80] Choi S, Baudys M, Kim SW. Control of blood glucose by novel GLP-1 delivery using biodegradable triblock copolymer of PLGA–PEG–PLGA in type 2 diabetic rats. Pharm Res 2004;21(5):827–31.

[81] Yannic B, Schuetz YB, Gurny R, et al. A novel thermoresponsive hydrogel based on chitosan. Eur J Pharm Biopharm 2008;68(1):19–25.

[82] Li L, Lim LH, Wang Q, et al. Thermoreversible micellization and gelation of a blend of pluronic polymers. Polymer 2008;49(7):1952–60.

[83] Hoare TR, Kohane DS. Hydrogels in drug delivery: progress and challenges. Polymer 2008;49 (8):1993–2007.

[84] Ward MA, Georgiou TK. Thermoresponsive polymers for biomedical applications. Polymers 2011;3:1215–42.

[85] Cooperstein MA, Canavan HE. Assessment of cytotoxicity of (N-isopropyl acrylamide) and poly(N-isopropyl acrylamide) coated surfaces. Biointerphases 2013;8(1):19.

[86] Galaev I, Mattiasson B. Smart polymers: applications in biotechnology and biomedicine. 2nd ed. Boca Raton, FL: CRC Press; 2007. p. 144–6.

[87] De Oliveira AM, De Oliveira PC, Dos Santos AM, et al. Synthesis and characterization of thermo-responsive particles of poly(hydroxybutirate-co-hydroxyvalerate)-b-poly(nisopropylacrylamide). Braz J Phys 2009;39(1a):217–22.

[88] Ci T, Chen L, Yu L, et al. Tumor regression achieved by encapsulating a moderately soluble drug into a polymeric thermogel. Sci Rep 2014;4:5473.

[89] Yu L, Hu H, Chen L, et al. Comparative studies of thermogels in preventing post-operative adhesions and corresponding mechanisms. Biomater Sci 2014;2(8):1100–9.

[90] Banerjee S, Chaurasia G, Ghosh A. Smart polymers: around the cosmos. Asian J Pharm Clin Res 2010;3 (3):135–41.

[91] You J, Shao RP, Wei X, et al. Near-infrared light triggers release of paclitaxel from biodegradable microspheres: photothermal effect and enhanced antitumor activity. Small 2010;6(9):1022–31 94 = 80 (previously given-1 ref will reduce finally).

[92] Lee SY, Tae G. Formulation and in vitro characterization of an in situ gelable, photo-polymerizable pluronic hydrogel suitable for injection. J Control Release 2007;119(3):313–19.

[93] Yin R, Tong Z, Yang D, et al. Glucose and pH dual-responsive concanavalin a based microhydrogels for insulin delivery. Int J Biol Macromol 2011;49(5):1137–42.

[94] Lal S, Clare SE, Halas NJ. Nanoshell-enabled photothermal cancer therapy: impending clinical impact. Acc Chem Res 2008;41(12):1842–51.

[95] Chen Z, Yin JJ, Zhou YT, et al. Dual enzyme-like activities of iron oxide nanoparticles and their implication for diminishing cytotoxicity. ACS Nano 2012;6(5):4001–12.

[96] Liu J, Zhang Y, Wang C, et al. Magnetically sensitive alginate-templated polyelectrolyte multilayer microcapsules for controlled release of doxorubicin. J Phys Chem C 2010;114(17):7673–9.

[97] Sun J, Zhang Y, Chen Z, et al. Fibrous aggregation of magnetite nanoparticles induced by a time-varied magnetic field. Angew Chem Int Ed Engl 2007;46(25):4767–70.

[98] Wang F, Kim DK, Yoshitake T, et al. Diffusion and clearance of superparamagnetic iron oxide nanoparticles infused into the rat striatum studied by MRI and histochemical techniques. Nanotechnology 2010;22 015103.

[99] Torchilin VP. Multifunctional, stimuli-sensitive nanoparticulate systems for drug delivery. Nat Rev Drug Discov 2014;13:813–27.

[100] Hu K, Sun J, Guo Z, et al. A novel magnetic hydrogel with aligned magnetic colloidal assemblies showing controllable enhancement of magnetothermal effect in the presence of alternating magnetic field. Adv Mater 2015;27(15):2507–14.

[101] Fang K, Song L, Gu Z, et al. Magnetic field activated drug release system based on magnetic PLGA microspheres for chemo-thermal therapy. Colloids Surf B Biointerf 2015;136:712–20.

[102] Yang F, Zhang X, Song L, et al. Controlled drug release and hydrolysis mechanism of polymer–magnetic nanoparticle composite. ACS Appl Mater Interf 2015;7(18):9410–19.

[103] Liu D, Wu W, Chen X, et al. Conjugation of paclitaxel to iron oxide nanoparticles for tumor imaging and therapy. Nanoscale 2012;4(7):2306–10.

[104] Xiong F, Chen Y, Chen J, et al. Rubik-like magnetic nanoassemblies as an efficient drug multifunctional carrier for cancer theranostics. J Control Release 2013;172(3):993–1001.

[105] Hayashi K, Nakamura M, Sakamoto W, et al. Superparamagnetic nanoparticle clusters for cancer theranostics combining magnetic resonance imaging and hyperthermia treatment. Theranostics 2013;3 (6):366–76.

[106] Xie J, Zhang Y, Yan C, et al. High-performance PEGylated Mn–Zn ferrite nanocrystals as a passive-targeted agent for magnetically induced cancer theranostics. Biomaterials 2014;35(33):9126–36.

[107] Song L, Zang F, Song M, et al. Effective PEGylation of Fe3O4 nanomicelles for in vivo MR imaging. J Nanosci Nanotechnol 2015;15(6):4111–18.

[108] Tanaka IN, Nishio I, Sun ST, et al. Collapse of gels in an electric field. Science 1982;218(4571):467–9.

[109] Murdan S. Electro-responsive drug delivery from hydrogels. J Control Release 2003;92(1–2):1–17.

[110] Paris JL, Cabañas MV, Manzano M, et al. Polymer-grafted mesoporous silica nanoparticles as ultrasound-responsive drug carriers. ACS Nano 2015;9(11):11023–33.

[111] Tong R, Tang L, Ma L, et al. Smart chemistry in polymeric nanomedicine. Chem Soc Rev 2014;43 (20):6982–7012.

[112] Moghimi SM. Recent developments in polymeric nanoparticle engineering and their applications in experimental and clinical oncology. Anticancer Agents Med Chem 2006;6(6):553–61.

[113] Zhu D, Liu F, Ma L, et al. Nanoparticle based systems for T1-weighted magnetic resonance imaging contrast agents. Int J Mol Sci 2013;14(5):10591–607.

[114] Wang YX. Superparamagnetic iron oxide based MRI contrast agents: current status of clinical application. Quant Imaging Med Surg 2011;1(1):35–40.

[115] Stephen ZR, Kievit FM, Zhang M. Magnetite nanoparticles for medical MR imaging. Mater Today 2011;14(7–8):330–8.

[116] Khemtong C, Kessinger CW, Gao J. Polymeric nanomedicine for cancer MR imaging and drug delivery. Chem Commun 2009;28(24):3497–510.

[117] Thomas R, Park IK, Jeong YY. Magnetic iron oxide nanoparticles for multimodal imaging and therapy of cancer. Int J Mol Sci 2013;14(8):15910–30.

[118] Muthiah M, Park IK, Cho CS. Surface modification of iron oxide nanoparticles by biocompatible polymers for tissue imaging and targeting. Biotechnol Adv 2013;31(8):1224–36.

[119] Liu Y, Chen Z, Liu C, et al. Gadolinium-loaded polymeric nanoparticles modified with anti-VEGF as multifunctional MRI contrast agents for the diagnosis of liver cancer. Biomaterials 2011;32 (22):5167–76.

[120] Huang CH, Tsourkas A. Gd-based macromolecules and nanoparticles as magnetic resonance contrast agents for molecular imaging. Curr Top Med Chem 2013;13(4):411–21.

[121] Huang CH, Nwe K, Al Zaki A, et al. Biodegradable polydisulfide dendrimer nanoclusters as MRI contrast agents. ACS Nano 2012;6(11):9416–24.

[122] Jeong SY, Kim HJ, Kwak BK, et al. Biocompatible polyhydroxyethylaspartamide-based micelles with gadolinium for MRI contrast agents. Nanoscale Res Lett 2010;5(12):1970–6.

[123] Lee HY, Jee HW, Seo SM, et al. Diethylenetriaminepentaacetic acid-gadolinium (DTPA-Gd)-conjugated polysuccinimide derivatives as magnetic resonance imaging contrast agents. Bioconjug Chem 2006;17 (3):700–6.

[124] Li Y, Qian Y, Liu T, et al. Light-triggered concomitant enhancement of magnetic resonance imaging contrast performance and drug release rate of functionalized amphiphilic diblock copolymer micelles. Biomacromolecules 2012;13(11):3877–86.

[125] Lee SM, Song Y, Hong BJ, et al. Modular polymer-caged nanobins as a theranostic platform with enhanced magnetic resonance relaxivity and pH-responsive drug release. Angew Chem Int Ed Engl 2010;49(51):9960–4.

[126] Zhang HW, Wang LQ, Xiang QF, et al. Specific lipase-responsive polymer-coated gadolinium nanoparticles for MR imaging of early acute pancreatitis. Biomaterials 2013;35(1):356–67.

[127] Cheng Z, Thorek DLJ, Tsourkas A. Porous polymersomes with encapsulated Gd-labeled dendrimers as highly efficient MRI contrast agents. Adv Funct Mater 2009;19(23):3753–9.

[128] Doiron AL, Homan KA, Emelianov S, et al. Poly(lactic-co-glycolic) acid as a carrier for imaging contrast agents. Pharm Res 2009;26(3):674–82.

[129] Broz P, Ben-Haim N, Santini F, et al. Polymer vesicles loaded with precipitated gadolinium nanoparticles: a novel target-specific contrast agent for magnetic resonance imaging. Eur J Nanomed 2009;2 (2):43–8.

[130] Peng XH, Qian X, Mao H, et al. Targeted magnetic iron oxide nanoparticles for tumor imaging and therapy. Int J Nanomed 2008;3(3):311–21.

[131] Frey NA, Peng S, Cheng K, et al. Magnetic nanoparticles: synthesis, functionalization, and applications in bioimaging and magnetic energy storage. Chem Soc Rev 2009;38(9):2532–42.

[132] Hong GB, Zhou JX, Yuan RX. Folate-targeted polymeric micelles loaded with ultrasmall superparamagnetic iron oxide: combined small size and high MRI sensitivity. Int J Nanomed 2012;7:2863–72.

[133] Huang H, Li Y, Li C, et al. A novel anti-VEGF targeting and MRI-visible smart drug delivery system for specific diagnosis and therapy of liver cancer. Macromol Biosci 2013;13(10):1358–68.

[134] Chiang WH, Huang WC, Chang CW, et al. Functionalized polymersomes with outlayered polyelectrolyte gels for potential tumor-targeted delivery of multimodal therapies and MR imaging. J Control Release 2013;168(3):280–8.

[135] Sivakumar B, Aswathy RG, Romero-Aburto R, et al. Highly versatile SPION encapsulated PLGA nanoparticles as photothermal ablators of cancer cells and as multimodal imaging agents. Biomater Sci 2017;5(3):432–43.

[136] Yoon HY, Saravanakumar G, Heo R, et al. Hydrotropic magnetic micelles for combined magnetic resonance imaging and cancer therapy. J Control Release 2012;160(3):692–8.

[137] Bourin M, Jolliet P, Ballereau F. An overview of the clinical pharmacokinetics of x-ray contrast media. Clin Pharmacokinet 1997;32(3):180–93.

[138] Rabin O, Manuel Perez J, Grimm J, et al. An X-ray computed tomography imaging agent based on long-circulating bismuth sulphide nanoparticles. Nat Mater 2006;5(2):118–22.

[139] Boote E, Fent G, Kattumuri V, et al. Gold nanoparticle contrast in a phantom and juvenile swine: models for molecular imaging of human organs using x-ray computed tomography. Acad Radiol 2010;17 (4):410–17.

[140] Aydogan B, Li J, Rajh T, et al. AuNP-DG: deoxyglucose-labeled gold nanoparticles as X-ray computed tomography contrast agents for cancer imaging. Mol Imaging Biol 2010;12(5):463–7.

[141] Ai K, Liu Y, Liu J, et al. Large-scale synthesis of Bi(2)S(3) nanodots as a contrast agent for in vivo X-ray computed tomography imaging. Adv Mater 2011;23(42):4886–91.

[142] Oh MH, Lee N, Kim H, et al. Large-scale synthesis of bioinert tantalum oxide nanoparticles for X-ray computed tomography imaging and bimodal image guided sentinel lymph node mapping. J Am Chem Soc 2011;133(14):5508–15.

[143] Liu Y, Ai K, Lu L. Nanoparticulate X-ray computed tomography contrast agents: from design validation to in vivo applications. Acc Chem Res 2012;45(10):1817–27.

[144] Christiansen C. X-ray contrast media-an overview. Toxicology 2005;209(2):185–7.

[145] Elrod DB, Partha R, Danila D, et al. An iodinated liposomal computed tomographic contrast agent prepared from a diiodophosphatidylcholine lipid. Nanomedicine 2009;5(1):42–5.

[146] Skajaa T, Cormode DP, Falk E, et al. High-density lipoprotein-based contrast agents for multimodal imaging of atherosclerosis. Arterioscler Thromb VascBiol 2010;30(2):169–76.

[147] de Vries A, Custers E, Lub J, et al. Block-copolymer-stabilized iodinated emulsions for use as CT contrast agents. Biomaterials 2010;31(25):6537–44.

[148] Kong WH, Lee WJ, Cui ZY, et al. Nanoparticulate carrier containing water-insoluble iodinated oil as a multifunctional contrast agent for computed tomography imaging. Biomaterials 2007;28(36):5555–61.

[149] Hyafil F, Cornily JC, Feig JE, et al. Noninvasive detection of macrophages using a nanoparticulate contrast agent for computed tomography. Nat Med 2007;13(5):636–41.

[150] Galperin A, Margel D, Baniel J, et al. Radiopaque iodinated polymeric nanoparticles for X-ray imaging applications. Biomaterials 2007;28(30):4461–8.

[151] Aviv H, Bartling S, Kieslling F, et al. Radiopaque iodinated copolymeric nanoparticles for X-ray imaging applications. Biomaterials 2009;30(29):5610–16.

[152] Yin Q, Yap FY, Yin L, et al. Poly(iohexol) nanoparticles as contrast agents for in vivo X-ray computed tomography imaging. J Am Chem Soc 2013;135(37):13620–3.

[153] Mieszawska AJ, Kim Y, Gianella A, et al. Synthesis of polymer lipid nanoparticles for image-guided delivery of dual modality therapy. Bioconjug Chem 2013;24(9):1429–34.

[154] Pan D, Williams TA, Senpan A, et al. Detecting vascular biosignatures with a colloidal, radio-opaque polymeric nanoparticle. J Am Chem Soc 2009;131(42):15522–7.

[155] Weissleder R, Tung CH, Mahmood U, et al. In vivo imaging of tumors with protease-activated near-infrared fluorescent probes. Nat Biotechnol 1999;17(4):375–8.

[156] Simpson CR, Kohl M, Essenpreis M, et al. Near-infrared optical properties of ex vivo human skin and subcutaneous tissues measured using the Monte Carlo inversion technique. Phys Med Bio 1998;43 (9):2465–78.

[157] Yaseen MA, Yu J, Wong MS, et al. In-vivo fluorescence imaging of mammalian organs using chargeassembled mesocapsule constructs containing indocyanine green. Opt Express 2008;16(25):20577–87.

[158] Sevick-Muraca EM, Houston JP, Gurfinkel M. Fluorescence-enhanced, near infrared diagnostic imaging with contrast agents. Curr Opin Chem Biol 2002;6(5):642–50.

[159] Chaney EJ, Tang L, Tong R, et al. Lymphatic biodistribution of polylactide nanoparticles. Mol Imaging 2010;9(3):153–62.

[160] Kim K, Lee M, Park H, et al. Cell permeable and biocompatible polymeric nanoparticles for apoptosis imaging. J Am Chem Soc 2006;128(11):3490–1.

[161] Veiseh O, Sun C, Gunn J, et al. Optical and MRI multifunctional nanoprobe for targeting gliomas. Nano Lett 2005;5(6):1003–8.

[162] Park K, Kim JH, Nam YS, et al. Effect of polymer molecular weight on the tumor targeting characteristics of self-assembled glycol chitosan nanoparticles. J Control Release 2007;122(3):305–14.

[163] Lee C-M, Jeong H-J, Cheong S-J, et al. Prostate cancer-targeted imaging using magnetofluorescent polymeric nanoparticles functionalized with bombesin. Pharm Res 2010;27(4):712–21.

[164] Lee S, Ryu JH, Park K, et al. Polymeric nanoparticle-based activatable near infrared nanosensor for protease determination in vivo. Nano Lett 2009;9(12):4412–16.

[165] Yang Z, Zheng S, Harrison WJ, et al. Long circulating near-infrared fluorescence core-crosslinked polymeric micelles: synthesis, characterization, and dual nuclear/optical imaging. Biomacromolecules 2007;8(11):3422–8.

[166] Lee J-J, White AG, Rice DR, et al. In vivo imaging using polymeric nanoparticles stained with near-infrared chemiluminescent and fluorescent squaraine catenane endoperoxide. Chem Commun 2013;49 (29):3016–18.

[167] Zhang Y, Yue X, Kim B, et al. Bovine serum albumin nanoparticles with fluorogenic near-IR-emitting squaraine dyes. ACS Appl Mater Interfaces 2013;5(17):8710–17.

[168] Li FQ, Su H, Wang J, et al. Preparation and characterization of sodium ferulate entrapped bovine serum albumin nanoparticles for liver targeting. Int J Pharm 2008;349(1–2):274–82.

[169] Ding D, Li K, Qin W, et al. Conjugated polymer amplified far-red/near-infrared fluorescence from nanoparticles with aggregation-induced emission characteristics for targeted in vivo imaging. Adv Healthc Mater 2013;2(3):500–7.

[170] Yen SK, Janczewski D, Lakshmi JL, et al. Design and synthesis of polymer-functionalized NIR fluorescent dyes-magnetic nanoparticles for bioimaging. ACS Nano 2013;7(8):6796–805.

[171] Kohl Y, Kaiser C, Bost W, et al. Preparation and biological evaluation of multifunctional PLGA-nanoparticles designed for photoacoustic imaging. Nanomedicine 2011;7(2):228–37.

[172] Liu P, Yue C, Shi B, et al. Dextran based sensitive theranostic nanoparticles for near-infrared imaging and photothermal therapy in vitro. Chem Commun 2013;49(55):6143–5.

[173] Manchanda R, Fernandez-Fernandez A, Nagesetti A, et al. Preparation and characterization of a polymeric (PLGA) nanoparticulate drug delivery system with simultaneous incorporation of chemotherapeutic and thermo-optical agents. Colloids Surf B Biointerf 2010;75(1):260–7.

[174] Yaseen MA, Yu J, Jung B, et al. Biodistribution of encapsulated indocyanine green in healthy mice. Mol Pharm 2009;6(5):1321–32.

[175] El-Daly SM, Gamal-Eldeen AM, Abo-Zeid MAM, et al. Photodynamic therapeutic activity of indocyanine green entrapped in polymeric nanoparticles. Photodiagn Photodyn Ther 2013;10(2):173–85.

[176] Lim YT, Noh YW, Han JH, Cai QY, Yoon KH, Chung BH. Biocompatible polymer-nanoparticle based bimodal imaging contrast agents for the labeling and tracking of dendritic cells. Small 2008;4 (10):1640–5.

[177] Saxena V, Sadoqi M, Shao J. Indocyanine green loaded biodegradable nanoparticles: preparation, physicochemical characterization and in vitro release. Int J Pharm 2004;278(2):293–301.

[178] Larush L, Magdassi S. Formation of near-infrared fluorescent nanoparticles for medical imaging. Nanomedicine 2011;6(2):233–40.

[179] Gamal-Eldeen AM, El-Daly SM, Borai IH, et al. Photodynamic therapeutic effect of indocyanine green entrapped in polymeric nanoparticles and their anti-EGFR-conjugate in skin cancer in CD1 mice. Photodiagn Photodyn Ther 2013;10(4):446–59.

[180] Ranjan AP, Zeglam K, Mukerjee A, et al. A sustained release formulation of chitosan modified PLCL: poloxamer blend nanoparticles loaded with optical agent for animal imaging. Nanotechnology 2011;22 (29):295104.

[181] Kim TH, Chen Y, Mount CW, et al. Evaluation of temperature-sensitive, indocyanine green-encapsulating micelles for noninvasive near-infrared tumor imaging. Pharm Res 2010;27(9):1900–13.

[182] Konan YN, Berton M, Gurny R, et al. Enhanced photodynamic activity of meso-tetra(4- hydroxyphenyl) porphyrin by incorporation into sub- 200nm nanoparticles. Eur J Pharm Sci 2003;18(3–4):241–9.

[183] Konan YN, Cerny R, Favet J, et al. Preparation and characterization of sterile sub-200nm meso-tetra(4-hydroxylphenyl)porphyrin-loaded nanoparticles for photodynamic therapy. Eur J Pharm Biopharm 2003;55(1):115–24.

13

Thermal Decomposition of Polymer Nanocomposites With Functionalized Nanoparticles

Klaudia Król-Morkisz, Kinga Pielichowska

AGH UNIVERSITY OF SCIENCE AND TECHNOLOGY, FACULTY OF MATERIALS SCIENCE AND CERAMICS, DEPARTMENT OF BIOMATERIALS AND COMPOSITES, KRAKÓW, POLAND

13.1 Introduction

The relatively low thermal stability of polymers strongly limits their range of applications [1]. Decomposition of traditional organic polymers starts usually at temperatures ranging from 150°C to 300°C. As a result, degradation reactions may already occur during the processing, but also during the service life of polymers [2]. In both cases, the material properties such as mechanical strength, thermal properties, color, material homogeneity, etc., may be changed [3]. Therefore, the enhancement of thermal stability of polymers is the key issue for further development of polymer products used in various applications. One possible approach to improve the thermal resistance of polymers is to incorporate nanofiller into the polymer matrix. Functionalized nanoparticles show great potential in this area. In recent years, a considerable number of studies on the influence of different kinds of functionalized nanoparticles on the decomposition behavior of polymer composites has been performed.

13.1.1 Thermal Decomposition Mechanisms of Polymers

Decomposition of polymer chains due to excessive heating may occur according to different mechanisms. Depending on heating temperature, heating rate, and the atmosphere different chemical processes occur in polymeric material [4]. The degradation of polymer may be induced both by heat (thermal degradation) and by oxygen (oxidative or thermooxidative degradation) [5]. Decomposition reactions may proceed via radical, ionic, or molecular mechanisms. Moreover, the ease of degradation of polymer is strongly related to its chemical structure. Generally, the weaker links (e.g., C–O) and some end groups (as hydroxyl groups) contribute to decrease in the thermal stability of polymers. Polymers containing strong bonds, like Si–O, Si–N, but also C–X bond (where X is halogen atom), are usually characterized by much higher thermal stability [6].

Polymer Composites with Functionalized Nanoparticles. DOI: https://doi.org/10.1016/B978-0-12-814064-2.00013-5

13.1.1.1 Thermal Degradation

Three main reactions during nonoxidative thermal degradation of polymers are depolymerization, random chains scission, and side-group elimination. In some cases, a recombination process that leads to cross-linking or cyclization may occur during degradation [7]. Usually, more than one degradation mechanism take place simultaneously.

Depolymerization (unzipping) is a free radical process reverse to polymerization, in which polymer chains are degraded to monomers, dimers, and/or oligomers at high temperature. This reaction generally starts at chain ends where the free radical is formed (initiation). Then, the monomers are sequentially detached from the main chain in the depropagation stage. Depolymerization is an unstoppable process and it lasts until the polymer chain is completely depolymerized (termination). Thermal depolymerization is characteristic for some polymers obtained by chain polymerization (e.g., poly(methyl methacrylate) (PMMA), polystyrene, and polyoxymethylene (POM)) as opposed to polycondensation polymers that do not undergo the depolymerization process [8,9].

Random chain scission is the second mechanism in the polymer thermal decomposition. In this multistep process the free radicals are randomly created along the main polymer chains, which causes the fragmentation of polymer into smaller molecules. These molecules differ in chain lengths and some of them are small enough to be volatile. Except for random scission, the end-chain scission process, in which monomers are effectively removed from the chain ends, is another possible pathway for the polymer thermal degradation. Chain scission processes drastically decrease the molecular weight and mechanical strength of polymer. Random chain scission is a typical degradation mechanism for polyolefins [10–12].

The third degradation mechanism is side-group elimination. In this reaction, the bonds between side groups and the main chain are broken. The detached side groups react with each other. In some cases the free side groups react with each other and convert into cyclic structures in the cyclization process. The molecules, formed in the elimination process, are usually volatile products of degradation [13].

13.1.1.2 Oxidative Degradation

Thermal degradation is usually accelerated by the oxidants, and as a result, the onset decomposition temperature of polymer is decreased in an atmosphere containing oxygen. The presence of oxidants may drastically change both the degradation rate and its mechanism. Oxidative degradation of polymers is a free radical chain reaction initiated by free radicals formed from peroxides and involves chain scission and cross-linking [14,15]. Some polymers, like PMMA, are less susceptible to the presence of oxidants at elevated temperatures, whereas, for example, polypropylene (PP) belongs to the oxidation-sensitive polymers. To protect polymers against oxidative degradation, antioxidants such as organic phosphines, amines, or sulfides are applied. They are responsible for the reduction of the free radicals formation during processing and service life of the polymer. Apart from antioxidants, heat stabilizers (based on epoxides, calcium, tin, zinc, and other metals) are used to protect the polymer against thermal decomposition both in presence or absence of oxygen [16,17].

13.1.1.3 Physical and Chemical Changes During the Thermal Decomposition of Polymers

Regardless of the thermal degradation mechanism, this process always contributes to deterioration of properties of macromolecular materials. Both irreversible physical and chemical changes occur in polymer due to the exposure to high temperatures. As macromolecules degrade, the molar mass of polymer decreases and the chemical structure of polymer also changes. Consequently, the mechanical properties of polymers, including tensile strength, are impaired. Moreover, the degradation process leads to viscosity loss, embrittlement, and changes in hardness of polymer. When the degradation degree is high enough, some visible changes on the polymer surface, such as changes in color (yellowing, browning) and texture, cracks, pores, or reduction of polymer transparency or gloss, are observed [18–21].

13.1.2 Thermal Stability and Heat Resistance of Polymers

Thermal stability of polymer is defined as the ability of the polymeric material to resist the action of heat [22] and to maintain its properties, such as strength, toughness, or elasticity at given temperature [23]. The thermal stability of polymers is usually determined by thermogravimetric analysis (TGA). The weight loss, as a function of temperature, is monitored and some characteristic temperatures as $T_{x\%}$ (where x is a mass loss, e.g., $x = 1\%$, 3%, 5%, 10%, 20%, 50%) are determined from the thermogravimetric (TG) profile [24]. Thermal stability of polymer depends very much on its chemical structure, degree of crystallinity, and molecular weight. It is well known that aromatic structures in the polymer backbone and cross-linking processes improve the thermal stability of polymers. On the other hand, double bonds or oxygen-containing structures in the main chain make polymers less resistant to high temperatures [25].

13.2 Effect of Functionalized Nanoparticles on the Polymer Nanocomposites' Stability

As it is commonly known, inorganic nanoparticles (nanoclays, metal oxides, carbon-based nanostructures, etc.) can improve the thermal stability of polymers. However, the compatibility between inorganic nanoadditives and polymer matrix in nanocomposites is often quite poor. The nanoparticles are characterized by high surface to volume ratio, therefore the intermolecular forces (such as electrostatic interactions) between nanoparticles are much stronger than between micron-size particles [26]. This can lead to nanoparticles aggregation and, in consequence, the dispersion of nanoadditive in polymer matrix is very difficult [27]. Consequently, the influence of nanoadditive on the improvement of composite material thermal stability is limited.

To enhance the dispersibility and the compatibility between nanoparticles and polymer, chemical modifications of nanoparticle surface are performed. The inorganic–organic hybrid materials are obtained in the grafting process. This procedure allows to graft polymer chains or other compounds onto the nanoparticles' surface via covalent bonds [28]. There are two main strategies to create grafted nanoparticles: "grafting to" and "grafting from." In the grafting-to procedure, the end-functionalized polymer chains are directly bonded to the nanoparticles via reactive groups, as hydroxyl groups, that are present on the surface of nanoparticles [29]. In the grafting-from method the initiator or chain transfer agent is attached to the surface of nanoparticles at first, and then the polymer chains propagate from the nanoparticles [30,31].

13.3 Decomposition of Polymer Nanocomposites With Functionalized Nanoparticles

Nowadays, functionalized nanoparticles and polymer nanocomposites containing such moieties, characterized by improved thermal stability, are the subject of intensive research in both academia and industry. The influence of functionalized nanoparticles on the decomposition process and thermal stability of polymer matrices, both thermoplastics and thermosets, as well as elastomers is extensively investigated. This chapter reviews the current state of the art in the area of polymer nanocomposites with functionalized nanoparticles and their impact on the decomposition behavior of selected polymers.

13.3.1 Inorganic Fillers

13.3.1.1 Nanoclays

Nanoclays are nanoparticles in the form of thin sheets of mineral silicates with dimension of 50–200 nm and thickness of 1 nm [32,33]. The common types of the layered silicates are mica, fluoromica, hectorite, fluorohectorite, saponite, etc., but in the nanocomposites field considerable attention is paid to montmorillonite (MMT), which belongs to the structural family known as the 2:1 phyllosilicates [34]. These silicates are widely used to improve mechanical, barrier, and thermal properties of composite materials [35]. In general, most polymers possess hydrophobic character and to improve adhesion between nanoclays and polymer matrix, as well as to facilitate intercalation and exfoliation of clay layers, the clay has to be modified with a surfactant to make the space between galleries sufficiently hydrophobic to allow it to interact with the polymer chains. As organic modifiers, ammonium, imidazolium, phosphonium, and stibonium compounds are mainly used [34,36]. In this section, the influence of different organoclays on the thermal decomposition behavior of polymer/nanoclay nanocomposite will be discussed.

MONTMORILLONITE

MMT is the most common and the best studied clay, which has been used in polymer nanocomposites for almost three decades. MMT belongs to the smectite family of clays and it has 2:1 layer structure, which means that the structure unit of MMT consists of two tetrahedral silica layers separated by an octahedral alumina layer [37]. The chemical structure of MMT is defined as $M_x(Al_{4-x}Mg_x)Si_8O_{20}(OH)_4$, where M is a monovalent cation, x is the degree of isomorphous substitution ranging from 0.5 to 1.3 [38]. As MMT is a hydrophilic material, its compatibility with most of the organic polymers is very poor, therefore it needs to be modified to be better dispersed in the polymer matrix [39]. Due to hydroxyl group presence in MMT's chemical structure it can be functionalized by grafting of polymer chains. Moreover, MMT is able to adsorb and exchange some cations such as Na^+ and K^+ in its interlayer spacing. The cation exchange capacity (CEC), which is the total number of cations that can be exchanged per unit mass or surface of the clay particles, ranges from 80 to 100 mEq/100 g for MMT [40]. Therefore, the cation exchange process is also used for the modification of MMT with organic compounds [41].

Cheng et al. [42] studied the effect of 4,4′-diphenylmethane diisocyanate (MDI)-grafted-organic montmorillonite (MOMMT) on selected properties of polyurethane (PU)/MOMMT nanocomposites. Ca-MMT was previously modified with cetyltrimethyl ammonium bromide to obtain organic montmorillonite (OMMT) and then the OMMT was grafted with MDI MOMMT. The matrix of the nanocomposite was composed of elastomeric PU made of MDI/poly(propylene glycol)/1,4-butanediol (1,4-BDO) with molar ratio of 3/1/2. The good dispersion of the nanoadditive in the PU matrix TEM microphotograph was confirmed in TEM observations. TGA of PU, PU/OMMT, and PU/MOMMT nanocomposites confirmed PU decomposition as a two-step process. Hard segments of PU degrade at first, and then the degradation step from soft segments is observed. The results from TGA are listed in Table 13–1. The initial degradation temperature for PU/MOMMT composites was more than 10°C higher in the first step (T_{I}^{i}) and 37°C higher in the second degradation step (T_{II}^{i}) in comparison with pristine PU. This improvement is most likely due to barrier effect of the silicate layers. In the first step, PU/OMMT nanocomposites were more stable against PU/MOMMT because of the acid catalytic effect of protonated silicate layers and thermal degradation of the swelling agent. At higher temperatures, PU/MOMMT nanocomposites were more thermally stable, because of the increasing barrier effect ("labyrinth" effect), which hinders diffusion of volatile degradation products. Moreover, the interactions between the surface of silicate layers and PU macrochains lower the molecular mobility.

Table 13–1 TGA Results for PU, PU/OMMT, and PU/MOMMT Nanocomposites

Sample	T_{I}^{i}(°C)	T_{I}^{max}(°C)	T_{I}^{i}(°C)	T_{II}^{max}(°C)
PU	277	292	347	376
PU//MMT(98/2)	301	321	382	399
PU//MMT(95/5)	300	323	382	399
PU//MMT(92/8)	302	322	383	400
PU//MMT(90/10)	302	322	383	401
PU/MOMMT(98/2)	290	314	385	401
PU/MOMMT(95/5)	289	311	385	401
PU/MOMMT(92/8)	288	309	389	403
PU/MOMMT(90/10)	288	309	389	404

T_{I}^{i} and T_{II}^{i} are the initial temperatures of weight loss in the first and second step, T_{I}^{max} and T_{II}^{max} correspond to the maximum rate of weight loss.

Source: Reprinted from Cheng A, Wu S, Jiang D, Wu F, Shen J. Study of elastomeric polyurethane nanocomposites prepared from grafted organic–montmorillonite. Colloid Polym Sci 2006;284(9):1057–61. With permission from Springer.

Mansoori et al. [43] successfully grafted the PMMA onto organophilic MMT. It was a two-step grafting process: MMT was silylated with vinyltrichlorosilane at first, and then it was free radically copolymerized with methyl methacrylate (MMA). The PMMA chains were grafted to the edges of the layered silicate via a strong Si—C linkage (Fig. 13—1) and polymer—clay nanocomposites (PCN) were obtained with 1, 3, and 5 wt.% PMMA-grafted MMT.

Thermal degradation behavior of pristine PMMA and PCN was investigated using the TGA method. Thermal decomposition of pure PMMA (A) is a three-step process (Fig. 13—2). The first step occurs at around 100°C and is related to decomposition of head-to-head configuration and evaporation of impurities and solvent residue. The second degradation step is due to end-chain saturation and it starts at c. 280°C. The final stage of degradation starts at c. 350°C and it is decomposition of the main chains of PMMA.

An additional step at about 166°C for PMMA nanocomposites was observed. This step has been attributed to the decomposition of grafted PMMA chains and the quaternary alkylammonium ions incorporated in MMT. The maximum weight loss in this step was observed for PCN 1% (by 18%) and the extent of weight loss decreases with increasing the PMMA-grafted MMT content (about 10.6 and 9.6 wt.% for PCNs 3% and 5%, respectively). This phenomenon has been explained by massive grafting of polymer chains onto inorganic substrate at lower clay loadings due to better dispersion. The temperature for end-chain saturation (T_{II}) was not significantly affected by the nanoadditive content. However, the temperature at the beginning of the random scission of the polymer chains (T_{III}) and the temperature of the maximum rate of weight loss of the third step (T_{III}^{max}) for this stage increased with an increase of the nanoadditive content.

Leszczyńska et al. [44] investigated the effect of a PU on the thermal properties of POM/OMMT nanocomposites. They found that the thermal stability of POM/MMT nanocomposites was improved by PU both in oxidative and inert atmosphere. Thermogravimetry-Fourier Transform Infrared Spectroscopy (TG-FTIR) and Thermogravimetry-Mass Spectrometry (TG-MS) studies revealed less intensive evolution of formaldehyde and formic acid during the thermal degradation of the obtained systems. Moreover, in case of thermooxidative degradation T_{DTGmax} has been moved to higher temperature more than it could be predicted from the TGA results obtained for neat POM, POM/thermoplastic polyurethane blend, and POM/OMMT. The observed effect was ascribed to the physical barrier effect.

In another work sodium MMT and organophilized MMT were modified by adsorption of disodium H-phosphonate (DHP). It has been found that the adsorption of DHP on MMT and OMMT delays the onset of the degradation temperature of PA6 by 8°C and 23°C, respectively and additionally, the flame resistance of nanocomposite with 0.75 wt.% of DHP stabilizer adsorbed on OMMT surface was significantly improved [45].

KAOLINITE

Kaolinite is aluminosilicate clay and its layer structure is of the 1:1 type. The basic structural unit of kaolinite consists of one tetrahedral (Si—O) sheet and one octahedral (Al—O) layer; the stoichiometric formula is $Al_2Si_2O_5(OH)_4$ [46]. It has quite low CEC of 3—15 mEq/100 g, compared with MMT [47].

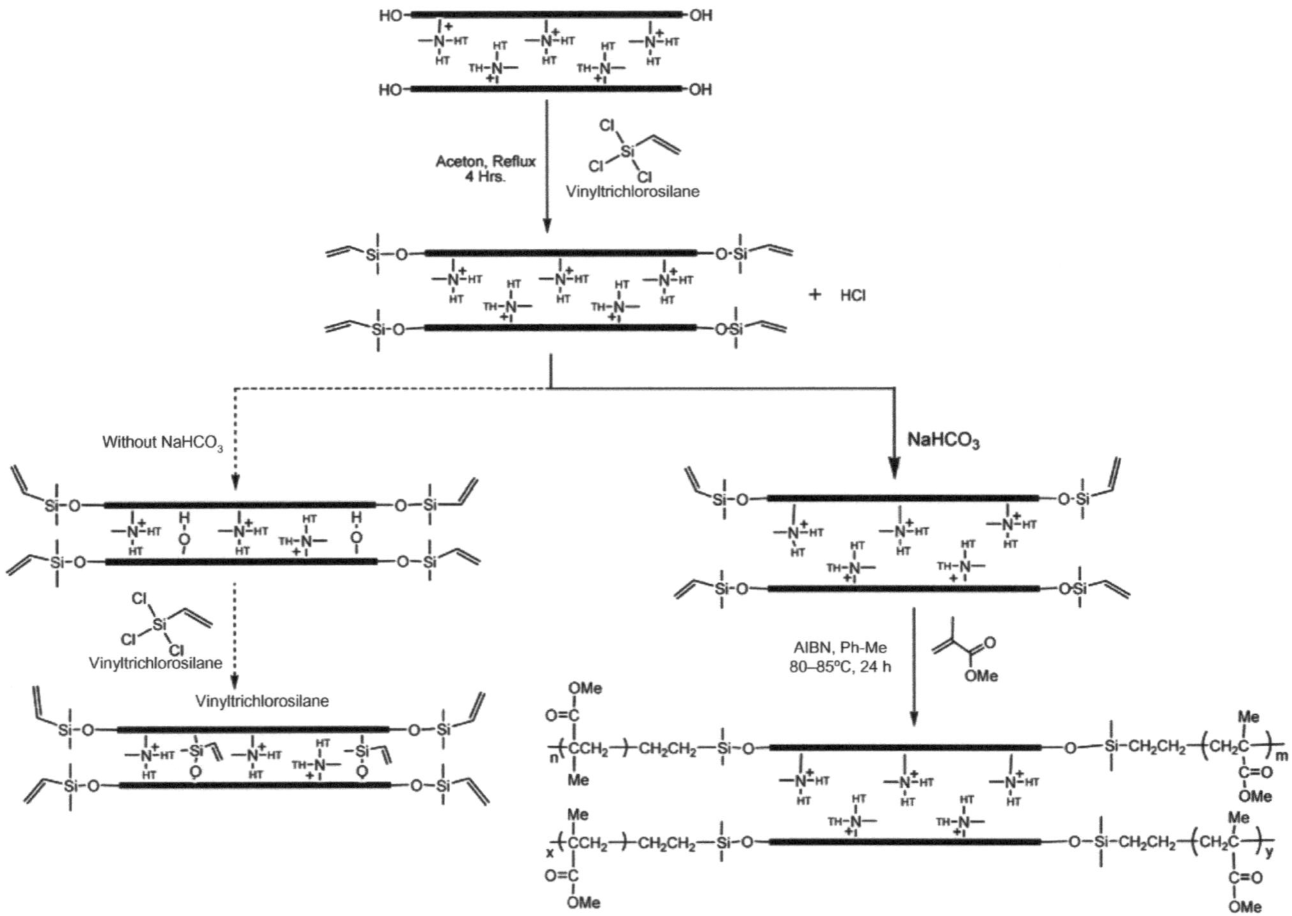

FIGURE 13–1 Free-radical grafting of PMMA onto organophilic MMT. *Reprinted from Mansoori Y, Atghia SV, Sanaei SS, Zamanloo MR, Imanzadeh G. PMMA-clay nanocomposite materials: free-radically grafting of PMMA onto organophilic montmorillonite (20A). Macromol Res 2010;18(12):1174–81. With permission from Springer.*

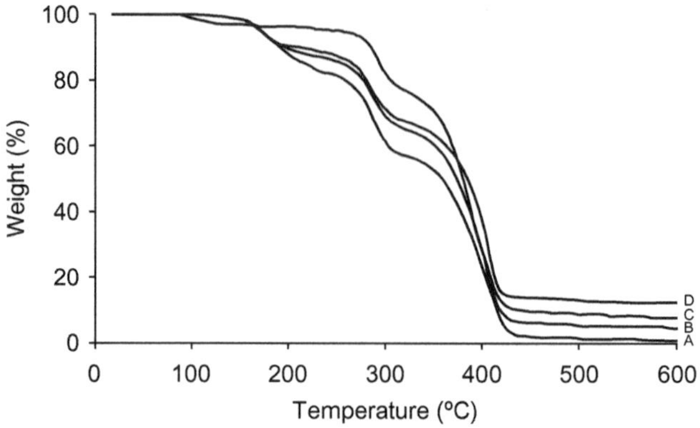

FIGURE 13–2 TGA thermograms (heating rate of 10°C/min under nitrogen flow) of (A) PMMA, (B) PCN 1%, (C) PCN 3%, and (D) PCN 5%. *Reprinted from Mansoori Y, Atghia SV, Sanaei SS, Zamanloo MR, Imanzadeh G. PMMA-clay nanocomposite materials: free-radically grafting of PMMA onto organophilic montmorillonite (20A). Macromol Res 2010;18(12):1174–81. With permission from Springer.*

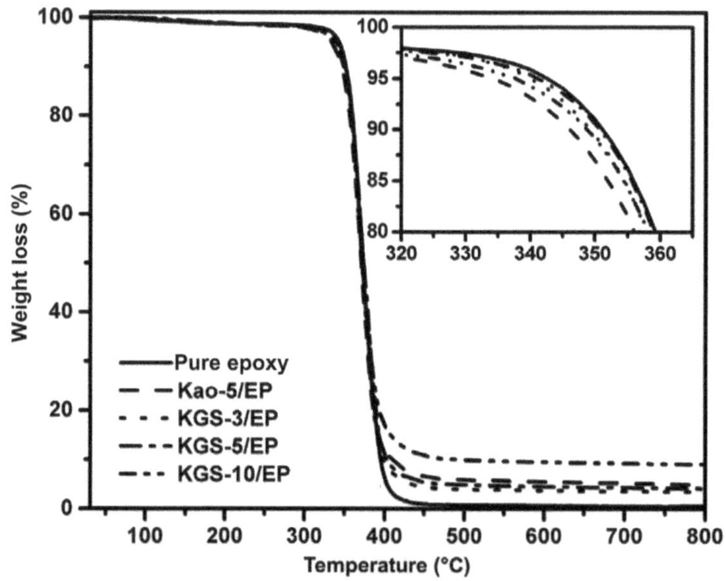

FIGURE 13–3 TG results for pure epoxy resin, 5.0 wt.% unmodified kaolinite/epoxy resin (Kao-5/EP) and 3, 5, 10 wt.% functionalized kaolinite/epoxy resin nanocomposites (KGS-3,5 or 10/EP). *Reprinted from Su L, Zeng X, He H, Tao Q, Komarneni S. Preparation of functionalized kaolinite/epoxy resin nanocomposites with enhanced thermal properties. Appl Clay Sci 2017;148:103–8.With permission from Elsevier.*

Functionalized kaolinite (KGS)/epoxy resin nanocomposites were prepared by Su et al. [48], who modified kaolinite with 3-aminopropyltriethoxysilane (APTES) in a silylation reaction. The KGS loadings in nanocomposites were 3, 5, and 10 wt.%, respectively. TG analysis results are shown in Fig. 13–3.

TG analysis revealed that for unmodified epoxy resin, the thermal decomposition that is attributed to the degradation of methylene groups in the epoxy resin network starts at c. 340°C. Along with temperature increase, a rapid mass loss corresponding to the degradation of alkyl chains was observed, and at temperatures above 400°C the thermal decomposition of epoxy resin was nearly completed. The addition of 5% of unmodified or functionalized kaolinite accelerates slightly the thermal decomposition of epoxy resin. It starts at the temperature of 335°C and 340°C, respectively. This undesired effect was possibly the result of the decreased cross-linking density of epoxy resins caused by the addition of kaolinite. Nevertheless, the functionalized kaolinite/epoxy nanocomposites showed still higher thermal stability than raw kaolinite/epoxy materials.

Albach et al. [49] prepared *n*-octyltriethoxysilane (OCTEO)-grafted kaolinite/isotactic polypropylene (iPP) nanocomposites. The raw kaolinite was functionalized with 10 wt.% (KO10) and 20 wt.% (KO20) of OCTEO and its possible structure is shown in Fig. 13–4. The iPP was modified with 5, 10, and 15 wt.% (K5, K10, and K15) of functionalized kaolinite.

Thermal decomposition behavior of iPP and its nanocomposites was investigated using TGA; based on the obtained results kinetic analysis was performed by Friedman method to determine the activation energy (E_a). It has been found that the initial thermal decomposition temperature of iPP/kaolinite composites and iPP, as well as the temperature of mass loss in T50% and T80%, decreased. Activation energy (E_a) of iPP degradation increases after introducing kaolinite and functionalized kaolinite. It was postulated that the improvement in the thermal stability and E_a can be attributed to the good dispersion of kaolinite particles in the polymer matrix that restrict the easy diffusion of volatiles from the bulk.

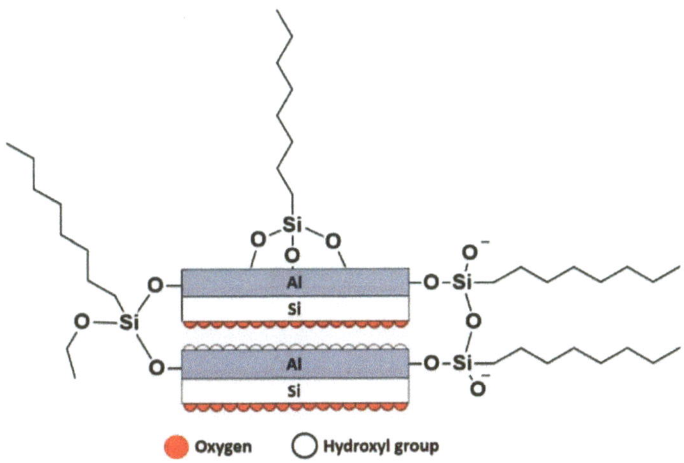

FIGURE 13–4 The possible structure of the silanized kaolinite showing the grafts at the edges and on the surface of the kaolinite layers. *Reprinted from Albach B, Munaro M, Santos PH, Zawadzki SF, Schreiner WH, Rampon DS, et al. Thermal, mechanical, and water vapor barrier behavior of polypropylene composite containing modified kaolinite. J Appl Polym Sci 2018;135(5):45785–45. With permission from Wiley.*

13.3.1.2 Silicon-Based Nanomaterials

Silicon-based nanoparticles are broadly used in polymer nanocomposites to increase their mechanical strength, melting temperature, tribological properties, bioactivity, etc. But only well-dispersed reinforcement in polymer matrix gives a considerable improvement in properties of composite even at very low filler volume content (1%–5%) [50]. Apart from high dispersion degree of nanoadditive in polymer matrix, the good interfacial interaction between inorganic nanofiller and organic matrix is a crucial issue [51]. To achieve both challenges, the surface of ceramic nanoparticles is modified with organic compounds.

SILICON AND SILICA

The silicon-containing compounds are covalently functionalized with various organic chains and then incorporated into different polymer matrices.

Baig et al. [52] obtained polyamide (PA) nanocomposite membranes modified by functionalized silicon nanoparticles (~Ø 15 nm). First, the amorphous silicon nanoparticles were synthesized using inductively coupled plasma method and next, the Si nanoparticles were functionalized with hydroxyl groups by dispersion in ethanol–water (ET) and isopropyl alcohol–water (IPA) solvents. The as-synthetized nanoparticles were incorporated into the PA thin layer, which was obtained in the interfacial polymerization reaction between the *m*-phenylenediamine, aqueous phase monomer containing nanoparticles, and the organic phase 1,3,5-benzenetricarbonyl trichloride monomer, on the inner surface of the polysulfone (PSU) hollow fiber membrane, as presented in Fig. 13–5.

The thermal stability of unmodified PSU and the PSU/PA-0.5% Si-OH (ET 0.5 and IPA 0.5) membranes were investigated using TGA method and results are given in Table 13–2.

Three-step degradation profiles for all investigated membranes were observed. The first step occurs in the temperature range up to 150°C and it was related to evaporation of the volatile organic compounds and water. In the second step (200–500°C), the polymer chains degraded. In the third step, the products of degradation carbonized. For unmodified

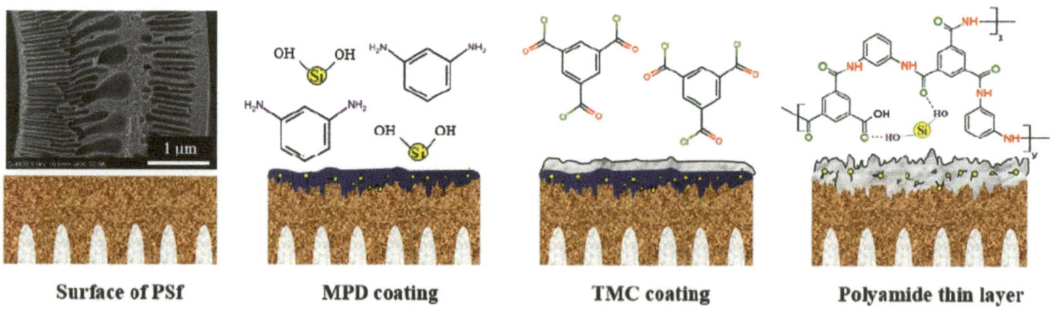

FIGURE 13–5 Schematic illustration of the formation of PA thin film containing nanoparticles on the PSU hollow fiber membrane. *Reprinted from Baig MI, Ingole PG, Jeon J-d, Hong SU, Choi WK, Jang B, et al. Water vapor selective thin film nanocomposite membranes prepared by functionalized silicon nanoparticles. Desalination 2017. With permission from Elsevier.*

Table 13–2 TGA Results for PSU and PSU/PA-Si-OH Nanocomposites

Sample	T_{I}^{i} (°C)	T_{II}^{i} (°C)	T_{III}^{i} (°C)	Char Residue at 800°C (%)
PSU	65.68 (5.66%)	234.88 (31.32%)	485.11 (39.59%)	24.37
PSU/PA-ET (0.5)	124.81 (2.17%)	208.57 (12.14%)	492.11 (24.19%)	31.70
PSU-PA-IPA (0.5)	125.26 (2.45%)	215.83 (12.26%)	496.32 (23.89%)	32.90

T_{I}^{i}, T_{II}^{i}, and T_{III}^{i} are the initial temperatures of weight loss in the first, second, and the third degradation step [52].

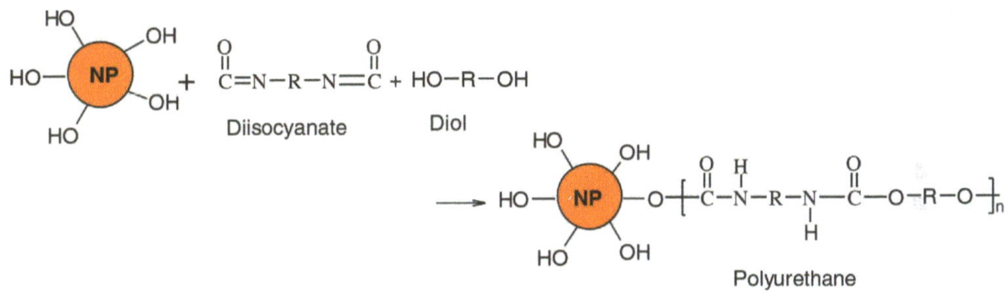

FIGURE 13–6 Scheme of the nanocomposite formation by the SIP method. *Reprinted from Guo Z, Kim TY, Lei K, Pereira T, Sugar JG, Hahn HT. Strengthening and thermal stabilization of polyurethane nanocomposites with silicon carbide nanoparticles by a surface-initiated-polymerization approach. Compos Sci Technol 2008;68(1):164–70. With permission from Elsevier.*

PSU, the polymer chains start breaking above 200°C due to the degradation of sulfonic groups. The biggest weight loss occurs in the temperature range 450–600°C and it corresponds to the polymer combustion. In the case of PSU/PA-Si-OH nanocomposites a strong increase in the stability of the PSU substrate was observed. The weight loss at 234°C in raw PSU is around 31% and it was reduced by approximately 12% for PSU covered with PA nanocomposite membranes. Moreover, the residual mass at 800°C was higher by 7.3% and 8.5% for PSU/PA-ET (0.5) and PSU/PA-IPA (0.5), respectively, in comparison with PSU. It indicates that the PA/Si-OH membranes improved the thermal stability of the PSU membrane. It is attributed to the silicon nanoparticles and their interaction with the PA membrane. However, the IPA silicon nanoparticles worked more effectively, as they were better dispersed in the polymer matrix.

One of the most common silicon-based nanomaterials is nanosilica. The surface of silica nanoparticles (SN) can be modified using various strategies while their basic geometry and mechanical strength are maintained. Many efforts have been made to use modified silica particles such as SiO_2-SO_3H and SiO_2-PO_3H_2 in the polymeric nanocomposites [53,54].

Silicon carbide/PU nanocomposites were obtained by a facile surface-initiated polymerization (SIP) method by Guo et al. [55] (Fig. 13–6).

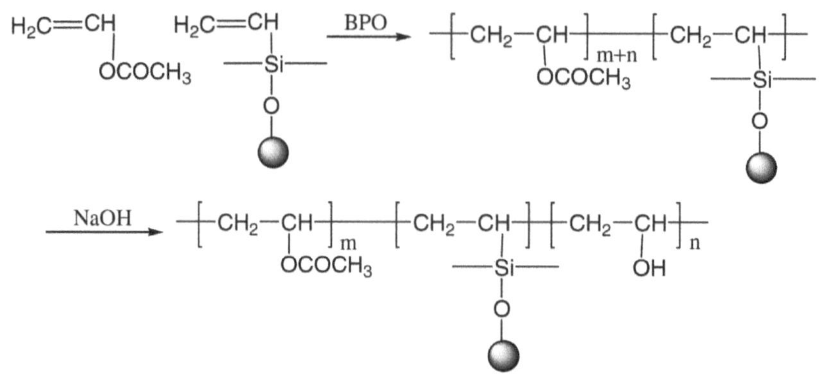

FIGURE 13–7 The process of preparation of PVA–SNs nanocomposites, where the gray circle represents the SNs. *Reprinted from Jia X, Li Y, Cheng Q, Zhang S, Zhang B. Preparation and properties of poly(vinyl alcohol)/silica nanocomposites derived from copolymerization of vinyl silica nanoparticles and vinyl acetate. Eur Polym J 2007;43 (4):1123–31. With permission from Elsevier.*

It was observed that the addition of silicon carbide nanoparticles leads to increasing of thermal stability of the PU matrix with a higher initial decomposition and complete decomposition temperature.

Another approach to silica functionalization has been presented in the work of Jia et al. [56]. The SN were functionalized with vinyltriethoxysilane. Then, the poly(vinyl alcohol) (PVA)/SN composites were obtained by in situ radical polymerization, as shown in Fig. 13–7. The concentration of vinyl SN in nanocomposites was 1, 2, 3, and 4 wt.%. The PVA/SNs nanocomposites were obtained in the form of hybrid films that were casted from the solution.

Decomposition behavior of PVA/SNs nanocomposites was investigated using TGA method (25–800°C, 20°/min, N_2 atmosphere) and the results are presented in Fig. 13–8.

Thermal degradation of PVA and its composites is a three-step process. A strong improvement in the thermal stability of PVA was observed with the increasing concentration of SNs attached to PVA main chains. The initial step of degradation corresponds to the vaporization of moisture and trapped solvent. The biggest mass loss in this region was observed for unmodified PVA. The second degradation step occurs from 250°C to 450°C and it was due to the dehydration reactions of the polymer chains. For nanocomposites containing 2 wt.% of Si nanoparticles the thermal degradation temperature was shifted toward higher temperature most of all samples. Additionally, the second mass loss decreased with the increasing content of SNs incorporated into PVA. It is most likely due to restricted mobility of PVA chains and suppressed decomposition by addition of SNs. The third step, which starts at the temperature of about 450°C, is caused by polyene residues degradation yielding carbon and hydrocarbons.

Another interesting silicon-based nanofiller is mesoporous silica characterized by high surface area thanks to ordered nanoporous structure [57]. In polymer/mesoporous silica nanocomposites macromolecules can locate in nanopores that lead to formation of materials

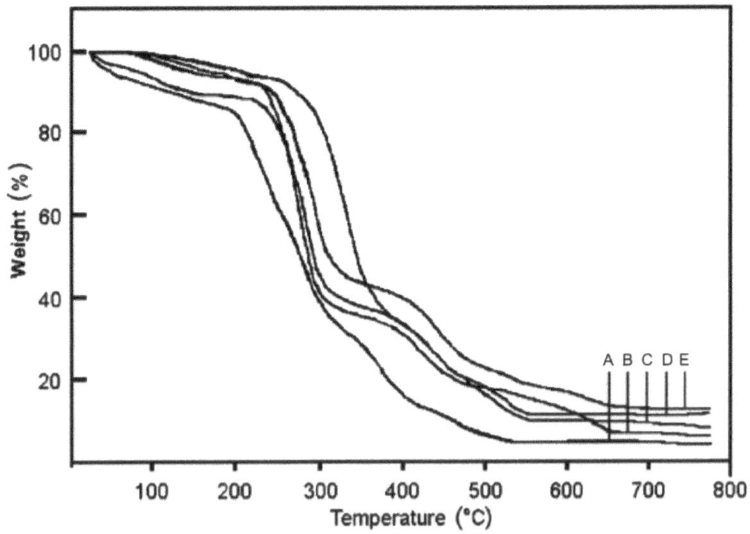

FIGURE 13–8 TGA curves of PVA and PVA–SNs: (A) PVA, (B) PVA/SNs1, (C) PVA/SNs2, (D) PVA/SNs3, and (E) PVA/SNs4. *Reprinted from Jia X, Li Y, Cheng Q, Zhang S, Zhang B. Preparation and properties of poly(vinyl alcohol)/silica nanocomposites derived from copolymerization of vinyl silica nanoparticles and vinyl acetate. Eur Polym J 2007;43 (4):1123–31. With permission from Elsevier.*

with novel properties as compared with nonporous fillers. Silica may show a hexagonal or cubic arrangement of mesopores, as well as a laminar structure [58]. For functionalization of mesoporous silica and preparation of porous hybrid materials three pathways are available: grafting, cocondensation, and production of periodic mesoporous organosilicas (Fig. 13–9).

Mesoporous silica/phenolic resin hybrid materials have been fabricated by Yu et al. [59] through in situ condensation polymerization of phenol and formaldehyde in the presence of silica. High thermal stability of the materials obtained has been attributed to the strong bonding between the components. Similar effect was observed when silica was modified using γ-glycidyloxypropyltrimethoxysilane as a coupling agent and nanocomposites were prepared via in situ polymerization [60].

Ji et al. [61] used organic-modified mesoporous silica particles as fillers to form organic/inorganic poly(3-trimethoxysilyl)propyl methacrylate (TMSPMA)-based nanocomposites. Mesoporous nanosilica was functionalized by coassembly of surfactant and silicate species prepared by hydrolysis and condensation reactions of tetraethoxysilane and 3-TMSPMA through an aerosol process. Next, infiltration and subsequent in situ polymerization of 3-TMSPMA in the presence of mesoporous silica particles were performed. Chemical bonding and strong interactions between the polymer and filler, confined segmental motion of the polymer chains within the mesoporous channels, and the use of the silica particles as pseudo-cross-linking agents lead to improvement of thermal and mechanical properties.

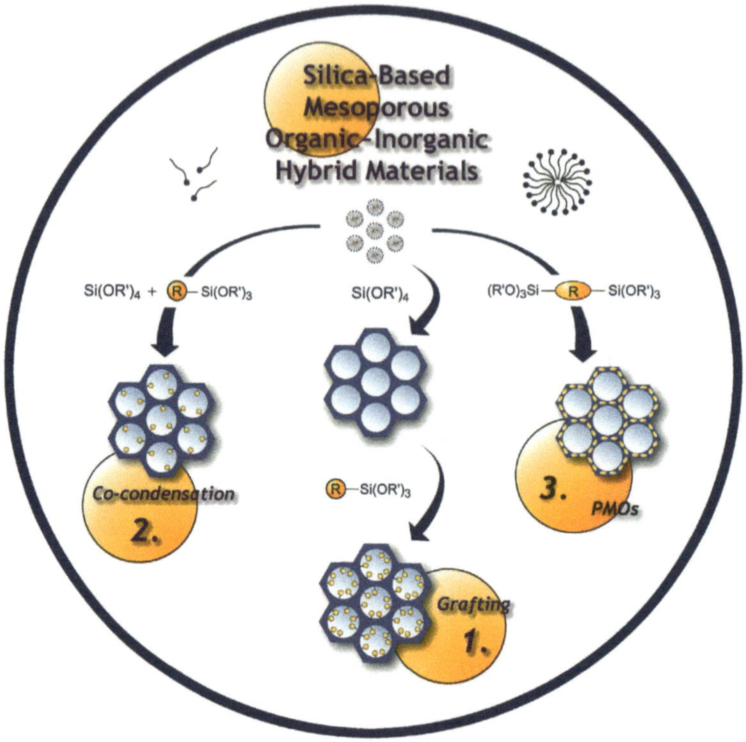

FIGURE 13–9 Available pathways for the synthesis of porous hybrid materials based on organosilica units. *Reprinted from Hoffmann F, Cornelius M, Morell J, Fröba M. Silica-based mesoporous organic–inorganic hybrid materials. Angew Chem Int Ed 2006;45(20):3216–51. With permission from Wiley.*

POLYHEDRAL OLIGOSILSESQUIOXANES

Polyhedral oligomeric silsesquioxanes (POSS) are nanostructures having the empirical formula $R_n(\mathrm{SiO1.5})_n$, where R is a hydrogen atom or an organic functional group such as an alkyl, alkylene, acrylate, hydroxyl, or epoxide unit with nanosized cage structures having dimensions comparable with those of most polymeric segments or coils [62,63]. Polymer/POSS systems have been widely investigated, since POSS molecules can be easily introduced into polymer matrix both by physical mixing and by chemical bonding through, for example, copolymerization or grafting. Different substituents may be fixed onto the Si atoms at the corners of the cages to optimize polymer–POSS interactions and to improve solubility of the POSS moieties in common solvents for synthesis purposes [64]. It has been reported that the incorporation of POSS to polymer matrix may lead to enhancements in the thermal stability.

Hence, Lin and Song [65] investigated cyanate ester resin (CY)/POSS nanocomposites. Generally, the air atmosphere accelerated degradation of CY, but after incorporation of 1 wt. % POSS, the end decomposition temperature increased by 146°C. Moreover, the heat generated by the thermal degradation of the CY/POSS nanocomposites was around four times less than that of the neat CY. Different behavior has been observed under nitrogen atmosphere, where the char yield increased up to 15 wt.% with the addition of POSS and the heat and degradation activation energy required for the degradation of the CY/POSS nanocomposites were much higher than that of pure CY.

Lee et al. [66] obtained polybenzoxazine/POSS nanocomposites by ring-opening polymerization of benzoxazine. Decomposition temperature was found to increase with increasing POSS content from 0 to 5 wt.%. This effect has been attributed to the reduced mobility of the molecular chains at high temperature due to the presence of POSS moieties. Similar trend was observed for the amount of char yield, indicating formation of a protective layer on the polymer surface [67].

The effects of reactive and nonreactive POSS and their loading level on the properties of polylactide (PLA) composites were studied by Sirin et al. [68]. It has been found that under nitrogen atmosphere the initial decomposition temperature, defined as the temperature at 5% weight loss (Td,5), and maximum weight loss (T_{DTGmax}) increased with the addition of POSS particles except glycidylisobutyl-POSS. However, it should be noted that the T_{DTGmax} value of all PLA/POSS composites was higher than pure PLA.

Fina et al. [69] investigated the thermal stability of PP metal functionalized POSS nanocomposites (Al-POSS and Zn-POSS) both under inert and oxidative atmosphere. It has been revealed that the incorporation of metal containing POSS strongly enhances the thermooxidative stability of PP (Fig. 13–10) by retarding maximum weight loss rate temperature by 50–60°C for composites containing 10 wt.% of metal POSS.

Significant increases in the thermal stability were observed for PU/POSS nanocomposites [70]. PU elastomers have been modified with varying levels of 1,2-propanediol-heptaisobutyl-POSS (PHIPOSS) as a chain extender unit, replacing 1,4-BDO. The thermal stability of PU/PHIPOSS indicated that covalent bonding of POSS to the PU chains network increases its nonoxidative thermal stability. Thermal volatilization analysis results have demonstrated that PU/PHIPOSS are more thermally stable than the unmodified PU matrix with evolving significantly lower levels of volatile degradation products and significant increase (c. 30°C) in onset degradation temperature [71]. They also revealed that the inclusion of POSS in the PU network changes the degradation mechanism of both the soft and hard segments of the PU elastomers. In other works of the same research group [72,73] the thermal degradation behavior of nanohybrid PU/functionalized POSS elastomers was studied. Three different types of functionalized POSS: 1,2-propanediolizobutyl POSS (PDI-POSS), disilanollsobutyl POSS (DSI-POSS), and octa(3-hydroxy-3-methylbutyldimethylsiloxy) POSS (OCTA-POSS) were used to obtain PU with POSS as pendant group (PDI-POSS), part of the main chain (DSI-POSS) and cross-link (OCTA-POSS). PU elastomers under O_2 atmosphere decompose in three separate steps with the maximum rate of weight loss of each step at the temperatures of 330°C, 420°C, and 536–595°C, respectively. At temperature of about 300°C the urethane bonds are broken and aromatic (primary) amines and CO_2 are formed. The next stage (350–450°C) is related to the degradation process of soft segments. The final step is due to oxidation of condensed aromatic residue. The initial decomposition temperature of PU/POSS materials is hardly affected by the presence of functionalized POSS. However, the char residue of hybrid materials was slightly higher than for pure PU, which is most probably an effect of formation of silicate oxidized char residue that acts as an insulating barrier limiting heat and mass

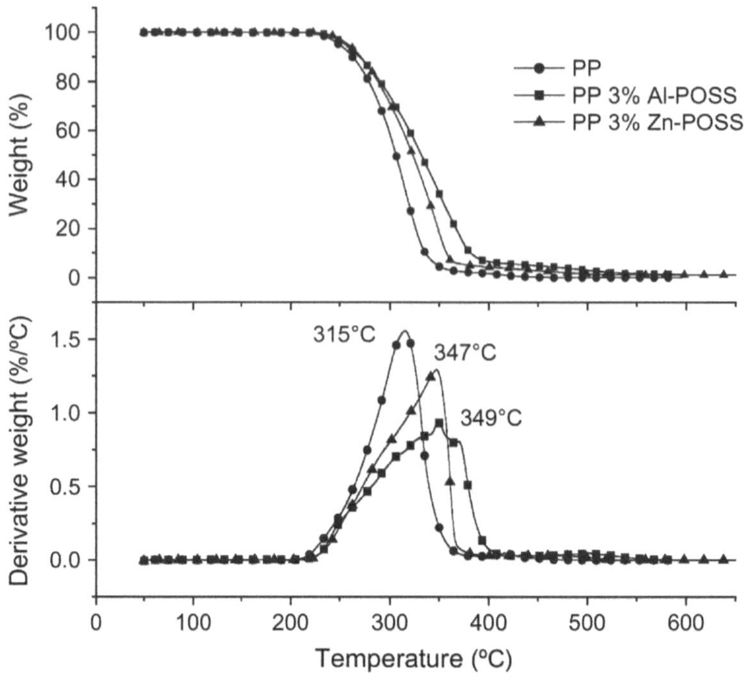

FIGURE 13–10 TGA in air for PP-3% metal-isobutyl POSS. *Reprinted from Fina A, Abbenhuis HCL, Tabuani D, Frache A, Camino G. Polypropylene metal functionalised POSS nanocomposites: a study by thermogravimetric analysis. Polym Degrad Stab 2006;91(5):1064–70. With permission from Elsevier.*

transfer. It could be also assumed that PU chains are immobilized by the covalently bonded POSS substituents and thermally induced motions are limited.

Sanchez-Soto et al. [74] investigated the influence of the functionalization of fully condensed POSS cages with different organic substituents such as glycidylethyl, aminopropylisobutyl, and poly(ethylene glycol) (PEG) on the properties of POM based nanocomposites. The temperatures at 5 wt.% weight loss ($T_{5\%}$) and of the maximum weight loss rate (T_{DTGmax}), as well as char residue at 600°C for POM/POSS composites decomposition under nitrogen atmosphere are displayed in Table 13–3.

The authors found that the thermal stability of POM under inert gas atmosphere significantly increased with the addition of any of the three POSS systems tested. The highest enhancement in thermal stability, up to 50°C higher than that of POM, was observed after addition of apib-POSS.

13.3.1.3 Other

HYDROXYAPATITE

Bioactive nanoparticles, such as hydroxyapatite (HAp), tricalcium phosphate, bioactive glass, or calcium silicate, are broadly used in polymer/bioceramics nanocomposites for bone tissue engineering. The major challenge is to increase the compatibility of inorganic bioceramics and organic polymer matrix [75,76] through properly designed functionalization.

Table 13–3 Temperatures of 5 wt.% Weight Loss ($T_{5\%}$), of Maximum Weight Loss Rate (T_{DTGmax}), and the Fraction of the Solid Residue at 600°C for POM/POSS Composites

Material	[Substituent] wt.%	$T_{5\%}$ °C	T_{max} °C	Solid Residue wt.%
POM		268	321	0
ge-POSS	–	239	296	15.4
	2.5	286	341	0.1
	5	291	344	1.4
	10	285	350	2.4
apib-POSS		261	306	19.5
	–	321	369	2.3
	521.5	322	376	3.5
	10	323	379	3.8
Peg-POSS	–	244	299	10.6
	2.5	279	320	2.6
	5	281	328	2.8
	10	284	330	3.6

Source: Reprinted from Sanchez-Soto M, Illescas S, Milliman H, Schiraldi DA, Arostegui A, Morphology and thermomechanical properties of melt-mixed polyoxymethylene/polyhedral oligomeric silsesquioxane nanocomposites. Macromol Mater Eng 2010;295 (9):846–58. With permission from Wiley.

For instance, hydroxyapatite possesses in its structure free hydroxyl groups that can be used for functionalization and grafting of other low molecular weight molecules or polymer chains. Mirhosseini et al. [77] functionalized hydroxyapatite nanoparticles (HAp) by immobilizing Pluronic F127 (a symmetric triblock copolymer composed of poly(ethylene oxide) and poly(propylene oxide) chains with molar mass of 12,600 g/mol) by using maleic anhydride. Next, HAp and HA-F127 were introduced to PCL/Pluronic P123 ($M = 5600$ g/mol) solution, to obtain HAp/PCL/P123 and HApF127/PCL/P123 solutions, respectively, and then fibers were obtained by electrospinning. Immobilization of Pluronic F127 chains on the HAp surface decreased the agglomeration of modified nanoparticles and improved their dispersion in polycaprolactone. Moreover, from TGA results it was found that, compared with pristine copolymer degradation temperature, the mass loss of attached Pluronic P127 chains in HApF127/PCL/P123 shifts to higher temperature and the thermal stability remarkably increases from 245°C to 300°C. This effect was attributed to a decrease in mobility of copolymer chains grafted to HAp nanoparticles.

Waterborne polyurethane (WBPU)/HAp nanocomposites were obtained through in situ polymerization by Zhao et al. [78]. In the first step HAp nanoparticles (HAp NPs) were reacted with 3-isocyanatemethyl-3,5,5-trimethyl-cyclohexylisocyanate (isophorone diisocyanate) to obtain grafted HAp NPs containing free isocyanate groups (HAp-g-NCO). In the next step HAp-g-NCO and IPDI were polymerized with polyester diol component to form the WBPU/HAp nanocomposites, as presented in Fig. 13–11.

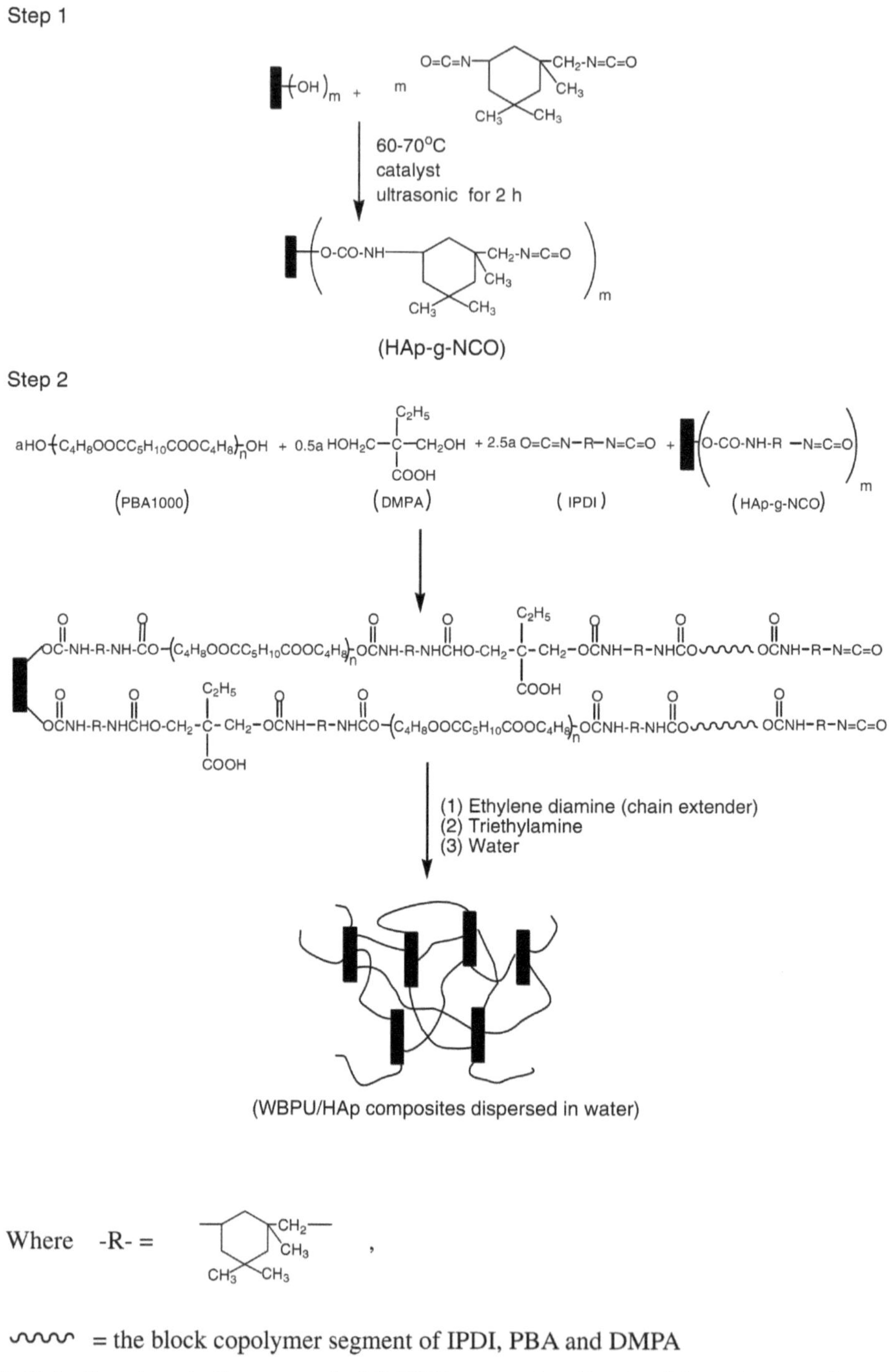

FIGURE 13–11 In situ polymerization process for WBPU/HAp composites. *Reprinted from Zhao C-X, Zhang W-D. Preparation of waterborne polyurethane nanocomposites: polymerization from functionalized hydroxyapatite. Eur Polym J 2008;44(7):1988–95.With permission from Elsevier.*

TG measurement results show that the thermal degradation mainly proceeded in two distinctive steps, corresponding to the soft and hard segments of WBPU decomposition, however, WBPU/HAp nanocomposites exhibited increases in thermal stability with the increase of HAp NPs content compared with unmodified WBPU. The onset decomposition temperature of WBPU was 292 while for WBPU/HAp nanocomposites 310°C. It has been postulated that this effect may be due to the formation of hydrogen bonds between the urethane groups of WBPU and unreacted free hydroxyl groups of HAp NPs, as well as cross-linking reactions.

In POM/HAp nanocomposites it was observed that the introduction of higher amounts of HAp causes decomposition of POM matrix, and changes the mechanism of POM degradation, including its kinetics as kinetic model of autocatalytic reaction (pristine POM) was replaced by phase-boundary or nth order reaction models for POM/HAp nanocomposites [79,80]. Moreover, thermoanalytical data revealed that the main degradation product of POM/HAp nanocomposites reinforced by higher amounts of HAp is formaldehyde, and formation of other degradation products is significantly reduced [80]. In the next work, HAp was functionalized by grafting poly(ethylene glycol) with diisocyanate as a coupling agent (HAp-*g*-PEG) [81]. Then HAp-*g*-PEG was incorporated to POM matrix using melt processing methods. The thermal stability of POM/HA-*g*-PEG composites was evaluated by TGA and the results are given in Fig. 13–12.

As it was reported previously [80], the incorporation of unmodified HAp nanoparticles into POM matrix significantly decreased the thermal stability of POM copolymer by c. 30°C (POM/10.0% HAp). On the other hand, functionalized HAp-*g*-PEG additive significantly improved the POM thermal stability from 296°C for pure POM to 326°C for POM/10% of HAp-*g*-PEG. This advantageous effect is most probably the result of the reduction of the number of basic hydroxyl groups in Hap, which are active sites involved in the POM decomposition reactions. Replacement of hydroxyl groups in HAp by urethane groups and grafting of PEG chains effectively reduced the catalytic effect of HAp on the polyacetal thermal depolymerization process. The thermal stability increase of POM/HAp-*g*-PEG nanocomposites can also be supported by the nitrogen atoms' presence in HAp-*g*-PEG nanoadditive as it is the case in other heat stabilizers for POM, such as PA, aromatic amines, and dicyandiamines [81].

13.3.2 Carbonaceous Nanoadditives

Carbon-based nanomaterials (CBNs) are a wide group of compounds including graphene, single- and multiwalled carbon nanotubes (MWCNTs), carbon nanofibers, fullerenes, nanodiamonds, etc. [82]. Owing to the outstanding combination of physical and chemical properties of CBNs such as thermal and electrical conductivity, high mechanical strength, and optical properties, they are vastly used in polymer nanocomposites [83]. However, chemical functionalization of carbonaceous nanoparticles significantly enhances the nanoadditive–polymer interface interactions and enables formation of nanoparticle networks, which improves the mechanical and thermal properties of nanocomposites [84]. Moreover, functionalized carbon nanoparticles are also easier to disperse in polymer matrix, which is of technological importance [85].

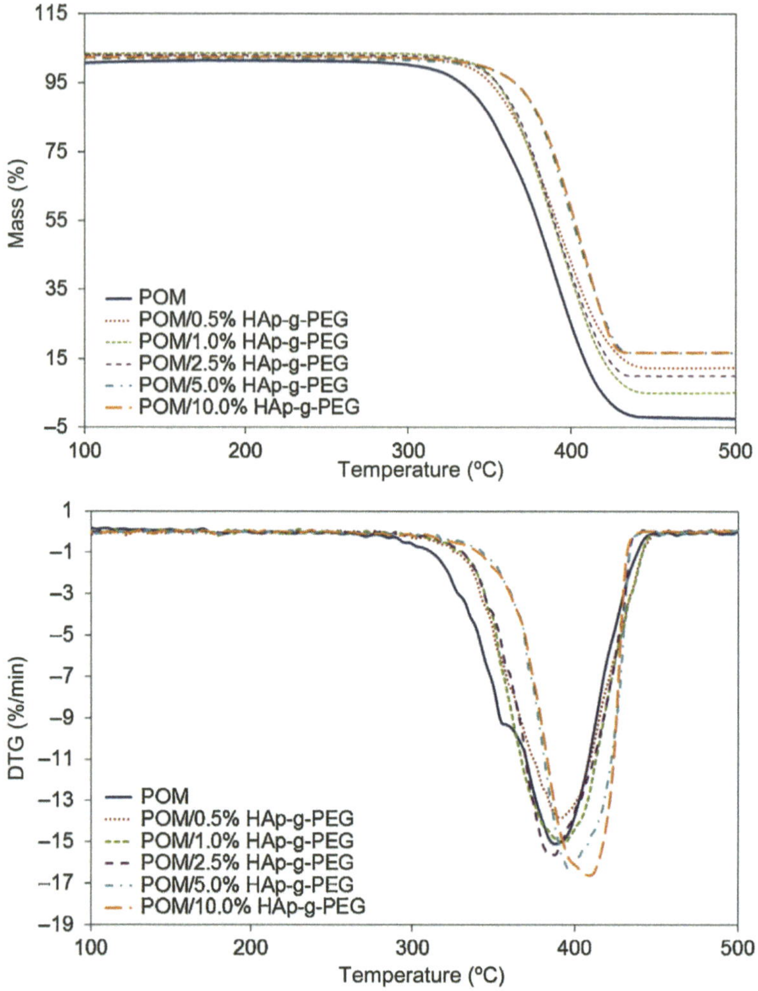

FIGURE 13–12 TG and DTG curves of POM and POM/HAp-*g*-PEG composites. *Reprinted from Pielichowska K, Król K, Majka TM. Polyoxymethylene-copolymer based composites with PEG-grafted hydroxyapatite with improved thermal stability. Thermochimica Acta 2016;633:98–107. With permission from Elsevier.*

13.3.2.1 Graphene

Functionalized graphene (FG) belongs to the new generation of nanofillers for polymer nanocomposites due to its physical properties and lower cost of fabrication in comparison with carbon nanotubes. In the last 10 years the rapid development of polymer/FG has been observed. The issue of graphene-based nanofillers in improving the thermal stability of polymers has been widely studied and it has been postulated that graphene-based nanofillers can act as barriers to prevent the propagation of heat generated from the external environment in polymeric matrices [84,86–88]. Enhancement of thermal properties in polymer nanocomposites with graphene depends on nanofiller amount and dispersion degree of nanoparticles in the polymer matrix as well as the possibility of formation of a nanofiller network [86,88].

Hence, Zhang et al. [89] investigated the influence of FG nanoplatelets (GnPs) on the thermal properties of silicone rubber (SR) composites. GnPs were modified using two different approaches: covalent bonding formation between the oxygen functional groups of GnPs and the modifiers APTES and vinyltrimethoxysilane (VTMS), and noncovalent attachment of PEG *p*-(1,1,3,3-tetramethylbutyl)-phenyl ether (Triton X-100). The functionalization process of GnPs is presented in Fig. 13–13.

Thermal decomposition of the neat SR starts at the temperature of c. 460°C and then a rapid, one-step weight loss, with its maximum rate at 528°C, corresponding to the decomposition of the siloxane main chains, occurs. For composites containing GnPs, the $T_{0.5}$ and T_{max} values are moved toward higher temperatures, probably due to the formation of a platelet barrier effect, slowing the escape of volatile decomposition products. Further, GnPs are characterized by high thermal conductivity so they might facilitate heat transport within the system. Therefore, they protect the polymer matrix against excessive heat concentration, which also enhances the thermal stability of SR nanocomposites. Moreover, the VTMS

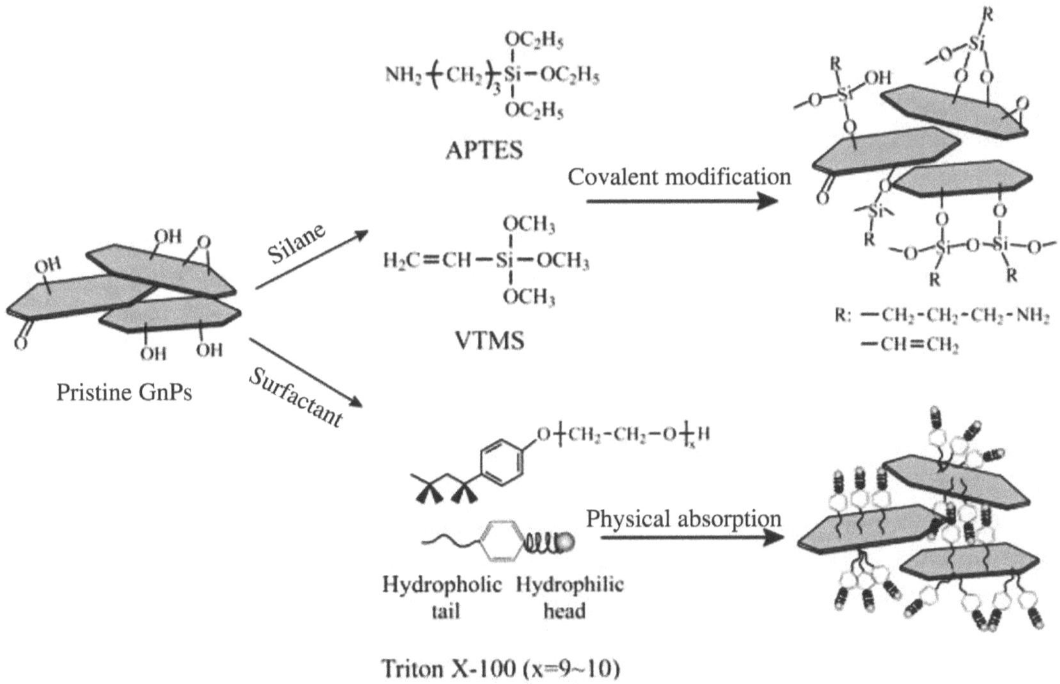

FIGURE 13–13 Schematic of the surface treatment of graphene nanoplatelets particles with different modifiers. *Reprinted from Zhang G, Wang F, Dai J, Huang Z. Effect of functionalization of graphene nanoplatelets on the mechanical and thermal properties of silicone rubber composites. Materials. 2016;9(2). With permission from MDPI.*

treated GnPs enable formation of chemical bonds with silicone and increase the rigidity of siloxane chains, which causes substantial improvements in the thermal stability of SR nanocomposites.

Nam et al. [90] performed edge-selective functionalization of graphene to obtain polyimide (PI)/edge-selectively functionalized graphene (EFG) nanocomposites. It has been observed that the decomposition temperatures increased with filler load increase. The authors attributed enhancement of the nanocomposite thermal stability to high thermal stability of EFG and to the interaction between EFG and PI chains.

Thermal stability improvement was also observed in PVA/graphene oxide nanocomposites. Both for unmodified PVA and its nanocomposite, a two-step decomposition process has been found, however, in nanocomposite the decomposition onset was shifted toward a higher temperature to that of pure PVA. The minimum on the DTG curve (T_{DTG}) represents the temperature at which the highest weight loss rate is reached. It was found that for the nanocomposite T_{DTG} appeared at c. 316.5°C and 428.6°C, which was higher than for PVA at about 5.5°C and 2°C, respectively [91].

Interestingly, an increase in the onset and the degradation temperatures by 16°C and 55°C, respectively, was observed in functionalized graphene sheet (FGS)/silicone porous nanocomposites. This effect was explained in terms of increased rigidity of the siloxane chains in the presence of FGS [92].

13.3.2.2 Carbon Nanotubes

Carbon nanotubes usually agglomerate in polymer matrix due to van der Waals forces and they are difficult to disperse. To prevent nanotube aggregation, the functionalization of nanotube (CNT) is often applied to better disperse and stabilize the dispersion of CNTs within a polymer matrix. The main methods of CNTs functionalization include defect functionalization, covalent functionalization, and noncovalent functionalization [93,94].

Along this line of interest, Wu et al. [95] investigated PLA nanocomposites containing carboxylic (c), hydroxy (h), functionalized, and purified (p) MWCNTs. It was found that the presence of all these three MWCNTs slightly increases the thermal stability of PLA at the initial stage of degradation as it can be seen in Fig. 13–14.

MWCNTs show a small effective barrier effect at the initial stage of PLA decomposition; composite with purified MWCNTs shows higher T5 wt.% compared with functionalized MWCNTs as the former is characterized by better thermal conductivity than that of functionalized MWCNTs. For higher decomposition level (e.g., at T_{DTGmax}), where the thermal conductivity and barrier effects of the MWCNTs become more effective, T_{DTGmax} of nanocomposites with MWCNTs increases by about 8–10°C compared with that of neat PLA. It can suggest that after initial stage of the decomposition, the MWCNTs are able to absorb heat and hinder transport of PLA degradation products more effectively, thus delaying the progress of decomposition process [95].

Different effect was observed when PLA chains were chemically bonded to side walls of MWCNTs, as investigated by Amirian et al. [96]. Pristine MWCNTs were functionalized by Friedel–Crafts acylation to introduce the aromatic amine groups on the sidewall of

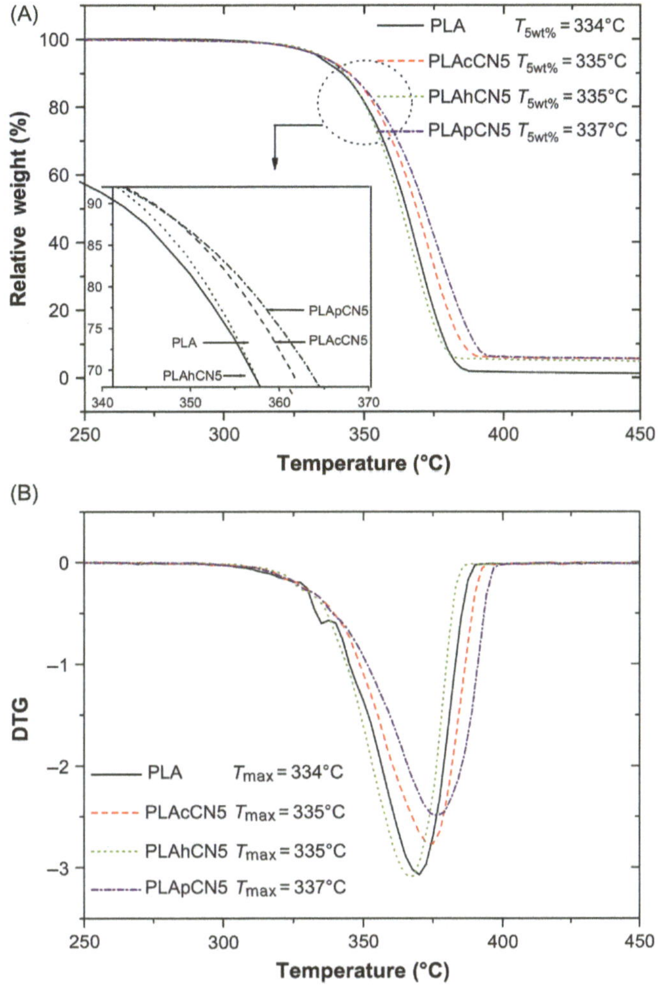

FIGURE 13–14 (A) TGA and (B) DTG curves for the neat PLA and its composites with various functionalized MWCNTs at identical loading levels of 5 wt.%. *Reprinted from Wu D, Wu L, Zhang M, Zhao Y. Viscoelasticity and thermal stability of polylactide composites with various functionalized carbon nanotubes. Polym Degrad Stab 2008;93(8):1577–84. With permission from Elsevier.*

MWCNTs and then, the PLA chains were covalently bonded to the sidewall carbon nanotubes by in situ ring-opening polymerization of L-lactide oligomer. Thermogravimetric results show a significant improvement in the thermal stability of composites with increasing the concentration of MWCNT-*g*-PLLAs in composites.

Ferreira et al. [97] presented the influence of dodecylamine (DDA) functionalized carbon CNTs on the thermal stability of high-density polyethylene (HDPE)-based composites. Multiwalled carbon CNTs have been functionalized according to a two-stage procedure: at first, the unmodified CNTs (CNT-H) were functionalized with carboxylic groups, by treating with sulfuric acid and nitric acid (H_2SO_4:HNO_3 = 3:1; the CNTs-COOH were labeled as CNT-Ac). Then, the CNT-Ac was mixed with DDA for 4 days to obtain CNT-DDA carbon nanotubes functionalized with alkyl groups. Finally, the CNT-DDA/HDPE composites, containing 0.8 wt.% of functionalized CNTs, were fabricated by melt processing method using a preprepared HDPE/CNT-DDA master batch. The addition of CNTs into HDPE matrix increased significantly both T_{onset} and T_{max}; however, the CNT-DDA/HDPE composite exhibited the highest increase in both temperatures. This phenomenon may be due to the consistent dispersion and improved interfacial interaction between the CNT-DDA nanoadditive and polyethylene matrix. The nanoadditive that is homogenously dispersed in polymer matrix can act as a strong barrier to the oxygen atoms supply during thermal decomposition of the nanocomposites according to the "tortuous path" mechanism.

13.3.3 Metal Fillers

Incorporation of metal nanoparticles into polymer matrix may impart the magnetic, optical, semiconductor, catalytic and sensing properties of polymer [98]. Some metals, such as silver, gold, zinc, and copper, exhibit antibacterial activity, therefore they are used in polymer composites for biomedical applications [99]. The extent of property improvement depends on several variables including the size of the metal filler, its degree of dispersion, and the adhesion at the filler–matrix interface. By coating the surface of metal nanoparticles with polymer, the van der Waals interactions can be reduced and the compatibility between the organic polymer and inorganic nanoparticles can be improved [26,100].

13.3.3.1 Metals

Chatterjee et al. [101] prepared silver/PMMA nanocomposites by mixing a colloidal solution of functionalized Ag nanoparticles with solution of PMMA polymer. The expected surface functionality of silver nanoparticles was achieved by using amphiphilic poly(2-(*N*,*N*-dimethylamino)ethyl methacrylate-b-(methyl methacrylate)-b-2-(*N*,*N*-dimethylamino) ethyl methacrylate) P(DMA-b-MMA-b-DMA), poly(styrene-b-poly(2-(*N*,*N*-dimethylamino)ethyl methacrylate), P(St-b-DMA), and poly(t-butyl methacrylate)-b-(2-(*N*,*N*-dimethylamino) ethyl methacrylate) P(tBMA-b-PDMA) block copolymers as stabilizers. Ag nanoparticles with hydrophobic PMMA brushes were obtained by reducing $AgNO_3$ using sodium borohydride as a reducing agent and di- or triblock copolymer of poly(2-(*N*,*N*-dimethylamino)ethyl methacrylate) (PDMA) P(DMA-b-MMA-b-DMA-Ag)/PMMA nanocomposite films with 0.32 and 0.64 wt.% of Ag were prepared by mixing a colloidal solution of P(DMA-b-MMA-b-DMA-Ag) nanoparticles with PMMA solution followed by slow evaporation of solvent.

The TGA results show that functionalized silver nanoparticles contribute to decrease in the weight loss of Ag/PMMA composites in comparison with pure PMMA. At the temperature of 360°C the weight loss for PMMA/Ag0.64, PMMA/Ag0.32 nanocomposite, and the PMMA samples were about 17%, 25%, and 30%, respectively. In addition, the thermal stability of PMMA/Ag0.64 composite increases by about 20°C at the inflection point. PMMA matrix is completely degraded above 425°C unlike PMMA/Ag nanocomposite, which remains

Table 13–4 The Data Obtained From TGA/DTG Curves of Pure PVA and PVA/TiO_2 Nanocomposites

Sample	T_5 (°C)[a]	T_{10} (°C)[b]	Char yield (%)[c]	LOI[d]
PVA	243.0	255.4	7.1	20.4
PVA/TiO_2 5 wt.%	262.1	272.7	11.7	22.1
PVA/TiO_2 10 wt.%	266.4	276.2	19.3	25.2
PVA/TiO_2 15 wt.%	267.6	279.3	21.7	26.1
PVA/TiO_2 20 wt.%	269.3	284.1	28.0	28.7

[a,b]$T_{5\%}$ and $T_{10\%}$ are the temperatures of 5% and 10% mass loss respectively.
[c]Char yield at 800°C [%].
[d]LOI - Limiting oxygen index.
Source: Reprinted from Mallakpour S, Barati A. Efficient preparation of hybrid nanocomposite coatings based on poly(vinyl alcohol) and silane coupling agent modified TiO_2 nanoparticles. Prog Org Coat. 2011;71(4):391–8. With permission from Elsevier.

undecomposed with 7 wt.% of mass remaining. These observations indicate some improvement in the thermal stability of the PMMA/Ag nanocomposites, attributed to the physical interaction of nanoadditive with the PMMA chains. Simultaneously, the thermal degradation pattern of PMMA and P(DMA-b-MMA-b-DMA)/PMMA (without Ag nanoparticle) is quite similar.

13.3.3.2 Metal Oxides

Mallakpour et al. [102] modified the surface of titanium dioxide (TiO_2) nanoparticles with γ-aminopropyltriethoxy silane used as a coupling agent. Then, PVA/TiO_2 nanocomposite coatings with different loading of modified TiO_2 were obtained under sonication. Next, the nanocomposite coatings were used for fabrication of PVA/TiO_2 films via solution casting method. In case of pure PVA, three steps of thermal decomposition process are observed.

In the first stage, in the range of 50–170°C, residual solvents evaporate, in particular the trapped water. The second weight loss, between 200°C and 350°C, corresponds to the elimination reactions of water. The final stage (400–550°C) is connected with further decomposition of polyene residues to yield the carbon and hydrocarbons. The TG and DTG curves of PVA/TiO_2 nanocomposites are almost identical to the pure PVA degradation curve. Nonetheless, the PVA/TiO_2 nanocomposites are characterized by higher thermal stability than pristine PVA (Table 13–4).

The decomposition temperature of the PVA/TiO_2 nanocomposites was slightly shifted toward higher temperatures with TiO_2 content increase. This enhancement in the thermal stability is due to the presence of nanofiller, which acts as a barrier for volatile degradation products.

13.4 Conclusions

The influence of covalently functionalized nanoparticles on the thermal decomposition behavior of polymer matrix in composite materials has been extensively investigated by many authors. Functionalized nanoparticles were found to affect the polymer behavior during heating, and in most cases they increase the thermal stability of polymer matrix due to the presence of specific interactions, barrier mechanism, or restricted chain mobility. The inorganic nanoparticles can be covalently functionalized with organic chains and the hybrid nanoadditives show a reduced tendency toward agglomeration in the polymer matrix compared with the unmodified nanoparticles. Functionalized MMT, hydroxyapatite, graphene, carbon nanotubes, or titanium dioxide can significantly improve the thermal stability of polymers as they create a barrier both to volatile degradation products and to oxygen supply. Factors such as the size and shape of the nanofiller, kind of organic decoration, degree of dispersion, and the adhesion forces at the polymer–filler interface govern the extent of thermal stability improvement. Interestingly, modified nanoparticles may add new properties to the polymer matrix, such as magnetic, catalytic, or antibacterial. They can be considered as tailor-made nanohybrid thermal stabilizers for polymer materials, which can be used together with traditional stabilizers applied in the polymer technology.

Acknowledgments

The authors are grateful to the Polish National Science Center for financial support under grant no. 2016/21/B/ST8/00449.

References

[1] Lu D, Xing Z. Study of the thermal stability of an organic polymer low-k material. Int J Mod Phys B 2002;16(28n29):4441–4.

[2] Kholodovych V, Welsh WJ. Thermal-oxidative stability and degradation of polymers. In: Mark JE, editor. Physical properties of polymers handbook. New York: Springer New York; 2007. p. 927–38.

[3] Harwood HJ, Hawkins WL. Polymer degradation and stabilization:. Berlin, Heidelberg: Springer; 2012.

[4] Moldoveanu SC. Chapter 2 Thermal decomposition of polymers. In: Techniques and instrumentation in analytical chemistry, 25. Amsterdam, The Netherlands: Elsevier; 2005. p. 31–107.

[5] Polymer Properties Database. Thermal-oxidative degradation of polymers. from: <http://polymerdatabase.com/polymer%20chemistry/Thermal%20Degradation.html>.

[6] Jail MA, Kulkarni AP Applied science—I: Technical Publications; 2009.

[7] Kaczmarek H, Chylińska M, Ziegler-Borowska M. Thermal properties of novel polymers based on poly (hydantoin-methyl-*p*-styrene) and their substrates. J Therm Anal Calorim 2012;110(3):1315–26.

[8] Pielichowski K, Njuguna J, Rapra technology L. Thermal degradation of polymeric materials: Rapra Technology; 2005.

[9] Crompton TR Thermal stability of polymers: Smithers Information Limited; 2012.

[10] Chu CC, von Fraunhofer JA, Greisler HP. Wound closure biomaterials and devices. Boca Raton, FL: Taylor & Francis; 1996.

[11] Grand AF, Wilkie CA. Fire retardancy of polymeric materials. Boca Raton, FL: CRC Press; 2000.

[12] Ueno T, Nakashima E, Takeda K. Quantitative analysis of random scission and chain-end scission in the thermal degradation of polyethylene. Polym Degrad Stab 2010;95(9):1862–9.

[13] Visakh PM, Arao Y. Thermal degradation of polymer blends, composites and nanocomposites. Switzerland: Springer International Publishing; 2015.

[14] Rabek JF. Chapter 4—Oxidative degradation of polymers. In: Bamford CH, Tipper CFH, editors. Comprehensive chemical kinetics, 14. Amsterdam, The Netherlands: Elsevier; 1975. p. 425–538.

[15] Zweifel H. Principles of oxidative degradation. In: Zweifel H, editor. Stabilization of polymeric materials. Berlin, Heidelberg: Springer; 1998. p. 1–40.

[16] Al-Malaika S. Antioxidants: an overview. In: Pritchard G, editor. Plastics additives: an A-Z reference. Dordrecht, The Netherlands: Springer; 1998. p. 55–72.

[17] Štěpek J, Daoust H. Antioxidants and metal ion deactivating agents. In: Štěpek J, Daoust H, editors. Additives for plastics. New York: Springer; 1983. p. 167–85.

[18] Izdebska J, Thomas S. Printing on polymers: fundamentals and applications. Oxford, UK: Elsevier Science; 2015.

[19] van Krevelen DW, te Nijenhuis K. Properties of polymers: their correlation with chemical structure; their numerical estimation and prediction from additive group contributions. Oxford, UK: Elsevier Science; 2009.

[20] Lu X, Jiang H, Li J, Zhao L, Pei Y, Zhao Y, et al. Polymer thermal degradation in high-temperature reservoirs. Petrol Sci Technol 2015;33(17–18):1571–9.

[21] International ASM, Lampman S Characterization and failure analysis of plastics. A S M International; 2003.

[22] International union of pure and applied chemistry. Commission on macromolecular N, Jones RG. Compendium of polymer terminology and nomenclature: IUPAC Recommendations, 2008: RSC Pub.; 2009.

[23] Charles J, Ramkumaar GR, Azhagiri S, Gunasekaran S. FTIR and thermal studies on nylon-66 and 30% glass fibre reinforced nylon-66. E-J Chem 2009;6(1):23–33.

[24] Sinha Ray S. 10—Thermal stability and flammability of environmentally friendly polymer nanocomposites using biodegradable polymer matrices and clay/carbon nanotube (CNT) reinforcements. Environmentally friendly polymer nanocomposites. Cambridge, UK: Woodhead Publishing; 2013. p. 295–327.

[25] Witkowski A, Stec AA, Hull TR. Thermal decomposition of polymeric materials. In: Hurley MJ, Gottuk D, Hall JR, Harada K, Kuligowski E, Puchovsky M, et al., editors. SFPE handbook of fire protection engineering. New York: Springer; 2016. p. 167–254.

[26] Jeon I-Y, Baek J-B. Nanocomposites derived from polymers and inorganic nanoparticles. Materials 2010;3(6).

[27] Khrenov V, Schwager F, Klapper M, Koch M, Müllen K. Compatibilization of inorganic particles for polymeric nanocomposites. Optimization of the size and the compatibility of ZnO particles. Polym Bull 2007;58(5):799–807.

[28] Merhari L. Hybrid nanocomposites for nanotechnology: electronic, optical, magnetic and biomedical applications. New York: Springer; 2009.

[29] Kumar SK, Jouault N, Benicewicz B, Neely T. Nanocomposites with polymer grafted nanoparticles. Macromolecules 2013;46(9):3199–214.

[30] Li S, Meng Lin M, Toprak MS, Kim DK, Muhammed M. Nanocomposites of polymer and inorganic nanoparticles for optical and magnetic applications. Nano Rev 2010;1. Available from: https://doi.org/10.3402/nano.v1i0.5214.

[31] Wu L, Baghdachi J. Functional polymer. coatings: principles, methods, and applications. Hoboken, NJ, USA: Wiley; 2015.

[32] Hosseini H, Shojaee-Aliabadi S, Hosseini SM, Mirmoghtadaie L. Chapter 11—nanoantimicrobials in food industry A2—Oprea, Alexandra Elena. In: Grumezescu AM, editor. Nanotechnology applications in food. Cambridge: Academic Press; 2017. p. 223–43.

[33] Hayles J, Johnson L, Worthley C, Losic D. 5—Nanopesticides: a review of current research and perspectives A2—Grumezescu, Alexandru Mihai. New pesticides and soil sensors. Boston, MA: Academic Press; 2017. p. 193–225.

[34] Leszczyńska A, Njuguna J, Pielichowski K, Banerjee JR. Polymer/montmorillonite nanocomposites with improved thermal properties: Part I. Factors influencing thermal stability and mechanisms of thermal stability improvement. Thermochim Acta 2007;453(2):75–96.

[35] Sun-Waterhouse D, Waterhouse GIN. 2—Recent advances in the application of nanomaterials and nanotechnology in food research A2—Grumezescu, Alexandru Mihai. Novel approaches of nanotechnology in food. Boston, MA: Academic Press; 2016. p. 21–66.

[36] Leszczyńska A, Njuguna J, Pielichowski K, Banerjee JR. Polymer/montmorillonite nanocomposites with improved thermal properties: Part II. thermal stability of montmorillonite nanocomposites based on different polymeric matrixes. Thermochim Acta. 2007;454(1):1–22.

[37] Jayrajsinh S, Shankar G, Agrawal YK, Bakre L. Montmorillonite nanoclay as a multifaceted drug-delivery carrier: a review. J Drug Deliv Sci Technol 2017;39:200–9.

[38] Shunmugasamy VC, Xiang C, Gupta N. Clay/polymer nanocomposites: processing, properties, and applications. In: Kim C-S, Randow C, Sano T, editors. Hybrid and hierarchical composite materials. Cham: Springer International Publishing; 2015. p. 161–200.

[39] Müller K, Bugnicourt E, Latorre M, Jorda M, Echegoyen Sanz Y, Lagaron JM, et al. Review on the processing and properties of polymer nanocomposites and nanocoatings and their applications in the packaging, automotive and solar energy fields. Nanomaterials 2017;7(4):74.

[40] Reible DD. Fundamentals of environmental engineering. New York: Taylor & Francis; 1998.

[41] Mohanty S, Nayak SK, Kaith BS, Kalia S. Polymer nanocomposites based on inorganic and organic nanomaterials. Hoboken, NJ: Wiley; 2015.

[42] Cheng A, Wu S, Jiang D, Wu F, Shen J. Study of elastomeric polyurethane nanocomposites prepared from grafted organic–montmorillonite. Colloid Polym Sci 2006;284(9):1057–61.

[43] Mansoori Y, Atghia SV, Sanaei SS, Zamanloo MR, Imanzadeh G. PMMA-clay nanocomposite materials: free-radically grafting of PMMA onto organophilic montmorillonite (20A). Macromol Res 2010;18 (12):1174–81.

[44] Leszczynska A, Pielichowski K. The mechanical and thermal properties of polyoxymethylene (POM)/ organically modified montmorillonite (OMMT) engineering nanocomposites modified with thermoplastic polyurethane (TPU) compatibilizer. In: Boudenne A, editor. Polymer composite materials: from macro, micro to nanoscale. Materials science forum, 714. Stafa-Zurich: Trans Tech Publications Ltd; 2012. p. 201–9.

[45] Majka TM, Leszczynska A, Kandola BK, Pornwannachai W, Pielichowski K. Modification of organo-montmorillonite with disodium H-phosphonate to develop flame retarded polyamide 6 nanocomposites. Appl Clay Sci 2017;139:28–39.

[46] Cronan CS. Ecosystem biogeochemistry: element cycling in the forest landscape. New York: Springer International Publishing; 2017.

[47] Ghosh D, Bhattacharyya KG. Adsorption of methylene blue on kaolinite. Appl Clay Sci 2002;20 (6):295–300.

[48] Su L, Zeng X, He H, Tao Q, Komarneni S. Preparation of functionalized kaolinite/epoxy resin nanocomposites with enhanced thermal properties. Appl Clay Sci 2017;148:103–8.

[49] Albach B, Munaro M, Santos PH, Zawadzki SF, Schreiner WH, Rampon DS, et al. Thermal, mechanical, and water vapor barrier behavior of polypropylene composite containing modified kaolinite. J Appl Polym Sci 2018;135(5) 45785-45.

[50] Cannillo V, Bondioli F, Lusvarghi L, Montorsi M, Avella M, Errico ME, et al. Modeling of ceramic particles filled polymer–matrix nanocomposites. Compos Sci Technol 2006;66(7):1030–7.

[51] Xiong M, Wu L, Zhou S, You B. Preparation and characterization of acrylic latex/nano-SiO_2 composites. Polym Int 2002;51(8):693–8.

[52] Baig MI, Ingole PG, Jeon J-d, Hong SU, Choi WK, Jang B, et al. Water vapor selective thin film nanocomposite membranes prepared by functionalized silicon nanoparticles. Desalination. 2017;.

[53] Moghadasi M, Mortaheb Hamid R. Incorporating functionalized silica nanoparticles in polyethersulfone-based anion exchange nanocomposite membranes. J Appl Polym Sci 2016;134(11).

[54] Shahi VK. Highly charged proton-exchange membrane: sulfonated poly(ether sulfone)-silica polyelectrolyte composite membranes for fuel cells. Solid State Ionics 2007;177(39):3395–404.

[55] Guo Z, Kim TY, Lei K, Pereira T, Sugar JG, Hahn HT. Strengthening and thermal stabilization of polyurethane nanocomposites with silicon carbide nanoparticles by a surface-initiated-polymerization approach. Compos Sci Technol 2008;68(1):164–70.

[56] Jia X, Li Y, Cheng Q, Zhang S, Zhang B. Preparation and properties of poly(vinyl alcohol)/silica nanocomposites derived from copolymerization of vinyl silica nanoparticles and vinyl acetate. Eur Polym J 2007;43(4):1123–31.

[57] Wei L, Hu N, Zhang Y. Synthesis of polymer—mesoporous silica nanocomposites. Materials. 2010;3(7).

[58] Hoffmann F, Cornelius M, Morell J, Fröba M. Silica-based mesoporous organic–inorganic hybrid materials. Angew Chem Int Ed 2006;45(20):3216–51.

[59] Yu C, Gao M, Feng J, Liu Y, Lv J, Liu H, et al. Thermal and frictional properties of mesoporous silica SBA-15/phenolic resin nanocomposites. Polym Compos 2016;38(S1):E351–8.

[60] Yu CB, Wei C, Lv J, Liu HX, Meng LT. Preparation and thermal properties of mesoporous silica/phenolic resin nanocomposites via in situ polymerization. Express Polym Lett 2012;6:783–93.

[61] Ji X, Hampsey JE, Hu Q, He J, Yang Z, Lu Y. Mesoporous silica-reinforced polymer nanocomposites. Chem Mater 2003;15(19):3656–62.

[62] Kuo S-W, Chang F-C. POSS related polymer nanocomposites. Prog Polym Sci 2011;36(12):1649–96.

[63] Pielichowski K, Njuguna J, Janowski B, Pielichowski J. Polyhedral oligomeric silsesquioxanes (POSS)-containing nanohybrid polymers. supramolecular polymers polymeric betains oligomers. Adv Polym Sci, 201. Berlin, Heidelberg: Springer; 2006. p. 225–96.

[64] Raftopoulos KN, Pielichowski K. Segmental dynamics in hybrid polymer/POSS nanomaterials. Prog Polym Sci 2016;52:136–87.

[65] Lin Y, Song M. Effect of polyhedral oligomeric silsesquioxane nanoparticles on thermal decomposition of cyanate ester resin. React Funct Polym 2018;129:58–63.

[66] Lee Y-J, Kuo S-W, Su Y-C, Chen J-K, Tu C-W, Chang F-C. Syntheses, thermal properties, and phase morphologies of novel benzoxazines functionalized with polyhedral oligomeric silsesquioxane (POSS) nanocomposites. Polymer 2004;45(18):6321–31.

[67] Lee Y-J, Kuo S-W, Huang C-F, Chang F-C. Synthesis and characterization of polybenzoxazine networks nanocomposites containing multifunctional polyhedral oligomeric silsesquioxane (POSS). Polymer 2006;47(12):4378–86.

[68] Sirin H, Kodal M, Ozkoc G. The influence of POSS type on the properties of PLA. Polym Compos 2014;37(5):1497–506.

[69] Fina A, Abbenhuis HCL, Tabuani D, Frache A, Camino G. Polypropylene metal functionalised POSS nanocomposites: a study by thermogravimetric analysis. Polym Degrad Stab 2006;91 (5):1064–70.

[70] Lewicki JP, Pielichowski K, Jancia M, Hebda E, Albo RLF, Maxwell RS. Degradative and morphological characterization of POSS modified nanohybrid polyurethane elastomers. Polym Degrad Stab 2014;104:50–6.

[71] Lewicki JP, Pielichowski K, De La Croix PT, Janowski B, Todd D, Liggat JJ. Thermal degradation studies of polyurethane/POSS nanohybrid elastomers. Polym Degrad Stab 2010;95(6):1099–105.

[72] Pagacz J, Hebda E, Janowski B, Sternik D, Jancia M, Pielichowski K. Thermal decomposition studies on polyurethane elastomers reinforced with polyhedral silsesquioxanes by evolved gas analysis. Polym Degrad Stab 2018;149:129–42.

[73] Pagacz J, Hebda E, Michalowski S, Ozirnek J, Sternik D, Pielichowski K. Polyurethane foams chemically reinforced with POSS-thermal degradation studies. Thermochim Acta 2016;642:95–104.

[74] Sanchez-Soto M, Illescas S, Milliman H, Schiraldi DA, Arostegui A. Morphology and thermomechanical properties of melt-mixed polyoxymethylene/polyhedral oligomeric silsesquioxane nanocomposites. Macromol Mater Eng 2010;295(9):846–58.

[75] Ramalingam M, Vallittu P, Ripamonti U, Li WJ. Tissue engineering and regenerative medicine: a nano approach. Boca Raton, FL: CRC Press; 2012.

[76] Pielichowska K, Blazewicz S. Bioactive polymer/hydroxyapatite (nano)composites for bone tissue regeneration. In: Abe A, Dusek K, Kobayashi S, editors. Biopolymers. Advances in polymer science, 232. Berlin, Heidelberg: Springer; 2010. p. 97–207.

[77] Mirhosseini MM, Haddadi-Asl V, Zargarian SS. Fabrication and characterization of polymer-ceramic nanocomposites containing pluronic F127 immobilized on hydroxyapatite nanoparticles. RSC Adv 2016;6(84):80564–75.

[78] Zhao C-X, Zhang W-D. Preparation of waterborne polyurethane nanocomposites: polymerization from functionalized hydroxyapatite. Eur Polym J 2008;44(7):1988–95.

[79] Pielichowska K. The influence of polyoxymethylene molar mass on the oxidative thermal degradation of its nanocomposites with hydroxyapatite. J Therm Anal Calorim 2015;1–15.

[80] Pielichowska K. Thermooxidative degradation of polyoxymethylene homo- and copolymer nanocomposites with hydroxyapatite: kinetic and thermoanalytical study. Thermochim Acta. 2015;600:7–19.

[81] Pielichowska K, Król K, Majka TM. Polyoxymethylene-copolymer based composites with PEG-grafted hydroxyapatite with improved thermal stability. Thermochim Acta. 2016;633:98–107.

[82] Bhattacharya K, Mukherjee SP, Gallud A, Burkert SC, Bistarelli S, Bellucci S, et al. Biological interactions of carbon-based nanomaterials: from coronation to degradation. Nanomed Nanotechnol Biol Med 2016;12(2):333–51.

[83] Cha C, Shin SR, Annabi N, Dokmeci MR, Khademhosseini A. Carbon-based nanomaterials: multifunctional materials for biomedical engineering. ACS Nano 2013;7(4):2891–7.

[84] Kashiwagi T, Du F, Douglas JF, Winey KI, Harris Jr RH, et al. Nanoparticle networks reduce the flammability of polymer nanocomposites. Nat Mater 2005;4:928.

[85] Ramanathan T, Fisher FT, Ruoff RS, Brinson LC. Amino-functionalized carbon nanotubes for binding to polymers and biological systems. Chem Mater 2005;17(6):1290–5.

[86] Cai D, Song M. Recent advance in functionalized graphene/polymer nanocomposites. J Mater Chem 2010;20(37):7906–15.

[87] Silva M, Alves Natália M, Paiva Maria C. Graphene-polymer nanocomposites for biomedical applications. Polym Adv Technol 2017;29(2):687–700.

[88] Ramanathan T, Abdala AA, Stankovich S, Dikin DA, Herrera-Alonso M, Piner RD, et al. Functionalized graphene sheets for polymer nanocomposites. Nat Nanotechnol 2008;3:327.

[89] Zhang G, Wang F, Dai J, Huang Z. Effect of functionalization of graphene nanoplatelets on the mechanical and thermal properties of silicone rubber composites. Materials 2016;9(2).

[90] Nam K-H, Cho J, Yeo H. Thermomechanical behavior of polymer composites based on edge-selectively functionalized graphene nanosheets. Polymers 2018;10(1).

[91] Liang J, Huang Y, Zhang L, Wang Y, Ma Y, Guo T, et al. Molecular-level dispersion of graphene into poly(vinyl alcohol) and effective reinforcement of their nanocomposites. Adv Funct Mater 2009;19 (14):2297–302.

[92] Verdejo R, Barroso-Bujans F, Rodriguez-Perez MA, Antonio de Saja J, Lopez-Manchado MA. Functionalized graphene sheet filled silicone foam nanocomposites. J Mater Chem 2008;18 (19):2221–6.

[93] Hirsch A. Functionalization of single-walled carbon nanotubes. Angew Chem Int Ed 2002;41 (11):1853–9.

[94] Sahoo NG, Rana S, Cho JW, Li L, Chan SH. Polymer nanocomposites based on functionalized carbon nanotubes. Prog Polym Sci 2010;35(7):837–67.

[95] Wu D, Wu L, Zhang M, Zhao Y. Viscoelasticity and thermal stability of polylactide composites with various functionalized carbon nanotubes. Polym Degrad Stab 2008;93(8):1577–84.

[96] Amirian M, Nabipour Chakoli A, Cai W, Sui J. Effect of functionalized multiwalled carbon nanotubes on thermal stability of poly (L-LACTIDE) biodegradable polymer. Sci Iran 2013;20(3):1023–7.

[97] Ferreira FV, Franceschi W, Menezes BRC, Brito FS, Lozano K, Coutinho AR, et al. Dodecylamine functionalization of carbon nanotubes to improve dispersion, thermal and mechanical properties of polyethylene based nanocomposites. Appl Surf Sci 2017;410:267–77.

[98] Nicolais L, Carotenuto G. Metal-polymer nanocomposites. New York: Wiley; 2004.

[99] Hernández-Sierra JF, Ruiz F, Cruz Pena DC, Martínez-Gutiérrez F, Martínez AE, de Jesús Pozos Guillén A, et al. The antimicrobial sensitivity of streptococcus mutans to nanoparticles of silver, zinc oxide, and gold. Nanomed Nanotechnol Biol Med 2008;4(3):237–40.

[100] Guo Z, Pereira T, Choi O, Wang Y, Hahn HT. Surface functionalized alumina nanoparticle filled polymeric nanocomposites with enhanced mechanical properties. J Mater Chem 2006;16(27):2800–8.

[101] Chatterjee U, Jewrajka SK, Guha S. Dispersion of functionalized silver nanoparticles in polymer matrices: stability, characterization, and physical properties. Polym Compos 2009;30(6):827–34.

[102] Mallakpour S, Barati A. Efficient preparation of hybrid nanocomposite coatings based on poly(vinyl alcohol) and silane coupling agent modified TiO_2 nanoparticles. Prog Org Coat. 2011;71(4):391–8.

14

Polymer Composites Containing Functionalized Nanoparticles and the Environment

Jamal Seyyed Monfared Zanjani[1,2], Oğuzhan Oğuz[1,2], Burcu Saner Okan[1,2], Mehmet Yildiz[1,2], Yusuf Ziya Menceloğlu[1,2]

[1]FACULTY OF ENGINEERING AND NATURAL SCIENCES, MATERIALS SCIENCE AND NANO ENGINEERING, SABANCI UNIVERSITY, TUZLA, ISTANBUL, TURKEY [2]SABANCI UNIVERSITY INTEGRATED MANUFACTURING TECHNOLOGIES RESEARCH AND APPLICATION CENTER & COMPOSITE TECHNOLOGIES CENTER OF EXCELLENCE, ISTANBUL, TURKEY

14.1 Introduction

Nanoparticles in general and functional nanoparticles in particular are of considerable interest for various industrial and technological applications due to their small dimensions, enhanced functional properties and features they offer that are not available in particles with larger dimensions or bulk of same materials [1–3]. The European Commission (EC) defines a nanomaterial as [4]: "A natural, incidental or manufactured material containing particles, in an unbound state or as an aggregate or as an agglomerate and where, for 50 % or more of the particles in the number size distribution, one or more external dimensions is in the size range 1–100 nm." Nanoparticles can be manufactured from various materials including metals, ceramics/glass, polymers, and semiconductors [5–11]. These particles can be produced in various shapes such as sphere [12], cube [13], rod [14], nanoplate [15], and others [16]. The properties of nanoparticles are closely linked to their size, shape, and surface functionalities. These variations and tailorable properties have led to a wide range of applications of this class of materials in the development of state-of-the-art materials to be used in structural applications, electronics, biomedicine, cosmetics, energy, catalysts, filters, and sensors [17]. As a result of their unique properties and a wide range of applications, the production of various nanoparticles has dramatically increased in recent years, which brings about the dilemma of their positive and negative effect on the environment and human health. These materials can be used in biological applications, water treatment, biosensors, filters, and energy technologies to save the environment and provide higher life quality for humankind. However, even nanoparticles synthesized from biologically inert materials can be harmful to

Polymer Composites with Functionalized Nanoparticles. DOI: https://doi.org/10.1016/B978-0-12-814064-2.00014-7

the cell by damaging cell membrane, disrupting cell function, or altering the genetics of the cell by attaching to its DNA [18].

Nanoparticle toxicology is a science of studying the ability of nanoparticles to damage or change the function of cells, genes, and living organs and determine the mechanism of that damage [19]. However, nanotechnology boundaries are always being pushed to create extremely small size particles with dimensions smaller than that of cellular structures [20,21]. The extremely small size of particles induces a significant impact on nanoparticles' properties and behavior in the biological environment due to their ultrahigh surface area and shape factors such as extreme sharp edges, which have led to the intrinsic toxicity of such materials [22–24]. The nanoscale dimension renders the chemical/physical properties of the particles in a way that toxicological behavior of nanoparticles cannot be extrapolated from information and available data of larger particles of the same composition [24,25].

One may say that in polymeric nanocomposites, nanoparticles are fundamentally bound up in the structure of polymeric nanocomposites, but there is the potential for exposure to nanoparticles and nanomaterials throughout the product chain during manufacture, application, and waste management [22]. One issue making particle toxicology more complex than other toxicology branches is the variation in behavior of nanoparticles of interest by several structural and environmental factors that enable the unfavorable effect of nanoparticles in diverse scenarios [22].

This chapter initially discusses the recent progress in the understanding of the toxic behavior of various nanoparticles that are used commercially and intensively in nanocomposite production. This section reviews these nanoparticles' effect on living organs by elaboration on mechanisms and parameters contributing to such behavior. In the second part of this chapter, different applications of nanocomposites to promote human quality of life and contributing to the wellbeing of the environment are discussed in selected areas to emphasize the importance of nanocomposites in technological advancements.

14.2 Toxicological and Exposure

The nanoparticles with dimensions in the range below 100 nm are prone to have a range of unexpected effects on biological systems. These particles are so small that they can pass through the phagocytic defense system without being detected, giving them access to blood flow and the nervous system, and they can also accumulate in the vital organs. Furthermore, these nanoparticles have the ability to interact with proteins and inactivate their function or create an autoimmune effect in living organs. Humans are exposed to nanoparticles by lung exposure through environmental air pollution, skin exposure through cosmetic products, and inhalation and consequent redistribution to other organs [21,26]. There are many knowledge gaps in our understanding on the toxicity of nanoparticles and their relations with particle size, roughness, shape, charge, composition, surface coating, and physiological mechanisms [24]. Therefore, a comprehensive understanding of the potentially harmful effects of nanoparticles is necessary to provide a safe facilitation of these materials to our lives and ensure future wellbeing. In this section, a summary of nanotoxicological studies of some important nanoparticles with high volume use such as carbon nanotubes (CNTs), graphene, nanoclays, and various metallic nanoparticles is discussed and their toxicity mechanisms as well as the effect of nanoparticles' physical and chemical structure in the environment are deliberated when available.

14.2.1 Carbon Nanotubes

CNTs are the thinnest tubes humans have ever made. They are chemically and thermally very stable with diverse properties and broad application range [27–29]. Some of CNTs' distinguished properties are high strength, high toughness, extremely high surface area, and excellent thermal and electric conductivity. CNTs' diameter ranges from ~0.4 to 3 nm for single wall and from ~1.4 to 100 nm for multiwall nanotubes [30–32]. CNTs are long thin structures that can have length up to many several tens of micrometers. Fig. 14–1 exhibits the molecular representations of single-walled CNTs and multiwalled CNTs (MWCNTs) and their typical transmission electron micrographs.

CNTs are usually synthesized by one of the following techniques: Carbon-arc discharge, laser ablation of carbon, or chemical vapor deposition (CVD) [34]. The first two methods employ solid-state carbon precursors as carbon sources by vaporization at high temperature. On the other hand, CVD method utilizes hydrocarbon gases as carbon sources and metal particles as "seeds" for nanotube growth [35]. These CNT production methods require a high amount of energy and generate large amounts of byproducts, which limits their production in large scale and raises several environmental concerns. CNTs are one of the least biodegradable manmade materials. The insolubility in solvents and lipophilic behavior of CNTs make their removal from environment and organs almost impossible [36,37]. These CNTs can enter the environment at all stages of synthesis, processing, application, and disposal,

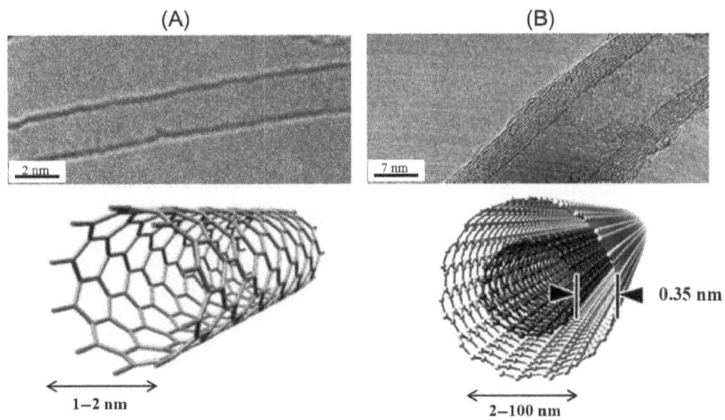

FIGURE 14–1 TEM images and schematic illustrations of CNT types showing typical diameters and separation between the graphene layers in MWCNT: (A) single-walled carbon nanotube (SWCNT) and (B) MWCNT. *Reprinted from Kaseem M, Hamad K, Ko YG. Fabrication and materials properties of polystyrene/carbon nanotube (PS/CNT) composites: a review. Eur Polym J 2016;79:36–62 [33], by permission of Elsevier.*

and exhibit their toxic properties. The toxicity and polluting effect of CNTs differ depending on their structural properties such as length, surface chemistry, and oxidative potential as well as the types of cells they are interacting with [38].

The health risk and environmental effect of CNTs strongly depend on their physical properties, shape, and surface functional groups. The toxicity of CNTs is a result of both their physical and chemical properties. The intensive interactions between CNTs and cell membranes may cause physical damage and consequent cell death. Moreover, CNTs are able to chemically interact with cells and promote cellular oxidative stress, which can lead to cell malfunction or even cell death [39]. Size is one of the most important parameters determining the toxicity of CNTs, for example, CNTs longer than 20 μm cause the same types of pathology that long biopersistent fibers cause such as fibrosis, cancer, pleural changes, and mesothelioma. On the other hand, CNTs with shorter length of 15–20 μm act like nanoparticles in which adverse effect and toxic behavior are driven by the small size and large surface area of the CNTs [26]. The surface functionalities of CNTs play a critical role in their behavior, toxicity, and state of aggregation in the environment. The surface functionalization of CNTs through acid treatments or any other methods can change their dispersion behavior significantly [40]. The surface functionalized CNTs displayed high mobility in water, which assesses the potential migration in natural porous media [41]. In addition, the high surface area and consequent high surface energy of CNTs lead to attachment of other potential pollutants and their transport throughout the environment [42]. Furthermore, the surface functional groups determine CNTs' interaction with cells and consequent toxic effects. The surface charge of CNTs is another parameter influenced by surface functionalization, which plays a critical role in the toxicity of CNTs. Shen et al. [43] showed that in vitro cytotoxicity and the biocompatibility of functionalized CNTs is largely dependent on their surface potential. They applied polyethyleneimine functionalization on the acid-treated CNT, which resulted in water-soluble and stable CNTs. The mentioned study showed that neutral and negatively charged CNTs are nontoxic to targeted cells at a concentration up to 100 μg/mL, whereas positively charged MWCNTs showed toxicity at 10 μg/mL [43]. Nevertheless, Pulskamp et al. [44] did not observe acute toxicity for three commercial grades of nonfunctionalized single-walled CNTs, whereas Jos et al. [45] confirmed the cytotoxicity of carboxylic functionalized single-walled CNTs on differentiated and nondifferentiated Caco-2 cells derived from a human intestinal adenocarcinoma at concentrations higher than 100 μg/mL. In another work, Saxena et al. [46] observed that acid functionalized single-walled CNT exerted stronger toxic effects in vitro (cytotoxicity, cell cycling inhibition, and apoptosis) and in vivo in comparison with unmodified single-walled CNT, and they showed that those toxic effects could be reversed by neutralizing the negative surface charges. Magrez et al. [29] showed that the toxicity of CNTs increase significantly when carbonyl (C–O), carboxyl (COOH), and/or hydroxyl (OH) groups are present on their surface without any clarification on the exact mechanisms. In contrast, Sayes et al. [47] observed that an increase in the degree of sidewall functionalization of single-walled CNTs decreases their cytotoxic property.

In general, size, shape, and surface charge of nanoparticles are believed to be the three most important factors affecting the nanoparticles' toxicity. However, the toxic behavior of

CNTs is also related to environmental parameters such as pH, temperature, and light. In fact, the toxicity of nanoparticles in general and CNTs, in particular, is a complicated process related to the whole effect of their physical properties interacted with specific biomolecules or the releasing contents of dissolved toxic ions in biological media. Therefore, due to the contribution of numerous effective parameters in CNT polluting and toxicity, the investigation, control, and prediction of its behavior and effects on organs and environments is a challenging task. However, a high level of fundamental understanding of biocompatibility and effects of CNTs on human health and environment should be achieved to ensure the safe integration of CNT-based products in our lives. Considering this information, extra precautions in manipulation and functionalization of CNTs are needed to be taken in manufacturing and applications of functionalized CNTs where CNTs are in direct contact with living organisms.

14.2.2 Graphene

Graphene with its unique 2D structure and exceptional physical and chemical properties has attracted the attention of scientists and engineers in various advanced applications such as development of novel polymeric composites with unique properties [15,48–50]. Graphene is one of the strongest materials, having a theoretical Young's modulus of 1060 GPa and an ultimate strength of 130 GPa, and also high specific surface area, which makes it a suitable candidate for nanocomposite production with extended physical and chemical properties [48]. These novel nanocomposite materials have great potential for various applications, ranging from high-performance structures, energy storage devices, electrically conductive polymeric materials, as well as antibacterial materials [51].

Several graphene grades can be found in the graphene family materials such as few-layer graphene, graphene nanosheets, graphene oxide (GO), and reduced graphene oxide (rGO) and graphite as shown in Fig. 14–2 [52]. Each member of the graphene family has unique behavior and sometimes diverse properties compared with others. Graphene family products are growing at a rapid rate and they are commercialized in various applications [53]. This increases the graphene production rate and consequently their release to the environment. The graphene/polymer nanocomposites may release graphene during manufacturing, degradation of the polymeric matrix, or direct environmental applications such as water or soil treatment [54–56]. The released graphene into the environment has the high potential to have significant adverse impacts at different levels from bacteria and mammalian cells to animals.

Herein, we will try to briefly review the available literature to provide an understanding on the effect of graphene-based materials on various organisms, the mechanisms involved in their toxicity, and also compare their toxicity with other similar materials. It is well known that GO and rGO as two renowned members of the graphene family show higher antibacterial activity over graphite oxide and graphite. This stems from their finer size and thinner layers, which give them the sharp edges to damage the plasma membranes on bacterial cells and kill the cell [39,57]. The sharp edges of GO and rGO are not available in other carbon

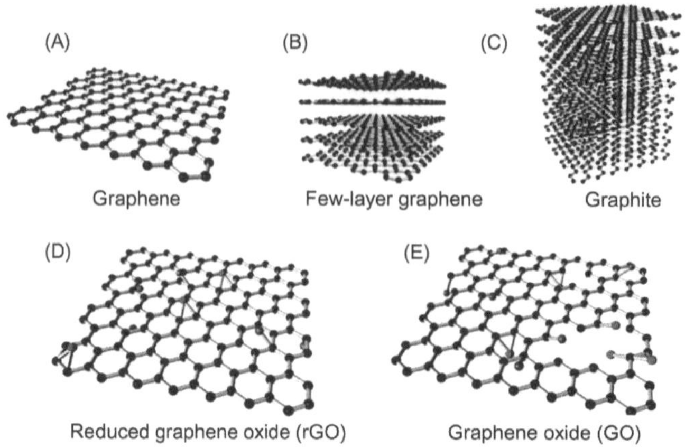

FIGURE 14–2 Graphene family nanomaterials. (a) Graphene, (b) few-layer graphene, (c) graphite, (d) reduced graphene oxide (rGO), and (e) graphene oxide (GO) *Reprinted from Kiew SF, Kiew LV, Lee HB, Imae T, Chung LY. Assessing biocompatibility of graphene oxide-based nanocarriers: a review. J Control Release 2016;226:217–28, by permission of Elsevier.*

nanomaterials such as carbon black, which restricts their antimicrobial activities through the mentioned mechanism. In addition, GO shows higher cytotoxicity compared with rGO; this is because of its higher dispersibility in aqueous media, and presence of active surface functional groups, which promote the possibility of direct contact with cells and inducing higher intracellular oxidative stress [49]. Contrarily, some studies showed higher toxicity of rGO is correlated with its thinner structure in the absence of oxygen functional groups (0.34 nm) compared with that of GO (1 nm), which creates sharper edges to damage the cell membrane easily [58,59]. Furthermore, it is observed that higher electrical connectivity of rGO increases the oxidative stress on intracellular glutathione by enhancing the electron transfer between the edges to membranes which promotes toxic effect on cells [60–62]. On the other hand, the hydrophobicity of rGO upholds its interaction with the cells' outer wall, which can hinder the nutrient absorption and gas exchange across the cell membrane [63,64].

The environment and cell type have a disputable effect on the toxicity of graphene. A small change in the environments may change the graphene's structure, transport, aggregation, and toxicity [65]. GO can go through a reduction process under light exposure or through exposure to chemical compounds [66–68]. Furthermore, biodegradability is another issue that should be addressed while discussing the concerns around the environmental effects of graphene and its derivatives. GO has shown biodegradability potential in presence of enzymes such as myeloperoxidase [69]. Surface functionalization is another parameter that influences the toxicity of graphene family materials and alters their biodegradability and dispersibility on the environment [70]. GO has the ability to be degraded by low concentrations of enzymes while rGO shows resistance to enzyme degradation [71]. The point is oxidation of graphite creates defect sites on the structure promoting the biodegradation of graphene. However, excessive functionalization with larger functional groups may prevent their biodegradation [72,73]. Nevertheless, the number of effective enzymes to be used in biodegradation of nanomaterial is very restricted and more biodegradation studies are needed to characterize better and more effective enzymes to provide biodegradability of graphene and minimize its effect on the ecosystem [70].

14.2.3 Nanoclays

The first polymer nanocomposite was a polymer–clay hybrid that was invented at Toyota Central R&D Labs, Inc. (Toyota) [74]. In the mentioned hybrid polymer composite system a small amount of silicate layer of clay mineral is randomly and homogeneously dispersed on a molecular level in a polymer matrix [75]. Nanoclays are layered mineral silicates, which are one of the common additives for polymer nanocomposite production with increased strength, modulus, and toughness and improved barrier and flame-retardant properties [76,77]. Clay-based polymeric nanocomposites possess a large portion of the nanocomposite market in various application areas.

Clays are abundant, highly stable, inexpensive, and believed to be an environmentally friendly nanoadditive [73]. Montmorillonite is a natural clay and the most frequently used nanoclay in the preparation of polymer nanocomposites with plate-like particles called platelets [78]. Montmorillonite platelets have an average thickness of only 1 nm, while their dimensions in length and width can be measured up to 1 μm [78]. Fig. 14–3 shows the structure of montmorillonite clay, which consists of two tetrahedral sheets covered with one octahedral sheet in between [79]. Despite the several useful biological and environmental applications of nanoclays, their nanoscale size enables them to penetrate to the cell membranes, and interfere with cellular processes [78,80]. Several mechanisms contribute to toxicity of nanoclays including decrease in both cellular growing rate, morphological changes of cell, cellular proliferation, mitochondrial damage, reactive oxygen species generation, and

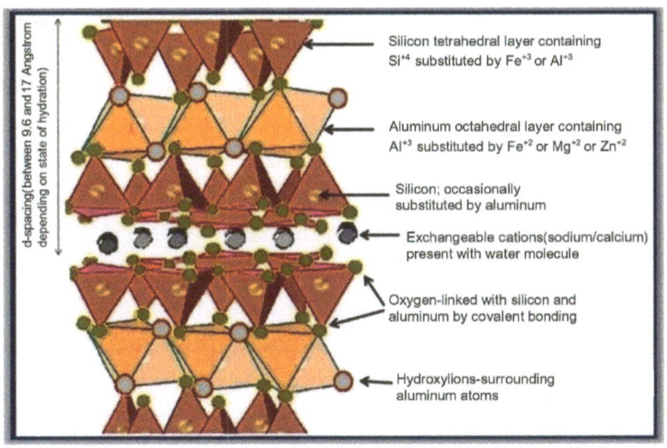

FIGURE 14–3 Structural unit of montmorillonite nanoclay. *Reprinted from Jayrajsinh S, Shankar G, Agrawal YK, Bakre L. Montmorillonite nanoclay as a multifaceted drug-delivery carrier: a review. J Drug Deliv Sci Technol 2017;39(Suppl. C):200–9, by permission of Elsevier.*

DNA alteration [81–84]. It worth noting that the prudency of each damage type is closely related to studied cell, dosage, and surface functionalization of clays.

In the clay structure, the interlayer space (known as the clay gallery) contains exchangeable cations such as Na^+ or K^+ [85]. Organomodification of nanoclays is a method to replace these cations with organic cationic surfactants. This organomodification increases gallery spacing, improves the compatibility, makes clay hydrophobic, and enhances its compatibility with most of the polymeric matrices [86]. The organomodification or functionalization of clays with organic chains plays a critical role in their cytotoxic response. Some studies find that organomodified clays have greater toxicity compared with unmodified ones [78,87] whereas other findings express lower toxicity of modified clays compared with their unmodified counterparts [88,89]. These studies also express the importance of functional group type used for organomodification on the toxicity of nanoclays [90–92].

The clay's physical structure and geometry (platelet or tubular) have significant effect on its toxicity [93]. Verma et al. [81] detected a noticeable difference in the number of detached and deformed cells exposed to platelet-type nanoclays compared with cells in contact with tubular clays, which indicates that platelet structured nanoclays are more cytotoxic than tubular type. Their study showed that tubular clays, known as halloysite nanotubes, do not show any toxic effect until doses of 250 μg/mL, while platelet nanoclays cause toxicity at 25 μg/mL [81].

Given the complex and uncertain effects of nanoclays on biological environments, systemic management, analysis, and precautions are required to assess and minimize the exposures during their manipulation, handling, and disposal. In addition, more research efforts are necessary to evaluate and understand the mechanisms of toxicity of nanoclays in cells and find a proper solution to design safe products of nanoclay-based polymeric nanocomposites for future applications.

14.2.4 Metal Nanoparticles

The incorporation of metal nanoparticles into polymeric matrices is an emerging area to further extend the applications of polymeric nanocomposites. The quantum size effects on metal nanoparticles give them great promise in many technical and also medical applications [94]. These unique properties can be tailored by controlling these metal nanoparticles' size, shape, and preparation methods. Metal nanoparticle-based polymeric nanocomposites provide enhanced performance and multifunctionalities as well as promising potential applications in electronics, optics, environmental, and biotechnological applications, which may not be possible with larger metal particles. Metal nanoparticles can be categorized into metals, metal oxides, and other metal-containing nanoparticles [95,96]. The wide range of application of metallic nanoparticles and high volume production connect this class of materials with some environmental effects uncertainties and serious risks. These materials can enter living organisms including the human body through active or passive pathways and have the high potential to create unrecoverable damage because of their activated toxicity due to their intrinsic properties and also their nanoscale size [97]. In this part, the environmental effects of selected metal nanoparticles, which are used intensively in polymer nanocomposite productions such as titanium dioxide (TiO_2), zinc oxide (ZnO), and silver (Ag) nanoparticles, are discussed in detail including the effective parameters and corresponding mechanisms in their toxicity.

TiO_2 is a naturally occurring mineral that can be categorized into anatase, rutile, or brookite types based on its crystal structure [98]. TiO_2 nanoparticles have one of the highest production rates worldwide among other nanoparticles [99]. TiO_2 is an effective reinforcing agent for several polymeric matrices and it also provides polymeric materials with several unique properties. TiO_2 nanoparticle is a large band gap semiconductor, which gives it exceptional electronic, optical, thermal, and photocatalytic properties. These unique functionalities of TiO_2 nanoparticles are of interest for the development of several functional polymeric nanocomposites [100–102]. TiO_2 nanoparticles show both beneficial and toxic effects on nature depending on their intrinsic properties and also the environmental factors [103]. TiO_2 have been accepted as safe materials and their cytotoxicity is often neglected whereas there are reports on their toxicity through generation of reactive oxygen species, increased cell adhesion, and alteration on carbohydrate metabolism [104–106]. Furthermore, TiO_2 nanoparticles have genotoxicity properties and they can interact with DNA of cells [107]. However, there are several conflicting results on toxicities of TiO_2 nanoparticles, some claiming no toxicity effect and some high toxic effect of these particles in the living organs. These conflicts stem from various morphology, crystal structure, size, purity levels surface areas, and also conducted test conditions [108]. To resolve these conflicts, a clear understanding of the molecular mechanisms behind the toxic and nontoxic responses of TiO_2 nanoparticles is needed [104].

The crystal structure of TiO_2 nanoparticles plays a critical rule in their toxic behavior [109]. In nature, TiO_2 has three common crystalline polymorphs, namely brookite, anatase, and rutile as shown in Fig. 14–4 [110]. However, only rutile and anatase TiO_2 are generally manufactured in titanium dioxide commercial applications. The relative toxicity of each

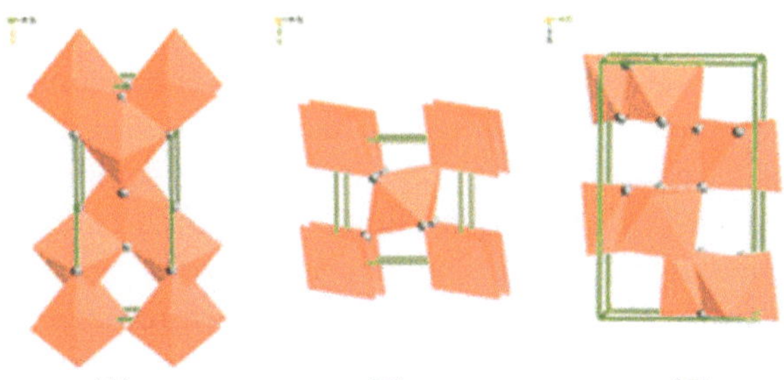

FIGURE 14–4 Representation of three TiO_2 polymorphs: (A) anatase, (B) rutile, and (C) brookite forms. *Reprinted from Liao Y, Que W, Jia Q, He Y, Zhang J, Zhong P. Controllable synthesis of brookite/anatase/rutile TiO_2 nanocomposites and single-crystalline rutile nanorods array. J Mater Chem 2012;22(16):7937–44, by permission of Royal Society of Chemistry.*

crystal structure of TiO_2 nanoparticles with respect to biological systems is broadly debated; no solid conclusion is available. In general, it is accepted that anatase TiO_2 nanoparticles are more active under light exposure or dark environment and generate higher reactive oxygen species and consequent DNA damages compared with other counterparts with different crystal structures [111,112]. The anatase TiO_2 nanoparticles are able to spontaneously generate reactive oxygen species whereas rutile TiO_2 nanoparticles do not exhibit similar behavior [112]. It is worth noting that anatase nanoparticles have a high affinity to proteins, mainly impairing mitochondrial function whereas the rutile TiO_2 nanoparticles have higher affinity to phospholipids, mainly targeting the plasma membrane and the lysosome membrane, and causing reactive oxygen species–independent cell death [109]. Furthermore, the rutile TiO_2 nanoparticles are able to induce hydrogen peroxide and oxidative DNA damage in the absence of light while anatase TiO_2 nanoparticles do not have this effect [113]. Additionally, some studies showed the higher toxicity of a mixture of anatase and rutile particles and its higher abilities to create DNA damage in the absence of light compared with individual anatase or rutile TiO_2 due to contribution of multiple toxicity mechanisms [113].

TiO_2 nanoparticles can be distributed to all organs or tissues from the original site upon entering the blood and deposited in the liver, lung, kidney, and other organs, which can cause severe damage to and malfunction of that organ [99,114,115]. Nanoparticles in general and TiO_2 nanoparticles, in particular, can enter the human/animal body through various routes such as oral, dermal, or injection and cause various toxicities in different organs [99]. Studies show that oral exposure to TiO_2 nanoparticles induces hepatic histopathologic damage, with alterations in serum lactate dehydrogenase and alpha-hydroxybutyrate dehydrogenase levels indicating myocardial damage as well as neurotoxicity and brain damage [116,117]. The high amount of application for TiO_2 nanoparticles in cosmetic products makes dermal exposure an important issue. However, there are conflicting findings in the literature in which some groups claimed that TiO_2 nanoparticles are unable to penetrate human skin whereas others present opposite findings [118–122]. Despite several recent studies on the toxicity of TiO_2, its toxicity mechanisms and its effect on the ecosystem are not completely understood. Therefore, further investigations and detailed research activities are required to understand the potential health effects and plan proper risk assessment procedures to pave the way for this useful material implementation in future applications.

Zinc oxide (ZnO) nanoparticle is another metallic nanoparticle widely used in the polymeric nanocomposite manufacturing. ZnO is an n-type semiconductor with wide band gap energy of 3.3 eV making it a suitable choice for a broad range of applications. The incorporation of ZnO nanoparticle into polymeric matrices provides them with various functionalities including enhanced mechanical performance, permeability, flame resistance, UV-shielding, antimicrobial activities, and unique thermal and electrical properties [73,123–127]. Even though ZnO is listed as a Generally Recognized as Safe material by the US Food and Drug Administration (21CFR182.8991) and it is widely used in products that have direct contact with humans such food packaging and coatings applications, there is a myriad of research reporting the toxic behavior of ZnO nanoparticles [128–131]. The ZnO nanoparticles have the ability to induce morphological modifications, oxidative stress, lipid peroxidation, cytotoxicity, genotoxicity, and chromosomal breakage forms of toxicities [132–137]. The toxicity of ZnO nanoparticles can be influenced by ZnO nanoparticle surface composition, size, and shape [133,138,139]. It is an accepted fact that antibacterial activity of ZnO nanoparticle increases with a reduction in particle size [139]. In addition, the shape of ZnO nanoparticles plays an important role in their toxic behavior [140]. At a fixed ZnO nanoparticle size and surface area, nanorod ZnO nanoparticles are more toxic than the corresponding spherical ones [140]. In addition to the toxicity mechanism correlated with the nanoscale size of the ZnO nanoparticles, it is shown that release of zinc ions is another toxicity mechanism in which these ions show higher toxicity compared with ZnO nanoparticles themselves [141].

The surface functionalization is another important parameter influencing the toxicity of ZnO nanoparticles. In addition, the correlation between surface functionality and toxicity of ZnO nanoparticles can be an effective way to modify the toxicity of ZnO nanoparticles to have specific cellular interactions with the biological molecules and preserve their functional properties, which will open up several novel applications for this group of nanomaterials [142–146]. In this regard, Ramasamy et al. [147] fabricated SiO_2-coated ZnO nanoparticles in which coated ZnO nanoparticles exhibit less cytotoxicity compared with uncoated ZnO nanoparticles. The mentioned work correlates that the decreases in cytotoxicity with fewer surface interactions of ZnO nanoparticles with cells after coating by SiO_2, decreased in the release rate of zinc ion, and changes on the hydrophilicity of the surface after coating of ZnO nanoparticles [147]. In another study, Hsiao et al. [148] reduced the cytotoxicity of ZnO nanoparticles by coating them with a TiO_2 layer applying a sol – gel method and they observed moderate toxicity on the ZnO/TiO_2 core/shell structure compared with uncoated ZnO nanoparticles. The TiO_2 coating reduces toxicity by decreasing the release of zinc ions, also restricting the contact area of the ZnO cores with the cell. Luo et al. [149] applied polyethylene glycol coating on ZnO nanoparticles and reduced the cytotoxicity of ZnO nanoparticles. These findings bespeak the important role of the surface properties of ZnO nanoparticles on their cytotoxicity. Despite these studies, the toxicity mechanism of ZnO particles is still not well understood and further investigations are required to understand the mechanisms involved in the toxicity of ZnO nanoparticles and develop the safety rules regarding exposure to ZnO nanoparticles.

Silver (Ag) nanoparticles with extraordinary optical, electronic, catalytic, and more importantly antibacterial properties have found great applications in the development of functional polymer nanocomposites [150–152]. The widespread use of Ag nanoparticles in toys, clothing, and food packaging products raises serious concerns about their effects on human health and ecosystems. These concerns bespeak the necessity of detailed safety evaluations and obtaining a clear understanding of the impact of Ag nanoparticles on human health and the environment [152,153]. Ag nanoparticles are one of the most effective and frequently used antimicrobial agents in polymeric products [154,155]. The broad range of Ag nanoparticles' applications increases the possibility of exposure and consequent severe damage risks to the ecosystem. Despite several proposed mechanisms for the toxicity of Ag nanoparticle, its exact mechanisms are not well understood yet [156]. Some of the proposed mechanisms are physical damage and penetration that cause cell malfunction [157], dissolution of Ag + from the Ag [158], and stimulation of reactive oxygen species [159]. Various factors that can influence the level of toxicity and its mechanism are particle size, surface properties, and shape [160,161]. Nevertheless, as silver in its metallic state is considered as an inert metal, it can easily react with moisture and get ionized, and is highly reactive and able to bind to proteins, DNA, and RNA [162].

It is suggested that Ag nanoparticle morphological properties indirectly affect antimicrobial activity and toxicity by influencing the Ag + release rate [163]. Helmlinger et al. [160] correlated the dissolution kinetics with the particle size in which particles with a higher specific surface area (finer particles) show more toxicity than particles with smaller specific surface areas (larger particles). Considering the Ag^+ release as the main toxicity source for Ag nanoparticles antibacterial and toxic activity, these activities can be manipulated by controlling the Ag^+ release through tailoring the particle size, shape, applied coating, and the environmental parameters [163]. However, there are studies suggesting morphological factors of Ag nanoparticles as responsible for its toxic behavior rather than its ionized state. The nanoscale size of Ag nanoparticles is able to produce reactive oxygen species and oxidative stress, which can damage the cell [164–166]. On the other hand, some research proposes a combined effect in which Ag nanoparticles enter the cells and act as a source of Ag^+, leading to cytotoxic and genotoxic damage known as the "Trojan horse" effect [73,167]. The high production in the increasing utilization of Ag nanoparticles in commercial products raises serious concerns about its environmental effects and their impact on the ecosystem in recent years. Therefore, more research and investigations are needed to enhance current understanding of their toxic behavior and address these issues for their further development.

14.3 Environmental Benefits and Application of Polymeric Nanocomposites

Despite several studies mentioning the adverse effect of nanoparticles and nanocomposites on the environment and the ecosystem, this class of materials brings about several novel and unique beneficial applications to serve nature and the ecosystem. In this section, some of the well-established environmental applications of polymeric nanocomposites that ease and protect human life, save the environment, and provide better future ecosystems are discussed. Packaging, membranes, and energy are the selected application areas, among others that nanotechnologies and especially polymeric nanocomposites have revolutionized.

14.3.1 Food Packaging by Nanotechnology and Nanocomposites

Advances in the food packaging industries as an essential part of today's economy have enabled effective transportation, distribution, and preservation of food and ensure its proper physical and physicochemical conditions at the delivery point. The desired packaging material should possess mechanical, optical, and thermal properties that prevent the transfer of unwanted substances such as microbial contamination, moisture, and oxygen into/from the package. Over the last few decades, the utilization of polymers for food packaging has increased greatly because of their superior properties such as functionality, light weight, ease of processing, variety, and low cost over other traditional materials (metals, ceramics, and paper). However, pristine polymer-based packaging does not satisfy industry needs for a packaging that provides the capability to keep the packaged materials for a longer time. To address the mentioned limitations, polymeric nanocomposites prepared by functionalized nanoparticles provide an opportunity to produce the next-generation food packaging. Polymer nanocomposite–based packaging material shows several unique and enhanced properties such as high barrier, antimicrobial activities, and the capability to act as a sensor for the detection of food-relevant gases, or other molecules. In addition, the possibility to implement low amounts of nanoparticles and achieve higher enhanced properties compared with microcomposites has led to a reduction in cost. One of the novel packaging technologies enabled by functionalized nanoparticles is active and intelligent packaging in which the package materials are designed to interact with the food or the food environment. These interactions can be releasing, absorbing, or modifying substances into or from the packaged food or the environment surrounding. Fig. 14–5 represents some of the nanocomposite-based active packaging technologies [168]. The developments in this type of packaging materials are mostly focused on low permeable and antimicrobial films [168–170].

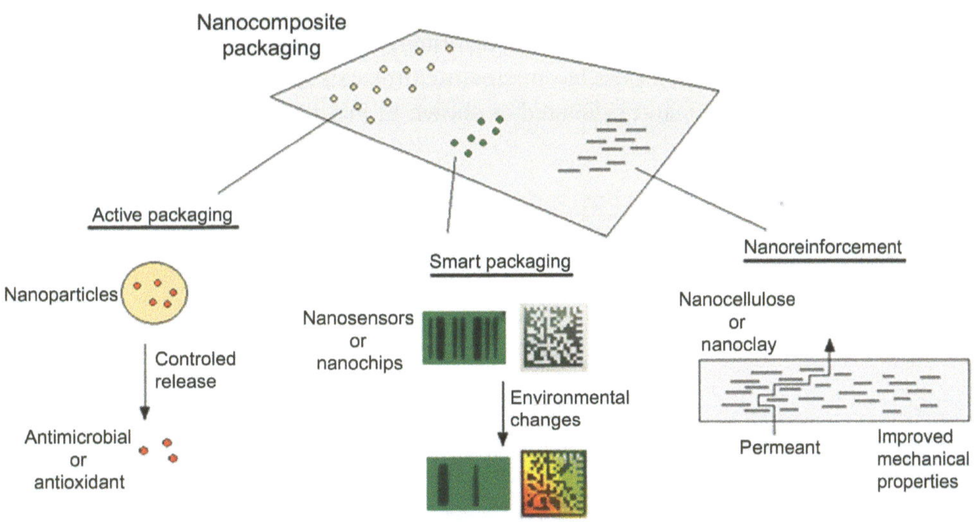

FIGURE 14–5 Applications of nanostructures in nanocomposite packaging. Active packaging can be developed through incorporation of nanoparticles encapsulating bioactive compounds with antimicrobial or antioxidant activity. The use of nanosensors or nanochips for development of smart/intelligent packaging may provide information about the condition of packaged food during transport and storage. Nanoreinforcement can be achieved by including nanofillers into biopolymer matrixes intended for packaging purposes, providing significant improvements in mechanical and barrier properties. *Reprinted from Brandelli A, Brum LFW, dos Santos JHZ. Nanostructured bioactive compounds for ecological food packaging. Environ Chem Lett 2017;15(2):193–204, by permission of Springer Nature.*

As mentioned earlier, one of the main limitations restricting the widespread application of polymeric materials in food packaging is their relatively high permeability to small molecules and moisture, which has an adverse effect on the lifetime of the packaged product. One of the effective methods to enhance the barrier properties of the polymeric materials is blending them with specific nanoparticles. In nanocomposites, the gas barrier performance is determined by nanoparticle properties, matrix permeability, nanoparticle–matrix interactions, and the orientation of the nanoparticles. Clay and graphene are the most common nanoparticles used for enhancing the barrier properties of the polymeric composites because of their high surface area to thickness ratio, and impermeability to most of the gases and water vapor. Even relatively low additions of nanoclay to a polymer matrix stems a reduction of transport cross section, and also increase of transport paths of penetrant molecules and causes extreme reductions in permeability [171]. Moreover, the addition of nanoplatelets results in modification of polymer chain mobility *and* provides lower available free volume to diffusing gas molecules and changes the solubility parameters [172,173]. The state of the dispersion and interface quality are also important parameters on nanocomposites' permeabilities, and play a critical role in the barrier performance of nanocomposites, which can be enhanced by proper surface modification of particles. However, low compatibility between polymer and nanoparticles results in a poor interfacial quality, which could lead to the presence of narrow gap around the nanofiller [174]. These gaps facilitate the flow and penetration of gas molecules and increase the permeability of the nanocomposite, which can have an adverse effect on the packaged material [174]. In addition, the polymer and nanoparticle compatibility state determines the dispersion state of the nanoparticles in the polymer matrix whose possible microstructures can be classified as phase-separated (microcomposite), intercalated, and exfoliated as shown in Fig. 14–6 [175]. In the case of phase-

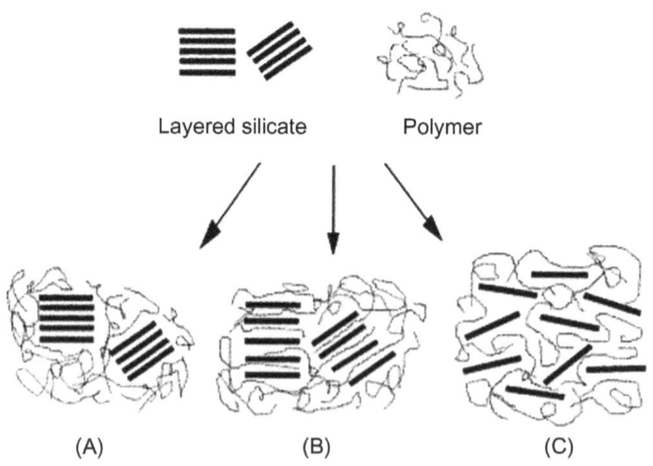

FIGURE 14–6 Scheme of different types of composite arising from the interaction of layered silicates and polymers: (A) phase-separated microcomposite, (B) intercalated nanocomposite, and (C) exfoliated nanocomposite. *Reprinted from Alexandre M, Dubois P. Polymer-layered silicate nanocomposites: preparation, properties and uses of a new class of materials. Mater Sci Eng Rep 2000;28(1):1–63, by permission of Elsevier.*

separated nanocomposites, nanoplatelet tactoids are formed in the matrix and no separation of nanoplatelets occurs and polymer chains surround clay nanoplatelets without penetration between the layers. In intercalated nanocomposites, some of the molecular polymer penetrates between nanoplates while the overall order of the nanoplatelets is preserved. In the exfoliated nanocomposites, the layers are completely separated and dispersed individually within the continuous polymer matrix. The highest barrier improvements are expected from fully exfoliated polymer nanocomposites because in this state the nanoparticle platelets will have the highest surface area [176].

Antibacterial activity is another desired property for food packaging materials that extends the shelf life and safety of food products by reducing the growth rate of microorganisms [177,178]. The polymeric nanocomposites based on metal nanoparticles such as Ag TiO_2 and ZnO as antimicrobial agents showed high antimicrobial performance for a broad spectrum of microorganisms [179,180]. The presence of antibacterial nanoparticles in polymeric nanocomposite structure not only inhibits initial undesirable microorganisms but also averts the residual activity over time by controlled migration of the compound into the food [181]. The antibacterial activity of nanoparticles stems from various mechanisms including direct interaction with the microbial cells and causing damage in the cells' structure, oxidizing cell components, and producing reactive oxygen species or dissolved heavy metal ions [182,183]. Fig. 14–7 shows various mechanisms of antimicrobial activities exerted by nanomaterials [182].

14.3.2 Nanocomposites in Membrane Technology

Membrane technology is widely applied in various industries such as water treatment, molecular separations, biomolecule purification, environmental remediation, gas separation,

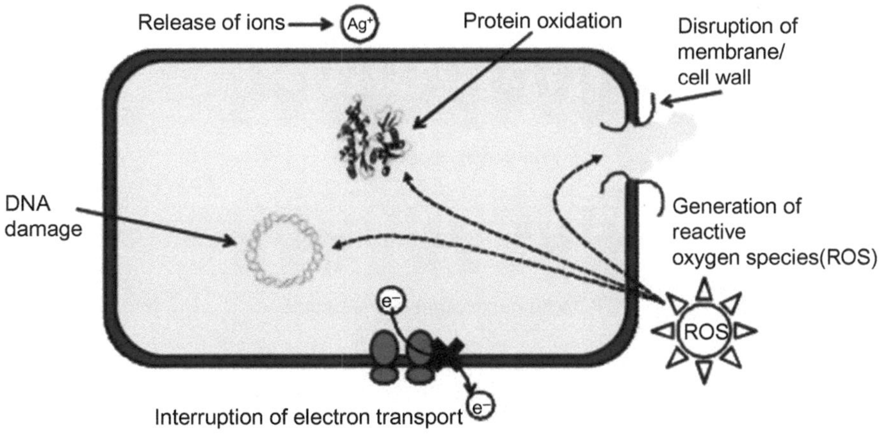

FIGURE 14–7 Various mechanisms of antimicrobial activities exerted by nanomaterials. *Reprinted from Li Q, Mahendra S, Lyon DY, Brunet L, Liga MV, Li D, et al. Antimicrobial nanomaterials for water disinfection and microbial control: potential applications and implications. Water Res 2008;42(18):4591–602, by permission of Elsevier.*

petroleum chemicals, and fuel production [184,185]. Polymeric membranes technology is an attractive alternative to inorganic membrane equivalents because of their high flexibility, low energy requirements, relatively low preparation cost, and environmental friendliness [186,187]. However, application of polymeric membranes is facing several challenges regarding their mechanical performance, thermal stability, permeability, and selectivity [188]. Nanomodifications and utilization of nanoparticles on the polymeric structures are considered as an effective way to address the mentioned challenges to produce membranes with higher reliability. These nanoparticle modified membranes are categorized into four distinct categories, namely, conventional nanocomposites, thin-film nanocomposites, thin-film composites with nanocomposite substrate, and surface located nanocomposites; these are schematically shown in Fig. 14–8 [186]. These nanocomposites can enhance the process efficiency and cost for various applications and have gained considerable attention in the cutting edge applications and high-performance membrane production in fuel cells, proton exchange membrane, sensors, batteries, solvents, and water treatment [189].

The nanocomposite structure combines the properties of each component and offers novel and unique properties that were not available previously. One of the important examples of nanocomposite technology development's effect on membrane application is the direct methanol fuel cell (DMFC). The conventional polymeric DMFC membranes' efficiency is constrained by the high rate of methanol permeation through a polymeric membrane, which causes the chemical reaction of the fuel with oxygen and depolarization of the cathode [190]. Therefore, it is highly desirable to have a membrane with reduced methanol

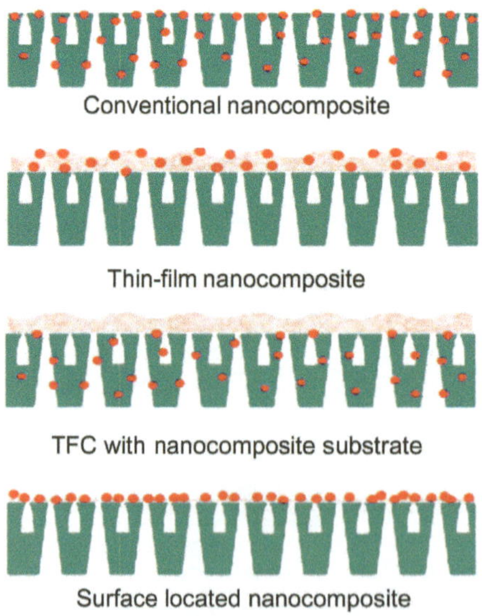

FIGURE 14–8 Typical types of nanocomposite membranes *TFC*, Thin-film composite. *Reprinted from Yin J, Deng B. Polymer–matrix nanocomposite membranes for water treatment. J Memb Sci 2015;479:256–75, by permission of Elsevier.*

permeation. One of the examples of utilization of nanocomposites in this area of application is proposed by Song et al. [190] who used Nafion polymer-layered silicate nanocomposite as a membrane in which the strong interaction of Nafion polymer chains with exfoliated clays significantly decreased methanol crossover through nanocomposite membranes by only 1 wt.% organoclay fillers. In addition, nanoclay improved the thermal decomposition temperature and mechanical properties of the Nafion-based membrane [190].

Another example of the enhanced performance of membranes through the incorporation of nanoparticles is membranes used in water and wastewater treatment in which contamination of membrane reduces the flux by the time [187,191]. To address this issue, Bae et al. [192] fabricated TiO_2 entrapped membranes from three different polymers of polysulfone, polyvinylidene fluoride, and polyacrylonitrile to be used in filtration of the activated sludge. The TiO_2 entrapped membrane showed lower flux decline compared with neat polymeric membrane regardless of the polymeric matrix. In another work, Cao et al. [193] prepared a poly(ethylene oxide)/graphene oxide proton conductive membrane with enhanced performance compared with commercially available ones. The presence of GO in the poly(ethylene oxide) matrix increases the conductivity through the proton released from the COOH groups on the GO sheets as an ion conductivity mechanism. There are several other examples of nanocomposite-based membrane designs in the literature that can further improve the efficiency and generate a new generation of membranes for future applications in energy and various environmental applications, which can boost the future view of our ecosystem.

14.3.3 Nanocomposites in Energy Applications

Sustainable, efficient, and clean energy storage is another area that is taken to a whole new level by nanocomposite materials. Hydrogen is one of the renewable, convenient, safe, versatile, and clean energy fuel sources, which provides a significant reduction in the greenhouse gas emissions and fossil fuels consumption, and has high energy conversion efficiency [194,195]. The hydrogen storage systems are regarded as the key challenge in large-scale and commercial applications of hydrogen energy [196]. As pressurized tank and cryogenic liquid hydrogen techniques fail to satisfy the primary requirements for a hydrogen storage system such as being inexpensive, safe, delivering low efficiency, and material-based storage systems are expected to provide a final solution for this issue [196,197].

However, simple material systems with only one component failed to operate in ambient temperature ranges, and they suffer from low efficiency of fueling and serious concerns about their safety and slow kinetics [198]. The polymeric nanocomposites are considered to have potential promise for hydrogen storage among other counterparts [199]. Saner Okan et al. [27] combined the beneficial features of polypyrrole (PPy) with those of CNT and considerably improved the hydrogen storage capacity of the nanocomposite structure compared with the neat polymer and pure CNT. In the mentioned work, CNT and PPy adsorbed 0.46 wt.% and 0.14 wt.% hydrogen while their composite adsorbed 1.66 wt.% hydrogen [27]. Fig. 14–9A–C exhibit the nanostructures of PPy, CNT, PPy:CNT composite, respectively. Furthermore, Fig. 14–9D shows hydrogen sorption curves of pristine CNT, PPy, oxCNT, and

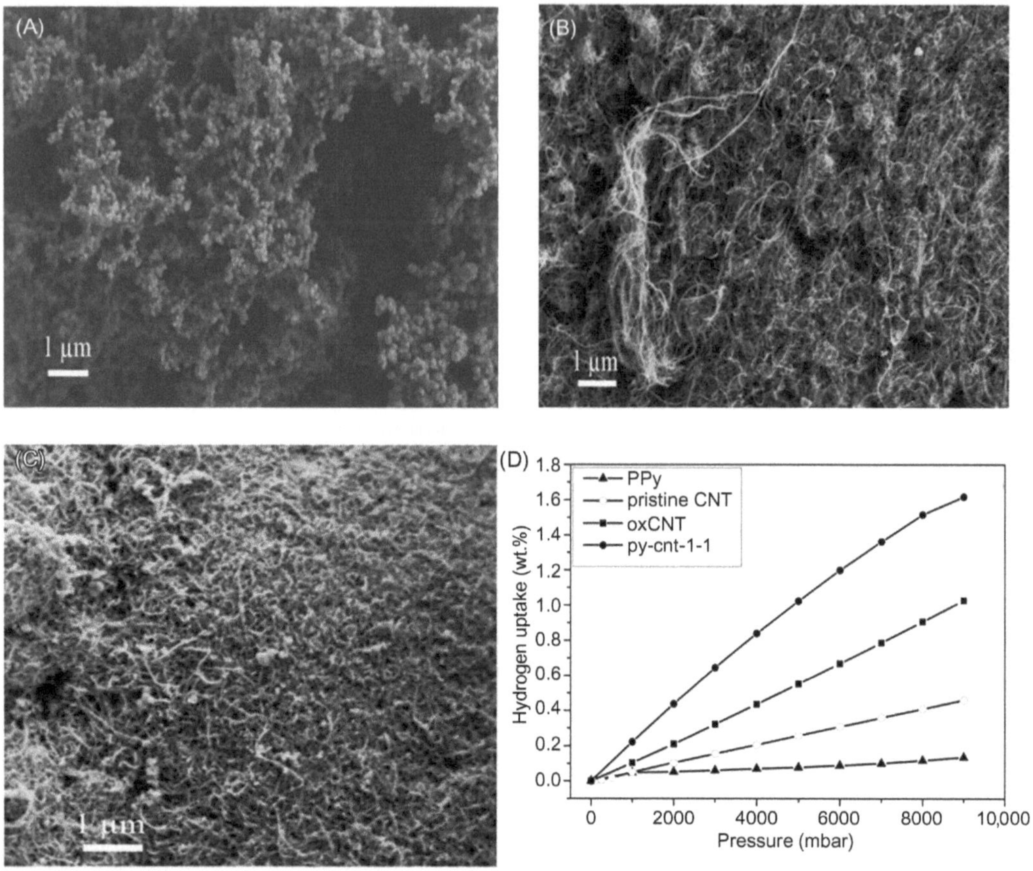

FIGURE 14–9 SEM images of (A) pristine PPy, (B) pristine CNT, (C) Py:CNT = 1:1 composite, and (D) the comparison of adsorption isotherms of pristine CNT, PPy, oxCNT, and Py:CNT = 1:1. *Reprinted from Okan BS, Zanjani JSM, Letofsky-Papst I, Cebeci FÇ, Menceloglu YZ. Morphology-controllable synthesis and characterization of carbon nanotube/polypyrrole composites and their hydrogen storage capacities. Mater Chem Phys 2015;167(Suppl. C):171–80, by permission of Elsevier.*

Py:CNT composite in the range of 0–9000 mbar pressure at room temperature indicating that pristine CNT, PPy, oxCNT, and Py:CNT adsorb 0.46, 0.14, 1.03, and 1.66 wt.% hydrogen, respectively bespeaking that the hydrogen storage properties of the produced nanocomposites are considerably higher than the individual components.

In another study Muthu et al. [200] employed sulfonated poly(ether-ether-ketone) (SPEEK) and hexagonal boron nitride (h-BN) nanoparticles, to obtain high-performance hydrogen storage material and they enhanced the hydrogen storage from 0.66 wt.% for neat SPEEK to 2.98 wt.% in SPEEK-BN nanocomposite. Kim et al. [201] reported hydrogen adsorption in the polyaniline–vanadium oxide nanocomposites of up to ~1.8 wt% at 77 K and ~0.16 wt% at 298 K whereas neither polyaniline (~0.2 wt% at 77 K) nor pristine vanadium pentoxide powder ($\sim$0.2 wt% at 77 K) were able to adsorb similar amount of hydrogen. Makridis et al. [202] used Mg-nanoparticles produced by laser ablation technique as shown in Fig. 14–10 and observed that Mg-nanoparticles in a polymer matrix exhibit more rapid hydrogen uptake compared with Mg-pure types (<20 minutes at 250°C) with a high capacity of 6 wt.% with excellent reversibility under vacuum at 250°C, which is a relatively low temperature with regard to the necessary $\sim$330°C for Mg-bulk materials. These studies bespeak the high potential of nanocomposites for hydrogen storage in the near future.

14.4 Conclusion and Outlook

The rapid development of nanoparticle-based nanocomposites has a huge impact on many technological areas. There are a vast variety of nanoparticles that are used in the preparation of polymeric nanocomposites in which a small change in their physical or chemical structure leads to a completely different property and a set of novel applications. Engineers and researchers are in a constant attempt to create novel nanomaterials or modify current ones to achieve many advantageous properties that were not available before. However, some of these properties that are desired from a technical point of view are undesirable and harmful to the environment and ecosystem. Therefore, there is a need for extensive toxicological research for this class of materials to avoid any possible risk and facilitate their safe implementation in future products. The first part of this chapter provides a short review on the

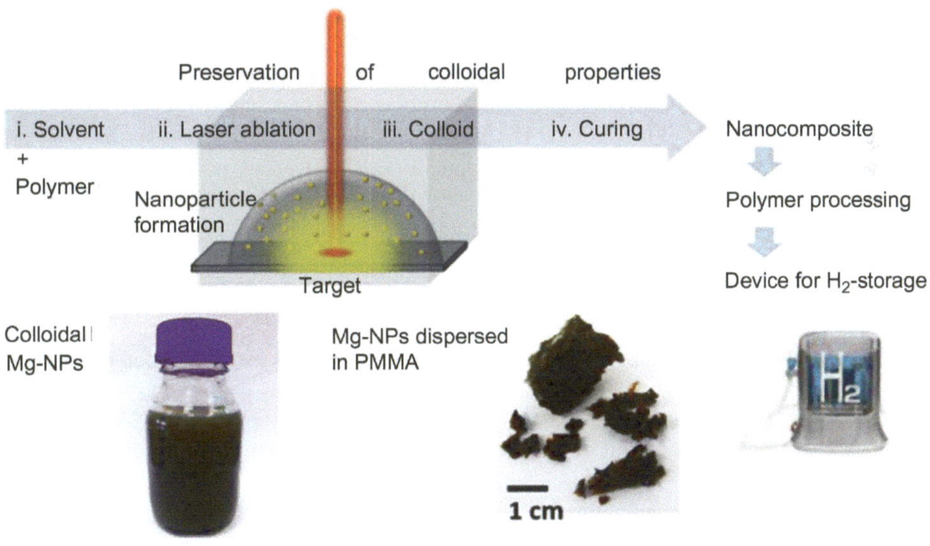

FIGURE 14–10 Laser-generated nanoparticles for nanocomposites. *Reprinted from Makridis SS, Gkanas EI, Panagakos G, Kikkinides ES, Stubos AK, Wagener P, et al. Polymer-stable magnesium nanocomposites prepared by laser ablation for efficient hydrogen storage. Int J Hydrogen Energy 2013;38(26):11530–5, by permission of Elsevier.*

toxicity of some of the most important and commercially used nanoparticles utilized in polymeric nanocomposite production. Despite several research attempts focusing on the toxic behavior of nanomaterials, it has been seen there is no clear and solid understanding of their molecular scale mechanisms and proper procedures to prevent their adverse effect on the environment. It has been discussed that functionalization can increase or decrease nanomaterials' toxic effect depending on the nanomaterials' nature and surface functionalization type. The complex and in several cases the unknown toxic behavior of nanoparticles raises serious concern about their effect on human health and ecosystems and signifies the necessity of further scientific exploration and governmental regulations on their use. In addition, nanoparticle-based polymeric nanocomposites are a useful material class for a wide variety of applications that can promote human life and also serve our ecosystem. The replacement of conventional materials with their nanocomposite counterparts has increased the efficiency, lifetime, and reliability, and enabled several novel applications in various areas such as food packaging, membranes, and energy storage systems, among others. To reiterate, nanocomposites with their unique properties are key for further technological advancement in many areas. However, further multidisciplinary research is needed to evolve the understanding of their environmental effects and produce nanocomposites with minimum adverse effect on the ecosystem.

References

[1] Lordan S, Higginbotham CL. Effect of serum concentration on the cytotoxicity of clay particles. Cell Biol Int 2012;36(1):57–61.

[2] del Pino P, Mitchell SG, Pelaz B. Design and characterization of functional nanoparticles for enhanced bio-performance. In: Guisan JM, editor. Immobilization of enzymes and cells. 3rd ed. Totowa, NJ: Humana Press; 2013. p. 165–207.

[3] Tritschler U, Pearce S, Gwyther J, Whittell GR, Manners I. 50th anniversary perspective: functional nanoparticles from the solution self-assembly of block copolymers. Macromolecules 2017;50(9):3439–63.

[4] The EU Commission Recommendation on the Definition of Nanomaterial E, European Commission, Brussels; 2011.

[5] Sreeprasad TS, Pradeep T. Noble metal nanoparticles. In: Vajtai R, editor. Springer handbook of nanomaterials. Berlin, Heidelberg; 2013. p. 303–388.

[6] Shindu CT, Harshita, Pawan Kumar M, Sushama T. Ceramic nanoparticles: fabrication methods and applications in drug delivery. Curr Pharm Des 2015;21(42):6165–88.

[7] Seyyed Monfared Zanjani J, Saner Okan B, Letofsky-Papst I, Yildiz M, Menceloglu YZ. Rational design and direct fabrication of multi-walled hollow electrospun fibers with controllable structure and surface properties. Eur Polym J 2015;62:66–76.

[8] Bangal M, Ashtaputre S, Marathe S, Ethiraj A, Hebalkar N, Gosavi SW, et al. Semiconductor nanoparticles. Hyperfine Interact 2005;160(1):81–94.

[9] Self-healing thermosetting composites: concepts, chemistry, and future advances In: Zanjani JSM, Okan Burcu S, Menceloglu Y, Yildiz M, Delville M, Taubert A, editors. Hybrid organic-inorganic interfaces: towards advanced functional materials. Wiley; 2017[chapter 3] . Available from: https://doi.org/10.1002/9783527807130.ch3.

[10] Ullah M, Seyyed Monfared Zanjani J, Haghighi Poudeh L, Siddiq M, Saner Okan B, Yildiz M, et al. Manufacturing functionalized mono-crystalline diamond containing electrospun fibers reinforced epoxy composites with improved mechanical characteristics. Diam Relat Mater 2017;76:90–6.

[11] Monfared Zanjani JS, Okan BS, Yildiz M, Menceloglu Y. Fabrication and morphological investigation of multi-walled electrospun polymeric nanofibers. MRS Proc 2014;1621:119–26.

[12] Haghighi Poudeh L, Saner Okan B, Seyyed Monfared Zanjani J, Yildiz M, Menceloglu Y. Design and fabrication of hollow and filled graphene-based polymeric spheres via core-shell electrospraying. RSC Adv 2015;5(111):91147–57.

[13] Salazar-Alvarez G, Qin J, Šepelák V, Bergmann I, Vasilakaki M, Trohidou KN, et al. Cubic versus spherical magnetic nanoparticles: the role of surface anisotropy. J Am Chem Soc 2008;130(40):13234–9.

[14] Zarghami Z, Ramezani M, Motevalli K. ZnO nanorods/nanoparticles: novel hydrothermal synthesis, characterization and formation mechanism for increasing the efficiency of dye-sensitized solar cells. J Clust Sci 2016;27(4):1451–62.

[15] Monfared Zanjani JS, Okan BS, Menceloglu YZ, Yildiz M. Nano-engineered design and manufacturing of high-performance epoxy matrix composites with carbon fiber/selectively integrated graphene as multi-scale reinforcements. RSC Adv 2016;6(12):9495–506.

[16] Jo DH, Kim JH, Lee TG, Kim JH. Size, surface charge, and shape determine therapeutic effects of nanoparticles on brain and retinal diseases. Nanomedicine 2015;11(7):1603–11.

[17] Liu W-T. Nanoparticles and their biological and environmental applications. J Biosci Bioeng 2006;102 (1):1–7.

[18] Tsoli M, Kuhn H, Brandau W, Esche H, Schmid G. Cellular uptake and toxicity of Au55 clusters. Small 2005;1(8–9):841–4.

[19] Donaldson K, Seaton A. A short history of the toxicology of inhaled particles. Part Fibre Toxicol 2012;9:13.

[20] Donaldson K, Stone V, Clouter A, Renwick L, MacNee W. Ultrafine particles. Occup Environ Med 2001;58(3):211.

[21] Shvedova AA, Kisin E, Murray AR, Johnson VJ, Gorelik O, Arepalli S, et al. Inhalation vs. aspiration of single-walled carbon nanotubes in C57BL/6 mice: inflammation, fibrosis, oxidative stress, and mutagenesis. Am J Physiol Lung Cell Mol Physiol 2008;295(4):L552–65.

[22] Donaldson K, Stone V, Tran CL, Kreyling W, Borm PJA. Nanotoxicology. Occup Environ Med 2004;61 (9):727.

[23] Tran CL, DBRTCASADJKD. Inhalation of poorly soluble particles. II. influence of particle surface area on inflammation and clearance. Inhal Toxicol 2000;12(12):1113–26.

[24] Ogden LE. Nanoparticles in the environment: tiny size, large consequences? Bioscience 2013;63(3):236.

[25] Knaapen AM, Borm PJA, Albrecht C, Schins RPF. Inhaled particles and lung cancer. Part A: Mechanisms. Int J Cancer 2004;109(6):799–809.

[26] Donaldson K, Aitken R, Tran L, Stone V, Duffin R, Forrest G, et al. Carbon nanotubes: a review of their properties in relation to pulmonary toxicology and workplace safety. Toxicol Sci 2006;92(1):5–22.

[27] Okan BS, Zanjani JSM, Letofsky-Papst I, Cebeci FÇ, Menceloglu YZ. Morphology-controllable synthesis and characterization of carbon nanotube/polypyrrole composites and their hydrogen storage capacities. Mater Chem Phys 2015;167(Suppl. C):171–80.

[28] Yumura M. Carbon nanotube industrial applications. AIST Today (Int Ed) 2003;10:8–9.

[29] Lu F, Gu L, Meziani MJ, Wang X, Luo PG, Veca LM, et al. Advances in bioapplications of carbon nanotubes. Adv Mater 2009;21(2):139–52.

[30] Ding R, Lu G, Yan Z, Wilson M. Recent advances in the preparation and utilization of carbon nanotubes for hydrogen storage. J Nanosci Nanotechnol 2001;1(1):7–29.

[31] Tang Z, Zhang L, Wang N, Zhang X, Wen G, Li G, et al. Superconductivity in 4 angstrom single-walled carbon nanotubes. Science 2001;292(5526):2462–5.

[32] Baughman RH, Zakhidov AA, de Heer WA. Carbon nanotubes—the route toward applications. Science 2002;297(5582):787.

[33] Kaseem M, Hamad K, Ko YG. Fabrication and materials properties of polystyrene/carbon nanotube (PS/CNT) composites: a review. Eur Polym J 2016;79:36–62.

[34] Dai H. Carbon nanotubes: synthesis, integration, and properties. Acc Chem Res 2002;35(12):1035–44.

[35] Dai H. Nanotube growth and characterization. In: Dresselhaus MS, Dresselhaus G, Avouris P, editors. Carbon nanotubes: synthesis, structure, properties, and applications. Berlin, Heidelberg: Springer; 2001. p. 29–53.

[36] Lam C-W, James JT, McCluskey R, Hunter RL. Pulmonary toxicity of single-wall carbon nanotubes in mice 7 and 90 days after intratracheal instillation. Toxicol Sci 2004;77(1):126–34.

[37] Wu Y, Hudson JS, Lu Q, Moore JM, Mount AS, Rao AM, et al. Coating single-walled carbon nanotubes with phospholipids. J Phys Chem B 2006;110(6):2475–8.

[38] Helland A, Wick P, Koehler A, Schmid K, Som C. Reviewing the environmental and human health knowledge base of carbon nanotubes. Ciência Saúde Coletiva 2008;13(2):441–52.

[39] Liu S, Zeng TH, Hofmann M, Burcombe E, Wei J, Jiang R, et al. Antibacterial activity of graphite, graphite oxide, graphene oxide, and reduced graphene oxide: Membrane and oxidative stress. ACS Nano 2011;5(9):6971–80.

[40] Kharissova OV, Kharisov BI, de Casas Ortiz EG. Dispersion of carbon nanotubes in water and non-aqueous solvents. RSC Adv 2013;3(47):24812–52.

[41] Lecoanet HF, Bottero J-Y, Wiesner MR. Laboratory assessment of the mobility of nanomaterials in porous media. Environ Sci Technol 2004;38(19):5164–9.

[42] Díaz-Flores PE, Arcibar-Orozco JA, Perez-Aguilar NV, Rangel-Mendez JR, Ovando Medina VM, Alcalá-Jáuegui JA. Adsorption of organic compounds onto multiwall and nitrogen-doped carbon nanotubes: insights into the adsorption mechanisms. Water Air Soil Pollut 2017;228(4):133.

[43] Shen M, Wang SH, Shi X, Chen X, Huang Q, Petersen EJ, et al. Polyethyleneimine-mediated functionalization of multiwalled carbon nanotubes: synthesis, characterization, and in vitro toxicity assay. J Phys Chem C 2009;113(8):3150–6.

[44] Pulskamp K, Diabaté S, Krug HF. Carbon nanotubes show no sign of acute toxicity but induce intracellular reactive oxygen species in dependence on contaminants. Toxicol Lett 2007;168(1):58–74.

[45] Jos A, Pichardo S, Puerto M, Sánchez E, Grilo A, Cameán AM. Cytotoxicity of carboxylic acid functionalized single wall carbon nanotubes on the human intestinal cell line Caco-2. Toxicol In Vitro 2009;23(8):1491–6.

[46] Saxena RK, Williams W, McGee JK, Daniels MJ, Boykin E, Ian Gilmour M. Enhanced in vitro and in vivo toxicity of poly-dispersed acid-functionalized single-wall carbon nanotubes. Nanotoxicology 2007;1(4):291–300.

[47] Sayes CM, Liang F, Hudson JL, Mendez J, Guo W, Beach JM, et al. Functionalization density dependence of single-walled carbon nanotubes cytotoxicity in vitro. Toxicol Lett 2006;161(2):135–42.

[48] Seyyed Monfared Zanjani J, Saner Okan B, Menceloglu Y. Manufacturing of multilayer graphene oxide/poly(ethylene terephthalate) nanocomposites with tunable crystallinity, chain orientations and thermal transitions. Mater Chem Phys 2016;176(Suppl. C):58–67.

[49] Okan BS, Marset A, Seyyed Monfared Zanjani J, Sut PA, Sen O, Çulha M, et al. Thermally exfoliated graphene oxide reinforced fluorinated pentablock poly(L-lactide-co-ε-caprolactone) electrospun scaffolds: insight into antimicrobial activity and biodegradation. J Appl Polym Sci 2016;133(22) n/a-n/a.

[50] Seyyed Monfared Zanjani J, Saner Okan B, Pappas P-N, Galiotis C, Ziya Menceloglu Y, Yildiz M. Tailoring viscoelastic response, self-heating and deicing properties of carbon-fibre reinforced epoxy composites by graphene modification. Compos Part A Appl S.**106**, 2018, 1–10.

[51] Dhand V, Rhee KY, Ju Kim H, Ho Jung D. A comprehensive review of graphene nanocomposites: research status and trends. J Nanomater 2013;2013:14.

[52] Kiew SF, Kiew LV, Lee HB, Imae T, Chung LY. Assessing biocompatibility of graphene oxide-based nanocarriers: a review. J Control Release 2016;226:217–28.

[53] Segal M. Selling graphene by the ton. Nat Nanotechnol 2009;4:612.

[54] Mas B, Fernández-Blázquez JP, Duval J, Bunyan H, Vilatela JJ. Thermoset curing through Joule heating of nanocarbons for composite manufacture, repair and soldering. Carbon NY 2013;63(Suppl. C):523–9.

[55] Zhang K, Kemp KC, Chandra V. Homogeneous anchoring of TiO_2 nanoparticles on graphene sheets for waste water treatment. Mater Lett 2012;81(Suppl. C):127–30.

[56] Chung H, Kim MJ, Ko K, Kim JH, Kwon H-A, Hong I, et al. Effects of graphene oxides on soil enzyme activity and microbial biomass. Sci Total Environ 2015;514(Suppl. C):307–13.

[57] Begum P, Fugetsu B. Induction of cell death by graphene in *Arabidopsis thaliana* (Columbia ecotype) T87 cell suspensions. J Hazard Mater 2013;260(Suppl. C):1032–41.

[58] Eda G, Fanchini G, Chhowalla M. Large-area ultrathin films of reduced graphene oxide as a transparent and flexible electronic material. Nat Nanotechnol 2008;3:270.

[59] Jasim DA, Boutin H, Fairclough M, Ménard-Moyon C, Prenant C, Bianco A, et al. Thickness of functionalized graphene oxide sheets plays critical role in tissue accumulation and urinary excretion: a pilot PET/CT study. Appl Mater Today 2016;4(Suppl. C):24–30.

[60] Akhavan O, Ghaderi E. Toxicity of graphene and graphene oxide nanowalls against bacteria. ACS Nano 2010;4(10):5731–6.

[61] Venza M, Visalli M, Beninati C, De Gaetano GV, Teti D, Venza I. Cellular mechanisms of oxidative stress and action in melanoma. Oxid Med Cell Longev 2015;2015:481782.

[62] Hirtz M, Oikonomou A, Georgiou T, Fuchs H, Vijayaraghavan A. Multiplexed biomimetic lipid membranes on graphene by dip-pen nanolithography. Nat Commun 2013;4:2591.

[63] Liu S, Hu M, Zeng TH, Wu R, Jiang R, Wei J, et al. Lateral dimension-dependent antibacterial activity of graphene oxide sheets. Langmuir 2012;28(33):12364–72.

[64] Zhang B, Wei P, Zhou Z, Wei T. Interactions of graphene with mammalian cells: molecular mechanisms and biomedical insights. Adv Drug Deliv Rev 2016;105(Part B):145–62.

[65] Zhao J, Wang Z, White JC, Xing B. Graphene in the aquatic environment: adsorption, dispersion, toxicity and transformation. Environ Sci Technol 2014;48(17):9995–10009.

[66] Guardia L, Villar-Rodil S, Paredes JI, Rozada R, Martínez-Alonso A, Tascón JMD. UV light exposure of aqueous graphene oxide suspensions to promote their direct reduction, formation of graphene–metal nanoparticle hybrids and dye degradation. Carbon NY 2012;50(3):1014–24.

[67] Akhavan O, Ghaderi E. Photocatalytic reduction of graphene oxide nanosheets on TiO_2 thin film for photoinactivation of bacteria in solar light irradiation. J Phys Chem C 2009;113(47):20214–20.

[68] Ye X, Zhou Q, Jia C, Tang Z, Zhu Y, Wan Z. Producing large-area, foldable graphene paper from graphite oxide suspensions by in-situ chemical reduction process. Carbon N Y 2017;114(Suppl. C):424–34.

[69] Kurapati R, Russier J, Squillaci MA, Treossi E, Ménard-Moyon C, Del Rio-Castillo AE, et al. Dispersibility-dependent biodegradation of graphene oxide by myeloperoxidase. Small 2015;11 (32):3985–94.

[70] Chen M, Qin X, Zeng G. Biodegradation of carbon nanotubes, graphene, and their derivatives. Trends Biotechnol 2017;35(9):836–46.

[71] Kotchey GP, Allen BL, Vedala H, Yanamala N, Kapralov AA, Tyurina YY, et al. The enzymatic oxidation of graphene oxide. ACS Nano 2011;5(3):2098–108.

[72] Modugno G, Ksar F, Battigelli A, Russier J, Lonchambon P, Eleto da Silva E, et al. A comparative study on the enzymatic biodegradability of covalently functionalized double- and multi-walled carbon nanotubes. Carbon NY 2016;100(Suppl. C):367–74.

[73] Duncan TV. Applications of nanotechnology in food packaging and food safety: barrier materials, antimicrobials and sensors. J Colloid Interface Sci 2011;363(1):1–24.

[74] Okada A, Kawasumi M, Kurauchi T, Kamigaito O. Synthesis and characterization of a nylon 6-clay hybrid. Abstracts of papers of the American Chemical Society, Washington, DC; 1987. p. 10-MACR.

[75] Okada A, Usuki A. Twenty years of polymer-clay nanocomposites. Macromol Mater Eng 2006;291 (12):1449–76.

[76] Sinha Ray S, Okamoto M. Polymer/layered silicate nanocomposites: a review from preparation to processing. Prog Polym Sci 2003;28(11):1539–641.

[77] Manias E, Touny A, Wu L, Strawhecker K, Lu B, Chung TC. Polypropylene/montmorillonite nanocomposites. review of the synthetic routes and materials properties. Chem Mater 2001;13(10):3516–23.

[78] Lordan S, Kennedy JE, Higginbotham CL. Cytotoxic effects induced by unmodified and organically modified nanoclays in the human hepatic HepG2 cell line. J Appl Toxicol 2011;31(1):27–35.

[79] Jayrajsinh S, Shankar G, Agrawal YK, Bakre L. Montmorillonite nanoclay as a multifaceted drug-delivery carrier: a review. J Drug Deliv Sci Technol 2017;39(Suppl. C):200–9.

[80] Hong S, Hessler JA, Banaszak Holl MM, Leroueil P, Mecke A, Orr BG. Physical interactions of nanoparticles with biological membranes: the observation of nanoscale hole formation. J Chem Health Saf 2006;13(3):16–20.

[81] Verma NK, Moore E, Blau W, Volkov Y, Ramesh Babu P. Cytotoxicity evaluation of nanoclays in human epithelial cell line A549 using high content screening and real-time impedance analysis. J Nanoparticle Res 2012;14(9):1137.

[82] Baek M, Lee J-A, Choi S-J. Toxicological effects of a cationic clay, montmorillonite in vitro and in vivo. Mol Cell Toxicol 2012;8(1):95–101.

[83] Murphy EJ, Roberts E, Anderson DK, Horrocks LA. Cytotoxicity of aluminum silicates in primary neuronal cultures. Neuroscience 1993;57(2):483–90.

[84] Wagner A, Eldawud R, White A, Agarwal S, Stueckle TA, Sierros KA, et al. Toxicity evaluations of nanoclays and thermally degraded byproducts through spectroscopical and microscopical approaches. Biochim Biophys Acta 2017;1861(1 Pt A):3406–15.

[85] Xia Y, Rubino M, Auras R. Release of nanoclay and surfactant from polymer–clay nanocomposites into a food simulant. Environ Sci Technol 2014;48(23):13617–24.

[86] Prado LASdA, Karthikeyan CS, Schulte K, Nunes SP, de Torriani IL. Organic modification of layered silicates: structural and thermal characterizations. J Non-Cryst Solids 2005;351(12):970–5.

[87] Zhang M, Lu Y, Li X, Chen Q, Lu L, Xing M, et al. Studying the cytotoxicity and oxidative stress induced by two kinds of bentonite particles on human B lymphoblast cells in vitro. Chem Biol Interact 2010;183 (3):390–6.

[88] Yoshida T, Yoshioka Y, Matsuyama K, Nakazato Y, Tochigi S, Hirai T, et al. Surface modification of amorphous nanosilica particles suppresses nanosilica-induced cytotoxicity, ROS generation, and DNA damage in various mammalian cells. Biochem Biophys Res Commun 2012;427(4):748–52.

[89] Liu Q, Yu H, Tan Z, Cai H, Ye W, Zhang M, et al. In vitro assessing the risk of drug-induced cardiotoxicity by embryonic stem cell-based biosensor. Sens Actuators B Chem 2011;155(1):214–19.

[90] Janer G, Fernández-Rosas E, Mas del Molino E, González-Gálvez D, Vilar G, López-Iglesias C, et al. In vitro toxicity of functionalised nanoclays is mainly driven by the presence of organic modifiers. Nanotoxicology 2014;8(3):279–94.

[91] Sachin K, Anupam M, Kaushik C. Effect of organically modified clay on mechanical properties, cytotoxicity and bactericidal properties of poly(ϵ-caprolactone) nanocomposites. Mater Res Express 2014;1(4):045302.

[92] Maisanaba S, Ortuño N, Jordá-Beneyto M, Aucejo S, Jos Á. Development, characterization and cytotoxicity of novel silane-modified clay minerals and nanocomposites intended for food packaging. Appl Clay Sci 2017;138(Suppl. C):40–7.

[93] Maisanaba S, Pichardo S, Puerto M, Gutiérrez-Praena D, Cameán AM, Jos A. Toxicological evaluation of clay minerals and derived nanocomposites: a review. Environ Res 2015;138(Suppl. C):233–54.

[94] Abdizadeh H, Atilgan AR, Atilgan C, Dedeoglu B. Computational approaches for deciphering the equilibrium and kinetic properties of iron transport proteins. Metallomics 2017;9(11):1513–33.

[95] Wang Z, Zhang L, Zhao J, Xing B. Environmental processes and toxicity of metallic nanoparticles in aquatic systems as affected by natural organic matter. Environ Sci Nano 2016;3(2):240–55.

[96] Li Y-F, Chen C. Fate and toxicity of metallic and metal-containing nanoparticles for biomedical applications. Small 2011;7(21):2965–80.

[97] Abdizadeh H, Atilgan AR, Atilgan C. Detailed molecular dynamics simulations of human transferrin provide insights into iron release dynamics at serum and endosomal pH. J Biol Inorg Chem 2015;20(4):705–18.

[98] Cox A, Venkatachalam P, Sahi S, Sharma N. Reprint of: silver and titanium dioxide nanoparticle toxicity in plants: a review of current research. Plant Physiol Biochem 2017;110(Suppl. C):33–49.

[99] Hong F, Yu X, Wu N, Zhang Y-Q. Progress of in vivo studies on the systemic toxicities induced by titanium dioxide nanoparticles. Toxicol Res 2017;6(2):115–33.

[100] Mallakpour S. Production, characterization, and surface morphology of novel aromatic poly(amide-ester-imide)/functionalized TiO_2 nanocomposites via ultrasonication assisted process. Polym Bull 2017;74(7):2465–77.

[101] Kubacka A, Serrano C, Ferrer M, Lünsdorf H, Bielecki P, Cerrada ML, et al. High-performance dual-action polymer – TiO_2 nanocomposite films via melting processing. Nano Lett 2007;7(8):2529–34.

[102] Fu G, Vary PS, Lin C-T. Anatase TiO_2 nanocomposites for antimicrobial coatings. J Phys Chem B 2005;109(18):8889–98.

[103] Dutilleul P, Han L, Valladares F, Messier C. Crown traits of coniferous trees and their relation to shade tolerance can differ with leaf type: a biophysical demonstration using computed tomography scanning data. Front Plant Sci 2015;6:172.

[104] Kuku G, Culha M. Investigating the origins of toxic response in TiO_2 nanoparticle-treated cells. Nanomaterials 2017;7(4):83.

[105] Jin C, Liu Y, Sun L, Chen T, Zhang Y, Zhao A, et al. Metabolic profiling reveals disorder of carbohydrate metabolism in mouse fibroblast cells induced by titanium dioxide nanoparticles. J Appl Toxicol 2013;33(12):1442–50.

[106] Jin C-Y, Zhu B-S, Wang X-F, Lu Q-H. Cytotoxicity of titanium dioxide nanoparticles in mouse fibroblast cells. Chem Res Toxicol 2008;21(9):1871–7.

[107] Demir E, Kaya N, Kaya B. Genotoxic effects of zinc oxide and titanium dioxide nanoparticles on root meristem cells of *Allium cepa* by comet assay. Turk J Biol 2014;38(1):31–9.

[108] Golbamaki N, Rasulev B, Cassano A, Robinson RLM, Benfenati E, Leszczynski J, et al. Genotoxicity of metal oxide nanomaterials: review of recent data and discussion of possible mechanisms. Nanoscale 2015;7(6):2154–98.

[109] Yu Q, Wang H, Peng Q, Li Y, Liu Z, Li M. Different toxicity of anatase and rutile TiO_2 nanoparticles on macrophages: involvement of difference in affinity to proteins and phospholipids. J Hazard Mater 2017;335:125–34.

[110] Liao Y, Que W, Jia Q, He Y, Zhang J, Zhong P. Controllable synthesis of brookite/anatase/rutile TiO_2 nanocomposites and single-crystalline rutile nanorods array. J Mater Chem 2012;22(16):7937–44.

[111] Johnston HJ, Hutchison GR, Christensen FM, Peters S, Hankin S, Stone V. Identification of the mechanisms that drive the toxicity of TiO(2)particulates: the contribution of physicochemical characteristics. Part Fibre Toxicol 2009;6 33-.

[112] Jin C, Tang Y, Yang FG, Li XL, Xu S, Fan XY, et al. Cellular toxicity of TiO_2 nanoparticles in anatase and rutile crystal phase. Biol Trace Elem Res 2011;141(1):3–15.

[113] Gurr J-R, Wang ASS, Chen C-H, Jan K-Y. Ultrafine titanium dioxide particles in the absence of photoactivation can induce oxidative damage to human bronchial epithelial cells. Toxicology 2005;213 (1):66–73.

[114] Fabian E, Landsiedel R, Ma-Hock L, Wiench K, Wohlleben W, van Ravenzwaay B. Tissue distribution and toxicity of intravenously administered titanium dioxide nanoparticles in rats. Arch Toxicol 2008;82 (3):151–7.

[115] Chen J, Dong X, Zhao J, Tang G. In vivo acute toxicity of titanium dioxide nanoparticles to mice after intraperitioneal injection. J Appl Toxicol 2009;29(4):330–7.

[116] Wang J, Zhou G, Chen C, Yu H, Wang T, Ma Y, et al. Acute toxicity and biodistribution of different sized titanium dioxide particles in mice after oral administration. Toxicol Lett 2007;168(2):176–85.

[117] Ze Y, Hu R, Wang X, Sang X, Ze X, Li B, et al. Neurotoxicity and gene-expressed profile in brain-injured mice caused by exposure to titanium dioxide nanoparticles. J Biomed Mater Res A 2014;102 (2):470–8.

[118] Pflücker F, Wendel V, Hohenberg H, Gärtner E, Will T, Pfeiffer S, et al. The human stratum corneum layer: an effective barrier against dermal uptake of different forms of topically applied micronised titanium dioxide. Skin Pharmacol Physiol 2001;14(Suppl. 1):92–7.

[119] Gamer AO, Leibold E, van Ravenzwaay B. The in vitro absorption of microfine zinc oxide and titanium dioxide through porcine skin. Toxicol In Vitro 2006;20(3):301–7.

[120] Sadrieh N, Wokovich AM, Gopee NV, Zheng J, Haines D, Parmiter D, et al. Lack of significant dermal penetration of titanium dioxide from sunscreen formulations containing nano- and submicron-size TiO_2 particles. Toxicol Sci 2010;115(1):156–66.

[121] Bennat Müller G. Skin penetration and stabilization of formulations containing microfine titanium dioxide as physical UV filter. Int J Cosmet Sci 2000;22(4):271–83.

[122] Rouse JG, Yang J, Ryman-Rasmussen JP, Barron AR, Monteiro-Riviere NA. Effects of mechanical flexion on the penetration of fullerene amino acid-derivatized peptide nanoparticles through skin. Nano Lett 2007;7(1):155–60.

[123] Ramesan MT, Jayakrishnan P, Sampreeth T, Pradyumnan PP. Temperature-dependent AC electrical conductivity, thermal stability and different DC conductivity modelling of novel poly(vinyl cinnamate)/ zinc oxide nanocomposites. J Therm Anal Calorim 2017;129(1):135–45.

[124] Qi Y, Zhang J, Qiu S, Sun L, Xu F, Zhu M, et al. Thermal stability, decomposition and glass transition behavior of PANI/NiO composites. J Therm Analy Calorim 2009;98(2):533.

[125] Castro-Mayorga JL, Fabra MJ, Pourrahimi AM, Olsson RT, Lagaron JM. The impact of zinc oxide particle morphology as an antimicrobial and when incorporated in poly(3-hydroxybutyrate-co-3-hydroxyvalerate) films for food packaging and food contact surfaces applications. Food Bioprod Process 2017;101 (Suppl. C):32–44.

[126] Zhang Y, Zhuang S, Xu X, Hu J. Transparent and UV-shielding ZnO@PMMA nanocomposite films. Opt Mater 2013;36(2):169–72.

[127] Anderson J, Chris GVdW. Fundamentals of zinc oxide as a semiconductor. Rep Prog Phys 2009;72 (12):126501.

[128] Choudhury SR, Ordaz J, Lo C-L, Damayanti NP, Zhou F, Irudayaraj J. From the cover: zinc oxide nanoparticles-induced reactive oxygen species promotes multimodal cyto- and epigenetic toxicity. Toxicol Sci 2017;156(1):261–74.

[129] Ng CT, Yong LQ, Hande MP, Ong CN, Yu LE, Bay BH, et al. Zinc oxide nanoparticles exhibit cytotoxicity and genotoxicity through oxidative stress responses in human lung fibroblasts and *Drosophila melanogaster*. Int J Nanomed 2017;12:1621–37.

[130] Venkatachalam P, Jayaraj M, Manikandan R, Geetha N, Rene ER, Sharma NC, et al. Zinc oxide nanoparticles (ZnONPs) alleviate heavy metal-induced toxicity in *Leucaena leucocephala* seedlings: a physiochemical analysis. Plant Physiol Biochem 2017;110(Suppl. C):59–69.

[131] Hou J, Wu Y, Li X, Wei B, Li S, Wang X. Toxic effects of different types of zinc oxide nanoparticles on algae, plants, invertebrates, vertebrates and microorganisms. Chemosphere 2018;193(Suppl. C):852–60.

[132] Guan R, Kang T, Lu F, Zhang Z, Shen H, Liu M. Cytotoxicity, oxidative stress, and genotoxicity in human hepatocyte and embryonic kidney cells exposed to ZnO nanoparticles. Nanoscale Res Lett 2012;7(1):602.

[133] Król A, Pomastowski P, Rafińska K, Railean-Plugaru V, Buszewski B. Zinc oxide nanoparticles: synthesis, antiseptic activity and toxicity mechanism. Adv Colloid Interface Sci 2017;249(Suppl. C):37–52.

[134] Heinlaan M, Ivask A, Blinova I, Dubourguier H-C, Kahru A. Toxicity of nanosized and bulk ZnO, CuO and TiO_2 to bacteria *Vibrio fischeri* and crustaceans *Daphnia magna* and *Thamnocephalus platyurus*. Chemosphere 2008;71(7):1308–16.

[135] Premanathan M, Karthikeyan K, Jeyasubramanian K, Manivannan G. Selective toxicity of ZnO nanoparticles toward gram-positive bacteria and cancer cells by apoptosis through lipid peroxidation. Nanomedicine 2011;7(2):184–92.

[136] Sharma V, Shukla RK, Saxena N, Parmar D, Das M, Dhawan A. DNA damaging potential of zinc oxide nanoparticles in human epidermal cells. Toxicol Lett 2009;185(3):211–18.

[137] Anand AS, Prasad DN, Singh SB, Kohli E. Chronic exposure of zinc oxide nanoparticles causes deviant phenotype in *Drosophila melanogaster*. J Hazard Mater 2017;327(Suppl. C):180–6.

[138] Guo D, Wu C, Jiang H, Li Q, Wang X, Chen B. Synergistic cytotoxic effect of different sized ZnO nanoparticles and daunorubicin against leukemia cancer cells under UV irradiation. J Photochem Photobio B Bio 2008;93(3):119–26.

[139] Nair S, Sasidharan A, Divya Rani VV, Menon D, Nair S, Manzoor K, et al. Role of size scale of ZnO nanoparticles and microparticles on toxicity toward bacteria and osteoblast cancer cells. J Mater Sci Mater Med 2008;20(1):235.

[140] Hsiao IL, Huang Y-J. Effects of various physicochemical characteristics on the toxicities of ZnO and TiO_2 nanoparticles toward human lung epithelial cells. Sci Total Environ 2011;409(7):1219–28.

[141] Hackenberg S, Scherzed A, Technau A, Kessler M, Froelich K, Ginzkey C, et al. Cytotoxic, genotoxic and pro-inflammatory effects of zinc oxide nanoparticles in human nasal mucosa cells in vitro. Toxicol In Vitro 2011;25(3):657–63.

[142] Zhou Z, Son J, Harper B, Zhou Z, Harper S. Influence of surface chemical properties on the toxicity of engineered zinc oxide nanoparticles to embryonic zebrafish. Beilstein J Nanotechnol 2015;6:1568–79.

[143] Bartczak D, Baradez MO, Merson S, Goenaga-Infante H, Marshall D. Surface ligand dependent toxicity of zinc oxide nanoparticles in HepG2 cell model. J Phys Conf Ser 2013;429(1):012015.

[144] Nguyen VS, Rouxel D, Vincent B, Badie L, Santos FDD, Lamouroux E, et al. Influence of cluster size and surface functionalization of ZnO nanoparticles on the morphology, thermomechanical and piezoelectric properties of P(VDF-TrFE) nanocomposite films. Appl Surf Sci 2013;279(Suppl. C):204–11.

[145] Jaramillo AF, Baez-Cruz R, Montoya LF, Medinam C, Pérez-Tijerina E, Salazar F, et al. Estimation of the surface interaction mechanism of ZnO nanoparticles modified with organosilane groups by Raman spectroscopy. Ceram Int 2017;43(15):11838–47.

[146] Abdelhay A, Carmen-Mihaela T, Christophe M, Cédric M, Hélène G, Ghouti M, et al. Physicochemical properties and cellular toxicity of (poly)aminoalkoxysilanes-functionalized ZnO quantum dots. Nanotechnology 2012;23(33):335101.

[147] Ramasamy M, Das M, An SSA, Yi DK. Role of surface modification in zinc oxide nanoparticles and its toxicity assessment toward human dermal fibroblast cells. Int J Nanomed 2014;9:3707–18.

[148] Hsiao IL, Huang Y-J. Titanium oxide shell coatings decrease the cytotoxicity of ZnO nanoparticles. Chem Res Toxicol 2011;24(3):303–13.

[149] Luo M, Shen C, Feltis BN, Martin LL, Hughes AE, Wright PFA, et al. Reducing ZnO nanoparticle cytotoxicity by surface modification. Nanoscale 2014;6(11):5791–8.

[150] Krystosiak P, Tomaszewski W, Megiel E. High-density polystyrene-grafted silver nanoparticles and their use in the preparation of nanocomposites with antibacterial properties. J Colloid Interface Sci 2017;498 (Suppl. C):9–21.

[151] Huang T, Xu X-HN. Synthesis and characterization of tunable rainbow colored colloidal silver nanoparticles using single-nanoparticle plasmonic microscopy and spectroscopy. J Mater Chem 2010;20 (44):9867–76.

[152] Oliani WL, Parra DF, Komatsu LGH, Lincopan N, Rangari VK, Lugao AB. Fabrication of polypropylene/silver nanocomposites for biocidal applications. Mater Sci Eng, C 2017;75(Suppl. C):845–53.

[153] Kawata K, Osawa M, Okabe S. In vitro toxicity of silver nanoparticles at noncytotoxic doses to HepG2 human hepatoma cells. Environ Sci Technol 2009;43(15):6046–51.

[154] Vaquero C, Gutierrez-Cañas C, Galarza N, López de Ipiña JL. Exposure assessment to engineered nanoparticles handled in industrial workplaces: the case of alloying nano-TiO_2 in new steel formulations. J Aerosol Sci 2016;102(Suppl. C):1–15.

[155] Kim S, Choi JE, Choi J, Chung K-H, Park K, Yi J, et al. Oxidative stress-dependent toxicity of silver nanoparticles in human hepatoma cells. Toxicol In Vitro 2009;23(6):1076–84.

[156] El Badawy AM, Silva RG, Morris B, Scheckel KG, Suidan MT, Tolaymat TM. surface charge-dependent toxicity of silver nanoparticles. Environ Sci Technol 2011;45(1):283–7.

[157] Fabrega J, Fawcett SR, Renshaw JC, Lead JR. Silver nanoparticle impact on bacterial growth: effect of pH, concentration, and organic matter. Environ Sci Technol 2009;43(19):7285–90.

[158] Choi O, Clevenger TE, Deng B, Surampalli RY, Ross L, Hu Z. Role of sulfide and ligand strength in controlling nanosilver toxicity. Water Res 2009;43(7):1879–86.

[159] Choi O, Deng KK, Kim N-J, Ross L, Surampalli RY, Hu Z. The inhibitory effects of silver nanoparticles, silver ions, and silver chloride colloids on microbial growth. Water Res 2008;42(12):3066–74.

[160] Helmlinger J, Sengstock C, Gro, Mayer C, Schildhauer TA, Koller M, et al. Silver nanoparticles with different size and shape: equal cytotoxicity, but different antibacterial effects. RSC Adv 2016;6 (22):18490–501.

[161] Ahamed M, Karns M, Goodson M, Rowe J, Hussain SM, Schlager JJ, et al. DNA damage response to different surface chemistry of silver nanoparticles in mammalian cells. Toxicol Appl Pharmacol 2008;233 (3):404–10.

[162] Rai M, Yadav A, Gade A. Silver nanoparticles as a new generation of antimicrobials. Biotechnol Adv 2009;27(1):76–83.

[163] Xiu Z-M, Zhang Q-B, Puppala HL, Colvin VL, Alvarez PJJ. Negligible particle-specific antibacterial activity of silver nanoparticles. Nano Lett 2012;12(8):4271–5.

[164] Asharani PV, Yi Lian W, Zhiyuan G, Suresh V. Toxicity of silver nanoparticles in zebrafish models. Nanotechnology 2008;19(25):255102.

[165] Choi JE, Kim S, Ahn JH, Youn P, Kang JS, Park K, et al. Induction of oxidative stress and apoptosis by silver nanoparticles in the liver of adult zebrafish. Aquat Toxicol 2010;100(2):151–9.

[166] Newton KM, Puppala HL, Kitchens CL, Colvin VL, Klaine SJ. Silver nanoparticle toxicity to *Daphnia magna* is a function of dissolved silver concentration. Environ Toxicol Chem 2013;32(10):2356–64.

[167] Navarro E, Piccapietra F, Wagner B, Marconi F, Kaegi R, Odzak N, et al. Toxicity of silver nanoparticles to *Chlamydomonas reinhardtii*. Environ Sci Technol 2008;42(23):8959–64.

[168] Brandelli A, Brum LFW, dos Santos JHZ. Nanostructured bioactive compounds for ecological food packaging. Environ Chem Lett 2017;15(2):193–204.

[169] Dainelli D, Gontard N, Spyropoulos D, Zondervan-van den Beuken E, Tobback P. Active and intelligent food packaging: legal aspects and safety concerns. Trends Food Sci Technol 2008;19(Suppl. 1): S103–12.

[170] Vanderroost M, Ragaert P, Devlieghere F, De Meulenaer B. Intelligent food packaging: the next generation. Trends Food Sci Technol 2014;39(1):47–62.

[171] Tan B, Thomas NL. A review of the water barrier properties of polymer/clay and polymer/graphene nanocomposites. J Memb Sci 2016;514(Suppl. C):595–612.

[172] Cui Y, Kumar S, Rao Kona B, van Houcke D. Gas barrier properties of polymer/clay nanocomposites. RSC Adv 2015;5(78):63669–90.

[173] Huang H-D, Ren P-G, Xu J-Z, Xu L, Zhong G-J, Hsiao BS, et al. Improved barrier properties of poly(lactic acid) with randomly dispersed graphene oxide nanosheets. J Memb Sci 2014;464(Suppl. C):110–18.

[174] Ren P-G, Wang H, Huang H-D, Yan D-X, Li Z-M. Characterization and performance of dodecyl amine functionalized graphene oxide and dodecyl amine functionalized graphene/high-density polyethylene nanocomposites: a comparative study. J Appl Polym Sci 2014;131(2) n/a-n/a.

[175] Alexandre M, Dubois P. Polymer-layered silicate nanocomposites: preparation, properties and uses of a new class of materials. Mater Sci Eng R 2000;28(1):1–63.

[176] Kalendova A, Merinska D, Gerard JF, Slouf M. Polymer/clay nanocomposites and their gas barrier properties. Polym Compos 2013;34(9):1418–24.

[177] Emamifar A, Kadivar M, Shahedi M, Soleimanian-Zad S. Evaluation of nanocomposite packaging containing Ag and ZnO on shelf life of fresh orange juice. Innov Food Sci Emerg Technol 2010;11 (4):742–8.

[178] Emamifar A, Kadivar M, Shahedi M, Soleimanian-Zad S. Effect of nanocomposite packaging containing Ag and ZnO on inactivation of *Lactobacillus plantarum* in orange juice. Food Control 2011;22 (3):408–13.

[179] Damm C, Neumann M, Münstedt H. Properties of nanosilver coatings on polymethyl methacrylate. Soft Mater 2005;3(2–3):71–88.

[180] Chaudhry Q, Scotter M, Blackburn J, Ross B, Boxall A, Castle L, et al. Applications and implications of nanotechnologies for the food sector. Food Add Contam Part A 2008;25(3):241–58.

[181] Mauriello G, De Luca E, La Storia A, Villani F, Ercolini D. Antimicrobial activity of a nisin-activated plastic film for food packaging. Lett Appl Microbiol 2005;41(6):464–9.

[182] Li Q, Mahendra S, Lyon DY, Brunet L, Liga MV, Li D, et al. Antimicrobial nanomaterials for water disinfection and microbial control: potential applications and implications. Water Res 2008;42 (18):4591–602.

[183] Emamifar A. Applications of antimicrobial polymer nanocomposites in food packaging. In: Hashim A, editor. Advances in nanocomposite technology. Rijeka: InTech; 2011 [chapter 13].

[184] Shiflett MB, Foley HC. Ultrasonic deposition of high-selectivity nanoporous carbon membranes. Science 1999;285(5435):1902–5.

[185] Merkel TC, Freeman BD, Spontak RJ, He Z, Pinnau I, Meakin P, et al. Ultrapermeable, reverse-selective nanocomposite membranes. Science 2002;296(5567):519–22.

[186] Yin J, Deng B. Polymer–matrix nanocomposite membranes for water treatment. J Memb Sci 2015;479:256–75.

[187] Ng LY, Mohammad AW, Leo CP, Hilal N. Polymeric membranes incorporated with metal/metal oxide nanoparticles: a comprehensive review. Desalination 2013;308:15–33.

[188] Cong H, Radosz M, Towler BF, Shen Y. Polymer–inorganic nanocomposite membranes for gas separation. Sep Purif Technol 2007;55(3):281–91.

[189] Ulbricht M. Advanced functional polymer membranes. Polymer (Guildf) 2006;47(7):2217–62.

[190] Song M-K, Park S-B, Kim Y-T, Kim K-H, Min S-K, Rhee H-W. Characterization of polymer-layered silicate nanocomposite membranes for direct methanol fuel cells. Electrochim Acta 2004;50(2):639–43.

[191] Jian K, Pintauro PN, Ponangi R. Separation of dilute organic/water mixtures with asymmetric poly (vinylidene fluoride) membranes. J Memb Sci 1996;117(1):117–33.

[192] Bae T-H, Tak T-M. Effect of TiO_2 nanoparticles on fouling mitigation of ultrafiltration membranes for activated sludge filtration. J Memb Sci 2005;249(1):1–8.

[193] Cao Y-C, Xu C, Wu X, Wang X, Xing L, Scott K. A poly (ethylene oxide)/graphene oxide electrolyte membrane for low temperature polymer fuel cells. J Power Sources 2011;196(20):8377–82.

[194] Pumera M. Graphene-based nanomaterials for energy storage. Energy Environ Sci 2011;4(3):668–74.

[195] Niemann MU, Srinivasan SS, Phani AR, Kumar A, Goswami DY, Stefanakos EK. Nanomaterials for hydrogen storage applications: a review. J Nanomater 2008;2008:9.

[196] Liu C, Li F, Ma L-P, Cheng H-M. Advanced materials for energy storage. Adv Mater 2010;22(8):E28–62.

[197] Aricò AS, Bruce P, Scrosati B, Tarascon J-M, van Schalkwijk W. Nanostructured materials for advanced energy conversion and storage devices. Nat Mater 2005;4:366.

[198] Dincă M, Long JR. Hydrogen storage in microporous metal–organic frameworks with exposed metal sites. Angew Chem Int Ed 2008;47(36):6766–79.

[199] Zhang Q, Uchaker E, Candelaria SL, Cao G. Nanomaterials for energy conversion and storage. Chem Soc Rev 2013;42(7):3127–71.

[200] Naresh Muthu R, Rajashabala S, Kannan R. Synthesis and characterization of polymer (sulfonated poly-ether-ether-ketone) based nanocomposite (h-boron nitride) membrane for hydrogen storage. Int J Hydrogen Energy 2015;40(4):1836–45.

[201] Kim BH, Hong WG, Lee SM, Yun YJ, Yu HY, Oh S-Y, et al. Enhancement of hydrogen storage capacity in polyaniline–vanadium pentoxide nanocomposites. Int J Hydrogen Energy 2010;35(3):1300–4.

[202] Makridis SS, Gkanas EI, Panagakos G, Kikkinides ES, Stubos AK, Wagener P, et al. Polymer-stable magnesium nanocomposites prepared by laser ablation for efficient hydrogen storage. Int J Hydrogen Energy 2013;38(26):11530–5.

15

Future Perspectives

Tomasz M. Majka, Krzysztof Pielichowski
DEPARTMENT OF CHEMISTRY AND TECHNOLOGY OF POLYMERS, CRACOW UNIVERSITY OF TECHNOLOGY, KRAKÓW, POLAND

15.1 Trends in the Development of Functionalized Hybrid Materials

Polymer composites with functionalized nanoparticles have attracted considerable interest in recent years as a group of promising materials with properties that can be controlled not only through the kind and size of nanomaterial but also via its tailor-made functionalization. Surface functionalization of nanoparticles helps to increase their compatibility with polymer matrix and limits aggregation effects. The surface modification may include functionalization of the nanoparticles' surface by using low molecular weight compounds with functional groups needed for further postfunctionalization with low molecular weight compounds or polymers. Generally, nanofiller surface modification protocols depend on the structure and reactivity of the nanoparticles. For instance, thiol and disulfide functional groups are used for modification of noble metal surfaces, while silane, phosphonate, acidic, or hydroxyl groups are applied for metal oxide surfaces. The presence of some functional groups could be limited due to their interference with the catalysts or polymer active centers during the polymerization process, when functional polymer is synthesized for subsequent grafting process. Hence, suitability of the functional groups for the modification of a nanofiller surface depends not only on the chemical structure of the particle but also on the method used for polymer synthesis and type of the modification process.

Nanocomposites containing functionalized nanoparticles form an important subgroup of polymer-based nanocomposites. Specific interactions and self-organization play an important role in polymer/functionalized nanoparticle systems, and through surface modification of inorganic nanoparticles these phenomena can be at least partially controlled and utilized in the design of novel hybrid materials with adjustable properties.

Techniques relying on the in situ synthesis allow for production of nanocomposites with high mass or volume fractions of functionalized nanoparticles that can be utilized in various applications, such as photocatalysis, dye-sensitized solar cells, and metamaterials fabrication. Some functionality can be cross-linked with polymers or take part in polymerization reactions. In situ synthesized silica nanoparticles in polymeric (or a monomer mix) media may be applied to produce hybrid nanocomposites with tunable properties. Differently

Polymer Composites with Functionalized Nanoparticles. DOI: https://doi.org/10.1016/B978-0-12-814064-2.00015-9

functionalized silica nanoparticles can be well distributed in the polymer matrices that provide good properties of composites with linear, branched or cross-linked polymers. The sol–gel method is a good alternative for silica nanoparticle synthesis, being able to tailor the particles' geometry and size. Surface functionality can be tailored by grafting reactions of the silanol groups in the surface of the silica particle with different compounds, such as chlorosilanes, to get better interactions and dispersion of the silica particles in the polymeric matrix. On the other hand, some silane-based functionalities can be used to cross-link macromolecules to improve the mechanical properties of the ultimate polymeric materials.

Polyhedral oligomeric silsesquioxanes (POSS), consisting of cage-like molecules with silicon in their vertices and oxygen on the edges, can be functionalized through introduction of reactive organic groups to the vertex positions via various chemical pathways. Consequently, functionalized POSS could be chemically incorporated into polymer structures or physically blended with existing polymer matrices, resulting in polymer–POSS organic–inorganic hybrid composites. Incorporation of suitable POSS derivatives into the polymer systems leads to improvement in thermal stability, mechanical strength, electric conductivity, antiatomic oxidation, and antiphotodegradation. With the advanced developments of organic chemistry, a large number of POSS compounds with a variety of functional groups have been prepared successfully, allowing them to be covalently attached onto polymeric networks through different polymerization or grafting reactions. On the other hand, modern polymer chemistry has offered many polymerization approaches, including atom transfer radical polymerization (ATRP), reversible addition-fragmentation chain transfer (RAFT), etc., some of which have shown great potential to precisely introduce POSS into polymeric architectures and therefore tailor-make the material's performance for emerging applications such as biomaterials for tissue engineering, drug delivery, etc.

The functionalization process of layered silicate nanofillers is relatively easy to perform and control, however, for some applications special requirements, for example, stability during high-temperature processing, require special functionalization procedures and/or modifiers. Therefore there is still a search for novel methods of clay-containing composites preparation and functionalization. Although the design and fabrication of this kind of nanocomposite have been successfully developed since the 1990s, the long-term behavior of clay embedded in the polymer matrix, including aging effects, is still the subject of investigation. Functionalized clay-containing polymer nanocomposites show enhancements in mechanical and barrier properties ("labyrinth effect"), as well as in thermal, optical, and physicochemical properties, compared with pristine polymers or classical microcomposites. However, structure-processing-properties physicochemical relationships still need to be formulated to better answer the question on how the presence of clay influences the nanocomposite characteristics.

Methods comprising the use of preformed nanoparticles pave the way to produce multidimensional nanoparticle arrays in block copolymer–based nanocomposites. In this context, tuning particle size, shape, and surface chemistry is of significance to tailor the thermodynamic driving forces taking place during the self-assembly processes. Along these lines, particular emphasis has to be devoted to clearly understand the kinetics of the coassemblies to further optimize critical design principles and processing parameters to provide a precise morphological control over the systems, not only at equilibrium, but also kinetically trapped states.

Many changes are proposed in the carbon materials field. For instance, while covalent functionalization would modify the carbon nanomaterials with oxygen, nitrogen, phosphorus, and halogen containing heteroatoms, noncovalent functionalization is used to physically disentangle carbon nanotubes and unstack graphene sheets with the assistance of ionic, nonionic, or polymeric surfactants.

By embedding gold and silver nanoparticles into polymer matrices a novel family of multifunctional materials can be developed taking advantage of both the intrinsic properties of the metal at the nanoscale and the features of organic macromolecules. These nanocomposites allow the technological application of unique properties of metals emerging at the nanoscale.

It is noteworthy that the processing method, morphology, size, and surface characteristics of the nanoparticles are key aspects to consider in the preparation of the polymer nanocomposites and the final properties that are desired. For instance, optical properties and bioactivity depend on morphology and surface charge, respectively.

Numerous research efforts are focused on nanostructured hybrid composites for electronic applications—the conjugated polymer-based nanohybrids with inorganic metallic nanoparticles and/or nanostructured carbon fillers have been found to exhibit intriguing multifunctional properties for various potential applications in energy storage, energy conversion, and electrochemical sensors.

In parallel with the remarkable efforts in synthesis, modification, and functionalization of inorganic nanoparticles, recent developments in polymer science include novel strategies toward production of high-performance nanohybrid materials with advanced functional properties as the next-generation materials that can be implemented in emerging fields of science and technology.

Development of high-performance hybrid nanocomposites requires a thorough design, material selection, and synthesis path. It involves the selection of polymers and fillers out of the vast number of polymers and fillers available today in search of the required property. It also requires determination of the suitable production process, considering the end use requirements, and finally fabrication of the material. Future developments will need to deal with the recycling issues of functionalized nanoparticles and composites made therefrom, with special attention paid to health and environmental safety issues.

Progress in the synthesis and processing techniques of polymer nanocomposites, including ultrasonication assisted solution mixing, shear mixing, three roll milling, ball milling, double-screw extrusion, as well as other ex situ and in situ processing methods, is also expected in the context of their applicability to different nanofillers and matrices, control of the process parameters, and the up-scaling for industrial applications.

In the design of polymer nanocomposites with functionalized nanoparticles it is of crucial importance to understand the influence of particle size, shape, and orientation of decorated filler on the composite (micro)structure, as well as on its compatibility with polymer matrix.

In the synthesis of polymer nanocomposites, proper dispersion of the nanofillers to prevent agglomeration effects is the most important consideration that should be taken into account when choosing the synthesis technique and/or processing parameters. Understanding the relationships between these factors is required for successful design of polymer nanocomposites. Considering the versatility of the nanocomposites together with the characteristic features of their organic and inorganic components, it can be expected that such complex systems shall not only offer new perspectives in the materials engineering field, but also bring new challenges of an interdisciplinary nature. However, it should be emphasized that recent developments in computational tools and modeling/simulation approaches, as well as progress in supramolecular and interface chemistry, look promising for the design and production of next-generation multifunctional nanocomposite materials with advanced physical properties. Moreover, imaging analysis will continue to play a pivotal role in understanding the structural properties of polymer composites with functionalized nanoparticles, especially carbon-based nanomaterials.

The application of nanoscience in medicine will probably employ more and more engineered nanoscale materials for novel therapeutic and diagnostic purposes. Polymeric nanoparticles have been widely studied as efficient nanocarriers in the medical field, owing to their biocompatibility, efficient drug delivery abilities, controlled and sustained release properties, and biodegradability. Nanoparticle surface structures can be tailored by functionalization and novel bioconjugation strategies are required for more advanced drug delivery, imaging, and sensing applications.

From the synthetic perspective, surface functionalities should be endowed with properties including stability in complex biological media, biocompatibility, and specificity for active targeting for precise drug delivery. The proper understanding of the mechanism of functionalized polymeric nanoparticles and biointerface phenomena offers superior features for development of advanced functional nanomaterials.

Therefore, there is a need for extensive toxicological research for this class of materials to avoid any possible risk and facilitate their safe implementation in future products. Despite several research attempts focusing on the toxic behavior of nanomaterials, it has been seen there is no clear and solid understanding of their molecular scale mechanisms and proper procedures to prevent their adverse effect on the environment. It has been discussed that functionalization can increase or decrease nanomaterials' toxic effect depending on the nanomaterials' nature and surface functionalization type. However, further multidisciplinary investigations are needed to evolve the understanding of their environmental effects and produce nanocomposites with minimum adverse effect on the ecosystem.

Index

Note: Page numbers followed by "*f*" and "*t*" refer to figures and tables, respectively.

A

ABS copolymer. *See* Acrylonitrile-butadiene-styrene copolymer (ABS copolymer)
Absorption of silane coupling agents, 90
ACNT. *See* Aligned-CNT (ACNT)
ACPA. *See* 4,4′-Azobis(4-cyanopentanoic acid) (ACPA)
Acrylonitrile (AN), 306
Acrylonitrile-butadiene-styrene copolymer (ABS copolymer), 155–156
Acryloxies (−O−C(H or CH_3) = CH_2), 287–288
Activation energy, 413
Activator regenerated by electron transfer ATRP (ARGET-ATRP), 17, 23
Active clay, 169
Active food packaging systems, 266
ADDL. *See* Amyloid β-derived diffusible ligands (ADDL)
Adverse effect, 440, 450–451
AEAPS. *See* *N*-(2-Aminoethyl) 3-aminopropyl-trimethoxysilane (AEAPS)
Agglomeration effects, 469–470
Aggregation-induced emission (AIE), 198
AIE. *See* Aggregation-induced emission (AIE)
Alginate, 390–391
Aligned-CNT (ACNT), 235–237
Alkylphenolethoxylates, 295*f*
α-Fe_2O_3 nanoparticles, 298
Alpha-hydroxybutyrate dehydrogenase levels, 446
Alumina (Al_2O_3), 312
Aluminum, 168
Amines (−NH_2), 287–288
p-Amino diphenylamine (PADPA), 363
Amino-POSS/PLLA, 198–200
N-(2-Aminoethyl) 3-aminopropyl-trimethoxysilane (AEAPS), 302
p-Aminophenyl trimethoxysilane (APhTMS), 95, 96*f*
3-Aminopropylmethylsilyl groups (APMS), 129–130
3-Aminopropyltriethoxysilane (APS), 137, 302, 412
Aminopropyltrimethoxysilane, 92–94
Ammonia, 160
Ammonium hydroxide (NH_4OH), 88–89
Amphiphilic Au NPs, 8–9
Amyloid β-derived diffusible ligands (ADDL), 272
AN. *See* Acrylonitrile (AN)
Anatase, 445–446
Aniline octamers
  polymerization mechanism, 363
  structures, 362*f*
Anilinium, 306–307
Anionic polymerization, 11–14
Anionic ring opening polymerization (AROP), 11–12
Anti-photodegradation, 192–193
Antiatomic oxidation (Anti-AO), 182, 192
  materials, 192
Antimicrobial materials, 259
Antimicrobial nanocomposite systems, 267
AOT. *See* Dioctyl sodium sulfosuccinate (AOT)
APhTMS. *See* p-Aminophenyl trimethoxysilane (APhTMS)
APMS. *See* 3-Aminopropylmethylsilyl groups (APMS)
APS. *See* 3-Aminopropyltriethoxysilane (APS)
APTES. *See* 3-Aminopropyltriethoxysilane (APS)
AR. *See* Aspect ratio (AR)
ARGET-ATRP. *See* Activator regenerated by electron transfer ATRP (ARGET-ATRP)
Argon ions, 350
Aromatic PA dendrimers, 184–186
AROP. *See* Anionic ring opening polymerization (AROP)
Aryl diazonium salts, 220
Ascorbic acid, 5

Aspect ratio (AR), 56–59
Aspergillus terreus, 253
Atom transfer radical polymerization (ATRP), 10, 15–19, 15*f*, 137–138, 180–181, 287–288
ATRP. *See* Atom transfer radical polymerization (ATRP)
Au-NWs. *See* Gold nanowires (Au-NWs)
AuNP/BC. *See* Gold nanoparticle/bacterial cellulose (AuNP/BC)
4,4′-Azobis(4-cyanopentanoic acid) (ACPA), 21

B

BA. *See* *n*-Butyl acrylate (BA)
Bacillus licheniformis, 252
Bacterial cellulose fibers (BC fibers), 257
Bacterial nanocellulose (BNC), 24–25
Ball milling, 71–72, 288
Basal spacing, 151–152
BC fibers. *See* Bacterial cellulose fibers (BC fibers)
BCB-EO. *See* 2-Hydroxyethyl benzocyclobutene (BCB-EO)
BCPs. *See* Block copolymers (BCPs)
BDGE. *See* 1,4-Butanediol diglycidyl ether (BDGE)
1,4-BDO. *See* 1,4-Butanediol (1,4-BDO)
Bentonite, 166–167
β-naphthalene sulfonic acid (β-NSA), 306–307
BHJ. *See* Bulk heterojunction (BHJ)
Bidentate ligands, 384
Bilayer structure, 341
Bioconjugation, 385–387
Biodegradability, 442–443
Biological applications, 437
Biological environment, 438
Biomedical applications of composites, 138–141
Biomolecule, 385–386
3,3′,4,4′-Biphenyltetracarboxylic dianhydride (BPDA), 135
1,3-Bis(4-aminophenoxy)benzene (APB), 135
Blending process, 306–307
Block copolymers (BCPs), 104
  BCP-based nanocomposites, 87–88
  BCP-directed nanoparticle hybrids, 121
BN tubes. *See* Boron nitride tubes (BN tubes)
BNC. *See* Bacterial nanocellulose (BNC)
Boehmite (AlOOH), 312
Boron nitride tubes (BN tubes), 50
Bottom-up approach, 288
Bovine serum albumin (BSA), 265
BPDA. *See* 3,3′,4,4′-Biphenyltetracarboxylic dianhydride (BPDA)
2-Bromoisobutyrate-based siloxane, 16
Brookite, 445–446
Brust–Schiffrin method, 250, 254*f*
BSA. *See* Bovine serum albumin (BSA)
Bulk heterojunction (BHJ), 222
1,4-Butanediol diglycidyl ether (BDGE), 184–186
1,4-Butanediol (1,4-BDO), 409
n-Butanol, 290
n-Butyl acrylate (BA), 9
n-Butyl glycidyl ether (nBGE), 184–186

C

Caco-2 cells, 169
Cadmium selenide/cadmium(II) sulfide (CdSe/CdS), 55
Cadmium sulfide nanoparticles (CdS NPs), 372–375
Caged silsesquioxanes, 179–180
Calcium carbonate ($CaCO_3$), 23
Calcium-iron complex oxides ($CaFe_2O_4$), 301
Calendering. *See* Roll milling
Capacitive Faradaic storage (CFS), 231–232
Carbohydrates, 250, 267
Carbon black (CB), 11
Carbon fiber (CF), 186–187
Carbon materials, 214*t*
Carbon monoxide (CO), 262–263
Carbon nanofiber (CNF), 309
Carbon nanomaterials (CNMs), 212–213, 235*f*, 239*t*
  polymer composites, 212–216
    composites of polymer with functionalized CNMs, 215–216
    functionalization of CNMs and effects, 216–228
    properties and applications, 213–215
  properties and applications of polymer composites, 228–238
    electrical properties and applications, 229–230

electrochemical properties and applications, 230–238
mechanical and structural enhancements, 238
Carbon nanotubes (CNTs), 2, 50, 119–120, 186–187, 212–213, 213*f*, 367, 426–428, 438–441, 439*f*
Carbon-based nanomaterials (CBNs), 423
Carbon-based nanoparticle functionalization, 5–6
Carbonaceous materials, 213
Carbonaceous nanoadditives, 423–428
CNTs, 426–428
graphene, 424–426
Catalytic processes, 88–89
Cation exchange capacity (CEC), 151, 408–409
Cationic polymerization, 10–11
Cause-and-effect relationship, 166–167
CB. *See* Carbon black (CB)
CBNs. *See* Carbon-based nanomaterials (CBNs)
CdS NPs. *See* Cadmium sulfide nanoparticles (CdS NPs)
CdSe/CdS. *See* Cadmium selenide/cadmium(II) sulfide (CdSe/CdS)
CE. *See* Composite electrolytes (CE); Counterelectrode (CE)
CEC. *See* Cation exchange capacity (CEC)
Cell adhesion, 445
Cellular oxidative stress, 440
Cellulose nanofibrils (CNFs), 2
nanocomposites, 53
Celluloses, 24–25
Cerium-doped titanium oxide (Ce-TiO_2), 310–311
Cetylpyridinium chloride (CPC), 23–24, 295*f*
Cetyltrimethyl ammonium bromide (CTAB), 99, 298
CF. *See* Carbon fiber (CF)
CFS. *See* Capacitive Faradaic storage (CFS)
Chain transfer agents (CTAs), 19–20
agent, 20–21
Charge storage mechanism, 231
Chemical functionalization methods, 224–225
Chemical in situ methods, 74
Chemical modification, 95
Chemical polymerization, 361–363, 366
Chemical vapor deposition (CVD), 150, 213–214, 288, 299–300, 439–440
Chitosan (CS), 259–260, 268, 390–391
with AgNPs, 267
Clay, 23–24, 166–167, 267, 450–451
clay-containing polymer nanocomposites, 155–156, 161
functionalization, 153–155
influence on human health and environment, 166–170
gallery, 444
and methods of functionalization, 151–155
silicates functionalization for fire retardant applications, 155
minerals, 160
swelling, 152–153
Cloisites, 97
CNF. *See* Carbon nanofiber (CNF)
CNFs. *See* Cellulose nanofibrils (CNFs)
CNMs. *See* Carbon nanomaterials (CNMs)
CNTs. *See* Carbon nanotubes (CNTs)
Cobalt oxide (Co_3O_4), 301
Coefficient of thermal expansion (CTE), 312
Compatibility, 3–4, 7
Complex fillers, 120–121, 136–138
Composite electrolytes (CE), 188–189
Composites
biomedical applications, 138–141
with polar polymers, 127–136
Computed tomography (CT), 393–394
Conducting polyaniline-based hybrid nanostructured composite, 367–368
potential applications, 368–375
electrochemical energy storage devices, 368–372
energy conversion systems based on polyaniline nanohybrids, 372–375
Conductive polymer nanocomposites (CPCs), 263
Controlled/living polymerization techniques, 10
Conventional flame retardants, 159
Conventional nanocomposites, 451–452
Conventional polymerization process, 365
Coprecipitation method, 291–294
Core–shell NPs, 20, 23

Core–shell strategy, 367
Counterelectrode (CE), 369
CPC. *See* Cetylpyridinium chloride (CPC)
CPCs. *See* Conductive polymer nanocomposites (CPCs)
CPENTYLPOSS, 194–195
Creep rate spectroscopy (CRS), 132–133, 132*f*
CRS. *See* Creep rate spectroscopy (CRS)
Crystal structure of TiO_2 nanoparticles, 445–446
Crystalline polymorphs, 445–446
Crystallinity, 195–196
CS. *See* Chitosan (CS)
CT. *See* Computed tomography (CT)
CTAB. *See* Cetyltrimethyl ammonium bromide (CTAB)
CTAs. *See* Chain transfer agents (CTAs)
CTE. *See* Coefficient of thermal expansion (CTE)
$CuFeMnO_4$, 292–294
Current–voltage measurement (I–V measurement), 372–375
CV. *See* Cyclic voltammogram (CV)
CVD. *See* Chemical vapor deposition (CVD)
Cyanate ester resin (CY), 418–419
Cyclic voltammogram (CV), 217, 369

D

d-spacing (d_{001}), 151–152
DBS. *See* Dielectric breakdown strength (DBS); Sodium dodecylebenzene sulfonate (DBS)
DBSA. *See* Dodecylbenzenesulfonic acid (DBSA)
DDA. *See* Dodecylamine (DDA)
Decorated metal oxides, polymer nanocomposites with
  polymer–metal oxide nanocomposite materials, 304–312
  surface modification of metal oxide nanoparticles, 301–304
  synthesis of metal oxide nanoparticles, 288–301
Defect functionalization processes, 224–225, 224*f*
Delaminated structure. *See* Exfoliated structure
Delocalized π–electrons system, 359
Dendritic aminofunctionalization, 136–137
Density functional theories, 1
Deoxyribonucleic acid (DNA), 260–261
  sensors, 269
Depolymerization, 406
DGEBA. *See* Diglycidyl ether bisphenol A (DGEBA)
DHP. *See* Disodium H-phosphonate (DHP)
"Diagnostic window", 394
Dielectric breakdown strength (DBS), 134
Dielectric matrix
  extinction of implanting silver nanoparticles in, 331–335
  ion synthesis of MNs in, 327–331
Diethylphosphatotriethoxysilane (SiP), 126
Differential scanning calorimetry (DSC), 194–195
Diglycidyl ether bisphenol A (DGEBA), 184–186
3,4-Dihydro-3-(prop-2-ynyl)-2H-benzoxazine (P-pa), 183–184
Dihygrolipoic acid, 384
Dimethacrylate (UDMA), 194–195
  UDMA-POSS, 194–195
2,2-Dimethoxy-2-phenylacetophenone (DMPA), 258
Dimethyl benzyl hydrogenated tallow ammonium (Dellite), 169
Dimethyl carbonate (DMC), 119–120
(2-Dimethylamino)ethyl methacrylate (DMAEMA), 17–18
Dimethyldichlorosilane (DMS), 119–120
Dinitrotoluene (DNT), 198
Dioctyl sodium sulfosuccinate (AOT), 294, 295*f*
1,1-Diphenylenethylene (DPE), 11
4,4′-Diphenylmethane diisocyanate (MDI), 409
Direct functionalization, 2–3, 224–225
  of carbon-based nanoparticles, 5–6
  of noble metals and metal oxides, 3–5
  of NPs surface, 4*f*
Direct melt intercalation, 157
Direct methanol fuel cell (DMFC), 452–453
Disilanollsobutyl POSS (DSIPOSS), 419–420
Disodium H-phosphonate (DHP), 155, 410
Dispersibility, 1, 5
Dispersion swelling emulsion techniques, 360
Disputable effect, 442–443
Ditelechelic POSS–PCL–POSS, 196–198

Dithiocarbamate ligands, 384
DMAEMA. *See* (2-Dimethylamino)ethyl methacrylate (DMAEMA)
DMC. *See* Dimethyl carbonate (DMC)
DMFC. *See* Direct methanol fuel cell (DMFC)
DMPA. *See* 2,2-Dimethoxy-2-phenylacetophenone (DMPA)
DMS. *See* Dimethyldichlorosilane (DMS)
DNA. *See* Deoxyribonucleic acid (DNA)
DNT. *See* Dinitrotoluene (DNT)
Dodecylamine (DDA), 427–428
Dodecylbenzenesulfonic acid (DBSA), 364
n-Dodecyltrimethoxysilane (DTES), 302
Dopamine, 5
Double-screw extrusion, 72–73
DPE. *See* 1,1-Diphenylenethylene (DPE)
Drug delivery, 387–391
DSC. *See* Differential scanning calorimetry (DSC)
DSIPOSS. *See* Disilanollsobutyl POSS (DSIPOSS)
DSSCs. *See* Dye-sensitized solar cells (DSSCs)
DTES. *See* *n*-Dodecyltrimethoxysilane (DTES)
Dye-sensitized solar cells (DSSCs), 237, 372–375, 467–468
Dynamic laser light scattering measurements, 137–138

E

EB. *See* Emeraldine (EB)
EC. *See* European Commission (EC)
ECIS. *See* Electric cell-substrate impedance sensing (ECIS)
ECPs. *See* Electronically conducting polymers (ECPs)
EDL. *See* Electrochemical double layer (EDL)
EDLCs. *See* Electric double layer capacitors (EDLCs)
EDNB. *See* Ethylene-dinitro-benzoyl (EDNB)
EDOT. *See* 3,4-Ethylene-dioxythiophene (EDOT)
EDTA. *See* Ethylenediaminetetraacetic acid (EDTA)
EEC. *See* 3,4-Epoxycyclohexylmethyl-3′,4′-epoxycyclohexanecarboxylate (EEC)
EEI. *See* Electrode/electrolyte interface (EEI)
EGDMA. *See* Ethylene glycol dimethacrylate (EGDMA)
EIS. *See* Electrochemical impedance spectroscopy (EIS)
Electric cell-substrate impedance sensing (ECIS), 169
Electric conductivity, 188–192
Electric double layer capacitors (EDLCs), 368–369
Electric field–induced alignment, 65
Electrical properties and applications of polymer–CNM composites, 229–230
Electrically conducting organic conjugated polymers, 361–375
  conducting polyaniline-based hybrid nanostructured composite, 367–368
    potential applications, 368–375
  nanostructured hybrid organic conjugated polymers, 366–367
  organic conjugated polymers, 361–364
  polyaniline-based nanomaterials, 364–366
Electrochemical
  behaviors, 369
  capacitors, 368–369
  codeposition, 234
  energy storage devices based on polyaniline nanohybrids, 368–372
  properties and applications, 230–238
    composites of electronically conducting polymers, 233–237
    composites of polymers and CNMs for, 237–238
  sensors, 269
  synthesis, 250
Electrochemical double layer (EDL), 231
Electrochemical impedance spectroscopy (EIS), 369
Electrode/electrolyte interface (EEI), 218–219
Electrodeposition approach, 268–269
Electrolytes, 188–189
Electron antennae, 268–269
Electronic technology, 357
Electronically conducting polymers (ECPs), 215, 358
  composites, 233–237
Electrostatic force, 152–153
Electrostatic interaction, 59–61, 386
Emeraldine (EB), 361

Encapsulation, 3
Endohedral functionalization, 63
Energy applications, nanocomposites in, 453–455
  laser-generated nanoparticles, 455*f*
Energy conversion, 229
  systems based on polyaniline nanohybrids, 372–375
Energy storage applications, CNMs for, 233–237
Engineering polymer, 155–156
Enthalpic interactions, 87–88
Entropy force, 345–346
Environmental air pollution, 438–439
Environmental benefits and application of polymeric nanocomposites, 448–455
  food packaging by nanotechnology and nanocomposites, 448–451
  nanocomposites
    in energy applications, 453–455
    in membrane technology, 451–453
EPDM. *See* Ethylene–propylene–diene monomer (EPDM)
Epoxide copolymers, 12
Epoxies (–$CHOCH_2$), 287–288
  epoxy ternary blends, 133–134
  epoxy-based nanocomposites, 91–92
  epoxy-functionalized polystyrene/silica NP, 127–128
  epoxy/silica nanocomposites, 91–92
  epoxy–matrix curing process, 135–136
Epoxy resins (ERs), 127–128, 155–156, 339
3,4-Epoxycyclohexylmethyl-3′,4′-epoxycyclohexanecarboxylate (EEC), 258
ε-caprolactam, 23
ERs. *See* Epoxy resins (ERs)
Eschericha coli (*E. coli*), 259–260
ET. *See* Ethanol–water (ET)
Ethanol, 290
Ethanol–water (ET), 414
Ethylene glycol dimethacrylate (EGDMA), 188–189
Ethylene-dinitro-benzoyl (EDNB), 237–238
3,4-Ethylene-dioxythiophene (EDOT), 235–237
Ethylene-vinyl acetate copolymer (EVA), 155–156
Ethylenediaminetetraacetic acid (EDTA), 302
Ethylene–propylene–diene monomer (EPDM), 183–184
European Commission (EC), 437
EVA. *See* Ethylene-vinyl acetate copolymer (EVA)
Ex situ method, 258–259
Ex situ processing methods, 469
Exfoliated structure, 159
Expanding clays, 151

F

F-POSS. *See* Fluorinated decyl polyhedral oligomeric silsesquioxane (F-POSS)
Fabricating graphene-type flower structures (f-GE structures), 251*f*
Fabrication methods of polyaniline-based nanomaterials, 365
Faradaic storage (FS), 231
FAS. *See* Fluoroalkyl silane (FAS)
Fatty acid coated gadolinium diethylenetriamine pentaacetic acid (GA-DTPA-FA), 392–393
FD-POSS. *See* Fluorodecyl-POSS (FD-POSS)
FDA. *See* US Food and Drug Administration (FDA)
FF. *See* Fill factor (FF)
FG. *See* Functionalized graphene (FG)
FG nanoplatelets (GnPs), 425
FGS. *See* Functionalized graphene sheet (FGS)
Fill factor (FF), 372–375
Fillers, structure of unmodified and modified silicas as, 122–125
Fire retardants
  functionalized clays as, 159–161
  silicate functionalization for fire retardant applications, 155
Florida test, 163–164
Flory theories, 1
Flow-induced alignment, 64–65
Fluorinated decyl polyhedral oligomeric silsesquioxane (F-POSS), 268
Fluorine, 5
Fluoroalkyl silane (FAS), 94–95, 135–136
Fluorodecyl-POSS (FD-POSS), 189–192
FMOs. *See* Fumed metal or metalloid oxides (FMOs)

FNPs. *See* Functionalized nanoparticles (FNPs)
Food packaging applications, 266–267
Food packaging by nanotechnology and nanocomposites, 448–451, 450*f*, 451*f*
Fourier transform infrared (FTIR), 164
  spectra analysis, 165
Free radical copolymerization, 101
Free radical polymerization (FRP), 1, 14, 21
Freeze-drying process, 291
FRP. *See* Free radical polymerization (FRP)
FS. *See* Faradaic storage (FS)
Fully oxidized graphene, 232
Fumed metal or metalloid oxides (FMOs), 119–120
Functional groups for NPs surface modification, 2–6, 2*f*
  direct functionalization of NPs surface, 3–6, 4*f*
  postfunctionalization reactions at NPs surface, 6, 7*f*
Functional surfaces, 196–198
Functionality-based designs, 53–54
Functionalization, 88–91
  of AgNPs and AuNPs, 253–256
  of carbon-based NPs, 5–6
  of clay, 169
  of CNMs and effects on polymer composites, 216–228
    halogenation of CNMs, 222
    nitrogenation of CNMs, 221–222
    oxygenation of CNMs, 216–220
    preparation of polymer–CNM composites, 225–228
    sulfonation of CNMs, 220–221
  methods, 212–213
  noncovalent, 222–225
Functionalized clay-containing composites, 149–150
  barrier properties of functionalized layered silicates, 161–163
  clay functionalization influence on human health and environment, 166–170
  clays and methods of functionalization, 151–155
  functional clays influence on aging of polymer composites, 163–166
  functionalized clay-containing polymer nanocomposites formation, 155–157
    BET adsorption isotherms, 156*f*
    composite structures, 158*f*
  functionalized clays as fire retardants, 159–161
  synthesis of nanoparticles, 150–151
  types of nanocomposites structures, 158–159
Functionalized clays as fire retardants, 159–161
Functionalized CNMs, polymer composites with, 215–216
Functionalized graphene (FG), 424–425
Functionalized graphene sheet (FGS), 426
Functionalized hybrid materials, 467–470
Functionalized layered silicates, barrier properties of, 161–163
Functionalized nanoparticles (FNPs), 86–87, 119–120, 467
  modern functionalized NPs for advanced applications
    functionalized CNFs, 24–25
    functionalized graphenes, 25–26
    functionalized silicates, 23–24
    Si/G nanocomposite fabricated by electrostatic attraction, 27*f*
    stimuli-responsive nanomaterials, 27–28
  polymer nanocomposites
    decomposition, 408–429
    effect on stability, 407–408
    environmental benefits and application, 448–455
  synthesis routes
    functional groups for NPs surface modification, 2–6, 2*f*
    NPs modification by polymer chains, 7–23, 8*f*
  toxicological and exposure, 438–448
    CNTs, 439–441
    graphene, 441–443
    metal nanoparticles, 444–448
    nanoclays, 443–444
Functionalized POSS-based hybrid composites, 182, 183*f*
  for performance enhancement of materials, 183–206
Functionalized PP (PPOH), 127
Functionalizers, 250

G

G-PIL. *See* Graphene-modified protic ionic liquid (G-PIL)
Galvanostatic charge–discharge (GCD), 369
Gas-to-solid transformation, 299–301
Gastrointestinal tract (GI tract), 388–389
GCD. *See* Galvanostatic charge–discharge (GCD)
Gd-based chelates, 392
GI tract. *See* Gastrointestinal tract (GI tract)
Glass transition dynamics (GTD), 129–130
Glass transition temperature, 103, 133–134
Glassy carbon electrodes, 250
Glycerol (Gy), 97–98
Glycidyl methacrylate (GMA), 127–128
Glycidyl-POSS membrane, 196–198
3-Glycidyloxypropyltrimethoxysilane (GPTMS), 105, 106*f*, 268, 302, 309–310
GLYMO. *See* 3-Glycidyloxypropyltrimethoxysilane (GPTMS)
GMA. *See* Glycidyl methacrylate (GMA)
GN. *See* Graphene (GN)
GNP. *See* Graphite nanoplatelets (GNP)
GnPs. *See* FG nanoplatelets (GnPs)
GO. *See* Graphene oxide (GO)
Gold nanocomposites
  catalytic activity of, 262–263
  electronic properties of, 263–264
  for medical application, 270–273
  for medical devices, 270–271
Gold nanoparticle/bacterial cellulose (AuNP/BC), 265
Gold nanoparticle–polyaniline hybrid composite (AuPANI), 270
Gold nanoparticles (AuNPs), 137–138, 249, 251
  antibacterial properties of, 259–260
  functionalization, 253–255
  for medical application, 270–271
  water treatment using, 264–265
Gold nanowires (Au-NWs), 264
GPTMS. *See* 3-Glycidyloxypropyltrimethoxysilane (GPTMS)
GQDs. *See* Graphene and quantum dots (GQDs)
Grafted oxide nanoparticles, 141
Grafting density, 302–303
"Grafting from" method, 91, 302
  modification using, 10–21
    anionic polymerization, 11–14
    ATRP, 15–19, 15*f*
    cationic polymerization, 10–11
    FRP, 21
    NMP, 14–15, 14*f*
    RAFT polymerization, 19–21, 19*f*
"Grafting to" strategy, 91, 302
  modification using, 7–10
"Grafting-through" technique, modification using, 21–23
Graphene (GN), 71–72, 212–213, 213*f*, 358, 424–426, 441–443, 442*f*, 450–451
  modified, 5
  nanosheets, 268–269
  oxygenation, 219–220
Graphene and quantum dots (GQDs), 25–26, 26*f*
Graphene oxide (GO), 6, 26, 119–120, 183, 214–215, 441
Graphene-modified protic ionic liquid (G-PIL), 25–26
Graphite nanoplatelets (GNP), 49–50
Green synthetic approach, 267
GTD. *See* Glass transition dynamics (GTD)
Guar gum polysaccharide, 269–270
Guest and host interactions, 367

H

h-BN. *See* Hexagonal boron nitride (h-BN)
Halloysite nanotubes, 444
Halogenation of carbon nanomaterials, 222
HAp. *See* Hydroxyapatite (HAp)
Hard template methods, 365–366
Hazardous components, 170
HBP. *See* Hyperbranched polyurethane (HBP)
HDPE. *See* High-density polyethylene (HDPE)
HDPE/OM-POSS. *See* High density polyethylene-based nanocomposites containing octamethyl-POSS (HDPE/OM-POSS)
Heat resistance of polymers, 407
Heat resistance of polymers, thermal stability and, 407
Heat stabilizers, 406

HEMA. *See* 2-Hydroxyethyl methacrylate (HEMA)
HepG-2 cells, 169
Hexadecyipyridinium chloride (HPC), 298
Hexagonal boron nitride (h-BN), 50, 454–455
Hexylamine, 303–304
High density polyethylene-based nanocomposites containing octamethyl-POSS (HDPE/OM-POSS), 100–101
High-density polyethylene (HDPE), 427–428
High-energy ball milling process, 150
High-temperature CVD method, 300
HISO-POSS, 194–195
HNCs. *See* Hybrid nanostructured composites (HNCs)
HOF. *See* Hydrogen-bonded organic framework (HOF)
HPC. *See* Hexadecyipyridinium chloride (HPC)
Hyaluronic acid, 390–391
Hybrid composites, 179
  organic conjugated polymer, 360
Hybrid nanostructured composites (HNCs), 119–120
  synthetic strategies for HNC preparation, 125
Hybrid PS composite, 183–184
Hydrochloric acid (HCl), 364
Hydrogen storage systems, 453
Hydrogen-bonded organic framework (HOF), 198
Hydrolyzable group, 301–302
Hydrophobic nonpolar polymers, 126–127
Hydrothermal synthesis, 250–251
Hydroxyapatite (HAp), 420–423
2-Hydroxyethyl benzocyclobutene (BCB-EO), 13
2-Hydroxyethyl methacrylate (HEMA), 12, 18
Hydroxyl groups (OH groups), 6, 440
Hydroxypropylmethylcellulose, 259–260
N-Hydroxysuccinimide (NHS), 6, 386–387
Hyperbranched polyurethane (HBP), 97–98, 184–186
Hyphenated techniques, 164

I

ICG. *See* Indocyanine green (ICG)
ILSS. *See* Interlaminar shear strength (ILSS)
Imaging, 391–395
  CT, 393–394
  MRI, 391–393
  optical imaging, 394–395
Immersion route, 104–105
In situ
  intercalative polymerization technique, 157
  method, 74, 257–258, 469
  nanocomposite formation, 307
  polymerization, 85, 95
Indocyanine green (ICG), 395
Infrared spectra (IR spectra), 123, 124*f*
Ink-jet printing, 230–231
Inorganic clay, 158
Inorganic dopants for polyaniline, 363–364
Inorganic fillers
  HAp, 420–423
  nanoclays, 408–413
  silicon-based nanomaterials, 414–420
Inorganic nanofiller and organic matrix, 414
Inorganic nanoparticles, 304–305
Inorganic nanostructured fillers, 121
Inorganic silica oxide–conducting polymer composite, 366
Intercalated structure, 158–159
Intercalation process of montmorillonite, 152*f*
Interfacial conjugation, 218–219
Interlaminar shear strength (ILSS), 137
Interlayer, 151–152
  space, 444
Interpenetrating network (IPN), 128
Intrinsically conducting polymers (ICPs). *See* Electronically conducting polymers (ECPs)
Intrinsically electrically conjugated polymers, 358–360, 359*f*
Ion exchange process, 153
Ion implantation, 327, 346
  MNs synthesis in PMMA by
    experimental procedure, 326
    extinction of implanting AgNPs in dielectric matrix, 331–335
    interface effects in silver core–carbon shell structure, 336–338
    ion synthesis of MNs in dielectric matrix, 327–331
    physical stages of nanoparticle synthesis, 327*f*
  synthesis of silver nanoparticles in viscous epoxy resin, 346–352

Ionically organic conducting polymers, 360
IPA. *See* Isopropanol (IPA); Isopropyl alcohol–water (IPA)
IPN. *See* Interpenetrating network (IPN)
iPP. *See* Isotactic polypropylene (iPP)
IR spectra. *See* Infrared spectra (IR spectra)
Iron oxide (Fe_2O_4), 303–304
Isodimensional nanoparticles, 50
Isopropanol (IPA), 88–89, 290
Isopropyl alcohol–water (IPA), 414
N-Isopropylacrylamide (NIPAM), 9, 137–138
Isotactic polypropylene (iPP), 413
I–V measurement. *See* Current–voltage measurement (I–V measurement)

J

Jsc. *See* Short-circuit current density (Jsc)

K

Kaolin, 151–152, 166–167
Kaolinite, 410–413, 412*f*, 413*f*
Kaolinosis, 167

L

L-lactide (L-LA), 11
"Labyrinth" effect, 409, 468
Laponite, 99
Layer morphology, 151
Layer-by-layer assembly (LBLA), 225, 228
Layer-by-layer method, 270
Layered silicates crystal structure, 151–152
LBLA. *See* Layer-by-layer assembly (LBLA)
LC. *See* Liquid crystalline (LC)
LDP. *See* Liquid-phase deposition (LDP)
LDPE. *See* Low-density polyethylene (LDPE)
Leucomeraldine base (LEB), 361
Ligand
  addition, 385
  exchange, 303–304, 304*f*, 384
    technique, 3, 287–288
  modification, 384–385
Lipophilic behavior, 439–440
Liquid crystalline (LC), 195–196
Liquid electrolytes, 188–189
Liquid exfoliation technique, 85
Liquid-phase deposition (LDP), 298–299
Listeria monocytogenes, 261–262, 262*f*
LiTFSI, 188–189
Lithium-ion batteries, 103
Living organisms, 252
Living polymerizations, 27
Localized surface plasmon resonance (LSPR), 259
Low-density polyethylene (LDPE), 302
LSPR. *See* Localized surface plasmon resonance (LSPR)

M

MA. *See* Methyl acrylate (MA)
MA-POSS, 183–184, 188–189, 194–198, 197*f*
Magnesium oxide (MgO), 310
Magnetic field–induced alignment, 65
Magnetic resonance imaging (MRI), 391–393
Magnusiomyces ingens, 253
Maxwell equations, 331–332
MB. *See* Methylene blue (MB)
MCA. *See* Mechanochemical activation (MCA)
MDI. *See* 4,4′-Diphenylmethane diisocyanate (MDI)
MDI-grafted-organic montmorillonite (MOMMT), 409
Mechanical alloying, 150
Mechanical milling, 150
Mechanical properties, 120
Mechanochemical activation (MCA), 120–121
Mechanochemical synthesis, 150
Mechanoresponsive polymeric scaffolds, 139–140
Medical applications, 387–395
  drug delivery, 387–391
  imaging, 391–395
Medical devices
  gold nanocomposites for, 270–271
  silver nanocomposites for, 272–273
Melt mixing technique, 95
Melt-compounding technique, 85
Membrane technology, nanocomposites in, 451–453, 452*f*
Mesoporous silica, 136
  mesoporous silica/phenolic resin hybrid materials, 417
  NPs, 139

Metal fillers, 428–429
metal oxides, 429
metals, 428–429
Metal nanoparticles (MNs), 325, 444–448, 451
fabrication of viscous polymer composites, 338–352
formation of silver nanoparticles during deposition, 339–346
synthesis of silver nanoparticles in viscous epoxy resin, 346–352
ion synthesis in dielectric matrix, 327–331
micrographs of epoxy layers with, 348*f*
synthesis, 249–253
biological methods, 252–253
chemical methods, 250–251
physical methods, 251–252
synthesis in PMMA, 326–338
Metal oxide nanoparticles
polymer–metal oxide nanocomposite materials, 304–312
surface modification, 301–304
synthesis, 288–301
physical methods, 299–301
wet chemical methods, 288–299
Metal oxide–based nanostructured composites, 372
Metal oxides, 429
direct functionalization, 3–5
nanocomposites, 287–288
nanostructures, 303
Metal/polymer nanocomposites. *See also* Polymer nanocomposites
AgNPs and AuNPs
antibacterial properties, 259–262
catalytic activity, 262–263
electronic properties, 263–264
for medical application, 270–273
water treatment using, 264–266
food packaging applications, 266–267
polymer composites with metal nanoparticles for sensors, 268–270
preparation method, 256–259
ex situ method, 258–259
in situ method, 257–258
textile industry, 267–268
Metallic silver, 261
Metals, 428–429
Metamaterials fabrication, 467–468
3-Methacryloxypropyl-trimethoxysilane (MPS), 22, 90*f*, 91, 138–139, 302
Methanol, 290
Methyl acrylate (MA), 9
Methyl methacrylate (MMA), 9, 410
Methyl orange (MO), 136
Methylene blue (MB), 263
Methylphenylsilicone resin (MPSR), 137
Methyltriethoxysilane, 92–94
MFC/NFC. *See* Micro/nanofibrillated cellulose (MFC/NFC)
MFCs. *See* Microbial fuel cells (MFCs)
Mica, 151–152
Micro or miniemulsion polymerization process, 365
Micro/nanofibrillated cellulose (MFC/NFC), 24–25
Microbial fuel cells (MFCs), 139
Microemulsion, 294–296
Microwave synthesis, 150
Mie electromagnetic theory, 326
Mie resonance bands, 332
Mie theory, 331–332
MMA. *See* Methyl methacrylate (MMA)
MMT. *See* Montmorillonite (MMT)
MNs. *See* Metal nanoparticles (MNs)
MO. *See* Methyl orange (MO)
Modified silicas as fillers, 122–125
Molecular dynamics methods, 1
Molecular electronics, 357
Molybdenum disulfide (MoS_2), 50, 71–72
MOMMT. *See* MDI-grafted-organic montmorillonite (MOMMT)
Monoclinic gallium oxide (β-Ga_2O_3), 301
Monofunctional alkoxy silanes, 99
Monolayer graphene, 66
Montmorillonite (MMT), 95–97, 151, 159–160, 166–167, 408–410, 443–444
MPS. *See* 3-Methacryloxypropyl-trimethoxysilane (MPS)
MPSR. *See* Methylphenylsilicone resin (MPSR)

MPTMS. *See* 3-Methacryloxypropyl-trimethoxysilane (MPS)
MRI. *See* Magnetic resonance imaging (MRI)
Multihydroxyl polyglycerol, 12
Multiwalled carbon nanotubes (MWCNTs), 137, 213–214, 423, 439
  MWCNTs-PMMA, 18
Myeloperoxidase, 442–443

N

Nafion polymer chains, 452–453
Nafion resin solution, 135
Nanoclays, 149, 408–413, 443–444, 443*f*
  kaolinite, 410–413
  MMT, 408–410
Nanocomposites, 127. *See also* Polymer nanocomposites
  based on block copolymers, 87–88
  containing functionalized nanoparticles
    polyhedral oligomeric silsesquioxane hybrids, 99–103
    systems, 103
  in energy applications, 453–455
  fabrication techniques, 73*t*
  food packaging by, 448–451, 449*f*
  in membrane technology, 451–453, 452*f*
  production, 441
  reinforced by modified silica nanoparticles, 91–95
  reinforced by organically modified layered silicates, 95–99
  self-organization in, 103–107
  structures, 158–159
    exfoliated structure, 159
    intercalated structure, 158–159
    phase-separated microcomposite structure, 158
"Nanoconfined geometries" effect, 133
Nanocrystal cellulose/cellulose nanowhiskers (NCC/CNWs), 24–25
Nanocrystals. *See* Three-dimensional nanofillers (3D nanofillers)
Nanofillers, 51*f*, 56, 57*f*, 169–170
  classification, 49–50, 49*f*
    1D nanofillers, 49–50
    2D nanofillers, 50
    3D nanofillers, 50
  surface modification protocols, 467
Nanogranules. *See* Three-dimensional nanofillers (3D nanofillers)
Nanomedicine, 381–382
Nanoparticles (NPs), 1, 50, 86, 139, 385–386, 407, 437
  functional groups for surface modification, 2–6, 2*f*
  functionalization and bioconjugation, 382–387
    bioconjugation, 385–387
    ligand addition, 385
    ligand exchange, 384
    ligand modification, 384–385
  modification by polymer chains, 7–23, 8*f*
    using "grafting from" technique, 10–21
    using "grafting to" technique, 7–10
    using "grafting-through" technique, 21–23
  synthesis, 150–151
  toxicology, 438
Nanoplatelet-type of structured filler particles (NSFPs), 151
Nanoplatelets, 66
Nanosilicas, 121
Nanospheres. *See* Three-dimensional nanofillers (3D nanofillers)
Nanostructured hybrid organic conjugated polymers, 366–367
  core–shell strategy, 367
  organic conjugated polymer-based hybrid structured nanocomposites, 367
Nanostructured starch-based film, 267
Nanotechnology, 149, 381–382
  food packaging by, 448–451
Nanotubes, 66
Nanowires, 66
Natural polymers, 262
  biosensor, 269
nBGE. *See* *n*-Butyl glycidyl ether (nBGE)
NCC/CNWs. *See* Nanocrystal cellulose/cellulose nanowhiskers (NCC/CNWs)
NCFS. *See* Noncapacitive Faradaic storage (NCFS)
NCRP. *See* Nonconjugated redox polymers (NCRP)

Near-infrared imaging probes (NIRF), 394
Neat infrared (NIR), 309–310
Nervous system, 438–439
NHS. *See* *N*-Hydroxysuccinimide (NHS)
Nickel oxide (NiO), 301
NIPAM. *See* *N*-Isopropylacrylamide (NIPAM)
NIR. *See* Neat infrared (NIR)
NIRF. *See* Near-infrared imaging probes (NIRF)
Nitrogenation of CNMs, 221–222
Nitrosonium tetrafluoroborate ($NOBF_4$), 303–304
Nitroxide mediated polymerization (NMP), 10, 14–15, 14*f*, 287–288
Noble metal
  direct functionalization of, 3–5
  NPs, 3–4
Nonaqueous sol–gel technique, 150
Noncaged silsesquioxanes, 179–180
Noncapacitive Faradaic storage (NCFS), 231–232
Nonconducting polymers, 226
Nonconjugated redox polymers (NCRP), 232–233
Noncovalent functionalization, 63, 222–225, 223*f*
Nonexpanding clays, 151
Nonporous NPs (NPNPs), 122
Nonylphenol, 295*f*
NPNPs. *See* Nonporous NPs (NPNPs)
NPs. *See* Nanoparticles (NPs)
NSFPs. *See* Nanoplatelet-type of structured filler particles (NSFPs)
Nylon 6/POSS, 184–186

O

OA. *See* Oleic acid (OA)
OAm. *See* Oleylamine (OAm)
OapPOSS-rGO. *See* Octaaminophenyl-POSS-rGO (OapPOSS-rGO)
OBZ-POSS, 183–184
Octa-azido functionalized POSS (OVBN3-POSS), 183–184
Octa(3-hydroxy-3-methylbutyldimethylsiloxy) POSS (OCTA-POSS), 183–184, 419–420
Octaaminophenyl-POSS-rGO (OapPOSS-rGO), 183–184
Octakis(dimethylsiloxypropylglycidylether) silsesquioxane (OGPOSS), 183–184
  n-Octyltriethoxysilane ΨΨ (OCTEO), 413
OCV. *See* Open circuit voltage (OCV)
OD. *See* Optical density (OD)
OGPOSS. *See* Octakis (dimethylsiloxypropylglycidylether) silsesquioxane (OGPOSS)
Oleic acid (OA), 303–304
Oleylamine (OAm), 303–304
OMMT. *See* Organic montmorillonite (OMMT)
One-dimensional nanofillers (1D nanofillers), 49–50
One-step in situ polymerization technique, 310–311
ONQ Silver Soaker, 272
Open circuit voltage (OCV), 237, 372–375
Optical density (OD), 326
Optical extinction, silver core–carbon shell structure on, 336–338
Optical imaging, 394–395
Organ-specific drug release, 389
Organic conjugated polymers, 358–364
  hybrid structured nanocomposites, 367
  polyaniline, 361
    advantages, 361–362
    inorganic and organic dopants, 363–364
    synthesis strategy, 362–363
  polymerization mechanism of aniline monomer, 363
  types, 358–360
    hybrid composites, 360
    intrinsically electrically conjugated polymers, 358–360
    ionically organic conducting polymers, 360
Organic dendrimers, 270
Organic molecules, 250
Organic montmorillonite (OMMT), 409
Organic polymer, 158
Organic solar cell (OSC), 372–375
Organofunctional silanes, 301–302
Organometallic precursor, 307
Organomodification of nanoclays, 444
Organophilic character, 95–97
Orientation, 63–65
  electric field–induced alignment, 65
  flow-induced alignment, 64–65
  magnetic field–induced alignment, 65

Orientation (*Continued*)
shear-induced alignment, 63–64
OSC. *See* Organic solar cell (OSC)
OSE. *See* Overall thermal stabilization effect (OSE)
OVBN3-POSS. *See* Octa-azido functionalized POSS (OVBN3-POSS)
Overall thermal stabilization effect (OSE), 155
Oxidative degradation, 406
Oxidative stress, 448
Oxoaminium perchlorate groups, 10
Oxygenation of CNMs, 216–220
deconvolution of C1 in X-ray spectrograph, 218*f*
oxygen-containing groups on surface of CNTs, 217*f*
oxygenation of graphene, 219–220

P

P-pa. *See* 3,4-Dihydro-3-(prop-2-ynyl)-2H-benzoxazine (P-pa)
P(NIPAM-co-A). *See* Poly(N-isopropylacrylamide-co-acrylamide) (P(NIPAM-co-A))
P3HT. *See* Poly(3-hexylthiophene) (P3HT)
PA. *See* Polyamide (PA)
PAA. *See* Poly(acrylic acid) (PAA)
PADPA. *See* *p*-Amino diphenylamine (PADPA)
PAHMA. *See* Poly(6-azidohexyl methacrylate) (PAHMA)
PAI. *See* Polyamide-imide (PAI)
PAM. *See* Polyacrylamide (PAM)
PAMAM. *See* Poly(amidoamine) (PAMAM)
PAN. *See* Polyacrylonitrile (PAN)
Panax ginseng, 252
PANI. *See* Polyaniline (PANI)
Partial-caged POSS with triols, 192–193
Partially oxidized graphene, 232
Particle size distribution (PaSD), 119–120
PAs. *See* Polyacrylates (PAs)
PaSD. *See* Particle size distribution (PaSD)
PBA. *See* Poly(n-butyl acrylate) (PBA)
PBFNP. *See* Polymer brush FNP (PBFNP)
PBMA. *See* Poly(butyl methacrylate) (PBMA)
PBS. *See* Poly(butylene succinate) (PBS)
PBZ. *See* Polybenzoxazines (PBZ)
PC. *See* Polycarbonate (PC)
PCGA. *See* Poly(″-caprolactone-co-D,L-lactic acid) (PCGA)
PCL. *See* Polycaprolactone (PCL)
PCL-diol. *See* Poly(ε-caprolactone)-diol (PCL-diol)
PCLA. *See* Poly(″-caprolactone-co-D,L-lactic acid) (PCLA)
PCN. *See* Polymer–clay nanocomposites (PCN)
PD. *See* Pentadecane (PD)
PDI-POSS. *See* 1,2-Propanediolizobutyl POSS (PDI-POSS)
PDMA. *See* Poly(2-(*N,N*-dimethylamino)ethyl methacrylate) (PDMA)
PDMS. *See* Poly(dimethylsiloxane) (PDMS)
PDPA. *See* Poly(diphenylamine) (PDPA)
PE. *See* Polyethylene (PE)
PEA. *See* Poly(ethyl acrylate) (PEA)
Peak heat release rate (PHRR), 155, 312
Pearson's theory, 255
Pebax/POSS, 196–198
PEDOT. *See* Poly(3,4-ethylenedioxythiophene) (PEDOT)
PEDOT:PSS. *See* Poly(3,4-ethylenedioxythiophene) polystyrene sulfonate (PEDOT:PSS)
PEEK. *See* Polyetheretherketone (PEEK)
PEG. *See* Poly(ethylene glycol) (PEG)
PEGylated silicon NPs, 9
PEI. *See* Polyethyleneimine (PEI)
PEM. *See* Polymer electrolyte membrane (PEM)
Pentadecane (PD), 294
N,N,N′,N″,N″-Pentamethyldiethylenetriamine (PMDETA), 26
PEO. *See* Poly(ethylene oxide) (PEO)
PEPS. *See* Poly(PEGMA–PPGMA–POSSMA) (PEPS)
Perchloric acid ($HClO_4$), 364
Percolation threshold, 64–65
Perfluorocyclopentenyl aryl ether polymers (PFCP), 189–192
Perfluorophenylazides (PFPA), 9
Pernigraniline (PNB), 361
PES. *See* Polyester (PES)
PET. *See* Poly(ethylene terephthalate) (PET)
PFCP. *See* Perfluorocyclopentenyl aryl ether polymers (PFCP)

PFDBT-POSS-Affibody, 194–195
PFPA. *See* Perfluorophenylazides (PFPA)
PGMA. *See* Poly(glycidyl methacrylate) (PGMA)
PH-POSS. *See* Poly(hydroxylether of bisphenol A) (PH-POSS)
Phagocytic defense system, 438–439
Phase-separated microcomposite structure, 158
Phase-transfer agent, 250
PHB. *See* Poly(3-hydroxybutyrate) (PHB)
PHEMA glass transition, 129–133, 131*f*, 132*f*
Phenolic resin, 155–156
N-Phenylaminomethyl-POSS/PU composites, 200–201
PHIPOSS. *See* 1,2-Propanediol-heptaisobutyl-POSS (PHIPOSS)
Phosphoric acid (H_3PO_4), 364
Photocatalysis, 467–468
PHRR. *See* Peak heat release rate (PHRR)
Phyllosilicate, 151–152
Physical adsorption, 95
Physical methods, 299–301
  CVD, 299–300
  PLD, 300–301
  in situ methods, 74
PI. *See* Polyimide (PI)
PI-b-PEO. *See* Poly(isoprene-b-ethylene oxide) (PI-b-PEO)
PIL. *See* Polymeric ionic liquid (PIL)
PIL-SNP. *See* Polymeric ionic liquid-functionalized mesoporous silica nanoplates (PIL-SNP)
PLA. *See* Polylactide (PLA)
Plasma-assisted CVD, 300
PLD. *See* Pulse laser deposition (PLD)
PLGA. *See* Poly(D,L-lactic acid-co-glycolic acid) (PLGA)
PMAA. *See* Poly(methacrylic acid) (PMAA)
PMAA-PEA. *See* Poly(methacrylic acid) polyethylacrylate (PMAA-PEA)
PMAA-PMA. *See* Poly(methacrylic acid) polymethacrylate (PMAA-PMA)
PMDETA. *See N,N,N′,N″,N″*-Pentamethyldiethylenetriamine (PMDETA)
PMMA. *See* Poly(methylmethacrylate) (PMMA)
PNB. *See* Pernigraniline (PNB)
PNIPAM. *See* Poly(*N*-isopropylacrylamide) (PNIPAM)
PNVK. *See* Poly(*N*-vinylcarbazole) (PNVK)
Polar polymers, 127–136
Polar surface groups, 218–219
Poly (propyleneimine) (PPI), 263
Poly-ε-caprolactone. *See* Polycaprolactone (PCL)
Poly(PEGMA). *See* Poly(poly(ethylene glycol) methylether methacrylate) (Poly(PEGMA))
Poly(″-caprolactone-co-D,L-lactic acid) (PCGA), 389–390
Poly(″-caprolactone-co-D,L-lactic acid) (PCLA), 389–390
Poly(″-caprolactone). *See* Polycaprolactone (PCL)
Poly(2-(*N,N*-dimethylamino)ethyl methacrylate-b-(methyl methacrylate)-b-2-(*N,N*-dimethylamino) ethyl methacrylate) P(DMA-b-MMA-b-DMA), 428
Poly(2-(*N,N*-dimethylamino)ethyl methacrylate) (PDMA), 428
Poly(3-hexylthiophene) (P3HT), 237
Poly(3-hydroxybutyrate) (PHB), 261
Poly(3,4-ethylenedioxythiophene) (PEDOT), 54, 215
Poly(3,4-ethylenedioxythiophene) polystyrene sulfonate (PEDOT:PSS), 263–264
Poly(4-vinylpyridine) (PVP), 155–156, 257–258, 303–304
Poly(6-azidohexyl methacrylate) (PAHMA), 20
Poly(acrylic acid) (PAA), 9–10, 385
Poly(amidoamine) (PAMAM), 136–137, 255, 263, 392–393
Poly(butyl methacrylate) (PBMA), 16, 18
Poly(butylene succinate) (PBS), 98–99
Poly(caprolactone-based urea-urethane), 188–189
Poly(D,L-lactic acid-co-glycolic acid) (PLGA), 389–390
Poly(dimethylsiloxane) (PDMS), 92, 119–120
Poly(diphenylamine) (PDPA), 359
Poly(ethyl acrylate) (PEA), 14–15
Poly(ethylene glycol) (PEG), 188–189, 270–271, 384, 420
Poly(ethylene oxide) (PEO), 95–97, 151, 155–156, 188–189

Poly(ethylene terephthalate) (PET), 127–128, 155–156, 312
Poly(glycidyl methacrylate) (PGMA), 127–128
Poly(hydroxylether of bisphenol A) (PH-POSS), 183–184, 196–198
Poly(isoprene-b-ethylene oxide) (PI-b-PEO), 104–105
Poly(methacrylic acid) (PMAA), 389
Poly(methacrylic acid) polyethylacrylate (PMAA-PEA), 389
Poly(methacrylic acid) polymethacrylate (PMAA-PMA), 389
Poly(methyl methacrylate)-*b*-poly(butyl acrylate)-*b*-poly(methyl methacrylate) (PMMA-*b*-PBA-*b*-PMMA), 27–28
Poly(methylmethacrylate) (PMMA), 55, 92, 101, 155–156, 219, 259–260, 312, 326, 406, 411*f*
  experimental optical density spectra, 331*f*
  MNs synthesis in, 326–338
Poly(n-butyl acrylate) (PBA), 14–15
Poly(N-isopropylacrylamide-co-acrylamide) (P(NIPAM-co-A)), 136
Poly(*N*-isopropylacrylamide) (PNIPAM), 16, 137–138, 263, 389–390
Poly(*N*-vinylcarbazole) (PNVK), 27
Poly(PEGMA–PPGMA–POSSMA) (PEPS), 183–184
Poly(poly(ethylene glycol)methylether methacrylate) (poly(PEGMA)), 17
Poly(styrene sulfonates) (PSS), 223
Poly(styrene-b-poly(2-(*N,N*-dimethylamino)ethyl methacrylate) P(St-b-DMA), 428
Poly(styrene-co-maleic anhydride) (PSMA), 196–198
  PSMA/Ti-POSS-NH_2, 198–200
Poly(t-butyl methacrylate)-b-(2-(*N,N*-dimethylamino) ethyl methacrylate) P(tBMA-b-PDMA), 428
Poly(vinyl chloride) (PVC), 155–156, 267
Poly(vinyl pyrrolidone). *See* Poly(4-vinylpyridine) (PVP)
Poly(vinylbenzyl chloride) (PVBC), 20
Poly(vinylidene fluoride-hexafluoropropylene) (PVDF-HFP), 94–95, 94*f*
Poly(vinylidene fluoride) (PVDF), 192, 268
Poly(ε-caprolactone)-diol (PCL-diol), 184–186
Polyacrylamide (PAM), 9
Polyacrylates (PAs), 258
Polyacrylonitrile (PAN), 155–156
Polyalcohols, 5
Polyamide (PA), 127–128, 261, 309
  6/clay system, 149
  6/silica HNCs, 128
Polyamide-imide (PAI), 95
Polyaniline (PANI), 215, 258, 306–307, 357–358, 369–370, 371*f*
Polyaniline, 361
  advantages, 361–362
  inorganic and organic dopants for, 363–364
  nanohybrids
    electrochemical energy storage devices based on, 368–372
    energy conversion systems based on, 372–375
  synthesis strategy, 362–363
Polyaniline-based hybrid nanostructured composite, 367–368
Polyaniline-based nanomaterials, 364–366
  fabrication methods, 365
  hard template methods, 365–366
  soft template methods, 365
  template-free methods, 366
Polybenzoxazines (PBZ), 183
  PBZ matrix, 183–184
  PBZ-OGPOSS, 183–184
  PBZ-POSS nanocomposites, 189–192
Polycaprolactam, 259–260
Polycaprolactone (PCL), 10, 59, 140, 155–156, 259–260, 389–390
Polycarbonate (PC), 155–156, 226, 259–260, 309–310
Polyelectrolyte membranes, 135
Polyester (PES), 261
Polyetheretherketone (PEEK), 264
Polyethylene (PE), 9, 63–64, 155–156, 255
Polyethylene glycol sorbitan trioleate (Tween 85), 295*f*
Polyethyleneimine (PEI), 257, 263, 268
  functionalization, 440
Polyfluorene, 54

Polyhedral oligomeric silsesquioxane (POSS), 23–24, 86, 179–180, 181*f*, 418–420, 468
hybrids, 99–103, 100*f*, 101*f*
POSS-capped PEO nanocomposite, 196–198
POSS–based hybrid composites for performance enhancement, 183–206
anti-photodegradation, 192–193
antiatomic oxidation, 192
biochemistry application, 200–201
catalytic activity, 198–200
crystallinity, 195–196
electric conductivity, 188–192
mechanical strength, 184–187
optical application, 194–195
porosity, 198
self-assembly, 202–206
surface properties, 196–198
thermal stability, 183–184
POSS–PCL/graphene, 188–189
reinforced PBZ/EP, 192–193
Polyhedral oligomeric silsesquioxanes-poly (carbonate-urea) urethane (POSS-PCU), 139–140
Polyhedral oligosilsesquioxane. *See* Polyhedral oligomeric silsesquioxane (POSS)
Polyimide (PI), 155–156, 188–189, 302, 426
Polylactic acid. *See* Polylactide (PLA)
Polylactide (PLA), 155–156, 183, 311–312, 419
PLA-g-MA/POSS-NH_2, 195–196
PLA-g-MA/POSS-OH, 195–196
Polymer brush FNP (PBFNP), 135
Polymer chains, NP modification by, 7–23, 8*f*
Polymer composites, 467
with carbon nanomaterials, 212–216
functional clays influence on aging, 163–166
morphological developments, 165–166
structural changes of polymer/clay composites, 164–165
functionalization of CNMs and effects on, 216–228
with functionalized silica
filled by silicas, 125–141
structure of unmodified and modified silicas as fillers, 122–125
with metal nanoparticles
functionalization of AgNPs and AuNPs, 253–256
preparation method, 256–259
properties, 259–273
for sensors, 268–270
synthesis of metal nanoparticles, 249–253
properties and applications of polymer composites with CNMs, 228–238
Polymer electrolyte membrane (PEM), 221
Polymer nanocomposites, 47–65, 52*f*, 85–86, 264, 287–288, 438. *See also* Metal/polymer nanocomposites
classification of nanofillers, 49–50, 49*f*
design parameters, 56–65, 57*f*
AR, 56–59
interface, 59–63
orientation, 63–65
environmental benefits and application of, 448–455
functionality-based designs, 53–54
functionalized nanoparticles effect on stability, 407–408
material, 47–48
carbonaceous nanoadditives, 423–428
decomposition with functionalized nanoparticles, 408–429
inorganic fillers, 408–423
metal fillers, 428–429
optical properties
fabrication of viscous polymer composites with MNs, 338–352
MNs synthesis in PMMA, 326–338
rationality-based design, 53
synthesis, 66–74
ball milling, 71–72
double-screw extrusion, 72–73
roll milling, 70–71
shear mixing, 68–70
ultrasonication-assisted solution mixing, 67–68
tailored property-based designs, 54–56
Polymer solar cells (PSCs), 372–375
Polymer-based nanocomposites, 86
Polymer-based restorative dental composite materials, 140

Polymer-coated cobalt NPs, 9–10
Polymer-layered silicate nanocomposites, 95, 97*f*
Polymer/AgNPs/cotton nanocomposites, 268
Polymer/AuNPs tissue engineering, 271
Polymer/functionalized nanoparticle systems, 103, 467
  functionalization and bioconjugation of nanoparticles, 382–387
  medical applications, 387–395
  nanocomposites containing functionalized nanoparticles, 91–103
  self-organization in nanocomposites, 103–107
  synthesis, modification, and functionalization, 88–91
Polymer–clay hybrid, 443
Polymer–clay nanocomposites (PCN), 410
Polymer–CNM composites, 225–228, 227*f*, 229*t*
  electrical properties and applications, 229–230
  electrochemical properties and applications, 230–238
  mechanical and structural enhancements in, 238
Polymeric ionic liquid (PIL), 137
Polymeric ionic liquid-functionalized mesoporous silica nanoplates (PIL-SNP), 137
Polymeric/polymer, 103, 136
  degradation, 405
  grafting, 302–303
  materials, 183–186
  matrix, 89, 95, 287–288
  microspheres, 263
  nanohybrids, 159–161
  nanomaterials, 149
  nanoparticles, 387
  physical and chemical changes, 407
  polymer-stabilized Au NPs, 8–9
  thermal decomposition mechanisms, 405–407
    oxidative degradation, 406
    thermal degradation, 406
  thermal stability and heat resistance, 407
Polymerization, 1, 7–8, 10–11, 15, 219–220, 287–288, 306–307, 467
  mechanism of aniline monomer, 363
  methods, 302
  polymerized complex method, 150–151
Polymer–metal oxide nanocomposite materials, 304–312
  applications, 312, 313*t*
    TTI, pHRR, and THR of PMMA nanocomposites, 313*f*
  methods for nanocomposites preparation
    blending, 306–307
    in situ nanocomposite formation, 307
    sol–gel processing, 307
  properties, 307–312
    electrical properties, 310–311
    mechanical properties, 307–309
    optical properties, 309–310
    thermal properties, 311–312
Polymer–particle interactions, 87–88
Polyoxyethylene(4) lauryl ether (Brij 30), 295*f*
Polyoxymethylene (POM), 406, 421*t*, 424*f*
Polypropylene (PP), 126–127, 155–156, 222, 258–259, 306, 406
Polypyrrole (PPy), 215, 263–264, 357, 453–454, 454*f*
Polysiloxane elastomers [SiRR–O–]$_x$, 127
Polystyrene (PS), 8–9, 67, 101, 155–156, 183, 223, 302–303
Polystyrene-*b*-polyisoprene-*b*-polystyrene (SIS), 27–28
Polystyrene–butadiene–polystyrene (SBS), 186–187
Polystyrenesulfonic acid (PSSA), 364
Polysulfone (PSU), 414, 414*f*
Polythiophene, 263–264
Polyurethane (PU), 98*f*, 155–156, 183–184, 224–225, 259–260, 272, 409, 409*t*
Polyurethane-poly(2-hydroxyethyl methacrylate) (PU-PHEMA), 129–130, 130*f*
Polyurethane–silica nanocomposites, 93*f*
Polyurethaneurea, 308
Polyvinyl alcohol (PVA), 54, 151, 226, 259–260, 360, 416, 417*f*
Polyvinyl-ferrocene (PVFc), 232–233
POM. *See* Polyoxymethylene (POM)
Pore size distribution (PSD), 123, 123*f*, 124*f*
Porosity, 198
POSS. *See* Polyhedral oligomeric silsesquioxane (POSS)

POSS-PCU. *See* Polyhedral oligomeric silsesquioxanes-poly(carbonate-urea) urethane (POSS-PCU)
Postfunctionalization, 2–3
  reactions at NPs surface, 6, 7*f*
Power spectral density analysis (PSD analysis), 165
PP. *See* Polypropylene (PP)
PPI. *See* Poly (propyleneimine) (PPI)
PPOH. *See* Functionalized PP (PPOH)
PPSD. *See* Primary particle size distribution (PPSD)
PPy. *See* Polypyrrole (PPy)
Precipitation method, 291–292
Primary particle size distribution (PPSD), 122
Pristine resin, 163–164
1,2-Propanediol-heptaisobutyl-POSS (PHIPOSS), 419–420
1,2-Propanediolizobutyl POSS (PDI-POSS), 419–420
Propanoic acid, 20
n-Propanol, 290
PS. *See* Polystyrene (PS)
PSCs. *See* Polymer solar cells (PSCs)
PSD. *See* Pore size distribution (PSD)
PSD analysis. *See* Power spectral density analysis (PSD analysis)
Pseudocapacitance electrode, 372
PSMA. *See* Poly(styrene-co-maleic anhydride) (PSMA)
PSS. *See* Poly(styrene sulfonates) (PSS)
PSSA. *See* Polystyrenesulfonic acid (PSSA)
PSU. *See* Polysulfone (PSU)
PU. *See* Polyurethane (PU)
PU-PHEMA. *See* Polyurethane-poly(2-hydroxyethyl methacrylate) (PU-PHEMA)
Pulmonary deposition, 168
Pulse laser deposition (PLD), 288, 300–301
PVA. *See* Polyvinyl alcohol (PVA)
PVBC. *See* Poly(vinylbenzyl chloride) (PVBC)
PVC. *See* Poly(vinyl chloride) (PVC)
PVDF. *See* Poly(vinylidene fluoride) (PVDF)
PVDF-HFP. *See* Poly(vinylidene fluoride-hexafluoropropylene) (PVDF-HFP)
PVFc. *See* Polyvinyl-ferrocene (PVFc)
PVP. *See* Poly(4-vinylpyridine) (PVP)
Pyrogenic synthesis, 121
Pyrophyllite, 151–152

Q

Quantum dots (QDs), 50
Quantum effects, 54–55
Quantum size effects, 444–445
Quaternary ammonium compounds (QACs), 168
Quaternary ammonium salt, 168

R

Radio frequency (RF), 300
RAFT. *See* Reversible addition–fragmentation chain transfer (RAFT)
Random chain scission, 406
Rationality-based design, 53
RDRPs. *See* Reversible-deactivation radical polymerizations (RDRPs)
RE. *See* Reference electrode (RE)
Reactive ligands, 303–304
Reactive oxygen species (ROS), 259
Redox polymers, 232–233
Reduced graphene oxides (rGO), 219, 441
  rGO-OapPOSS, 189–192
  rGO/PS nanocomposites, 67
Reference electrode (RE), 369
Relative humidity (RH), 135
Reverse osmosis process, 134
Reversible addition–fragmentation chain transfer (RAFT), 287–288
  polymerization, 8–9, 19–21, 19*f*, 180–181
Reversible-deactivation radical polymerizations (RDRPs), 1
RF. *See* Radio frequency (RF)
rGO. *See* Reduced graphene oxides (rGO)
RH. *See* Relative humidity (RH)
Ring opening polymerization (ROP), 183–184
RMS. *See* Root mean square (RMS)
Roll milling, 70–71
Root mean square (RMS), 166
ROP. *See* Ring opening polymerization (ROP)
ROS. *See* Reactive oxygen species (ROS)
Rutile, 445–446

S

S-amine. *See* SiO_2 nanocomposites with amine (S-amine)
S-epoxide. *See* SiO_2 nanocomposites with epoxide (S-epoxide)
S-NCO. *See* SiO_2 nanocomposites with isocyanate (S-NCO)
SA. *See* Sodium alginate (SA)
Sargassum tenerrimum, 253
SBR. *See* Styrene–butadiene rubber (SBR)
SBS. *See* Polystyrene–butadiene–polystyrene (SBS)
Scanning electron microscopy (SEM), 192–193
SDS. *See* Sodium dodecylsulfonate (SDS)
SEBS. *See* Styrene-ethylene/butylene-styrene (SEBS)
SEBS grafted maleic anhydride (SEBS-g-MA), 126–127
Self-assembly processes, 468–469
Self-organization in nanocomposites, 103–107
  electron micrograph, 107*f*
  hierarchical coassemblies of nanocrystals, polymers, 104*f*
SEM. *See* Scanning electron microscopy (SEM)
Sensors, polymer composites with metal nanoparticles for, 268–270
SERS. *See* Surface-enhanced Raman spectroscopy (SERS)
Serum lactate dehydrogenase, 446
Shear mixing, 68–70
Shear-induced alignment, 63–64
Sheet-like morphology, 162
Short-circuit current density (Jsc), 372–375
SI-ATRP. *See* Surface-initiated atom-transfer radical polymerization (SI-ATRP)
Silane coupling, 301–302
  agents, 287–288
Silane-based functionalities, 467–468
Silane-modified tungsten trioxide, 309–310
Silanization
  effect, 138–139
  of silica filler, 138–139
Silica nano/microspheres (SNS), 127
Silica nanoparticles (SN), 9, 14–16, 415, 416*f*
  aggregation effects, 139
  functionalized with vinyl isocyanate, 129
  nanocomposites reinforced by modified, 91–95
  sizes, 122*f*
Silica NPs. *See* Silica nanoparticles (SN)
Silica(s), 120–121, 414–417
  gel, 288–289
  polymer composites filled by, 125–141
    biomedical applications of composites, 138–141
    complex fillers, 136–138
    composites with polar polymers, 127–136
    hydrophobic nonpolar polymers, 126–127
  silica-based fillers, 119–120
Silicate nanotubes (SNT), 59
Silicates, 151
  functionalization for fire retardant applications, 155
Silicon, 414–417
  POSS, 418–420
  semiconductors, 357
  silicon-based nanomaterials, 414–420
Silicone rubber (SR), 425
Silsesquioxane, 179–180, 180*f*
Silver (Ag), 444–445, 447–448
Silver nanocomposites
  electronic properties of, 264
  for medical application, 270–273
  for medical devices, 272–273
Silver nanofibers (Ag-Nfbs), 262
Silver nanoparticles (AgNPs), 249, 251
  antibacterial properties, 260–262, 261*f*
  catalytic activity, 263
  in dielectric matrix, extinction of implanting, 331–335
  formation during deposition on viscous epoxy resin, 339–346
  functionalization, 253–256
  granulometric characteristics, 348*t*
  histogram of size distribution, 343*f*
  for medical application, 272–273
  number of ultrafine, 344*f*
  synthesis in viscous epoxy resin, 346–352
  water treatment using, 264–266
Silver-doped zeolites, 267
Silverline, 272

Single-walled carbon nanotubes (SWCNTs), 18, 213–214, 214*f*, 439
Singlet oxygen, 259
SiO_2 nanocomposites with amine (S-amine), 312
SiO_2 nanocomposites with epoxide (S-epoxide), 312
SiO_2 nanocomposites with isocyanate (S-NCO), 312
SiO_2 nanoparticles, 289–290
SiP. *See* Diethylphosphatotriethoxysilane (SiP)
SIP method. *See* Surface-initiated polymerization method (SIP method)
SIS. *See* Polystyrene-*b*-polyisoprene-*b*-polystyrene (SIS)
Smectites, 151–152
SN. *See* Silica nanoparticles (SN)
SNS. *See* Silica nano/microspheres (SNS)
SNT. *See* Silicate nanotubes (SNT)
Sodium alginate (SA), 259–260
Sodium dodecylebenzene sulfonate (DBS), 298
Sodium dodecylsulfonate (SDS), 298
Soft template methods, 365
Solar energy conversion, 372–375
Sol–gel
  method, 467–468
  processing, 288–291, 307
  synthesis, 151, 156–157
Solid electrolytes, 188–189
Solution blending technique, 85
Solution intercalation, 157
Solvent exchange technique, 85
Solvothermal synthesis, 250–251
Solvothermal/hydrothermal technique, 296–299
Sorbitan laurate (Span 20), 295*f*
Sorbitane monooleate (Span 80), 295*f*
sp^2-bonded carbon atom, 214–215
SPEEK. *See* Sulfonated poly(ether-ether-ketone) (SPEEK)
Spherical silica particles, 133–134
SPIs. *See* Sulfonated polyimides (SPIs)
SPP13 polymer, 188–189, 190*f*
SPR. *See* Surface plasma resonance (SPR)
Spray coating technique, 259
SPS/CS. *See* Sulfonated polystyrene/chitosan (SPS/CS)
SR. *See* Silicone rubber (SR)
SRMs. *See* Stimuli-responsive materials (SRMs)
St-b-DMA. *See* Poly(styrene-b-poly(2-(*N,N*-dimethylamino)ethyl methacrylate) P(St-b-DMA)
Stannic oxide (SnO_2), 301
Stimuli-responsive materials (SRMs), 27
Stimuli-responsive nanomaterials, 27–28
  synthesis and properties of photoactuating
    nanocomposites, 28*f*
Stöber method, 288–289
Structure–morphology–property behavior, 85
Styrene-ethylene/butylene-styrene (SEBS), 126–127
Styrene–butadiene rubber (SBR), 309
Submicron silica particles surface-capped with diglycidyl ether, 135–136
Sulfo NHS, 386–387
Sulfonated poly(ether-ether-ketone) (SPEEK), 454–455
Sulfonated polyimides (SPIs), 188–189
Sulfonated polystyrene/chitosan (SPS/CS), 263
Sulfonation of CNMs, 220–221
Sulfuric acid (H_2SO_4), 364
Sunset yellow (SY), 369–370
Supercapacitors, 368–369
  application, 372, 374*t*
Supercritical hydrothermal synthesis, 151
Surface functional groups, 230
Surface functionalization, 153, 442–443
  of CNMs, 216
  CNTs, 440
  of nanoparticles, 467
  NPs, 105
  silica NPs, 92
Surface located nanocomposites, 451–452
Surface modification, 1, 89, 91, 467
  of metal oxide nanoparticles, 301–304
    chemical treatments, 301–302
    ligand exchange, 303–304, 304*f*
    polymer grafting, 302–303
Surface morphology, 135–136
Surface plasma resonance (SPR), 325
Surface-enhanced Raman spectroscopy (SERS), 265

Surface-initiated atom-transfer radical polymerization (SI-ATRP), 257–258
Surface-initiated polymerization method (SIP method), 10, 415
Surfactants, 294, 298
SWCNTs. *See* Single-walled carbon nanotubes (SWCNTs)
SY. *See* Sunset yellow (SY)
Synergetic effects, 121
Synergistic effect, 371–372
Synthetic metal, 358

T

t-BuLi. *See* *Tert*-butyllithium (t-BuLi)
Tailored property-based designs, 54–56
tBA. *See* *Tert*-butyl acrylate (tBA)
tBMA. *See* *Tert*-butyl methacrylate (tBMA)
tBMA-b-PDMA. *See* Poly(t-butyl methacrylate)-b-(2-(*N,N*-dimethylamino) ethyl methacrylate) P (tBMA-b-PDMA)
Telechelic POSS–PLLA, 196–198
TEM. *See* Transmission electron microscopy (TEM)
Template-free methods, 366
TEMPO based alkoxyamine. *See* 2,2,6,6-Tetramethyl-1-piperidinyloxy based alkoxyamine (TEMPO based alkoxyamine)
TEMPO–oxidized cellulose nanofibers (TOCNs), 24–25, 25*f*
TEOS. *See* Tetraethoxysilane (TEOS); Tetraethylortosilicate (TEOS)
TERM. *See* Tissue engineering and regenerative medicine (TERM)
Ternary transition metal oxide, 292–294
Tert-butyl acrylate (tBA), 12
Tert-butyl methacrylate (tBMA), 17
Tert-butyllithium (t-BuLi), 11–12
Tetradecylphosphonic acid, 303–304
Tetraethoxysilane (TEOS), 127–128, 289–290
Tetraethyl orthotitanate [Ti(OEt)$_4$], 307
Tetraethylortosilicate (TEOS), 88–89, 89*f*
2,2,6,6-Tetramethyl-1-piperidinyloxy based alkoxyamine (TEMPO based alkoxyamine), 15, 24–25
Tetramethylethylenediamine, 12
Tetraoctylammonium bromide, 250
Textile industry, metal/polymer nanocomposites in, 267–268
TG-FTIR techniques, 164, 410
TGA. *See* Thermogravimetric analysis (TGA)
Thermal decomposition of polymer nanocomposites
  decomposition with functionalized nanoparticles, 408–429
  functionalized nanoparticles effect on stability, 407–408
  mechanisms of polymers, 405–407
  thermal stability and heat resistance of polymers, 407
Thermal degradation, 406
Thermal depolymerization, 406
Thermal stability of polymers, 183–184, 407
Thermogravimetric analysis (TGA), 407, 412*f*, 415*t*, 420*f*, 427*f*, 429*t*
Thermooxidative reaction, 165
Thermoplastic copolymers (TPE), 27–28
Thermoplastic elastomeric composites, 186–187
Thermoplastic polyurethane (TPU), 99
Thermoresponsive polymers, 389–390
Thin-film composites with nanocomposite substrate, 451–452
Thin-film nanocomposites, 451–452
Thiol–disulfide exchange reaction, 200–201
Thiols (–SH), 287–288
3-Thiopropyltriethoxysilane (TPTES), 128
THR. *See* Total heat release (THR)
Three-dimensional nanofillers (3D nanofillers), 50
Time to ignition (TTI), 161, 312
TIPNO. *See* 2,2,5-Trimethyl-4-phenyl-3-azahexane-3-nitroxide (TIPNO)
Tissue engineering and regenerative medicine (TERM), 271
Titanium dioxide (TiO_2), 298, 301, 303–304, 312, 429, 444–445, 445*f*
TMCS. *See* Trimethylchlorosilane (TMCS)
TMDs. *See* Transition metal dichalcogenides (TMDs)
TMP. *See* Trimethylolpropane (TMP)
TNT. *See* Trinitrotoluene (TNT)

TOCNs. *See* TEMPO–oxidized cellulose nanofibers (TOCNs)
Top-down approach, 288
Tortuous factor, 161–162
Tortuous path model, 162, 162*f*
Total heat release (THR), 161, 312
Toxic behavior, 440
TPE. *See* Thermoplastic copolymers (TPE)
TPTES. *See* 3-Thiopropyltriethoxysilane (TPTES)
TPU. *See* Thermoplastic polyurethane (TPU)
Tracheobronchial deposition, 168
Traditional fillers, 161
Traditional flame retardants, 160
Transition metal dichalcogenides (TMDs), 69
Transition metal oxides, 371
Transmission electron microscopy (TEM), 326
1-Trichlorosilyl-2-(*m-p*-chloromethylphenyl) ethane, 16
2,2,5-Trimethyl-4-phenyl-3-azahexane-3-nitroxide (TIPNO), 14–15
Trimethylchlorosilane (TMCS), 119–120, 125*f*
Trimethylolpropane (TMP), 97–98
Trinitrotoluene (TNT), 198
"Trojan horse" effect, 448
TTI. *See* Time to ignition (TTI)
Tungsten trioxide (WO_3), 301, 309–310
Turbinaria conoides, 253
Two-dimension (2D)
  nanofillers, 50
  sp^2-bonded carbon atom, 214–215

U

Ubiquicidin 29–21 (UBI), 255
UDMA. *See* Dimethacrylate (UDMA)
Ultrafine particles, 168
Ultrahigh electric insulator materials, 310
Ultrasonication, 68, 288
  ultrasonication-assisted solution mixing, 67–68
Ultrasound (US), 388
  micromolding technology, 138
Ultraviolet (UV)
  radiation, 192–193
  resistance, 50
Unmodified clay, 168–169
Unmodified silicas as fillers, 122–125
Unsaturated polyester nanocomposites, 307–308
Unsaturated polyester resin (UPR), 186–187
US. *See* Ultrasound (US)
US Food and Drug Administration (FDA), 387–388, 446–447

V

van der Waals force, 152–153, 291
VBa-POSS, 183–184
Vinylbenzyltrimethylammonium chloride, 23
Vinyls (–CH = CH2), 287–288
Vinyltriethoxysilane, 92–94
Vinyltrimethoxysilane (VTMS), 425
Vinyltris(2-methoxyethoxy)silane (VMES), 302
Viscous epoxy resin
  formation of silver nanoparticles during deposition, 339–346
  synthesis of silver nanoparticles, 346–352

W

Water treatment using AgNPs and AuNPs, 264–266
Water-in-oil microemulsion technique, 294
Waterborne polyurethane (WBPU), 421, 422*f*
Wet chemical methods, 288–299
  coprecipitation method, 291–294
  microemulsion technique, 294–296
  sol–gel processing, 288–291
  solvothermal/hydrothermal technique, 296–299

X

X-ray computed tomography, 393
X-ray diffraction (XRD), 292
X-ray photoelectron spectroscopy (XPS), 217, 220*f*

Y

Young's equation, 59–61

Z

Zero dimensional nanoparticles, 50
Zinc oxide (ZnO), 301, 444–447
  ZnO-modified SnO_2, 301

Kolibri 360 GmbH
Ferdinand-Jühlke-Straße 7
99095 Erfurt, Deutschland
produktsicherheit@kolibri360.de